Functional Vertebrate Morphology

Functional Vertebrate Morphology

Edited by

Milton Hildebrand
Dennis M. Bramble
Karel F. Liem
David B. Wake

The Belknap Press of
Harvard University Press
Cambridge, Massachusetts
and London, England
1985

Printed in the United States of America
10 9 8 7 6 5 4 3 2 1

This book is printed on acid-free paper, and its binding materials have been chosen for strength and durability.

Library of Congress Cataloging in Publication Data
Main entry under title:
Functional vertebrate morphology.
Bibliography: p.
Includes index.
1. Vertebrates—Morphology. 2. Vertebrates—Physiology. I. Hildebrand, Milton, 1918–
QL805.F86 1985 596′.01852 84-19175
ISBN 0-674-32775-6 (alk. paper)

Preface

In the course of revising my lectures and textbook, and in conversations with students and colleagues, I became convinced that the study of functional vertebrate morphology would be well served by a book that was more advanced than available texts, yet less technical than publications resulting from symposia. The former can only introduce topics, and the latter tend to be restricted in scope and too detailed for nonspecialists. Accordingly, I drafted a prospectus and invited the assistance of my three coeditors — selected because they are productive, respected morphologists and congenial friends, and because their research interests complement my own. Together we developed a topical outline and a list of proposed contributors, and the project was under way.

The resulting book is intended neither as a text for first-level courses nor as a guide for specialists, but instead as a text for second-level undergraduate or graduate courses and as a reference work for instructors and advanced students. More than 1100 references are given. The chapters, which summarize significant knowledge in their respective areas, are intended to offer balanced, though not exhaustive, coverage and to suggest areas for future study. The final chapter surveys the current state of the discipline of functional morphology, and evaluates present-day approaches and limitations. With minimum use of jargon the authors have endeavored to make their contributions interesting and readable without avoiding necessary detail.

We assume that the reader has a general background in vertebrate systematics, structure, and development, as well as in basic physics and physiology. We do not presuppose advanced training in mathematics, paleontology, biochemistry, or ultrastructure. Although topics like swimming and flying can hardly be presented at the secondary level without mathematics, technical sections can be skimmed without losing the progression of ideas.

A principal strength of this volume is its multiple authorship; indeed, the book could have been written in no other way. We have not tried to eliminate the individuality that results from diverse preferences and styles, and from the varied nature of the subject matter. Nevertheless, we recognize that multiauthored works can be flawed by insufficient balance and integration. Each editor reviewed many chapters, and the contributors exchanged drafts of various chapters or portions of chapters.

All of the coeditors gratefully acknowledge the expertise of the contributions and the patience and good humor of the contributors. We are also deeply grateful to the many distinguished colleagues who reviewed early drafts of the chapters; they are acknowledged individually at the end of the book.

M.H.

Contents

Functional Vertebrate Morphology

Chapter 1

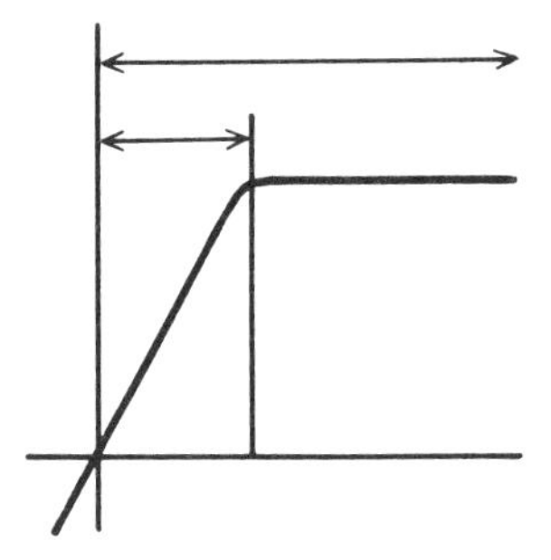

Functional Adaptation in Skeletal Structures

Lance E. Lanyon
Clinton T. Rubin

One of the most distinctive characteristics of vertebrates is that they move. Movement requires that the tissues generate and also withstand the forces involved in locomotion. The generation of force is an active process performed solely by cells. Where it is necessary only to withstand forces applied from some external source, however, this can be accomplished by an extracellular material. Nevertheless, like all components of the body, these load-bearing materials are elaborated, organized, and maintained by cells, and it is only as a result of continuing cellular activity that their structural competence is maintained. In this chapter we consider how cellular activity adjusts the structural properties of load-bearing tissues according to their functional requirements.

The development of an organism from an embryo into a normal adult is a complex process requiring both appropriate genetic instructions and adequate nutrition. However, this input alone is not sufficient to form a functional skeleton, since development occurring under these conditions (as in a paralyzed limb) produces skeletal structures that lack the detailed shape, mass, and arrangement of tissue necessary for load bearing. "Normalcy" of architecture, and the structural competence that it reflects, is achieved and maintained only as a result of an adaptive response of the cells to load bearing. This response is functional adaptation.

The degree of functional adaptability varies with both the structures and the functions concerned. For instance, whereas the mass and architecture of the monkey mandible are profoundly affected by the nature and intensity of the chewing performed (Bouvier and Hylander, 1981a), the presence and form of bony brow ridges cannot be directly influenced by the degree of intimidation — or admiration — that they induce in other monkeys. In the latter case there has been evolutionary selection for an increase in the size of a structure composed of tissue whose sensitivity to load bearing is now superfluous. Thus, the size of structures such as brow ridges is primarily the result of genetic determination. The shape and number of points in deer antlers is also determined genetically, but in this case full expression occurs in response to levels of circulating testosterone (Lincoln, 1975). The specific mechanism or mechanisms that match structure to function within the various tissues are unknown. However, tissue architecture is profoundly affected by mechanical use in structures whose primary role is load bearing. The characteristic conformation of these structures is attained as a result of functionally adaptive modification of the form, mass, and material properties of a basic genetically determined template (Fig. 1-1). Once achieved, this "functional" architecture is maintained only so long as normal activity continues.

The importance of training for athletic achievement is well recognized. A person's potential to be a particular sort of athlete may be genetic, but the ability to excel can only be attained, and retained, as a result of continued appropriate training. It is not so widely appreciated that everyone engaged in functional activity is constantly in a certain state of training and that the "normalcy" of structure is a reflection of the

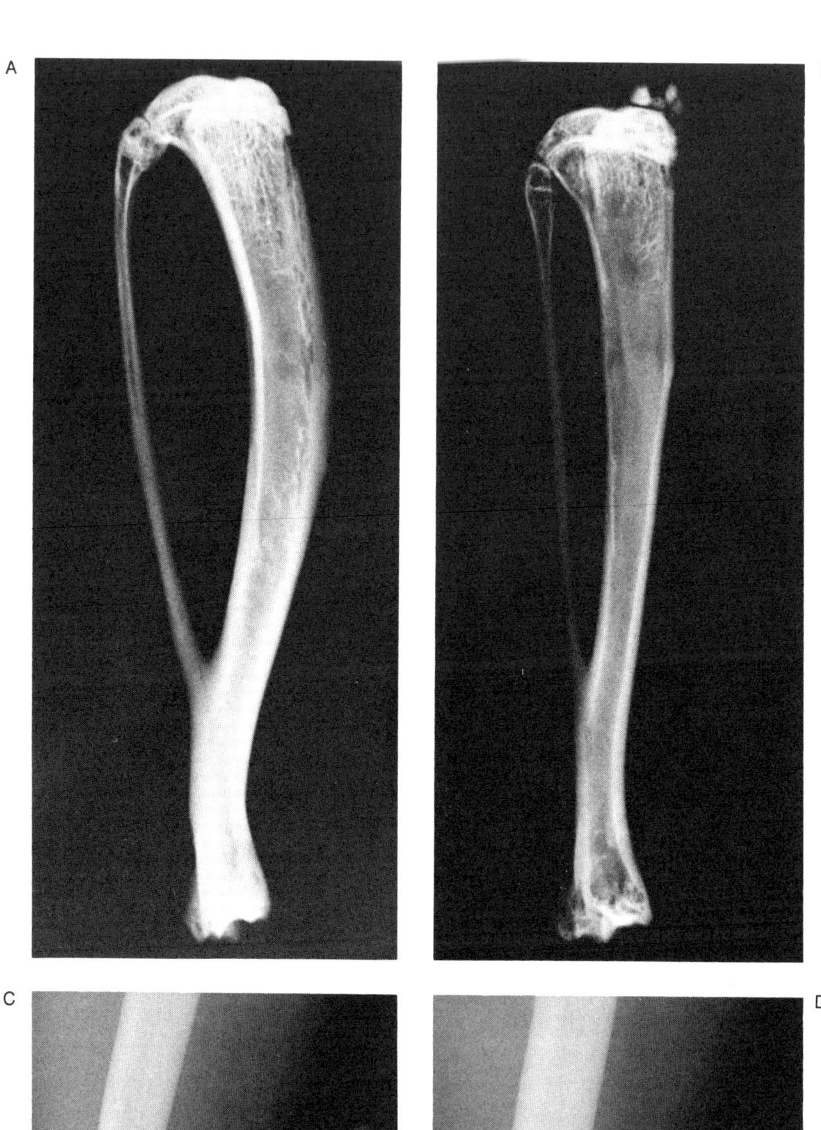

Figure 1-1 Lateral radiographs of tibias and fibulas taken from (A) a normal rat and (B) a rat that had had a unilateral sciatic neurectomy on that side during its growing period. (From Lanyon, 1980.) Relative blackening of the film reflects respective mineral content. The bone lengths are similar, but the functional bone is wider, its cortices are thicker, it contains more cancellous bone, and its curvature is greater. Lateral radiographs of the distal humerus of a male professional tennis player: compare the arm that simply throws the ball into the air (C) with the serving arm (D). The latter shows a 35% increase in cortical thickness. (From Jones et al., 1977, in *Journal of Bone and Joint Surgery.*)

"normal" intensity of physical activity. The achievements of individuals can be seen, therefore, as the cumulative result of functional adaptation within all their relevant tissues and organs. Bilateral symmetry reflects symmetrical use of the body, whereas vigorous use of the limbs on one side results in larger muscles and bones in those limbs (Jones et al., 1977; Montoye et al., 1980). Conversely, relative or complete underuse causes both the bones and the muscles of that side to become smaller (Fig. 1-1).

Structural Requirements of Load-Bearing Tissues

The loads that the structural elements of the body must withstand can be divided into (1) longitudinal tension and (2) everything else, namely, compression, bending, and torsion. Muscles, along with the tendons responsible for transmitting the force of their contraction, are self-aligning under load and hence are required to withstand only longitudinal tension. Apart from the anatomical requirement that they be a certain length and attach in specific locations, there is little need for them to be of any particular shape. Provided that they do not stretch unduly and that they have enough fibers to bear the peak loads applied to them without damage, there is also no requirement for them to retain their shape under load. By contrast, the requirements for the rigid, compression-bearing elements of the skeleton are not only that they be able to bear the diverse loads imposed upon them, but also that they retain their shape while doing so.

The principal responsibility of any structural component of the body is that it bear the loads of coordinated activity without breaking, failing, or sustaining so much damage that its continued load-bearing capacity is impaired. Although the tension-bearing extensor appodeme in the locust appears to be loaded to within 15% of its breaking strength (Vincent, 1980), the margin of safety between the working range and the danger level in vertebrate skeletons appears to be at least two times (Wainwright et al., 1976; Alexander, 1981; Rubin and Lanyon, 1982b, 1983c). Thus, all the loads incurred during coordinated activity and most incurred during uncoordinated accidents can be withstood without immediate damage. The majority of bone fractures occur as sudden traumatic events brought about by collisions with trees, rocks, automobiles, and so on. The more robust the skeleton, the greater the proportion of these accidents that can be sustained without damage. However, tissue is metabolically expensive both to produce and subsequently to transport, and so there is a trade-off between building structures that are massive and essentially indestructible and building those that are economical to produce and move but in greater danger of failure. The elements of this compromise underlie the design of all the load-bearing organs in the body.

Consequences of Loading Body Tissues

The immediate consequence of loading any structure is that it deforms. The amount of deformation depends on the magnitude and direction of the load applied and on the mass, arrangement, and physical properties of the material that is being loaded. The overall deformation can be resolved into strains. Strain is the ratio of the change in any dimension to the original dimension. Thus a linear strain of 1 means a doubling of a linear dimension, whereas a strain of 0.001 (1000 microstrain) means a 0.001-mm change of length for every mm of original length (Fig. 1-2).

The deformation caused by external loading is resisted by forces generated within the tissue (the stresses). At equilibrium, these stresses equal the external load applied per unit area of the load-bearing tissue. A typical stress : strain curve for a biological material is shown in Figure 1-3. The safest level of loading is within the linear portion of the curve where stress is directly proportional to strain. This is called the elastic region, since by definition (although not by popular usage) *elastic* means that the deformation sustained on loading will be completely recovered on unloading. The slope of this linear portion of the curve (or the tangent to it if it is not quite linear) is the elastic — or Young's — modulus. This modulus represents one of the most important physical properties of the material. The steeper the slope of the

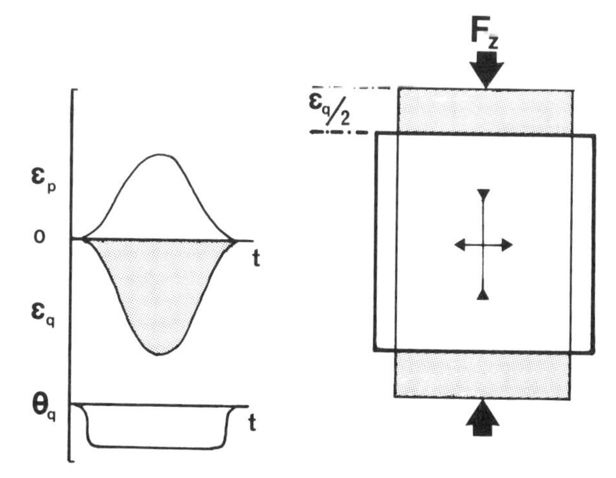

Figure 1-2 Load:strain relationships in a simple rectangular structure. When loaded by an axial force F_z, it compresses in the longitudinal direction and stretches in the transverse direction. The compressive strain (ϵ_q) is the decrease in length *(shaded area)* compared with the original length. The tensile strain (ϵ_P) is the increase in width over the original width. The graph *(left)* shows these strain changes during the application and removal of load: tensile strain positive, compressive strain negative *(shaded)*, and the orientation of the compressive strain (θ_q) to an arbitrary axis *(bottom)*. These variables are the same as those shown for the dog during walking in Fig. 1-8.

stress: strain curve, the greater the resistance to deformation and thus the greater the material's stiffness.

Loading and unloading curves usually differ slightly and enclose an area known as a hysteresis loop (Fig. 1-4A), which is proportional to the energy dissipated in the structure during the load-unload process. When loading is confined within the elastic region, the unloading curve ends at the point from which the loading curve started, indicating that no damage was caused and that the energy was lost as heat. As long as functional loading is confined to this elastic region, the tissues are not subjected to loads from which they cannot recover.

Coral skeleton is a very brittle material that behaves elastically until it breaks. Tougher materials, such as bone, behave in an (essentially) elastic manner up to a certain point, beyond which substantial deformation occurs without proportional increase in stress. This nonlinear behavior implies that nonrecoverable damage is being caused within the tissue (Fig. 1-3). The strain at which this disruption starts to occur is the point of yield, and the stress at this stage is the yield strength of the material. Loading beyond the material's yield strength causes plastic deformation until ultimate failure occurs and the material actually separates into fragments.

Obviously, it is safer for a material to be loaded within the plastic yield region than to complete failure, since internal repair can occur more easily within a structure whose gross integrity has been retained. In bone considerable energy can be absorbed and dissipated by causing structural damage without causing actual separation into fragments (Burstein et al., 1972). This is because bone has a hierarchy of structural arrangements that tend to prevent cracks from propagating and thus make the material tough as opposed to brittle. The ease with which a crack propagates depends very much on the radius of curvature of its tip. The stress at the tip of sharp cracks is very high and may exceed the yield strength of the material; blunt cracks, on the other hand, require considerable energy input before they can propagate. Should a crack start to propagate in bone, it will soon, depending upon its size, encounter either a collagen-mineral interface, a canaliculus, a lacuna, a vascular channel, or the cement line surrounding a haversian system (Fig. 1-5). If the crack encounters a void, its tip will be blunted; if it encounters one of these tissue interfaces, its course will be diverted and its energy dissipated. A measure of the amount of energy required to propagate a crack through a material is an indication of its toughness. Although crack-arresting features make it safer for bone to be occasionally loaded into its yield range, each such event nevertheless causes some irreversible change in the bone's shape and some disruption of its material.

Fatigue Failure

The skeleton is at risk not only from immediate failure during single loading incidents and from damage accumulated during repeated loading into its yield range, but also from fatigue. In some materials, specifically ferrous metals, there is a

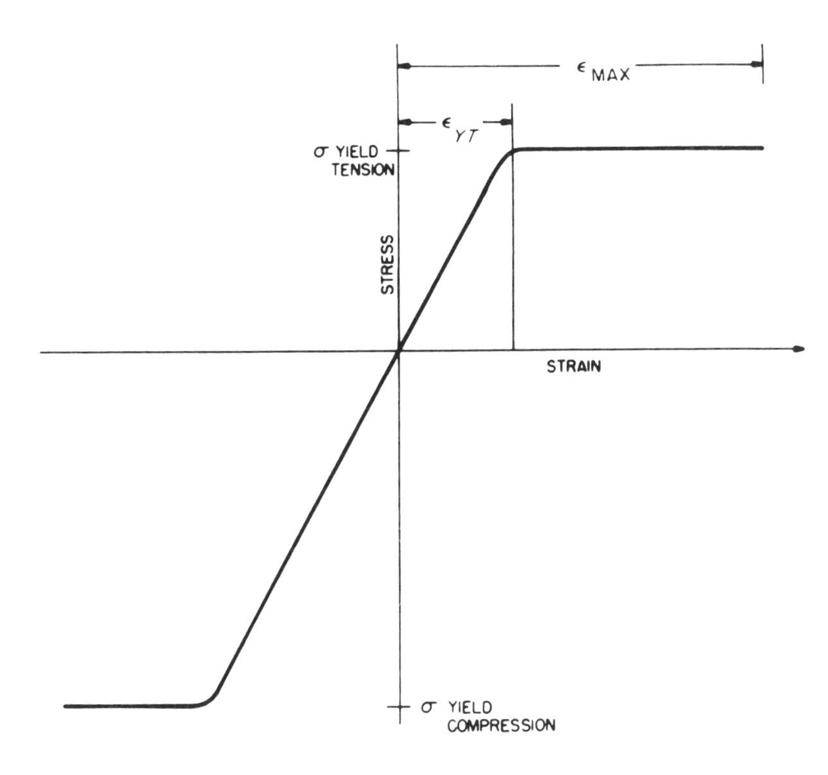

Figure 1-3 Idealized stress : strain curve for bone loaded in tension and compression. The modulus of elasticity in this tissue is the same in tension and compression (slope of the linear portion of the curve). The area beneath the linear portion represents the amount of energy that can be absorbed and recovered without permanent damage. The area beneath the curve in the yielding region shows the much larger amount of energy dissipated by causing damage while still retaining the gross integrity of the structure. The strain at which the curve flattens out (great increase in strain for small increase in stress) is the yield strain of the material (6800 μE for bone). At a certain strain (15,700 μE), the structure fragments; this is the ultimate strain of the material. The yield strength of the material is the stress at the yield point (129 MPa in tension, 284 MPa in compression). σ = stress; ϵ = strain; ϵ_{YT} = strain at yield. (Values from Burstein et al., 1972; Carter et al., 1981a,b.)

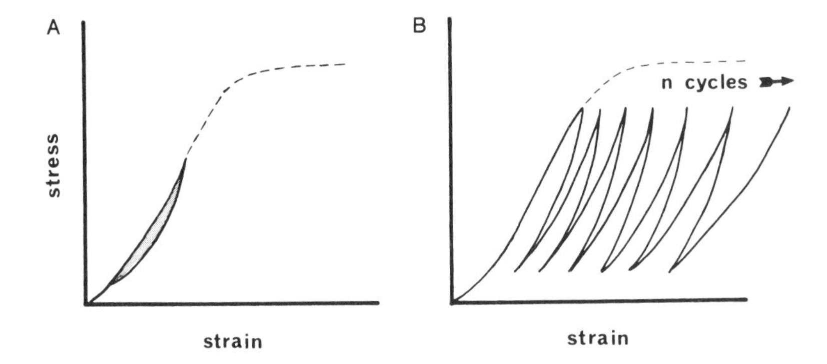

Figure 1-4 Monotonic stress : strain curves *(dotted lines)* and load : deformation curves *(solid lines)* of two specimens being cyclically loaded. A specimen loaded at the lower end of the stress : strain scale (A) sustains no internal damage; the repeated loading curves are superimposed and the hysteresis loop returns to the starting point after each cycle. A specimen being deformed to a greater extent but still below the yield point (B) sustains fatigue damage; the hysteresis loops are open and do not return to their starting point.

Figure 1-5 A sharp crack propagates easily through a homogenous material *(bottom)*, but less easily through an inhomogenous material (*middle* and *top*). Propagation is halted if the crack penetrates a void, which effectively blunts its tip, or if it encounters a discontinuity in the tissue. These arrangements increase the material's toughness by increasing the energy required for cracks to propagate.

strain level below which an infinite number of strain cycles can be sustained without damage. This strain level is termed the endurance limit. If the material is intermittently strained above the endurance limit or if it is a material, like aluminum, that has no endurance limit, then even strains that never reach the yield point can produce dislocations and microcracks that propagate until, eventually, a fracture is generated across the whole section. This type of damage is called fatigue fracture. As the amount of fatigue damage accumulates, the stress : strain curve for each unloading cycle does not return to its origin and the hysteresis loops open (Fig. 1-4B). Permanent deformation is thus produced even though the monotonic yield strain is never reached. The amount of damage accumulates and increases with each additional loading cycle as the effective load-carrying area is decreased. The larger the applied strain, the further the cracks will advance with each load application. The fatigue life of a structure is expressed as the number of strain reversals it can sustain without causing failure, and this is usually expressed as a curve showing the number of cycles to failure. This is an s-n curve (Fig. 1-6).

Because the relationship between strain and number of cycles is exponential, a rapid increase in damage can be caused by a comparatively slight cyclical overload (a doubling of the applied load causes a hundredfold loss of fatigue life). Recent studies on fatigue in bone suggest that it has no endurance limit and that its fatigue properties are very poor (Carter et al., 1981a,b). Indeed, if the logarithmic relationship between strain reversals and the fatigue life found in these experiments is extrapolated to the normal physiological levels, a strain range of 3000 microstrain would result in a fatigue life of only 100,000 cycles for the humerus in flying geese and the limb bones of galloping horses (Rubin and Lanyon, 1982a,b). This is dangerously short for these migratory and cursorial animals, particularly those flying for long distances over water. However, other experiments suggest that bone is more resistant to fatigue failure than is indicated by Carter's pessimistic predictions (King and Evans, 1967; Swanson, Freeman, and Day, 1971; Lafferty and Raju, 1979) and that possibly this tissue has an endurance limit in vivo (Seireg and Kempke, 1969).

Although the extent of the threat of fatigue failure for biological tissues is as yet undefined, it

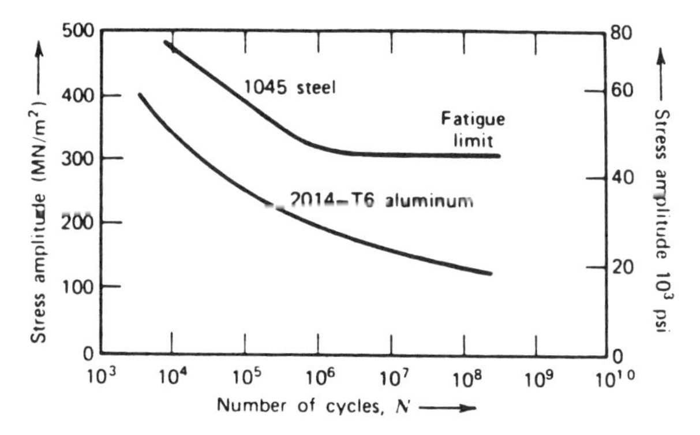

Figure 1-6 Cycles to failure (s-n) curves for plain carbon steel and age-hardened aluminum alloy. At 300 megapascalls (MN/m^2) the steel shows a fatigue or "endurance" limit, that is, a stress below which it will not fail regardless of the number of cycles. The aluminum alloy shows no such limit, but the slope of stress versus log N becomes less negative as the stress decreases. (From Radin et al., 1979; reprinted by permission of John Wiley and Sons, Inc.)

appears to be a real danger. It may be countered in one of three ways: (1) by limiting the strain range during functional loading so that the fatigue life of the tissue is long in relation to the animal's lifespan; (2) by adjusting the properties of the material so that it becomes stiffer and/or tougher; and (3) by constantly renewing the structure's constituent material so that the fatigue life of the structure exceeds that of the tissue. Options (1) and (2) are relevant to both fatigue and monotonic failure and will be considered later. Option (3) is relevant primarily to fatigue failure and is considered first.

Mechanisms of Bone Remodeling

One disadvantage of using a tissue that by nature has a large resistance to deformation is that when structural alterations become necessary, the tissue cannot be molded or expanded from within. Cartilage is sufficiently soft that growth and change in form can be achieved by cellular division and the elaboration of new matrix within the depths of the tissue, which then expands in volume. This interstitial growth is impossible for bone, the extracellular matrix of which is calcified and rigid. Bone can thus grow only appositionally by adding to one of its surfaces. Consequently, change in the structure's shape is normally achieved by a coordinated process of removal of tissue from places where it is no longer needed and deposit of new tissue on surfaces where it is now required.

Formation of bone is essentially a twofold process. First the osteoblasts elaborate a collagenous matrix called osteoid, and then this matrix is mineralized. A number of osteoblasts become trapped within the matrix before it is calcified, and they are then called osteocytes. However, the contribution of these resident cells to subsequent structural remodeling is negligible. Where bone has to be removed, this is performed by large multinucleated cells called osteoclasts. These cells, derived from blood-borne monocytes, attach to tissue that is to be eroded. It is the coordinated activity of osteoblasts and osteoclasts that is responsible for the adjustments of the form and size of bone. Renewal of bone tissue in cancellous regions is achieved by removal of a "packet" of bone from the surface and its replacement with a similar "packet" of new tissue. Within cortices or larger trabeculae, tissue renewal is accomplished by populations of osteoclasts that bore tunnels through the tissue that is to be removed (Fig. 1-7). In long bones these "cutting cones" of cells usually start from the endosteal surface and run obliquely in a longitudinal direction toward the periosteum. Each tunnel is rapidly filled in with circumferential layers of new bone by a much larger team of osteoblasts. In this way the existing tissue becomes secondarily remodeled with cores of new tissue, each one of which is called a haversian system, or secondary osteone (Jaworski, 1981; Currey, 1982).

There is considerable debate as to the requirement for secondary remodeling and the structural significance of the direction of the haversian systems. On the basis of ultimate strength, yield strength, and fatigue resistance, the remodeled haversian bone is inferior to undamaged primary bone (Currey, 1959; Hert et al., 1965; Carter, Hayes, and Schuman, 1976). Nevertheless, it is probably superior to primary bone that has accumulated microdamage. The tissue renewal involved by this remodeling extends the fatigue life of the whole bone to many times longer than that of its constituent tissue.

In addition to its structural role, the skeleton is important as a mineral reservoir. During periods of calcium drain, such as lactation, pregnancy, and antler growth (Hillman, Davies, and Abdelbaki, 1973; Rasmussen, 1977), the skeleton is raided for its calcium regardless of the structural damage this may cause. The bone's cortex usually becomes thinner (as a result of endosteal resorption) and more porotic (as a consequence of incomplete infilling of resorption cavities; see Fig. 1-14B). If calcium deprivation is temporary, restitution of bone mass usually occurs once the period of deprivation is over. In some animals, including man, it is probable that a certain amount of remodeling is constantly necessary to "even out" daily dietary variations in available calcium. Reduction of bone mass is not achieved by resorption alone, nor is its subsequent restitution achieved solely by formation. Rather, each adjustment is achieved by the balance between re-

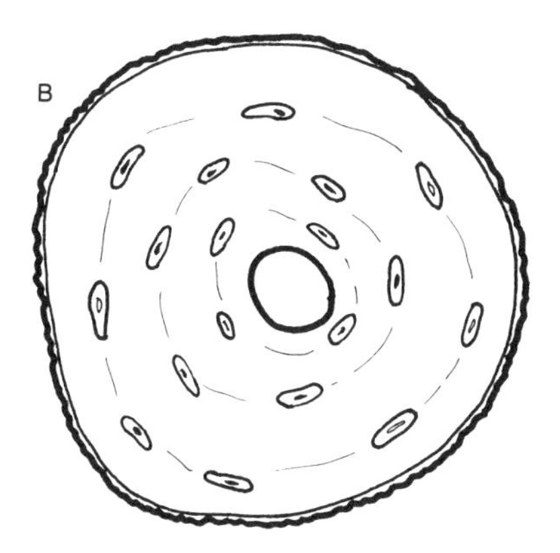

Figure 1-7 A: The formation of a haversian system, shown in longitudinal section. Osteoclasts are eroding a circular tunnel through bone tissue; a larger population of osteoblasts, which line the walls of the eroded cavity, subsequently fill it with successive circumferential lamellae of new bone. The innermost circle is therefore the most recent addition. B: The appearance of this cylindrical core of bone tissue when cut in transverse section.

sorption and formation. Thus, paradoxically, a feature of bone loss may be an actual increase in the absolute amount of bone formation accompanied by an even larger amount of resorption. Hence, adjustments of mass may be achieved by changing the balance of remodeling at either normal or augmented levels of turnover.

Design Strategies for Load-Bearing Tissues

In any structure the product of load, geometry, mass, and material properties is the strain produced within the tissue (Fig. 1-2). It is desirable to hold the strains below levels that cause substantial fatigue damage. For an organism to be sparing with the amount of tissue used in its skeleton and yet not incur dangerously high functional strains, the options include: (1) regulation of the manner and magnitude of skeletal loading (the load); (2) adjustment of the material properties of the load-bearing tissues (the material properties); (3) alignment of tissue in relation to the direction of loading (the geometry); and (4) adjustment of the mass of tissue (the mass).

Regulation of Applied Loading

Self-aligning tensile elements, such as ligaments and tendons, are required to resist loads applied to them from only one direction. Accordingly, the difference between operational strains and those likely to cause damage can be small. The rigidity of the hard tissue components of the skeleton correlates with a far wider range of loading conditions. Limb bones derive most of their loading directly or indirectly through articular surfaces and the insertions of tendons and ligaments. Joint surfaces directly load a restricted region (predominantly in compression), and tendons and ligaments attach to fairly localized insertions (loading them predominantly in tension). These loads are then distributed throughout the structure. Such complexity in the manner of loading is not intrinsically uneconomical. Lack of economy arises only if the tissue is not advantageously placed with regard to load distribution or if the patterns of loading are diverse. The more diverse each structure's manner of loading, the less it can achieve a favorable orientation with respect to each load and thus the more massive and less "tuned" the bone must become in order to preserve an adequate margin of safety. It follows that reducing the diversity of loading is an effective strategy for achieving economy of tissue.

In order to investigate the mechanical behavior of bones during functional loading, strain gauges have been applied to the limb bones of a number of species to record changes in strain during walking, running, or flying (Lanyon, 1973; Lanyon et al., 1975; Lanyon and Bourne, 1979; Lanyon, Magee, and Baggott, 1979; Rubin and Lanyon, 1981, 1982a,b, 1983b; Biewener et al., 1983). In terrestrial locomotion each stride typically involves a sharp increase in strain as the limb is placed on the ground and a return to zero as the leg again enters the swing phase. During each stance phase, however, the orientation and the relative magnitude of the strains on the different cortices of the bone remain remarkably constant (Fig. 1-8). It appears, therefore, that the forces generated within the bones of the limb and those transmitted through them interact throughout the load-bearing period to provide a restricted load distribution (Rubin and Lanyon, 1982b). This constancy of loading is maintained not only throughout each stance phase at one speed but also throughout the animal's entire speed range. This restricted manner of loading should allow for a unique (and optimal) structural solution at each skeletal location.

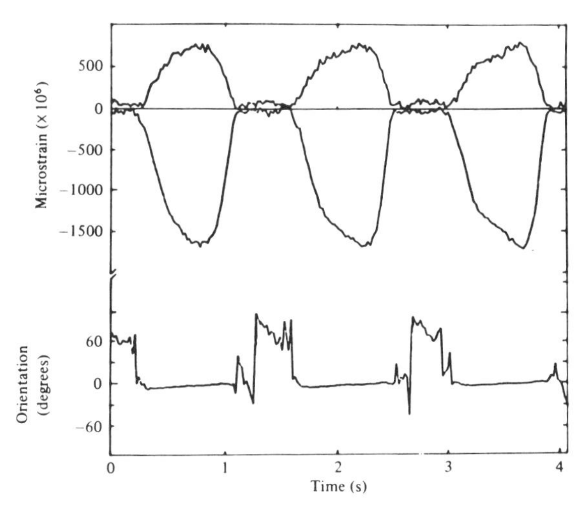

Figure 1-8 Peak principal tensile *(upper trace)* and compressive *(middle trace)* strains on the caudal surface of the radius of a dog, together with the orientation of the principal compressive strain to the long axis of the bone *(lower trace)*, taken while the animal walked at 4.2 kph (see Fig. 1-2). Although both the tensile and compressive strains change quickly during the stance phase of the limb, the strain orientation remains remarkably constant. (From Rubin and Lanyon, 1982b.)

Most of these strain gauge experiments were conducted with the subjects walking or running in a straight line on a flat surface, usually a treadmill. Turning, accelerating, and decelerating would certainly induce more diverse strain patterns. However, it appears to be at least a fortuitous circumstance that during steady-state locomotion, bone loading is fairly constant. For maximum economy, however, not only should the diversity of loading be restricted, but the magnitude of loading should also be kept low. Measurements of strain from the limb bones of dogs and horses indicate that behavioral mechanisms for the regulation of load do exist. During walking, peak strains within the limb bones increase with increasing speed and then jump incrementally when the animals change to a trot. However, when the change is from a trot to a canter, the peak strains in the bones (and the peak force on the ground) are reduced by as much as 40% (Fig. 1-9; Rubin and Lanyon, 1982b). Only at the highest speeds (fast gallop) are the peak strains again as high as those engendered in the trot. This regulation of limb loading (limiting the stress developed in tendons, ligaments, muscles, and bones) may be associated with the recruitment of the muscles of the back and trunk that occurs when the animal breaks into a canter. Although the primary benefit from gait change may be muscular or metabolic, change of gait is also a behavioral adaptation by which increased speed can be attained without placing additional requirements on either the material or structural properties of the load-bearing elements of the limbs.

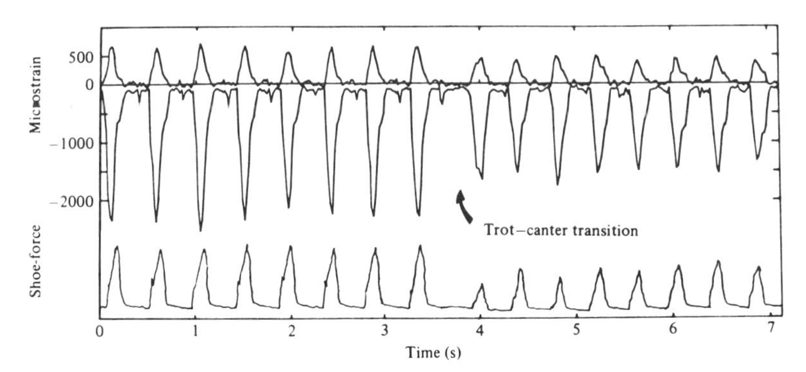

Figure 1-9 The tensile *(upper trace)* and compressive *(middle trace)* strains at the caudal surface of the radius of a horse during transition from trot to canter at constant speed on a treadmill. The lower trace is from a load-sensitive shoe attached to the same forefoot one year later. The 42% decrease in peak compressive strain at the transition is reflected by a similar decrease in peak force recorded from the shoe. (From Rubin and Lanyon, 1982b.)

Adjustment of Material Properties

Bone is a composite material consisting predominantly of collagen and mineral. The mineral component imparts stiffness, but by itself is very brittle, requiring little input of energy to propagate a crack and instigate failure. Collagen, on the other hand, is extremely tough, and it is practically impossible to generate a crack through it. Bone has a requirement to be both stiff (to resist deformation) and tough (to prevent crack propagation). However, there is a trade-off between the degree of these qualities that can be attained in one material because the final properties of the tissue are a consequence of the proportions of its organic and inorganic components. Comparatively small changes in the mineral content of bone tissue have substantial effects on its properties as a material (Currey, 1969, 1975). These have best been demonstrated in a comparison of bone tissue from the bovine femur, the deer antler, and the whale tympanic bulla (Currey, 1979). Of these three bones, only the femur retains traditional load-bearing responsibilities. The particular roles of the antler and tympanic bulla are now best served by rather different material properties (Table 1-1).

Antlers probably produce their greatest effect by their imposing appearance. However, antlers are also used in battle where they must first resist impact and then not be broken by the head twisting that follows. The stiffness required for this is only moderate, but the amount of work they can absorb without breaking should be as high as possible. This requirement for toughness can be achieved at the price of reduced stiffness by a low mineral content (59% compared with the 67% of the femur). The tympanic bulla of the whale carries practically no load but acts instead to isolate the middle ear from sound reaching it by any route other than the chain of auditory ossicles. To do this its acoustic impedance must be high, which can best be accomplished by high density and a high modulus. Its mineral content (86%) causes its modulus to exceed that of the femur by 220%. As a result, the bone is brittle, but this is of no consequence since it carries no load. It is the traditional load-bearing responsibilities of the femur that require the compromise between extremes of toughness and stiffness. The result is a desirably high resistance to bending (Table 1-1).

The large changes in physical properties that can be achieved with relatively small adjustments in mineral content suggest a possible means for the adaptive response. It will be remembered that the internal remodeling process in bone operates by physically removing old tissue and replacing it with new. The first material deposited is collagen, which subsequently becomes mineralized. This mineralization commences almost immediately, but it normally takes several weeks—or months—to complete, and its rate and extent are under cellular control. Accordingly, it seems to be possible for the cellular population to adjust the mineral content of the tissue as an adaptive response to changes in load bearing. There is, however, no evidence to suggest that this option is exercised.

Table 1-1 Mechanical properties and mineral content of three diverse types of bone.

Feature	Antler	Femur	Bulla
Work of fracture (% of antler):	100%	28%	3%
Bending strength (% of femur):	73%	100%	13%
Modulus of elasticity (% of bulla):	24%	43%	100%
Mineral content (%)	59%	67%	86%

Source: Adapted from Currey, 1979.

Instead, where differences in mineral content do exist, they appear to be related primarily to the rate of tissue turnover. This remodeling rate is high in areas where the functional strain level (and thus the degree of cortical damage) are also high (Lanyon and Baggott, 1976). The cumulative result of high tissue turnover is that the mineral content (and therefore the elastic modulus) of the tissue is lower than it is where turnover is less active. However, this does not appear to be part of an adaptive response to load bearing. Woo et al. (1981a,b) found that after young swine had exercised for a period of twelve months, the improved structural properties of the femur resulted solely from an increase in the amount of tissue present rather than from a change in its material properties. Similarly, the decrease in structural properties that follows reduced loading results primarily from a decrease in the amount of tissue present, with little change in its material properties. It was once thought that even though the modulus of fully mineralized haversian bone is no greater than that of primary bone, the different arrangement of tissue might produce better fatigue properties. It has since been shown (Carter et al., 1976) that this is not so, and that the fatigue life of haversian bone is inferior to that of primary bone. Thus, internal bone remodeling is not an adaptive response to improve the material properties of the bone tissue. Nevertheless, although such remodeling improves neither the strength nor the fatigue resistance of the material, it does increase the fatigue life of the structure.

In bones bearing usual loads, the tissue's material properties are fairly constant. These properties are similar not only throughout the skeleton of each individual animal but also throughout the skeletons of animals of widely varying size (Table 1-2).

The appropriateness of the material properties of bone tissue, both within "traditional" bones such as the femur and within specialized ones such as deer antlers and the whale bulla, appears to be the result of natural selection rather than of functional adaptation. Adaptive alterations in the structural properties of the "traditional" components of the skeleton are achieved primarily by

Table 1-2 Bending strength of cortical bone from the femurs of selected small and large animals.

Animal	Mass (kg)	Bending strength (MPa)	Source
Chipmunk	0.09	263	Biewener, 1982
Squirrel	0.14	219	Biewener, 1982
Rat	0.52	182	Engesaeter et al., 1979
Human	70.00	186	Sedlin and Hirsch, 1966
Horse	400.00	186	Yamada, 1970
Cow	700.00	264	Burstein et al., 1972

the regulation of tissue mass and by the adjustment of tissue architecture.

Functional Loading

Alignment of Bone Tissue The strategic alignment of the rigid elements of the skeleton in relation to their direction of customary loading is achieved by progressive remodeling of the tissue until it conforms to the appropriate shape. The more diverse the loading, the more blurred and less evident the structural solution. Accordingly, this effect is easiest to see where the manner of loading is restricted, as, for instance, in the trabeculae of cancellous bone.

Trabeculae occupy the expanded extremities of long bones and provide the majority of the tissue in the carpus, tarsus, and vertebral column. Their function is exclusively concerned with load bearing and, unlike the whole bones of which they are part, they have no individual requirement to be any particular shape or to provide joint surfaces or anchorages for adjacent tissues. In the same way that a complex shape can be resolved into a great many short straight lines, the complex loading of a block of cancellous bone tissue can be resolved at the level of each trabecula. It is possible, therefore, for each element within the cancellous lattice to be placed so as to encounter loads that are either predominantly compressive or predominantly tensile.

In young animals, before the onset of functional loading, cancellous bone has a random honeycomb architecture. As functional loads are applied, tracts of trabeculae become discernible that are arranged in the same directions as the principal stresses within the tissue (Fig. 1-10A). Even when the loading of a structure is simple, the pattern of strain induced within it is complex. In response to a simple compressive load (Fig. 1-2), a block of material compresses in the direction in which the load is applied and expands in the direction at 90° to it. The compressive and tensile directions in the deformed material are called the principal strain directions. In any direction other than these, there is a shear strain component which is at a maximum at 45° to them. If this cube of material is not a solid block but instead consists of a lattice of struts, then those aligned with the principal tensile strain are stretched, those aligned with compression are compressed, and those aligned obliquely are bent (Fig. 1-10). The significance of this arrangement is that columns and struts can carry the greatest load with the least strain when they are loaded axially.

Since material can withstand far less load in bending than in axial loading, the danger for slender compressive elements is not that they will fail in compression but that they will buckle and fail in bending. Once buckling has occurred, dangerously high strains can be engendered by comparatively small loads. To prevent a column or sheet from buckling under compression, cross braces are introduced, thus making a lattice. For these cross braces to be most effective and not themselves subject to bending, they are aligned at right angles to the primary, compression-bearing trabeculae and are thus in line with the directions of principal tension within the tissue. The coincidence of trabecular directions with those of the principal strains has been called the trajectorial arrangement of bone architecture, since (in an isotropic material at least) the directions of the principal strains coincide with those of the principal stresses or stress trajectories (Fig. 1-10C).

Although the requirement for cross braces appears straightforward, the actual nature of load distribution within cancellous tissue is far from clear. When a compressive member starts to buckle, the cross braces will be loaded in tension or in compression, depending on the direction of buckling. In some locations, particularly where there are muscular attachments, the cancellous bone tissue must also be subjected to substantial tensile, bending, and torsional loads. The manner in which loading of the volume of tissue becomes resolved into loads at the trabecular level remains to be determined. However, despite this uncertainty, the original suggestion by Wolff (1870) that trabeculae follow the lines of "functional stress" appears to be borne out by subsequent measurement (Lanyon, 1974), correlation (Oxnard, 1972), and analysis (Koch, 1917; Hayes and Snyder, 1981). It is likely that this orientation of trabeculae enables cancellous bone to withstand large loads with great economy of material.

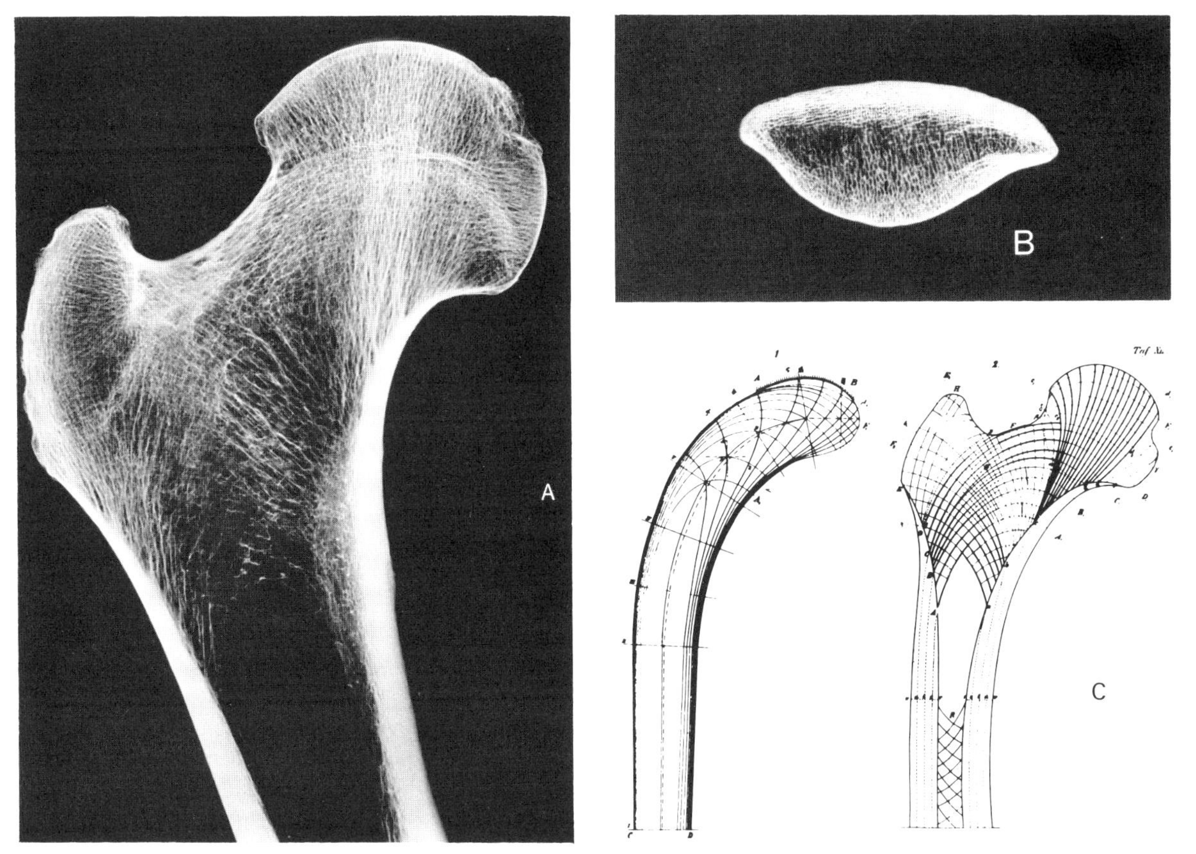

Figure 1-10 The lattice orientation of trabeculae. A: A radiograph of the proximal end of a human femur; B: a transverse microradiograph of a human patella. C: The distribution of loads and stresses in Culman's crane, which are similar in their course and distribution to the trabeculae in the human femur (reproduced from Wolff, 1870, with original symbols).

The structural appropriateness of the orientation of trabecular bone has led to investigations of the orientation of structural features within cortical bone. For haversian systems to have a functionally relevant orientation it is necessary for the osteoclastic cutting cones, which determine their direction, to pursue a functionally relevant course. Although there is a general feeling that this is so, there is scant hard evidence. In the only study known to us where the direction of functional strain was compared with that of secondary osteones (Lanyon and Bourne, 1979), there appeared to be some agreement between off-axis locomotor strains and off-axis haversian systems, but a causal relationship was not demonstrated.

Another method used in attempts to relate the "grain" of the tissue to directions of presumed functional strain was the split-line technique. This involved demineralizing the bone surface and then inducing it to split by inserting a round needle. The direction of the split was supposed to follow the direction of the collagen in the bone. Bouvier and Hylander (1981b) have demonstrated, however, that the split-line pattern is unrelated to the direction of functional strain.

Attempts have also been made to correlate the orientation of the crystalline components of the hydroxyapatite with that of strain (Bacon, Bacon, and Griffiths, 1980). This study seemed to show an agreement between the orientation of apatite crystals and the lines of assumed muscle pull. However, since the actual directions of strain within the tissue were not determined, the significance of this pattern is not clear.

Curvature of Whole Bones The shape and architecture of each bone is a compromise among requirements for shape (such as length and appropriately contoured joint surfaces), the need for accurately located muscle attachments, and an arrangement and mass of tissue consistent with both economy of material and structural competence as a load bearer.

The position and loading of whole bones (unlike that of their internal cancellous architecture) assures that, in addition to axial loads, they will also be subject to bending and torsion. The basic form of each bone is genetically determined, and even in individuals paralyzed during development the bones assume approximately normal length, tubular shape, and characteristic anatomical relationships. Thus, with adequate nutrition, the genetic program is capable of producing a skeleton sufficiently well shaped to allow developmental organization to occur. However, there are many other features, particularly those associated with load-bearing capacity, that develop normally only under the influence of normal functional activity. In long bones these features include the presence and size of various processes and tuberosities (Howell, 1917; Murray, 1936; Stinchfield et al., 1949; Moore, 1965), the presence, size, orientation, contiguity, and areal density of trabeculae within cancellous regions (Hayes and Snyder, 1981), and the girth, cortical thickness, cross-sectional shape, and longitudinal curvature of the shaft (Ralis et al., 1976; Lanyon, 1980).

A long bone is similar to a trabecula in that it can carry the greatest load with the least strain if it is loaded axially. However, a long bone has the shape constraints just mentioned, it is loaded from various angles, and it cannot be cross-braced. It is inevitable, therefore, that some bending is imposed on the bone's shaft. The strain developed within the bone depends not only on the degree of external bending applied, but also on bending moments engendered internally as a result of applying an axial load to a structure with the bone's longitudinal shape and cross-sectional geometry.

Figure 1-11 illustrates the components that combine to produce the final strain at the midshaft of the radius in a quadruped. At midstance (and peak strain) there is little or no bending as the result of limb angulation because the front limb is then nearly perpendicular to both the ground and the trunk (Biewener et al., 1983). Despite this, the greatest proportion of the strain at midshaft (85%) is due to bending, two thirds of which results from the bone's longitudinal curvature. One of the consequences of long bones' having a longitudinal curvature is that it determines the manner in which the bone deforms in response to end-to-end loading. If the bones were straight and axially loaded they would be capable of carrying large loads with little strain right up to the point of buckling, when they would bend and fail in a practically random direction. When there is a curvature the degree of strain for small loads is comparatively high, and it gets progressively greater with increasing loads up to the point of failure. Although this situation appears dangerous, it has the potential advantage that it is easier to adjust bone form in response to a progressively increasing danger than in anticipation of a potential catastrophe that gives little indication of its coming.

Whether or not this hypothesis is valid, there must be some advantage in having a curvature that outweighs the apparent disadvantage of increased functional strains. Such curvatures are characteristic of many long bones, and since they appear not to develop in paralyzed limbs (Fig. 1-1), they seem to be a specifically engendered response to functional activity (Lanyon, 1980). Frost (1973), among others, following the reasoning that bone form is primarily adjusted to reduce tissue strain, postulated that curvatures develop so that the bending moments they engender cancel those due to eccentric muscle loading and limb angulation to the ground. However, as we have seen, this does not seem to be the case. On the contrary, it could be postulated that the curvatures of some bones are developed specifically to generate strain. If continued intermittent strain were to provide some (as yet unidentified) benefit to bone tissue, then this strain could be engendered by curvatures in regions where other requirements (such as the need to have sufficient surface area for muscle attachments) dictate structures of sufficient size that customary loading would otherwise induce only small strains.

In locations where such requirements are not

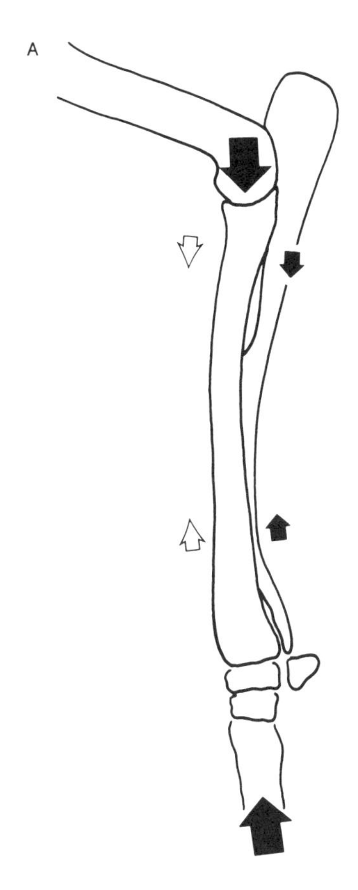

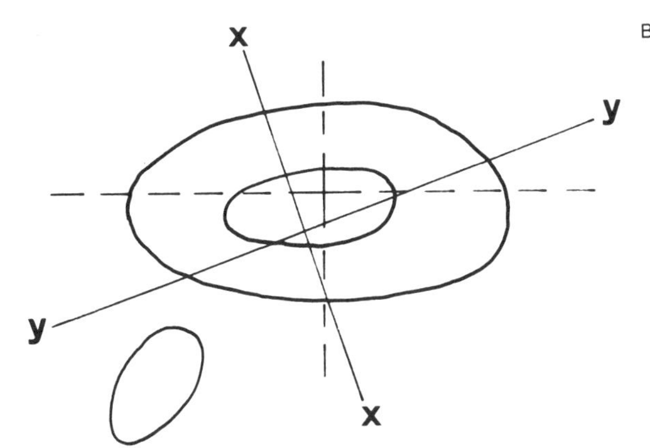

Figure 1-11 The radius and ulna of a sheep (other ungulates are similar). The lateral view (A) shows the bones' slight craniocaudal curvature. Axial loading *(large black arrows)* causes bending, which places the cranial surface in tension and the caudal surface in compression. Tension in the flexor muscles *(small black arrows)*, which are active during weight bearing, contributes to this bending rather than opposing it. In B, a transverse section of the radius and ulna shows the principal axes. The greatest resistance to bending is in the *y-y* direction; the least is in the *x-x* direction, which is close to the bending direction engendered by locomotor loading. (From Lanyon et al., 1982.)

operative or where customary loading is sufficient so that strains are already at an appropriately high level, curvatures would not be necessary. This situation is illustrated by the metacarpus of the horse, which is practically straight and at midstance during steady-state locomotion maintains essentially the same perpendicular alignment to the ground as the radius. However, because it does not have a curvature, the longitudinal strains produced within it are compressive all the way around its circumference, thus indicating that it is subjected to almost axial compression (Biewener et al., 1983). Since there is little or no bending, these strains are also much lower than they would be if there were a longitudinal curvature. This situation changes, however, during acceleration and deceleration when external bending moments in the metacarpus increase, thus increasing the strains within the bone until they equal those in the radius (whose strain level is unaffected by acceleration). If the metacarpus possessed a curvature or were a less substantial structure, then the strains within it would be about equal to those in the radius during steady-state locomotion but would become damagingly high during acceleration and deceleration. Thus, when the activities that cause the highest strains at these two locations are considered, it appears that both bones are designed to be in similar danger of failure. The implications of uniform peak functional strains throughout the skeleton are considered further below. Before that, however, we consider the other architectural means by which regulation of strain may be achieved.

Cross Section of Whole Bones The basic tubular shape and circular cross section of long bones is the most economical arrangement to resist both torsion and bending from any direction. This simple, genetically determined geometry is modified by adaptive remodeling so that in a functional limb, midshaft girth, cross-sectional shape, and cortical thickness all reflect the manner and magnitude of loading. The wider the diameter of a tube (and thus the farther away from the centroid the tissue is placed), the larger the structure's resistance to bending. In paralyzed limbs, the bones have narrow midshaft diameters, the "normal" diameter being achieved only under the influence of load bearing. Cross-sectional shape may also be influenced by functional load bearing. The classic case is the cross section of the tibia, which in most species is triangular in the bone's proximal portion. Although the functional significance of this shape is not clear, it is plainly influenced by the presence of the tibial crest anteriorly and the concave bed that is necessary to accommodate the anterior tibial musculature. In the absence of load bearing, the triangular shape does not develop and the bone's cross-sectional shape remains circular.

It is important to distinguish among the various influences on bone remodeling produced by load-bearing activity since such activity involves strain within bone tissue, intermittent tension in adjacent musculature, and pressure on the periosteum. Compressive loads are generally applied to bone tissue through articular cartilage without any intervening cellular membrane. Similarly, when orthopedic devices such as screws and pins are inserted directly into bone, large loads are transferred directly to the tissue because there is no interposed layer of cells. In contrast, comparatively light pressure applied to periosteum produces local resorption. This indicates that blood vessels and muscle masses can mold the surface of the bones they contact. It is by this process that teeth can be moved through the jaws in response to the pressure induced by orthodontic braces. This pressure, acting through the alveolar ligament, causes the tooth to migrate by means of coordinated resorption on one side of the alveolus and deposition on the other.

We can distinguish between the effects of functional adaptation owing to strains acting within the bone tissue and bone remodeling in response to direct pressure on periosteum. Both effects influence bone morphology, and both may be related to the nature of physical activity. However, only the former is an adaptive response to the size and distribution of strains throughout the bone tissue. The cross-sectional shape of bone is influenced by the conformation of the adjacent joint surface, the influence of pressure from local musculature, and requirements for particular geometric properties, such as area to resist axial loads, resistance to bending or second moment of area, and surface area to provide muscle attachment. The best place to assess shape in relation to strain regulation alone is in the midshaft of a long bone, remote both from joint surfaces and from muscle insertions. In order to resist bending, it is necessary to place material as far away from the neutral axis as possible. If bending occurs in only one direction, then a high second moment of area can most economically be achieved by an I beam. The material in the top and bottom cross-pieces provides the resistance to bending and the central element of the beam keeps these cross-pieces apart. A circular cross section is effective because it is essentially an I beam oriented to all directions, and since its continuous shell maintains the opposing cortices apart, no central element is necessary.

However, when bending is confined to one direction, then economy of material can be achieved without reduction of the second moment of area in that direction if the bone's diameter is reduced at right angles to the direction of bending. The bone's cross section would then be an ellipse with its larger diameter oriented to provide the greatest resistance to the bending normally imposed. This interpretation can be applied to the case of the phalanges in bats (see Chapter 8), the zygomatic arch of carnivores, the neural spine of ungulates, and so on.

Surprisingly, however, observation and strain gauge measurement from the radius in dogs, horses, and sheep demonstrate that the reverse may also be true, since the orientation of the elliptical cross section in these bones is directed to provide the least resistance in the direction in which bending normally occurs and the greatest resistance at right angles to it. Strain gauge instrumentation of the sheep radius (Lanyon and Baggott, 1976; Lanyon et al., 1982) suggests that at peak deformation during the stance phase of a stride, the flexor musculature on the bone's caudal surface is active, and that this tension increases the bending moment established by axially loading the bone's longitudinal curvature (Fig. 1-11). Thus, even though we have traditionally believed that most features of bone morphology reduce functional strain, both of these particular features (curvature and shape of cross section) actually appear to increase it. In the sheep, horse, and dog radius and tibia, bending accounts for approximately 85% of the total strain engendered (Rubin and Lanyon, 1982b), and most of that strain is caused by the bone's curvature. It appears, therefore, that orientation of the elliptical cross section acts to confine the direction of bending to the one plane in which the bone's resistance to it is not great, and at the same time provide greater resistance to bending in the plane at right angles to it.

In light of this evidence, it is difficult to preserve the idea that reduction of strain is the primary objective in the architecture of long bones. It is too simplistic to assume that *all* structural strains are detrimental just because high strains can be dangerous. Provided that strains are within the elastic range and do not induce an unacceptable degree of fatigue damage, they do not in themselves constitute a danger to the bone structure. Perhaps intermittent deformation benefits the bone by, for example, increasing the perfusion of the tissue or maintaining a flux of strain-generated electrical charge. The *potential* for damage caused by strains produced during coordinated functional activity must, therefore, be distinguished from the *actual* damage caused by strains produced by loading accidents.

Although safety from fatigue and monotonic failure can each be achieved by regulating functional strains, neither requires that these strains be minimized. Indeed, it seems likely that, in addition to the advantages of tissue economy, there may be other benefits from accepting strains that are above certain critical levels. Functional adaptation is thus required to ensure that (1) strains are sufficiently low so that fatigue life is not unacceptably short (the upper strain limit in relation to fatigue is perhaps 3000 microstrain, probably equivalent to a fatigue life, uncomplicated by remodeling or repair, of over 1 million cycles); and (2) bones are sufficiently robust to withstand a reasonable proportion of loading accidents.

Skeletal fractures usually result from single, massive, aberrant loadings, not from exaggerated versions of the strains produced by normal activity. Thus, one does not break one's leg by running so fast that the bones will no longer stand the increasingly high strains. Instead, one breaks it by accidental collision or uncoordinated activity leading to unnatural loading and disastrously high strains. Naturally, if functional adaptation results in bones that are massive (and functional strains that are low), then the general possibility of fracture is reduced. However, the precise matching of the bone's architecture to its customary strain environment is almost irrelevant in this context. Thus, the relationship between the customary strain pattern and the general robustness of the skeleton is not clear. At least in the radius of the quadrupeds studied, the elliptical cross section of the bone is far more resistant to lateral bending (which is more likely to occur during accidental loading or dangerous cornering) than it is to bending in the normal, craniocaudal (locomotor) direction. In this bone both the curvature and cross-sectional area can be interpreted as providing a high structural stiffness to resist loads applied from an abnormal direction and a lower stiffness to permit or induce high strains owing to bending in the customary direction.

Although superficial examination suggests that the particular load-bearing requirements within different animals are very different (from digging in the ground to flying through the air), at the level of the tissue these requirements are essentially similar. Since the properties of bone tissue (and thus the strain level at which damage occurs) are also essentially similar, it should come as no surprise that in those bones whose primary role is load bearing, the peak functional strains are remarkably similar (2000 to 3000 microstrain, Rubin and Lanyon, 1982b), as shown in Table 1-3.

This constancy in both tissue properties and peak strains implies that regardless of a bone's location, if its principal function is load bearing, selection pressure ensures a similar safety factor between the strains produced during peak physical activity and those likely to cause yield or fatigue failure. This relationship has been called dynamic strain similarity (Rubin and Lanyon,

Table 1-3 Peak compressive strains measured from bone-bonded strain gauges in various animals during the customary activity that elicited the highest recorded strains, along with safety factors to yield and ultimate failure.

Bone	Activity	Peak strain (microstrain)	Factor to yield[a]	Factor to ultimate failure[a]	Source
Horse radius	Trotting	−2800	2.4	5.6	Rubin and Lanyon, 1982b
Horse tibia	Galloping	−3200	2.1	4.9	Rubin and Lanyon, 1982b
Horse metacarpus	Accelerating	−3000	2.3	5.2	Biewener et al., 1982
Dog radius	Trotting	−2400	2.8	6.5	Rubin and Lanyon, 1982b
Dog tibia	Galloping	−2100	3.2	7.4	Rubin and Lanyon, 1982b
Goose humerus	Flying	−2800	2.4	5.6	Rubin and Lanyon, 1981
Cockerel ulna	Flapping	−2100	3.2	7.4	Rubin and Lanyon, 1982a
Sheep femur	Trotting	−2200	3.1	7.1	Rubin et al., 1983
Sheep humerus	Trotting	−2200	3.1	7.1	Lanyon and Rubin, unpublished data
Sheep radius	Galloping	−2300	3.0	6.8	O'Connor et al., 1982
Sheep tibia	Trotting	−2100	3.2	7.4	Lanyon et al., 1982
Pig radius	Trotting	−2400	2.8	6.5	Goodship et al., 1979
Fish hypural	Swimming	−3000	2.3	5.2	Lauder, 1981
Macaca mandible	Biting	−2200	3.1	7.1	Hylander, 1979
Turkey tibia	Running	−2600	2.5	6.0	Rubin and Lanyon, 1983c

a. Safety factors to yield and failure calculated using values of 6,800 and 15,700 microstrain, respectively, taken from Carter et al. (1981a,b).

1983b). It does not follow, of course, that all animals are in equal danger of accidentally breaking their bones. Just because a monkey's humerus is strained to the same extent during swinging from tree to tree as an elephant's femur is while stomping through the undergrowth, it does not mean that the inherent dangers of slipping or accidental loading in the two situations are equal. However, in each individual, the cellular population responsible for functional adaptation can only respond to actual and not potential strains. Natural selection reflects the associated risks and determines the scope of the adaptive response.

In each individual, regulation of peak functional strains is achieved by the behavioral and structural means just discussed. These involve regulation of the manner and magnitude of skeletal loading and adjustment of the mass, alignment, and material properties of the load-bearing tissue (Fig. 1-12). The architectural changes in bone dimensions necessary to maintain strain similarity in animals of different sizes are outlined in Chapter 2, which deals with scaling. If geometrically similar animals of different body sizes were to load their bones to a similar extent in relation to size, this would put large animals in considerably greater danger of failure than small ones (Biewener, 1982).

Indeed, it can be predicted that animals much larger than horses would destroy themselves if the weight-related loads they employed to run at full speed were similar to those used by a chipmunk for the same activity. However, the reason large animals do not self-destruct is that, in addition to the structural changes they make to their bones (including adjustment of cortical thickness and curvature, both of which are ignored in the classic scaling arguments), they also allometrically scale the forces that their peak functional activity produces (Rubin and Lanyon, 1983b). Thus, although an elephant may move over the ground faster than a chipmunk, its relative mode of locomotion is far more sedate and it imposes far smaller weight-related loads on its bones. We suggest that similar safety factors are maintained not by allometrically scaling bone dimensions, but rather by allometrically scaling the magnitude of the peak forces applied to them during vigorous locomotion. Although this is an effective strategy for maintaining the structural competence of the

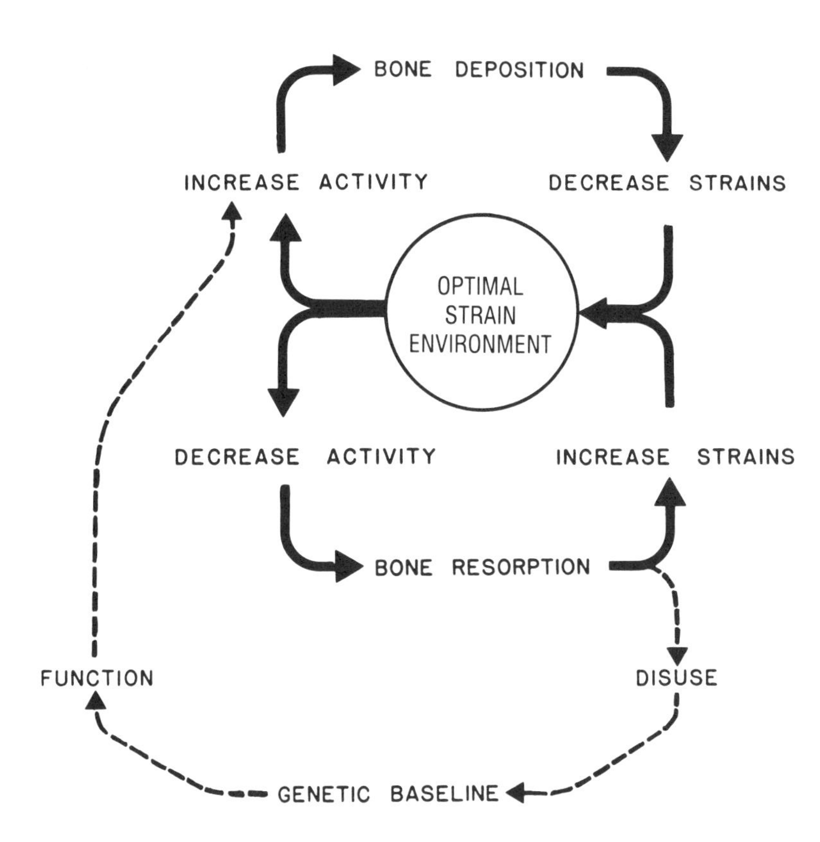

Figure 1-12 The effects of functional strains within bone tissue on remodeling to achieve an optimal strain environment. In any individual, activity that causes increased (above optimal) strains stimulates bone deposition, thus increasing skeletal mass and restoring optimal strain level. The resorption that accompanies decreased functional loading tends to restore optimal functional strains by reducing bone mass.

skeleton of larger animals, it is achieved by reducing their locomotor agility. Perhaps a major benefit of large size is that one need not be so active.

Control of Bone Mass The adaptive process in bone tissue operates by modifying the shape and adjusting the mass of a basic template to produce bone tissue that is adequate to withstand functional loads. The points of loading (at insertions and articulations) are points of potentially high stress, which can be avoided only by local increases in mass. Many crests, lines, tuberosities, local cortical thickenings, and the expanded joint surfaces can be interpreted as local reactions to avoid stress concentrations. The fact that many of these features do not develop in the absence of normal functional activity supports this relationship. Thus, tensional loading at an underdeveloped muscle insertion will cause unacceptably high strains in that vicinity, resulting in local bone hypertrophy and the development of an insertion process (tuberosity, trocanter, and so on) that will become substantial enough to distribute the load and maintain the strains below the level likely to cause damage. Even in completely paralyzed appendages bones do not disappear completely, so their mass and architecture presumably represent the basic, genetically determined template from which "normal" bone form is achieved by functional adaptation. No corresponding upper limit of bone mass has been established. Woo et al. (1981a,b) showed that by running pigs for a one-year period, they could induce a 17% increase in the cortical thickness of the femur. Jones et al. (1977) demonstrated that the cortical thickness of the humerus of professional tennis players is 35% greater in men, and 28% greater in women, in their serving arm than in the arm that simply throws the ball into the air. Goodship, Lanyon, and MacFie (1979) showed that, following removal of the ulna in skeletally immature pigs, a 50% increase in cross-sectional area of the overloaded radius occurs within three months. In none of these studies is there evidence that an additional overload could not have increased bone mass even further. However, experimental evidence from studies with bed rest (Donaldson et al., 1970), immobilization (Uhthoff and Jaworski, 1978), and space flight (Morey and Baylink, 1978; Tilton, Degioanni, and Schneider, 1980) has shown that once any functional level of bone mass has been achieved, it is maintained only as a result of the continuing stimulus of load bearing.

In trabecular bone there also appears to be a quantitative relationship between trabecular density (that is, the amount of the space available that is occupied by bone tissue) and the deformation within the tissue (Hayes and Snyder, 1981). The maximum shear strain is the numerical difference between the principal tensile strain (positive) and the principal compressive strain (negative). When this difference is large, the spatial density of trabeculae is great, leaving little space between the elements of the lattice. When this strain difference is small, then less of the available space is occupied by bone.

To achieve a mass and architecture appropriate to prevailing mechanical circumstances, the cellular population responsible for remodeling must be able to "sense" the bone's structural competence in relation to its current load-bearing requirement. This can be most readily achieved if the cells respond directly or indirectly to the strains induced within the tissue by functional activity (Fig. 1-12). Not only is strain the most relevant variable to control, but it is difficult to see how the cell population could sense either load or stress directly. We have seen that although the peak strains at any point around a bone's circumference are very different (ranging from tensile strain on one cortex, through zero at the neutral axis, to compressive strain on another surface), the maximum peak strain appears fairly constant, even in different bones and different species (Rubin and Lanyon, 1982b). However, adaptive remodeling is influenced not only by strains engendered during peak performance, but also by those resulting from customary activity. Indeed, under normal circumstances it is the customary strains, often well below the peak level (and thus also well below the level that would cause damage), that are responsible for maintaining skeletal architecture.

The sensitivity of adaptive remodeling to changes in strain well within the physiological range is illustrated in Figure 1-13. In this experiment, a portion of the ulna in the sheep forelimb was removed in order to cause a functional overload on the radius (Lanyon et al., 1982). The sheep ulna is such a small structure that its removal produced only a small change in the strains on the radius. Nevertheless, the absence of the ulna's structural contribution was somehow "sensed" by the radius, which compensated for it by increasing the thickness of its own cortex. This increase was achieved by the deposition of an amount of bone exactly equivalent in cross-sectional area to that of the removed ulna and located precisely in the spot where it could best imitate the ulna's structural contribution. During the one-year period in which bone hypertrophy was allowed to occur, the animals were walked around a circular track for one hour each day. However, they were never given the opportunity to run or jump and thus probably never engendered the peak strains of which they were capable. The fact that appropriate hypertrophy still occurred suggests that bone tissue is highly tuned to its customary loading pattern and is responsive to even minor changes in it. Thus, it seems that each bone is an organ inhabited or surrounded by cells capable not only of detecting small changes in customary strain, but also of mounting a coordinated response throughout a sufficient volume of tissue that the bone can be remodeled, and maintained, as an "organized strain-sensitive structure." This means that a strain distribution that is normal in the midshaft of one bone would cause adaptive remodeling if applied to the midshaft of another.

For each structure there must be a genetic "blueprint" not only for anatomical shape and position, but also for appropriate magnitude and distribution of functional strain. The bones' cell populations must be constantly comparing the specifications of this "blueprint" with the features of their prevailing strain environment and directing adaptive remodeling to remove any discrepancy. These capabilities require a considerable sensory and computational capacity from the osteocyte/osteoblast population.

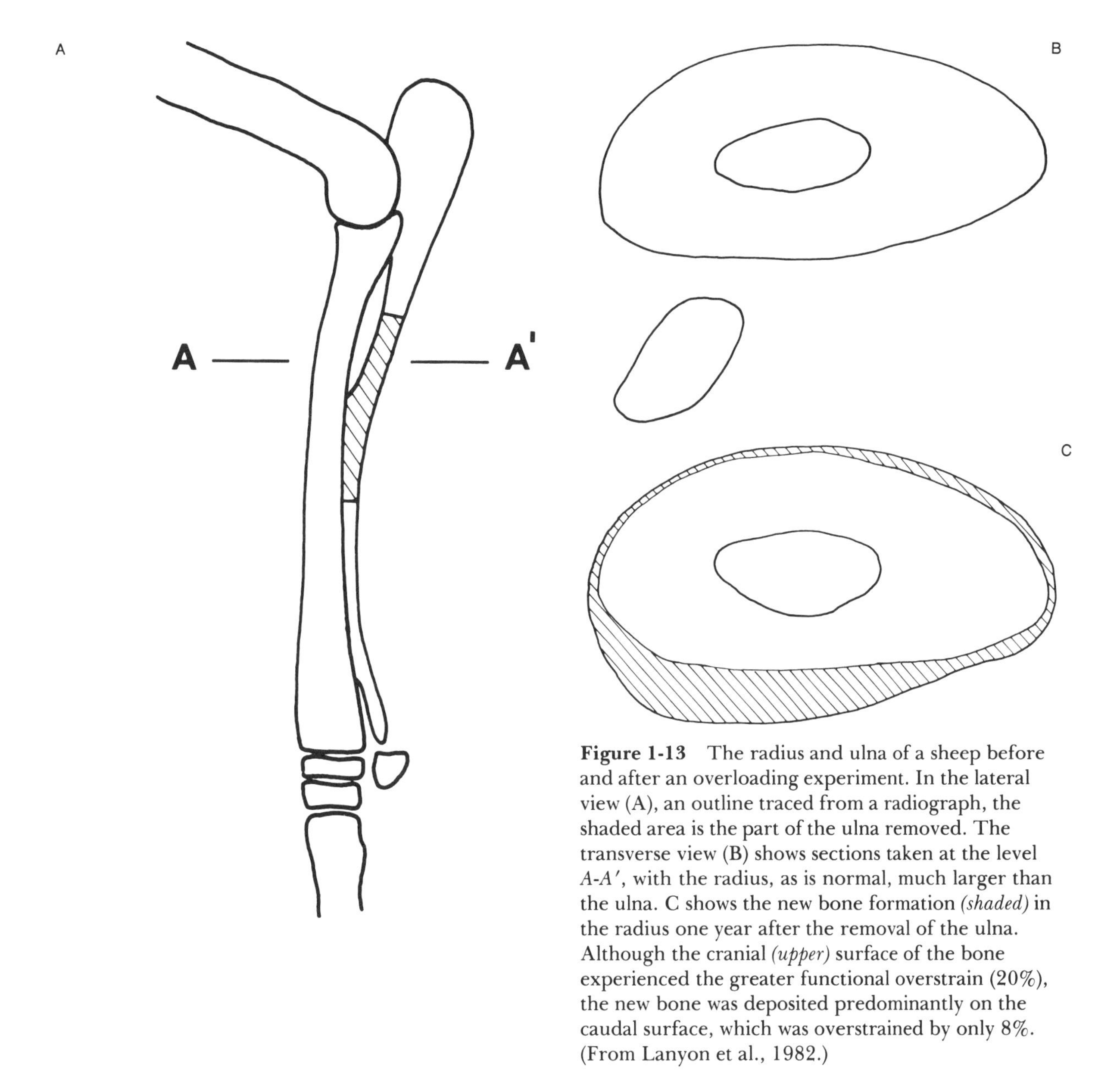

Figure 1-13 The radius and ulna of a sheep before and after an overloading experiment. In the lateral view (A), an outline traced from a radiograph, the shaded area is the part of the ulna removed. The transverse view (B) shows sections taken at the level *A-A′*, with the radius, as is normal, much larger than the ulna. C shows the new bone formation *(shaded)* in the radius one year after the removal of the ulna. Although the cranial *(upper)* surface of the bone experienced the greater functional overstrain (20%), the new bone was deposited predominantly on the caudal surface, which was overstrained by only 8%. (From Lanyon et al., 1982.)

Stimulus for Adaptation of Bone Tissue

The means by which cell populations assess the need for remodeling and control its course are unknown. However, this question is being investigated by applying loads from an external source to bones in situ and observing the response. This approach was pioneered by Hert, who applied static and sinusoidally varying dynamic loads to the tibia of rabbits. He demonstrated that the adaptive response was inherent in the bone itself and needed no central nervous connection (Hert, Skelenska, and Liskova, 1971). He also demonstrated that an osteogenic response was achieved only by intermittent, and not by static, loading

(Hert, Liskova, and Landa, 1971). Churches et al. (1979, 1980, 1981) applied external loads to the sheep metacarpus and varied both the magnitude of strain and the proportion of strain that was due to bending and compression. Their results suggested not only a relationship between the amount of stress induced and the quantity of the bone formed, but also that the remodeling response was induced by compression alone and not by bending. If this is true, it is a singular finding, since it has already been shown that the largest (and most potentially dangerous) type of bone strain is that due to bending (Rubin and Lanyon, 1982b; Biewener et al., 1983). Thus, it would be strange if bending were to have no effect on adaptive remodeling.

In addition to a selective sensitivity to various components of the strain distribution, it seems that bone cells respond differently according to the waveform of the strains produced within the material surrounding them. O'Connor et al. (1981) applied external loads to a sheep's radius for one hour per day and varied both the peak strains achieved and the rate at which the strains were applied during loading and release. The rate at which the strains changed correlated far better with the remodeling response than the peak strains achieved.

One disadvantage with the above experiments is that the effects of the artificial strain regime were superimposed onto those arising from normal load bearing. The introduction of the bird ulna preparation (Rubin and Lanyon, 1981, 1983a,c) was a significant advance in this respect since the bone shaft is isolated from normal load bearing by subarticular osteotomy, but it remains in situ with its normal physical and nutritive connections. Any remodeling at the bone's midshaft can thus be ascribed to the difference between the natural mechanical situation prior to surgery and the artificial one engendered by the loading apparatus afterward. With an appropriately con-

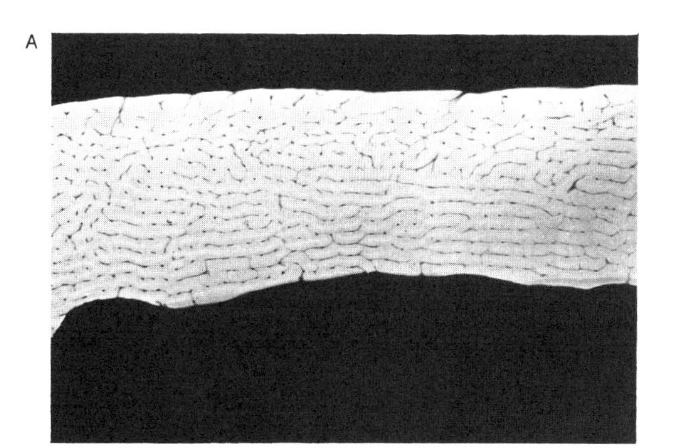

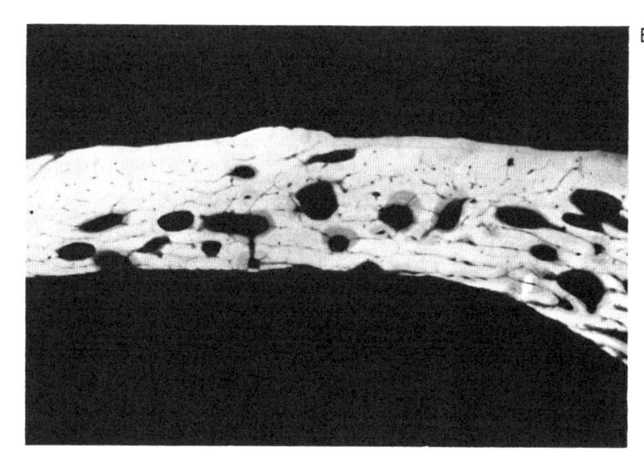

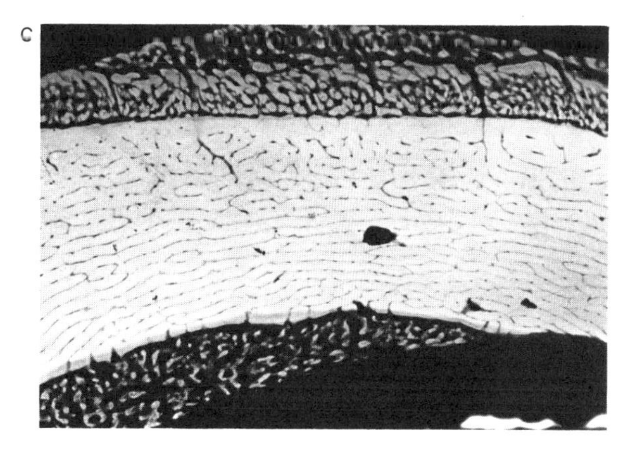

Figure 1.14 Transverse microradiographs of the rooster ulna: normal, underexercised, and exercised. In A, the cortical bone is normal. In B, the bone has become osteoporotic and its cortex has thinned following six weeks without functional load bearing. The low mineralization of this new bone is evident from its lower mineral density. The similar bone in C had received only 36 consecutive (0.5-Hz) applications of a compressive load every day. It shows no resorption and a periosteal layer of new bone being deposited. (From Rubin and Lanyon, 1981, 1983c.)

trolled mechanism for external loading, this allows separation of the parameters of the dynamic loading regime. For instance, in a series where peak strain was varied and strain rate kept constant, it was shown that the adaptive remodeling produced was indeed proportional to peak strain (Rubin and Lanyon, 1983a). Interestingly, however, large increases in bone mass can be produced by strain levels that are smaller, but have a different distribution, than those induced naturally during wing flapping. This suggests that the bone cells are capable of detecting and responding to changes not only in magnitude of peak strain (which reflects the structural danger that the bone is in), but also in strain distribution (which reflects the bone's particular sensitivity to the prevailing pattern of loading).

A remarkable finding from using the bird wing preparation was how few loading cycles are necessary to elicit an adaptive response. When the number of 0.5-Hz load cycles was reduced from 1800 a day to 360 or 36, there was no difference in the character or the amount of new bone deposited (Rubin and Lanyon, 1981, 1983c). Indeed, only 4 cycles per day of a potentially osteogenic stimulus, involving only eight seconds of loading per day, although not producing any increase in bone mass, were sufficient to prevent the resorptive remodeling that occurs with functional deprivation alone (Fig. 1-14). This reinforces the concept that it is not the damage caused by deformation that engenders the remodeling response; instead, bone cells control adaptive remodeling by virtue of some sensory capacity that enables them, after only a very few cycles, to recognize that they are being exposed to a strain pattern repeated often enough to demand "attention" and different enough from the pattern to which the bone has previously adapted to require a remodeling response.

Cartilage

The success of bone tissue as the building material for rigid skeletal structures is evidenced by its comparatively uniform mechanical properties over the large number of species in which it is found. The price to be paid to achieve these properties is the need for a complicated mineral metabolism and the requirement that both structural modification and repair involve physical destruction and removal of existing tissue. In cartilage some of these problems are avoided. Cartilage, like bone, is a connective tissue consisting of a collagenous matrix reinforced to make it rigid. Rather than using a mineral phase, however, cartilage achieves its stiffness by restraining water within a fibrous net. Articular cartilage consists of 75% water, 15–20% collagen, and 2–10% proteoglycans. Some of the water is free and can be expelled with physical pressure, but much of it is combined with the proteoglycans to form a gel held captive by the collagenous net. The large osmotic pressure exerted by the proteoglycan molecules causes the gel to expand against the collagen fiber framework, which is thus placed in tension. It is this captive water that supports the external loads applied to the tissue and accounts for its mechanical behavior.

Naturally, the shape of the tissue can be affected by the osmolarity of the fluid with which it is in contact, as well as by the size and character of external physical forces applied to it. Thus, under a constant static load, cartilage initially deforms not unlike an elastic material. However, instead of then maintaining its shape, it will continue to deform slowly with time. This continued, time-related deformation under constant load is termed *creep*. At the initial instant of load application, the cartilage reacts as a constant volume of trapped water. However, as the load continues to be applied, the free water moves through the tissue and is expelled to its surface, resulting in additional deformation. At the same time, some of the bound water becomes free as the local mechanical pressure exceeds the osmotic swelling pressure of the proteoglycans. This water also then moves through and abandons the tissue, thus diminishing its hydration. When loading is released, there is an initial rapid elastic recovery, followed by a slow rehydration of the gel until the structure's original shape is gradually reestablished. It is because of this difference in the degree of hydration of intervertebral disc cartilage under load that people are about a quarter of an inch taller at the beginning of each day than at the end and some one to two inches taller at the end of a prolonged space mission than on leaving earth's gravity.

In some regions, the proportion of fibers in the gel matrix is so small that fibers are not evident on gross examination and the tissue appears clear. This is called hyaline cartilage, which is seen covering the articular surfaces in synovial joints that are subjected solely to compressive loads. In locations where it might be subjected to tensile or bending loads, the gel is substantially reinforced by abundant fibers, which are obvious macroscopically, and the tissue is called fibrocartilage. In regions such as the intervertebral discs where resistance to exploding is required, the fibers are collagenous, adding tensile strength to the tissue, whereas in ear pinnae, which must be flexible as well as strong, the fibers are elastic.

In many ways, cartilage is a much simpler tissue than bone. It is avascular, requiring no capillary exchange for the nutrition of the chondrocytes deep within it. Instead, these cells exchange with extracellular fluid, which is assisted in moving through the tissue by mechanical loading. Because this tissue is less rigid than bone, it can grow by interstitial expansion from within as well as by appositional recruitment of its perichondrium. During growth, many of the bones of the skeleton begin as cartilage. Through a process called endochondral ossification, the cartilage is gradually replaced with bone. The cartilage that first develops in the embryo is sufficiently rigid to allow other structures to organize around it. It grows rapidly, and because it is avascular, it does not require a complicated mineral metabolism or nutrient exchange system. The extent to which these cartilaginous precursors have been replaced by bone in utero varies with the degree of activity of the newborn. In altritial animals, such as humans and marsupials, the degree of ossification of the skeleton at birth is poor and these animals are very dependent on their parents to defend, feed, and move them around. In precocial animals, like sheep, lizards, and chickens, the only cartilage remaining at birth is that covering the articular surfaces and separating primary and secondary ossification centers (the epiphyseal growth plate), thus allowing these animals to ambulate almost immediately after birth. These plates are usually located toward the ends of long bones, separating the metaphysis (shaft) and epiphysis, and are associated with joints or the points of insertion of large muscles. Since the cartilage covering articular surfaces is specialized for an articular rather than a proliferative role, it is only as a result of interstitial growth and osseous replacement of the cartilage at the growth plates that substantial increase in length of the bone is actually accomplished.

In short bones (such as those in the carpus and tarsus), there is only one ossification center, and in these bones increase in length is achieved by osseous encroachment into the single layer of cartilage, which provides a growth zone on its deep surface and an articular surface superficially. When growth of cartilage ceases, longitudinal growth of the bone ceases also. However, the covering of cartilage over the bone's articular extremities is retained throughout life, since appropriately lubricated cartilage provides an admirably low-friction surface.

Although some animals, such as sharks, have developed cartilage to an extensive degree, this tissue has disadvantages. Its avascularity makes it difficult to heal—hence the intractability of osteoarthritis. Cartilage is also not particularly strong. In synovial joints it is protected from shearing loads by the film of lubricant over its surface, and is thus primarily subjected to compression, which is the manner of loading that it is best able to withstand. However, at muscle insertions it is subjected to substantial amounts of tension and may separate, with consequent avulsion of the tuberosity concerned. In its position as a growth plate in the epiphyses of long bones, it is subjected to all the same components of load as the rest of the bone, including shear. To compensate for this weakness, growth plates are generally thin and tend to align themselves perpendicular to the direction of compressive loading. They also develop irregular contours so that the two ossification centers are joined by interdigitating bony processes, separated only by a narrow zone of cartilage. Nevertheless, despite these adaptations, growth plates always present a plane of weakness that is eliminated only at maturity when the cartilage cells cease to divide and bony continuity is established between the primary and secondary ossification centers.

Functional adaptation has been studied far less in cartilage than in bone. The physical properties of the tissue are profoundly affected by the degree of hydration and the immediate loading history of the tissue. The nutrition of articular cartilage also relies to some extent on continued intermittent loading, which forces fluid through the tissue. Under conditions of disuse the thickness of articular cartilage is generally reduced, and there are reports (Holmdahl and Ingelmark, 1948) that during training its thickness increases. The proliferative activity of cartilage cells in growth plates is also affected by the degree of loading. If the functional loading across the growth plates is confined to very high or very low loads, longitudinal growth of the bones is reduced. Maximum growth appears to occur at slightly subnormal loading levels (Hert, 1969). It is difficult to determine to what extent this modulation of the growth plate's activity is an adaptive response or a direct consequence of load bearing. Whereas the primary influences on longitudinal growth are almost certainly genetic and hormonal, the load-related modification of cartilage may be more a direct physical effect of load bearing on cell division, cell nutrition, or the physical properties of cartilage tissue itself than an adaptive modification of cellular activity in response to functional use.

Chapter 2

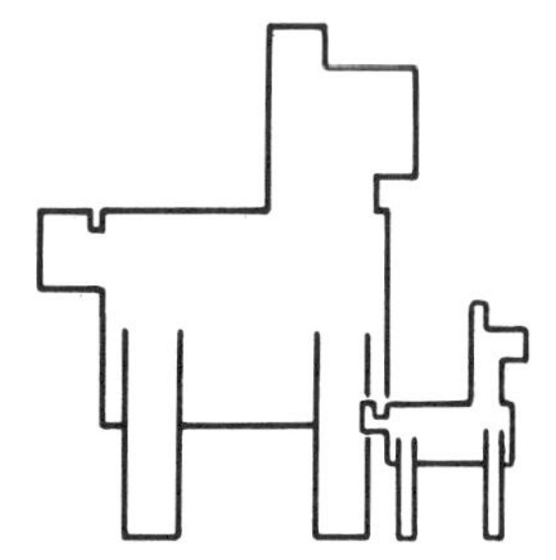

Body Support, Scaling, and Allometry

R. McNeill Alexander

Terrestrial animals rely on legs and flying animals on wings to support the weight of their bodies. This contrasts with fishes and other aquatic animals, which are supported largely by buoyancy. In this chapter we investigate how the muscles and skeletons of nonaquatic vertebrates perform their support function.

Any general discussion of body support in vertebrates must take account of differences of body size. Pygmy shrews *(Sorex minutus)* have masses of about 3 g, whereas African elephants *(Loxodonta africana)* are about a million times heavier with masses of about 3 tons. Suppose a shrew were somehow enlarged to 100 times its linear dimensions without any change of shape. It would be 100 times longer, 100 times wider, and 100 times higher, and its volume would be increased by a factor of 1 million (100^3). If the densities of its parts were unchanged it would be as heavy as an elephant. However, the cross-sectional areas, and hence the strengths, of the bones and muscles that support it would be increased by a factor of only 10,000 (100^2) It seems unlikely that the giant shrew could stand. How, then, are elephants possible?

The answer, as we shall see, is that elephants are not scaled-up shrews. Some of their linear dimensions are more than 100 times those of shrews and others are less, which results in many differences in body proportion between them and shrews. Elephants stand with their legs straighter than shrews and are not very agile. Differences like these that can be related to differences of body size between animals are described as scale effects. This chapter is about the scaling problems associated with body support.

The theories about scaling generally produce equations of the form

$$y = ax^b, \tag{2-1}$$

where x and y are dimensions or other quantitative properties of animals of different sizes, and a and b are constants. We have already had a hint of this in the expressions 100^3 and 100^2 in an earlier paragraph. Furthermore, scientists who have made measurements on related animals of different sizes have usually found that their results are described fairly accurately by equations like (2-1). For instance, in Figure 2-1, the points near the lower line show the diameters of the femurs d (in mm) of mammals of different masses m (in kg). All the points lie close to the line, which represents the equation

$$d = 5.2m^{0.36}. \tag{2-2}$$

Equations like this one are called allometric equations.

One of the properties of this form of equation is that it gives straight lines when plotted (as in Fig. 2-1) on logarithmic coordinates. It can be shown that it must do so by taking logarithms of Eq. (2-1):

$$\log y = \log (ax^b) = \log a + b \log x. \tag{2-3}$$

Hence, a graph of $\log y$ against $\log x$ (or a graph of y against x on logarithmic coordinates) is a straight line of slope b.

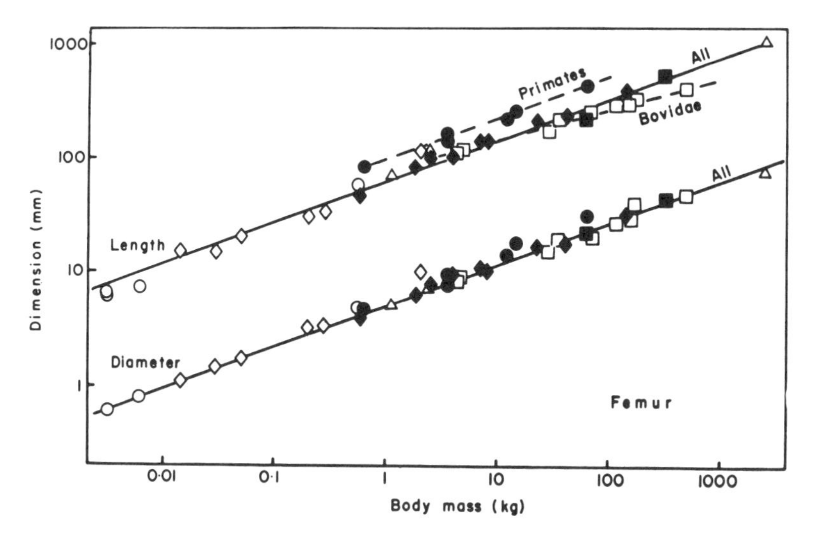

Figure 2-1 Lengths and diameters of the femurs of mammals plotted against body mass on logarithmic coordinates. Mammals included are insectivores *(clear circles)*, primates *(black circles)*, rodents *(clear diamonds)*, carnivores *(black diamonds)*, and bovids *(clear squares)*. (From Alexander et al., 1979.)

Limitations of a Leg

Scaling can be understood on the basis of a simplified model of a leg (Fig. 2-2). This model looks like the hind leg of a mammal from the knee down, but the conclusions derived from it will apply to legs in general. The two leg segments that are shown are inclined at angles α_1 and α_2 to the force F that is acting on the foot. The upper segment contains a bone of length l_b and diameter d_b. Alongside the bone runs a muscle of cross-sectional area A_m whose muscle fibers have length l_m when the leg is in this particular position. The tendon of the muscle has length l_t and cross-sectional area A_t.

We will calculate the force F that the muscle must exert. The perpendicular distance from the force to the joint between the two segments is $l_f \sin \alpha_2$, so the force has a moment (torque) $Fl_f \sin$

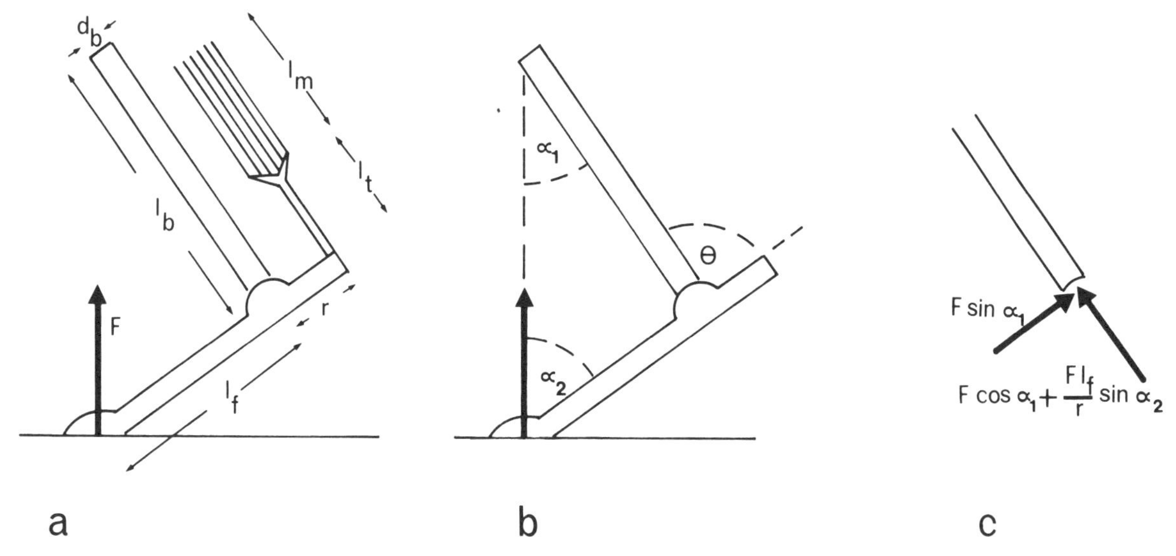

Figure 2-2 A simplified model of a leg showing (a) dimensions, (b) angles, and (c) forces acting on the distal end of the upper bone.

α_2 about the joint. This moment must be balanced by the muscle, which has a moment arm r at the joint. Hence the muscle must exert a force $(Fl_f/r) \sin \alpha_2$. This is equal to $\sigma_m A_m$, where σ_m is the stress in the muscle, and so F is $\sigma_m A_m r/(l_f \sin \alpha_2)$.

This force F is limited because the stress σ_m is limited. If $\sigma_{m,max}$ is the greatest stress that the muscle can exert, then

$$F \leq \frac{\sigma_{m,max} A_m r}{l_f \sin \alpha_2}. \tag{2-4}$$

The force exerted by the muscle is transmitted by the tendon and must not break it; that is, the stress in the tendon must not exceed its tensile strength $\sigma_{t,max}$. Hence,

$$F \leq \frac{\sigma_{t,max} A_t r}{l_f \sin \alpha_2}. \tag{2-5}$$

Furthermore, the bone must not be broken. The force F has components $F \cos \alpha_1$ along the axis of the upper bone and $F \sin \alpha_1$ at right angles to it. The force exerted by the muscle $(Fl_f/r) \sin \alpha_2$ acts parallel to the axis of the bone. Hence, forces must act on the distal end of the bone as shown in Figure 2-2c. The axial component of force sets up compressive stresses throughout the bone, whereas the component at right angles tends to bend the bone, extending the anterior face and compressing the posterior one.

The bone has diameter d_b. Assume for simplicity that it is solid and cylindrical, so that its cross-sectional area is $\pi d_b^2/4$. The stress in it, owing to the axial component of force, is

$$\sigma_{b,axial} = -\frac{4F}{\pi d_b^2}\left[\cos \alpha_1 + \frac{l_f}{r} \sin \alpha_2\right].$$

(The negative sign indicates a compressive stress.)

The component of force at right angles to the bone exerts bending moments, which increase from zero at the distal end of the bone to $Fl_b \sin \alpha_1$ at the proximal end. The resulting stresses, in each cross section, range from a maximum compressive stress at one face of the bone to a maximum tensile stress at the other. By applying a standard engineering equation (Alexander, 1968), we find the stresses $\sigma_{b,bending}$ throughout the bone lie in the range:

$$-\frac{32Fl_b}{\pi d_b^3} \sin \alpha_1 \leq \sigma_{b,bending} \leq \frac{32Fl_b}{\pi d_b^3} \sin \alpha_1.$$

The total stress at any point in the bone is $\sigma_{b,axial} + \sigma_{b,bending}$, but in the parts where the stresses are largest and fracture most likely to start, the bending component is likely to dominate. For instance, calculations concerning the tibia of a dog making a large jump showed that the extreme value of $\sigma_{b,axial}$ was about -20 MPa, whereas the range of $\sigma_{b,bending}$ was ± 80 MPa. We shall therefore ignore the axial component.

Let the tensile strength of bone be $\sigma_{b,ult}$. If the bone is not to break, then

$$F \leq \frac{\pi d_b^3 \sigma_{b,ult}}{32 l_b \sin \alpha_1}. \tag{2-6}$$

Expressions (2-4), (2-5), and (2-6) set limits to the forces the leg can exert. There are also limits to the movements it can make because muscle fibers can work only over restricted ranges of length. The maximum length at which an extended muscle fiber can exert force is generally about twice the minimum length at which it can still exert force when it shortens. Let the range of length over which force can be exerted be ϵl_m, where l_m is the resting length and ϵ is some constant fraction. If the muscle has a constant moment arm r and can move the joint through a range of angles $\Delta\theta$, then

$$\Delta\theta \leq \frac{\epsilon l_m}{r}. \tag{2-7}$$

The limitations identified here are used in subsequent sections to show the implications of different possible scaling rules.

Geometric Similarity

Two objects are geometrically similar if their corresponding angles are equal and one could be made identical to the other simply by multiplying all linear dimensions by some factor λ (Fig. 2-3). Areas are multiplied by a factor λ^2 and volumes by λ^3. If the objects have the same density, masses are

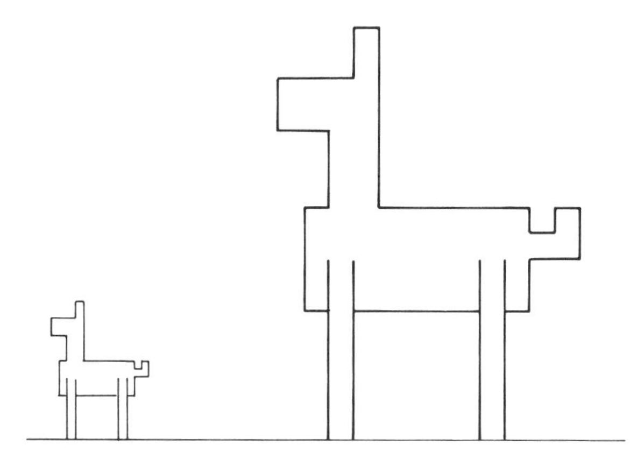

Figure 2-3 Two geometrically similar animals. All linear dimensions of the larger one are 3 times the corresponding dimensions of the smaller one.

also multiplied by λ^3. Thus, geometrically similar objects made of the same materials have corresponding lengths proportional to mass$^{1/3}$ and areas to mass$^{2/3}$.

Suppose that related animals of different sizes are geometrically similar and that they adopt geometrically similar positions when exerting maximum forces on the ground. Suppose also that they are made of the same materials, and apply (2-4) to them. They have equal values of $\sigma_{m,max}$ and α_2, whatever their size. The linear dimensions l_f and r are proportional to $m^{1/3}$ (m is body mass), whereas the area A_m is proportional to $m^{2/3}$. With these proportionalities, the maximum possible force F_{max} obeys the rule

$$F_{max} \propto m^{2/3}. \quad (2\text{-}8)$$

Expressions (2-5) and (2-6) lead to exactly the same conclusion, which is indeed the conclusion reached by a much simpler argument in the beginning of this chapter: an elephant-sized animal, 1 million times as heavy as a geometrically similar shrew, could exert only 10,000 times as much force as the shrew. The more complicated argument is unnecessary in this case, but it will be needed when patterns of scaling other than geometric similarity are discussed.

Hill (1950) pointed out other consequences that would follow from geometric similarity. Muscle cross-sectional areas A_m would be proportional to $m^{2/3}$, and so the maximum forces $A_m\sigma_{m,max}$ that the muscles could exert would also be proportional to $m^{2/3}$. The lengths l_m of muscle fibers would be proportional to $m^{1/3}$, and so the amount each muscle could shorten would also be proportional to $m^{1/3}$. Thus the work each muscle could do (force times shortening distance) would be proportional to m. This work could be converted to kinetic energy ($\frac{1}{2}mv^2$), and so animals of different sizes could accelerate themselves to the same speed v when taking off in a standing jump. They could all jump to the same height (in a high jump) or over the same distance (in a long jump). Alternatively, the work could be used to accelerate a leg, moving it relative to the body. Animals of different sizes could accelerate their feet to the same speed and could therefore run equally fast. Hill pointed out that rat kangaroos (*Bettongia*, about 1.5 kg) can jump almost as high as the large kangaroos (*Macropus*, about 50 kg) and that whippets run almost as fast as greyhounds. However, the predictions of Hill's simple theory do not hold when very large or very small mammals are considered. Neither very small nor very large mammals have athletic performance as good as that of intermediate-sized ones. Neither shrews nor elephants jump as well as gazelles, or run as fast.

Elastic Deformations

All solids are elastic to some extent. Tendons can be stretched by 5% to 10% before they break, and bone by about 1%. Muscle can elongate elastically by about 2%, but it can also make much larger changes of length by inelastic processes involving detachment and reattachment of cross-bridges.

Consider the elastic deformations of the leg illustrated in Figure 2-2 that result from the force F on the foot. It has been shown that the force in the muscle and its tendon is $(Fl_f/r)\sin\alpha_2$. The stress in the tendon can be obtained by dividing this by the cross-sectional area A_t. If the tendon has unstretched length l_t and is made of material of Young's modulus E_t (Young's modulus is the tensile stress divided by the resulting strain), it will be stretched by an amount

$$\delta l_t = \frac{F l_f l_t}{r A_t E_t} \sin \alpha_2 .$$

Since the moment arm is r, this will allow the angle θ of the joint to increase by

$$\delta\theta = \frac{F l_f l_t}{r^2 A_t E_t} \sin \alpha_2 . \qquad (2\text{-}9)$$

Furthermore, the bone will bend. The component of force $F \sin \alpha_1$ (Fig. 2-2c) will deflect the end of the bone by a distance s that can be calculated from Eq. (15) in Alexander (1968). Expressed as a fraction of the length of the bone, this deflection is

$$\frac{s}{l_b} = \frac{64 F l_b^2}{3\pi E_b d_b^4} \sin \alpha_1 , \qquad (2\text{-}10)$$

where E_b is Young's modulus for bone. For geometrically similar animals made of the same materials, Eq. (2-9) gives

$$\delta\theta \propto F m^{-2/3}, \qquad (2\text{-}11)$$

and Eq. (2-10) gives

$$\frac{s}{l_b} \propto F m^{-2/3}. \qquad (2\text{-}12)$$

It has already been shown in (2-8) that the maximum possible force on the foot F_{max} would be proportional to $m^{2/3}$, and so when it acted, $\delta\theta$ and s/l_b would be independent of body size: animals of all sizes would deform elastically in geometrically similar fashion.

Elastic Similarity

Elastic deformations have figured prominently in theories of scaling. Rashevsky (1962) and McMahon (1973, 1975) built a theory around the assumption that animals of different sizes have evolved so as to sag under their weight in geometrically similar fashion. This is the theory of elastic similarity. Notice that the deforming forces it deals with are the weights of the animals, not the maximum forces discussed in the preceding section. When an animal runs or jumps, forces are required to accelerate it as well as to support its weight. When it stands still, these additional forces do not act. Thus, the theory of elastic similarity concerns the deformation of stationary animals.

Consider geometrically similar animals of different sizes, standing still. The forces F on their feet would be proportional to m. The angular deflections $\delta\theta$ of their joints owing to tendon elasticity (2-11) would be proportional to $m^{1/3}$. The degree of bending of their bones s/l_b (2-12) would also be proportional to $m^{1/3}$. Elastic similarity requires $\delta\theta$ and s/l_b to be independent of body mass, and so geometrically similar animals made of the same materials are not elastically similar.

Elastic similarity requires systematic deviation from geometric similarity. Linear dimensions have to be classified into two groups, lengths, such as l_f, l_b, l_m, and l_t, and diameters, such as d_b and r. (It is perhaps not immediately obvious that the moment arm r should be classed as a diameter, but the theory requires it.) Cross-sectional areas such as A_m and A_t are, naturally, proportional to the square of diameter. It will be shown that the theory of elastic similarity requires all lengths to be proportional to $m^{1/4}$ and all diameters to $m^{3/8}$. Thus, diameters are proportional to $l^{3/2}$, and larger animals are required to have relatively stouter legs and bodies than smaller ones (Fig. 2-4). Notice that volumes must be proportional to ld^2, and that these rules make ld^2 proportional to m, as consistency requires. By making lengths proportional to $m^{1/4}$, diameters proportional to $m^{3/8}$, angles independent of m, and the force F proportional to m, in Eqs. (2-9) and (2-10), we find that $\delta\theta$ and s/l_b would both be independent of body mass. Animals built according to the rule of elastic similarity would deform under their own weight in geometrically similar fashion.

By making the same substitutions in (2-4) we find, for the maximum force consistent with the strength of the muscle, that

$$F_{max} \propto \frac{d^3}{l} \propto m^{7/8}. \qquad (2\text{-}13)$$

The same conclusion would be reached by considering tendon strength (2-5) or bone strength (2-6). Contrast this with (2-8), which shows that for geometrically similar animals F_{max} would be proportional to $m^{2/3}$. The difference can be em-

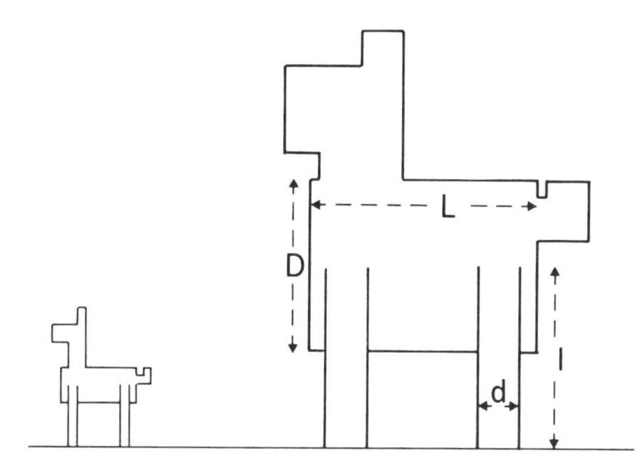

Figure 2-4 Two elastically similar animals. The larger animal has lengths *(l,L)* that are 3 times the corresponding lengths of the smaller one, but its diameters *(d,D)* are $3^{3/2} \cong 5$ times those of the smaller animal.

phasized by returning to the example of the shrew and the elephant-sized animal of 1 million times its mass. If the large animal were geometrically similar to the shrew, it could exert only 10,000 ($1{,}000{,}000^{2/3}$) times as much force as the shrew, but if it were elastically similar it could exert 178,000 ($1{,}000{,}000^{7/8}$) times as much force as the shrew. This would make it much more likely to be able to stand.

Geometrically similar animals of different sizes would be able to move their joints through equal angles, but elastically similar ones would not. Since the lengths of muscle fibers in elastically similar animals scale as lengths and moment arms scale as diameters, (2-7) gives

$$\Delta\theta_{max} \propto \frac{l}{d} \propto m^{-1/8}. \qquad (2\text{-}14)$$

An elephant-sized animal that was elastically similar to a shrew, therefore, could move its joints through angles only 0.178 ($1{,}000{,}000^{-1/8}$) times as large. If the shrew could move a particular joint through 180°, the elephant-sized animal could move the homologous joint through only 32°. This argument assumes that shrews do not have muscle fibers so long that their whole range of contraction cannot be used.

Elastic similarity might give large animals very restricted joint mobility, but it would not alter Hill's conclusions about jumping ability and running speed because the work each muscle could do would be proportional to ld^2 and hence to body mass.

There is no obvious reason why evolution should favor elastic similarity. Terrestrial vertebrates are constructed so that they do not sag much under their own weight, and there is no obvious advantage to sagging in geometrically similar fashion. The inertial forces that act on the bodies of small animals when they run and jump are often many times their body weight. Expression (2-13) points to a possible advantage of elastic over geometric similarity, but a scaling rule that deviated even further from geometric similarity could make F_{max} proportional to an even higher power of m.

There is a snag about any scaling rule that involves systematic deviation from geometric similarity: it cannot be maintained over more than a limited range of animal sizes. For instance, an elephant-sized animal that was elastically similar to a shrew would have a body 32 ($1{,}000{,}000^{1/4}$) times as long as those of shrews, but it would have legs 178 ($1{,}000{,}000^{3/8}$) times as thick. The bases of the legs would be too thick to fit onto the body. In contrast, there would be no difficulty in fitting together the parts of an elephant-sized animal that was *geometrically* similar to a shrew, although such an animal might be too weak to stand.

Data

Do the legs of mammals and birds scale according to geometric similarity, elastic similarity, or some other rule?

Figure 2-1 and a great many similar graphs show that the scaling of the parts of legs can generally be described quite closely by allometric equations. The situation is particularly simple for bone diameters, for instance, for the diameters of femurs shown by the lower line in Figure 2-1. All the points lie very close to the line, showing that Eq. (2-2) fits the data well. The constants in the equation can be determined fairly precisely. They would have to be revised if more data were collected, but statistical analysis shows that with 95% probability, the "true" value of the factor (based on all possible data) lies in the range 5.0 to

Table 2-1 Mean exponents b of allometric equations of the form $y = am^b$, where m is body mass, for various dimensions of the legs of mammals in general, of bovids, and of flying birds, in comparison with the exponents required for elastic and geometric similarity.

Dimension	Mammals	Bovids	Birds	Geometric similarity	Elastic similarity
Bone length, l_b	0.35	0.26	0.37	0.333	0.250
Bone diameter, d_b	0.36	0.36	0.41	0.333	0.375
Moment arm, r	0.40	0.34	0.39	0.333	0.375
Muscle fiber length, l_m	0.24	0.25	0.24	0.333	0.250
Muscle fiber area, A_m[a]	0.81	0.77	0.79	0.667	0.750
Muscle mass	1.05	1.02	1.01	1.000	1.000
Tendon cross-sectional area, A_t	0.69	0.75	0.47	0.667	0.750

Sources: Alexander et al., 1979, 1981; Maloiy et al., 1979; Prange, Anderson, and Rahn, 1979.

a. Muscle volume divided by fiber length.

5.4, and (also with 95% probability) the "true" value of the exponent lies in the range 0.35 to 0.37. Such precision in allometric equations is generally only possible when data are available for animals of an extremely wide range of sizes. In this instance the range is from tiny shrews to an elephant almost a million times heavier.

Differences between related groups of animals are often made apparent by allometric data. This is illustrated by the upper graph in Figure 2-1, which shows relationships between femur length and body mass. The points for primates lie above those for other mammals, showing that primates have unusually long femurs for their body masses, but the slope of the primate line is the same as the slope of the line for mammals in general. The points for bovids (antelopes, cattle, sheep) lie along a line of smaller slope than those for other mammals, indicating a lower allometric exponent. For mammals in general the exponent is 0.36 ± 0.02 (mean and 95% confidence limits), but for bovids it is 0.27 ± 0.03. The bovids are the only mammals among those included in Figure 2-1 that differ significantly from the rest in their allometric exponents for leg bone length, a peculiarity that has not been explained. Because of their lower exponents for all leg bones, bovids will be treated separately in this discussion.

To find whether geometric or elastic similarity or some other allometric rule holds widely, we need to investigate many different dimensions of several different groups of animals. Some data for the widest possible range of body sizes are given in Table 2-1. As in Figure 2-1, the mammals cover the whole terrestrial range from shrews to elephants. The bovids range from one of the smallest species of antelope (4 kg) to buffalo (*Syncerus,* 500 kg). The birds whose skeletons were measured range from a 4-g hummingbird to one of the largest flying birds, a 16-kg Kori bustard (*Ardeotis*). Because birds whose muscles were measured range only from about 100 g upward, the exponents for bird muscles are known less precisely than those for bird bones.

The values given in Table 2-1 are in most instances means of exponents for several different bones or several different muscles, but the exponents whose means have been taken are generally very close to one another. For instance, the exponent given for the diameters of mammal bones is the mean of five exponents for individual bones, ranging only from 0.32 (a metatarsal bone) to 0.38 (humerus). The exponents for mammals and birds are in most instances almost identical, suggesting that the same scaling rule applies to both groups.

All the exponents for bovids are very close to the predictions of elastic similarity, but the mammals in general and the birds show discrepancies. In striking contrast to the requirement of elastic similarity, these have exponents for bone length of 0.35 and 0.37. The exponent for the cross-sectional areas of bird tendons also seems to deviate from elastic similarity, but this particular exponent is not known at all precisely: it is the mean of values with very wide confidence limits.

The departure of bone lengths from the rule of elastic similarity has a profound effect on the external appearance of animals of different sizes because it affects leg length. Large mammals and birds have much longer legs than they would if they were elastically similar to their small relatives. The manner in which leg bone dimensions scale also implies that leg skeletons make up larger fractions of body mass in larger animals. If leg bone length l_b is proportional to $m^{0.35}$ and diameter d_b to $m^{0.36}$, then leg bone mass should be proportional to $l_b d_b^2$ or to $m^{1.07}$. I have no data specifically on leg bone mass, but the mass of the complete skeleton is proportional to $m^{1.09}$ in mammals and $m^{1.07}$ in birds (Prange, Anderson, and Rahn, 1979). The skeleton is 5% of body mass in a shrew and more than 20% in an elephant.

Expression (2-7), with the exponents for l_m and r given in Table 2-1, shows that mammals in general should be able to move their joints through angles proportional to $m^{-0.16}$, bovids to $m^{-0.09}$, and birds to $m^{-0.14}$; larger animals should be restricted to smaller angular movements of their joints. The situation is actually a little more complicated because (at least in mammals) l_m tends to have a larger exponent for proximal leg muscles than for distal ones. Proximal muscles such as the biceps femoris (Fig. 2-5) tend to be parallel-fibered, and their fiber lengths scale approximately in proportion to the lengths of the bones they run alongside. Distal muscles such as the gastrocnemius (Fig. 2-5) tend to be pinnate, with muscle fibers that are much shorter than the neighboring bones, especially in large mammals. In extreme cases, such as the interosseous and plantaris muscles of camels, the muscle fibers have disappeared almost completely, leaving only a tendon in place of the muscle.

It might be supposed that the movements of the distal joints of the legs of large mammals would be very severely restricted. This is not the case for two reasons. First, many of the muscles cross

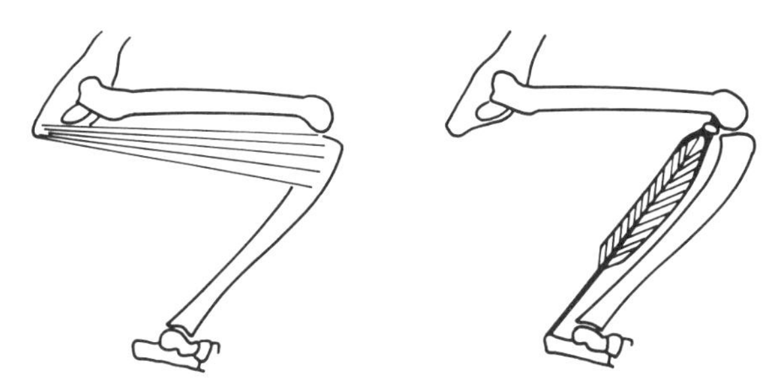

Figure 2-5 Skeleton of a mammal leg with the biceps femoris muscle *(left)* and gastrocnemius muscle *(right)*.

more than one joint so that, for instance, any degree of bending of the ankle is possible without changing the length of the gastrocnemius, provided the knee bends as well. Second, the tendons may be stretched quite a lot, allowing extra movement at the joint when large forces act on the feet.

The tension in a tendon is limited by the force its muscle can exert. If the fiber area of the muscle is proportional to $m^{0.81}$ (as for mammals in general, Table 2-1) and the cross-sectional area of the tendon to $m^{0.69}$, the maximum stress that the muscle can impose on the tendon and the resulting elastic strain should be proportional to $m^{0.12}$. This allometric trend could hardly apply over an indefinitely large range of sizes of animals because as size increased, a point would eventually be reached at which the muscle was strong enough to break its own tendon.

If tendon strain is proportional to $m^{0.12}$, it should be twice as great in a 50-kg mammal (such as a large kangaroo, *Macropus*) as in a 100-g mammal (such as a kangaroo rat, *Dipodomys*). Kangaroos and kangaroo rats both hop on their hind legs, and mechanical analysis confirms that the distal leg tendons of kangaroos suffer much larger strains than those of kangaroo rats (Biewener, Alexander, and Heglund, 1981). As explained in Chapter 3, tendon elasticity saves quite a lot of energy for large running and hopping mammals but not for small ones.

Scaling of Forces

Expressions (2-4) to (2-6) set limits to the forces that feet can exert on the ground. The exponents from Table 2-1 can be used to predict how these limits should scale. For mammals in general, A_m should be proportional to $m^{0.81}$, r to $m^{0.40}$, and l_f (like other bone lengths) to $m^{0.35}$, and so by (2-4) the limit to ground force, set by muscle strength, should be proportional to $m^{0.86}/\sin \alpha_2$. The corresponding expressions for bovids and birds are $m^{0.85}/\sin \alpha_2$ and $m^{0.80}/\sin \alpha_2$. Since large mammals tend to hold their legs rather straighter than small ones, α_2 may be proportional to some small negative power of m and ground forces (if limited by muscle strength) should perhaps be about proportional to $m^{0.9}$.

The limit to the forces the bones can withstand is set by (2-6). With the exponents given in Table 2-1, it should be proportional to $m^{0.73}/\sin \alpha_1$ for mammals in general, to $m^{0.82}/\sin \alpha_1$ for bovids, and to $m^{0.85}/\sin \alpha_1$ for birds. This implies that if $\sin \alpha_1$ is proportional to $\sin \alpha_2$, bone strength in mammals fails to keep pace with muscle strength as body size increases. However, a small error in the exponent for d_b could be responsible for a misleading conclusion, as d_b is cubed in (2-6). There is a fair amount of evidence that leg bones of dogs and larger mammals are stressed in fast locomotion to between about a quarter and a half of their tensile strength, but there is little comparable information for smaller vertebrates.

So far, exponents for the scaling of ground forces have been inferred from anatomical evidence. There is little more direct information. It has been calculated that a 0.25-kg galago that jumped to a height of 2.26 m must have exerted peak forces on the floor of at least 7.5 times body weight with each hind foot. A 42-kg kangaroo on a force platform exerted a maximum of 3.0 times body weight with one hind foot. These forces are proportional to $m^{0.82}$. The forces exerted on the ground by the feet of running animals can be estimated from the fraction of the time that each foot remains on the ground. This has been done for ungulates ranging from small gazelles (about 20 kg) to giraffes (about 1000 kg) galloping at top speed (Alexander, Langman, and Jayes, 1977). The maximum forces seemed to be proportional to $m^{0.86}$ for hind feet and to $m^{0.89}$ for forefeet, agreeing well with the scaling rule inferred from anatomical data.

Kinematic Similarity

The concept of geometric similarity refers to shapes. A related concept that applies to movements is kinematic similarity. Two movements are kinematically similar if one could be made identical to the other by multiplying all linear dimensions by some constant factor λ and all time intervals by some constant factor τ. An example of two kinematically similar motions is provided by two pendulums of different lengths swinging through the same angle.

Kinematic similarity is possible only under strictly defined conditions. For motions in which gravity is important (for instance, the swinging of pendulums or the running of animals), it is possible only if the motions have equal Froude numbers, v^2/gh. Here v is some characteristic speed (defined in the same way for both motions), g is the acceleration of free fall, and h is a characteristic length. In comparisons of pendulums, v might be defined as the speed of the bob at midswing and h as the length of the pendulum. In comparisons of running animals v is defined as the speed of running and h as leg length.

Animals cannot run in kinematically similar fashion except at speeds that give them equal Froude numbers. There is no mechanical necessity for them to run in kinematically similar fashion even when their Froude numbers are equal, but there seems to be a strong tendency for them to do so (see Alexander, 1977). Animals as diverse as men, horses, small rodents, and ostriches, when running with equal Froude numbers, take strides that are approximately equal multiples of leg length. Mammals running with the same Froude number generally use the same gait. For instance, horses, gazelles, cats, and mice all trot at Froude numbers between about 0.6 and 3 (Alexander, 1977).

Running with equal Froude numbers requires v to be proportional to $h^{1/2}$ (since g is constant). Running in kinematically similar fashion requires stride length to be proportional to h. Stride frequency, the number of strides taken in unit time, is equal to speed divided by stride length and is therefore proportional to v/h or to $h^{-1/2}$ for animals running in kinematically similar fashion. Pennycuick (1975) observed mammals ranging from gazelles to elephants moving at their chosen speeds through their natural habitat. He found that their stride frequencies were proportional to $h^{-0.6}$ at the walk, to $h^{-0.5}$ at the trot, and also to $h^{-0.5}$ at the gallop.

Among the evidence cited by McMahon (1975) for the wide scope of the theory of elastic similarity was the relationship between the size of mammals and the stride frequency at which they change from trotting to galloping. This frequency is about 7 Hz for mice and 2 Hz for horses. For the series mouse, rat, dog, horse it is proportional to $m^{-0.14}$. Elastically similar structures have frequencies of free vibration proportional to $m^{-0.125}$. This may seem to be good agreement, but a running mammal is not very closely comparable to an elastic structure in free vibration. Although the animal loses kinetic energy at some stage of the stride and gains it at others, only a portion of this energy is stored in the interval and returned by an elastic recoil. The rest is degraded to heat as active muscles are stretched and supplied afresh from chemical energy as they contract (see Chapter 3). Furthermore, the animals presumably had leg lengths h proportional to some power of m between about 0.25 and 0.35, and so a stride frequency proportional to $m^{-0.14}$ was also approximately proportional to $h^{-0.5}$. This is as expected if animals tend to run in kinematically similar fashion, whether or not they are constructed according to elastic similarity.

Scaling of Wings

So far, this discussion has dealt exclusively with legs. An alternative support for the weight of birds and bats is their wings, which are in turn supported by the aerodynamic force called lift. Lift is given by

$$\text{lift} = \tfrac{1}{2}\rho S v^2 C_L. \qquad (2\text{-}15)$$

Here ρ is the density of the air, S is the plan area of the wings, v is speed, and C_L is a quantity called the lift coefficient. This coefficient varies according to the angle at which the wings are held, but it has an upper limit $C_{L,max}$, which is approximately constant in the ranges of size and speed that concern us. The minimum speed v_{min} at which the wings can support the weight mg of the body is given by

$$mg = \tfrac{1}{2}\rho S v_{min}^2 C_{L,max}, \text{ and}$$

$$v_{min} = \left(\frac{2mg}{\rho S C_{L,max}}\right)^{\frac{1}{2}} \propto \left(\frac{m}{S}\right)^{\frac{1}{2}}. \qquad (2\text{-}16)$$

The speed at which flight should theoretically need least power is also proportional to $(m/S)^{1/2}$.

Wing areas have been measured and allometric exponents calculated for various groups of birds and bats (Greenwalt, 1975; Norberg, 1981). The general rule seems to be that wing areas are about proportional to $m^{0.7}$, which would be approximately true if flying animals were geometrically similar to one another. With (2-16) this implies that flying speeds should be proportional to $m^{0.15}$: a typical 10-kg bird (the size of a swan) should fly 2.8 times as fast as a typical 10-g bird (the size of a warbler). There is, unfortunately, little information about the speeds of birds and bats of different sizes.

Large birds could fly as slowly as small ones if S increased in proportion to m instead of to $m^{0.7}$. This would, however, give very large birds extraordinarily large wings (Fig. 2-6).

The lengths of the wings of bats are fairly nearly proportional to $m^{0.33}$, but those of birds are about proportional to $m^{0.41}$. This is a marked deviation from geometric similarity: the areas of bird wings scale about as required for geometric similarity, but larger birds have relatively longer, narrower wings. This tendency is exaggerated in Figure 2-6: albatrosses have unusually long, narrow wings, even for birds of their size.

The wing bones of birds deviate even more markedly from geometric similarity. For flying birds ranging from a 4-g hummingbird *(Phaethornis)* to an 11-kg swan *(Cygnus)*, humerus length was found to be proportional to $m^{0.48\pm0.04}$ and ulna length to $m^{0.43\pm0.06}$ (95% confidence limits; Prange, Anderson, and Rahn, 1979). The diameters of these bones were proportional to $m^{0.38}$ and $m^{0.36}$. The exponents for diameters are close to the exponents for bird and mammal legs (Table 2-1), but the exponents for lengths seem to be

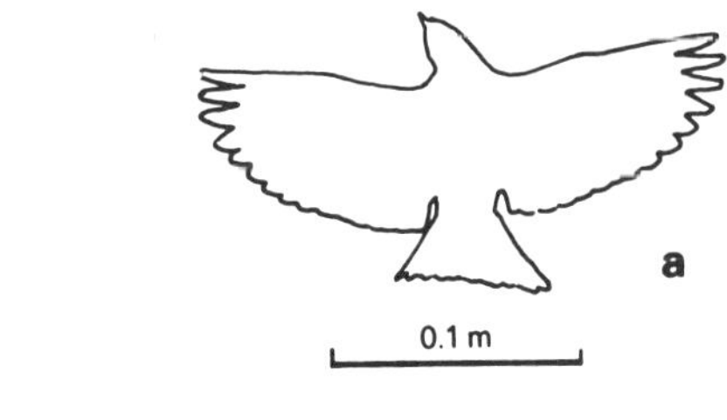

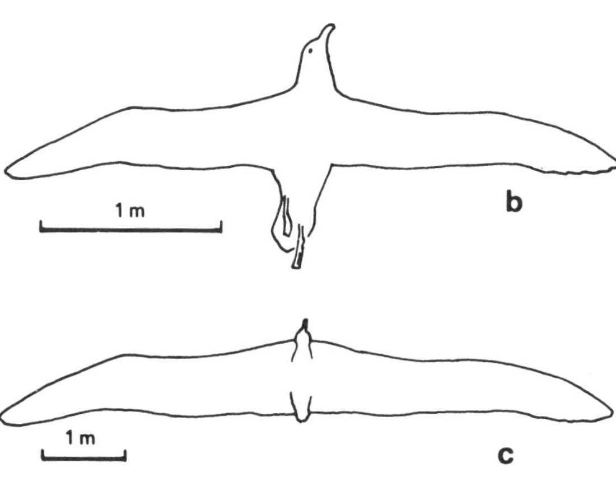

Figure 2-6 Comparison of body size with wing area for (a) a sparrow *(Passer domesticus)* having mass 30 g and wing area 100 cm^2; (b) an albatross *(Diomedia exulans)* having mass 8.5 kg and wing area 0.62 m^2; and (c) an imaginary "albatross" retaining mass 8.5 kg but with wing area adjusted to give the same mass-to-area ratio as the sparrow. (From Alexander, 1982.)

larger than for legs. The wing bones of large birds tend to be slimmer, relative to their length, than those of small ones.

If the diameter of the humerus is proportional to $m^{0.38}$ and if the thickness of the wall of this hollow bone is a constant fraction of its diameter, then the bending moment needed to break it is proportional to diameter^3, or to $m^{1.14}$. The assumption about thickness is doubtful, but we will nevertheless examine its consequences. If wing length is proportional to $m^{0.4}$, the distance from the base of the wing to the point where the lift effectively acts is probably also proportional to $m^{0.4}$. Hence the maximum lift that the humerus can bear (the breaking moment divided by the distance) may be proportional to $m^{0.74}$. In unaccelerated flight the lift required to support the bird equals its weight and is therefore proportional to m, but much more lift is needed for maneuvers such as fast cornering. This suggests that very large birds might not be able to withstand lift much greater than their weight, even when flying fast enough to generate such lift. However, the conclusion is tentative and I have no information about muscle strength to support it.

Scaling of the Trunk

The theory of elastic similarity requires that the trunk length of terrestrial animals, like their limb lengths, be proportional to $m^{1/4}$ and that trunk diameter, like limb diameter, be proportional to $m^{3/8}$. Large mammals should have relatively shorter, stouter bodies than small ones (Fig. 2-4). This seems to be true for ungulates, which have total lengths of head plus trunk proportional to $m^{0.29}$, but not for carnivores, which have them proportional to $m^{0.37}$ (Radinsky, 1978). It also seems to be true for primates in general, which have trunk lengths about proportional to $m^{0.25}$, but not for a more coherent group of arboreal monkeys, which have trunk lengths proportional to $m^{0.32}$ (Aiello, 1981). The exponent for primates in general is as required for elastic similarity, the exponent for ungulates lies between elastic and geometric similarity, and the exponent for carnivores is higher than for either elastic or geometric similarity. A paper by Economos (1983) seems to resolve this confusing situation. He shows that small mammals (under 20 kg) tend to have (head plus trunk) length proportional to $m^{0.34}$, but that large ones (over 20 kg) tend to have this length proportional to $m^{0.27}$. In this respect, small mammals scale about as predicted by geometric similarity and large ones about as predicted by elastic similarity.

In contrast, swimming vertebrates of all sizes tend to be geometrically similar. A sample of perch *(Perca fluviatilis)* with masses ranging from 0.2 g to 1.5 kg had lengths proportional to $m^{0.32}$ (Le Cren, 1951). It would be easy to distinguish large and small perch in equal-sized photographs because large ones have relatively smaller eyes, but body shape remains remarkably constant over the whole size range. Similarly, whales with masses ranging from 30 kg to 100 tons have lengths proportional to $m^{0.34}$ (Economos, 1983).

Fish and whales are supported largely or entirely by buoyancy and it seems unlikely that support is important in selection for body shape; swimming performance is probably much more important.

Conclusion

The principal question at present concerning the scaling of body support is whether the theory of elastic similarity remains useful. It seems doubtful that it does, even though it has played a valuable role in stimulating thought and research. It has always been open to the objection that it is based on a dubious assumption, and evidence is now accumulating that there are important discrepancies between the exponents it predicts and those that can be observed.

The dubious assumption is that the deformation of animals under their own weight has been predominantly important in natural selection for body proportions. Such elastic deformations as occur in bone and muscle seem trivial, and in muscle they are very much smaller than the changes of length that can be achieved by sliding of the myofilaments. Forces much larger than the body's own weight act on it in strenuous activities such as running and jumping. These forces do not scale in proportion to body weight; they tend to be larger multiples of weight in smaller animals.

Many of the exponents in Table 2-1 are close to the exponents required for elastic similarity, but there is a large, consistent discrepancy in the exponents for bone length in mammals (excluding bovids and probably other ungulates) and birds. This discrepancy affects leg length and therefore the scaling of external appearance. It tends to destroy the geometric similarity of elastic deformation on which the theory depends. The exponents for the lengths of muscle fibers shown in Table 2-1 seem consistent with elastic similarity, but they are the means of values that are significantly different for different groups of muscles. The scaling of trunk proportions may be consistent with elastic similarity in some groups of mammals, but it seems not to be in others.

McMahon (1975) tried to extend the theory of elastic similarity to explain gaits as well as body proportions, but the attempt involved an awkward and arbitrary decision to treat stride length as a diameter. The theory offers one possible explanation for the observed scaling of stride frequency, but another explanation based on the concept of kinematic similarity is equally successful. Norberg (1981) considered the possibility that wings might scale so as to maintain elastic similarity. They do not seem to do so.

Table 2-1 shows remarkable agreement between mammal and bird legs but it does not seem to suggest any universal theory. It seems unprofitable in any case to persist in looking for similarity principles. There is no reason to expect evolution to seek similarity as such. Rather it seeks some kind of optimality. Ultimately, it maximizes fitness, but this often involves minimizing mortality or energy expenditure or maximizing food intake or reproductive rate. Until we have a theory of scaling that explains body proportions in some such terms, our theories will be unsatisfactory.

Chapter 3

Walking and Running

Milton Hildebrand

This chapter deals with the functional morphology of animals that walk or run on the ground. Quadrupeds that commonly run and are structurally modified to benefit speed or endurance are referred to as cursors. Noncursors, on the other hand, include the many animals that walk freely but run infrequently, as well as those small species that do run but have few postural or structural modifications associated with speed (such as various lizards, insectivores, and rodents). The on-ground walking of animals that are primarily modified for other types of progression (for example, several fishes, moles, bats, sloths, pinnipeds) is omitted. Scaling, energetics, and neuromuscular activity are dealt with further in Chapters 2, 10, and 17.

General Requirements

The morphological, physiological, and behavioral demands imposed by walking and running increase markedly as speed, endurance, and body size increase. In general, a first requirement is for support and stability despite only intermittent contact with the substrate. Limb posture, gait, and relative length of limb are contributing factors. Second, propulsion is needed to accelerate the body and to compensate for internal and external resistance. This requires that the limbs, and often the body as well, make coordinated oscillations; the periodic activity of muscles, bony levers, and joints provides strides of adequate length, rate, and thrust. Third, there must be sufficient variability of direction of thrust in relation to mass to give needed maneuverability. Finally, the locomotor mechanism must be efficient enough to provide adequate endurance (if required) at an acceptable cost in energy. This may necessitate the avoidance of nonessential oscillations, design to maximize strength in relation to weight, a favorable distribution of mass, and the development of passive mechanisms to restrict extraneous motions, cushion impact, and store energy.

Limb Posture and Excursion

The limb posture of a quadruped, whether sprawling, upright, or intermediate, is important to its manner of locomotion. Amphibians, turtles, lizards, and some early squamates and ruling reptiles are (or were) sprawlers, a posture characterized by lateral placement of the feet and consequently a wide track. The propodium (humerus or femur) is held nearly horizontal and, when in a neutral position, at about right angles to the vertebral column. During locomotion (Fig. 3-1) the distal ends of propodium, epipodium (forearm or lower leg), and foot describe successively larger horizontal ellipses relative to the shoulder and hip (Snyder, 1962; Sukhanov, 1968; Walker, 1971; Brinkman, 1980). The musculature must be sufficiently forceful to support the body between the laterally placed feet. The elbow and knee remain bent. The feet may twist in their tracks so that at the end of the propulsive phase the thrust against the ground is delivered by the medial edge of the feet and medial digits. Motions of the spine and

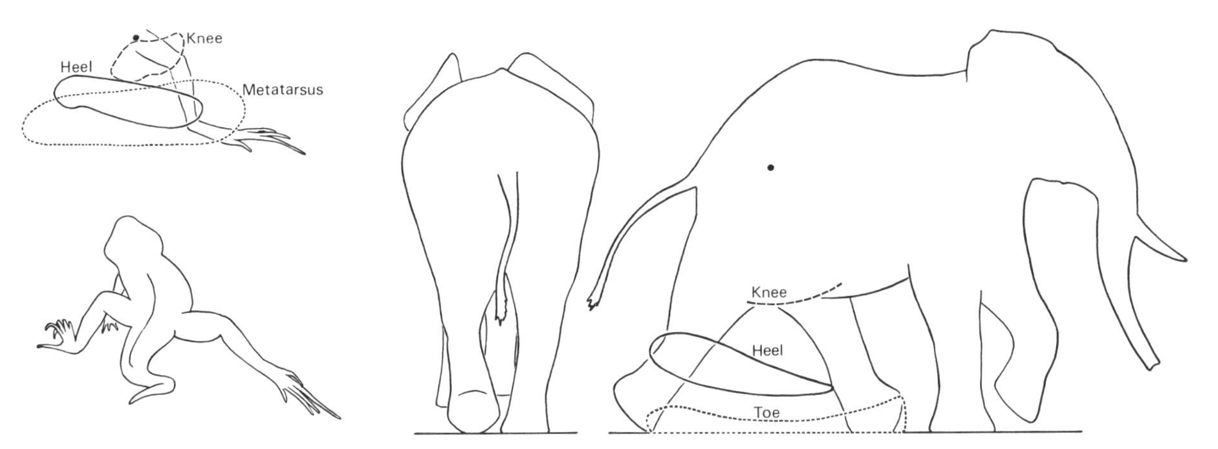

Figure 3-1 Posterior view of sprawling and upright postures and right lateral view of the arcs described by knee, heel, and toe (elephant) or distal metatarsus (lizard). (Lizard redrawn from Snyder, 1962.)

girdles that can contribute to limb excursion are in the horizontal plane (Figs. 3-4 and 3-12). See Chapter 9 for the nature of the motions and Rewcastle (1981) for an analysis of sprawling.

Cursorial and graviportal (very heavy-bodied) vertebrates use upright posture with limbs in relatively extended positions and little or no abduction of the propodium. All limb segments swing more or less in sagittal planes, and the feet are placed well under the body. Motions of the spine and girdles that can contribute to limb excursion are also in a sagittal plane (Fig. 3-13). All limb segments of graviportal quadrupeds are nearly vertical as the animal stands, although the propodium of cursors forms an acute angle with the horizontal (as viewed laterally).

Most quadrupeds have limb postures intermediate between sprawling and upright. This is true of fast-moving crocodilians, most of the earliest ruling reptiles (thecodonts, discussed by Charig, 1972), monotremes, infants of most mammals, and adult noncursorial mammals. The propodium is somewhat abducted, and elbow and knee are held, respectively, near the levels of the shoulder and hip joints (Jenkins, 1970, 1971). The feet are placed somewhat farther apart (transverse to the line of travel) than in animals with a fully upright posture.

Among striding bipeds the more proficient, such as many birds, man, and presumably ornithischian and theropodian dinosaurs (Coombs, 1978), have leg motions that resemble those of the hind legs of upright quadrupeds. Less skilled bipedal striding, as of the chimpanzee, more nearly resembles the hind-limb action described above as intermediate (Jenkins, 1972).

Quadrupeds having a sprawling or intermediate limb posture tend to have plantigrade feet (wrist and heel on the ground as the animal stands). Those with upright posture tend to have either digitigrade feet (only the digits on the ground), as seen in various dinosaurs, all birds, some dasyurid marsupials, cavies and agoutis among rodents, and cursorial carnivores, or to have unguligrade feet (only hoofs on the ground) as seen in ungulate mammals (but not in dinosaurs). Further correlates of foot posture are noted below.

Gaits and Gait Selection

A gait is a regularly repeated sequence and manner of moving the legs in walking or running. The gaits selected by animals correlate with their posture and excursion of limbs, stability and maneuverability, speed and endurance, body proportions, and cost of locomotion. Walking gaits have each foot on the ground half or more of the time; running gaits have each foot on the ground less than half the time. The time that a foot is on the ground is termed the contact interval or duty factor and is expressed as either a percentage or a decimal fraction of the complete cycle.

Gaits are classed as symmetrical if the footfalls (or any other given action) of the two feet of a pair are evenly spaced in time. They are asymmetrical if corresponding actions of the feet of a pair are unevenly spaced in time. The walks, pace, and trot are symmetrical; bounds and gallops are asymmetrical. For symmetrical gaits, the contact intervals of the four feet are usually equal, or nearly so. All such gaits can be identified by two numbers. The first is the contact interval, which varies with the rate of travel; the second relates the action of one pair of feet to that of the other, and is often expressed as the percentage of the cycle that a fore footfall follows the hind footfall on the same side of the body. The two numbers are conveniently plotted on a square graph, on which all symmetrical gaits can be represented (Fig. 3-2).

At the pace the two legs on the same side of the body move in phase. The camel family and some horses and large dogs pace. The gait has the advantage for long-legged animals that there can be no interference between fore and hind feet. Since it is relatively unstable at slow rates of travel, it is usually used at the run or near run.

When the two legs on the same side of the body are out of phase, then diagonally opposite legs swing in unison and the gait is the trot. Because the trot is more stable than the pace, the walking trot is employed by salamanders, reptiles, and some large mammals (although all more frequently use other, still more stable gaits when moving slowly). The running trot is used by most quadrupeds that run.

When consecutive footfalls of the four feet are about equally spaced in time, the gait is a singlefoot. If each fore footfall follows that of the hind on the same side of the body, the singlefoot is in lateral sequence. When walking, most tetrapods use either this very stable gait or gaits falling between this one and the trot (for the terminology of such intermediate gaits see Hildebrand, 1976). The running singlefoot in lateral sequence is a smooth gait used by elephants (when young, at least) and by horses trained to do the rack, running walk, or slow gait. The singlefoot in diagonal sequence (each fore footfall follows that of the hind on the opposite side of the body) is used at the walk by most primates and by a few other mammals (aardvark, kinkajou, and armadillo). It is used at the run only by several small artiodactyls. This gait is potentially a little less stable at slow speed than is the singlefoot in lateral sequence, and it is more likely to have interference at high speed. (For further descriptions and analyses of symmetrical gaits see Sukhanov, 1968, on lizards; Zug, 1971, on turtles; Edwards, 1976, on salamanders; Hildebrand, 1976, on mammals; and references they cite.)

With few exceptions asymmetrical gaits have been observed only in mammals. They may have nine variables, but it is practical to consider only five. These can be combined in ways that produce useful graphs of all possible performances, but the analysis is more complex than serves our purpose here (see Hildebrand, 1977). Since in this class of gaits the footfalls of a pair of feet are unevenly spaced in time, they occur as couplets (unless they are simultaneous). The foot of a pair that strikes the ground first in each couplet is the trailing foot; the other is the leading foot (it leads in position, not time). The longer the step taken between trailing and leading feet, the longer the lead. The hind lead tends to be shorter than the fore lead. Indeed, the hind legs often thrust together to initiate a jump (weasels, many rodents). The forefeet may have a moderate to long lead to prolong their role in support, or they may function virtually together to share the impact after a jump (impala, jumping mouse, squirrels). If the gait has at least a moderate fore lead but little or no hind lead, it is called a half bound (Fig. 3-3). If neither pair of feet has more than a slight lead, the gait is a bound. These gaits are most used by small, agile mammals that move by a series of leaps over terrain that is rough in relation to body size. Such locomotion is costly in energy, but it is used primarily for short dashes and quick acceleration. Rarely, all four footfalls are in unison, a bound called the pronk. This gait is used occasionally by certain rodents and artiodactyls.

If an animal runs with a lead both fore and hind, it is galloping. When fore and hind leads are on the same side of the body, the gait is a transverse gallop; when on opposite sides, a rotary gallop. The transverse gallop appears to be the more stable, whereas the rotary gallop has some advantage for maneuvering. The former is favored by

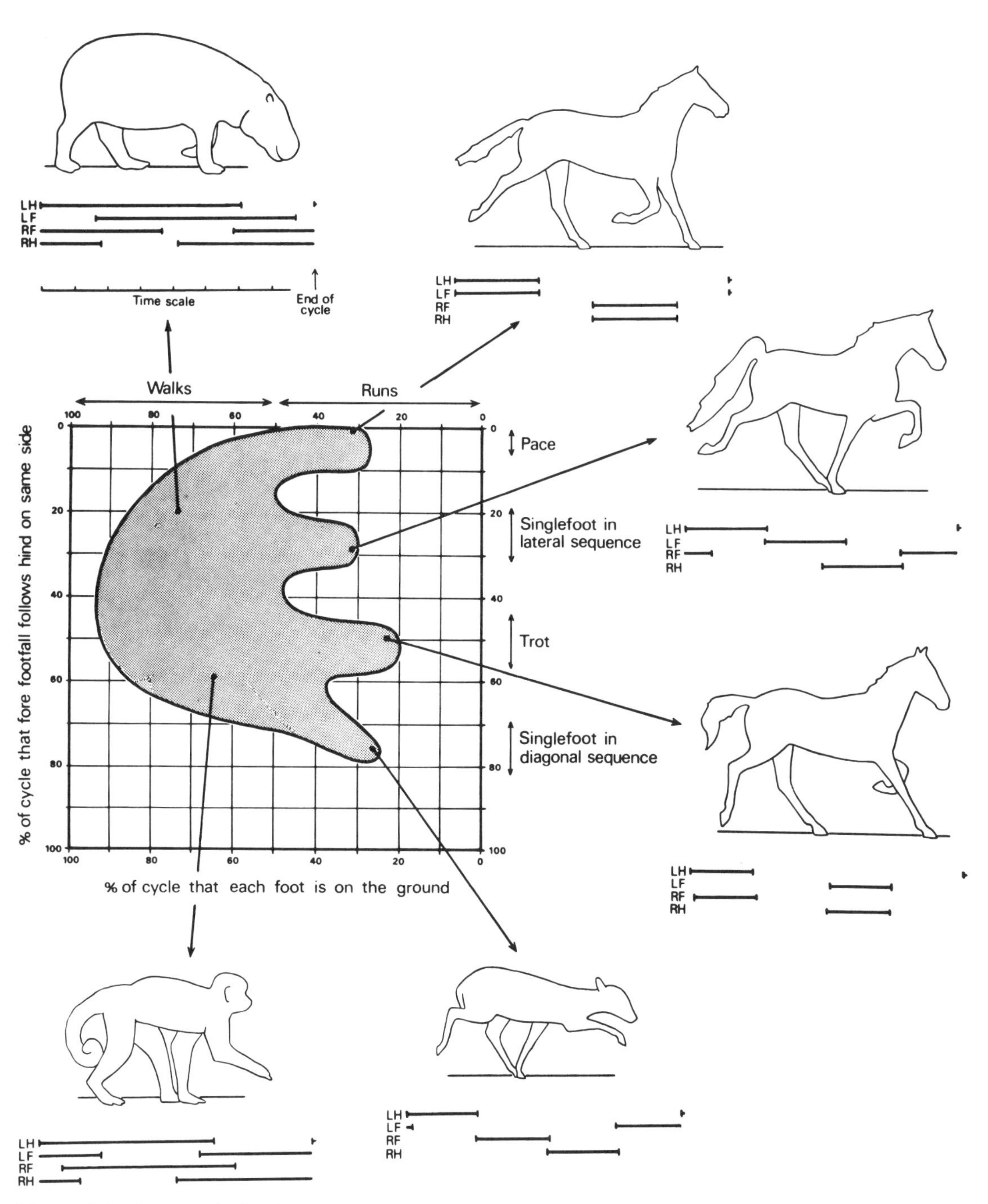

Figure 3-2 Symmetrical gaits as defined by the duration of contacts with the ground and the phase relationship of fore and hind feet. The outlined area encloses nearly 1200 plots for 156 genera, including amphibians, reptiles, and 16 orders of mammals. For the seven gaits recorded, a pigmy hippopotamus, three horses, a duiker, and a monkey are shown at the instant the left hind foot strikes the ground. Gait diagrams indicate the timing and durations of the respective contact intervals. Time scales *(upper left)* for the different animals are independent. *H* = hind; *F* = fore; *R* = right; *L* = left.

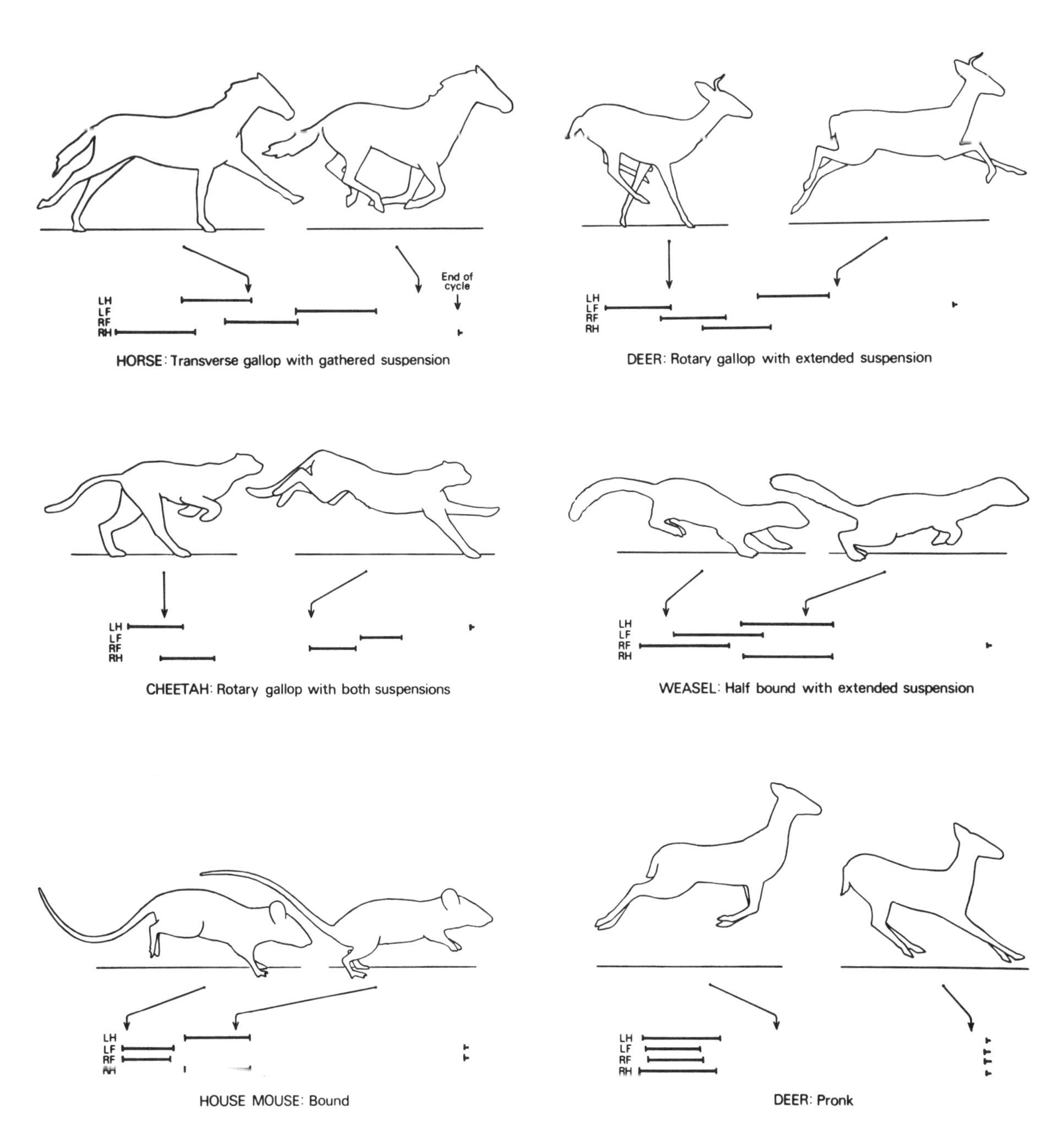

Figure 3-3 Six representative asymmetrical gaits. The gait diagrams start with the footfall that follows the suspension, or the longer suspension if there are two. Time scales and symbols are the same as in 3-2.

larger animals moving over open terrain, particularly when running slowly; the latter is used by smaller cursors, especially when moving fast.

As galloping mammals increase their speed, they increase the length of the stride more than its rate. This is done in part by introducing periods of suspension, when all feet are off of the ground. Suspensions that occur when the legs are flexed under the animal (that is, just after the leading forefoot leaves the ground) are termed gathered, whereas those occurring when fore and hind legs are angled forward and backward, respectively (that is, just after the leading hindfoot leaves the ground), are extended. Gathered suspensions tend to be short and are typical of large animals. Extended suspensions are part of a leap and are usual for small, agile mammals that do the bound or half bound. Many cursors use both suspensions when moving fast. Like the symmetrical gaits, the asymmetrical gaits form a continuum, merging into one another. (For a detailed analysis of asymmetrical gaits, see Hildebrand, 1977.)

Two of the most important factors in gait selection are stability and economy of effort. The first is interpreted further in the next section; the second is mentioned in several places elsewhere in the text. For a more detailed analysis of gait selection see Hildebrand (1980).

Stability

Principles of the mechanics of support for standing quadrupeds, including the loading of the spine and legs in relation to the distribution of mass, body proportions, and posture, are well illustrated and described by Gray (1944, 1968) and Kummer (1959). The stability of quadrupeds moving at slow to moderate rates of travel appears to follow the same principles, and it is hard to see how it could be otherwise. Nevertheless, little direct research has been done on this subject.

Stability increases as the relative size of the feet increases, as the area enclosed by those feet that are on the ground increases, and as the height of the center of mass above the ground decreases. For all of these reasons quadrupeds that sprawl are quite stable when three or four feet are on the ground.

In view of their need for maneuverability, most quadrupeds that move about at moderate speeds are better off to have a more upright stance and to derive stability largely from selection of the appropriate gait. A first general consideration is that the footfalls be timed so as to maximize the number of feet that are on the ground at a time and to favor those placements of three supporting feet that enclose the largest triangles. Because the triangles of support are larger for walking gaits in lateral sequence than in diagonal sequence, the former are selected by most quadrupeds (Fig. 3-4). The relative duration of support by three feet is maximized at the singlefoot, but at gaits intermediate between the singlefoot and trot, support by four feet increases a little (for any given contact interval) and support by two feet is always by the relatively stable diagonal opposites (Hildebrand, 1976). Accordingly, many animals walk using the lateral sequence with each forefoot striking the ground at 25 to 45 percentage points of the cycle after the hind foot on the same side (Fig. 3-2). Jayes and Alexander (1980) have shown that although the walking of turtles usually departs somewhat from expectation on the basis of mechanical considerations, their walking is the most stable for an animal that can alter only rather slowly the forces exerted at the feet. The transverse gallop has larger triangles of support than the rotary gallop, and this appears to be a reason that the former is selected by most large cursors running over open terrain (Fig. 3-4).

A second general consideration in maintaining stability is that the center of mass move favorably over the base of support. The duration of a stable condition (as for a walking salamander) would be maximal if the center of mass were near a margin of a triangle of support at the instant that that triangle is established, and if the center of mass then traveled over a relatively long chord of the triangle before it reached another margin.

When the body is supported by only two feet, as often occurs in fast walking and running, then the "area" of support approaches a line of support unless the feet are large. If the legs are not long and the feet of a pair straddle somewhat, as is usual for noncursorial quadrupeds, then the body tends to roll if supported only by the feet on one side of the body. Animals with such proportions select gaits that have, instead, bipedal support by diagonally opposite feet so that their line of support will fall more nearly under the center of mass. Long-legged cursors place their feet nearly under the midline of the body; the line of support established by, for instance, a forefoot with either hind foot then remains nearly under the center of mass as forward motion occurs.

The faster a quadruped moves, the smaller its base of support can be. This is because the center of mass cannot move far from a position over its support in a very short time, and the running animal quickly establishes a new base of support,

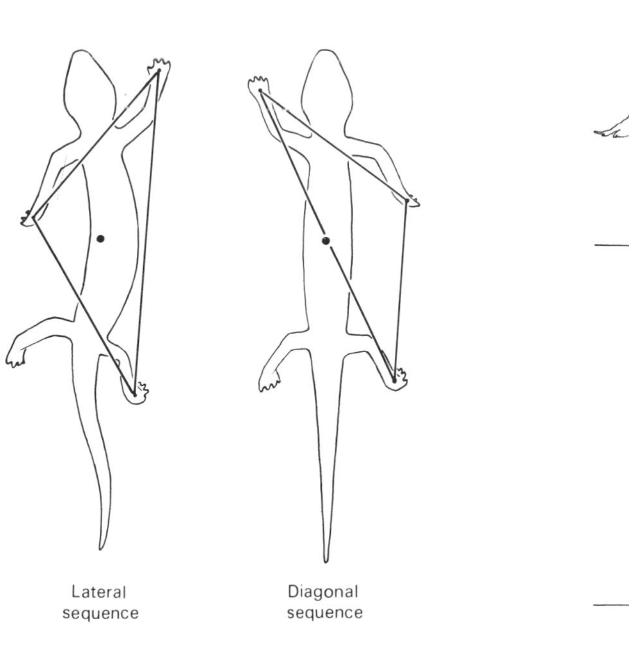

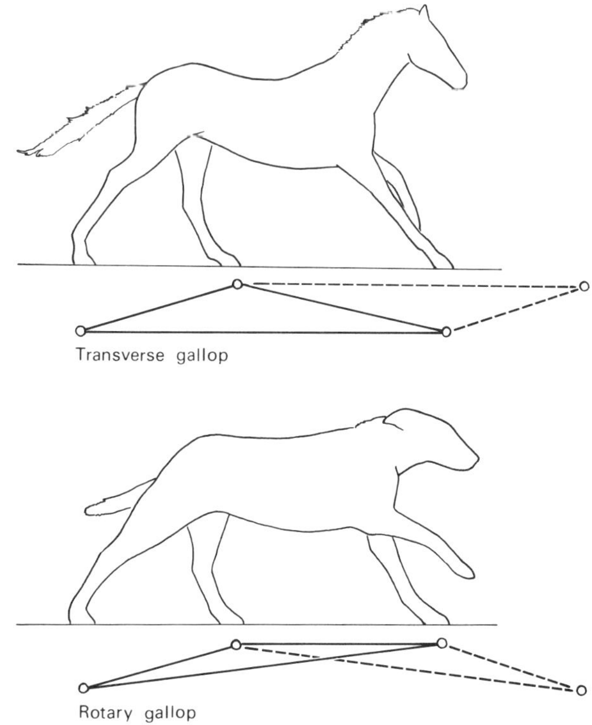

Figure 3-4 Factors affecting stability for a sprawling salamander and upright horse and dog. Using lateral sequence, the salamander benefits from a larger triangle of support that is better positioned under the center of mass *(dot)*, as well as from undulation that increases excursion for both pairs of legs. If diagonal sequence were used, undulation could not simultaneously benefit excursion for both pairs of legs. Triangles of support are larger for the transverse gallop than for the rotary gallop (width of track arbitrary and exaggerated).

slightly adjusted as necessary to provide stability. Indeed, the running quadruped, like the fast-walking or running human, can be falling much of the time provided that, overall, the successive thrusts of the feet counteract gravity.

Maneuverability

Maneuverability is the capacity for rapid and controlled change of speed and direction. Predator and prey often have high maneuverability according to tactics. Running animals must maneuver to place their feet safely among obstacles; herding animals must maneuver to coordinate their fast travel. The structural bases for maneuverability have scarcely been studied, but the anatomy of such seeming experts as large felids and small antelope does accord with expectations.

The requirements for maneuverability appear to be the opposite of those for stability. With relatively long legs, cursors have a relatively high center of mass. Being unguligrade or digitigrade, the supporting part of the foot is relatively small. Gaits are favored that provide relatively small areas of support (for instance, rotary gallops and running gaits that have no triangles of support). To alter the direction of travel, the long legs are readily brought down to one side or the other in relation to the center of mass, thus thrusting somewhat obliquely.

Although the trajectory of the center of mass cannot be altered while an animal is suspended, the orientation of the body can be changed in flight by flexing the spine, swinging the head and neck, or whipping the tail (see, for example,

Bartholomew and Caswell, 1951). Such behaviors prepare the animal for dodging in a new direction when the feet next strike the ground.

As an animal turns, centrifugal force causes it to tend to skid out of the turn. This force varies directly with the mass of the animal and the square of its velocity and inversely with the radius of the turn. To oppose this force, the animal leans into the turn. The reaction of the ground against the diagonally thrusting foot then has a horizontal component directed into the turn (that is, centripetal), which is expressed as friction between foot and ground. The animal must lean sufficiently (without loss of traction) so that the centripetal force equals the centrifugal force. Furthermore, while turning, the galloping quadruped leads with its inside forefoot. This enables successive footprints of the pair of feet to be closer to the curving line of travel than if the opposite lead were used. Animals that can dodge readily change lead with facility.

Since acceleration equals force divided by mass, and momentum equals mass times velocity, maneuverability is favored by small mass. Very large quadrupeds can have only limited agility. Most cursors can be agile by virtue of the structural and behavioral modifications noted above, but it appears that small quadrupeds that move quickly, such as many lizards and mice, attain high maneuverability without evident modifications of these kinds.

Lever Mechanics and Torque

The locomotor mechanisms of walkers and runners consist largely of bony levers activated by muscles. Accordingly, the principles of lever mechanics were frequently applied to studies of tetrapod locomotion following the publication of an influential paper by Maynard Smith and Savage in 1956. As we shall see, however, the interactions of moving limb segments can be exceedingly complex, monitoring loads and angles for parts of the functioning body is difficult in practice, and mechanics alone cannot be sufficient to interpret body form. Therefore, this method has recently been less stressed as other approaches to the analysis of locomotion have developed. Nevertheless, the principles are basic and do help to explain much observed structure. The terms used here, and some fundamental relationships, are shown in Figure 3-5.

Applications Consider a simplified model: a limb is rotated at a proximal pivot by a single one-joint muscle, joints distal to the pivot are stiff, and the muscle is unloaded (that is, the limb is weightless). The out-velocity (V_o) produced at the foot equals the in-velocity (V_i) at the muscle insertion, times the out-lever (or load arm) (L_o), divided by the in-lever (or power arm) (L_i). In-velocity varies over a rather limited range for muscles of similar architecture. Other factors being equal, evolution might be expected, there-

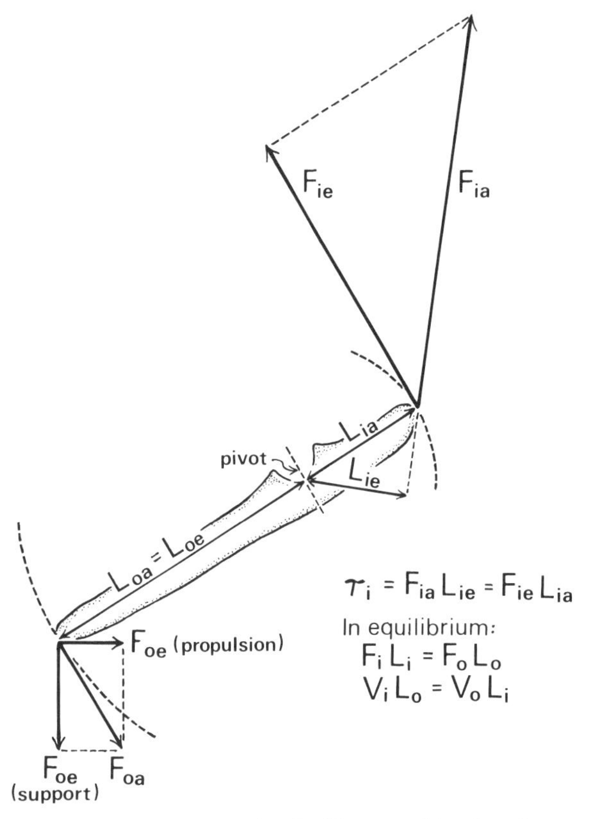

Figure 3-5 Lever mechanics illustrated by the ulna and triceps muscle. F = force; L = lever arm; i = in; o = out; a = actual; e = effective; τ = torque; V = velocity.

fore, to increase V_o at the foot by increasing L_o (lengthening the limb segments) and decreasing L_i (moving the muscle insertion closer to the pivot). See Figure 3-6.

As expected, among tetrapods that run quadrupedally, the faster an animal can move (compared with related forms), the longer are its legs in relation to body length. Among tetrapods that run bipedally (various mammals, birds, and ornithopodian and theropodian dinosaurs) the hind limbs lengthen even more, whereas the forelimbs are responsive to other behaviors. Among vertebrates that tend to bipedalism at maximum speed (various lizards: Rieser, 1977; small ceratopsian dinosaurs: Coombs, 1978) the hind limbs lengthen, but forelimbs are little modified.

Likewise, it is commonly recognized that many limb muscles of mammalian cursors do insert closer to the associated pivots than the same muscles of their less swift relatives. The scant data for other tetrapods (for instance, Rieser, 1977, on some lizards) are in agreement. As noted below, however, shortening L_i increases load, which modifies muscle physiology. Accordingly, the relative value of L_i must not be interpreted merely on the basis of lever mechanics.

When several muscles tend to turn a bony lever in the same plane and direction at the same time, the out-torque equals the sum of the several in-torques. However, if really contracting against no load, the in-velocities of several muscles acting about a single pivot could not be added to increase V_o. The muscle with the highest ratio V_i/L_i would determine V_o, because the others could not "keep up."

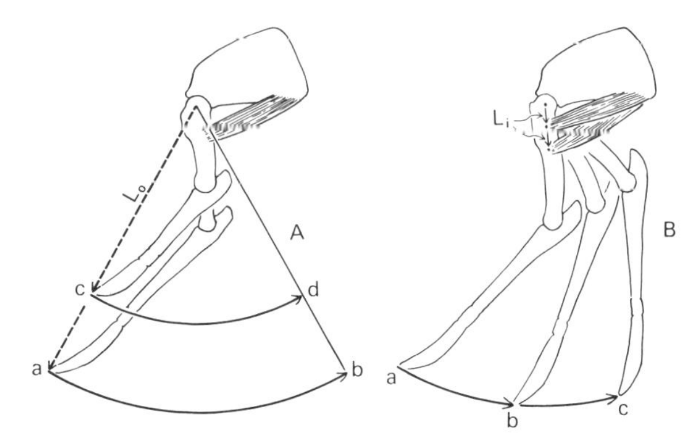

Figure 3-6 Some consequences of lever mechanics. In A, the foot of the longer leg travels from *a* to *b* in the same time that the foot of the shorter leg travels the shorter distance from *c* to *d*. In B, the muscle having the shorter in-lever moves the foot from *a* to *c* with the same shortening (hence time) that the other muscle needs to move it from *a* to *b*. For clarity of illustration the shoulder is shown as the pivot. The same principles apply as the muscle draws the shoulder over the stationary foot.

Complications Functional morphologists can usually measure or approximate the variables so far reviewed, and they customarily carry their analyses at least this far. We now complicate the model, first by recognizing that the lines of action of several muscles having similar function and inserting on one bony lever are unlikely to be in the same place. Furthermore, unless the limb is free to pivot in the plane of action of the resultant vector (as with a ball-and-socket joint), one may need to calculate the effective component of the vector that *is* in the plane of motion. Similarly, the actual F_o and V_o (as of a thrust against the ground) may not be in the plane of the effective force and velocity desired (as for support or propulsion). Three-dimensional analysis is difficult in practice because angles are often hard to measure and represent, so a two-dimensional approximation is frequently used.

We now add the reality that this lever has weight. When the system is static, the component of the weight that is in the direction of motion (according to the constraints of the pivot), times the distance from the center of mass to the pivot, is a torque that augments or (more often) counters the action of the muscles. Similarly, forces resulting from the actions of other parts of the skeleton on the given lever, reaction of the substrate, and wind resistance all load the muscles.

Often one can approximate the mass of a limb (or limb segment), but how much of, for instance, the weight of proximal, extrinsic limb musculature actually loads a given limb muscle at any instant in time is virtually impossible to determine. The center of mass of a segment is easy to find only if the part can be isolated, and its location depends on the positions of any contained joints. Wind resistance, which must be substantial for the faster cursors, has so far not been measured for any quadrupedal runner, let alone for its segments.

Understanding that the load on limb muscles increases as the center of mass of the lever moves away from the pivot (a fact that is underscored below when our model is allowed to oscillate) provides further insight into body proportions. As noted above, runners have relatively longer legs than nonrunners. For mammals it is always the metapodial and epipodial segments, not the propodial segment, that lengthen. This appears to be because the proximal segment is heaviest. Keeping it relatively short holds the center of mass of the entire limb closer to the girdle and allows it to move more slowly. Compactions and fusions of foot bones, reductions and loss of foot bones and muscles, and change of foot posture all help to keep the distal limb segments of cursors light. A study I have recently concluded with J. P. Hurley verifies that there is significantly more energy in the fast-oscillating leg of the cheetah than in the more tapered leg of the pronghorn, even when adjustment is made for weight, length, or joint angles. (These mechanical considerations notwithstanding, the relative lengthening of distal limb segments is less pronounced in cursorial dinosaurs and is not found in cursorial lizards. Furthermore, Taylor et al., 1974, did not find differences in length of limb relative to body weight or in distance from center of mass of the extended limb to proximal pivot to correlate with the energy cost of slow running for the cheetah, gazelle, and goat. Perhaps differences in the recycling of spring energy and other functional factors compensate, at least in part, for the structural differences among the limbs of these animals.)

The concept of several muscles acting on one loaded lever has long suggested to various authors (for example, Maynard Smith and Savage, 1956) the analogy of high and low gears. Since a long L_i favors F_o at the expense of V_o, if L_i differs for synergists (as for the gluteal muscles versus the muscles at the back of the thigh), then those with the shorter L_i (here the gluteals) have "higher gears" than their counterparts. At usual velocities all can contribute to F_o, although the high-gear muscles translate F_i less effectively. Possibly, however, at maximum instantaneous velocities of a lever, only the high-gear muscles are effective, the others not being able to "keep up." Electromyography cannot distinguish among muscles on this basis. (See mention of alternative approaches for future study in Chapter 17 and further discussion of the concept of gears below and in Chapter 4.)

Difficulties The model is now extended to include levers that (like all limbs) are jointed, and muscles that (like many limb muscles) cross, by their tendons, more than one joint. Bock (1968, 1974) used the free body diagram method to show that an isolated, multiple-joint, bone-muscle system is nonstatic, unstable, and indeterminate. The consequences of a muscle force cannot be ascertained from a knowledge of morphology. Any of the joints crossed by the muscle may be moved, in any combination and to various degrees. Furthermore, since one cannot calculate motions at joints spanned by a multiple-joint muscle, one also cannot calculate changes the muscle force will produce in the position of the center of mass, either of the entire unit or of any of its segments.

Far from causing "difficulties" to the animal, multiple-joint systems have potential advantages (Bock, 1968, 1974). If not pinnate (that is, having short fibers that slant to insert on one or more long tendons that can be central or superficial to the belly of the muscle), they tend to have longer fibers and longer lever arms. Motions they cause at the successive joints they cross may compensate in such a way that an active muscle is isometric or nearly so. As noted below, there are physiological advantages to muscle activity that is nearly isometric. Bock reasoned that multiple-joint muscles tend to be more versatile in their actions than one-joint muscles. This is potentially true on mechanical grounds, but Gambaryan et al. (1971) found that a multiple-joint muscle (biceps femoris) of the cheetah tends to link hip, knee, and ankle in a relatively limited way. The comparative function of single-joint versus multiple-joint systems invites further study.

If several joints of a limb are independently turned in the same direction at the same time, the V_o produced in the segments can be added to derive the combined V_o at the foot. This is a reason that cursors free as many joints as possible to rotate in unison: lizards, carnivores, and rabbits rotate the spine in the plane of the leg swing, the scapula of mammals rotates with the forelimb, and digitigrade and unguligrade foot posture allows rotation of the foot, as in the wrist of a greyhound and the fetlock of a horse.

The model is no longer simple for the human analyst (although the cerebellum of the active animal does manage), and changing from a static to a dynamic model will further complicate the mechanics. Before we consider oscillation, however, some relevant aspects of muscle function should be noted.

Some Properties of Muscles

Most striated muscles of most of the tetrapods so far tested have more than one type of fiber, the ratio of the types varying by muscle and by species. The characteristics of the types are fully presented in Chapter 17. It appears that as demands on the muscles increase, the three kinds of twitch fibers are recruited in sequence. Slow oxidative (SO) fibers are activated to maintain posture and engage in slow repetitive activities; fast oxidative glycolytic (FOG) fibers are added as activity increases (walk, trot); and fast glycolytic (FG) fibers come in at the moderate trot and gallop. This sequence may even apply within one action cycle of one muscle, thus establishing a gear system at the tissue level (Goldspink, 1981). One might predict that sprinters would have relatively many FG fibers in certain limb and back muscles. There is some suggestion of this, but the data base remains too small to draw conclusions.

The rate of shortening of an unloaded muscle increases with temperature and is proportional to intrinsic rate of shortening and the number of sarcomeres in series. The intrinsic rate differs from species to species, muscle to muscle, and even fiber to fiber, yet it varies only about one order of magnitude for tetrapods of very different habits and sizes. It is not yet known if intrinsic rate of shortening tends to be faster for runners than nonrunners. The number of sarcomeres in series is about proportional to fiber length, but not necessarily to muscle length. The more fleshy limb segments of cursors are relatively shorter, not longer, than those of noncursors.

Velocity of shortening falls off rapidly as load increases, dropping to less than half the unloaded rate when load increases to only one quarter of the muscle's capacity at resting length (see discussion and curves in Chapter 17). Increasing load also decreases the amount of shortening. In discussing the mechanics of unweighted levers we noted that F_o is favored when the ratio L_i/L_o is high, and V_o is favored when it is low. However, reducing L_i/L_o increases load, which *decreases* both the rate and distance of shortening. Since loads may be considerable, very high "gear ratios" may not benefit cursors after all. Seemingly, the body must seek optimum compromise.

Muscles generate maximum force when contracting at near their resting lengths, as is characteristic of isometric contractions. Furthermore, such contractions are relatively efficient (that is, they use minimal energy to accomplish their function) because they need generate little or no heat of shortening. Again, the benefit falls with the short L_i.

Motion, Energy, and Oscillation

Our model has so far been static, or, if moving we have considered only instantaneous torques, lever arms, and vectors. We proceed to dynamics. As an animal accelerates, as when a predator breaks from ambush to overtake its prey, equations for rectilinear motion apply reasonably well for the body as a whole: the propulsive force equals the body's mass m times its acceleration, the velocity v equals its acceleration times time, its momentum equals force times time, and its kinetic energy equals $\frac{1}{2}mv^2$. These relationships explain the observations that very heavy animals can alter velocity only relatively slowly, whether by acceleration, deceleration, or turning, and that graviportal quadrupeds can be ambulatory but not truly cursorial.

Oscillation Nearly all motions of moving quadrupeds, being associated with lever systems, are curvilinear. The limbs always oscillate, and spine and girdles often do so. Start with a simplified model consisting of a rigid limb that oscillates in one plane around a pivot at the hip or shoulder. The demands on its muscles are in proportion to the amount of inertia I of the limb, which is an expression of its resistance to acceleration and deceleration, and $I = mD^2$, where D is the radius of gyration. The radius of gyration varies with the length of the limb and the distribution of its mass. It is low when the weight is more proximal and high when it is more distal. Since D is squared in the formula for I, its reduction is a principal reason for the limb proportions and proximal weight distribution of cursors that were noted in a preceding section. The mechanics are simple, but applications are not: there is no way to measure D for either an irregularly shaped object or one that is not uniformly dense (and a limb is both of these) without isolating it and precisely timing its period of oscillation.

Now modify the model to a leg having a proximal segment that oscillates about the hip or shoulder, a middle segment that oscillates at knee or elbow, and a distal segment that oscillates at wrist or heel. Other oscillations in the feet, shoulder, and spine could be added, but already the problem of calculating the energy required to move the system at a given rate has escalated because every motion of each segment alters the kinetics of all the others. If the dimensions, masses, and centers of mass of the segments are known, and the trajectories of the joints are placed in time, then fluctuations in the energy of the system can be calculated (see Manter, 1938, for an early example, and Fedak, Heglund, and Taylor, 1982, for a more recent example).

Complexities of the mechanical approaches so far noted do not curtail progress in the analysis of walking and running, for there are other approaches. For example, animals moving on treadmills can be used to study energy-budget equations (for instance, Taylor et al., 1974; Chapter 10). Slow-motion cinematography (including cineradiography), force plate measurements taken at ground level, and strain gauge recordings from the moving skeleton offer productive means of investigating the dynamics of locomotion (for example, Jenkins and Camazine, 1971; Alexander, 1974; Jayes and Alexander, 1978, Jenkins and Weijs, 1979; Biewener, Alexander, and Heglund, 1981). Data obtained in these ways may be used to provide realistic assumptions for models of locomotion, but many models assume rather stylized form and behavior. Each model advances our understanding of some aspect of locomotion.

A Work Model and A Ballistics Model Alexander (1980) has presented a complex mathematical model of locomotion (bipedal and quadrupedal) that is a work theory. It evaluates the work done by the muscles (disregarding the various forms of mechanical energy and the total energy budget) to estimate the total metabolic power requirement for walking with various contact intervals, stated phase relationships between fore and hind limbs, and various values of a factor q. This factor describes the time course of the vertical component of force on the foot (for instance, showing if there are one or two maxima per cycle), thus introducing the power required to repeatedly lift the animal's center of mass. The model assumes that head and body are a rigid unit and that the limbs are weightless and are held straight. Contours on a graph, like isotherms on a weather map, show the power required (in relative terms) for man, a sheep, and dogs as they walk. At slow speeds man selects gaits that are close to the theoretical optimum for matching values of contact interval and q; sheep and dogs tend to select gaits with lower values of q than predicted.

Mochon and McMahon (1980) developed a ballistics model of human walking. This model, like those of subsequent sections, follows energy theory: it is concerned with the total mechanical energy of a system. The optimum relationships between step frequency and step length are calculated, assuming that the muscles impart initial velocities to thigh and shank at the start of the swing phase of the limb and that the leg then swings entirely under the influence of gravity until the foot strikes the ground. The pelvis is assumed not to rotate or tilt, the support leg is straight, and the ankle is fixed at a right angle. The knee of the swinging leg has sufficient initial flexure so the foot will clear the ground at midswing, but it must lock straight before the heel touches down. Natural walking matches the theory well.

Limbs as Pendulums The total amount of energy that would otherwise be required for locomotion may be much reduced by either or both of two mechanisms: an exchange in each locomotor cycle between gravitational potential energy and kinetic energy (as for a pendulum), and the storage and subsequent recovery of elastic strain energy in tendons and other tissues (as for a spring).

A pendulum is a body that is suspended from a point and exchanges potential and kinetic energy as it swings to and fro under the pull of gravity. Over the years many authors have considered that pendulumlike oscillation of limbs conserves work, but none has determined how apt the analogy is or has quantified such benefit as may accrue to the animal.

A simple pendulum has its mass concentrated at one point, its center of mass, and is suspended by a weightless cord. The period T of a pendulum swinging through a small angle is $2\pi\ \sqrt{(L/g)}$, where L is the length of the cord and g is the acceleration of gravity. The period is not affected by the mass. The amplitude θ is the angular displacement from the up position to the down position (not the full swing); it ranges to about 15° for the hind leg of a galloping horse and to 55° for that of a dashing cheetah. Although T increases as θ increases, the increment is negligible for values of θ below 60° and so can be ignored here.

A leg is not a simple pendulum; it has an irregular shape and uneven distribution of density. For such a pendulum T is not calculated using L (pivot to center of mass), but instead using the distance from pivot to the center of oscillation. This is the radius of gyration D and, as noted above, is difficult to measure for a limb. (For a simple pendulum only, $L = D$.)

Nevertheless, D can be approximated for limbs of representative proportions and foot postures by models that join a prism, representing the upper leg, and a rectangular solid, representing the lower leg (Fig. 3-7). The upper leg is here assumed to be $1\frac{1}{2}$ times as thick as the lower leg. Only relative measurements are needed, and these are easily determined from the dimensions and volumes of the shapes as drawn on graph paper.

If M is the mass of the prism, m is the mass of the rectangular solid, and other symbols are as keyed in the figure, the moment of inertia of such a "leg" is approximated by

$$(M + m)D^2 = M\left(\frac{h^2}{6} + \frac{a^2}{24}\right) + \frac{m}{3}(b^2 + c^2) + md^2. \quad (3\text{-}1)$$

Values of D for the five models shown are expressed as percentages of the length of the "leg," pivot to ground.

Figure 3-8 shows the natural periods of pendulums having radii of gyration within the range of those of the hind legs of tetrapods. In the left column are radii for various mammals as approximated from crude measurements of total leg length and interpolation from the models in Figure 3-7. On the right are actual periods (taken from moving pictures) for adults of the same animals walking at usual rates. The match is good; as expected, mammals do tend to walk at the natural periods of their limbs (see also the section on kinematic similarity in Chapter 2 and below for a comment on the accuracy of the method.)

During part of the swing phase of the step cycle the limb is flexed, which decreases D, but only very little for most walkers. The duration of the swing phase remains relatively constant as animals walk more and more slowly, whereas the support phase is retarded beyond the natural period of an equivalent pendulum.

If the stance of a tetrapod is intermediate between upright and sprawling, the leg may swing as a pendulum that is not vertical in the down position (for a raccoon or opossum it might be displaced to the side about 15°). This increases the period, but only slightly for displacements below 40°.

If the pivot of a swinging pendulum is raised on the downswing and lowered on the upswing, T remains nearly constant, but θ increases. The pivot of a limb (relative to the body) is moved in this manner during the support phase as reaction of the ground pushes the girdle up over the foot. This factor may help to compensate for damping resulting from deviation from the limb's natural period.

Since walking tetrapods vary the duration of their step cycles, their limbs often only approxi-

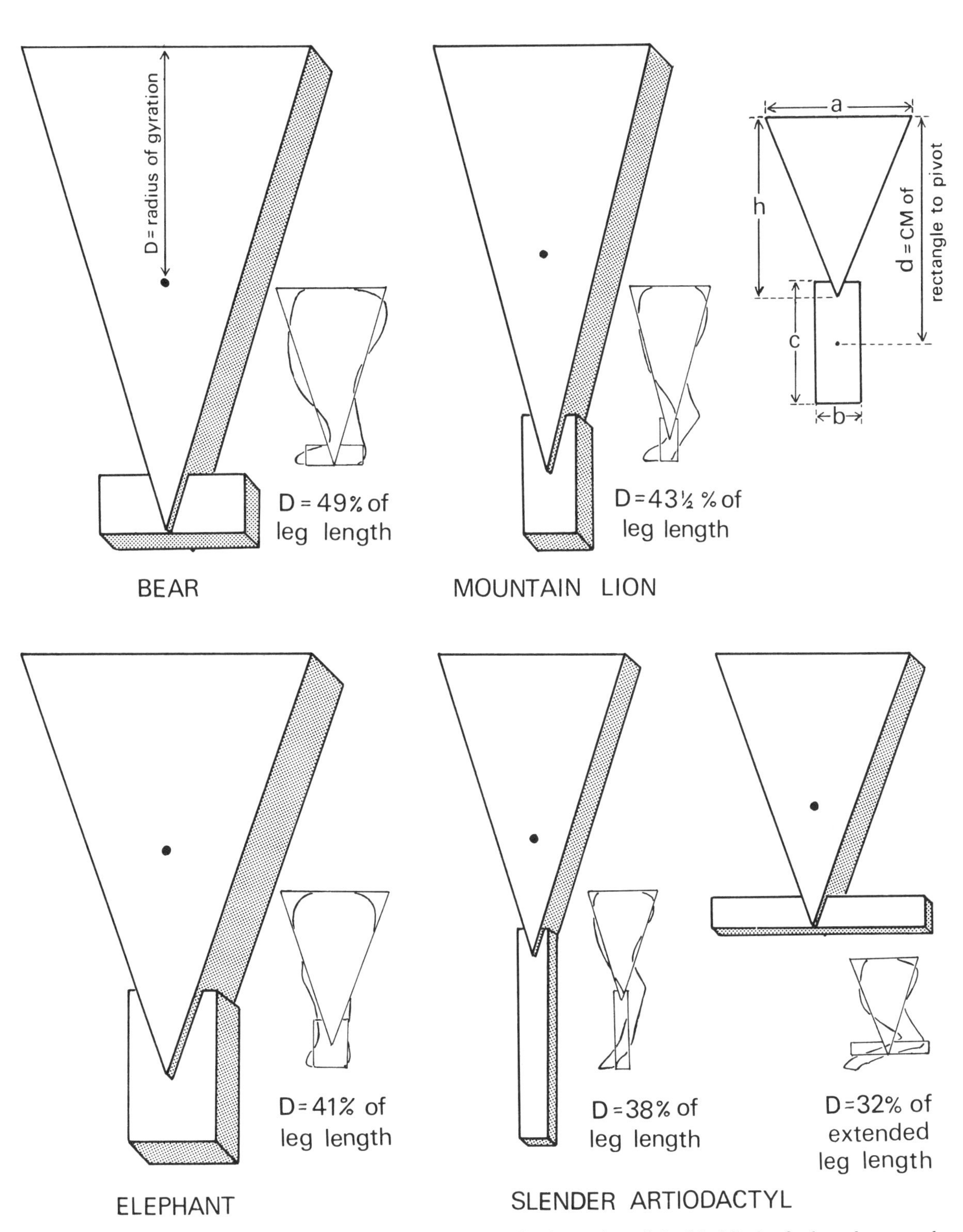

Figure 3-7 Models for the calculation of approximate radii of gyration of the hind limb of selected mammals using the formula in Eq. (3-1).

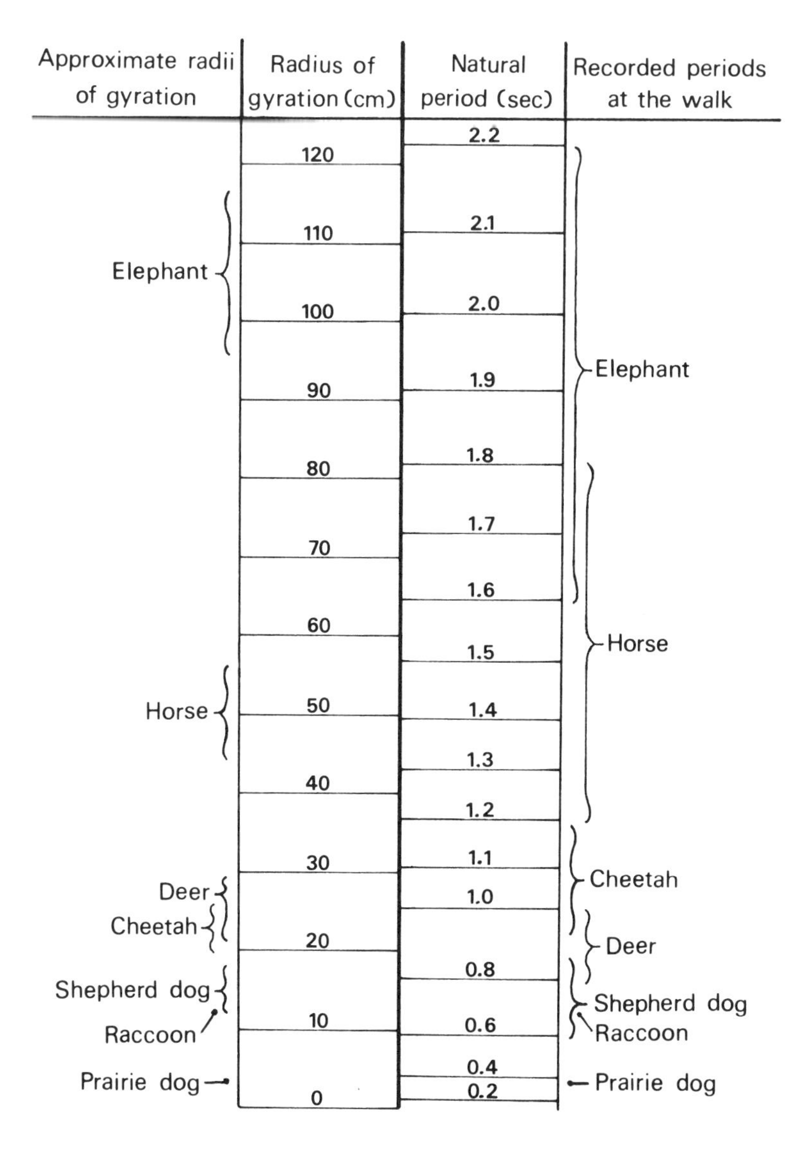

Figure 3-8 Observed periods of oscillation of the hind legs of selected walking mammals, compared with the natural periods of pendulums having equivalent radii of gyration. Sample sizes range from 3 to about 40.

mate their natural periods. At a fast gallop, the cycle of a horse drops to roughly $\frac{1}{3}$ of the natural period of the hind leg, and for a cheetah to about $\frac{1}{5}$ of the natural period. How much energy is stored by a driven pendulum? The work done by a pendulum is

$$2mgD\theta^2 \left|1 - \frac{T^2}{T_o^2}\right|,$$

where T_o is the natural period and T the driven period. Figure 3-9 shows work, where $2mgD\theta^2 = 1$, plotted against T/T_o. It is seen that although dashing cheetah and racing horse benefit little, savings are substantial where the driven period ranges from $\frac{2}{3}$ to $1\frac{1}{4}$ times the natural period. This encompasses most walking gaits, easy trots, and slow gallops (thus giving confidence in the general conclusion, even though the dimensions and distributions of mass assumed in Fig. 3-7 are inexact).

These data indicate theoretical maximum savings of energy attributable to the pendulumlike action of the limbs. Actual savings remain to be measured and doubtless are less.

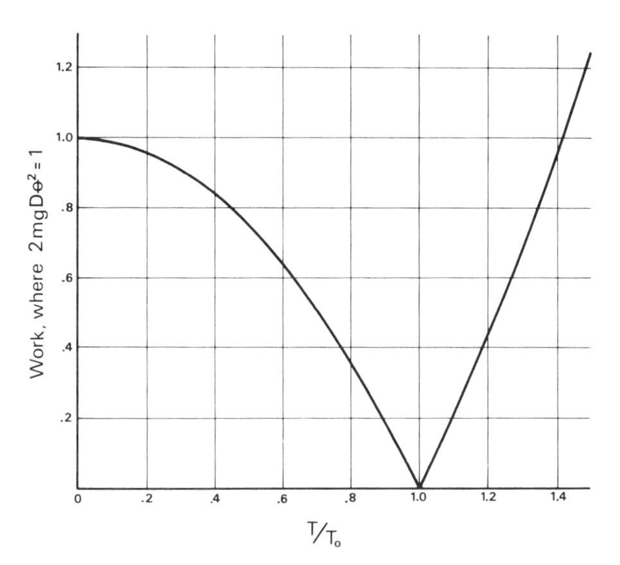

Figure 3-9 Work needed to drive a pendulum plotted against the ratio of the driven period *(T)* to the natural period *(T_o)*.

Limbs as Inverted Pendulums Several authors have equated the limb in the propulsive phases of successive locomotor cycles to an "inverted pendulum" to explain a method bipeds and quadrupeds use to conserve work output. Cavagna, Heglund, and Taylor (1977) have provided experimental analysis and are followed here.

Starting with the center of mass (CM) vertically over a supporting foot (position *A* in Fig. 3-10), the CM arcs forward and downward to position *B*, converting potential energy to kinetic energy. As CM then vaults up over the next supporting foot (to position *C*), it decelerates as it is lifted and the cycle is concluded. A push off by the rear foot exerts a force on CM that is normal to its downward arc and smoothly alters its path to the upward arc. An exchange of energy can occur only if there are actual changes in both kinetic and potential energy that are equal in magnitude, opposite in phase, and mirror images of each other. Cavagna and his associates used a force platform to measure the horizontal and vertical components of the resultant force applied by various animals to the ground. The components were integrated electronically to determine instantaneous kinetic energy and change in potential energy. They found remarkable similarity in force and velocity records for three bipedal striders (man, rhea, turkey), two bipedal hoppers (kangaroo, springhare), and three quadrupeds (dog, sheep, monkey). They reported the recovery of energy to be as much as 70% for the bipedal striders at intermediate walking speeds and 35% to 50% for the quadrupeds. At slower and faster walking speeds the savings declined. When trotting, hopping, or running there is scant exchange of kinetic and potential energy because of the phase relationship of their variations. Alexander (1980) believes that actual savings are less than reported because of the work done by one leg of a pair against the other. Mochon and McMahon (1980) find their ballistics model to be consistent with the inverted pendulum model.

Tendons and Ligaments as Springs Elastic storage and release is a second way that some animals recycle energy during locomotion. The capacity of muscles for storing elastic strain energy is limited, but the collagen of tendon and ligament can store 2000 to 9000 Jkg^{-1} (Alexander and Bennet-Clark, 1977). The most effective are the long tendons of muscles that are stretched before shortening as the feet contact the ground. Such muscles are often pinnate. For mammals, extensors of the ankle and knee and flexors of the wrist and digits are of this kind.

Quantification of elastic storage is relatively difficult. Savings appear to range from about 15% of the work output that would otherwise be required, as in kangaroo rats (Biewener, Alexander, and Heglund, 1981), to nearly 50%, as in

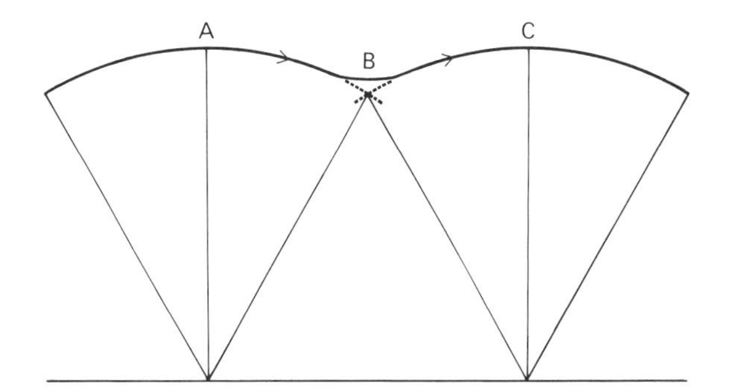

Figure 3-10 Schema of the inverted pendulum mechanism of a walking quadruped. (Redrawn from Cavagna et al., 1977.)

camels (Alexander et al., 1982). Maximum savings approximate 30% in dogs, monkeys, and rams (Cavagna, Heglund, and Taylor, 1977). Spring storage becomes significant only at gaits faster than the walk. At higher speeds the kangaroo's hop is more efficient than quadrupedal running; the animal even hops more economically at moderate than at slower rates (Dawson and Taylor, 1973).

Ungulates have long springing ligaments (evolved from muscles) that pass down behind the cannon bone, under the fetlock joint, and onto the phalanges (Fig. 3-11). These provide passive support during standing; during running they cushion impact and give substantial elastic storage. The check ligament of the front foot (between the tendon of the deep digital flexor and the cannon bone) of a 400-kg horse broke when the foot was experimentally loaded to 535 kg. The suspensory ligament (between phalanges and cannon bone) stretched 8% before breaking at 720 kg (Camp and Smith, 1942). Wentink (1979) believes that part of the energy stored on impact by the associated tendons of the digital flexor muscles stretch (when released) the tendinous peroneus tertius (in front of the shank), which in turn stores the energy and, like an echo, releases it to flex the hock (heel) when the foot is lifted.

Some mammals are more efficient when galloping slowly than when trotting at the same speed. Taylor (1978) believes that this is the reason the animals change gait and that at the gallop they recruit a spring in the back that is longer than those of the legs. English (1980) shows that all lumbar epaxial muscles are active in the galloping cat. In canids, felids, and rabbits (at least) the longissimus and iliopsoas muscles, which are pinnate and have long tendons, are in position to alternately extend and flex the back of the galloping animal and may function as reciprocal springs. The iliopsoas is particularly large in rabbits and the cheetah.

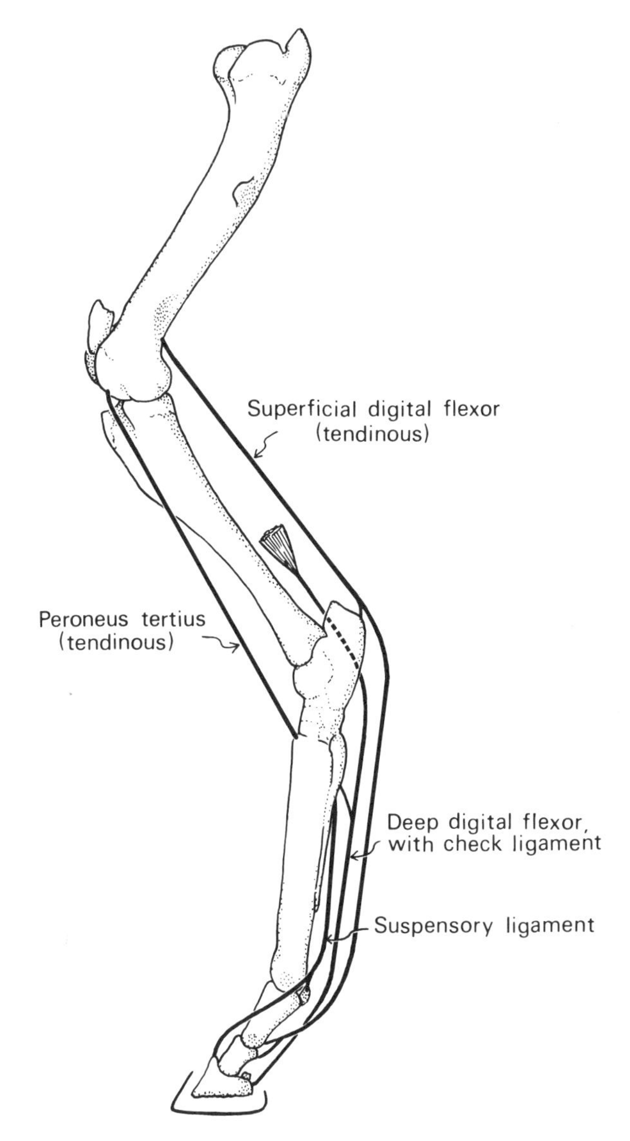

Figure 3-11 Springing ligaments and tendons in the hind leg of the horse.

Oscillations of the Back and Girdles Some rotation of the back, and with it of the girdles, is usual for most moving quadrupeds. For each gait the degree of rotation increases with speed. Rotation in the horizontal plane (Figs. 3-4, 3-12) ranges up to 49° in salamanders (Edwards, 1976), is about the same for the iguana (Brinkman, 1980), occurs in various birds, and reaches at least 14° in mammals of intermediate limb posture (Jenkins and Camazine, 1977). Rotation around the long axis of the spine is observed in erect and semierect bipeds. Rotation in the sagittal plane is typical of various running carnivores, rabbits, and rodents, reaching about 70° in the cheetah (Hildebrand, 1959; Fig. 3-13).

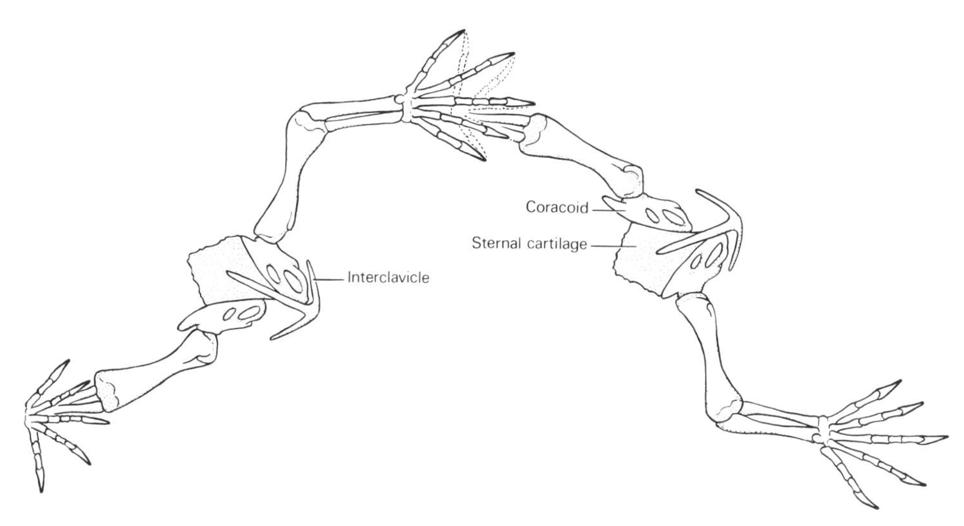

Figure 3-12 Ventral view of the pectoral girdle and forelimb skeleton of a walking savannah monitor lizard *(Varanus exanthematicus)* showing how a long step results from a combination of lateral undulation of the sternum and thorax (40–60°), translocation of the coracoid on the sternum (equivalent to 40% of the length of the coracoid), anteroposterior excursion (40–55°) and axial rotation (30–40°) of the humerus on the coracoid, and various motions of the forearm and foot. (Redrawn from Jenkins and Goslow, 1983. The data were derived using cineradiography.)

Furthermore, the pectoral girdle may oscillate on the thorax. The coracoid of one walking varanid lizard is displaced along the sternum by as much as 40% of the length of the coracoid (Jenkins and Goslow, 1983; Fig. 3-12). The shoulder joint of painted turtles rotates about 16° relative to the carapace (Walker, 1971). The clavicle of mammals serves as a spoke, which assures that the shoulder joint will move in an arc around the manubrium (Jenkins, 1974). Where the clavicle is rudimentary or absent (canids, felids, rabbits, elephants, ungulates), the scapula is free to rotate in the same plane as the swinging limb. The ex cursion may reach 20° and may be coupled with translocation (Hildebrand, 1961; English, 1978).

Rotation of the vertebral column in the same plane and direction as the swinging limb and the coordinated rotation of the scapula on the thorax (when that occurs) enable the animal to take a longer step in the same time interval. Because the angular velocities of independent joints rotating in the same direction at the same time can be added at the foot, the animal might also complete the same step in less time.

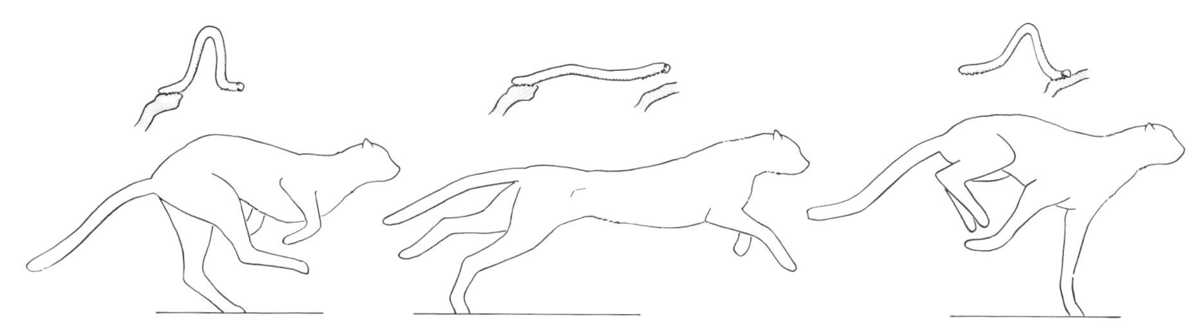

Figure 3-13 Measuring worm–like progression of the cheetah resulting from oscillation of the back coupled with appropriate timing of footfalls.

Two other benefits of spinal rotation may accrue to galloping mammals (Hildebrand, 1959): the distance between the girdles is greater when the vertebral column is extended than when it is ventroflexed, and the timing of limb action is such that the trunk of the animal progresses like a measuring worm between successive support phases by fore and hind limbs (Fig. 3-13). Also, during ventroflexion (when forefeet are on the ground) the animal's CM moves forward farther, and therefore faster, than the pectoral girdle, and during extension (when hind feet are thrusting) it moves forward faster than the pelvic girdle. Consequently, the animal (that is, its CM) travels a little faster than the active girdle.

Conservation of Angular Momentum As noted above, the moment of inertia of an oscillating limb equals mD^2. Angular momentum L equals $mD^2\omega$, where ω is angular velocity. Since L is conserved, if ω for a heavier, more proximal segment of an oscillating limb decreases, ω for the lighter, more distal segments tends to increase. Also, shortening D increases ω for the entire limb. This principle is exploited by divers and gymnasts (Frohlich, 1980), but has not yet been applied to walking and running animals. For many mammals, most of the extension of the lower leg at the knee during the recovery stroke occurs after the forward swing of the heavier thigh has decelerated or stopped. For some mammals the same is true of the forefoot relative to the forearm.

Some Passive Mechanisms

Large quadrupeds and cursors of all sizes have passive mechanisms to reduce loads and conserve energy.

Strength with Light Weight Runners sacrifice some strength to conserve weight. They tend to be more trim and less powerful than walkers and diggers of similar body size. However, several classes of features contribute adequate strength to their oscillating parts in spite of light weight. First, there may be reduction or loss of digits, ranging from one (most birds and various genera among dinosaurs, marsupials, carnivores, rodents, and rabbits), through two digits (some representatives among dinosaurs, birds, rodents, ungulates, and extinct notoungulates and litopterns), to loss of three or four digits (ostrich, some bandicoots, many ungulates, some litopterns). Among the remaining metapodials compaction is usual and fusion occasional (birds, artiodactyls). The result is more strength per unit weight of skeleton. The compressive load on the cannon bone (third metacarpal) of a horse reaches nearly five times body weight at the trot (Schryver et al., 1978).

Similarly, the fibula of birds and cursorial mammals is reduced, bound to the tibia, or vestigial, and the ulna of such mammals may be reduced distally and adherent or fused to the radius. Accompanying these changes are reduction or loss of pronators, supinators, and adductors and abductors of digits.

Cushioning of Impact The strength and weight that would otherwise be needed by runners may be greatly reduced by passive mechanisms that cushion impact. Springing mechanisms were described above that store energy for utilization on release. Importantly, the same mechanisms serve also to cushion impact. The hoofs of ungulates spread when heavily loaded (Butler, 1974), and a pad of fatty connective tissue under the foot (elephant) or terminal phalanx (horse) distributes the load evenly to more distal supportive tissues.

Restriction of Motion The joints of walkers more than climbers, and of runners more than walkers, restrict limb motion nearly to sagittal planes and passively resist dislocation. Unfortunately, joint mechanics of nonprimates is a neglected, though potentially rewarding, field (but see, for instance, Cracraft, 1971, for leg of a pigeon; Yalden, 1970, 1971, for carpus of carnivores and ungulates; Jenkins and Camazine, 1977, for hip of carnivores; Brinkman, 1980, for feet of lizards; Jenkins and Goslow, 1983, for shoulder of a lizard).

For running birds and mammals (though scarcely for running dinosaurs) limb joints distal to shoulder and hip become hingelike. The hinge may have a proximal cylinder (like the pin of a door hinge) that turns in a distal trough (as at elbow and interphalangeal joints). Lateral deviation is passively but strongly resisted (as in diggers, see Chapter 6) by splines or flanges that flare from the cylinder at right angles to its axis of rotation and fit into grooves in the trough. Alternatively, the hinge may consist of stacked blocks that pivot along their posterior edges to open away from one another (podials of ungulate and graviportal mammals). The range of flexion may be great (Yalden, 1971).

At the hip cursorial (as opposed to noncursorial) carnivores restrict axial rotation and abduction by having a deeper acetabulum, more restricted articulatory surface on the femoral head, and more horizontally oriented ischium (Jenkins and Camazine, 1977). Ungulates have been less studied in this regard, but in them (and kangaroos) the heads of femur and humerus have a markedly longer radius of curvature in the transverse plane of the body than in the longitudinal plane. This must limit axial rotation of the bones and restrict motions that are not in the sagittal plane. The head of the femur has a proximal shoulder and consequently little or no constriction of the neck. In birds and in dinosaurs with an upright stance, force is transmitted to the girdle primarily by the dorsal surface of the femoral head, and (for reasons inadequately explained) the acetabulum becomes perforate (Charig, 1972). The shoulder joint of mammals is intrinsically weak, yet is stabilized by the synchronous activity of subscapularis, supraspinatus, and infraspinatus muscles (Jenkins and Weijs, 1979).

Most hinge joints are bordered by collateral ligaments. If one end of such a ligament is anchored at the axis of rotation of the joint, then its length, like the radius of a circle, remains constant as the joint turns. If the ligament is anchored elsewhere (usually being a little longer than the above radius), then it is under more tension in some joint positions than others. Such "snap joints" have long been identified at the elbow and hock (heel) of ungulates, but their function is inadequately known. They are in equilibrium in the open and closed positions, and considerable force (not yet quantified) is needed to move the joint between these positions. Perhaps they provide some passive support against collapse of the leg, or perhaps they allow muscles to contract briefly isometrically before they shorten, thus (like those of the click beetle) increasing the velocity of their initial shortening.

Graviportal and large cursorial mammals passively restrict motion of the posterior trunk region, apparently to avoid expending the energy that would be needed either to cause controlled oscillation or actively to prevent oscillation (in reaction to the action of the limbs) in such massive parts. This is accomplished by having broad (lengthwise to the animal) neural spines and transverse processes that nearly abut against one another, by enveloping zygapophyses, by ligamentous ties, and by shortening the lumbar region (that is, by adding ribs).

Future Study

The functional morphology of overground locomotion has been as much studied as that of any locomotor habit, yet so much remains to be learned that we are again reminded that the moving animal is a wonderfully complex machine. Force analysis, now familiar for jaw and elbow, must be extended to feet, pelvis, and vertebral column. The physics of inertial systems and curvilinear motion has scarcely been applied to moving animals using one plane, let alone three dimensions. Acceleration has received scant attention. The ontogeny of form in relation to behavior has attracted only several pilot studies. The extreme (limiting) efforts of the master cursors have been recorded by the camera but not the oscillograph. The responses of heart, vessels, and respiratory and sensory structures to locomotor habit have only recently attracted attention (for example, Bramble and Carrier, 1983). Walking and running have been analyzed and modeled using metabolic output, lever mechanics, ballistics, pendulum and inverted pendulum, spring, joint angles and gait, force plate, sensory feedback, motor pathway, and other approaches. One hopes we may soon progress to better integration of these avenues. The moving animal integrates them all—perfectly, constantly, and almost instantaneously.

Chapter 4

Jumping and Leaping

Sharon B. Emerson

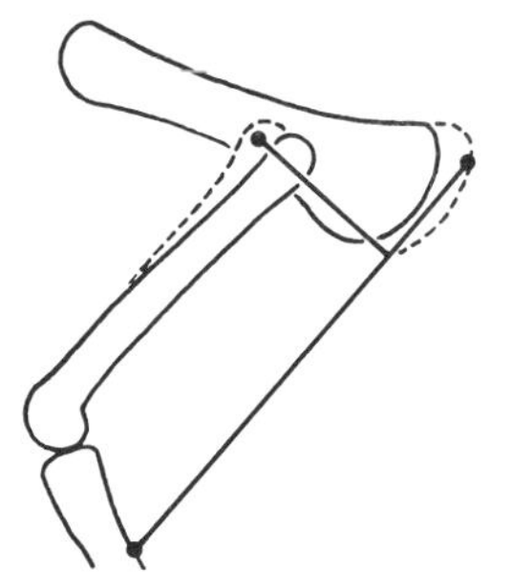

A number of aspects of jumping locomotion make it an interesting behavior to study from the perspective of the evolution of function. Jumping has evolved independently among vertebrates at least six times, and it occurs in species whose body weights vary from less than a gram to more than 50 kg. Jumpers are characterized by a diversity of body forms and saltatory behaviors. Kangaroos and springhares use jumping for sustained locomotion, whereas frogs and the vertical-clinging and leaping prosimians usually make only one or a few jumps at a time. Film tracings of jumps show considerable variation in the degree of flexion and extension of the limbs, angle of takeoff, positioning of the body, and bending of the vertebral column among jumping species (Gambaryan, 1974). For at least some species, the storage and release of spring energy makes sustained jumping more energy-efficient than the quadrupedal gallop, as explained in Chapter 3. Because the maximum stresses that act on limb bones and muscles appear to be greater for a standing jump than for running (Alexander, 1974), mechanical analyses of jumping provide insight into the general shape and architecture of bones and muscles. The small size of some vertebrate jumpers seems to require that they use stored energy in elastic tissues to provide a portion of the power during takeoff, but this has not been closely examined. Finally, jumping occurs in species that are ecologically and behaviorally diverse, which suggests that, as a mode of locomotion, jumping has several potential ecological advantages: quickness and unpredictability, as well as height and distance.

The term *jumping* has been used to describe the locomotion of a wide diversity of animals with a multitude of gaits. In this chapter it is restricted to the locomotor mode in which both hind limbs extend simultaneously to provide the total propulsive thrust. Studies on frogs, kangaroos, prosimian primates, and bipedal rodents are emphasized, but hopping birds, the Pliocene marsupial *Argyrolagus*, and (probably) some dinosaurs would also qualify as jumpers by this definition.

Theoretical Framework

Simple ballistics formulas provide a foundation for predicting the morphological specializations that maximize different jump parameters among animals of similar body shape but different body size. Assuming constant acceleration, the height (h) and the distance (d) that an animal can jump are

$$h = \frac{v^2 \sin^2 \theta}{2g}, \text{ and} \tag{4-1}$$

$$d = \frac{v^2 \sin^2 \theta}{g}, \tag{4-2}$$

where g is the acceleration due to gravity, v is the takeoff velocity, and θ is the angle of takeoff. The velocity at takeoff is given by

$$v^2 = 2as, \tag{4-3}$$

where a is acceleration and s is the distance through which the force acts.

Given these relationships, jump height is maximized at a takeoff angle of 90° and jump distance at 45°. Both height and distance increase with increasing takeoff velocity. Takeoff velocity, in turn, is dependent on the distance through which the force acts and on the acceleration.

If the takeoff velocity is determined by the kinetic energy imparted to the animal by its muscles, this energy (E) is

$$E = \frac{mv^2}{2}, \tag{4-4}$$

where m is the mass of the animal. Combining Eqs. (4-1), (4-2), and (4-4), height and distance of the jump can also be expressed as

$$h = \frac{E \sin^2 \theta}{mg}, \text{ and} \tag{4-5}$$

$$d = \frac{2E \sin^2 \theta}{mg}. \tag{4-6}$$

It follows that animals having the same ratio of muscle to body weight and capable of producing the same specific energy should all jump to the same height or distance, regardless of size (Hill, 1950; Bennet-Clark, 1977).

The power (p) that must be produced in a jump is

$$p = mav = Fv, \tag{4-7}$$

where F is force. It will become apparent in a later section that generation of power poses a special problem for very small jumpers.

Problems with the Model

Studies of the forces and accelerations during jumping show that, contrary to the assumptions of the ballistics formulas, acceleration and force are not constant throughout the takeoff. Figure 4-1 illustrates a force platform record from the jump of *Rana temporaria* (Calow and Alexander, 1973) and the acceleration profile of a *Rana ridibunda* (Hirsch, 1931). The force increases for the first 20 msec of takeoff, then is constant for about 40 msec before beginning to decline. The acceleration is highest during the first part of the takeoff, declining as the animal leaves the ground.

With force and acceleration not constant, the ballistics formulas will underestimate the peak mechanical stresses during the takeoff sequence. However, it is unclear whether this discrepancy is biologically significant. Calculations of maximum bone stresses taken from actual force plate records show that the stresses do not approach the yield stress of bone even though they may be higher than the stresses calculated from an assumption of constant force (Calow and Alexander, 1973; Alexander and Vernon, 1975).

Morphological Predictions for Jumping Species

The ballistics formulas suggest three structural modifications for jumping: longer hind limbs to

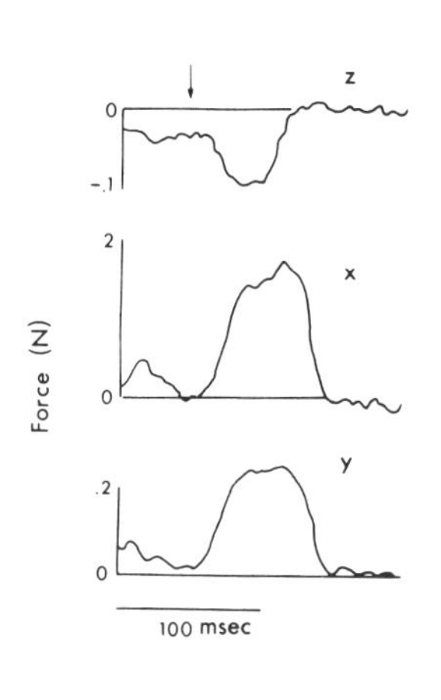

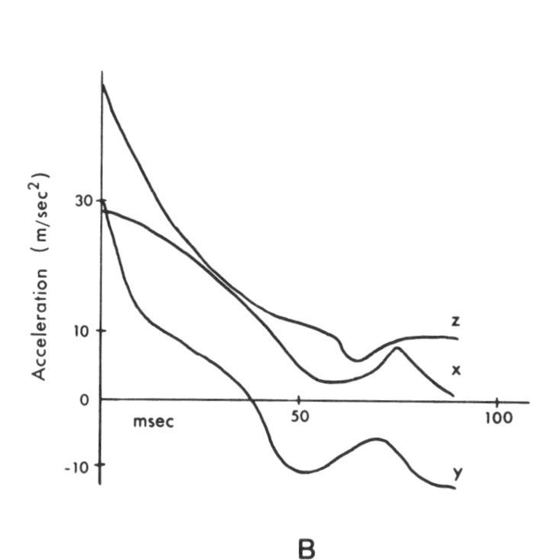

Figure 4-1 Measurements taken during a standing jump of a frog. A: force platform recordings in three planes (from Calow and Alexander, 1973); B: acceleration profile (from Hirsch, 1931).

increase the distance (or length of time) through which the force acts, larger propulsive hind-limb musculature to increase the energy of the jump, and shifts in the origins and insertions of the musculature to maximize the angular velocity of the limbs, the acceleration, and the takeoff velocity within usual load limits.

Problems of Comparison A number of studies have compared jumping and nonjumping species by contrasting limb proportions and body length. Unfortunately, most of the comparisons express the proportions of limb segments as ratios without including the actual measurements. If these ratios vary among species, differences in the numerator, the denominator, or both may be responsible. Without actual measurements, these possibilities cannot be distinguished. Second, some studies compare ratios among jumpers of different sizes, such as a kangaroo and a jerboa. The effects of allometry have not been considered in such studies.

Another problem results from the frequent use of body length as a measure of body size. A common specialization seen in jumpers is shortening of the vertebral column. Thus, relatively high ratios of hind limb to vertebral column in a jumper, as compared with a nonjumper, could result either from an elongation of the hind limb or from selective shortening of the vertebral column. Furthermore, for analysis of jumping, body length is not the most appropriate indicator of size. It is the mass of the animal that is important in terms of the energy of the jump.

Body weights are rarely available for skeletonized museum specimens, but they are available from the literature. Since the sizes of the jumpers in this study range over four orders of magnitude, intraspecific variation in body size is quite low compared with interspecific differences. Thus, the use of body weights from individuals other than those measured for limb lengths will not result in serious inaccuracies in a regression analysis. Consequently, comparisons in this chapter use body weight as a measure of animal size whenever possible.

Length of the Hind Limbs Methodological problems aside, jumpers do have longer hind limbs than do nonjumpers. As shown in Figure 4-2A, saltatory mammals share across taxonomic lines a generally similar relationship of hind-limb

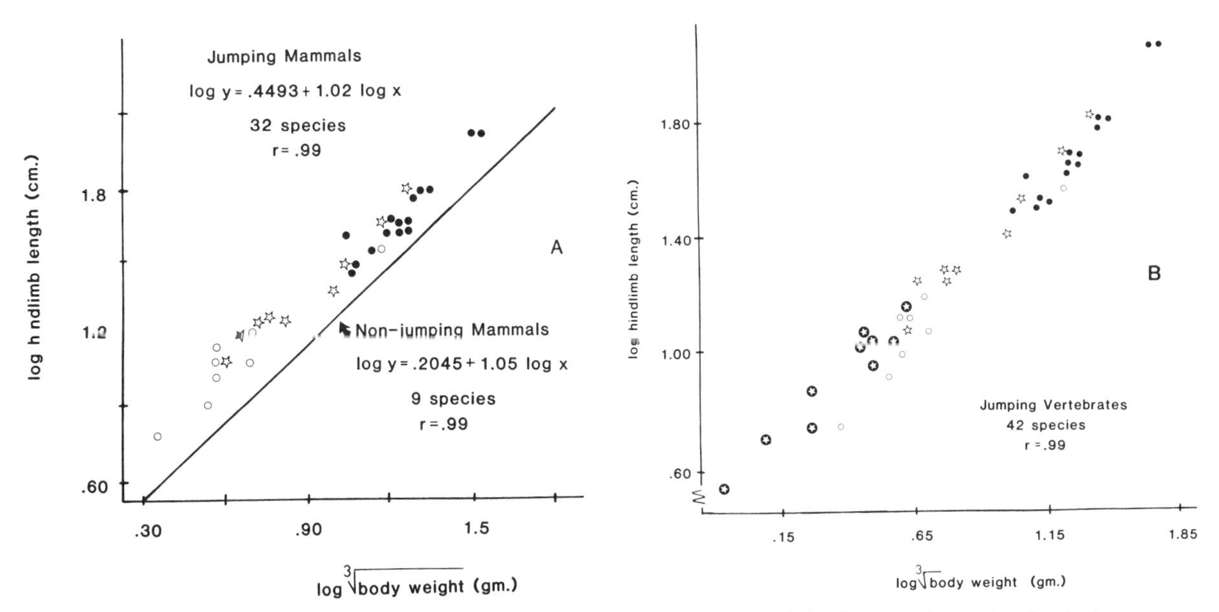

Figure 4-2 Plot of log hind-limb length versus log cube root of body weight for jumping animals. A shows mammals: kangaroos *(black circles)*, vertical clingers and leapers *(clear stars)*, and bipedal rodents *(clear circles)*. The straight line represents the same ratio in nonjumping mammals. (Data from Gambaryan, 1974.) B shows 42 species of jumping vertebrates: frogs *(circle-enclosed stars)*, other symbols as in A.

length to cube root of body weight. Thus, although lengthened hind limbs evolved independently in each group, the ratios of hind-limb length to body weight converge. The regression equation describing this relationship is significantly different (95% confidence level) from that describing the relationship of hind-limb length to body weight in nonjumping mammals. Furthermore, as shown in Figure 4-2B, hind-limb lengths of frogs are very similar to those of such small jumping mammals as tarsiers, galagos, and jerboas.

Examination of ratios of hind-limb length to body weight among prosimian primates illustrates the importance of considering allometric effects when comparing animals of different sizes. Among the jumping prosimians, there are two size classes of vertical clingers and leapers. The tarsiers and galagos are smaller than 1000 g, whereas the indriis (*Avahi, Indri,* and *Propithecus*), which include the largest living prosimians, have body weights over 1000 g. The total range in body size of the vertical clingers and leapers is more than two orders of magnitude, with the smallest, *Galago demidovii,* at 60 g and the largest, *Indri indri,* at 6250 g.

The two groups of vertical clingers and leapers have been distinguished from nonjumping prosimians primarily by differences in relative hind-limb length (Jouffroy and Lessertisseur, 1979, used ratios of hind-limb length to vertebral column length; Oxnard et al., 1981, studied limb proportions using multivariate techniques). If, however, one plots the relationship of hind-limb length to cube root of body weight for nonjumping prosimians, a strong positive allometry (slope of regression 1.45 for 11 species) is revealed (Fig. 4-3). This means that the hind limbs are relatively longer in larger nonjumpers than in smaller ones. The figure shows that small vertical clingers and leapers have relatively longer hind limbs than do nonjumpers of the same size, but that the two larger indriis, *Propithecus* and *Indri,* have the same hind-limb length as would a nonjumping prosimian of the same body size. Therefore, the relatively longer hind limbs of these indriis result from their larger size. The differences in ratios of hind-limb length to vertebral column length for *Propithecus* and *Indri* versus nonjumping prosimians (Jouffroy and Lessertisseur, 1979) are due to the fact that the nonjumping prosimians are smaller than the two indriis, not that *Propithecus* and *Indri* have a longer hind limb than would a nonjumping prosimian of the same body size. *Avahi,* the indri that overlaps the size range of the nonjumping prosimians, does have a longer hind limb than the nonjumping prosimians of the same body size (Fig. 4-3).

Although jumpers all have relatively long hind limbs, there are often differences among related

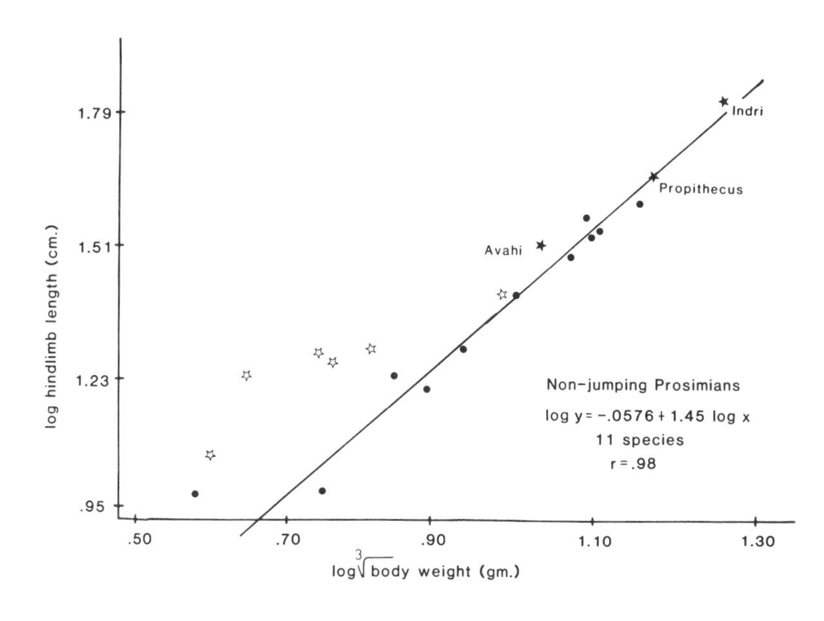

Figure 4-3 Plot of log hind-limb length versus log cube root of body weight for nonjumping prosimians *(black circles)* and the vertical clingers and leapers *(stars).*

Table 4-1 Hind-limb proportions of same-sized jumpers.

Parameter	Jerboa (*Jaculus jaculus*)	Frog (*Rana ridibunda*)	Galago (*Galago demidovii*)
Body weight (g):	60	66	60
Hind-limb length (mm):	122.1	137.7	118.9
Femur (% hind limb):	21.7	28.0	31.7
Tibia (% hind limb):	36.4	27.3	35.2
Tarsus (% hind limb):	7.0	14.6	15.3
Foot (% hind limb):	34.9	30.6	16.6

species of jumpers (Fig. 4-3). Such differences have generally been considered to reflect differences in jumping ability (see, for example, Gambaryan, 1974), although the relationship of hind-limb length to jump performance has been examined only in frogs. In a series of studies dating back to the 1930s (Wermel, 1934; Rand, 1952; Zug, 1972), jump lengths of frogs have been compared among species with different ratios of hind-limb length to body size. Not surprisingly, the results of most of this work show that species with relatively shorter hind limbs jump shorter distances. However, these studies did not take into account differences in body size, and they also compared species that varied morphologically in respects other than length of the hind limbs. Thus, it is not clear from these studies that differences in jump performance in frogs are due only to differences in relative hind-limb length. Two other studies (Stokey and Berberian, 1953; Emerson, 1978) compared two different populations of single species (*Rana pipiens* and *Pseudacris triseriata*, respectively). In these cases, although differences in ratios of hind-limb length to body size occurred between populations, other aspects of body morphology were similar. And although there were significant differences in the mean ratios between the two populations of each species, there was no significant difference in the mean maximum jump distance among animals of the paired populations. Hind-limb length is not, then, the only factor involved in determining jump distance.

Length of Hind-Limb Segments Different segments of the hind limb are disproportionately lengthened among the various jumpers. A comparison of three species of approximately the same size (Table 4-1) shows that although the frog, galago, and jerboa have roughly similar hind-limb lengths, the jerboa has a relatively short femur and tarsus, the frog a relatively short tibia, and the galago a relatively short foot.

Figure 4-4 plots tibia length against cube root of body weight for two groups of different-sized jumpers: the bipedal rodents and the kangaroos. Both groups have longer tibias than do their non-jumping relatives (*Rattus, Didelphis, Trichosaurus*), but the relationship between tibia and body weight is not the same for the two groups. Projecting the regression lines into overlapping size ranges suggests that a rodent the size of a large kangaroo would have a relatively shorter tibia and that a kangaroo the size of a small jumping rodent would also have the relatively shorter tibia.

Two groups of vertebrate jumpers are characterized by lengthened tarsal segments: the frogs and the small vertical-clinging and leaping primates. When tarsus length is plotted against body weight for members of these two groups, a striking example of convergence is revealed: the points fall on a straight line. Not only do galagos, tarsiers, and frogs lengthen the hind limb in an unusual area—the ankle region—but they all lengthen that segment in the same regular way with respect to body weight (Fig. 4-5).

Vertebral Column and the Center of Mass The propulsive thrust of the hind limbs produces a torque on the body of a jumper during takeoff if the center of mass is not aligned along the line of

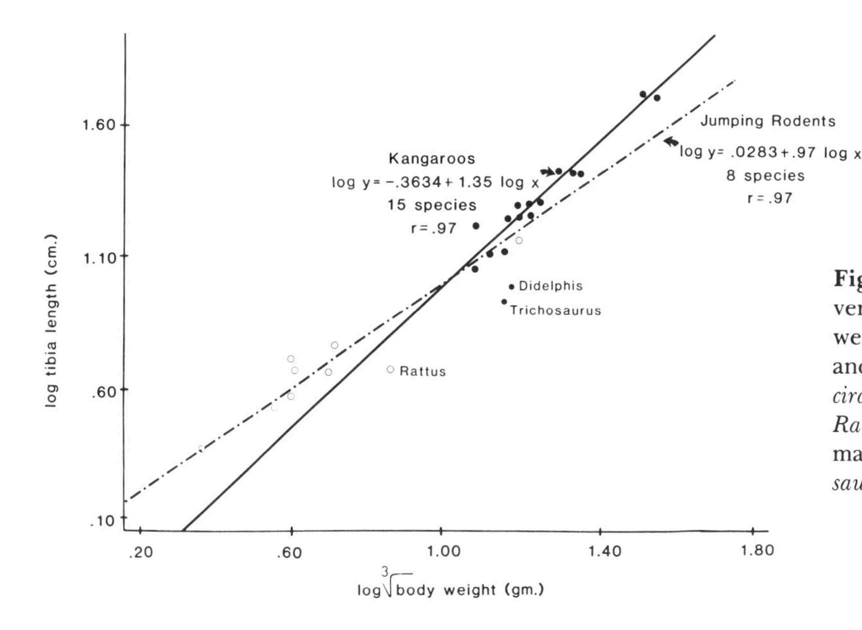

Figure 4-4 Plot of log tibia length versus log cube root of body weight for kangaroos *(black circles)* and bipedal jumping rodents *(clear circles)*. A nonjumping rodent, *Rattus*, and two nonjumping marsupials, *Didelphis* and *Trichosaurus*, are shown for comparison.

action of that propulsive force. This torque causes the body to rotate clockwise (when viewed from the right) if the center of mass is anterior to the line of action of the force and counterclockwise if the center of mass is posterior to the line of action of the force. Most mammalian jumpers minimize clockwise torques by having their centers of mass shifted backward to be more in line with the propulsive thrust of the hind limbs. This shift is accomplished in a number of ways: shorter trunks, reduced forelimbs, more massive hind-limb muscles, and/or longer, heavier tails. Additionally, the epaxial musculature, which plays an important role in the alignment of the body and in the maintenance of semierect posture, is enlarged in jumpers.

Loading of the vertebral column during terrestrial locomotion is poorly understood in general, and it has never been studied for saltatory locomotion. The functional significance of the mor-

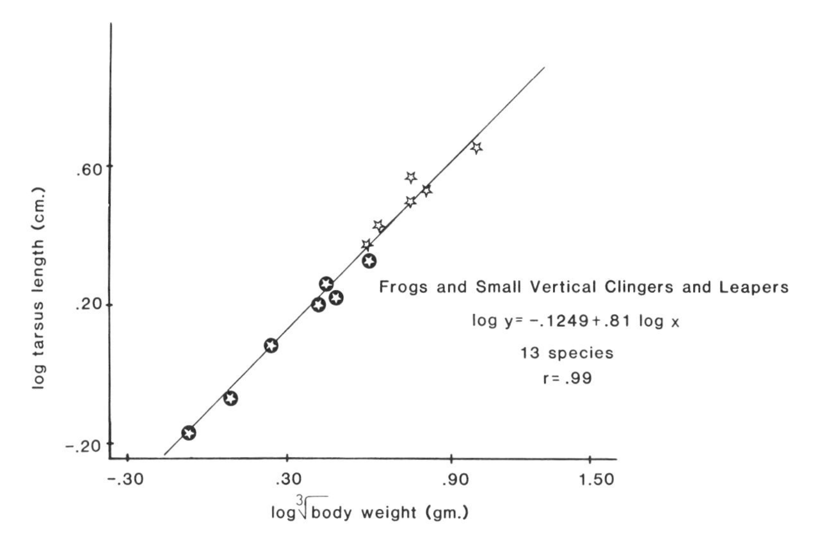

Figure 4-5 Plot of log tarsus length versus log cube root of body weight for frogs *(circle-enclosed stars)* and the small vertical clingers and leapers *(clear stars)*.

phological modifications described in the literature for jumpers is one of a number of questions waiting to be explored.

Modifications of the vertebral column in jumpers have been most extensively studied in the small bipedal rodents. Hatt (1932) found that fused cervical vertebrae and a shortened neck (less than 15% of thoracolumbar length versus greater than 15% in nonjumpers) characterize jumping rodents belonging to the families Pedetidae, Heteromyidae, and Dipodidae. The number of fused cervical vertebrae varies from two to six among species, and in all cases the fusion is correlated with the elongation of a single, large, anteriorly located neural spine. An important dorsal flexor muscle of the head, the spinalis dorsi, attaches to this spine. Hatt interpreted this feature as a means of reducing head bobbing during takeoff and reducing the work of the dorsal neck muscles. These functional hypotheses have not, however, been tested experimentally.

Another modification of the vertebral column in bipedal jumping rodents is the lengthening and anterior rotation of the neural spines of the posterior lumbar vertebrae. Hatt suggested that this shift provides better leverage for epaxial muscles such as the multifidus spinae, which presumably support the anterior part of the body during bipedal locomotion. Again, this explanation, although plausible, has never been confirmed experimentally.

Hatt also found that the bipedal rodents have longer tails than their nonjumping relatives. The longer tails result because the caudal vertebrae both lengthen and increase in number. The lengthened tail is important in reducing torques and pitching during takeoff and landing. When the animal is in the air, the position of the tail also influences the orientation of the body during the jump. On the ground, the animals use the tail as one "leg" in a support triangle, which frees the forelimbs for other activities.

Frogs are also characterized by a shortened presacral vertebral column. Whereas most nonjumping amphibians, such as salamanders, have at least 20 presacral vertebrae, frogs have fewer than 10. Unlike mammalian jumpers, frogs lack a tail for counterbalance, and the postsacral vertebrae are reduced and fused into a single bony element, the urostyle. Cineradiographic studies show that during the initial stages of the takeoff the urostyle and presacral vertebral column rotate counterclockwise (viewed from the right) relative to the ilia. The rotation results in an alignment of the anterior part of the frog's body along the axis of the pelvic girdle, paralleling the direction of the propulsive force. During landing the pelvic girdle rotates counterclockwise, relative to the sacrum, bringing the legs underneath the body. Vertical rotation of the presacral vertebral column is possible, in part, because the ligamentous attachments between the sacral diapophyses and the anterior ends of the ilia form a hinge joint. Precise movement at the hinge is controlled by four paired muscles surrounding the iliosacral joint (Emerson and de Jongh, 1980).

Hind-Limb Musculature Fokin (1978) compared the relative weight of hind-limb muscle (versus body weight) among similar-sized rodents, jumping and nonjumping. Whereas the hind-limb muscles are generally larger in the jumpers, it is the biceps femoris, an extensor of the hip, that shows the most dramatic enlargement — close to an order of magnitude — in the jumpers. In a recent study on the allometry of the hind-limb muscles of mammals, kangaroo rats, springhares, and kangaroos were found to have larger quadriceps, hamstrings, and ankle extensors than nonhoppers of the same body size (Alexander et al., 1981). Gambaryan (1974) compared the relative weights of the hind-limb muscles of rodents and kangaroos. However, since he presented the weight of each hind-limb muscle as a percentage of the total weight of the combined forelimb and hind-limb musculature of each species and since bipedal rodents and kangaroos have reduced forelimbs, his ratios do not allow interpretation of degree of enlargement of hind-limb muscles in jumpers.

Bauschulte (1972) compared relative weights of the individual hind-limb muscles of a kangaroo and a monkey with the total weight of hind-limb muscles expressed as a percentage of total body weight. She examined differences in the relative sizes of functional groups of hind-limb muscles and, interestingly, found few differences in the relative sizes of such groups between the two animals. In the most extreme case, the relative size of the knee extensors of the kangaroo was 15% larger than that of the monkey. Emerson (1978), in a study on comparative locomotor performance in frogs, found significant differences among species in the size of the hind-limb extensor muscles (compared with body weight), which correlated with differences in jumping ability.

These are most of the studies that have used a quantitative approach in studying differences in musculature among jumpers and nonjumpers. In all these cases, gross muscle size rather than the cross section was used as a predictor of force, and degree of pinnation was not a consideration. Most comparative myology is based on qualitative descriptions (Howell, 1932; Dunlap, 1960; Klingener, 1964; Badoux, 1965). Although such studies suggest that in all jumpers there is some enlargement of hind-limb musculature (specific muscles varying among species), they do not provide precise information on the degree of the differences, and they are difficult to interpret when they involve comparisons among species of different sizes.

Some studies have also examined positions of origin and insertion of the hind-limb muscles. Changes in position have been analyzed functionally on the basis of a lever-system model: shifts resulting in longer effective moment arms increase torque and out-force, and changes moving an insertion closer to a joint produce a higher out-velocity. For example, some of the hip extensor muscles have a more proximal insertion on the tibia in jumping rodents, kangaroos, and prosimians than in their nonjumping relatives (Gambaryan, 1974; McArdle, 1981). The shift in the insertion toward the joint has generally been interpreted as increasing the velocity of limb rotation (about the hip) during takeoff. Unfortunately, the biomechanics of the hind limb are probably more complex than this explanation would suggest.

Howell (1932) found a progressive shortening of the ilium in kangaroo rats. Maynard Smith and Savage (1956) suggest that the reduction in length of the ilium in jumpers implies that the main function of the hip extensor muscles of bipedal jumping mammals is to supply a large acceleration at the beginning of the propulsive stroke. This role would best be performed by the ischiopubic muscle group, which has a relatively large moment arm. But, many of the ischiopubic muscles that are hypothesized by Maynard Smith and Savage to be arranged for force production during hip extension are the same muscles that other authors (Gambaryan, 1974) have suggested are inserting more proximally on the tibia in jumpers to increase the angular velocity of the femur during takeoff. As it is impossible to maximize velocity and force simultaneously in a single-lever system, the biomechanical explanations present a paradox.

These contradictory hypotheses can be examined by measuring, in both jumping or nonjumping mammals, absolute length of the pelvis, as well as length of its ischial and ilial portions, and by constructing accurate proportional models of the pelvis and associated hip extensor musculature (Fig. 4-6).

In rodents the pelvic girdle is the same overall length in jumping and nonjumping species of the same body size. In the jumping rodents an anterior shift in the position of the acetabulum, as reflected by the relatively shorter ilium and longer ischium, results in the maintenance of the same moment arm for hip extensors as in nonjumpers, despite the proximal shift in the position of the insertion of these muscles on the tibia.

Kangaroos show both a shift in relative lengths of ilium and ischium and an increase in absolute size of the pelvic girdle, as can be seen by comparing a nonjumping marsupial, *Didelphis*, with a kangaroo of the same size (Fig. 4-6). This combination of change in shape and size results in the maintenance of the same length of moment arm for hip extensors despite a more proximal insertion for some of the muscles (*b* and *b′* in the figure). Furthermore, for other hip extensors that do not shift position of insertion, an actual increase in the moment arm of the muscle results from the increase in length of the pelvis and the shift in relative position of the acetabulum (*a* versus *b* in the figure).

	Sicista	Rattus	Jaculus
pelvis length	70 units	70	70
ilium : pelvis	.65	.61	.53
ischium : pelvis	.35	.39	.47

	Didelphis	Aepyprymnus
pelvis length	70 units	78.9
ilium : pelvis	.70	.57
ischium : pelvis	.30	.43

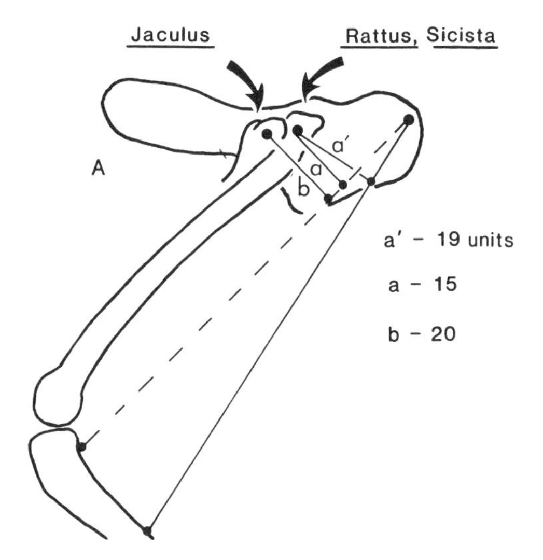

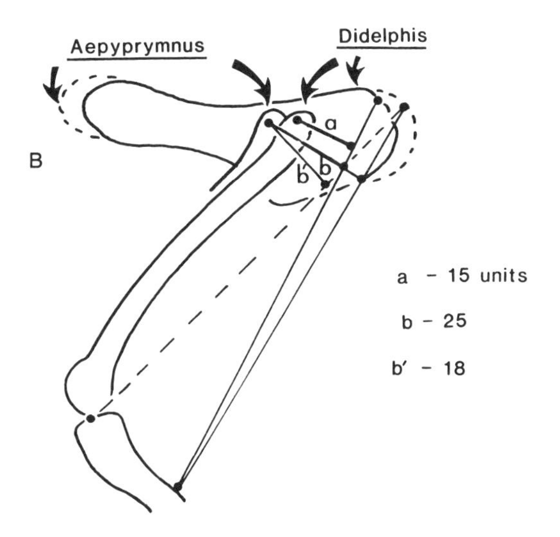

Figure 4-6 Scale models of pelvic girdles comparing moment arms (a, a′, b, b′) of some hip extensor muscles: A, the quadrupedal rodents (*Rattus* and *Sicista*) and the bipedal jumping rodent (*Jaculus*): B, the quadrupedal marsupial (*Didelphis*) and the bipedal kangaroo (*Aepyprymnus*).

Biomechanical Analyses

In many respects classical comparative studies of specializations for jumping have uncovered more questions than they have answered. They have verified the expected: jumpers elongate the hind limbs and modify their propulsive musculature. However, those studies have also shown a degree of morphological variation, both within and among groups, that does not sort into obvious patterns. This apparent lack of form-function patterns may have resulted from inferring the action of muscles from static analysis or from comparing animals of different size, of different jumping behaviors, or of diverse phylogenetic groups.

In any case, we must ask different kinds of questions about jumping to find insights into the complex morphological variation uncovered by the descriptive comparative work. Two types of studies stand out in this regard: those that have examined the problems of scaling in jumping animals, comparing parameters of jump performance and muscle energetics for animals of different size both within and among taxonomic groups, and those involving experiments designed to provide information on the biomechanics of hind-limb morphology at multiple levels of resolution, from the physiology of muscle contraction to gross muscle mechanics.

Muscle Forces, Bone Stress, and Energy Storage In the first of a series of papers examining different biomechanical aspects of a jump (Calow and Alexander, 1973; Alexander, 1974; Alexander and Vernon, 1975; Biewener, Alexander, and Heglund, 1981), Alexander used a force platform to obtain records of the magnitude and direction of forces exerted on the ground by a frog during jump takeoff. As the forces were recorded, jumps were filmed simultaneously in dorsal and lateral view for placement of the line of action of the ground reaction force and for calculation of moments of the ground reaction about the hip, knee, and ankle joints. In the strongest recorded jumps, a 28-g frog had a takeoff velocity of 1.8 m/sec at 55° to the horizontal. Maximum forces of about 0.35 N were exerted by each foot. Moment arms of the extensor muscles about the hip, knee, and ankle were determined from X rays and from measurements of dead frogs with their legs in a series of positions corresponding to the successive positions during takeoff. Forces developed in isometric contraction were measured in vitro for the same hind-limb extensor muscles. Both pinnate and parallel-fibered muscles and both one- and two-joint muscles were included in the study. In order to compare performance of different-sized extensors, the excised muscles were weighed and their lengths measured before the force recordings were made. Muscle stress (force/unit area) was calculated from the maximum measured isometric forces, and ranged from 150 to 350 kN/m^2 among the muscles. The amount each muscle must have shortened during the jump takeoff as the leg moved from the flexed to the extended position was obtained from the area under the graph of moment arm of the muscle against angle of the joint.

The information was used to address three questions: (1) How does the maximum thrust exerted on the water in swimming compare with the forces exerted in jumping? (2) What are the maximum bending moments that the femur and tibiofibula have to withstand during jump takeoff? and (3) What is the optimal arrangement of jumping muscles for the greatest takeoff velocity?

Calow and Alexander found that the thrust exerted on the water during swimming never approached the magnitude of the moments found to act in jumping. The maximum thrust per foot in swimming is about 0.1 N, compared with about 0.35 N per foot in jumping.

When leg bones were tested for stiffness and strength, the femur and tibiofibula had a stiffness of 2.6 and 2.1 Nm/degree, and mean bending moments at fracture of 63 and 70 mNm respectively. In the extreme case, when the ground reaction acts at right angles to the bone, the bending moment halfway along the bone equals that reaction multiplied by one half the bone length. Tibiofibulas of the frogs studied were about 28 mm in length and the femurs about 25 mm. The maximum ground reaction force recorded was 0.35 N. Hence, the greatest bending moments that forces at joints were likely to exert on the bones' midpoints during jumping were about 5 mNm, enough to bend them through only 2 to 3 degrees.

However, because long slender bones might be in danger of buckling under the axial loads imposed by the muscles acting alongside them, buckling loads (b) were estimated from Euler's formula: $b = EI/L^2$, where E is Young's modulus, I is the second moment of area of the cross section, and L is the length of the bone. The estimated value for the frog femur is about 60 N and for the tibiofibula about 40 N. These are much larger loads than the sums of the maximum isometric forces that the muscles beside the bones can exert. Even if all the extensor thigh muscles examined could develop 350 kN/m^2, their total force would only be 20 N, and the plantaris longus, which is the major calf muscle, can only exert 5 N. There is, therefore, a wide margin of safety.

The muscles that a frog, or any jumper, uses when it jumps accelerate a constant mass. Calow and Alexander predict, from theoretical considerations, that the highest velocity of that mass is achieved when the mechanical advantage is such that the final value of the muscle force is equal to about 0.20 of the maximum isometric force. Theoretically, the muscles used in jumping are most advantageously arranged if the force each exerts declines to about one fifth the maximum isometric force at takeoff. Calow and Alexander tested this hypothesis by examining the stresses in the hind-limb muscles of frogs during the initial and final stages of takeoff. These muscles developed, in vitro, maximum isometric stresses of about 300 kN/m^2; therefore, if their model is correct, the final stresses are about 60 kN/m^2. The stresses of 65 – 110 kN/m^2 estimated for the later stages of takeoff are close to this expectation, but, contrary to their predictions, they found no indication that substantially higher stresses acted in the earlier stages of takeoff when the angular velocities of the joints and the rates of shortening of the muscles were lower.

In a similar biomechanical analysis of the hopping of a kangaroo and a wallaby, Alexander examined the role of energy storage in the tendons of the gastrocnemius and plantaris muscles. He found that during a jump the leg muscles, by lengthening, do about 24 J negative work in the first (landing) phase of contact and 24 J positive work during the second (takeoff) phase as the animal gains energy. In the wallaby the cross-sectional areas of the gastrocnemius and plantaris tendons were 8.9 mm^2 and 12.4 mm^2, respectively; the forces exerted by the two muscles during a hop were, therefore, estimated to be 480 N and 370 N, respectively. From this information the stresses in the two tendons were calculated. Given that Young's modulus for fresh mammal tendon is about 1.2 GN/m^2, the stresses could be expected to stretch the gastrocnemius tendon by 4.5% and the plantaris by 2.5% of their length. By measuring the respective lengths of the two muscles from origin to insertion, it was established that both tendons extend about 11 mm during the first phase of contact. As the tendons taken together for the two legs exert a maximum total force of 1700 N, when they stretch elastically 11 mm they store 9 J of energy ($0.5 \times 0.11 \times 1700$). The total positive work of the second phase of contact is 23 J. If 9 J is supplied by elastic recoil, only 14 J must be supplied by actively contracting muscles. This represents an energy savings of close to 40%.

Strain energy can be stored in muscle fibers as well as in tendons. However, Alexander and Bennet-Clark (1977) have shown that for large animals, including kangaroo and wallaby, tendon storage is more important. Calculations show that only about one eighth of the 35% energy recovery is attributable to muscle storage (Biewener, Alexander, and Heglund, 1981).

In contrast, a study of elastic energy storage in hopping kangaroo rats (Biewener, Alexander and Heglund, 1981) suggests that energy storage in muscles may be important in these small animals. Depending on the force contributed by the individual ankle extensors (soleus, gastrocnemius, and plantaris), elastic energy savings range from 13% to 24%, with muscles storing from two ninths to one half the total. If all three muscles supply the required force, then about 4.4 mJ strain energy is stored in the tendons, and up to 5.4 mJ in the muscles. If the plantaris muscle supplies the entire force, the elastic elements in the muscle could be storing up to 4 mJ strain energy with each step, whereas the tendons store 14 mJ.

Unfortunately, the approach used in these studies on energy storage does not show how much of the force is actually contributed by each muscle. Equal stresses and similar firing patterns are assumed for the plantaris and gastrocnemius (and soleus in the case of the kangaroo rat). The studies also assume that the tendons of kangaroo rats and kangaroos have the same Young's modulus and that the modulus is identical to that of cat and human tendon. Possible viscoelastic behavior of the muscle fibers and tendons is also ignored. These assumptions may be reasonable, but they need to be tested before the data from the two studies are used to address broader questions concerning the energetics of terrestrial locomotion.

Scale Effects in Jumping Animals Hill (1950) suggested that animals that are geometrically similar should jump the same distance regardless of size (see Eqs. 4-5 and 4-6). Smaller animals have faster accelerations, whereas larger animals have longer distances through which the forces act. Emerson (1978) examined this hypothesis in frogs, which preserve a generally similar body shape ontogenetically over a wide size range, and are thus an interesting group in which to study the relationship of body geometry and size to locomotor function. Maintenance of geometric similarity imposes limitations on acceleration and jump distance with increasing size, but if geometric similarity is not maintained, acceleration should remain constant and jump distance should increase with increasing size. Accordingly, if the principal function of the jump is to provide a specific absolute distance and if Hill's model is correct, then one would predict that as size increases, body proportions and jump distance would remain constant and acceleration would decrease. However, if the principal function of jumping is quickness or maintenance of the same relative jumping distance, one would predict that as size increases, shape would change. Larger frogs would jump longer distances than smaller frogs, and acceleration would remain constant.

Analysis of three species of frogs (*Rana pipiens, Bufo americanus,* and *Pseudacris triseriata*) reveals that in all species absolute jump distance increases with increasing size, relative jump distances decrease as size increases, and the juveniles and adults of two of the species maintain constant accelerations despite size increases. Juvenile and adult frogs do maintain geometric similarity with ontogenetic size increase in body length, length of hind limbs, and weight of hind-limb extensor muscles. Only the size range from just metamorphosed to juvenile shows a positive allometry in body proportions.

There are two possible explanations for this paradox. Hill's model may be wrong or some aspect of locomotor morphology or physiology not measured above may be appropriately allometric. In other cases (see Chapter 2) Hill's predictions are not supported by empirical data—the basic assumption of geometric similarity is violated. A similar situation appears to be the case for frogs. Later work (Sperry, 1981) has revealed that as body size increases, frogs, unlike mammals, increase the number of muscle fibers per unit mass of muscle. Cross-sectional area of the hind-limb extensors scales with a positive allometry. This recruitment of additional muscle fibers with increasing size changes the relationship of energy output to unit mass of the muscles, and it may account for the constant acceleration and increasing jump distance seen within species of frogs (Sperry, 1981).

Also, the fact that in frogs acceleration remains constant with increasing size suggests that quickness of movement, in addition to jump distance, may be biologically important. This is in contrast to the assumptions of most earlier studies of jump performance, which have concentrated on distance as the critical parameter for frogs.

A paper by Bennet-Clark (1977) considers both quickness and distance while examining scale effects in jumping animals. The amount of specific energy and power required to produce jumps of varying heights by animals of different sizes can be obtained from the empirical data on stored energy and power output of contracting muscles and by consideration of the relationships among jump height, power, energy, and animal size as expressed in the ballistics formulas. From Eqs. (4-1), (4-3), and (4-6),

$$h = \left(\frac{2sp}{m}\right)^{2/3} \frac{g}{2}.$$

If s is proportional to an animal's length (l) and p is proportional to the mass of muscles (which range from 5% to 20% of an animal's total mass), then h and d are both proportional to energy/unit mass and are independent of size. Most of the available data for vertebrates compare lengths of jumps rather than heights, and Figure 4-7 is a modification of Bennet-Clark's original figure in that it plots standing jump length, rather than height, as a function of body length. The 100 W/kg line shows the upper limit for power produced by direct muscle action for an animal whose muscle mass is 10% of its total mass and whose specific power output is 1 kW/kg (from Bennet-Clark, 1977). Almost all the small vertebrate jumpers lie to the left of the 100 W/kg line. This indicates that small vertebrate jumpers appear to require more power (based on data from the literature) than can be produced by direct muscle action. This is logical when one considers that as animals become smaller their limbs become shorter and the time available for acceleration is reduced. Accelerations and forces, therefore, both increase with decreasing size, and at some point it is no longer possible for the peak power to be produced by direct muscle action; there is simply not enough time. Judging by the distances of their jumps, small saltatory vertebrates must be storing muscular energy before the jump and releasing it during takeoff.

Among mammalian jumpers there is much variation in the relative lengths of different limb segments (Table 4-1); however, in all mammalian jumpers the tibia is lengthened. The gastrocnemius and plantaris tendons are probably the major energy stores for most saltatory vertebrates, and elongation of the tibia is correlated with an increase in the length of these tendons. However, there are major differences in the shape of these tendons between large and small

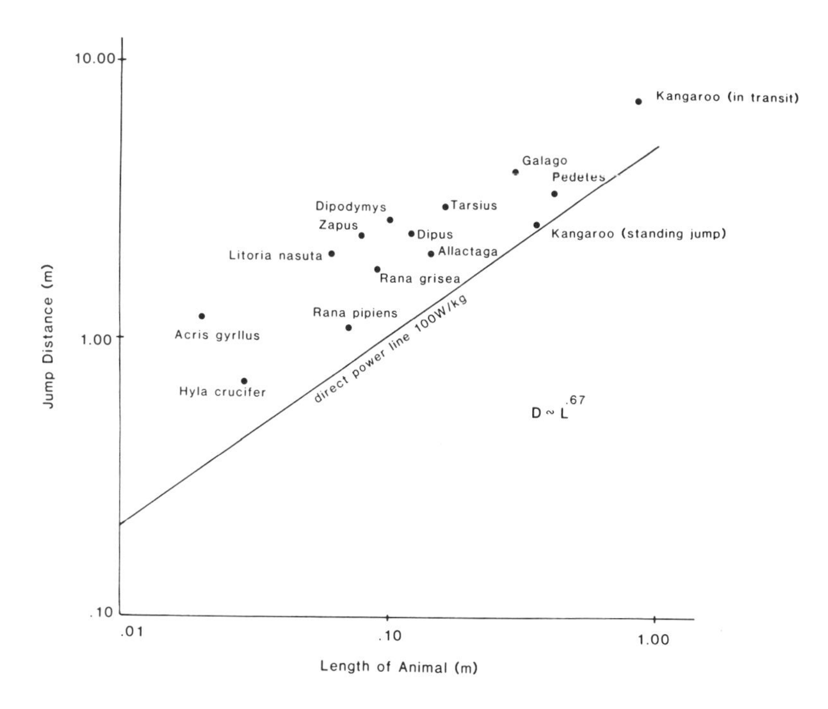

Figure 4-7 Log-log plot of jump distance versus body length for a diversity of saltatorial vertebrates.

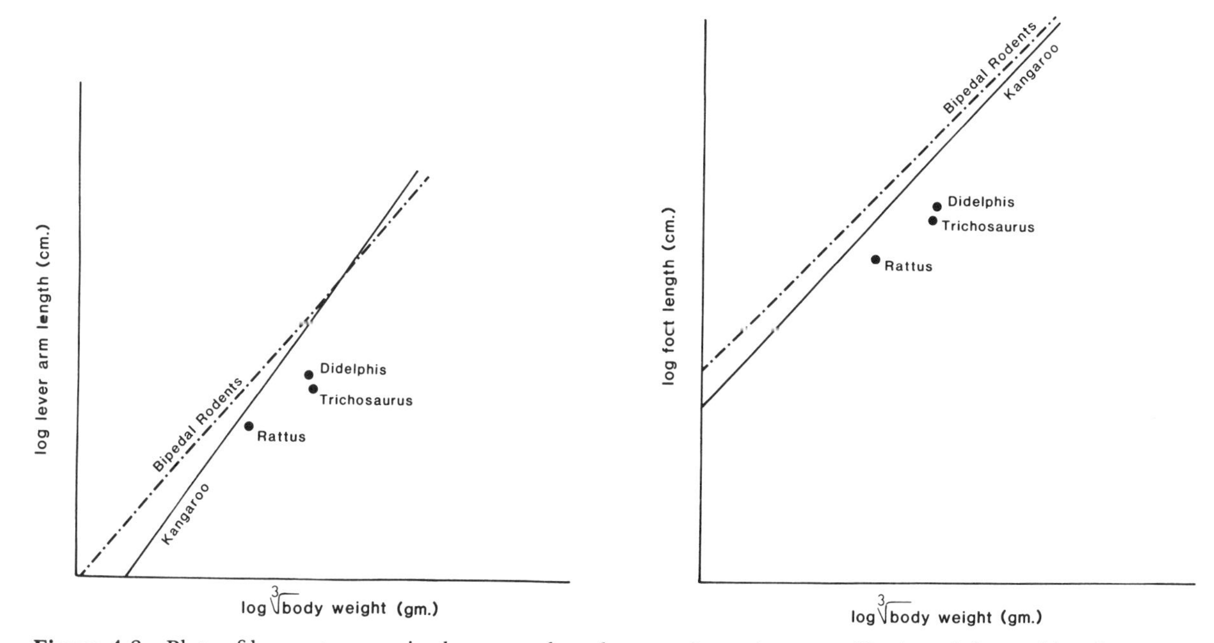

Figure 4-8 Plots of log gastrocnemius lever-arm length versus log cube root of body weight, and log foot length versus log cube root of body weight for bipedal rodents, kangaroos, and three nonjumping genera.

species. The tendons of the ankle extensors of kangaroo rats are thicker than would be those of a kangaroo of the same body size, and much thicker than would be required for an equal percentage of energy recovery (Biewener, Alexander, and Heglund, 1981). To match the 35% recovery rate of a wallaby, the plantaris and gastrocnemius tendons of kangaroo rats would need to have cross-sectional areas of about 0.16 and 0.11 mm², respectively; their actual cross-sectional areas are about 0.62 and 0.89 mm². From treadmill data it appears that the ankle extensor tendons of kangaroo rats are too thick to save much energy by elastic storage during sustained locomotion, and in fact they seem thicker than necessary to withstand the forces exerted by their muscles. However, long-distance, sustained jumping is not the characteristic locomotor pattern for kangaroo rats in the wild. Ecological and behavioral studies indicate that kangaroo rats are fairly sedentary and do not run for sustained periods (Kenagy, 1973). Rather, they use quick-starting, evasive locomotion over short distances (Bartholomew and Caswell, 1951). These observations suggest that energy storage may be important to kangaroo rats for increasing the quickness of initial jump takeoff rather than for efficiency during sustained locomotion. These small rodents fall into the size range where time limits the peak power produced by direct muscle action. Any energy storage would be advantageous since it would increase acceleration during takeoff. In addition, the energy storage of their tendons may be larger for a standing jump than that calculated for sustained locomotion. Pictures of standing jumps show ankle flexion of approximately 140°, whereas the maximum ankle flexion recorded during treadmill locomotion was 120°.

Morphological studies show that the tibialis anterior is enlarged in kangaroo rats. This muscle is a dorsi flexor of the foot, and its contraction stretches the gastrocnemius and plantaris tendons. The tibialis anterior could make a major contribution to energy storage before initial takeoff (in a standing jump) by actively preventing extension. Additional force from this muscle might also explain in part why the tendons of the ankle extensors of kangaroo rats are larger than would be expected from considering only the

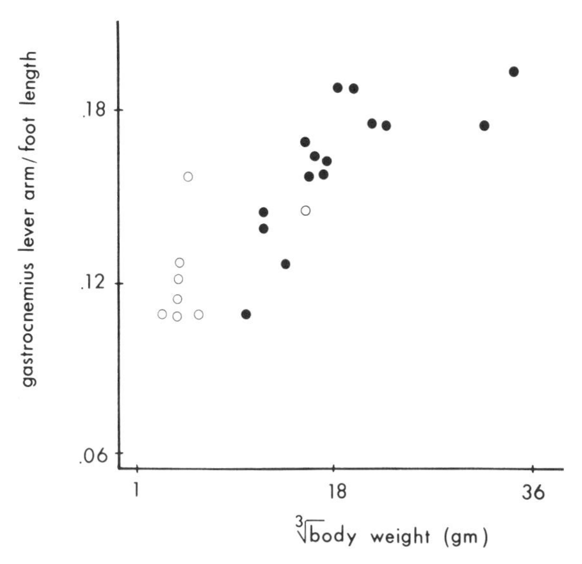

Figure 4-9 Plot of the ratio of gastrocnemius lever-arm/foot length versus cube root of body weight for bipedal jumping rodents *(open circles)* and kangaroos *(solid circles).*

sizes and forces of the gastrocnemius and plantaris. These possibilities have not been considered because investigators so far have examined only sustained locomotion, not standing jumps.

The distance from the articulation of the astragalus and tibia to the application point of the ground force is the in-lever in the energy storage system, and the lever arm of the gastrocnemius is the out-lever. This in-lever, which is basically the length of the foot, is lengthened in all jumpers, but so is the out-lever (Fig. 4-8). There are differences in the degree of lengthening of the two segments among kangaroos and among bipedal rodents, as well as between the two groups (Fig. 4-9). Attempts to understand the biomechanical significance of this variation must examine quantitative data on muscle action, muscle forces, tendon length, and jump performance, all in the context of the animals' actual locomotor behaviors in the field.

Integration of Morphological Differences and Biomechanical Analyses

The work on energy storage, muscle action, and scale effects in jumping indicates a complex inter-

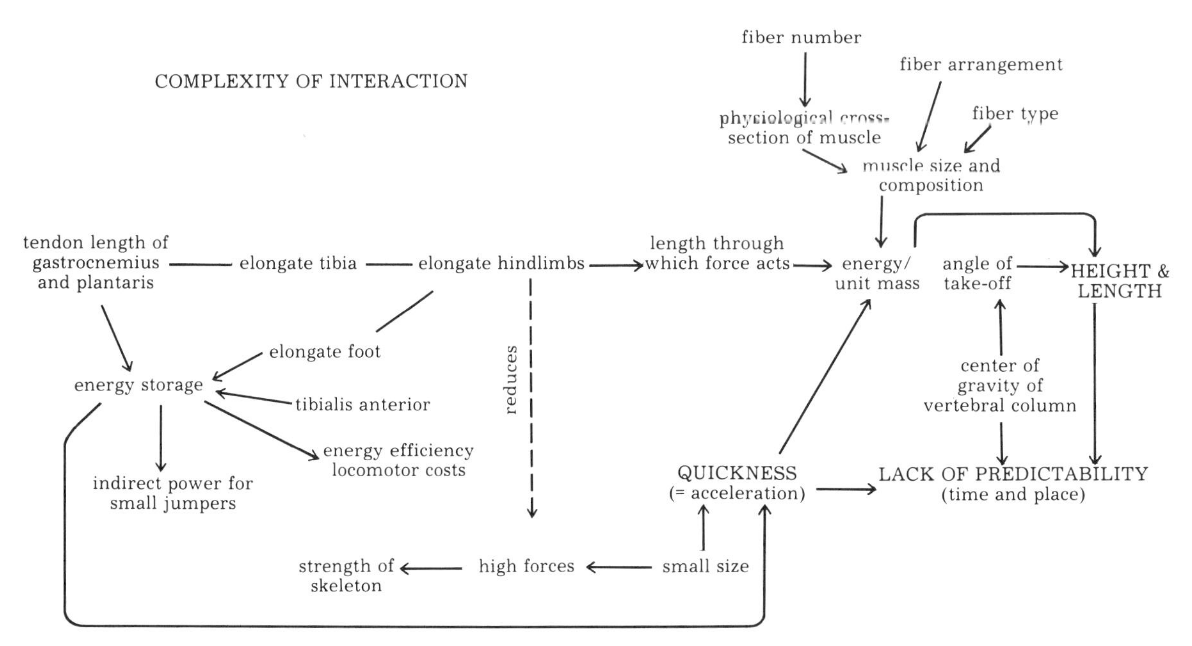

Figure 4-10 Interactions among morphology, performance, and size of jumpers.

action among behavior, morphology, and size, as diagramed in Figure 4-10. Depending on which jump parameters are considered, different aspects of morphology are emphasized; the size of the animal alone affects some properties of a jump (such as quickness where size is small), whereas other correlates of morphology (such as energy storage) may be dependent on both size and behavior.

The techniques are available to measure parameters of jump performance, biomechanics, size, and aspects of musculoskeletal morphology ranging from fiber types to gross anatomical arrangement and proportions. Integrated studies that examine comparative performance among related species in groups where jumping has evolved remain a necessary step before a meaningful general synthesis can be made about jumping among vertebrates.

Chapter 5

Climbing

Matt Cartmill

The term *climbing* refers to locomotion on supports with vertical or steeply sloping surfaces. Although some vertebrates have special adaptations for moving about on steep rock faces, most climbing vertebrates live in the branches of trees — or at any rate enter trees habitually to forage or seek shelter. Because the branches of trees are limited and variable in width, discontinuous, and oriented at all possible angles to the gravity vector, it is easy to fall from them; and because they grow high above the ground, a tree-dwelling animal that falls is likely to be injured when it lands. Even if the animal survives a fall unhurt, its involuntary trip to the forest floor may expose it to increased danger from predators, and it is compelled to waste time and energy in climbing back up to the level from which it fell. For all these reasons, natural selection favors the development of behavior and morphology that minimize an arboreal animal's chances of falling. Such animals therefore have developed a variety of solutions to the problems posed by discontinuous supports of varying size and angulation. Anyone who has ever climbed a tree will recognize that these are real problems, and that *Homo sapiens* is not very well equipped anatomically to deal with them.

There are no flat or level surfaces to stand on in a tree. Even horizontal branches have a more or less circular cross section, and so an animal standing on such a branch contacts a surface that is everywhere sloping except along the infinitely thin line connecting the uppermost points on the infinity of possible cross sections. Although standing or moving on a horizontal branch may not demand special adaptations if the animal is sufficiently small relative to the branch (a shrew atop a fallen sequoia trunk, for example, is in no danger of sliding off), vertical supports of any size whatever present unique problems. Because the unresisted force of gravity will make an animal on a vertical support fall at the same rate as if there were no support there, any animal capable of staying on such a support must generate a vertical reaction force between itself and the support that is equal and opposite to the gravitational vector. Such a force must (like the gravitational force) be tangential to the support's surface. There are only two ways of generating such a vertical force: (1) by interlocking the surface of the animal with that of its support in such a way as to generate a new, nonvertical contact surface between the two, or (2) by developing bonds between the animal and its support that are too strong to be broken by the animal's weight. The first mechanism is exemplified by animals that climb and cling using claws; the second, by animals that grip their supports using adhesion or suction. A friction grip involves both mechanisms.

The limited and variable width of tree branches presents arboreal animals with another set of problems that demand morphological solutions. A relatively large animal walking along a horizontal branch tends to topple to one side or the other because all its support points are effectively collinear (Fig. 5-1; Napier, 1967). There are four ways of diminishing or resisting this tendency: (1) having relatively short limbs, thus keeping the body's center of mass close to the support; (2) having prehensile hands and feet that can grip the branch and exert a torque that resists

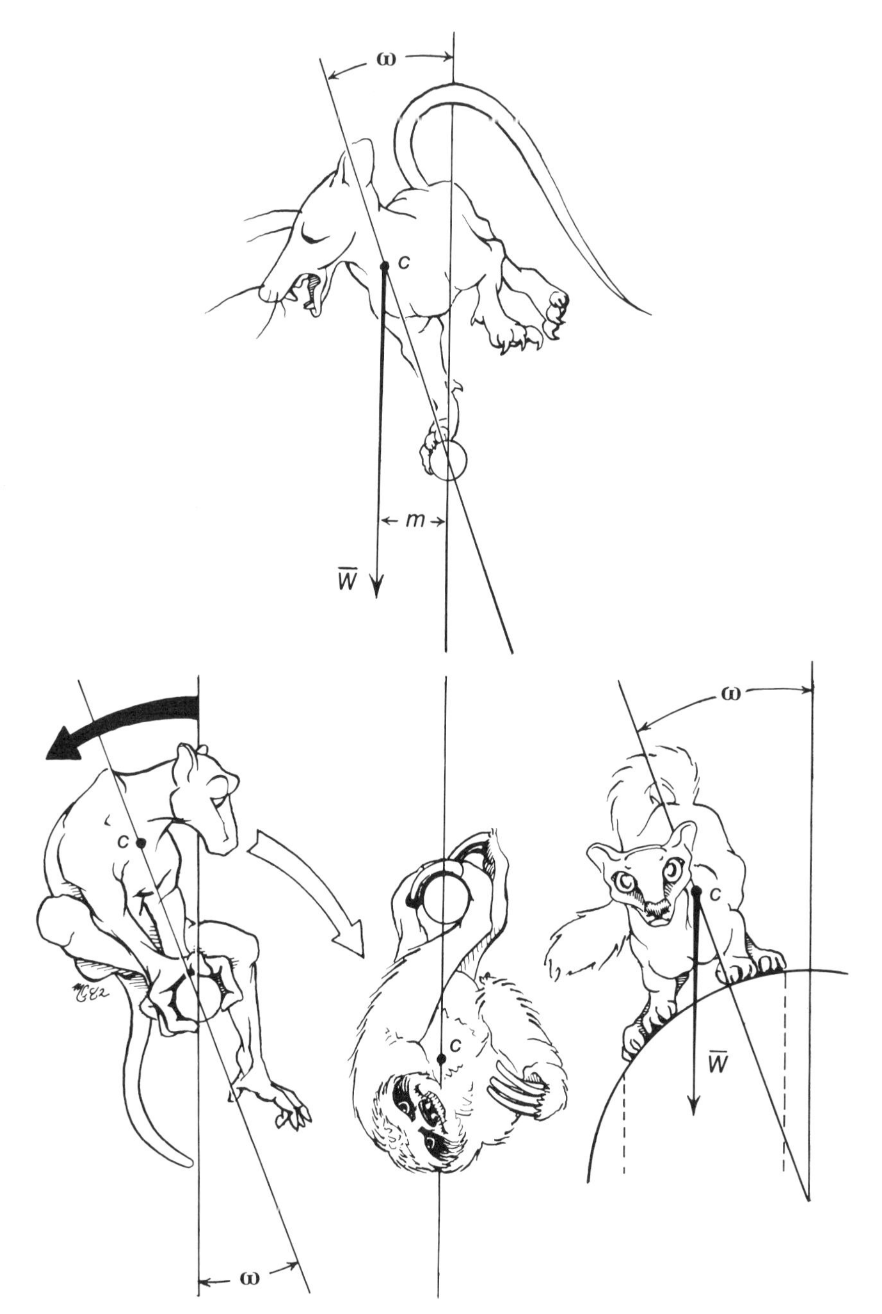

Figure 5-1 An animal standing on a relatively small horizontal cylinder *(top)* can easily roll or pitch into a position where its center of mass *(c)* no longer lies above its narrow base of support, whereupon gravity produces a rotatory moment of magnitude *Wm* (weight times horizontal distance between the support axis and the line of gravity) that topples the animal from the support. This can be resisted in several ways: by using grasping extremities *(bottom, left)* to generate a torque *(white arrow)* equal and opposite to *Wm (black arrow)*, or by hanging underneath the support *(bottom, middle)*, or by reducing body size or the height of *c* above the support *(bottom, right)* and so producing a relative increase in the size of the support polygon *(dashed lines)*.

the toppling moment; (3) hanging underneath the branch; or (4) reducing body size so that the support points are spread more widely on the surface of the branch. (The last solution, however, makes climbing on vertical supports more difficult: for a small animal, a thick vertical support becomes, in effect, a flat perpendicular surface.) A large animal, even if it has suspensory adaptations as effective as those of a sloth or an ape, has difficulty feeding in slender terminal branches, which tend to bend or break under the weight of its body. Large animals can address this problem to some extent by developing specializations of the joints and extremities that allow them to spread their weight out as widely as possible among multiple supports. However, the largest animals that habitually forage in trees today are male orangutans, which weigh no more than about 90 kg (Kay, 1973), and this is probably close to the maximum practical body weight for animals feeding in trees.

Because tree branches are discontinuous, arboreal vertebrates must sometimes cross gaps from one branch to the next. This can be done by leaping (or gliding or flying) across the gap, or by bridging it and transferring body weight more or less gradually from one support to the other. Here again, body weight partly determines the appropriate strategy; a 90-kg orangutan cannot afford to fling itself through the treetops like a 6-kg gibbon. This is partly because fewer branches will bear the shock of the orangutan's impact. However, the bending stress that a cylindrical beam can bear is proportional to the cube of its diameter (Jensen and Chenoweth, 1967), and so the branch that the orangutan lands on needs to be only 2.47 times as thick as the branch on which the gibbon lands, other things being equal. Of more importance is the fact that large animals are evidently more fragile than small animals. Although it seems intuitively obvious that a cat can sustain a longer fall than a horse can, and the work of Biewener (1982) provides some empirical support for this intuition, it is not clear why this should be the case. I suggest that it is due to the impossibility of scaling the cross-sectional areas of all supporting structures proportional to the $\frac{3}{2}$ power of body weight, which would be needed to withstand the tensile and compressive stresses of impact (assuming a constant terminal velocity). If safety factors were held constant while body weight increased, the entire weight of the animal would have to be given over to supporting tissues after a few doublings in mass. Whatever its explanation, the inability of large animals to withstand falls ensures that most of the more acrobatic climbing animals are relatively small and that the largest arborealists are cautious animals with relatively long extremities useful in bridging gaps. However, specializations for slow, cautious climbing also characterize several taxa of small vertebrates, probably because they allow such animals to move about silently and thus avoid detection by their prey or predators.

Every solution that an arboreal animal adopts to any one of the problems posed by arboreal locomotion limits its options in dealing with others. For example, an animal that has shortened its limbs to lower its center of mass on slim horizontal branches is thereby handicapped in trying to bridge gaps without leaping. The limited number of such solutions available, as well as the logical connections among them, has resulted in much convergent evolution among tree-dwelling vertebrates, so that we find the trees populated by vaguely kinkajoulike monkeys, lorislike lizards, apelike lemurs, and so on. Identifying the functional principles underlying these multiple convergences is a fascinating problem in itself, and it holds the additional interest of being directly relevant to our understanding of our own evolutionary origins.

The Role of Friction in Climbing

The forces of static friction (F) between two solid bodies are represented to a first approximation by the equation $F = \mu_s L$, where L is the load (that is, the force that is perpendicular or normal to the two bodies' area of contact) and μ_s is a constant called the coefficient of static friction. This constant is empirically determined for each combination of materials. For example, for wood on leather it is about 0.3 to 0.4, which means that the force needed to start a wooden block sliding across a horizontal leather surface equals 30% to 40% of the block's weight.

The equation $F = \mu_s L$ implies that the frictional force is proportional to the load. For an animal standing on a flat horizontal surface, the load or normal force is equal to its weight W. (Because static friction is ordinarily independent of the area of macroscopic contact, it makes no difference how that weight is distributed over various contact surfaces provided that they all exhibit the same coefficient of static friction against the substrate.) For an animal standing on a flat surface that deviates from the horizontal by some angle $\alpha < 90°$, the gravitational vector can be analyzed into a component tangential to the supporting surface and a component normal to that surface (Fig. 5-2), respectively equal to $W \sin \alpha$ and $W \cos \alpha$. The animal will slide down the incline when the tangential force exceeds the force of static friction—that is, when $W \sin \alpha > \mu_s W \cos \alpha$, or when $\tan \alpha \geqq \mu_s$. If $\alpha = 90°$, then static friction drops to zero no matter what μ_s is, because the gravitational force has no component normal to the supporting surface. Thus, when an arboreal animal clings to a vertical support, it must either use grasping mechanisms that do not rely on friction or else generate normal forces by its own muscular effort. This latter strategy is possible only on a nonplanar surface.

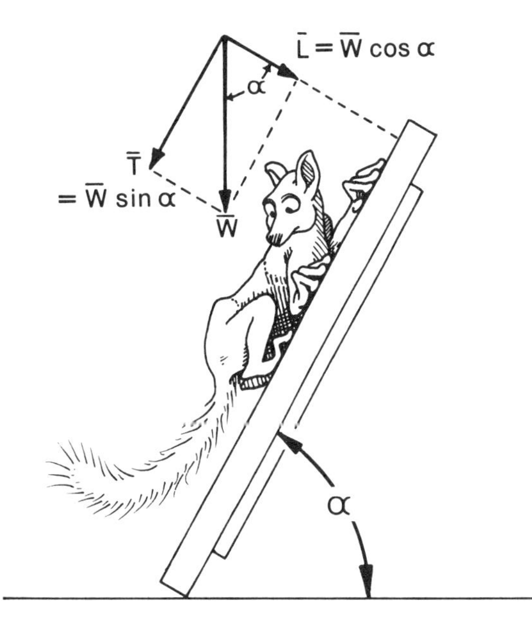

Figure 5-2 An animal's weight *(W)* can be analyzed into forces normal to its support (load, *L*) and tangential to its support *(T)*. The relative magnitude of the two on a planar support is a simple trigonometric function of support angulation (α). (From Cartmill, 1979.)

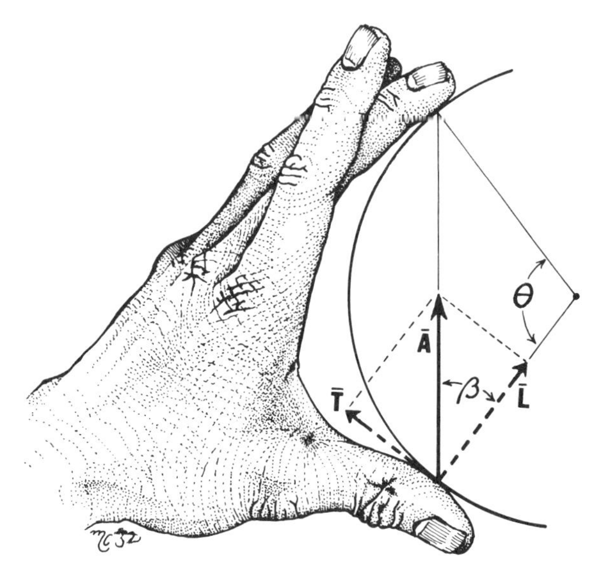

Figure 5-3 A clawless animal grasping an object with a circular cross section exerts adduction force *(A)* along the chord of the arc (θ) that the animal subtends. Adduction force can be analyzed into a component normal to the surface *(L)* and a tangential component *(T)*. Normal force equals $A \cos \beta$, and tangential force equals $A \sin \beta$, where $\beta = (180° - \theta)/2$. (Modified from Cartmill, 1979.)

Figure 5-3 illustrates an animal grasping a surface with a circular cross section. For simplicity's sake, it is assumed that the animal contacts the surface at only two points, which subtend an arc with a central angle θ. The animal generates an adduction force A by squeezing the surface—that is, by drawing each contact point toward the other. If $\theta = 180°$, A is normal to the support surface. If $\theta = 0°$, A is wholly tangential to that surface, and no frictional grip is possible. For intermediate values of θ, A has a normal component L and a tangential component T, which can be expressed, respectively, as $L = A \cos \beta$ and $T = A \sin \beta$, where $\beta = \frac{1}{2}(180° - \theta)$. The same reasoning that was applied in the case of the sloping plane (Fig. 5-2) shows that the animal's grip will fail if θ drops below the point at which $\tan \beta = \mu_s$. In practice, an animal that relies wholly on friction in clinging to a tree trunk or other vertical cylindrical surface must subtend a larger θ than the theoretical minimum that its coefficient of friction permits. This is because the gravitational vector is also tangential to the support's surface, making the total tangential force that must be overcome by friction equal to the vector sum of $A \sin \beta$ and some fraction of the animal's weight.

The central angle that such an animal needs to subtend depends (other things being equal) on the animal's weight; but it does so in a complicated way, because the coefficient of friction between two surfaces is not a constant if one of the two is a curved surface composed of a soft viscoelastic polymer like the skin on our own palms and soles (Bowden and Tabor, 1973). For such surfaces, static frictional forces (F) against a solid surface are not proportional to load (L), but to the load raised to a fractional exponent; that is, $F = KL^n$, where $n < 1$ and K and n are empirically determined constants. This means that as load goes up, the coefficient of static friction declines. Readers can prove this to their satisfaction by placing a glass or other cylindrical object on a table and grasping it between thumb and forefinger at a central angle of about 120°. Most materials can be grasped securely at that angle with light pressure; but squeezing harder causes a reduction in the coefficient of friction, and eventually a point is reached at which $A \sin \beta > \mu_s A \cos \beta$, causing the cylinder to slip from the grasping fingertips. Thus, if an animal clinging by adduction-generated friction to a vertical cylinder at a minimal value of θ increases its weight (thus increasing tangential forces), it cannot stay on the cylinder simply by gripping harder (increasing A); it must also increase its central angle, or fall. Conversely, reduction in body weight permits an animal to cling with a smaller central angle. Experimental data on primates clinging to inclined wooden planes (Cartmill, 1979) show that friction allows a 25-g animal to cling to a surface with a slope of 72.5°, whereas adult humans weighing 70–80 kg invariably slide at angles less than 45°. Equations derived from these data suggest that a diminutive primate like *Microcebus*, with a weight of 100 g and an armspan of 10.5 cm, should be unable to cling to a vertical wooden cylinder much larger than 13 cm in diameter if its grip depends wholly on friction. The available field and laboratory data for small prosimians roughly confirm this prediction (Cartmill, 1974; Charles-Dominique, 1977).

Nonfrictional Grip Mechanisms

We have been treating friction as if it were a single simple force. It is more realistic to express static frictional forces as the sum of an adhesion term and a plowing term (Bowden and Tabor, 1973), both of which ultimately reflect the strengths of various sorts of intermolecular forces. To start one surface sliding across another, it is necessary to disrupt any bonds or attractions that have formed between the molecules of the two apposed surfaces. The force needed to do this represents the adhesion term of static friction. The plowing term results from interlocking of the two surfaces; submicroscopic projections of the harder substance penetrate the softer substance when they are placed together, and these minute asperities must be sheared off or dragged along through the softer substance when one surface starts to slide across the other. Doing this involves breaking intermolecular bonds *within* the substance of one or both surfaces. The initial force needed to break these bonds is equal (but opposite) to the plowing term of the total static frictional force.

Some arboreal vertebrates can cling to a flat vertical surface, or even to the underside of a flat horizontal surface. Most of these animals rely either on interlocking or on adhesion to cling to such supports. In these cases, we no longer speak of the forces holding the animal against its support as frictional forces; doing so would complicate the mathematics too much (for example, we would have to define some finite frictional forces as the product of zero load and an infinite coefficient of friction). But the factors that permit adhesive or interlocking clinging are only aspects of ordinary friction that have been greatly enhanced by special morphological adaptations.

Interlocking Clinging by interlocking is the simplest and most familiar of the nonfrictional grip mechanisms. It is exemplified by any animal

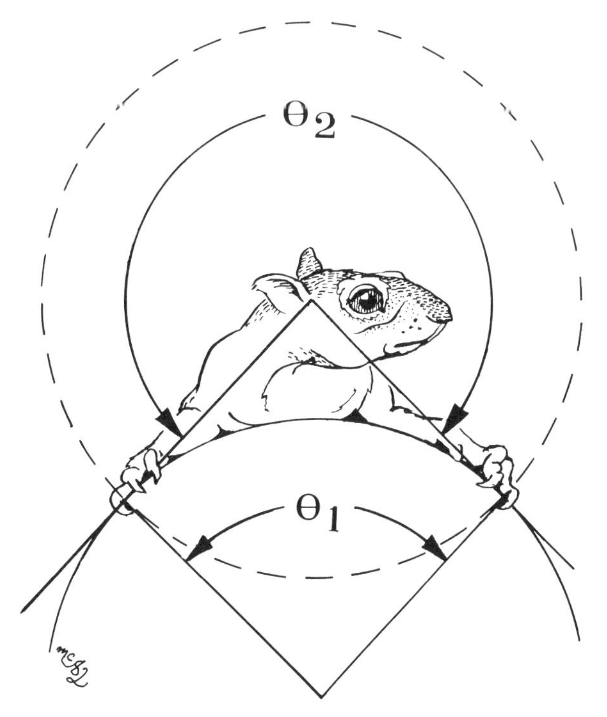

Figure 5-4 Clawed animal clinging to a cylindrical support, showing actual support surface tangent to the animal's volar skin *(solid circle)* and virtual support surface tangent to volar aspect of embedded claws *(dashed circle)*. The virtual or effective central angle (θ_2) is a function of the actual central angle (θ_1) that the animal subtends and of the angle at which the claw tips penetrate the support.

that clings to a support using claws. A clawed animal can climb a vertical cylinder of any diameter, or even a flat vertical surface (Cartmill, 1974), so long as it can dig the tips of its claws into its support. By doing so, it generates a new contact surface that is more nearly perpendicular to the adduction force. The result is an effective increase in the central angle that the animal subtends (Fig. 5-4). The amount of increase depends on the claws' curvature and the depth to which they penetrate. Mammals that rely on claws in climbing trees tend to have sharp claws that are abruptly recurved at the tip, whereas their close terrestrial relatives have more blunt, gently curved claws (Pocock, 1922; Peterka, 1937; Loveridge, 1956). Similar differences distinguish tree-climbing birds from their closest nonclimbing relatives (Bock and Miller, 1959).

Capillary Adhesion No mammals rely on adhesion in clinging, but several groups of reptiles and amphibians do. The specializations that allow them to do this include some of the most elaborate and poorly understood climbing adaptations found among vertebrates. Several sorts of adhesion can be distinguished (Emerson and Kiehl, 1980), but only two are known to be significant for tree-climbing vertebrates: dry adhesion and capillary adhesion. In capillary adhesion the animal is held to its support by the surface tension of a liquid that covers its contact surface. The force of this capillary attraction would be expected to vary with the area of the liquid film holding the two surfaces together, and it would therefore be enhanced by expansion and flattening of the animal's contact surface. Many arboreal frogs and salamanders have developed such specialized surfaces for capillary adhesion, including enlarged disc-shaped pads on the tips of the fingers and toes (Noble and Jaeckle, 1928; Green and Alberch, 1981). In tree frogs the surface of these pads is typically a pebbly mosaic of epidermal "tiles" about 0.01 mm across, separated by clefts through which mucus glands open onto the surface. When wetted by mucus, the smooth epidermal surfaces provide for effective capillary adhesion to smooth supports; the slight roughness produced by the gaps between "tiles" enhances interlocking effects for nonadhesive clinging to rough support surfaces (Emerson and Diehl, 1980; Green, 1981).

Dry Adhesion Since reptiles generally lack skin glands, they are not readily able to secrete the fluid film needed for capillary adhesion. However, several lineages of arboreal lizards have evolved mechanisms for dry adhesion. Dry adhesion, which is equivalent to the adhesion term of ordinary friction, is produced by the intermolecular forces that appear between two surfaces placed in extremely close contact. These forces are quantum electrodynamic phenomena that result from the coupling of electron motions in the electron "clouds" surrounding adjacent molecules. The strength of these forces depends in a complex way on the geometry of the apposed surfaces and the physical properties of their constituent molecules. They decline markedly with distance (theoretically, as the inverse sixth power of distance for two monatomic molecules) and are negligible for molecules more than about 5×10^{-10} m apart (Andrews and Kokes, 1962). Animals using dry adhesion in climbing therefore need to develop very specialized contact surfaces that ensure extremely intimate contact with the support at a submicroscopic level. The apparent or macroscopic area of contact is not directly relevant, as it is in the case of capillary adhesion.

Specializations for dry adhesion have been developed in tree-dwelling representatives of at least three lizard families (Williams and Peterson, 1982). These specializations are most pronounced and best understood in geckoes (Gekkonidae). Most digits of most geckoes bear claws, which function in the usual way in climbing; but each penultimate phalanx is covered ventrally by an adhesive pad as well. In the pes of *Gekko gecko*, each of these pads contains a blood-filled sinus that acts as a hydrostatic skeleton and ensures a broad, even transmission of digital pressure to the support surface (Russell, 1975). The surface of the pad is thrown into convolutions, forming transverse lamellae (Ruibal and Ernst, 1965) that can be raised and lowered by means of a complex system of internal tendons; those of the pes have indirect connections to the gastrocnemius. (The claw-flexing muscles are independent of the lamellar system.) The proximal-facing surface of each lamella is covered with thousands of setae roughly 0.1 mm in length, each resembling a complexly branching tree with a long "trunk" and a disclike expansion at the tip of each "branch" (Fig. 5-5). It is thought that when a gecko climbs using adhesion, it rolls the pad down onto the support surface (proximal lamellae first) and then exerts tension along the length of the digit, throwing the setae into S-shaped curves that press the terminal expansions into the support with sufficient force and intimacy to yield strong intermolecular attraction (Hiller, 1968). The process is reversed when lifting the foot, so that the setae are peeled gently off the support without tearing or breaking; damaged setae cannot be replaced until the skin of the pad is next shed, and the lizard's adhesive power depends on keeping its setae intact. (During climbing using claws, the setae can be held in reserve by keeping the lamellae erected.) Despite the delicacy of this procedure, a climbing gecko can move with astonishing speed and can run up a vertical sheet of glass using

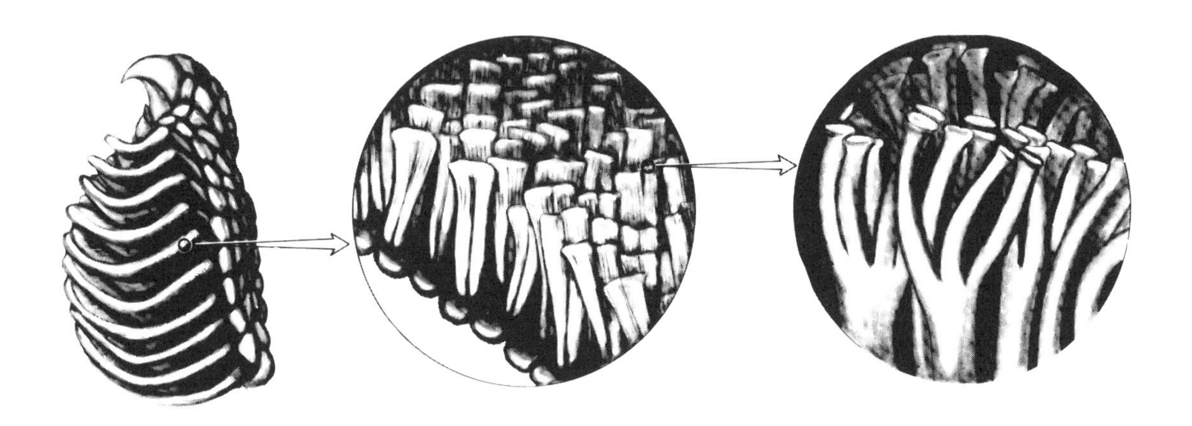

Figure 5-5 Adaptations for dry adhesion on the apical "pads" of a gecko *(Gekko)*: *(left)* digit with mobile lamellae and terminal claw; *(center)* enlarged view of a lamella, showing setae; *(right)* a further enlargement (diameter of field is 2.5μ), showing the terminal bristles and their endplates on each seta. (From Hildebrand, 1982; reprinted by permission of John Wiley and Sons, Inc.)

a fast gait in which three of its four feet are at times off the support, a pattern otherwise seen among lizards only in cursorial ground-dwellers (Russell, 1975). Similar but less thoroughly studied mechanisms of dry adhesion permit similar behavior in some anoline iguanids (which have shorter setae with spatulate tips) and scincids (Peterson and Williams, 1981; Williams and Peterson, 1982). The pads of certain rhacophorid tree frogs also bear setalike marginal projections, which may provide some potential for dry adhesion (Welsch, Storch, and Fuchs, 1974).

Capillary and dry adhesion differ in their relative strengths on different surfaces. For *capillary* adhesion the smoothness of the support surface is critical; rough surfaces preclude the formation of an extensive liquid film over the area of contact. The strength of *dry* adhesion varies directly with the so-called surface energies of the support surface, which are a direct function of the strength of the bonds between the molecules of that surface (Hiller, 1968). Cohesive materials have high surface energies and are easily wetted. Plant surfaces adapted for shedding water — such as the waxes that coat the cuticles of leaves — have low surface energies and are therefore poor supports for dry-adhesion climbers. However, since such surfaces are usually smooth and glossy, they are very well suited for capillary adhesion. Capillary-adhesion mechanisms are also seen in nonarboreal amphibians that use leaves as resting sites or climb on smooth, slippery rocks (Emerson and Diehl, 1980).

Suction A final nonfrictional grip mechanism of importance to arboreal vertebrates is suction, in which an animal is held against its support by air pressure. Such an animal must be separated from its support by a film of gas or liquid under lower pressure than the surrounding air, and must therefore have (1) a specialized contact surface that can enclose a sealed-off volume under negative pressure and (2) some means of producing that negative pressure. When this is accomplished, air pressure yields a force normal to the contact surface, pressing that surface against the support and increasing friction correspondingly. If the suction mechanism is efficient enough, the animal can cling to the underside of a horizontal surface, relying exclusively on air pressure to resist the force of gravity.

Early analyses attributed the clinging powers of geckoes and tree frogs to suction, but this interpretation is contravened by the fact that these animals go on adhering to smooth supports when circumambient air pressure is drastically reduced (Hiller, 1968; Emerson and Diehl, 1980). Certain hyracoids are said to climb rocks and trees using suction (Vaughan, 1972), but the evidence for this is wholly anecdotal (see, for example, Sclater, 1901). At present, the only air-breathing vertebrates known to generate suction forces in clinging are certain plethodontid salamanders and tropical bats.

The plethodontid genus *Bolitoglossa,* comprising over 70 species of Neotropical salamanders, includes several species that have adapted to life in lowland rain forests by utilizing bromeliads and other water-retaining plants as microhabitats. Rain forest *Bolitoglossa* exhibit specializations that permit them to cling to the smooth, waxy cuticles of these plants. The experiments of Alberch (1981), in which living *Bolitoglossa* were induced to cling to a flat surface with their feet placed over instrumented holes, show that these salamanders are able to generate negative pressure between their support and their volar surfaces. The species that cling most effectively to smooth surfaces have disclike hands and feet with subcircular outlines and extremely smooth epithelium on the volar margins, specializations which Alberch (1981; Green and Alberch, 1981) interprets as producing a tighter and more reliable seal around the edge of the volar "suction cup." However, since these salamanders can walk upside down on porous surfaces, do not lose their grip when circumambient air pressure is reduced, and can be washed off a sheet of glass by pouring water down it (Alberch, 1981), it is clear that capillary adhesion is their principal gripping mechanism, as it is for tree frogs.

The suction cups found on the wrists, ankles, or both of certain bats appear to have evolved from more typical mammalian volar friction pads, and various intermediate morphologies occur among living forms. In the rare Asian vespertilionid *Eudiscopus,* both wrist and ankle bear superficially similar dermal pads; but the carpal pad is a convex pressure-bearing cushion of the usual mammalian sort, whereas the tarsal pad has a concave center that lacks subcutaneous fat and is bound by strong collagenous bundles to the first four metatarsal bones. Although it is not known how the tarsal pad functions in living *Eudiscopus,* Schliemann and Rehn (1980) suggest that it serves as a suction disc, and that metatarsal movements can generate negative pressure by drawing the concave center of the pad away from a support surface. At the other extreme of the spectrum of specialization is the Neotropical bat *Thyroptera,* whose wrists and ankles bear stalked, rubbery suction discs containing a flexible fibrocartilaginous skeleton that can be tensed by certain extrinsic muscles to lift the center of the disc, thus producing suction, or to lift the edge, thus releasing suction (Wimsatt and Villa, 1970). The surface of the disc is richly supplied with sweat glands that secrete a watery fluid to moisten the disc and help produce a tight seal against the pressure of the surrounding air. Once established, negative pressure is maintained effortlessly by the elasticity of the fibrocartilage; a dead *Thyroptera* can be made to stick onto a vertical surface by a single thumb disc (Dunn, 1931). *Thyroptera* rests during the day inside the curled-up new leaves of large-leafed plants like bananas, using its suction discs to adhere to the inside of the leaf. The tiny vespertilionid *Tylonycteris,* which has expanded tarsal pads that may also function as suction cups, roosts inside bamboo stems and is thought to use its pads to adhere to the inner surface (Walker, 1975). The suction organs of these animals, and perhaps of other bats with similar specializations (such as *Myzopoda*), thus appear to be analogs of the specializations developed in various arboreal amphibians for capillary adhesion to plant surfaces with low surface energies.

Specializations Enhancing Skin Friction

Vertebrates that lack claws, suction cups, or adhesive surfaces must rely wholly on frictional forces generated by adduction in climbing. These forces can be enhanced to some extent by specialized sculpturing of the skin covering the volar contact surfaces. Some tree-climbing vertebrates with scaly volar skin exhibit imbrication of, or pointed projections on, the volar scales. These features can be plausibly interpreted as specializations for enhancing the plowing component of skin friction (Bock and Miller, 1959; Peterson and Williams, 1981).

Arboreal mammals lack volar scales, but some exhibit other friction-enhancing specializations of the volar skin. In primitive therian mammals the palms and soles bear a number of elevated pads composed of fat lobules bound to the underlying aponeuroses and deep fascia by a dense web of collagenous and elastic fibers. These fatty cushions are loaded under net compression during locomotion and accordingly lack the organized tracts of tension-bearing fibers seen in the superficially similar adhesive "pads" of tree frogs (Green, 1981). The volar epidermis of mammalian pads is thick, keratinized, and usually covered with small caruncles or conical papillae, and numerous sweat glands open onto its surface. The primitive arrangement of the pads (Fig. 5-6) comprises 11 pads per extremity: an apical pad under each distal interphalangeal joint, an interdigital pad under the distal end of each intermetapodial space, and a thenar or hypothenar pad beneath the base of each marginal metapodial (digits 1 and 5).

The dermal caruncles and sweat glands of the volar pads of friction-climbing mammals are commonly reorganized into a system of dermatoglyphic friction ridges or "fingerprints." Such ridges are found in all primates and are variably developed in arboreal marsupials, carnivorans, and insectivorans. It is generally thought that friction ridges enhance skin friction by producing surface interlocking that augments the plowing term of static friction, but this has been debated. Whipple (1904), who was the first to argue that mammalian friction ridges augment interlocking effects, noted in support of this thesis that such ridges are usually oriented "at right angles with the force that tends to produce slipping, or to the resultant of such forces when these forces vary in direction." But in most prosimian primates and arboreal marsupials, apical pad ridges, where present, run mostly parallel to the digital axis (Wood Jones, 1924; Dankmeijer, 1938; Biegert, 1961), and Kidd (1907) argued that this fact re-

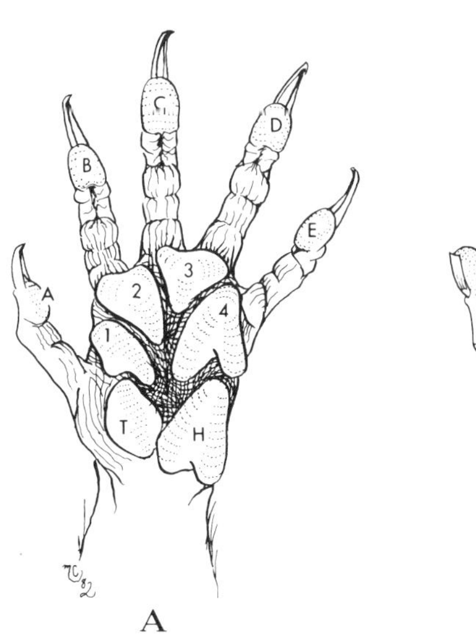

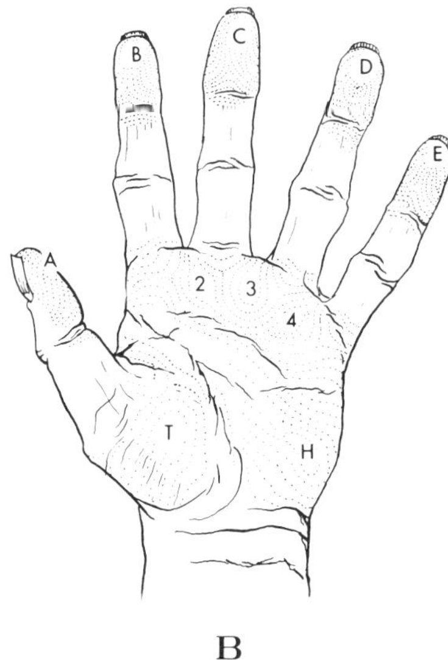

Figure 5-6 Ventral surface of the left hands of *Tupaia glis* (A) and *Homo sapiens* (B), showing pads and major patterns of dermatoglyphic "friction ridges" *(dotted lines).* A-E = apical pads; 1–4 = interdigital pads; T = thenar pad; H = hypothenar pad.

futes Whipple's thesis, since such axially oriented ridges cannot oppose the backward thrust of the digits in locomotion. However, primates and arboreal marsupials generally have grasping feet, in which some of the digits are widely divergent from the others and are used in opposition to them when clasping branches. When an animal with such hands or feet walks along a branch, some or all of its digits are placed roughly at right angles to the axis of the branch, and so its axially striated apical pads are oriented perpendicular to the direction of movement, as predicted by Whipple's analysis (Cartmill, 1974). In arboreal mammals with less specialized hands and feet—those in which the digits of a single extremity remain more nearly parallel to one another—the digits are put down roughly parallel to the branch's axis and apical pad striation, if present, is predominantly transverse. This is seen, for example, in *Potos* and tupaiids (Cummins and Midlo, 1943).

In many tree-climbing mammals (such as catarrhine primates, indriids, koalas, sloths) the volar pads have become fused into a broad, soft contact surface like those of our own hands and feet (Fig. 5-6). Among primates (the only arboreal mammal taxon for which this trend has been examined systematically) pad fusion is roughly correlated with body weight and tends to be accompanied by an expansion of dermatoglyphics over the entire volar surface (Biegert, 1961; Hershkovitz, 1977; Cartmill, 1979). This primate trend is often regarded as a manipulatory adaptation, as a tactile specialization, or as a generally progressive indication of perfected arboreal adaptation or "phylogenetically higher development" (Biegert, 1963; cf. Hill, 1953, p. 8; Le Gros Clark, 1971, pp. 187, 202; Hershkovitz, 1977, p. 35). However, a locomotor function for this flattening and fusion of volar pads has been claimed by Cartmill (1979), who notes that the exponent n in the frictional equation $F = KL^n$ would be expected to increase with decreasing curvature of the contact surfaces (Bowden and Tabor, 1973, pp. 79–81). If this is granted, it follows that small clawless tree-climbers would be better off with discrete, protuberant pads to enhance their coefficient of static friction in climbing large supports (where the normal component of the adduction force is reduced), whereas larger and heavier animals would be better off with fused pads and a coefficient of friction that remains more nearly constant under the high adduction forces needed to support their weight. More sophisticated theoretical analysis and additional comparative and experimental data are needed to evaluate this conjecture.

The friction grip of clawless arboreal mammals may be enhanced somewhat by pad fusion and friction ridges. It nevertheless remains unclear why these animals have lost their claws—or, in some cases, unrolled them from their primitive quill-like shape into flattened nails like those typical of primates. Claws that can be dug into a support surface would seem to be adaptively advantageous for most tree-climbers (Fig. 5-4), and their absence in many arboreal vertebrates demands special explanation. Some plant surfaces cannot be pierced by claw tips; and it is no surprise that vertebrates that climb principally or exclusively on such surfaces tend to have clawless digits, specialized for adhesion (as in leaf-climbing amphibians) or for a friction grip (as in bamboo-dwelling mice, discussed by Musser, 1972; Emmons, 1981). But this analysis is inapplicable to most clawless arboreal mammals. The "nails" of arboreal hyracoids are probably modified hooves, reflecting these animals' descent from terrestrial ungulatelike forms; as Kingdon (1971) notes, "the structure of the feet [in tree hyraxes] . . . appears to be a phylogenetically late makeshift." Primate clawlessness has been described as a primitive retention from the mammalian ancestry (Wood Jones, 1916), but this interpretation is contradicted by the wide distribution and morphological uniformity of claws among living mammals (Le Gros Clark, 1936) and by the presence of claw-shaped distal phalanges in such phylogenetically critical fossil groups as cynodont therapsids, Triassic nontherian mammals, and plesiadapids (Russell, 1964; Jenkins, 1971; Jenkins and Parrington, 1976). Apical pads supported by flattened nails may be superior to clawed fingertips as tactile and manipulatory organs (although the clawed fingers of raccoons are notorious for their sensitivity and dexterity); but in primate evolution the loss of claws seems to have preceded the emergence of a monkeylike interest and facility in handling things (Bishop, 1964; Cartmill, 1975), and so primate clawlessness cannot have *originated* as a manipulatory adaptation. Most recent accounts of claw loss in primates assume that claws get in the way in certain sorts of arboreal locomotion—for example, in grasping fine branches securely (Hershkovitz, 1970; Cartmill, 1974), in leaping from large vertical supports (Szalay, 1972), or in landing on long, flexible branches (Szalay and Dagosto, 1980). All these accounts run afoul of the counterexample of marmosets (Callitrichidae). Except for the hallux, the digits of these small Neotropical monkeys bear functional claws, apparently reevolved from typical primate nails as part of an adaptation for foraging on vertical tree trunks (Coimbra-Filho and Mittermeier, 1975; Rosenberger, 1979; Szalay, 1981). However, marmosets are nonetheless also adept at all the sorts of locomotion with which claws supposedly interfere (Thorington, 1968; Kinzey, Rosenberger, and Ramirez, 1975; Garber, 1980). The clawlessness of more typical primates and some other climbing mammals (such as burramyid marsupials) thus presents a persistent theoretical problem.

Claw Grip and Vertical Climbing

Most arboreal vertebrates descend vertical supports headfirst, which improves their chances of spotting dangers in their path. The biomechanical problems that this descent posture poses for a typical quadruped are obvious to anyone who has watched a domestic cat trying to climb down a tree. In headfirst descent the major propulsive muscles of the limbs—the retractors of the femur and humerus—cannot act as they do in headfirst ascent to help support the animal's weight. Animals relying on a claw grip find themselves especially handicapped in this position, since their claws tend to point the wrong way and cannot be dug in at the proper angle to resist the craniad tug of gravity.

Headfirst descent is facilitated by morphology that permits the hind limb to be inverted or supinated so that the sole of the foot can be applied to the support with the digits pointing laterally caudad. This allows the toe flexors to help counter the gravitational vector in headfirst descent. Placing the pes in this orientation presents no great problems for lizards and other vertebrates that preserve a more or less primitive tetrapod locomotor posture; but in mammals it demands additional specializations of the hind-limb joints. Most clawed arboreal mammals (and many clawless ones) have evolved such specializations, which permit the foot to be rotated into a position in which the sole faces ventrally but the toes point almost directly caudad. The mechanisms that afford this mobility are diverse; the approximately 170° of rotation needed occurs partly at the hip joint, supplemented by subtalar rotation (in squirrels, *Tupaia, Bassariscus,* and most primates), calcaneonavicular supination (in galagos and tarsiers), tibiofibular supination (in *Felis wiedi*), fibular translation (in marsupials), or some combination of these movements (Barnett and Napier, 1953; Leyhausen, 1963; Hall-Craggs, 1966; Trapp, 1972; Cartmill, 1974; Jenkins, 1974; Szalay and Drawhorn, 1980).

Although birds' ability to fly exempts them from the more serious dangers of arboreal locomotion, woodpeckers and other birds that forage on tree trunks experience the same sorts of problems that other vertebrates encounter in clinging to large vertical supports. Trunk-climbing birds have clawed digits that can be dug securely into bark; but since they have only two extremities that can be used in grasping, they cannot shinny up or down a tree trunk with the body's weight supported partly by the upper limbs (under tension) and partly by the lower limbs (under compression) as do typical clawed arboreal mammals (Fig. 5-7). Trunk-climbing birds have evolved two ways of dealing with this problem. In nuthatches (Sittidae, including Neosittidae) one hind limb grips the trunk above the bird's center of mass and functions as a tension member; the other grips below and is stressed under compression. With this limb deployment nuthatches can ascend or descend a trunk headfirst, holding the body axis somewhat obliquely to the trunk's vertical axis to enhance the backward rotation of the trailing limb (Richardson, 1942). Since each foot must subtend as large a central angle as possible so that it can maintain an independent grip on the trunk, the toes of nuthatches are elongated—especially the hallux, which opposes the other three toes and provides adduction force to keep the claws imbedded in the bark.

Other trunk-climbing birds—woodpeckers (Picidae), creepers (Certhiidae), and woodcreepers (Dendrocolaptidae)—use both hind limbs as tension members and have stiff, pointed tail feathers (borne on an enlarged pygostyle) that are braced against the trunk to support part of the bird's weight from below under compression (Richardson, 1942; Bock and Miller, 1959).

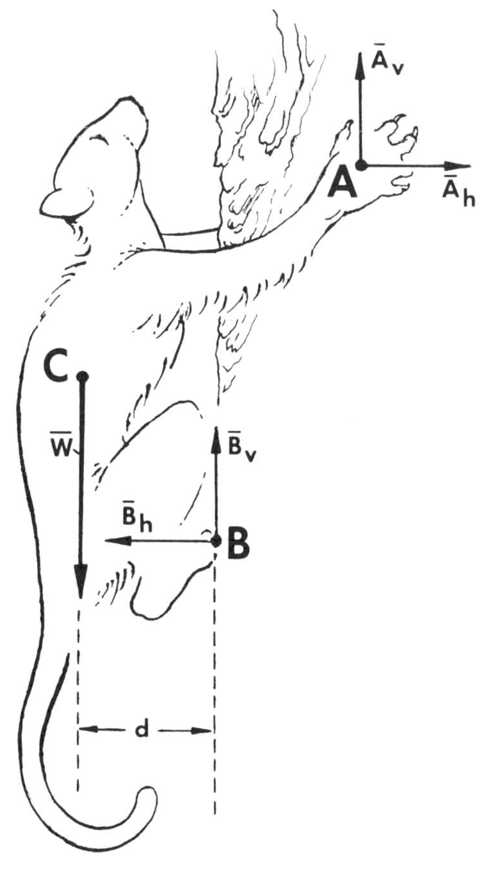

Figure 5-7 Mammal clinging quadrupedally to vertical support. Shown are C, center of mass of animal; W, weight of animal (gravitational-force vector); and horizontal (A_h, B_h) and vertical (A_v, B_v) components of reaction forces acting on animal through its contact points, A and B. The upper limb is stressed under tension ($A_h + A_v$); the lower, under compression ($B_h + B_v$). To reduce the gravitational moment around B, which tends to topple the animal away from its support, d is usually minimized by bringing the lower extremities close together (thus shifting B toward W). This means that a friction grip at B depends on B_h, with little help from adduction force. Increase in W for a clawless mammal thus demands increase in B_h, yielding an equal but opposite increase in A_h, and thus necessitating increase in the central angle subtended by the upper limbs. Therefore, upper-limb length in arboreal animals that rely on a friction grip is positively allometric. (Modified from Cartmill, 1974.)

Analogous specializations of the tail for clinging to vertical supports are seen in cliff-perching swifts (Apodidae) and in some primates and rodents (Tarsiidae, Anomaluridae). These specializations of the tail feathers make descent of a trunk awkward (and headfirst descent impossible), and so trunk-clinging birds that rely on the tail as a compression member tend to forage by working their way upward to the top of one trunk and then flying back down to start at the bottom of the next one. The most specialized trunk-feeders among the woodpeckers have modified the primitive picid foot by rotating digits 1 and 4 around from an opposed position (the zygodactylous condition) into a position parallel to digits 2 and 3 (the pamprodactylous condition). In these forms all four toes point upward in clinging tail-down to a tree trunk, and their claws can be dug in at the appropriate angle to resist the pull of gravity. The trunk-foraging parrot *Micropsitta* has a similarly modified zygodactylous foot, as well as woodpeckerlike specializations of the tail feathers (Bock and Miller, 1959).

Prehensile Extremities

The feet of many arboreal vertebrates have one or two digits that point in a markedly different direction from the rest of the digits and are separated from them by a cleft. Slender branches can be gripped in the cleft by adduction or flexion of the opposed sets of digits. This sort of grasping specialization of individual extremities is useful chiefly on relatively small supports on which the span of a single hand or foot can subtend a significant central angle and a secure claw grip cannot be obtained. It is perfectly possible to climb and forage among slender branches without such extremities; squirrels, for example, whose permanently adducted digital rays lack any sort of opposability, run and balance atop slim horizontal branches, shinny up sloping ones inchworm-fashion by alternately gripping with fore and hind pairs of limbs, and grasp and handle food items while hanging head-down by their clawed toes (Cartmill, 1974). But grasping hands and feet make these activities somewhat easier and safer, and permit others (such as standing bipedally on a branch to reach an overhead food item or reaching out horizontally away from a branch to secure an item barely within reach) that are difficult or dangerous for an animal that relies on balance to stay on a horizontal support. Because grasping extremities provide better control of pitching or rolling moments around the support axis (Fig. 5-1), they are useful in maneuvers on slender branches. (They are also indispensable for tree-roosting birds, although this has nothing to do with climbing as such.) In foraging activities among fine branches, prehensile extremities are probably most valuable for insect-eaters and other predators, whose food is not likely to stay put if they execute the sort of noisy and energetic approach that a squirrel uses to sneak up on a pecan.

When crossing from one branch to the next, an arboreal vertebrate must bridge the gap from a support of proven reliability to one that may break, bend precipitously, or prove unreachable. The danger of falling is reduced if the animal can keep a firm grip on the support behind while testing the one ahead. Thus, the grasping ability of the trailing end of the body is of more consequence than that of the forelimb, and so grasping specializations of the pes are usually more pronounced than, and seem invariably to be antecedent to, those of the manus. For example, a broadly divergent, grasping first toe is characteristic of all primates except ourselves, whereas there are at least 15 extant primate genera that lack a significantly divergent thumb.

With few exceptions, both hands and feet of primates are of the kind described by Haines (1955) as "clasping" or "opposable"—that is, they display enlarged contrahentes muscles of the first and fifth digits, which originate from a median raphe and run transversely across the palm or sole to insert on metapodials or proximal phalanges. This arrangement augments the adduction force that a single extremity can generate in squeezing a support between the marginal digits. In the grasping hind foot of typical primates the markedly divergent first digit (the hallux) is correspondingly furnished with a greatly enlarged contrahens, which may expand its attachments across the entire sole (Jouffroy and Lessertisseur, 1959). Most of these generalizations apply equally to the strikingly primatelike hind feet seen among arboreal marsupials (with the exception of tree kangaroos, which have secondarily reentered the trees after a period of adaptation to ricochetal hopping on the ground). Specialized grasping hind feet with divergent marginal digits are also found in chameleons, the bamboo-climbing rodents mentioned above, and certain tree frogs (such as *Phyllomedusa bicolor;* Noble, 1931). In primates and marsupials the first toe is opposed to the other four in grasping; but chameleons oppose toes 1 and 2 to the other three, and some climbing rodents (such as *Dendromus*) have a divergent, thumblike fifth toe (Jullien, 1968). Tree porcupines of the genus *Coendu* have developed a grasping pes by converting the tibial sesamoid and overlying pad into a surrogate hallux (Wood Jones, 1953; Hildebrand, 1978). In some arboreal mammals with enlarged claws, efficient single-extremity prehension of slender supports is made possible by enlargement of the proximal volar pads to form a cushion against which the clawed fingers can be flexed in a tight, sharp-edged grip; specializations of this sort are seen in arboreal pangolins, sloths, and xenarthran anteaters (Whipple, 1904; Böker, 1932).

A similar morphological diversity is seen in grasping hands. The primitive primate hand appears to have resembled those of living prosimians that lack markedly divergent thumbs — for instance, mouse lemurs *(Microcebus)* and small galagos. These animals tend to place the hand down on small supports with the support's axis indifferently falling through or to either side of the axis of the second digit (Bishop, 1964; Cartmill, 1974). More stereotyped and specialized patterns of manual prehension among primates (between digits 1 and 2 as in our own hands, or between digits 2 and 3 as in *Alouatta* and some other New World monkeys, or between digits 1 and 3, with digit 2 reduced to a stub, as in lorises) are derivable from the *Microcebus* pattern. Analogous grasping specializations of the manus have been evolved convergently in a few other tree-dwelling vertebrates; koalas and some other marsupials oppose digits 1 and 2 to the other digits of the hand, and chameleons oppose digits 1 – 3 to the other two fingers.

The prehensility of the trailing end of the body is greatly enhanced in chameleons, some arboreal snakes and salamanders, and a variety of arboreal mammals by the development of a grasping tail. A few Old World mammals (tree pangolins, some muroid rodents, and the viverrid *Arctictis*) are capable of using their tails to grasp a support, and many phalangerid marsupials of Australia and New Guinea have opossumlike prehensile tails; but the real homeland of the prehensile tail is South America, where for some reason no fewer than six lineages of mammals appear to have evolved such tails convergently: didelphid marsupials, the erethizontid rodent *Coendu,* the procyonid carnivoran *Potos,* the arboreal anteaters *Tamandua* and *Cyclopes,* and at least two groups of monkeys (*Cebus* and the atelines, including *Alouatta*). Prehensile-tailed arboreal vertebrates (other than snakes) usually also have grasping hind feet, which are used together with the tail to provide a stable three-point support base in feeding head-downward among terminal branches, as well as to enhance the animal's security in moving between supports. The extraordinarily specialized tails of spider monkeys and other atelines, which end in a fingertiplike apical pad covered with dermatoglyphics and are capable of fine manipulation, presumably evolved from a more typical "safety belt" prehensile tail of the sort seen in *Cebus.*

Gap-Bridging Adaptations and Hominoid Morphology

It was noted above that a large tree-dweller presumably cannot afford to be as acrobatic as a small one and should be more likely to negotiate a gap between supports by "transferring" or bridging behavior, reaching across the gap to grasp the next support and smoothly shifting its weight from one support to the next. The field studies of Fleagle and Mittermeier (1980) bear out this prediction: among seven sympatric species of New World monkeys, body weight is negatively correlated with frequency of leaping, and the heaviest animals (Atelinae) rarely leap from one support to the next. The great apes, the largest of all arboreal vertebrates, also rarely or never leap in the trees. Some arboreal mammals that are in theory small enough to be acrobatic leapers — lorises (Lorisinae), sloths (Bradypodidae), arboreal xenarthran anteaters *Tamandua, Cyclopes*), and some marsupials (for example, *Didelphis virginiana*)—exhibit a similar clambering mode of locomotion that avoids free flight between supports. In lorises this locomotor habit is usually interpreted as an adaptation for cryptic concealment (Cartmill, 1974; Charles-Dominique, 1977), an explanation that may hold for other small, cautious arboreal mammals as well. The slowness of sloths may serve the same end, but it also accords plausibly with the tight energy budget imposed by their diet and high population densities (Montgomery and Sunquist, 1978).

Such cautious, nonacrobatic tree-dwellers must be able to reach out precisely and get a secure grip on any support to which they wish to move. The parts of the body that are used in reaching across to the next support must therefore be flexible enough to adjust to the endless variability in the orientation of tree branches; but they must also be sufficiently strong and rigid to bridge the gap. In slow-climbing arboreal mammals these conflicting demands are met by specializations of the limbs and axial skeleton.

All nonleaping arboreal mammals have markedly specialized vertebral columns. Elongation of the trunk, particularly of its more rigid thoracic segment, is seen in lorises and sloths, which have increased the number of rib-bearing vertebrae from the primitive mammalian number of 13 or so to as many as 17 in lorises and 24 in sloths—representing maxima for primates and mammals respectively (Flower, 1885; Schultz, 1961). The peculiar specialized ribs of the loris *Arctocebus* and of the slow-climbing xenarthran anteaters appear to represent more advanced adaptations for this sort of bridging behavior (Jenkins, 1970), as do the specialized zygapophyses of certain tree snakes (Johnson, 1954). The great apes, which are so large and top-heavy that they cannot afford to elongate the trunk as a gap-bridging adaptation, have instead diminished the distance between rib cage and pelvis by reducing both the number and the height of individual lumbar vertebrae (Benton, 1967), thus gaining more effective control over movement between thorax and pelvis when crossing from one support to another. The large New World monkeys exhibit both lumbar shortening and thoracic elongation (Erikson, 1963). The net effect in all these animals is the same: the thorax contributes disproportionately to total trunk length. The tail of cautious climbers, no longer needed as a balancing organ in leaping or other acrobatic activity, either becomes a prehensile organ (as in *Tamandua, Cyclopes,* Atelinae, and *Didelphis*) or is reduced to a vestige, as in lorises, sloths, and apes.

Lorises, sloths, and apes also share a number of specializations of the forelimb, including dorsoventral flattening of the rib cage and associated laterad reorientation of the glenoid, reduction of the ulnocarpal contacts, and mobility-enhancing modifications of various joint surfaces (Cartmill and Milton, 1977; Mendel, 1979). The large New World monkeys are convergent with apes in most of these features as well (Erikson, 1963). These modifications increase the range of forelimb circumduction in reaching out to bridge gaps.

Most of these mammals also exhibit elongation of the forelimb relative to the hind limb. The degree of this elongation varies directly with body size, and so it is probably not simply an adaptation for bridging discontinuities. As noted above, a clawless animal clinging to a vertical support needs to subtend a large central angle on that support, and any increase in body weight demands an increase in the angle subtended (because of the exponential relationship between load and coefficient of friction). We might therefore expect increases in body size to be attended by a general increase in limb length. However, a clawless animal climbing up or down a tree trunk needs to retain a more secure grip with its upper pair of limbs than its lower pair, because failure of the upper grip will cause the animal to topple backward away from the tree and fall (whereas if the lower grip slips, the animal may only swing forward into the tree trunk). All these considerations (Fig. 5-7) imply that the ratio of total upper-limb length to lower-limb length should increase with body weight in clawless arboreal mammals, either by elongation of the forelimbs, shortening of the hind limbs, or both. This prediction (Cartmill, 1974) has been borne out by biometric studies of primates (Jungers, 1978; 1985) and tree kangaroos (Ganslosser, 1980). However, relative forelimb elongation does not vary with body size in tree squirrels (Thorington and Heaney, 1981) and other typical clawed tree-climbers. There are probably at least two reasons for this: first, reliance on claw grip exempts these animals from having to maintain a minimal central angle on a support, and second, they both ascend and descend vertical trunks headfirst, so that no pair of limbs is consistently uppermost.

A trend toward suspensory locomotion would also be predicted from an increase in body size. Because an animal walking along a relatively slender horizontal support is in continual danger of toppling (Fig. 5-1), it will benefit from any anatomical or behavioral change that reduces its rolling moment around the support axis. The simplest solution to this problem is to shorten the limbs. Some clawed tree-dwellers have adopted this solution; for example, arboreal viverrids have shorter limbs than their exclusively terrestrial relatives, and they carry the center of mass lower when walking along horizontal supports (Taylor, 1970, 1976). However, limb shortening is contraindicated for large, clawless animals like apes, which need to preserve relatively long forelimbs for crossing gaps and climbing large vertical supports. We would thus expect that any lineage of clawed arboreal mammals undergoing progressive increase in body size would increasingly tend to deal with the problem of toppling by hanging underneath horizontal supports rather than walking on top of them. This prediction (Napier, 1967; Hershkovitz, 1970; Cartmill and Milton, 1977) is supported by the observations of Fleagle and Mittermeier (1980), who find that the frequency of suspensory locomotion among New World monkeys is a direct function of their body weight.

In summary, body-size increase in a vertebrate that climbs using a friction grip would be expected on theoretical grounds to produce behavioral trends toward more frequent hanging and less frequent leaping, correlated with such anatomical features as elongated forelimbs, highly mobile limb joints, a disproportionately short lumbar segment of the vertebral column, and a dorsoventrally flattened thorax and correspondingly redirected scapula (to increase the range of humeral circumduction). All large, clawless arboreal vertebrates thus ought to look something like apes, at least from the neck down. This expectation is borne out by the presence of these apelike features in tree sloths, whose enlarged, blunt-tipped claws are used more like rigid fingers than like typical claws in grasping supports (Mendel, 1979). It is also supported by the parallel evolution of these features in at least four lineages of large arboreal primates: the apes, the atelines, the extinct palaeopropithecine lemurs of Madagascar (Tattersall, 1982), and the Miocene cercopithecoid *Oreopithecus* (Szalay and Delson, 1979).

The great apes, the lesser apes (gibbons and siamangs, Hylobatidae), and human beings (Hominidae) are distinguished from other extant Old World anthropoids by this suite of features and several other functionally related specializations (for example, visceral fixation). There is little doubt that this grouping represents a monophyletic clade, and it is accordingly recognized by primate systematists as a superfamily, Hominoidea. Since the pioneering work of Keith (1923), the synapomorphies of the living hominoids have generally been regarded as a heritage from an ancestor that practiced acrobatic, gibbonlike brachiation or "ricochetal arm-swinging" (Tuttle, 1969). However, the hylobatids may have developed apelike morphology at least in part independently of the other hominoids. The peculiar wrist morphology and allometrically disproportionate forelimb elongation of hylobatids (Lewis, 1972; Biegert and Maurer, 1972; Jenkins, 1981; Jungers, 1985) indicate that their acrobatic locomotor habits and adaptations are not inherited from the last common ancestor of the living hominoids; and the widespread distribution of apelike anatomical traits in slow-climbing arboreal mammals suggests that hominoid morphology need not have originated as an adaptation to brachiation, but may rather have been preadaptive to it. The relict apelike traits that are preserved in our own bodies, and that have contributed so much to the liberation and specialization of our forelimbs for making and using tools, appear to be best understood not as products of our ancestors' peculiar brachiating locomotion, but as consequences of the same elementary biomechanical considerations that have governed the evolution of climbing adaptations in other vertebrates.

Chapter 6

Digging of Quadrupeds

Milton Hildebrand

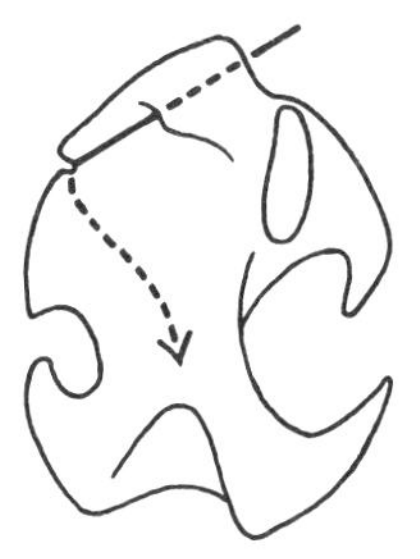

Vertebrates that spend all or nearly all their lives underground are said to be *subterranean* in terms of habitat and, if they maintain an open burrow system, *burrowers* in terms of interaction with the environment. Some burrowers venture above ground from time to time. Other vertebrates also dig effectively to secure food or shelter, yet spend most of their active lives above ground. All these animals are described as *fossorial* in terms of digging ability. This chapter explores the functional morphology of fossorial animals that are also quadrupeds. The diggers among amphibians and reptiles that have weak limbs, or none, are omitted here because their underground progression is included in Chapter 9.

Fossorial Quadrupeds and Their General Requirements

Because animals dig in many different ways, it will be necessary to refer to many taxa, some of which are not well-known (see Table 6-1). Skeletons, skins, preserved specimens, or a combination of these for all the genera listed were examined for this study. In a work such as this, there is danger of unwarranted generalization from species to genus to family; in this I have tried to be conservative, and the table shows the extent of my material. Extinct genera are not shown in the table and are identified to family or order where mentioned, as are several genera described from the literature but not seen by me. The orders and families having the most morphological adaptations for digging are included, but drawing the line is arbitrary. Many other quadrupeds have behavioral adaptations and adequate structure to enable them to dig well. Examples include numerous additional salamanders; most turtles; the tuatara; many lizards; crocodilians; some penguins, bee-eaters, swallows, and puffins; the platypus; some bandicoots; tenrecs and some shrews; several rabbits and pikas; the mountain beaver; marmots and prairie dogs; pocket mice and kangaroo rats; the springhare; mole voles; jerboas; and dogs and bears.

Most fossorial vertebrates are much modified for their mode of life. Their morphology relates, first, to the need to loosen and move resistant material. This requires a digging tool, capacity to produce and transmit much force, a transport mechanism for soil, passive resistance to various loads, and ability to sustain high or long-term activity. Second, burrowers cope in a confined, dark, often humid habitat. This requires adjustments of body form and behavior, changes in sense organs, and specializations related to ventilation, gas transport, and heat dissipation.

Digging Tools

Teeth Various rodent burrowers loosen soil primarily with the incisors, whereas others use both incisors and claws. In these species the incisors are large and strong and have roots extending close to or (in blesmols) into the condyle of the mandible and, in the upper jaw, to or (again in blesmols) even behind the maxillary cheekteeth (Figs. 6-1, 6-2).

Table 6-1 The living orders and families that include the most fossorial of quadrupeds, the genera seen for this study, and their common names.

Order and family	Genus	Common name
Anura		
Rhinophrynidae		
Pelobatidae		
Leptodactylidae		
Bufonidae		
Ranidae		
Microhylidae		
Urodela		
Ambystomatidae		
Plethodontidae		
Chelonia		
Testudinidae	*Gopherus*	Gopher tortoise
Monotremata		
Tachyglossidae	*Tachyglossus*	Echidna
Marsupialia		
Notoryctidae	*Notoryctes*	Marsupial mole
Phascolomidae	*Lasiorhinus*	Wombat
Insectivora		
Erinaceidae	*Erinaceus, Paraechinus*	Hedgehogs
Talpidae	*Dymecodon, Neurotrichus, Parascalops, Scalopus, Scapanus, Urotrichus, Talpa, Condylura*	Moles, shrew moles
Chrysochloridae	*Amblysomus, Chrysochloris, Chrysospalax, Eremitalpa*	Golden moles
Edentata		
Myrmecophagidae	*Myrmecophaga, Tamandua*	Anteaters
Dasypodidae	*Cabassous, Chaetophractus, Dasypus, Euphractus, Priodontes, Tolypeutes, Zaedyus*	Armadillos
Pholidota		
Manidae	*Manis*	Pangolin
Rodentia		
Sciuridae	*Spermophilopsis, Spermophilus*	Ground squirrels
Geomyidae	*Geomys, Thomomys, Cratogeomys*	Pocket gophers
Cricetidae	*Myospalax*	Zokor
Spalacidae	*Spalax*	Mole-rat
Rhizomyidae	*Cannomys, Rhizomys, Tachyoryctes*	Bamboo rats
Octodontidae	*Octodon*	Coruro
Ctenomyidae	*Ctenomys*	Tuco-tuco
Bathyergidae	*Bathyergus, Cryptomys, Georychus, Heterocephalus*	Blesmols
Carnivora		
Mustelidae	*Taxidea*	Badger
Tubulidentata		
Orycteropodidae	*Orycteropus*	Aardvark

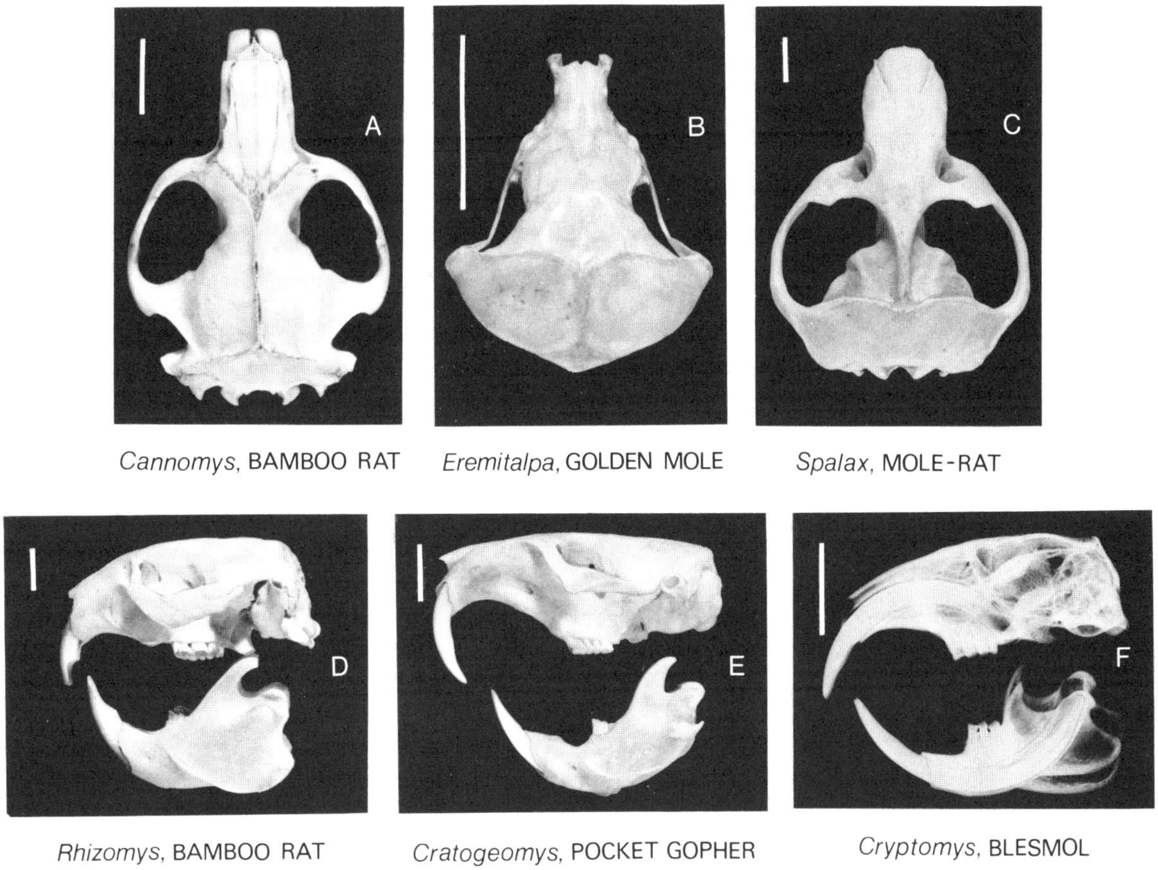

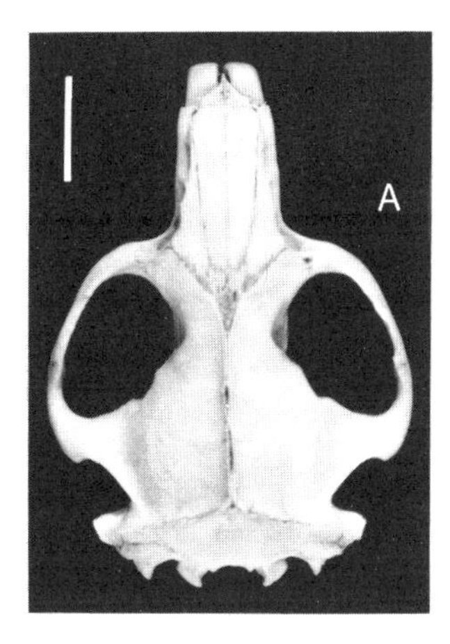

Cannomys, BAMBOO RAT

Eremitalpa, GOLDEN MOLE

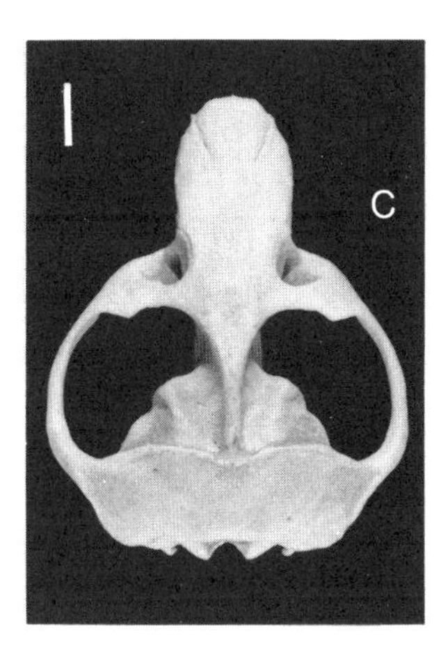

Spalax, MOLE-RAT

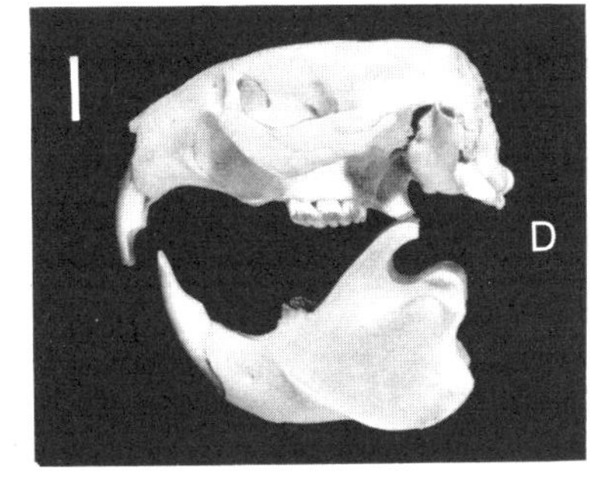

Rhizomys, BAMBOO RAT

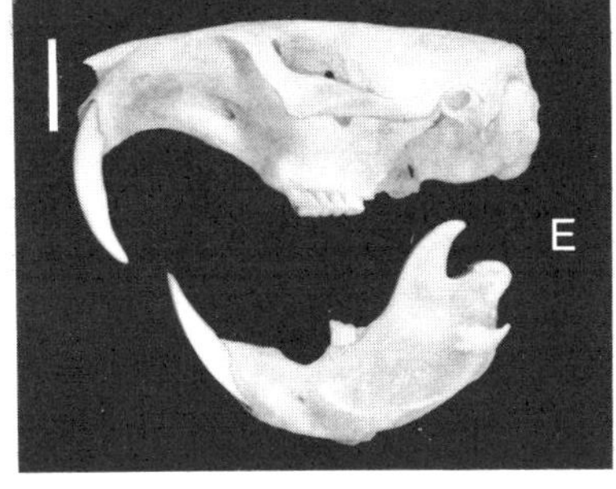

Cratogeomys, POCKET GOPHER

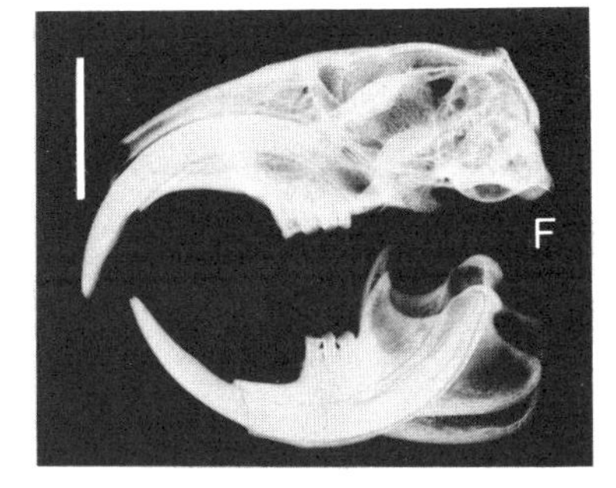

Cryptomys, BLESMOL

Figure 6-1 Skulls of burrowing mammals showing massive, firm construction; large occiput (particularly B,C); forward sloping occiput (B,C); sagittal and nuchal crests (A,C); strong, flaring zygomatic arches (A,C); deep zygomatic arches (D,E); base of upper incisor behind cheekteeth (F, which is an X-ray photo); base of lower incisor beside (D) or within (F) the mandibular condyle; upper incisors ranging from recurved (E) to procumbent (F); premaxillae flaring to support nose pad (B); rostrum massive to support nose pad (C). Animals B and C are head-lift diggers; A and C-F are chisel-tooth diggers; B and E are scratch diggers. The reference lines are 10 mm long.

Lower incisors are used to manipulate the earth more commonly than the uppers. Consequently, they wear and grow faster. Miller (1958) reported that each upper incisor of the pocket gopher *Thomomys* grows about 0.58 mm per day and that each lower tooth grows about twice that much. Although Manaro (1959) found somewhat lower rates for *Geomys,* the incisors of pocket gophers grow two to nearly four times as fast as those of selected nonfossorial rodents and rabbits. As in other rodents, the incisors of diggers are self-sharpening because enamel covers the softer dentine only on the anterior face of each tooth.

Tubercles and Claws Most Anura that dig do so hind-feet-first, and most of these species have a horny tarsal tubercle that serves as a cutting edge during sweeping motions of the leg. Reptilian diggers generally use their claws as tools, the tortoise *Gopherus* having large, spatulate ungual phalanges.

Many fossorial mammals break the soil with their claws, which are then large and fast growing (Figs. 6-2, 6-5). Gambarian (1960) noted that if the claw serves as a cutting tool, it tends to be laterally compressed and to have the bony phalanx surrounded over much of its length by the unguis, the softer sole plate being displaced

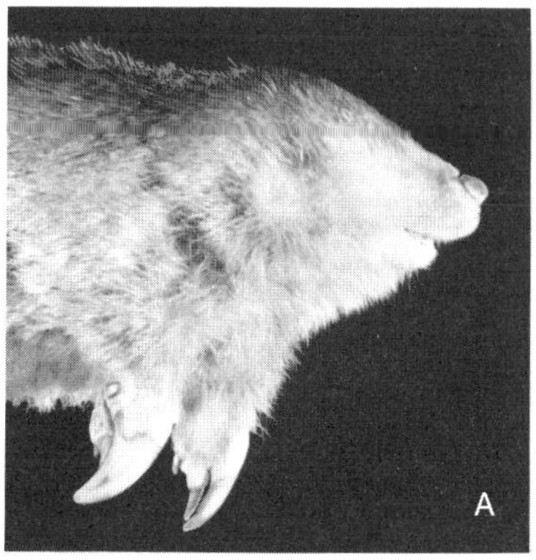

Amblysomus, GOLDEN MOLE
Scratch & head-lift digger

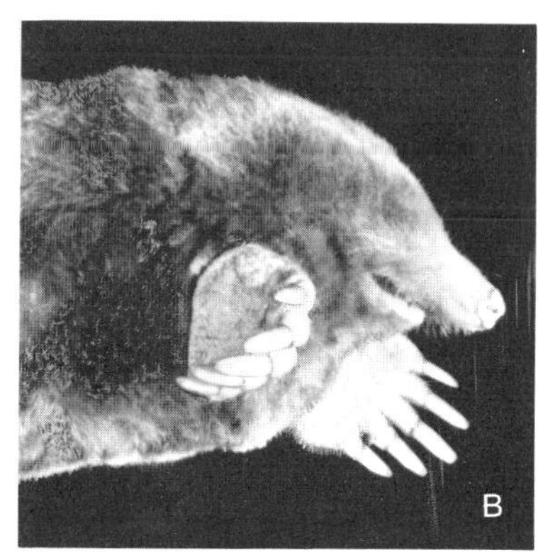

Talpa, MOLE
Humeral-rotation digger

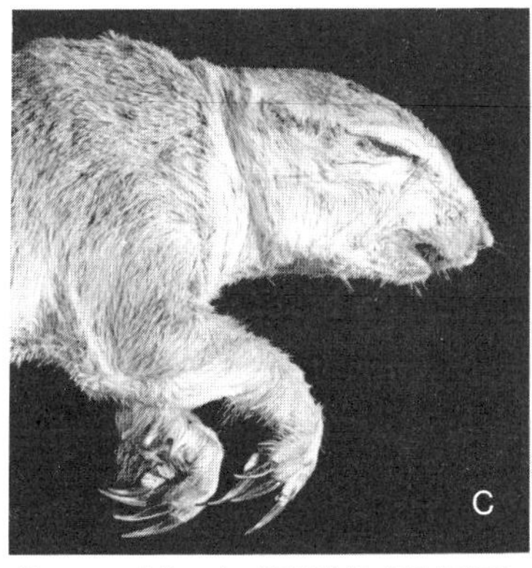

Spermophilopsis, GROUND SQUIRREL
Scratch digger

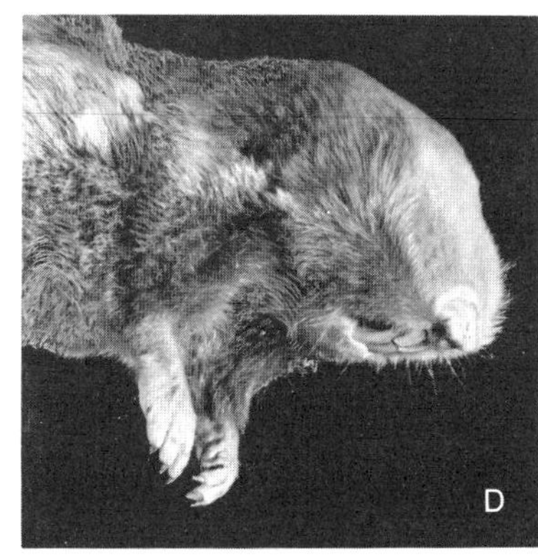

Spalax, MOLE-RAT
Chisel-tooth & head-lift digger

Bathyergus, BLESMOL
Scratch & chisel-tooth digger

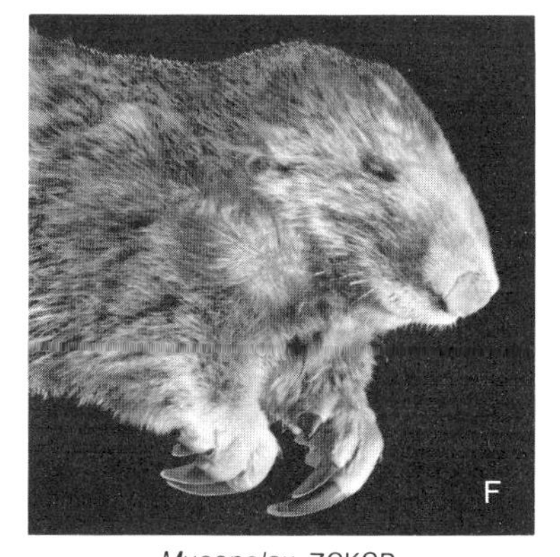

Myospalax, ZOKOR
Scratch, chisel-tooth & head-lift digger

Figure 6-2 Burrowing mammals showing claws adapted for scratch digging (A,C,E,F); incisors adapted for chisel-tooth digging (D,E); short forelimb (all except C); only the manus extending beyond the body contour (A,B); eyes reduced (E,F) or vestigial (A,B,D); external ear reduced or vestigial; nose pad (A,D,F); lips closing behind incisors (D,E); stiff hairs flanking a broad muzzle (D); broad manus (particularly B); neck virtually absent (A,B,D-F).

toward the base of the claw. For the rodent genera he studied (and also golden moles and some armadillos), wear is greater on the lateral side of the claw than on the medial side (because of the angle of strike or asymmetry of the phalanx). Accordingly, the horn of the lateral surface is thicker, grows faster, and extends under its counterpart to the midventral line, thus establishing a sharp edge. The hind claws of some of the same animals, and the foreclaws of others that use the forefoot to move earth but not to break hard soil, are broader and have a more extensive sole plate. In general, sharp claws exert several times more pressure but also several times less force against the substrate than do broad, blunt claws (Goldstein, 1968).

The horn-forming tissue of fast-growing claws is highly vascular. Some diggers have a narrow channel for vessels along the middorsal curve of the bony claw and extending to its tip (various pocket gophers, blesmols, anteaters). In some species the channel widens to a prominent groove that causes the tip of the bony claw to be slightly cleft (moles, see Reed, 1951). Such a groove is no longer entirely vascular because a thickened ridge on the inside of the horny claw fits into it. Gambarian (1960) noted that the bony claw of the pangolin is deeply grooved over most of its length and markedly bifid at the tip. The same is true of golden moles (Fig. 6-3). The terminal phalanx was similarly cleft in some extinct mammals (chalicotheres and at least some creodonts, mesony-

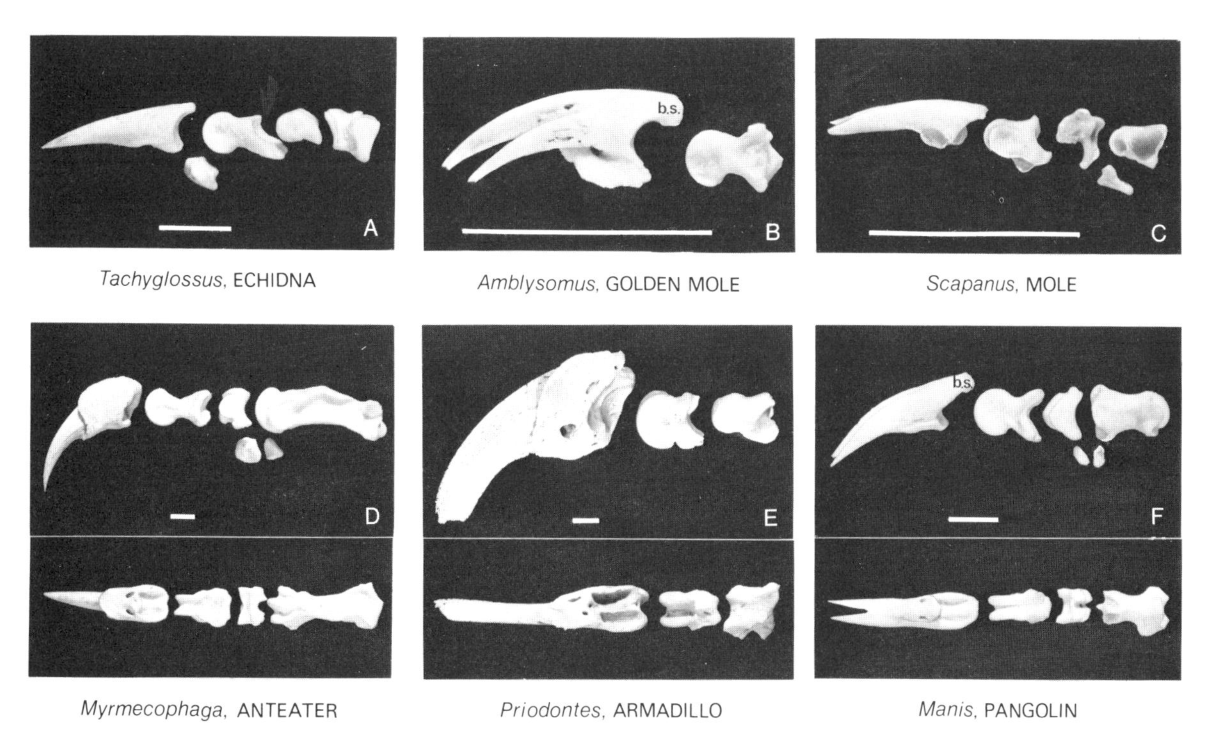

Figure 6-3 Third left (A-C) and right (D-F) metacarpal and digit of scratch diggers (all except C) and a humeral-rotation digger (C) shown in lateral views (A-C), medial views (D-F, *above*) and ventral views (D-F, *below*). Note shortening of metacarpal and all phalanges except the last; enlargement of the terminal phalanx; reduction of phalanges to two (E) or one (B); bifid terminal phalanx (B,C,F); achievement of rigidity between the metacarpal and first phalanx (except B and D) and between first and second phalanx by flat-ended or angled (not arcing) articulatory surfaces (best seen in F) or by a vertical spline and groove (F); bony stops *(b.s.)* that prevent hyperextension of terminal phalanx (best seen in B and F); strengthening of terminal joint by having a long radius of curvature, and by splines and grooves (best seen in the ventral views); large sesamoid bones (not all shown) that strengthen joints (A), brace joints against motion (C), guide tendons (C,F), or serve as attachments for tendons (A). Reference lines are 10 mm long.

chids, and toxodonts). All these large animals may have dug to secure part of their food, and the forelimb skeleton of some of them makes this seem probable. The cleft in a bifid phalanx is filled by a vertical partition within the horny claw, which apparently increases the strength of the latter. (The terminal phalanx was vascular but *not* cleft in the specimens seen of the marsupial mole, a bamboo rat, a wombat, armadillos, and the aardvark.)

Feet, Nose Pad, and Head If soil is friable, a broad digging tool that scoops or pushes can supplement, or substitute for, teeth and claws. Thus, moles excavate using their broad feet. Modifications of such feet for digging also function in transporting loose soil and are described below under that topic.

Four genera of microhylid frogs described by Menzies and Tyler (1977) have skin over their snouts that is about six times thicker than elsewhere. These authors think that such frogs may use the snout in burrowing, but behavioral study is needed to verify this possibility because many frogs have thickened skin over the snout. Large, tough nose pads are used as an entering wedge or to dislodge, move, and tamp dirt by the marsupial mole, golden moles, the myomorph rodents *Spalax* and *Myospalax,* and, to a lesser extent, by several other burrowers (Fig. 6-2). Possibly, the protuberances on the snouts of the Miocene mylagaulid rodent *Ceratogaulus* and the Miocene armadillo *Peltephilus* had similar functions, as did the extremely spatulate snout of Oligocene palaeanodonts. I found by X ray that bones of the rostrum of the marsupial mole are thick and cancellous under the nose pad. The premaxillae of golden moles flare under the pad to provide support (Fig. 6-1B). Histological sections I made of the pad of *Amblysomus* show it to resemble the integument of volar surfaces: the epidermis is about nine times thicker (to the base of its ridges) than that of adjacent skin, and it interdigitates with long dermal papillae. The reticular layer is thick, laced with collagenous fibers, and rich in fat cells.

Some frogs, the marsupial mole, golden moles, some rodents, and possibly some armadillos use the top of their broad, firm heads to displace or compact soil. A notable example is the mole-rat *Spalax.* Its muzzle is flanked by large fibrous cushions (Krapp, 1965) and functionally broadened by lateral fringes of deep-rooted, stiff, upcurled hairs (Figs. 6-2D, 6-6).

Production and Active Transmission of Force

Analysis of lever systems (see Chapter 3) shows that one way for diggers to exert adequate force against the earth (out-force) is to have high in-force. Accordingly, diggers have enlarged muscles compared with their nondigging relatives, and in several instances a muscle has been modified to enable it to supplement or replace its usual function with another that is of particular importance to the given manner of digging. Furthermore, out-force increases as the ratio of in-lever (power arm) to out-lever (load arm) increases. Hence, the muscles of diggers have their origins and insertions relatively far from the joints they turn, and the distance from pivot to digging tool (for example, length of radius, carpus, metacarpus, and all but distal phalanges) is short.

Muscle structure and some principal bone-muscle mechanisms of fossorial quadrupeds can be understood in terms of the animals' method of digging. I have used six categories of digging, but some of these are not sharply set off from others, and many burrowers dig in more than one way.

Muscle Structure Muscles used in burrowing tend to be conspicuously enlarged in cross section and volume, though usually not in length (quantification is provided in several of the references cited, including Gambarian, 1953, 1960). The comparative internal structure of the muscles of diggers is virtually unknown. Increase in pinnation and firmness of the collagenous framework may be predicted. In several instances a muscle has become partly or entirely ligamentous, thus creating a passive linkage between bones (see below). The comparative histology of the muscles of diggers is also little known. Goldstein (1971) found that in four forelimb muscles of the mole *Scapanus,* FG fibers predominated and SO fibers were absent, whereas three species of ground squirrels (genus *Citellus*) had many FOG and SO fibers (see Chapter 17 for a description of fiber types). He attributed the difference to greater diversity of function for the forelimbs of the squirrels, but more comparative studies are needed before conclusions can be drawn.

Scratch-Digging Most scratch-diggers extend the forefeet to the earth and then draw the claws more or less downward, toward, or under the body. Several of the diggers included here break the soil also, or instead, with the digits and palm, and some also excavate with the hind feet. In mammals action of the two feet of a pair usually alternates in rapid succession, the momentum of the feet at impact contributing to their effectiveness in breaking the soil. Scratch-diggers include some frogs, some tortoises and an (extinct) dicynodont reptile (Cox, 1972), fossorial marsupials, burrowing insectivores except moles (but including extinct palaeanodonts, Rose and Emry, 1983), armadillos, the pangolin, ground squirrels, pocket gophers, the coruro and tuco-tuco, the blesmol *Bathyergus* and cricetid *Myospalax*, extinct mylagaulid rodents, carnivores that dig, and the aardvark.

The relatively few frogs that dig with the forefeet sweep the limbs alternately in lateral arcs that first displace earth to the side and then propel the body forward. The forelimb retractors are enlarged, and the pectoral girdle is modified accordingly (Emerson, 1976). Digging tortoises make similar slow, lateral or ventrolateral strokes.

The major musculoskeletal modifications of mammalian scratch-diggers provide increased strength in flexing the larger digits, flexing the wrist, extending the elbow, flexing the humerus on the scapula, and stabilizing the shoulder. The medial epicondyle of the humerus, which is an origin of digital and carpal flexors and of the pronator, is very prominent, and the lateral epicondyle (an origin of extensors and the supinator) moderately so (Fig. 6-4). In Table 6-2, where the width of the humerus across the epicondyles is expressed as a percentage of the length of the bone from trochlea to head, the values for scratch-diggers range from the 30s (many examples) to the 70s and above (some golden moles). By contrast, the percentage for the nonfossorial raccoon is about 21. (In most instances only one or several skeletons were available, so the figures should be considered only approximate for the respective genera. Also, no compensation has been made for scaling. However, some of the largest, such as the giant anteater and giant armadillo, and smallest, such as the golden mole *Eremitalpa*, have typical or intermediate values for their families.)

The medial surface of the olecranon process is also an origin for digital and carpal flexors. It is usual for this process to curve medially in scratch-diggers. A flaring medial epicondyle and an inflected olecranon greatly increase the area of origin of these muscles and also much increase the effective force of the pronator by altering its line of action.

Various authors (such as Puttick and Jarvis, 1977) have noted that a sturdy bone replaces part of the muscle and tendon of the deep digital flexor of golden moles (Fig. 6-5A). It is about 65% of the length of the radius in an *Eremitalpa* and about 110% of the radius in an *Amblysomus.* Proximally the bone is anchored by a large ligament (replacing the humeral head of the muscle) to the distal margin of the long medial epicondyle. The bone is extended distally by a large tendon that branches to insert under the terminal phalanges of the third and fourth digits. The fleshy ulnar head of the deep digital flexor also inserts on this flexor bone.

The functional significance of this morphology has not been described heretofore. The median epicondyle of the humerus slopes down (away from the head of the humerus), and the ligament to the flexor bone originates from a prominent scar on the distal edge of the epicondyle. Thus, the origin of the ligament is slightly distal to the axis of rotation of the elbow. Accordingly, flexion of the elbow draws the ligament away from the carpus and, via the flexor bone and its tendons, passively flexes the claws. This can be verified by manipulation of preserved specimens. These coordinated motions of elbow and claws occur on the recovery stroke (Puttick and Jarvis, 1977) and probably also when the animal is forcing its head forward in the earth. (Contraction of the ulnar head of the deep digital flexor probably depresses the claws independently of flexion of the elbow. It

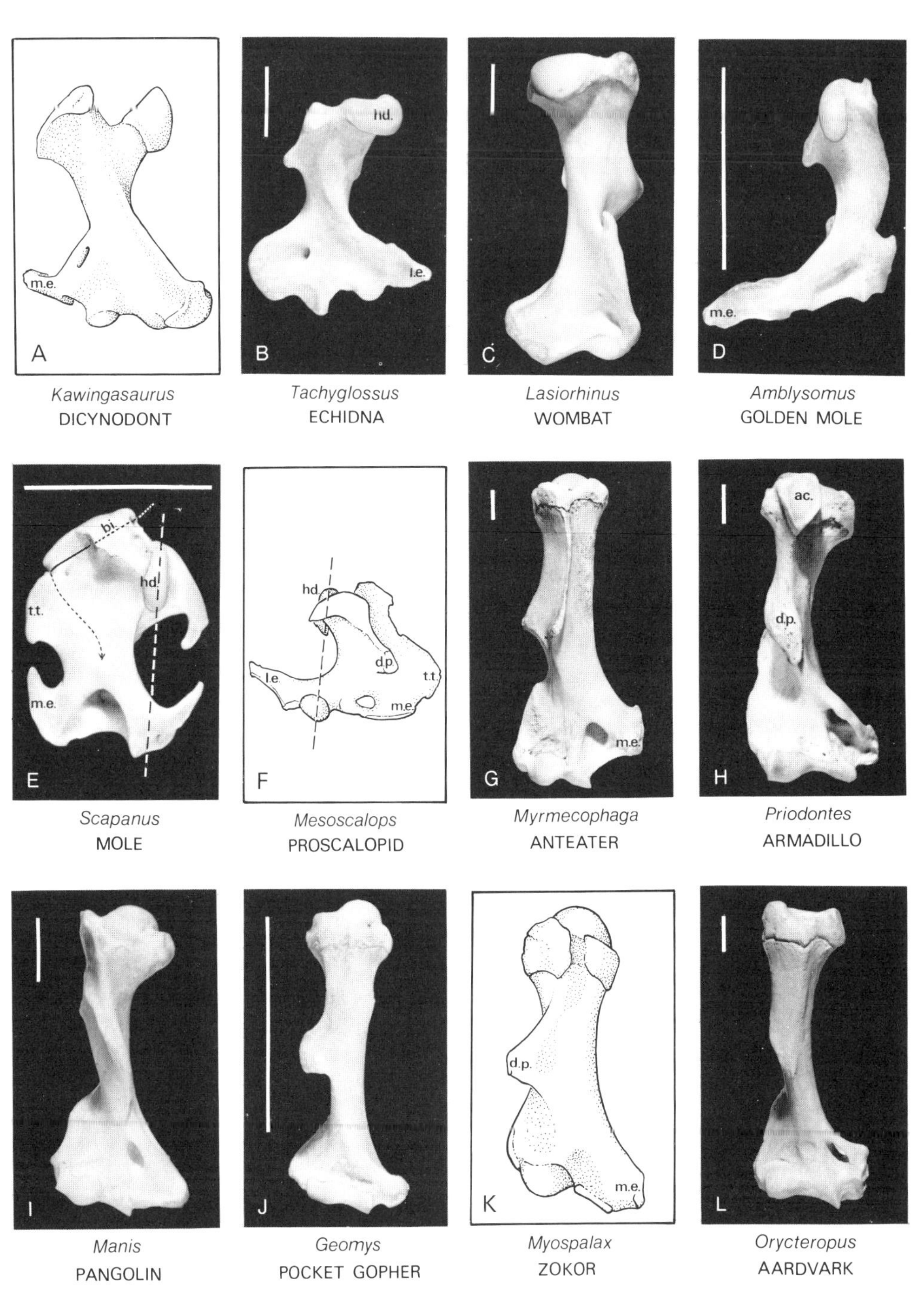

Kawingasaurus DICYNODONT

Tachyglossus ECHIDNA

Lasiorhinus WOMBAT

Amblysomus GOLDEN MOLE

Scapanus MOLE

Mesoscalops PROSCALOPID

Myrmecophaga ANTEATER

Priodontes ARMADILLO

Manis PANGOLIN

Geomys POCKET GOPHER

Myospalax ZOKOR

Orycteropus AARDVARK

Figure 6-4 Right humerus of selected diggers, in posterior (A-E) and anterior (F-L) views, showing generally rugged construction; large lateral epicondyle *(l.e.)* and very large medial epicondyle *(m.e.)*; prominent deltoid process *(d.p.)* extending for distad; oblong head *(hd.)* (D,E); flaring teres tubercle *(t.t.)* and offset axis of rotation *(dashed line)* of humeral-rotation diggers; path of biceps and its tendon *(bi.)* through bicipital tunnel, over bicipital groove, and onto anterior face of bone (E); surface that articulates with the acromion process *(ac.)* (H). Reference lines are 15 mm long. A redrawn from Cox, 1972; F redrawn from Barnosky, manuscript; K redrawn from Gambarian, 1953.

Table 6-2 Width of humerus across epicondyles, expressed as % of length from trochlea to head, and length of olecranon process, expressed as % of length of remainder of ulna, for single or a few specimens of selected fossorial mammals.

Mammal	Humeral width to length (%)[a]	Olecranon to ulnar length (%)[a]
Monotreme		
Tachyglossus	93	40
Marsupials		
Notoryctes	56	97
Lasiorhinus	48	37
Moles, shrew moles		
Dymecodon	54	30
Parascalops	77	46
Scapanus	84	57
Talpa	75	55
Condylura	69	35
Golden moles		
Amblysomus	97	75
Chrysochloris	80	—
Chrysospalax	71	77
Eremitalpa	69	68
Anteaters		
Myrmecophaga	44	28
Tamandua	48	29
Armadillos		
Cabassous	58	89
Chaetophractus	39	60
Dasypus	36	66
Euphractus	41	95
Priodontes	50	79
Tolypeutes	35	47
Zaedyus	39	56
Pangolin		
Manis	48	51
Sciuromorph rodents		
Spermophilopsis	30	23
Geomys	42	37
Thomomys	35	30
Myomorph rodents		
Myospalax	54	—
Spalax	35	60
Cannomys	30	34
Rhizomys	36	40
Tachyoryctes	30	42

continued

Table 6-2 *(continued)*

Mammal	Humeral width to length (%)[a]	Olecranon to ulnar length (%)[a]
Hystrichomorph rodents		
Ctenomys	33	31
Bathyergus	32	36
Cryptomys	29	39
Georychus	32	39
Heterocephalus	28	36
Carnivore		
Taxidea	34	35
Aardvark		
Orycteropus	43	50

a. Measured from trough of trochlear fossa.

is a unipinnate muscle, which possibly explains the advantage of the flexor bone over a tendon.)

Armadillos have under the pes a sesamoid that articulates with the navicular and cuboid. In *Dasypus* the sesamoid is larger by weight than either of those bones or the patella. The fibular flexor inserts on the bone, and its tendons run from the sesamoid to the terminal phalanges. Similarly, there is a sesamoid under the manus which, in *Dasypus,* weighs nearly half as much as the entire carpus. It distributes to the digits the force of the large deep digital flexor. A large palmar sesamoid also is present in the carpus of the marsupial mole (Wilson, 1894) and Oligocene palaeanodonts (Rose and Emry, 1983). The mechanism seems to assure that all claws flex as a unit, but the advantages need further study.

Scratch-diggers are highly modified for strong extension of the elbow and movement at the shoulders. The olecranon process is the lever arm of the triceps and dorsoepitrochlearis. The length of the process, expressed as a percentage of the length of the remainder of the ulna (both measured from the trough of the trochlear fossa), is shown in Table 6-2. Values for scratch-diggers range from the 20s (a ground squirrel) to 90s (marsupial mole, some armadillos). By contrast, the percentage for the nonfossorial raccoon is about 12.

The musculature of various scratch-diggers has been described by Orcutt (1940), Miles (1941), Gambarian (1960), Lehmann (1963), Puttick and Jarvis (1977), and by authors they cite. The large scapular head of the triceps extends along the posterior border of the scapula relatively far from the glenoid. The dorsoepitrochlearis is variable, but often large, and in the armadillo *Dasypus* appears like a massive thoracic head of the triceps (Fig. 6-5C). These two-joint muscles can move either the elbow or shoulder. The powerful latissimus dorsi, teres major, and deltoids insert relatively far from the glenoid, thus increasing their effective lever arms. The posterior angle of the scapula is often enlarged to provide greater area of origin for the teres major.

It is usually assumed that the large extrinsic muscles of the shoulder "fix" the girdle, but Gambarian (1960) finds them variable in size, and behavioral studies are needed. Functional studies are required to interpret the acromion process (usually large; articulating with the humerus in the giant armadillo *Priodontes;* wanting in pangolins), metacromion process (often large, often absent), clavicle (tending to be long and slender), "extra" small bone between the clavicle and acromion in blesmols and the unrelated mole-rat *Spalax,* variation of the fibula (ranging from large and free to small and fused), and other features.

Chisel-Tooth Digging As mentioned earlier, some rodents dig primarily with their incisors (bamboo rats, the octodontid *Spalacopus,* bles-

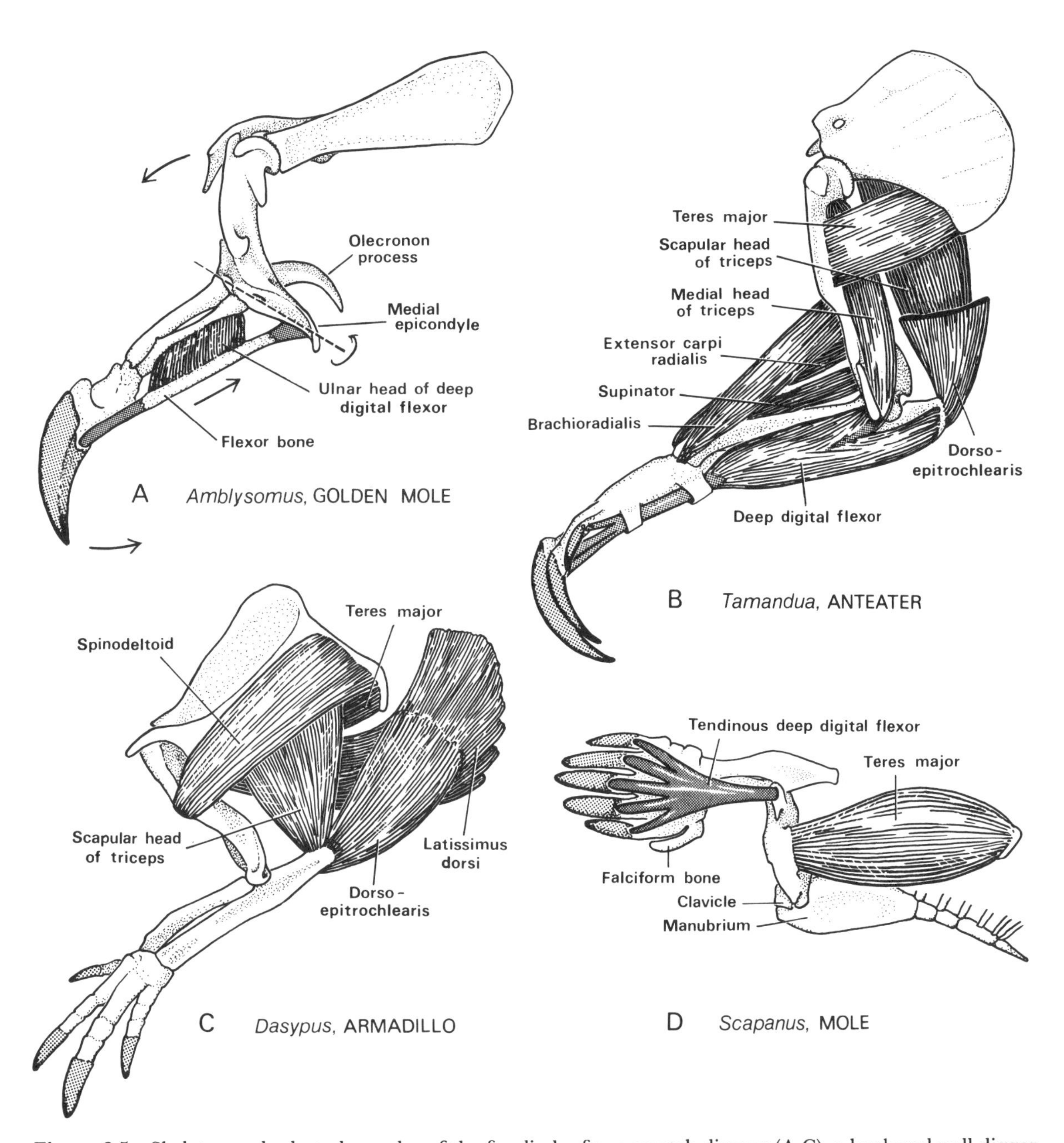

Figure 6-5 Skeleton and selected muscles of the forelimb of two scratch diggers (A,C), a hook-and-pull digger (B), and a humeral-rotation digger (D) shown in anteromedial (A), medial (B), and lateral (C,D) views. Tendons and ligaments (both in fine stipple) and muscles are semidiagrammatic. For A, the dashed line shows the approximate axis of rotation at the elbow, arrows show coordinated movement relative to a fixed ulna, and the first and second digits have been removed.

mols except *Bathyergus*). Others use both incisors and claws (pocket gophers, tuco-tuco, *Bathyergus*); or both incisors and head (the mole-rat *Spalax*); or incisors, claws, and head (the cricetid *Myospalax*).

The skull of the chisel-tooth digger is relatively large: its weight was 46% of the entire skeleton in a *Heterocephalus* and 51% in a *Georychus* (both blesmols), compared with 25% for a *Rattus* and 23% for a woodrat, *Neotoma*. The rostrum is strong, palatal foramina small, zygomatic arches strong and usually flaring, and sagittal and nuchal crests often evident (Fig. 6-1).

Rodents slide the mandible forward on the skull to engage the incisors and backward to engage the cheekteeth. This action is accentuated in chisel-tooth diggers. The mandibular symphysis is usually sufficiently flexible to allow the lower incisors to be separated. The mandibular fossa is broad and flat in some of these animals, notably in the bamboo rat *Rhizomys*. The bamboo rat *Cannomys* has a shallow fossa lateral to the mandibular fossa that appears (on dry skulls) to receive the projection that caps the root of the lower incisor. *Spalax* has a "second fossa" posterior and a little dorsal to the mandibular fossa. Krapp (1965) says that it does not receive the condyle, as has been claimed, but in order for the cheekteeth to be aligned, the condyle must fit into this pit—at least loosely. When the cheekteeth function, forces at the fossa are less than when the incisors function.

Clearly, the functions of the massive jaw musculature are many and complex. The muscles have been described for pocket gophers (Hill, 1937; Orcutt, 1940) and *Spalax* (Krapp, 1965; Fig. 6-6). Krapp provides functional interpretations but did not study living animals. Gambarian (1960) shows that rodents with relatively vertical exposed incisors have relatively vertical muscular forces. The jaw is shifted forward to engage the incisors for digging and then, remaining forward, moves largely in the vertical direction. He states that rodents with more procumbent exposed incisors (contrast E and F in Fig. 6-1) have more

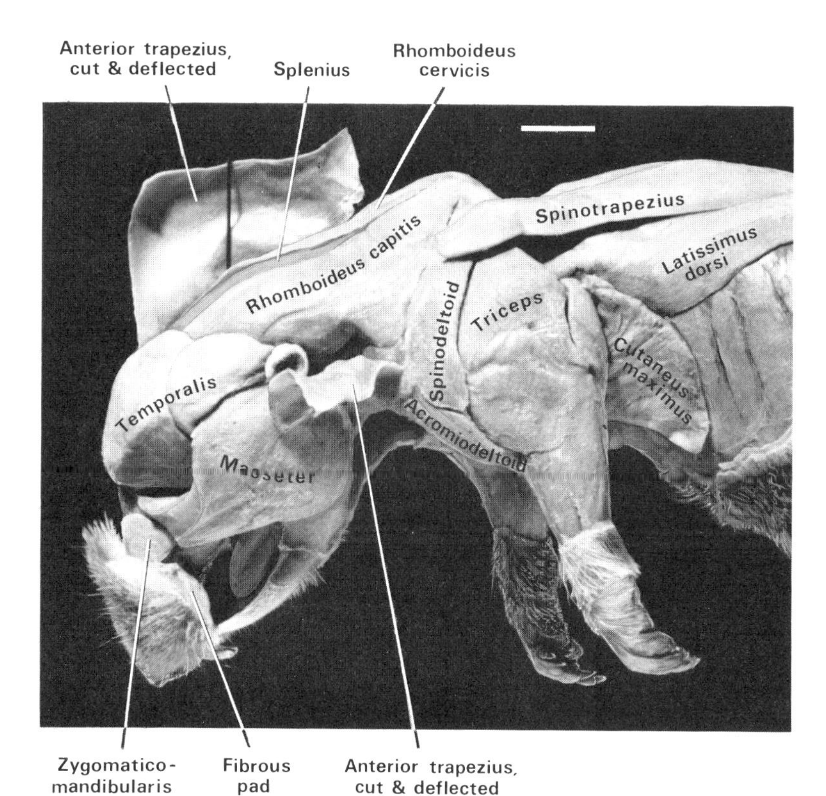

Figure 6-6 Dissection of the mole rat *Spalax* showing the large head and massive development of the muscles of the head, neck, and shoulder of this head-lift and chisel-tooth digger. The reference line is 15 mm long.

anteriorly directed muscle forces and, in gnawing, slide the mandible somewhat forward and backward with each bite.

The neck muscles of chisel-tooth diggers are much enlarged compared with those of nondiggers, as a large occiput and prominent nuchal crests attest. They are needed to stabilize the jaws on the body. Also, when jaws are open and fixed, dorsal neck muscles may be used to force the lower incisors up through the soil. The latter action necessitates a downward thrust with the forefeet. Accordingly, the modifications for extending the forelimb (digits excluded) are similar to those seen in scratch-diggers.

Head-Lift Digging Various quadrupeds dig with the head to a limited degree, and for the marsupial mole, golden moles (Puttick and Jarvis, 1977), mole-rat *Spalax,* and cricetid *Myospalax* (and probably for Oligocene palaeanodonts), head-lifting is a principal way of making shallow tunnels (the displaced earth then breaking the surface) or compacting soil in deeper runways. Gambarian (1953) found that, in a laboratory device, *Spalax* could repeatedly lift 15 to 20 times its body weight. A 57-g golden mole (genus *Amblysomus*) escaped from an earth-filled fish bowl by nudging aside a 9.5-kg cover of iron (Bateman, 1959).

Muscles that raise the head, neck, and body are developed in mammalian head-lift diggers to a degree equaling or exceeding that of chisel-tooth diggers. The rhomboideus capitus and serratus ventralis are said by Gambarian (1960) to be among the greatly enlarged muscles. The biventer cervicus, splenius, anterior trapezius, and cleido-occipitalis also contribute (Puttick and Jarvis, 1977). The occiput is remarkably high and wide, extending onto the bases of the zygomatic arches in *Spalax* and golden moles (Fig. 6-1). The occiput and nuchal crests of various armadillos are sufficiently developed to suggest head-lifting either to move soil or brace the body. Since head-lifting also requires that the animal push firmly down with the forelimbs, muscles and bones of the shoulder and brachium are comparable to those of scratch-diggers.

Hook-and-Pull Digging Anteaters dig primarily by engaging the large second and even larger third claws of the forefoot in a hole or crevice and then forcefully pulling the foot toward the body. The claws can be flexed to a marked degree to grip against the palm, and there is strong pronation and supination of the forearm to deflect the claws from side to side. The functional morphology of the forelimb of the collared anteater, *Tamandua,* has been well described by Taylor (1978), and the structure of the giant anteater, *Myrmecophaga,* is essentially the same.

The most striking feature of hook-and-pull diggers is the modification of the large medial head of the triceps. It does not insert on the ulna. Its tendon passes under the distal border of the large medial epicondyle of the humerus to become continuous with the tendon of the deep digital flexor (Fig. 6-5B). Since the pulley around the elbow joint is in line with the axis of rotation of that joint, contraction of the medial head of the triceps has scant effect on the joint. Instead, its force nearly matches that of the deep digital flexor alone in depressing the central claw. The tendon of the system is enormous compared with that of nondiggers; on one 30-kg *Myrmecophaga* its area of insertion on the distal phalanx is 1.6 cm^2.

The origins of the brachioradialis and extensor carpi radialis muscles are unusually proximal on the humerus, even for diggers. Thus, they strongly supplement the biceps and brachialis as flexors of the forearm. The brachioradialis also flexes the carpus and contributes to supination. The widely flaring medial epicondyle increases the effective force of the pronator.

Humeral-Rotation Digging The four ways of digging described thus far all require great strength for flexing or extending the shoulder, elbow, and wrist and sometimes of joints of the hind limb. Chisel-tooth digging and head-lift digging also require strength of dorsal neck muscles; scratch-digging and hook-and-pull digging also require strength in flexing the claws. We come now to a different, more unusual method of digging (which, however, may be linked to scratch-digging in the extinct insectivore family Proscalopidae, as discussed by Barnosky, 1981, 1982).

Moles and shrew moles dig by rotating the humerus around its own long axis. The functional anatomy of these animals has been described by Reed (1951) and Yalden (1966). Humeral rotation is the principal movement of the forearm of the echidna *Tachyglossus* during walking (Jenkins, 1970) and probably also in digging, although its fossorial behavior is inadequately known. This account is based on moles. (The most modified moles are *Scapanus, Scalopus,* and *Talpa*).

The long, nearly horizontal scapula and long, keeled manubrium carry the glenoid forward to the level of the occiput. This allows the short forearm and manus to reach beyond the nose when digging. The clavicle, which is shorter than broad, articulates with the humerus but not the scapula. The distinctive humerus is abducted to such a degree that its distal end is well above the glenoid and its anatomically medial surface is lateral in position (Fig. 6-5D). The elbow joint is held at close to 90° during the power stroke. The manus is constantly oriented halfway between the pronated and supinated positions, which makes it vertical, palm facing outward and backward, in place to sweep against the burrow wall as the humerus rotates. Action of the two feet alternates and is slow relative to that of scratch-diggers. The forefoot is never placed either under or above the body; the animal rotates its body in the tunnel to dig against all surfaces.

The effectiveness of this mechanism is attested by the facts that *Scalopus* can dig a surface tunnel in compact soil at the rate of 4 m/h, can lift 3 kg when making such a run, and, in a laboratory device, can exert against a constriction in an artificial tunnel a force equal to 32 times its body weight (Arlton, 1936).

Three aspects of the digging mechanism will be noted. First, the digging stroke is powered primarily by the teres major, which is the largest muscle of the body. It inserts on a large tubercle that is lateral in position (because of the elevated position of the distal humerus). Because the humerus is very broad and because its axis of rotation passes near the margin that is away from the teres tubercle, the lever arm of the muscle is remarkably long (Fig. 6-4E and F). (The latissimus dorsi, subscapularis, and part of the pectoralis assist in the power stroke.)

Second, as the long tendon of origin of the biceps leaves the scapula, it is deflected far to the side by a tunnel in the humerus (Fig. 6-4E) before it returns to insert near the axis of the forelimb. This devious course is somewhat lengthened by rotation of the humerus. Consequently, contraction of the biceps during the recovery stroke not only flexes the elbow but also counterrotates the humerus. (Other counterrotators are the supraspinatus, spinodeltoid, cleidohumeralis, and part of the pectoralis.)

Third, the deep digital flexor is entirely tendinous and originates from a pit on the medial epicondyle. As the humerus rotates, this tendon is drawn back (relative to the fixed ulna) as the radius is pushed forward. Thus, the manus, which was hyperextended during the recovery stroke, is automatically brought to the straight position during the next power stroke.

Hind-Feet-First Digging Ninety-five percent of Anura that dig enter the substrate hind feet first, moving backward into the ground. Turtles also excavate their nests with their hind feet. This brief account follows Emerson (1976) and is based primarily on the microhylid *Glyphoglossus.*

Morphological modifications for jumping appear to have preadapted frogs for digging, even though some fossorial frogs may no longer jump. The hind legs are positioned behind the body, the acetabula are lateral in position, the feet are large, and the legs fold sharply on themselves. Moving alternately, each foot is first positioned behind the body and inverted to orient the horny tarsal tubercle toward the soil. Next the leg is partially extended at the knee (not hip). The foot scoops dirt to the side creating a hole into which the body falls. From time to time the frog rotates its body so that it can dig on all sides of its earthen chamber.

Fossorial frogs differ from others in having relatively short hind legs, the tibiofibula being the most reduced segment. The tibialis anticus brevis, which inverts the foot, is relatively long, seemingly to increase its rate of shortening. Muscles that extend the lower leg and flex the tarsus (tibialis anticus longus, extensor cruris brevis) are relatively large.

Stabilization of Joints

Because of the relatively great forces developed by fossorial quadrupeds, they, more than other vertebrates, must have mechanisms to resist hyperextension, dislocation, and counterproductive deflection of certain joints. Most affected are the joints of the forelimb and neck, according to method of digging. Unfortunately, the joints and ligaments of diggers have scarcely been studied, and even the bones and muscles of the feet of small burrowers are virtually unknown. Original observations given here are based on only one or several specimens of each genus noted and are to be interpreted only as approximations.

First, motion may tend to become limited to one plane at joints that, in less specialized quadrupeds, allow motion in several planes. Thus, the wrist joint is hingelike, not elipsoid, permitting only flexion and extension in moles (Reed, 1951), golden moles, and probably echidnas, the aardvark, and others. Similarly, the head of the humerus of diggers often has a greater radius of curvature in the vertical plane than in the horizontal plane (Fig. 6-4D,E). This limits or, in moles and probably in golden moles, prevents adduction and abduction. The long acromion of some forms may also limit abduction (golden moles, aardvark, armadillos). Motion between the second through seventh cervical vertebrae tends to become limited to extension and flexion by virtue of very wide, flat-ended centra (the echidna *Tachyglossus,* the mole-rat *Spalax,* armadillos, pangolin).

Second, joints that provide for motion in one plane tend to become strengthened against dislocation. Thus, scratch-diggers and hook-and-pull diggers strengthen, in several ways, the joint between the second and terminal phalanges of the stronger digits: the distal end of the second phalanx may be enlarged, and its radius of curvature relatively great (golden moles, moles, anteaters, pangolin), thus increasing the surface contact between the bones (Fig. 6-3). A large sesamoid bone may be present under the joint; it is attached to the distal phalanx and functionally extends that bone's envelopment of the second phalanx (the echidna *Tachyglossus,* golden moles, moles, armadillos). Furthermore, the distal articular surface of the second phalanx may have a deep groove in the vertical plane (rather than the more usual shallow depression) that is flanked by arcing ridges. The terminal phalanx engages the groove and, in anteaters (Taylor, 1978), some armadillos, and the pangolin, also encloses the lateral ridges (Fig. 6-3D–F). The result is a hinge joint having enormous resistance to dislocation.

The joint between a metacarpal and first phalanx normally allows much motion in the vertical plane if foot posture is digitigrade. The spline (or low crest) that is common on the ventral part of the distal articulatory surface of each metacarpal then enlarges in diggers and extends around the end of the bone (Fig. 6-3D). A groove on the first phalanx receives this spline to give the joint great internal strength (anteaters, aardvark).

Third, diggers commonly have passive mechanisms to prevent hyperextension of hinge joints. An enlarged extensor tubercle of the terminal phalanx often seats against the second phalanx to form a bony stop. Examples are seen in the pangolin, golden moles (Fig. 6-3), and the pocket gopher *Geomys.* The joint between the metacarpal and first phalanx may also have a bony stop (aardvark).

Fourth, joints that permit motion in other quadrupeds may, in diggers, be modified to allow little or no motion. All joints of the digits of the tortoise *Gopherus* are rather flat-ended, and freedom of the mesocarpal joint is reduced (Bramble, 1982). The first and second phalanges of mammals that dig with their forefeet tend to be short and broad and are usually joined by an immovable joint. The articulatory surface of the more proximal bone is not larger than that of the other, thus preventing any sliding motion between them. The joint surfaces are somewhat V-shaped or multiangled (the echidna *Tachyglossus,* pangolin, moles, anteaters, Fig. 6-3) or are provided with a peg and socket (the armadillo *Dasypus*).

Similarly, the joint between metacarpal and first phalanx is more or less rigid in several diggers, thus establishing obligatory plantigrade posture. The large keel on the distal end of a metacarpal of the pangolin is not arced in the vertical plane (Fig. 6-3F). *Dasypus* has another peg and socket here that limits flexion; some moles wedge a single large "sesamoid" bone into the joint (derived from the usual pair of smaller bones), which appears to restrict flexion (Fig. 6-3C).

Finally, rigidity may be achieved by the loss or fusion of bones. Gopher tortoises have the reduced phalangeal formula 2–2–2–2–1 (but nonfossorial testudinids have the reduced formula 2–2–2–2–2). The third digit of the manus of golden moles has only one free phalanx (Fig. 6-3B), and the remaining three digits have two phalanges (the proximal elements having fused to the respective metacarpals). Similarly, the marsupial mole has only one free phalanx in digit four, two each in digits one through three, and none in digit five (Carlsson, 1904). The armadillo *Priodontes* has two free phalanges in digits three (Fig. 6-3E) and four.

Two cervical vertebrae fuse in the pocket gopher *Cratogeomys* (Gupta, 1966), two or more in armadillos, the second through fourth in adults of the mole *Talpa* (Dubost, 1968), third through seventh in the cricetid *Myospalax* (Dubost, 1968), second through sixth in the marsupial mole (Carlsson, 1904), and second through fifth in the extinct proscalopid *Mesoscalops* (Barnosky, 1981).

Bracing the Digging Mechanism

As diggers apply force to an earth face in order to crumble it, the earth applies to the burrower an equal force in the opposite direction. If the digger is not to be pushed momentarily in the direction away from the digging tool, thus negating its efforts, this tendency must be opposed.

Large diggers such as aardvarks (50 to 80 kg), giant armadillos (to 60 kg), wombats (15 to 35 kg), and echidnas, pangolins, and badgers (all usually more than 4 kg) hunch over their forelimbs as needed to brace the body and weight the claws by gravity. The weight of most subterranean quadrupeds, however, is not nearly sufficient. While keeping the digging tool free, they use other parts of their bodies to apply great muscular force to opposite sides of their burrows, thus bracing or jamming themselves firmly in their tunnels.

Mammals that dig with the forefeet usually brace with the hind feet, often supplemented by the tail serving as a prop. The hind limbs can be abducted to an unusual degree and the pes turned outward. Toes and claws grip the soil. Changes in the pelvis appear to relate to the forces, well in excess of body weight, that converge on the girdle. The sacral complex (including caudal vertebrae fused or sutured with the ancestral sacrum) enlarges to include 4 vertebrae (wombat, pangolin), 5 or 6 vertebrae (marsupial mole, moles, golden moles, pocket gophers, anteaters), or as many as 8 to 13 vertebrae (armadillos, where, however, rigidity of the shell is a factor). The innominate is fused (moles, pangolin) or firmly sutured to the sacral complex at one, two (some edentates), or three (some moles) levels. If at more than one level, one includes the ischium (Fig. 6-7). Compared with those of nondiggers, the ischium tends to be long, the ilium more nearly in line with the spine, and the acetabulum of the rodent and mole burrowers high, flanking or nearly flanking the spinal axis. Chapman (1919) believed that these features enable abducted limbs to stabilize the girdle or to deliver force to the spine in the direction of motion, or of resistance to motion.

Another distinctive feature of the pelvis of nearly all fossorial mammals, large and small, is weakness or loss of the pelvic symphysis. Nestling moles have a cartilaginous symphysis that is subsequently lost (Eadie, 1945). Adults of some moles (*Talpa*, some *Scalopus*) acquire a secondary symphysis (sometimes fused, sometimes not closed) that is more anterior and is dorsal to the digestive and urogenital organs (Fig. 6-7B). Golden moles also lack the symphysis. The union is short and weak in the marsupial mole, fossorial hedgehogs, rodents, edentates, and pangolin. The symphysis of female pocket gophers is permanently resorbed during pregnancy (Hisaw, 1924), and I find a narrow bony symphysis in two male mole-rats *(Spalax)* but none in a female. Weakening or loss of the symphysis appears to be associated with the firm union of girdle with sacrum and reduction of the adductors of the leg. When a bony symphysis is absent, the rectus abdominus muscles sometimes cross, or cross in part, to take origin from the pubis or pelvic ligament on the other side. Chapman (1919) believed

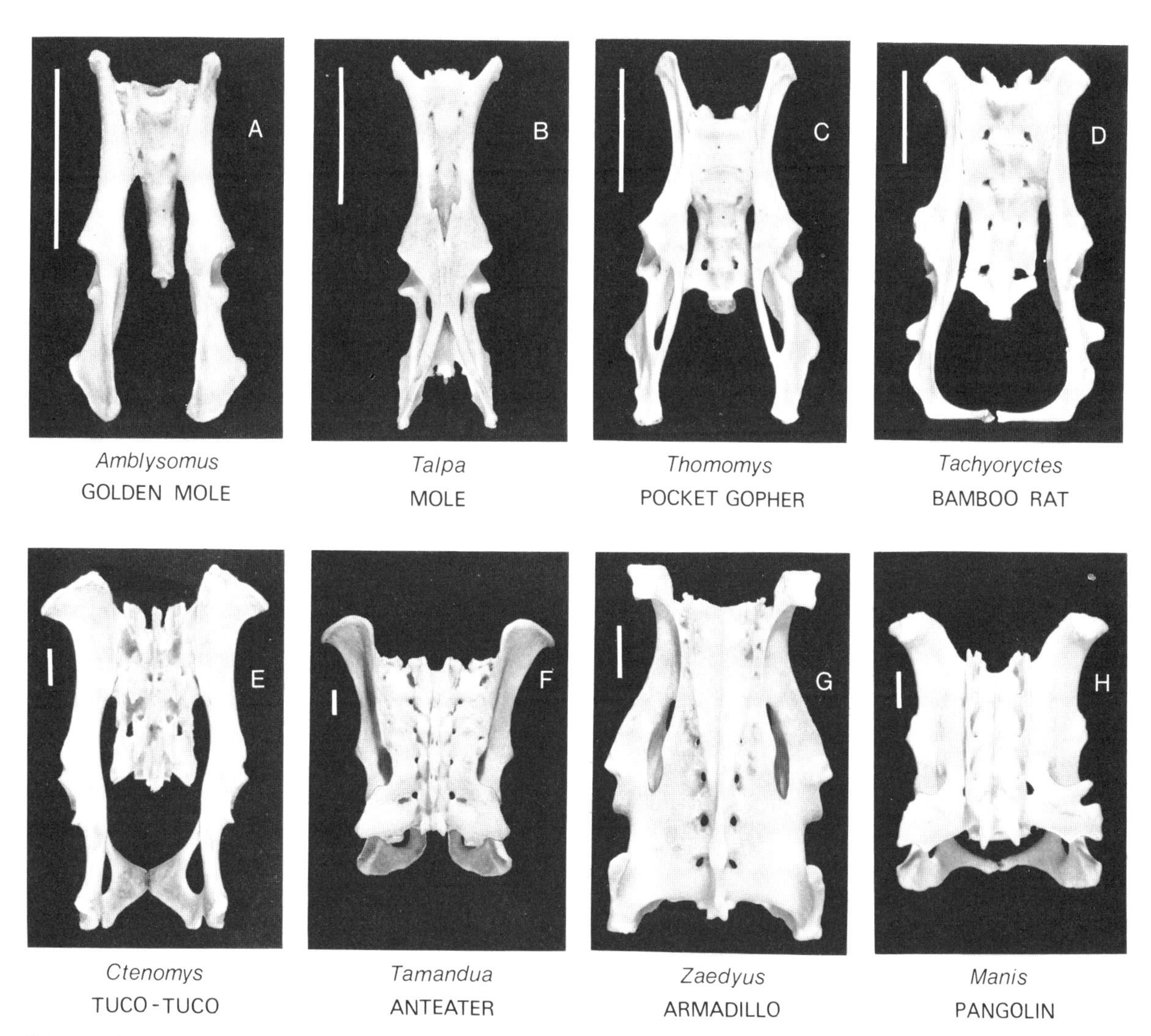

Figure 6-7 Pelvis of selected diggers in ventral (A-D) and dorsal (E-H) views showing sacral complex of many vertebrae; firm attachment of sacral complex to girdle, including (in B,F-H) the ischium; absence (A,C) or weak development (D-H) of the pelvic symphysis; secondary pelvic symphysis dorsal to gut (B); laterally facing acetabula (best seen in D and F). The reference lines are 10 mm long.

that in the absence of the bony arch, this arrangement increases support for abdominal viscera. A modern functional analysis of the pelvis of these animals would be welcome.

Other fossorial quadrupeds that dig with the forefeet brace their bodies in other ways. One of the anurans that dig headfirst is the ranid *Hemisus.* On entering the soil it flexes the head to an unusual degree and presses the snout into the substrate. Thus, the head supplements the feet in support as the forefeet alternately move against the soil (Emerson, 1976). The tortoise *Gopherus* forces its broad head against the wall of the burrow at the position that best resists forces generated by the digging manus. The cervical vertebrae are uniquely heavy, and the zygapophyses distinctive to adapt the neck for transmitting forces between head and trunk (Bramble, 1982). The marsupial mole has a patch of fur over the sacrum that is darker than the pelt elsewhere. Howe (1975) states that this patch is in almost constant contact with the soil during burrowing, and that the body is braced at the patch. Moles brace one broad manus against one wall of the burrow as the other digs against the opposite wall (Hisaw, 1923).

Mammalian diggers usually use the hind feet in unison to push loose dirt to the rear (see below). Moles then stabilize the body by bracing the forefeet against opposite walls of the tunnel. The mole-rat *Spalax* pushes down with the forefeet as it pushes up with the top of the head or nose pad (Gambarian, 1953). It is probable that other mammals with nose pads (marsupial mole, golden moles) do likewise. The blesmols *Heliophobius* and *Georychus* instead jam their relatively procumbent incisors into the top of the burrow (Jarvis and Sale, 1971).

Transport and Stabilization of Soil

Many quadrupedal diggers avoid, at least part of the time, the need to transport soil away from the place of digging; earth is disturbed but not moved far. Thus, a horned lizard submerges in loose sand by vibrating its flat body, and anurans, ambystomids, and lizards may wriggle into tractable soil without forming an open burrow. Some subterranean mammals (marsupial mole, moles, golden moles) merely displace sand or earth upward when they make foraging runs just below the surface. The runway remains open or collapses behind the digger, depending on soil characteristics. Finally, the various methods of bracing the body just described all compact soil. Even deep, open burrows may be made by compaction alone in soils that are damp and not too firm.

Nevertheless, most subterranean quadrupeds having open burrow systems must transport much soil to the surface from the places where tunnels are extended. It is usual for animals that dig with the teeth or forelimbs to push the loosened soil under the body with the forefeet. From time to time the hind feet kick the spoils behind the body. When there is a suitable accumulation of dirt, digging is interrupted to push the loose earth back along the tunnel to an exit.

Animals that turn around to do this job head-first include moles (Hisaw, 1923), golden moles (Bateman, 1959), bamboo rats (Jarvis and Sale, 1971), pocket gophers, and the mole-rat *Spalax*. The forefeet push the dirt, usually one at a time, and the side or top of the head, teeth (of some rodents), or chest may also be used. The tortoise *Gopherus* pushes some dirt with the upturned end of its plastron. Armadillos, blesmols (Bateman, 1959; Genelly, 1965; Jarvis and Sale, 1971), and the tuco-tuco (Pearson, 1959) instead back up, vigorously kicking the dirt behind them. The hind feet are usually used synchronously. The blesmol *Heliophobius* may instead place its hind feet against the dirt and hold them there as the forefeet push the body backward (Jarvis and Sale, 1971). The marsupial mole also moves dirt with its tail (Howe, 1975), which is short but robust, broad at the base, and naked and cornified below.

The loose dirt of a surface mound is sometimes stabilized by tamping, and loose earth within a tunnel may be compacted somewhat to facilitate its transport. This may be done with the nose pad. Some blesmols "vibrate" the hind feet against the dirt at more than 20 Hz. Some other rodents "bounce" on the loose soil with the forequarters (Lehman, 1963).

Numerous modifications of the feet adapt them for transporting loose earth. With few exceptions they are large, broad, thick, and firm (see good illustrations for the mammals in Dubost, 1968). Plantar tubercles are commonly very thick and tough (Fig. 6-2). The toes are more or less webbed, particularly in the marsupial mole, moles, and golden moles. The manus of the more specialized moles is further widened by the falciform bone, which is a flat "sesamoid" that lies medial to the carpus, first metacarpal, and the proximal phalanx of the pollex (Fig. 6-5D). Long, stiff, marginal hairs border fore and hind feet, and often the individual toes as well, thus increasing their functional breadth (Fig. 6-2). Claws are long if they are used to loosen soil, and they are broad in the gopher tortoise, echidnas, wombats, moles, armadillos, pangolins, and aardvark. The foreclaws of the marsupial mole and golden moles are instead thin blades, being deep in the sagittal plane. It is probable that these animals move loose earth with a sideways, sweeping action of the manus (Bateman, 1959).

External Form and Maneuverability

In order for burrowers to prosper within their confined quarters, it is desirable that their bodies be modified to eliminate impediments to free movement, minimize snagging on projecting roots or rocks, keep dirt out of orifices, allow easy backing up, and facilitate turning in a short radius.

The body is compact and stocky. In fossorial Anura, tortoises, echidnas, and hedgehogs, it is short and squat. Subterranean mammals have cylindrical bodies. The head may be large and blunt or smaller and pointed, but usually merges into the body without constriction at the short neck (Fig. 6-2). The legs are relatively very short, their proximal segments (and also the forearm of the marsupial mole, moles, and golden moles) being contained within the contour of the body.

The fur of mammalian burrowers tends to be either short and upright, thus brushing in either direction (moles), or longer but soft and lax. (The spine-covered echidnas and hedgehogs may avoid backing up within burrows.) Vigorous shaking of the pelage has been observed in some golden moles (Bateman, 1959), blesmols (Genelly, 1965), bamboo rats (Jarvis and Sale, 1971), and pocket gophers, and frequent grooming with teeth or claws has been described for several kinds of rodents.

The tail of burrowers may be as long as 40% of the remainder of the body (some moles), but is usually short and sometimes absent (golden moles, the mole-rat *Spalax*). Short tails may, nevertheless, be strong, and they are usually in frequent use as a brace, tactile organ, or radiator (see below).

The external ear (pinna) of mammalian burrowers ranges from small (cricetid voles, pocket gophers, bamboo rats) to virtually absent (marsupial mole, moles, golden moles, blesmols, *Spalax*). Most burrowers have valvular closure of the nostrils; possibly some can close off the auditory canals. The ear opening may be obscured by dense fur (moles, golden moles) or guarded by stiff hairs (marsupial mole). The tympanic membrane of fossorial anurans, lizards, and tortoises tends to be smaller than that of nondiggers.

The incisors of chisel-tooth diggers protrude through furry skin, and thus are excluded from the mouth, which can be sealed by the lips even when the incisors are in use. Regression of the eyes is noted below. The testes of many burrowers are seasonally or permanently abdominal.

Meager available information indicates that some burrowers execute a lateral turn to reverse direction (Anura, the bamboo rat *Tachyoryctes*, the blesmol *Heliophobius*), whereas others place the head between the forefeet and then do a half somersault and roll. Pocket gophers commonly tuck the head under one foreleg and instantly twist about in a plane that is neither vertical nor horizontal. For many rodents, such maneuvers are facilitated by a limber spine and the loose fit of their skins on the body.

Senses

Directional Sense and Touch Fossorial vertebrates that remain above ground, or retreat only into dens of simple structure, have senses that are not distinctive. Those of the subterranean diggers, however, are much modified—and inadequately known. An inherent sense of direction and the memory of an accurate map of the home range are of primary significance to at least some burrowers, yet are scarcely understood. When released on the surface, some of the animals quickly dig into their burrow systems. Kriszat (1940b) showed that if a portion of the burrow system of the mole *Talpa* is completely destroyed, the earth turned, and the animal then released underground at the far edge of the disturbed area, it constructs a pattern of runways similar to the original, new tunnels entering old ones end on, with close registration. Similar orientation of the bamboo rat *Tachyoryctes* is reported by Jarvis and Sale (1971). Eloff (1951) found that when a portion of a runway of the blesmol *Cryptomys* was blocked, the rodent dug a new tunnel exactly parallel to the old one until the obstruction was passed, then came into the original burrow. He postulated that the animal has a surface map that it monitors from below.

A mole released by Kriszat from a nest box

above ground found a food source seemingly by random search (until in close proximity, when it smelled the food). When released again, with the food in a different place, the mole went directly to the first location, then found the second location in a random way. The (blind) mole could always return straight to its box without retracing steps. Arlton (1936) reported similar observations.

Touch is highly developed in burrowers. For *Talpa* it is clearly the most significant of the familiar senses for orientation within tunnels, though not for finding food (Kriszat, 1940a). Vibrissae are prominent, particularly in rodent diggers (yet are seemingly absent in the marsupial mole). The stiff hairs that broaden the toes and feet are also sensory. Tactile hairs are present on the tails of most burrowers that have tails, and the tail may be moved about as a feeler while the animal backs. Sensory hairs are spaced over the entire body of some diggers (for example, blesmols, including otherwise naked *Heterocephalus*) or are concentrated on the rump (pocket gophers and some golden moles). The sensitive proboscis of moles has thousands of sensory papillae, each of which has 15 to 20 nerve fibers (see Dubost, 1968, for illustration).

Hearing Kriszat trained a *Talpa* to respond to sounds in the range of 250 to 3500 Hz. Bramble (1982) believes that the uniquely large inner ear, saccular otolith, and macula of the fossorial tortoise *Gopherus* enable it to detect weak, low-frequency ground motion produced by footfalls. The greatly hypertrophied malleus-incus complex of certain golden moles and extinct palaeanodonts (a remarkable example of convergence) is interpreted by Rose and Emry (1983) as an adaptation for the reception of sound of low frequency. Other authors have also emphasized that mammalian burrowers tend to be adapted for the detection of low-frequency and ground vibration, a subject that is included in Chapter 15.

Vision and Light Detection As vertebrates become more and more committed to subterranean life, vision becomes a less and less useful sense. Regression of the eyes has been relatively well documented (see Dubost, 1968, for a review, illustrations, and citations). The diameter of the eye is reduced nearly to 2 mm in pocket gophers and the tuco-tuco, although vision is retained. The eye measures about 1 mm in some moles, most blesmols, and the bamboo rat *Tachyoryctes.* It is a mere vestige under the skin in other moles, golden moles, the marsupial mole, and the mole-rat *Spalax.* The eye is insensitive to light in the last group and probably in most blesmols. As vision is lost the cornea thickens, the lens becomes cellular and finally amorphous, and the vitreous body regresses and becomes fibrous. The retina and optic nerve degenerate relatively slowly.

It has been shown that for both lizards (Gundy and Wurst, 1976) and mammals (Pevet, Kappers, and Nevo, 1976) parietal or pineal function is enhanced in subterranean vertebrates. In *Talpa* and *Spalax* the pinealocytes are unusual for their indications of intensive synthetic activity. Presumably, a dorsal organ is a more suitable photoreceptor for a burrower than are lateral eyes.

Energetics

In Chapter 10 the principles of energetics are presented with emphasis on overground locomotion, flying, and swimming. The energy requirements of fossorial quadrupeds, particularly of those living in closed burrows, are sufficiently unusual and critical to warrant review here. These animals must tolerate high work loads in atmospheres having relatively low O_2 tensions, high CO_2 tensions, high humidities, and limited temperature variation.

Digging requires the expenditure of much energy. Two species of fossorial anurans were shown by Seymour (1973) to maintain higher activity, as measured by O_2 consumption, and to fatigue more slowly than nonfossorial controls. Vleck (1979) developed a cost-of-burrowing model showing that, according to soil type and burrow characteristics, burrowing by the pocket gopher *Thomomys* can require 360 to 3400 times as much energy as moving the same distance on the surface. While digging in clay soil, O_2 consumption can increase sevenfold over resting levels. Thus, energy conservation may be important to diggers, particularly if they burrow to reach limited and dispersed food (Jarvis, 1978; Vleck, 1981).

Tolerance of low O_2 tension seems to be universal for burrowing quadrupeds. When covered by 8 inches of soil, a resting echidna did not become sufficiently uncomfortable to move to the surface until O_2 tension dropped to 7% or 8% (Augee et al., 1971), and the mole-rat *Spalax* remains active at an O_2 tension of only 5.3% (Ar, Arieli, and Shkolnik, 1977). The blood of burrowers has high affinity and high capacity for O_2: the red cell count is high, and the O_2 dissociation curve shifts to the left. The myoglobin content of selected skeletal muscles is high in *Thomomys* (Lechner, 1976) and *Spalax* (Ar, Arieli, and Shkolnik, 1977), yet Goldstein (1971) found relatively few red fibers in muscles of the mole *Scapanus.* Marked bradycardia seems to be usual for active burrowers. Associated with its O_2 metabolism is the ability of the armadillo *Dasypus,* and probably of other diggers, to hold the breath for long periods while active, evidently to avoid breathing dust (Scholander, Irving, and Grinnell, 1943). (Desert iguanids have complex narial passages to serve as dust traps.)

Burrowers are also tolerant of high CO_2 tension. Darden (1972) reported atmospheric concentrations to 3.8% for *Thomomys.* Augee et al. (1971) noted tensions of 10% to 12% for an echidna, and Ar, Arieli, and Shkolnik (1977) measured 13% for *Spalax.* The blood accepts high levels of CO_2, becomes relatively acidic, and has high noncarbonic buffering strength (Chapman and Bennett, 1975).

Since the humidity within closed burrows is commonly near saturation, evaporative cooling is not effective. Many mammalian diggers have relatively low body temperatures and low basal metabolic rates, although the converse is true of those weighing less than 60 g (McNab, 1979, 1980). Although ambient temperatures are relatively constant, the range of thermoneutrality is relatively great. Hypothermia lowers the basal metabolic rate and conserves O_2; tolerance of elevated body temperature during activity postpones heat stress. Thus, some burrowers are not strict homeotherms (Bradley and Yousef, 1975). Many fossorial rodents have reduced thermal conductance (Vleck, 1979), yet when body temperature is high, pocket gophers dissipate heat from their naked and vascular tails, and the blesmol *Heterocephalus* is entirely naked.

Future Study

Inadequacies in our knowledge have been noted throughout this chapter. Descriptions of musculoskeletal anatomy are available for some diggers (such as moles and pocket gophers), but are fragmentary or wanting for others (such as most nonmammals, and various rodents and armadillos). Owing to the small size of most burrowers and to the great difficulty of observing or monitoring them as they dig freely in their tunnels, behavioral and functional analyses lag behind those for cursors, scansors, and even swimmers and flyers. Slow-motion cinematography is fragmentary, electromyography absent, and behavioral studies rarely sufficient for testing functional interpretations of bone-muscle systems derived from work on preserved material. The forces and torques associated with the distinctive girdles, necks, and feet of diggers await analysis, and the ontogeny of the unique forms and functions of these structures are virtually unknown. Also, the paleontology and phylogeny of most fossorial quadrupeds is obscure. Convergence of the extreme kind demonstrated in burrowing provides outstanding opportunities for delimiting necessary conditions for various morphological traits using the comparative method.

Chapter 7

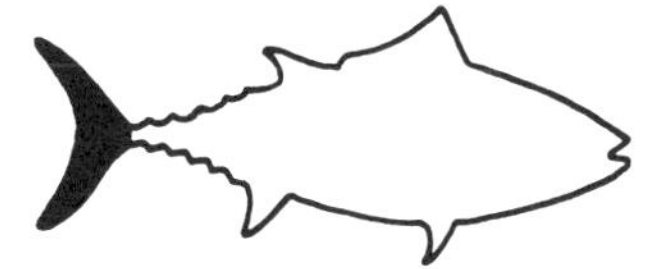

Swimming

Paul W. Webb
Robert W. Blake

Swimming animals are found in all groups of vertebrates, and all possible structures associated with the axial and appendicular skeletons are used for propulsion under or upon the water. Some aquatic vertebrates have evolved directly from aquatic ancestors. These are primary swimmers and include all the fish and the larval stages of most amphibians. Aquatic or semiaquatic reptiles, birds, and mammals are secondary swimmers, having returned to the water from terrestrial habitats. Primary swimmers show the greater diversity of locomotor mechanisms. For example, depending on their immediate needs for speed and maneuver, fish may swim using the body and caudal fin, long (or short) median dorsal and/or anal fins, or paired pectoral and pelvic fins (Fig. 7-1).

As a consequence of the different evolutionary histories of the various vertebrate groups, there is substantial diversity in the morphological structures that are used for locomotion. A summary of propulsor mechanisms by Lindsey (1978) lists 12 major categories and some subcategories to classify morphological variability in fish alone. However, from a mechanical point of view, this variation can be more simply subsumed under undulatory mechanisms and oscillatory mechanisms (Table 7-1). In the former category, waves are passed along the body or fin propulsor. Such waves are easily seen in long-bodied animals such as lampreys, eels, sea snakes, and crocodilians, and in long-finned fish such as bowfins *(Amia)*, electric eels *(Gymnotus)*, sea horses *(Hippocampus)*, and porcupine fish *(Diodon)*. At the other extreme, the tail of tunas (Thunnidae), whales and dolphins (Cetacea), and the boxfish *(Ostracion)* appear to the unaided eye to beat to and fro, although careful analysis of cine films shows that, in reality, these movements continue to reflect a basic undulatory motion. Oscillatory propulsors are usually paired appendages that beat backward and forward, and/or up and down, and function like a paddle or bird wing.

Undulatory propulsors are characteristic of primary swimmers, but they also occur among some secondary reptilian swimmers (such as crocodilians and snakes) and mammalian swimmers (such as cetaceans and sirenians). Oscillatory propulsors are most common among secondary swimmers, and they are the only mechanisms found among birds. Most mammals rely on oscillatory propulsors (legs), but they may use an undulating tail as an aid to controlling stability and turning, as do muskrats *(Ondatra)*.

The more detailed classifications of propulsion mechanisms, such as that described by Lindsey, recognize major patterns, or modes, of locomotion within continua of variation in each of the two groups defined here on the basis of mechanics. The detailed classification is useful when it is functionally oriented as, for example, in the description of undulatory modes of body or caudal fin (see Table 7-1, Fig. 7-1, and below), but it adds little that is conceptually useful where it is anatomically oriented, as for descriptions of many undulatory and oscillatory mechanisms of the appendages. The emphasis in this chapter is on the functional morphology of propulsion sys-

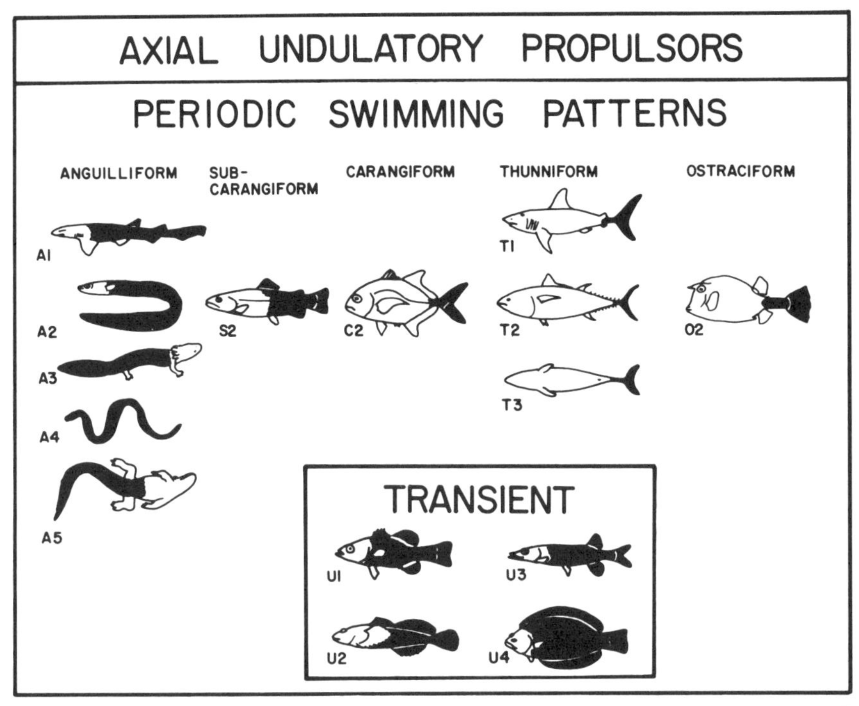

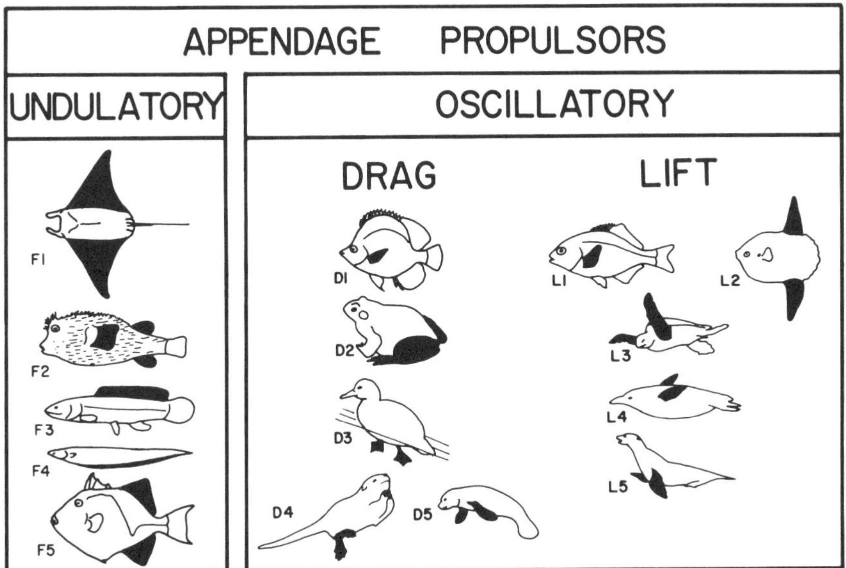

Figure 7-1 A diagrammatic summary of propulsion mechanisms in aquatic animals *(propulsors shaded).* A1 = *Scyliorhinus,* dogfish shark; A2 = *Anguilla,* eel; A3 = *Necturus,* a newt; A4 = *Pelamys,* sea snake; A5 = *Alligator;* S2 = *Salmo,* trout; C2 = *Caranx,* jack; T1 = *Lamna,* shark; T2 = *Thunnus,* tuna; T3 = *Cephalorhychus,* dolphin; O2 = *Ostracion,* box fish; U1 = *Micropterus,* bass; U2 = *Cottus,* sculpin; U3 = *Esox,* pike; U4 = *Psettodes,* a flat fish; F1 = *Manta,* a ray; F2 = *Diodon,* porcupine fish; F3 = *Amia,* bowfin; F4 = *Gymnotus,* a weakly electric fish; F5 = *Balistes,* trigger fish; D1 = *Chaetodon,* a butterfly fish; D2 = *Rana,* frog; D3 = *Anas,* duck; D4 = *Ondatra,* muskrat; D5 = *Manatus,* manatee; L1 = *Cymatogaster,* surf perch; L2 = *Mola,* ocean sunfish; L3 = *Chelone,* turtle; L4 = *Spheniscus,* jackass penguin; L5 = *Eumetopias,* sea lion.

Table 7-1 A functional classification of locomotor mechanisms in aquatic animals based on common kinematic patterns of the propulsor, along with terminology used in the original classification of fish locomotor mechanisms by Breder (1926).

Functional categories	Description	Breder's terminology[a]
Undulatory propulsion mechanisms		
Axial mechanisms (body and, where relevant, caudal fin)		
Anguilliform	Anguilliform to thunniform continuum of increasing length of the propulsive wave relative to the body length and of progressive concentration of major lateral movements at posterior portion of the body	Anguilliform
Subcarangiform		Subcarangiform
Carangiform		Carangiform
Thunniform		Thunniform
Ostraciiform	Propulsive wavelength long and concentration of major lateral movements caudally, but not associated with low drag or high Froude efficiency	Ostraciiform
Appendicular mechanisms		
Long-based	Parallels the anguilliform to thunniform continuum, but with increasing propulsive wavelength relative to the fin length; amplitude usually relatively constant over the whole appendage	Amiiform
Short-based		Gymotiform
		Balistiform
		Rajiform
		Diodontiform
		Tetraodontiform
Oscillatory propulsion mechanisms		
Drag-based		Labriform
Lift-based		Labriform

a. Breder's terms are included in the classification where they are based on function rather than on morphology alone.

tems based on mechanical principles, and earlier nomenclature will be used only where it is useful in clarifying function.

Undulatory Propulsion

Undulation, the most ubiquitous form of swimming, is found in all vertebrate classes except birds. It includes all swimming modes utilizing the axial skeleton. In addition, waves pass along the length of long-based fins (that is, fins that are long in relation to their depth) and along many shorter based median and paired fins as well. Short-based fins may undulate exclusively, or undulatory propulsion may be part of a range of mechanisms that includes oscillation.

In spite of the kinematic and mechanical similarities underlying the various undulatory mechanisms, the different major morphological systems used are functionally important because each generates thrust and control forces of different magnitudes. Axial mechanisms (the body and/or caudal fin) are associated with muscle masses that may exceed 50% of the body mass, and propulsive movements may convert more than 90% of the muscle energy into thrust energy. Therefore, the body or caudal fin can generate very large propulsive forces for fast swimming, rapid acceleration, and high-powered turns. In contrast, the other median and paired fins are associated with relatively little muscle (<20% of the body mass) and are usually used for slow swimming. However, the distribution of these appendages around the center of mass allows for a precision that has been central to the radiation of fish into many structurally complex habitats, such as coral reefs. Thus, two morphologically distinct and functionally and ecologically important swimming patterns emerge among undulatory propulsors: (1) the use of body and/or caudal fin for high-power activities, and (2) the use of appendages for low-power activities. However, these categories are modes defined within a continuum of function, and there are exceptions. For example, undulatory body movements by the eel *(Anguilla)* permit precise slow swimming; oscillatory motions of dorsal and anal fins by the ocean sunfish *(Mola)* permit fast cruising.

In addition to the general differences between low-power swimming using appendages and high-power swimming using axial structures, there are further mechanical and functional distinctions that apply primarily to the latter. The distinction is based on whether locomotor patterns are periodic or transient. In periodic locomotor patterns, swimming movements are regularly repeated, as in continuous cyclic beating of the body/caudal fin used to swim at fairly uniform speeds and two-phase swimming patterns (beat-and-glide or beat-and-coast propulsion) involving small accelerations (Weihs and Webb, 1983). Transient swimming, in contrast, describes nonrepeated locomotor activities, such as fast starts and turns that are characterized by brief but high rates of linear and angular acceleration. The distinction between periodic and transient swimming is necessary because they differ in regard to major sources of drag and the contributions that the different parts of the body make to thrust.

Most research has concentrated on periodic swimming at constant speed because of the availability of water and wind tunnels and the ease of replicating experiments. In reality, most free-swimming fish spend much of their time in transient motions, with frequent stops, starts, and turns as they defend their habitats or search for food, shelter, and mates. Furthermore, it appears that the perpetuation of transient swimming has been important in many evolutionary lines, especially in actinopterygians (Webb, 1982a) and probably also in many elasmobranchs and amphibians.

Transient Swimming Transient swimming is most important in fast starts and in turns having rapid acceleration, as animals evade predators or attack elusive prey. Transient swimming with time-dependent speed, or variable linear and angular acceleration at lower rates, also occurs during routine swimming. Powered turns and fast starts are mechanically similar (and distinct from periodic swimming).

The body movements of a rainbow trout *(Salmo gairdneri)* during a fast start are shown in Figure 7-2A. Three stages have been defined (Weihs, 1973): During stage 1 (0–67 msec in the figure)

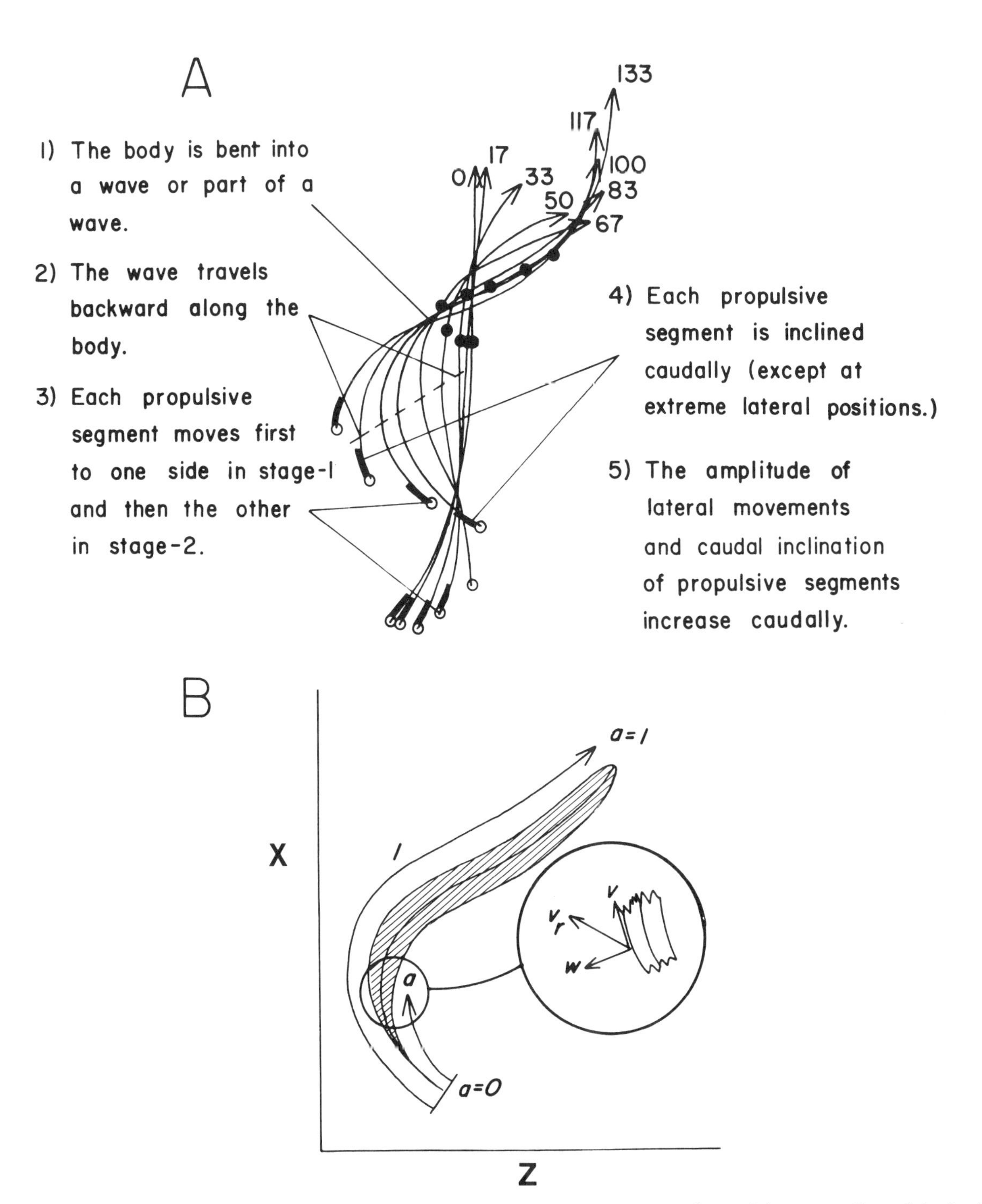

Figure 7-2 The kinematic features involved in thrust generation. A: Tracings of the center line of the body of a trout, *Salmo gairdneri,* taken from cine film of a fast start. Times between tracings are shown in msec. Shaded circles show the location of the center of mass. B: Convention for coordinates used to analyze locomotor movements such as those in A, based on Weihs (1973). Note the body is arbitrarily divided into infinite propulsive segments from the trailing edge ($a = 0$) to the tip of the nose ($a =$ length). One of these segments is shown at a, along with the directions of its forward, lateral, and resultant velocities (V, W, and V_r).

Table 7-2 Symbols used in text.

Symbol	Definition
C_D	Drag coefficient
C_n	Normal drag coefficient
c	Wave velocity
D	Drag force acting on a body or paddle
D_p	Drag during power phase of beat cycle
D_r	Drag during recovery phase of beat cycle
d	Body depth
dr	Spanwise length of an element of a paddle blade
E	Total work done during the power phase of the beat cycle
$\overline{E}$	Mean work done
E_D	Work done in overcoming drag
E_o	Total work done during a beat cycle
F	Force
F_a	Acceleration force
F_n	Normal force
L	Lift force
m	Added mass
N	Number of paddles
n	Normal coordinates of a propulsive segment
P	Total power output
P_k	Rate of loss of kinetic energy
P_T	Thrust power
R	Radius
S	Surface area
T	Thrust
$\overline{T}$	Mean thrust
t	Time
t_p	Duration of the power phase of the beat cycle
t_o	Duration of the beat cycle
u	Speed of a paddle
V	Forward velocity of animal, propulsive segment, or blade element
v_r	Resultant velocity of a propulsive segment or blade element
v_n	Normal velocity of an element of a paddle blade
v_s	Spanwise velocity of an element of a paddle blade
W	Lateral velocity of a propulsive segment
w	Velocity of water accelerated by a propulsive segment
β	Added mass correction for the effect of body
γ	Angle between a paddle and the body axis
η	Efficiency
θ	Angle between the trailing edge of a propulsive segment and the swimming axis
ρ	Density of water
ω	Angular velocity

the body bends from its initial relatively straight posture into the shape of a C or S. During stage 2 (70–130 msec) the body bends in the direction opposite to that in stage 1. This is due to the backward movement of the muscle contraction that initially bent the body in stage 1. Stage 3 is not shown because it is highly variable. During this stage the fish may continue to swim or start to glide. Most of the acceleration of the body is completed by the end of stage 2. Max mum acceleration rates (resolved about the center of mass) of 40 to 50 m/sec^2 have been recorded, and fish can reach speeds of the order of 1 to 1.5 m/sec (2.2 to 3.4 mph) in 60 to 120 msec. Furthermore, fish can turn at rates of the order of 60 rad/sec (3400 degrees per sec, or 9.5 full circles per sec!), and in radii as small as 11% of the body length. This performance is very impressive (Eaton et al., 1977; Webb, 1976, 1978a).

To analyze how the body movements generate thrust, the body (and fins as appropriate) is divided into arbitrarily defined "propulsive segments"; each point along the body, such as that at *a* in Figure 7-2B, then represents a propulsive segment (Webb, 1978b). As the body is bent into a wave, each propulsive segment is displaced laterally at some velocity W, and at the same time it moves forward at some velocity V as the fish begins to accelerate forward. Each propulsive segment, therefore, has some resultant velocity v_r. The component of v_r that is normal to the fish's centerline is w. If this has normal coordinates n and it accelerates a mass of water (called the added mass m) to w, then the normal force on a propulsive segment is mwn/t. (All symbols are redefined in Table 7-2.)

The force F_a acting on the body owing to the acceleration of water along the whole length is the sum of the contributions from all propulsive segments (Weihs, 1973). Using calculus, this can be formally stated as

$$F_a = \frac{d}{dt} \int_0^1 mwn(du), \tag{7 1}$$

where da is the length of a small segment of the body.

In addition, further thrust is generated in the form of lift by any free fin or portion of the body that has sharp edges and can act like "wings" (see Chapter 8). If the lift force is L and there are k such hydrofoils, then the total acceleration or turning force F acting on the body is

$$F = \frac{d}{dt} \int_0^1 mwn(da) + \sum_{i=1}^{k} L. \tag{7-2}$$

The integral of Eq. (7-2) generates the larger force.

The various propulsive segments do not contribute equally to the force acting on the body. More posterior segments move through greater distances, and hence move faster. Such segments are also directed more caudally than are the more anterior segments (Fig. 7-2). Therefore, they produce a relatively greater force that is more nearly aligned with the direction of motion of the fish. The magnitude of F_a also depends on the mass m of water affected by each propulsive segment, which in turn is affected by body shape. If the depth of the body is d and the density of water is ρ, then m is given by (Lighthill, 1975):

$$m = \frac{\beta \rho \pi d^2}{4}, \tag{7-3}$$

where β is a factor, usually close to unity, that takes into account the relative contributions of the body and fins to the total depth d.

From Figure 7-2, showing body motions, and Eqs. (7-2) and (7-3), it follows that the more posterior parts of the body contribute most to thrust in accelerations and turns and that the force acting on any segment increases as depth increases. Forces on the body are maximized when the greatest depth of body-plus-fin is at the posterior part of the body. These features are clearly shown by the form of fish specialized for acceleration, for example the cottid *Cottus,* esocid *Esox,* and pleuronectid *Psettodes* (Fig. 7-1), and by the fact that bony fish extend all their median fins before accelerating or turning (Eaton et al., 1977; Webb, 1978b).

Rapid side movements that generate thrust also create a problem because there is a component of the force acting on the body that tends to rotate the head. This tendency for the "tail to wag the head," or lateral recoil, wastes energy. The loss can be minimized by adding mass at the center of mass, thereby increasing the moment of inertia resisting recoil. Large girth, and hence inertia, is common in many fish. In addition, a dorsal fin is common over the center of mass, which adds to body depth, and hence to the added mass (Eq. 7-3) resisting lateral recoil. Thus the presence of an anterior medial fin is an important adaptation for stability when propulsion is by undulation of axial structures.

The performance of an aquatic animal depends not only on the magnitude of the thrust force propelling it forward, but also on the force resisting that motion. The origins of the resistance force are fairly complex and are merely suggested here; they are described in an especially readable form by Vogel (1981). The total resistance can be divided into several components. For rectilinear horizontal swimming, the resistance force is comprised of an inertial component, which is due to the mass of the animal plus entrained water, and a drag force attributable to the viscosity of water. When an animal moves vertically, an additional force must be considered to include any work done against the acceleration owing to gravity. The vertical force reduces the total resistance in negatively buoyant animals swimming toward the bottom, but adds to the total resistance of such animals rising in the water column.

When an animal turns, a centripetal acceleration must be applied, adding to the resistance (Weihs, 1981). This resistance force acts at right angles to those that resist continued forward motion.

The total resistance force can then be summarized as

$$\text{Total resistance} = \sqrt{[(\text{inertial force} + \text{drag force} + \text{vertical force})^2 + (\text{centrifugal force})^2]}. \tag{7-4}$$

The magnitudes of the various components of resistance obviously vary with the type of swimming activity. The drag force is the most complex. It arises primarily from viscous effects in a thin skin of water, the boundary layer, that surrounds any moving aquatic animal. This is the region of flow where water velocity declines from the speed of the animal, at that animal's surface, to the speed of the surrounding, undisturbed water. A velocity gradient is set up, and energy is dissipated in shear forces that resist distortion in the gradient, that is, in the boundary layer (see Chapter 8). Thus, skin friction is involved that varies with the total surface area of the animal. The magnitude of the drag force also depends on how the boundary layer interacts with the rest of the flow around the whole animal, but this total drag is nevertheless expressed in relation to the surface area that is in contact with the water.

During high rates of acceleration, the boundary layer gradually increases in thickness. Initially, although acceleration is high, speed is low. Thus, frictional drag is small enough to neglect during fast-starts, and resistance is dominated by the inertial force (Webb, 1982b). This is important. Performance depends on the balance between thrust and drag. For transient swimming it is advantageous for all propulsive segments along the body to be large to maximize thrust, and because frictional drag is so small, the large area carries little penalty in terms of drag and hence performance.

The principal source of resistance in transient swimming is thus the mass of the system, or more correctly, the nonmuscular "dead weight." Any adaptation that decreases dead weight increases the proportion of the body mass represented by muscle and thereby increases capacity for acceleration. Such adaptations are common because they also reduce density and thus reduce the problem of regulating buoyancy. Some specific acceleration-related adaptations are to be expected, but only one, reduced skin mass in some fish, has been described so far (Webb and Skadsen, 1979). The proportion of body muscle to dead weight is also important with regard to the trade-off between resistance to acceleration and the distribution of depth along the length of the body, which, as noted, influences the magnitude of thrust. Depth of body and muscle percentage among postjuvenile fish appear to be inversely related, so capacity for acceleration varies relatively little. Thus it seems probable that the evolution of predators and their prey has resulted in various strategies for balancing drag against thrust in order to preserve adequate acceleration performance. The resultant morphological variation allows for a range of locomotor options, thus permitting the exploitation of different habitats (Webb, 1982).

The various factors affecting thrust and resistance to motion in accelerations and turns can be summarized in terms of an optimal morphology, defined as that which would maximize performance. These features are: large body and caudal area, anterior stabilization of the body using median fins over the center of mass, flexible body to maximize amplitude, and a large ratio of muscle mass to body mass.

Although fast starts are spectacular, transient swimming also occurs at low and moderate speeds, and the same principles apply concerning the generation of thrust and resistance to motion. However, at moderate speeds drag becomes the dominant resistance component, comparable to that of periodic swimming as discussed below. Body area becomes a penalty rather than a benefit. Thrust is modulated mainly by small changes in the amplitude and frequency of the various propulsors and thrust and drag can be more easily estimated by applying mechanical principles developed for periodic swimming.

Periodic Swimming Historically, most research has concentrated on periodic swimming. The now classic work of Sir James Gray, in 1933, first enunciated the basic relationships between body movements and the generation of thrust that have been substantiated for all species studied. The fundamental undulatory movements during swimming at a uniform speed are illustrated in Figure 7-3 for the eel *(Anguilla)*. These movements underly most patterns of periodic swimming, including two-phase swimming because acceleration rates are low during the swimming phase. Mechanically, the same general principles apply to the swimming phases of periodic swimming as were seen for transient swimming (Fig.

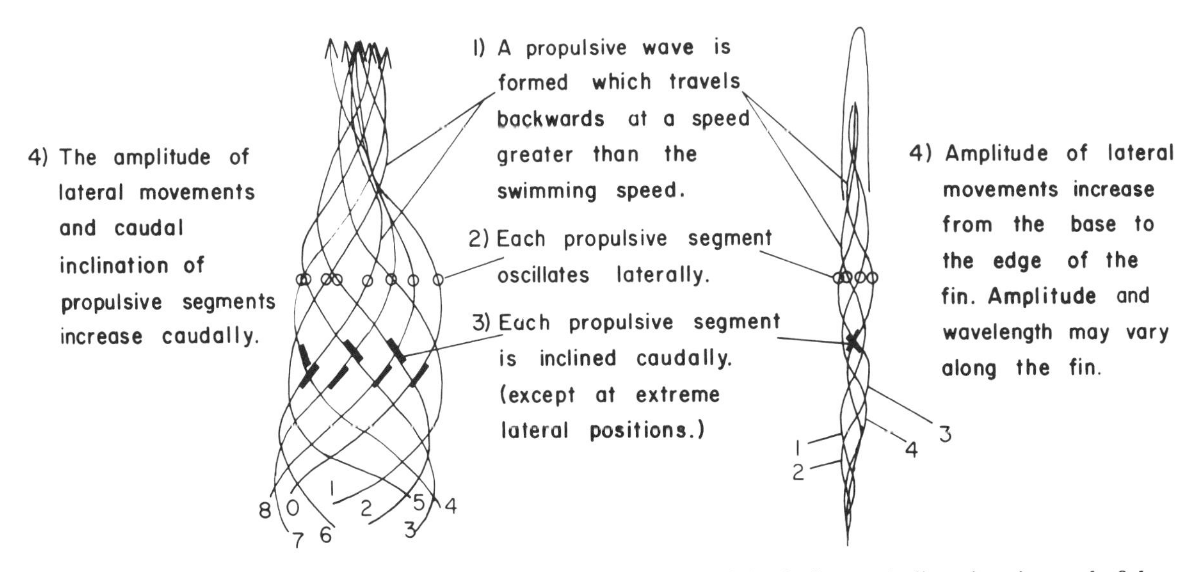

Figure 7-3 Tracings of the centerline of an eel, *Anguilla anguilla (left)*, during periodic swimming and of the undulatory movements of the long ventral fin of the weakly electric knife eel, *Gymnotus (right)*, with outline of the head. Frame numbers are shown for 90-msec intervals. The motion has been normalized relative to the head to illustrate body movements more clearly. Usually, each body position is displaced forward as the eel swims.

7-3, *left*). Thus, the body is bent into a wave that travels backward over the body as the animal swims forward. The backward speed of the wave exceeds the forward speed of the animal. Each propulsive segment oscillates relative to the head, with amplitude increasing caudally. Each propulsive segment is inclined backward, the more caudal segments inclining more posteriorly than the anterior segments. As with acceleration and turns, each propulsive segment accelerates the water in its vicinity, and thrust is proportional to the reaction on each propulsive segment.

However, there is a fundamental difference between the mechanics and kinematics of periodic and transient undulations. During periodic swimming movements, including the swimming phases of two-phase swimming, the various propulsive segments interact in space and time. To see this and its effect on calculating thrust, consider the water in the vicinity of the anterior part of the body. This water is accelerated laterally and caudally by a caudally inclined propulsive segment. Immediately, the water is affected by the next posterior propulsive segment because the propulsive wave travels backward faster than the animal moves forward. Furthermore, since the amplitude of the more posterior segment is greater than that of the first segment, the water is further accelerated. Finally, because the posterior segment is inclined more posteriorly, the motion of the water is inclined more posteriorly; that is, it is better aligned with the swimming axis. Eventually, after being influenced by all propulsive segments, the water is discharged into the wake beyond the tip of the body or caudal fin (the trailing edge). From Newton's Law, the mean rate at which water momentum is shed to the wake (thrust) is equal to the total resistance force on the fish body. As with transient swimming, the thrust acting on the fish at any instant in time is equal to the sum of all contributions from propulsive segments. However, for one tail beat the mean force can be determined from the bulk changes in momentum associated with motions of the trailing edge. Lighthill (1975) and Wu (see Wu et al., 1975; Pedley, 1977) used hydromechanical theories to describe these changes and to predict mean thrust power output and propulsive efficiency (that is, Froude efficiency, or total power input divided by useful power output).

The magnitude of the mean thrust force can be calculated by considering a slice of water approaching the extreme caudal, or trailing, edge of the body (Fig. 7-4). The fish swims forward at a

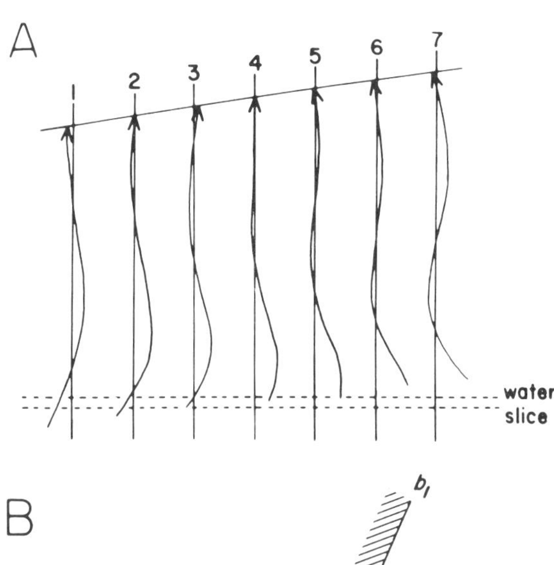

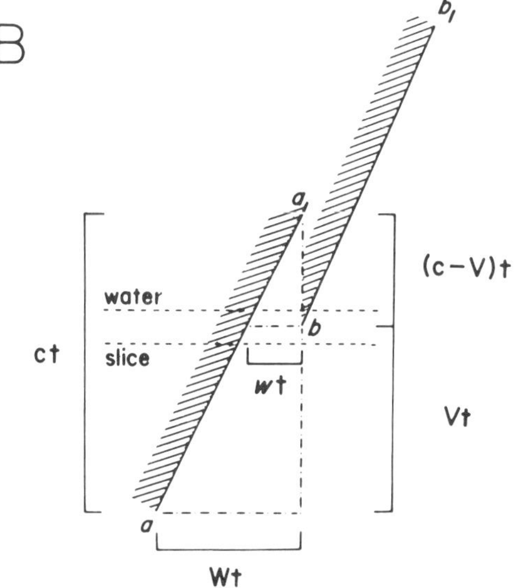

Figure 7-4 A simplified representation of body/caudal fin movements of a fish showing how thrust is generated. A: Tracings of the body centerline, made at 15-msec intervals, of a 30-cm trout swimming at 50 cm/sec; B: representation of a propulsive segment at the trailing edge at two successive instants in time. (The shading represents the body.) The segment moves from aa_1 to bb_1 in a small time t. It acts on a water slice, just leaving it behind after it. The trailing edge of the segment moves laterally at a velocity W and forward at a velocity V. The propulsive wave travels backward over the body at a velocity c. The slice of water is accelerated to a velocity w. (From Webb, 1978b).

mean velocity V while the propulsive wave travels backward at a velocity c. The trailing edge moves laterally with a mean speed W. Consider a propulsive segment just as it ejects the slice of water at the trailing edge. The length of the segment is defined for convenience in relation to the distance ct traveled by the propulsive wave in a short period of time t. Then, the water slice travels backward a distance Vt, but is displaced laterally a distance wt (where w = velocity of water accelerated by the propulsive segment), which can be calculated from the geometry of the system:

$$\frac{wt}{Wt} = \frac{(c - V)t}{ct} \quad \text{and} \quad w = W\left(1 - \frac{V}{c}\right). \tag{7-5}$$

The momentum gained by the water shed to the wake is mw, and it is shed to the wake at the swimming speed V. The trailing edge does work against the wake at a rate of W, so the mean total power output P is

$$P = mwWV. \tag{7-6}$$

However, loss of kinetic energy is associated with the acceleration of the water to w. If the propulsive segment is oriented at some angle θ to the direction of motion, then the rate of this energy loss P_k is

$$P_k = \frac{mw^2V}{2\cos\theta}. \tag{7-7}$$

Thrust power P_T is the difference:

$$P_T = P - P_k. \tag{7-8}$$

Froude efficiency η (defined above) is

$$\eta = \frac{P_T}{P} = \frac{V + c}{2c}. \tag{7-9}$$

Thrust at a given speed is maximum when w (the velocity given to the water) is small relative to W (the lateral velocity of the propulsive segment), providing, of course, that w is not too small. Efficiency is maximum when the swimming speed V is only a little smaller than the wave speed c.

Both thrust and efficiency vary with swimming speed. In periodic swimming, drag increases with the square of the swimming speed, and so thrust must similarly increase with speed. Increased thrust is achieved through increases in the frequency and/or amplitude of the tail beat, both of which increase both w and W. Increased frequency is the more important. In ray-finned fish amplitude usually varies only at low speeds of typically less than 1 to 2 body lengths per sec, but at higher speeds is more or less constant at about 20% of the body length. Accordingly, frequency increases linearly with speed. In some sharks amplitude may decline at higher speeds while frequency compensates by increasing with speed at a rate greater than that seen in teleosts.

The propulsive wavelength usually does not vary much with speed, so the wave velocity c increases with frequency of tail beat, and therefore also with increasing forward speed V. The difference between c and V becomes increasingly smaller as speed increases. As a result, Froude efficiency increases with speed and can reach values in excess of 0.9 at high speeds. There is only one adequately documented exception to this increase in efficiency with speed, and this was obtained for schooling sardine *(Sardinops)* rather than for solitary individuals (Hunter and Zweifel, 1971). It is not clear why Froude efficiency would decline with increasing swimming speed or why relationships should be different for individuals in a school compared with solitary fish. For fish that routinely swim using their appendages (such as the boxfish *Ostracion*) the Froude efficiency of the caudal fin is low, taking values of about 0.5. These fish reserve the caudal fin primarily for emergencies.

The magnitude of thrust is affected by a range of morphological features that influence both kinematics and the mass of water accelerated by body movements (Webb, 1978b). One feature of importance is body flexibility, which influences the length of the propulsive wave relative to body length. This is seen in specific modes distributed along a continuum ranging from long, thin, flexible eel-like animals with more than one wavelength within the length of the body (anguilliform mode) to short, stout, fusiform tunalike animals showing less than one wavelength (thunniform mode). The latter fish tend toward deeper, narrower caudal fins (tails having high aspect ratio), attached to a more or less rigid body by a narrow caudal peduncle (narrow necking). These trends increase thrust and Froude efficiency, and the thunniform swimmers (fish in the families Scombridae, Thunnidae, Lamnidae, and the order Cetacea) are the fastest swimmers of the periodic type. Swimming performance of the fossil reptilian ichthyosaurs, which had similar body forms, was probably equally impressive.

However, as the propulsive wavelength becomes longer relative to the body length, then (as for fast starts) a problem of lateral recoil arises. Exactly the same solution is used to minimize loss of energy owing to recoil: a large anterior mass, enhanced by large median fins, increases the inertia of the anterior part of the body. The narrow caudal peduncle also reduces the magnitude of the side force that would otherwise contribute to loss of energy owing to recoil.

The morphology of some sharks having large gaps between the dorsal fins (such as *Scyliorhynus* in Fig. 7-1) may influence thrust and performance while retaining body waves that are relatively short compared with the length of the fish. Motion of the water shed at the trailing edge of the more anterior dorsal fin may be out of phase with the motion of the caudal fin. The tail then pushes against the momentum of the water shed by the anterior fin, experiences a larger reaction force, and hence develops greater thrust. The different relationships among tail beat frequency, amplitude, and swimming speed that are seen in some sharks, as compared with teleosts, may ensure that the correct wave difference between the anterior median and caudal fins is maintained (Webb and Keyes, 1983).

In addition to these morphological features affecting thrust through kinematics, thrust is also enhanced when the added mass of a propulsive segment is large. This occurs when the depth of the trailing edge is large (Eq. 7-3). Species specialized for high-speed cruising have substantially deeper caudal fins or flukes than do more generalized swimmers.

Variation in body form clearly influences the magnitude of thrust, but performance (speed) depends on the balance of thrust and drag. Returning to Eq. (7-4), the only significant drag component during periodic swimming is the drag force owing to water viscosity. However, the magnitude of the drag force can be variable, and this is a topic of continuing debate. Minimum values for drag occur with well-designed, rigid, streamlined bodies, such as those of submarine or aircraft fusilages. Fish, however, are self-propelled, flexing bodies, and drag is substantially higher than for rigid bodies. Using the rigid body minimum as a reference, drag appears to increase by 5 to 10 times over the caudal portion of the flexing body where lateral amplitudes of the body are large. Over the entire body, mean drag is increased by 3 to 5 times by flexion. This increased drag of flexing bodies was first pointed out by Alexander (1967) using metabolic data in his classic study of functional design in fishes.

The magnitude of the drag is also affected by the nature of flow in the boundary layer (laminar versus turbulent) and by whether the boundary layer separates from the body (see Webb, 1978b). Most fish appear to have large areas of laminar flow that minimize frictional drag over the body. Exceptions are some elasmobranch and teleost fishes that have denticles or scales that induce turbulence. In these cases, the induced turbulence stabilizes the flow in such a way as to avoid separation, which might otherwise increase drag to values several times that of the frictional drag. Other fish, particularly in the more thunniform swimming modes, are believed to minimize separation with well-designed, streamlined shapes.

It follows from the nature of drag and the effect of swimming movements that drag is minimized by the same morphological adaptations that improve swimming thrust and efficiency in periodic swimming. For example, narrow necking of the caudal peduncle reduces body area just where drag is increased most by body movements. In addition, narrow necking permits the anterior parts of the body to be more rigid and to profit more from streamlining. Thus, the trend from anguilliform to thunniform swimming modes provides reduced drag as well as increased thrust and efficiency.

The various factors relating to thrust and drag can be combined to predict the optimum morphology for periodic swimming, defined as that maximizing speed. The optimal form is clearly shown by tunas, lamnid sharks, and cetaceans (Fig. 7-1). The morphological features are: tail with high aspect ratio, that is, large depth but small area; narrow caudal peduncle; small total caudal area; stabilization of the anterior of the body with an anterior median fin; and stiff, streamlined body. It is important to realize that these requirements are almost entirely opposite to those for transient swimming, and this conclusion has significant consequences for the radiation of various body forms among aquatic animals. These are discussed in the last section below.

The discussion so far has concentrated on undulations of the body. Among fish undulatory movements of fins are frequently used for slow routine swimming and precise maneuver (Breder, 1926; Gosline, 1971). The mechanics of these systems can be studied using a variety of analytical models, ranging from the hydromechanical theory regularly used for body movements to momentum jet models used for bird flight and derived from fixed-wing aircraft and helicopter theory. The latter models are discussed in Chapter 8.

In terms of the theory of hydromechanics used for body/caudal fin propulsion, there are some obvious differences in kinematics between undulatory fins and undulations of the body. This is mainly because fin movements occur about a relatively rigid fin-base axis. Amplitudes of fin movements do not increase continuously over the length of a fin, which may reduce efficiency as compared with that of body undulations. The amplitude of fin movements also varies along the span of the fin, being least at the fin base and increasing toward the distal edge. This substantially reduces the mean values of w and W and may generate mean forces normal to the axis of mean progression. The fins may interact with the body, particularly when the body has a large width relative to the depth of the fin (as in the electric eel, *Electrophorus*). Problems such as these are only beginning to be examined and may make momentum jet models useful in the study of many undulatory fin propulsors (Blake, 1983).

There are also similarities between undulation of appendages and of the body. There is a continuum from long-based to short-based appendages that is analogous to that from the anguilliform to thunniform modes, and is similarly associated with progressively longer propulsive wavelengths relative to length of fin base. These trends are apparently associated with enhanced thrust and/ or efficiency. In addition, other nonlocomotor factors may be important. For example, at one extreme the knife eel *(Gymnarchus)* swims by passing waves of long length, low frequency, and high amplitude along a dorsal fin that extends over most of the body length. At the other extreme, the sea horse (*Hippocampus)* has a short dorsal fin, along which it passes propulsive waves of short length, high frequency, and low amplitude. The Froude efficiency of *Gymnarchus* may be as great as 0.8, compared with 0.4 for *Hippocampus.* The lower efficiency is seen because more kinetic energy is lost in the acceleration of small masses of water to high velocities than in accelerating larger water masses to lower wake velocities. However, the high frequencies are probably above the frequency of flicker fusion of potential predators. Sea horses, and the related pipefishes, are small, diurnally active, and would be more susceptible to attack by visual predators; knife eels are often larger, nocturnal, and frequently inhabit turbid water (Blake, 1980).

Undulatory appendages have some advantages over undulatory bodies. The amplitude and wavelength of fin undulations can be varied locally, providing for very fine control over thrust. In many armored fishes each median and paired fin has such a narrow base that thrust can be oriented in any direction relative to the body. This provides for even finer control over maneuver. For example, sea horses and trunkfishes can turn about their own body axis without moving laterally or forward. This is important for probing structurally complex habitats.

All fish that are propelled by undulatory fins can reverse the direction of the wave propagated along their fins and thus can swim backward almost as well as they can swim forward. Fish propelled by long ribbon fins (for instance, the long dorsal fins of the Gymnarchidae or long anal fins of Gymnotidae) are particularly proficient at swimming backward.

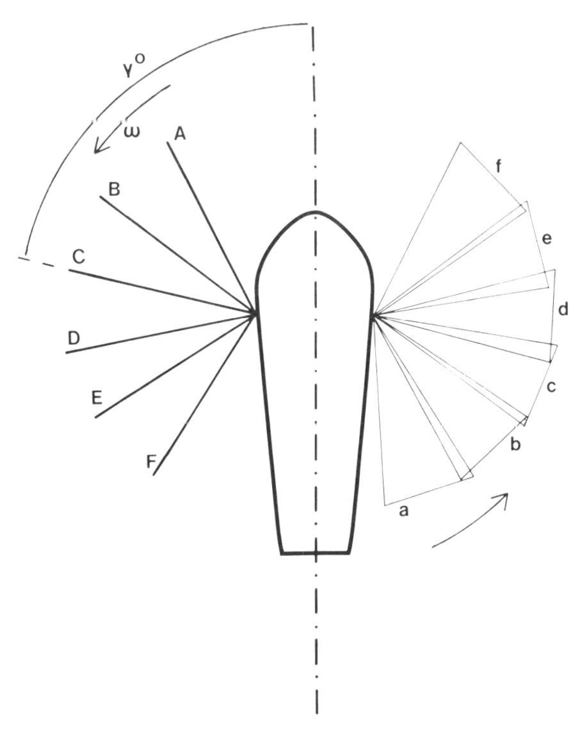

Figure 7-5 Diagram of a rowing animal seen from above. The positions of the fin are shown at six instances (*A* to *F*) during the power-stroke phase of the beat cycle. The fin moves at angular velocity ω through some angle measured about the body axis. The fin is also shown during the recovery stroke (*a* to *f*), feathered so that the fin area is parallel to the direction of motion of the fin.

Oscillatory Propulsion

Oscillatory propulsors can be divided into two basic types: those that are drag-based, operating like an oar or paddle, and those that are lift-based, operating like a helicopter rotor or a wing. A paddlelike appendage is pulled through water at an angle as close as possible to right angles to its direction of motion. The flow separates from the sharp edges and a large pressure difference occurs across the paddle. The pressure force (actually a drag force for the paddle) is oriented in the opposite direction to that in which the body moves, so that it is the thrust force for the whole organism. Flat plates oriented broadside to the flow are good paddles and are known hydrodynamically as bluff objects.

A paddle-propulsor always has a power stroke and recovery stroke (Fig. 7-5). During the power stroke phase of the paddle "beat cycle," the major force on the paddle is the pressure force, and frictional drag can be ignored. During the recovery stroke phase, drag must be minimal for good efficiency. This is achieved by collapsing or feathering the paddle so that it has a minimum area broadside to the flow. Pressure drag is minimal and frictional drag is the only major force on the paddle. The frictional drag during recovery is very small compared with the propulsive pressure force during the power stroke.

Lift-based propulsors operate on the principle of the lifting aerofoil or wing. The physical principles are outlined in Chapter 8. Briefly, the lift force acts on a surface (such as a fin) moving at a small angle of incidence to its axis of motion and acts in a direction that is normal to the direction of motion. The lift force is, therefore, also normal to the drag force, which requires a quite different orientation of an appendage acting as a wing as compared with one acting as a drag-based propulsor. The magnitude of the lift force depends on a lift coefficient (which varies according to the shape of the hydrofoil and angle of the incident flow), speed, projected area of the hydrofoil, and the density of the water.

Drag-Based Mechanisms Drag-based mechanisms are common among aquatic animals and are usually associated with slow swimming and precise maneuver. Among primary swimmers, especially fish, these complement or supplement high-powered caudal swimming. Some pinnipeds (such as seals), Sirenians (such as manatees), and the platypus *(Ornithorhynchus)* also may supplement paddling by their limbs with undulatory motions of the body. Adaptations for rowing are well developed among anurans with large webbed feet (such as the pipids *Pipa, Xenopus,* and *Hymenochirus* and ranids *Rana, Phytobatis,* and *Banina*). Some anurans are less well adapted than others; for example, the leptodactylids *Crinia, Heleioporus,* and *Myobuliabchus* have only slight webbing between the toes.

Many reptiles, both living and extinct, have paddlelike limbs. The latter include the euryapsid placodonts, plesiosaurs, and nothosaurs, the anapsid mesosaurs, and the lacertid mosasaurs. Many turtles (for example, among Cheloniidae, Chelyidae, Dermatemydidae) probably swim in the drag-based mode.

Paddling locomotion is common among living aquatic birds (as in cormorants, loons, grebes, and ducks), and also in certain extinct forms including flightless, diving *Hesperornis.*

The principles of drag-based swimming are the same for all paddling animals (Blake, 1981). To understand how the animal is propelled, consider a plate at right angles to the flow, moving backward at a constant speed. The drag D on the plate is given by the classical Newtonian equation:

$$D = \frac{\rho S(u - V)^2 C_D}{2}, \tag{7-10}$$

where ρ = density of water, S = surface area, u = speed of the paddle, and C_D = drag coefficient.

D is the thrust force as the paddle moves backward, pushing the animal forward during the power phase of the beat cycle. During the recovery phase, no thrust is developed and the drag of the feathered paddle adds to that of the body, resisting forward motion. If the drag of the body during the power phase is D_p and that of the body and feathered paddle during the recovery phase is D_r, then when the fish is swimming steadily, the total thrust in each beat from Eq. (7-10) must be

$$D = D_p + D_r. \tag{7-11}$$

Equation (7-10) applies only to paddles moving parallel to the body, or at best to paddles that have a large oar blade at the end of a very long oar pole. In addition, it assumes that the paddle has a constant velocity. Unfortunately, animals have very short paddles that move cyclically. In order to evaluate paddle morphology from a mechanical perspective a more complex model must be considered. Blade element theory has frequently been used and has been particularly successful for swimming animals.

Blade element theory simply means that the blade of a propulsor is divided along its length into a series of sections or elements, stretching crosswise from the axis of rotation to the tip of the paddle. These are analogous to the propulsive segments already encountered in undulatory propulsion. The properties of the paddle can then be determined by summing the contributions of each of the elements over the whole sur-

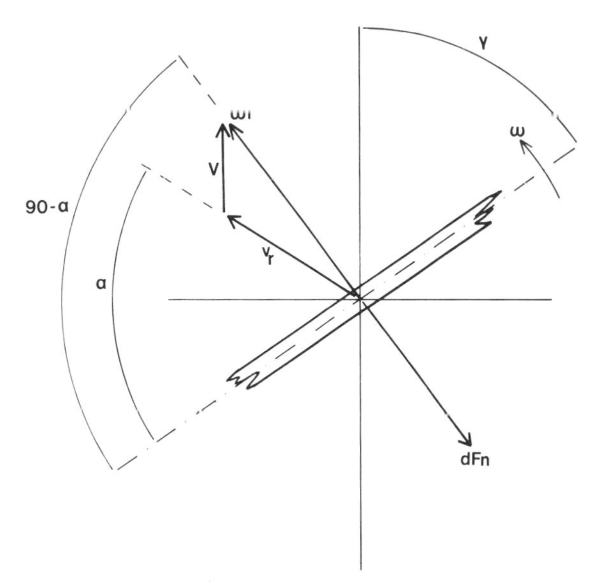

Figure 7-6 Representation of a single blade element of a fin, such as that seen from above in Figure 7-5. The element is rotating at angular velocity ωr as the fish moves forward at velocity V, so the resultant velocity is V_r (see Eq. 7-10). The element strikes the water at a hydrodynamic angle of attack α; ω = angular velocity; γ = angular excursion; dFn = normal or reaction force acting on the blade element.

face of the propulsor and throughout the power stroke.

Consider the blade element on a paddle that is at a distance r from the base (Fig. 7-6). The resultant velocity v_r of the segment depends on the speed of the animal V, on which is superimposed the angular velocity ω of the blade element. Because of the periodic nature of the motion, ω varies throughout the beat cycle (Fig. 7-7) but is related to the position of the fin as it oscillates between two points (A and F in Fig. 7-5). Therefore ω can be related to γ, the angle between the body and the position of blade element at any instant, and v_r can be calculated as:

$$v_r = \sqrt{(v_n^2 + v_s^2)} = \sqrt{[(\omega r - V \sin \gamma)^2 + (V \cos \gamma)^2]}, \tag{7-12}$$

where v_n = velocity normal to the surface of the blade element, including the component velocities owing to rotation of the paddle and to forward motion of the body, and v_s = spanwise velocity of an element.

The normal (reaction) force acting on a blade element, dF_n, is now given by the drag equation (7-10):

$$dF_n = \frac{\rho v_n^2}{2} C(dr)C_n. \tag{7-13}$$

In this equation $\frac{1}{2}\rho v_n^2$ = the dynamic pressure owing to the motion of the water, C = the chord or width of an element, and dr = the spanwise length of an element—and therefore $c(dr)$ = the area of an element, comparable to S in Eq. (7-10), and C_n = coefficient of normal force, similar to the drag coefficient in Eq. (7-10).

C_n depends on the Reynolds number (a nondimensional parameter that relates the magnitude of viscous and nonviscous forces in the water, as discussed further in Chapter 8), the angle of attack between the paddle element and the incident water, and the shape of the paddle. However, for most paddles these effects are not large, and C_n typically takes values of about 1.1.

In order to calculate the thrust generated by a paddle, one sums the forces on all elements throughout the power phase (lasting from zero time to t_p) and also sums all forces along the paddle length (varying from zero at the base to R at the tip). The mean thrust, $\overline{T}$, is found by dividing by the duration of the power stroke t_p. In formal terms:

$$\overline{T} = \frac{1}{t_p} \int_0^{t_p} \int_0^R \frac{\rho C_n v_n^2}{2} c(dr)(dt). \tag{7-14}$$

Each blade element generates a torque, so the total work done by the paddle is the sum of the torques for all elements throughout the power phase of the beat cycle. If the torque for each blade element is dE, then

$$dE = dF_n r. \tag{7-15}$$

The average work done by the whole paddle $\overline{E}$ is

$$\overline{E} = \frac{1}{t_p} \int_0^{t_p} \int_0^R \frac{\rho C_n v_n^2}{2} c(dr)(dr)(dt). \tag{7-16}$$

The work done by oscillating appendages is used to overcome the drag force D, which acts on

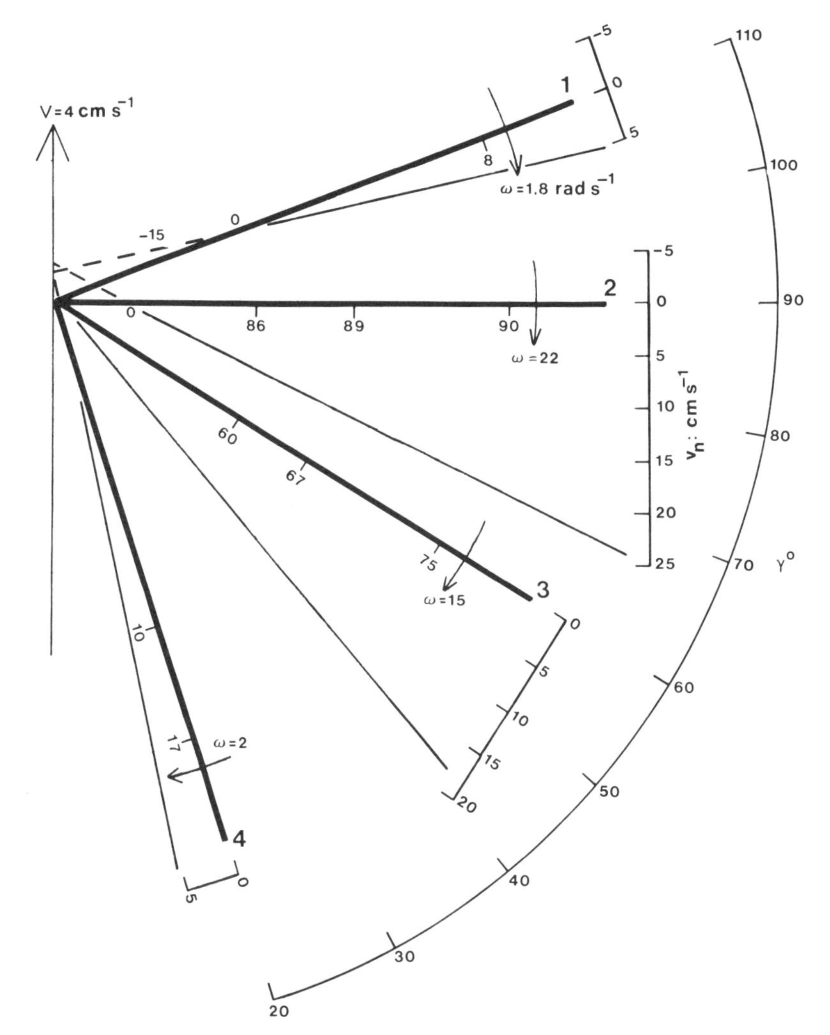

Figure 7-7 Diagram of a paddle, seen from above, at four instants (1 through 4) during the power phase of the beat cycle, illustrating how the kinematic parameters vary. The paddle is represented by solid lines at angles γ of 110°, 90°, 60°, and 20° for positions 1 through 4, respectively. The thin line shows the normal velocity v_n at each point along the paddle (scale shown at the distal end of the paddle oriented normal to the paddle). The angular velocity ω varies from small values at the beginning and end of the power phase to larger values through the middle of the beat.

the animal during the whole beat-cycle. If this cycle has a duration t_o, the work required to overcome this drag force is

$$E_D = \frac{\rho V^3}{2} SC_D t_o. \quad (7\text{-}17)$$

As with undulatory propulsion, the Froude efficiency η is the ratio of useful work done in overcoming drag E_D, compared with the total work performed by all paddles. If there are N paddles, then

$$\eta = \frac{E_D}{N\bar{E}}. \quad (7\text{-}18)$$

Thus, a variety of kinematic and morphological factors affect thrust and efficiency. Obviously, greater beat frequencies and amplitudes will increase ω (and hence v_n) and also the work done (Eq. 7-16). In addition, although the shape of a paddle has little effect on the drag of a blade element, it does effect the thrust force of the whole paddle. The shape, or geometry, is characterized by a "shape factor" that is calculated using the hydromechanical model. It is largest for paddles that maximize area away from the base of the appendage and minimize the area near the base. There are two reasons for this. First, the velocity of a blade element increases spanwise to reach maximal values at the distal edge of the paddle. Second, flow reversal and negative angles of attack occur at the base of the fin at the beginning and end of the power stroke (Fig. 7-7). These increase the total drag of the animal. Therefore, paddles of triangular shape that are narrow at the base produce the greatest propulsive force. Appendages of this shape are common in anurans, aquatic birds, and many mammals and fish. Among the latter, however, broad fin bases are also frequently found, making the fin more square than triangular, as for the pectoral fins of the angelfish *(Pterophyllum)*. This may allow individual fin rays greater freedom of motion to pass undulatory waves along the fin, thus enhancing the diversity of locomotor opportunities. For example, such fish can hover in addition to paddling, and when they are near the bottom, the wake from the fin interacts with the "ground" to improve swimming efficiency (Blake, 1979).

Another feature that affects thrust and efficiency is not developed by the simplified model above: the angle between any element and the flow varies, and it reaches 90° only during part of the stroke (usually the middle). When this angle is less than 90°, the contribution of the normal force to thrust is reduced. It follows that the greatest contribution to thrust occurs at the middle of the power stroke when angular velocity should be greatest (see Eqs. 7-14, 7-16), as is commonly observed (Fig. 7-7).

Animals with jointed limbs can orient their paddles normal to the body for longer periods of the power cycle than can animals lacking such joints, such as fish. This is achieved by controlling the angles between the limb elements. For example, the muskrat *(Odonatra)* varies the angles between femur and tibia and between tibia and foot to maintain hydrodynamic angles of attack of the feet that are close to 90° for much of the power stroke.

The efficiency of a drag-based propulsor depends on the ratio of drag to the mechanical work performed by the paddles. The drag of the body is minimized by streamlining, a common adaptation in most aquatic species. The drag of the fin during the recovery phase of the beat cycle is minimized by feathering the paddle and/or reducing its area. Values for the Froude efficiency of paddling animals are about 0.2 to 0.3, very much lower than typical values for body/caudal fin undulatory propulsion. However, the efficiency of these latter mechanisms decreases rapidly at lower speeds, and in the speed range where paddles are used, the efficiency of body undulation is probably lower than for paddling propulsion. This probably explains the near ubiquity of paddle mechanisms among animals that also have other well-developed propulsor mechanisms.

Lift-Based Mechanisms Oscillatory lift-based mechanisms of propulsion are not as common among aquatic vertebrates as are undulatory wave or paddling propulsors. However, they do occur in some teleost families (such as Serranidae, Chaetodontidae, Embiotocidae) and probably in many skates and rays. The fin motions of the surfperch (*Cymatogaster aggregata,* Embiotocidae) have been likened to the movements of a bird's wing. Studies on *Cymatogaster* indicate that lift-based mechanisms of oscillatory propulsion can have a mechanical efficiency of the same order as those using an undulatory wave. Among the birds, auklets, murres, and puffins (Alcidae), diving petrels (Pelecanoididae), and penguins (Spheniscidae) are propelled under the water by the action of their wings. Studies on the kinematics of swimming in the little blue penguin *(Eudyptula minor)* indicate that lift and thrust forces are produced on both the down stroke and the up stroke of the wing.

Some detailed hydromechanical models for oscillating lifting surfaces have been developed, but none has as yet been applied to swimming animals, partly because of their complexity (Blake, 1983). It might be simpler to use the same approach to lift-based propulsors as is used for drag-based mechanisms. Thus, a blade element theory could be applied, providing a detailed resolution of the function of an appendage. Such an analysis would parallel the model discussed above in first determining the incident velocity of a blade element. This velocity, together with the appropriate lift coefficients, could be used with equations similar to (7-14) to determine lift (thrust) forces. A more detailed account of lift mechanisms, which may also be applicable to aquatic animals, is given in Chapter 8.

Interactions among Swimming Mechanisms

The diversity of swimming mechanisms is large, as are the morphological requirements for different propulsors and activities (Gosline, 1971). All aquatic animals have variable requirements for speed and acceleration that make use of different morphologies, and many aquatic animals also walk or fly. Locomotion in more than one habitat is most prevalent in secondary swimmers. Different body forms affect locomotor performance in the various swimming modes and at various activity levels, and they also relate to the habitats occupied. The problem of swimming in many different ways and in different microhabitats is greatest for primary swimmers, which consequently show the greatest morphological and behavioral diversity. This discussion is, therefore, limited to fish.

First, there are differences in design principles for periodic swimming (cruising, sprinting) versus transient swimming (acceleration and turns) using the body and caudal fin. Acceleration is maximized by a large body area and a large muscle mass. Specialized accelerators include cottids, esocids, many deep-bodied species, and flat fish (see Fig. 7-1). Alternatively, specialized cruisers and sprint swimmers must maximize the depth of the tail and minimize the area anterior to the trailing edge of the tail. Typical examples of swimmers in the thunniform mode are shown in that column of Figure 7-1. In addition, sharks (Fig. 7-1, *A*1) are probably specialized for cruising in the anguilliform mode, and like other cruisers have reduced the surface area between their fins.

Some compromise is possible. Most commonly, bony fish have fins that are collapsible and erectile. This permits changes in the relationship between body shape and fin shape, the fins being erected to maximize area for acceleration and all but the caudal and anterior median fin being collapsed for cruising. Other fish, such as the dolphin fish *(Coryphaena),* may specialize separate fins for specific activities. In this case the caudal fin is used for cruising, with the dorsal and ventral fins collapsed. These latter fins are erected during accelerations and turns. In evolutionary terms, however, good transient performance appears to be a dominant trend, probably because of its importance in maneuver for evading predators (Webb, 1982b).

Second, some compromises appear necessary for fish specializing in undulatory and/or oscillatory fin propulsion. The key function of these modes appears to be efficient slow swimming and precise maneuver. This adds to the propulsive repertoire of fish, but when very specialized, it appears to impair high-speed, steady performance. Fish that routinely swim using noncaudal modes invariably retain acceleration abilities, presumably to avoid predation or to catch elusive prey. This is apparent from the body forms of *Chaetodon, Cymatogaster,* and *Ostracion* (Fig. 7-1, *D*1, *L*1, and *O*2).

Most bony fish have radiated about relatively short-bodied forms, with large caudal regions and anterior regions of variable area owing to collapsible fins. These body forms are known to be well designed for adequate cruising and acceleration, but the latter is the more important. Bony fish have also evolved numerous and variable specialist forms ranging from eels to tuna and the beautiful, maneuverable denizens of coral reefs (see Webb, 1982b; Webb and Weihs, 1983).

In contrast, elasmobranchs have radiated around the rather limited body forms of sharks and rays, and their history is predominantly that of benthic living. Their lack of diversity does not mean, however, that elasmobranchs are not sophisticated. The sharks in particular are highly specialized for cruising. In contrast to the well-known teleost cruisers, such as tuna, most have used another principle to enhance performance. This is flow interaction between discrete median fins, which increases thrust and efficiency. Such interactions are possible only with an elongate, flexible body and many vertebrae, as are characteristic of elasmobranchs.

Understanding of how the locomotor behavior and morphology of propulsors are related, however, remains in its infancy. Indeed, knowledge of most mechanisms is incomplete, and in all areas good comparative studies are lacking. This is particularly true for slow swimming in oscillatory modes, even though these mechanisms are the most common. There is no doubt that excellent research on functional morphology will continue to evaluate the performance of locomotor adaptations. But the most exciting area is that which directly couples functional morphology with testable hypotheses on behavior, such as foraging tactics that maximize net energy gain or maneuvers to catch elusive prey. Similarly, the manipulation of alternating periods of acceleration and deceleration, use of tides, schooling, and use of the bottom can influence both performance and the costs of moving around. Functional morphologists will probably come more and more to experimentation with animals indulging in normal behavior, including locomotion.

Chapter 8

Flying, Gliding, and Soaring

Ulla M. Norberg

Although active flight requires an extremely high output of power (work per unit time), it is very efficient in terms of the transport of a unit mass over a unit distance. For a given body size flying is a far cheaper way to move than is running, although more expensive than swimming (Schmidt-Nielsen, 1972). Flight also enables an animal to cross water, deserts, and other inhospitable environments, and to do so at heights where the temperature is suitable for cooling purposes.

Most birds, and all bats, can fly actively (that is, maintain flight in still air). Flexibility derived from the ability to fly explains why the various groups of flyers have undergone such dramatic adaptive radiations: insects are the most diverse and numerous class of animals (about 750,000 species); flying reptiles had enormous diversity of form and included the largest flying animal known; birds have more than 8,000 species; and bats are the second largest order of mammals (about 800 species).

The type of habitat a flying or gliding animal lives in and the way it exploits that habitat are closely related to its body size, wing form, flight style, and flight energetics. It is assumed that natural selection has provided near-optimal combinations among these variables. Some birds, such as albatrosses and vultures, can soar for hours with a minimum cost in energy, whereas hummingbirds must hover when foraging, hovering being the most energy-demanding type of locomotion. Very different wing morphology is required for such different flight activities. Many size-dependent factors also affect wing morphology, and birds span a size range from about 2 to 12,000 g (exclusive of nonflying species), and bats from about 3 to 1400 g.

The ancestors of animals with active flight were probably passive flyers, as are those living species of amphibians, reptiles, and mammals that have acquired the ability to glide or parachute. Birds probably evolved from small, warm-blooded coelurosaurian dinosaurs. The first known bird, *Archaeopteryx lithographica* (about 150 million years old), was long thought to be unable to fly actively, but to be a good glider. Several morphological features indicate, however, that *Archaeopteryx* was able to fly actively. It had, for instance, a well-developed furcula, which provided part of the origin for the wing-depressing pectoral muscle and may have served as an energy-saving device during wing flapping (Schaefer, 1975). It also had a large cerebellum, the center of neuromuscular control, and highly asymmetrical primary feathers which (as explained below) are associated with flight (Feduccia and Tordoff, 1979). The bird may have made short flights among trees, using the long tail for control and stability.

The oldest known fossil bat, *Icaronycteris index,* lived about 50 million years ago and was rather similar to modern bats. It may have been directly ancestral to all or some living bats (Jepsen, 1970).

The pterosaurs (subclass Archosauria) of the Jurassic and Cretaceous periods survived from about 180 until 65 million years ago. Some species were small and possibly tree-living, whereas others were enormous—the largest estimated span was 11–12 m for *Quetzalcoatlus northropi* (Langston, 1981). Pterosaurs were adapted for active flight, but the largest ones were probably mostly gliders and soarers, like the giant pleistocene condor *Teratornis* whose estimated mass was about 20 kg.

The aerodynamic principles governing animal flight are summarized in papers by Pennycuick (1972a, 1975) and Lighthill (1975, 1977). Earlier papers were concerned with flight kinematics (for instance, Brown's classic work on flight of the pigeon), whereas many recent ones deal with the energy cost of flight (for instance, Pennycuick, 1969, 1975; Tucker, 1973; U. M. Norberg, 1976a; Oehme, Dathe and Kitzler, 1977; Ellington, 1978; Rayner, 1979a,b,c). Ellington (1978) and Rayner (1979a,b,c) introduced a new theory of animal flight based on vorticity. The structure, form, and function of flying animals has been thoroughly treated by Hertel (1966). A new trend in research on animal flight is to relate flight energetics to both the morphology and the ecology of the animal (for example, Feinsinger et al., 1979; U. M. Norberg, 1979, 1981b; Rayner, 1981).

Basic Aerodynamics

Aerodynamics is not an exact science, as many believe. It is full of approximations, simplifications, and empirical coefficients. Flapping flight can, nevertheless, be understood by the application of theories of airplane wings (blade element theory) and of airscrews or propellers (momentum theory). The theory of animal flight has recently progressed to the vortex theory, which considers airflow around a wing.

Forces and Velocities The flow of air induced by an airfoil is such that negative pressure (and increased air speed) is generated on the upper surface of the airfoil and positive pressure (and decreased air speed) on the lower side (Fig. 8-1a). This pressure difference provides an aerodynamic force (R in Fig. 8-1b), which is conventionally resolved into a lift component L normal to the direction of the airflow and a drag component D parallel to it. A good airfoil maximizes the pressure difference and minimizes the drag. Drag is a retarding force, and the animal must either do mechanical work with its muscles to overcome it or must descend through the air at such an angle that a component of its weight balances the drag. (All symbols used in text are redefined in Table 8-1.)

The drag experienced by the animal has various components: frictional drag of the wings (conventionally termed profile drag) and body (termed parasite drag), and induced drag, which is caused by a flow of air around the wing tips (the trailing wing tip vortices) to equalize pressure between the upper and lower wing surfaces (Fig.

Table 8-1 Symbols used in text.

Symbol	Definition
b	Wingspan
C	Cost of transport
C_D	Drag coefficient
C_L	Lift coefficient
D	Drag force
e	Energy per unit mass
f	Flapping frequency
f_m	Mass flow
g	Gravitational acceleration
ind	Induced (subscript)
$inert$	Inertial (subscript)
L	Lift force
M	Body mass
m	Muscle mass
mp	Minimum power (subscript)
mr	Maximum range (subscript)
P	Output of mechanical power
P_a	Muscle power available
P_b	Basal metabolism
P_r	Muscle power required to fly
par	Parasite (subscript)
pro	Profile (subscript)
Q	Work per muscle contraction
R	Aerodynamic force; radius
r	Turn radius
S	Area of wing
S_d	Wing disk area
T	Thrust
V	Forward speed
V_r	Incident flow
V_w	Flapping speed
Y	Distance flown
α	Angle of attack
η	Mechanical efficiency
θ	Gliding angle
ρ	Air density

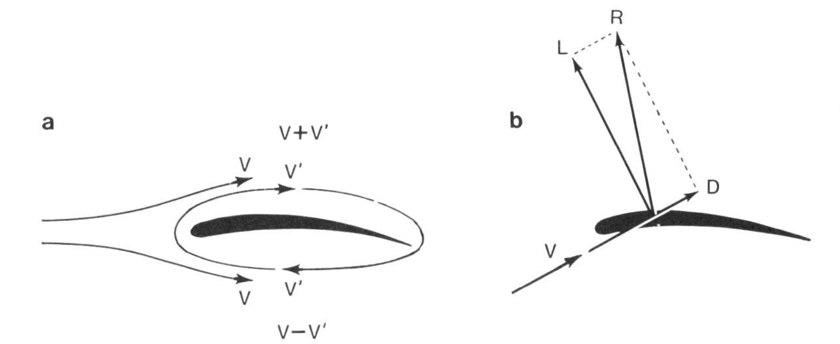

Figure 8-1 a: Schematic view of the airflow around a wing (seen in cross section) during flight; V is speed of the airstream meeting the wing and V' speed of circulation. b: The lift L and drag D force components of the resultant air force R experienced by the airfoil. c: Trailing wing tip vortices responsible for the induced drag.

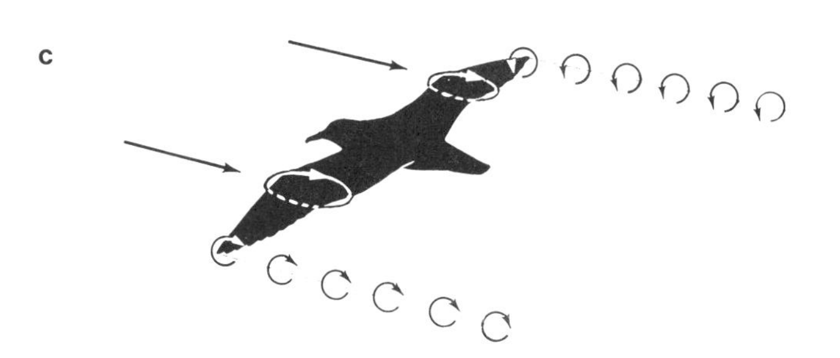

8-1c). A way to minimize the induced drag is to spread the wake, that is, to space the wing tips far apart by having a large wingspan.

In flapping flight the aerodynamic forces are generated by actively moving wings. The basic aerodynamic requirements for sustained horizontal forward flight are that the wings generate enough lift so that the mean vertical force balances body weight (Mg, where M is body mass and g acceleration owing to gravity), and the mean thrust T' balances the overall drag. The mechanical power required to fly is the product of mean thrust and speed.

The airflow V_r meeting the wings in forward flight is the vector sum of the forward speed V of the animal and the flapping speed V_w of the wing. Thus, the direction and magnitude of the resultant airflow, and hence of the local resultant air force, are different at each section of a flapping wing and during different phases of the wing stroke. The total force experienced by the flying animal is the vector sum of the forces on the different wing sections.

The above quantification of the action of an airfoil has been applied to natural flight by several authors using blade element theory, which assumes that the wings operate under steady-state airfoil conditions; however, it may be a good approximation also under quasi-steady conditions, as in powered flight with low wing beat frequencies. Lift and drag of a wing are then generated according to the formulas

$$L = \tfrac{1}{2}\rho V_r^2 S C_L \tag{8-1}$$

and $$D = \tfrac{1}{2}\rho V_r^2 S C_D, \tag{8-2}$$

where ρ is air density, S is area of the wing, and C_L and C_D are coefficients of lift and drag, respectively. During flight with no vertical acceleration, the average vertical component (L') of $R = Mg = L' \propto L$. Then, from Eq. (8-1), $Mg \propto V_r^2 S$, and $V_r \propto (Mg/S)^{1/2}$, where Mg/S is the wing loading. Thus, low wing loading enables the animal to fly at low speeds and still produce enough lift. The relationship applies also to other speeds. Figure 8-2 is a simplified diagram of horizontal flapping flight and the power components at different flight modes.

The coefficients of lift and drag are dimensionless numbers that indicate the capacity of an airfoil to generate lift and drag at a given angle of attack α (that is, the angle between wing chord

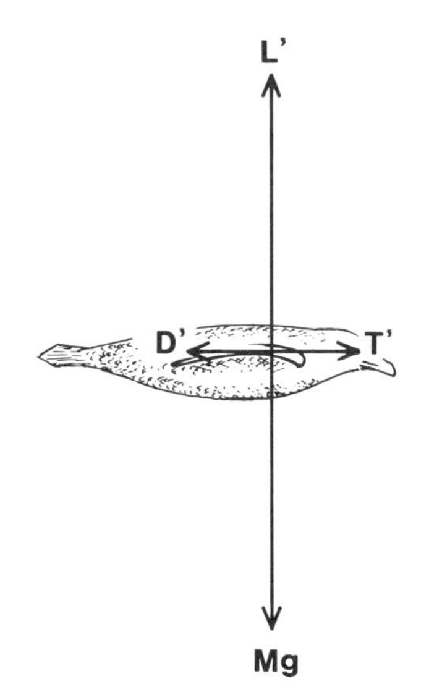

Horizontal flight at steady speed V:

$T' = D', \quad Mg = L', \quad P = T'V = D'V$

Climbing at angle θ:

$T' = D' + Mg \sin\theta, \quad P = V(D' + Mg \sin\theta)$
$L' = Mg \cos\theta$

Descending at angle θ:

$T' = D' - Mg \sin\theta, \quad P = V(D' - Mg \sin\theta)$
$L' = Mg \cos\theta$

$L = 1/2 \rho V^2 S C_L, \quad D = 1/2 \rho V^2 S C_L$

Forward flight: $D' = D'_{ind} + D'_{pro} + D_{par}$

$$D'_{ind} = \frac{k\, Mg^2}{2\rho V^2 S_d} = \frac{k\, Mg^2}{1/2 \rho \pi b^2 V^2}, \quad D'_{pro} = 1/2 \rho V^2 S C_{Dpro}, \quad D_{par} = 1/2 \rho V^2 S_b C_{Dpar} = 1/2 \rho V^2 A$$

$$P = D'V = \frac{2k\, Mg^2}{\rho \pi b^2 V} + 1/2 \rho V^3 S C_{Dpro} + 1/2 \rho V^3 A$$

Hovering flight: $D' = D'_{ind} + D'_{pro}$

$$D'_{ind} = \left(\frac{Mg}{2\rho S_d}\right)^{1/2}, \quad D'_{pro} = 1/2 \rho V^2 S C_{Dpro}, \quad P = D'V = \frac{2k\, Mg^{3/2}}{(2\rho \pi b^2)^{1/2}} + 1/2 \rho V^3 S C_{Dpro}$$

Figure 8-2 Simplified equilibrium diagram for horizontal flapping flight and formulas for lift, drag, and power according to classical aeronautical theory. The index ′ denotes the average, or effective, forces. L' is the average of the fluctuating lift force over an integral number of wing strokes. The effective drag force D' is a hypothetical average force the animal has to work against in powered flight. It includes three different drag components, of which the induced and profile drag components, D_{ind} and D_{pro}, have different direction and are dependent on a different airspeed than the parasite drag, D_{par}. The speed V for calculating wing profile drag in hovering depends only on induced speed and the wing's flapping speed, whereas in forward flight the forward speed also adds to V. D' can be estimated as (total mechanical power for flight)/(air speed), $D' = P/V$. The induced drag factor k would be equal to 1 in the ideal case of an elliptical spanwise lift distribution. In airplane wings it is 1.1 to 1.2. S_d is the wing disk area and equals $\pi b^2/4$ (area of a circle with span b as diameter). S_b is the projected frontal area of the body and A is the area of a flat plate with $C_{D_{par}} = 1$, which gives the same drag as the body.

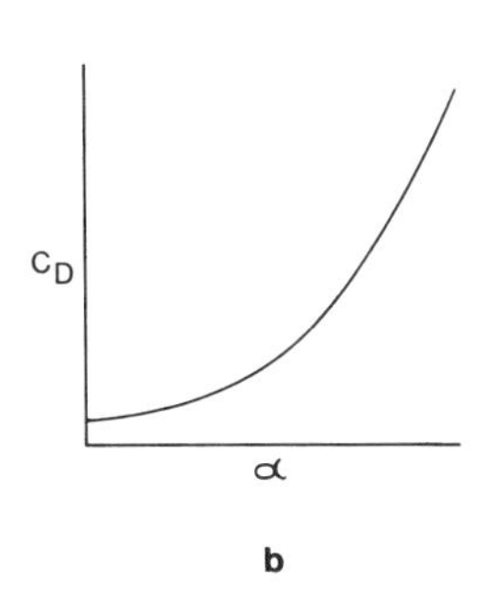

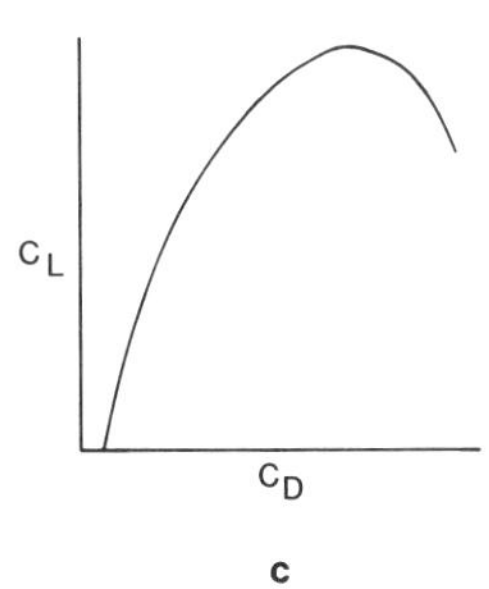

Figure 8-3 a: Lift coefficient C_L versus angle of attack α; b: drag coefficient C_D versus α; and c: C_L versus C_D.

and resultant airstream). They are dependent on the shape of the wing and the angle of attack. An efficient airfoil is designed so that C_L is large and C_D small, thus permitting weight support with minimum work. Figure 8-3 shows the relationships among C_L, C_D, and α.

As α is increased, C_L increases, up to a certain value, but at the cost of a higher C_D. When a critical α (stalling angle) is exceeded, the airstream separates from the wing's upper surface with a sudden fall of C_L and an increase of C_D.

In passive flight (gliding and parachuting), $T' = 0$, and the lift-to-drag ratio (L/D) establishes the gliding angle θ ($\tan \theta = D/L$; Fig. 8-4). The transition between gliding and parachuting is at the 45° inclination of the glide path, where lift and drag contribute equally to weight support; at shallower paths $L/D > 1$ and the animal glides, whereas at steeper paths $L/D < 1$ and the animal parachutes. The speed along the path increases as the angle steepens until the (now vertical) force R becomes equal to the weight Mg. In steady gliding at small angles and in level powered flight with no vertical acceleration, the time-averaged lift force equals the weight, so that the stalling speed is approximately

$$V_{min} = \left(\frac{Mg}{\frac{1}{2}\rho S C_{Lmax}} \right)^{1/2}, \tag{8-3}$$

where C_{Lmax} is the maximum obtainable lift coefficient. In birds it is usually between 1 and 2.

During the downstroke in forward flight the net aerodynamic force is directed upward and forward, thus supporting the weight as well as providing thrust (Fig. 8-5b). This can be seen in birds, for which the primary feathers are bent in the direction of the resultant force and curve upward and forward. During the upstroke the wings are partly flexed to avoid large retarding effects. On the upward stroke they may give negative lift, no lift at all, or small positive lift (Fig. 8-5a), depending on whether the resultant airstream

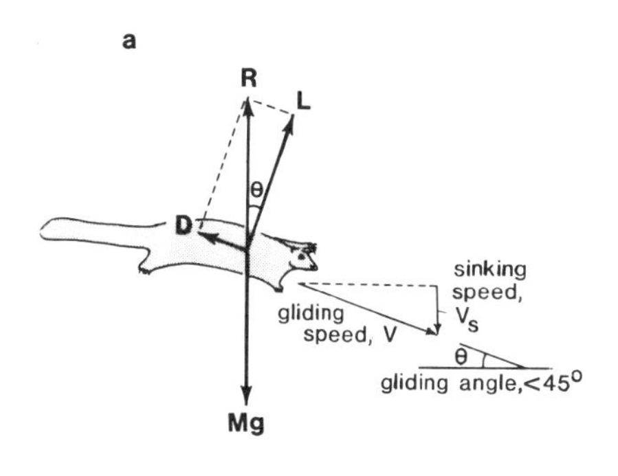

GLIDING:

In steady glide,

L=Mg cosθ, D=Mg sinθ

$V_s \propto V(D/L)$, $V \propto (Mg/S)$

L/D>1

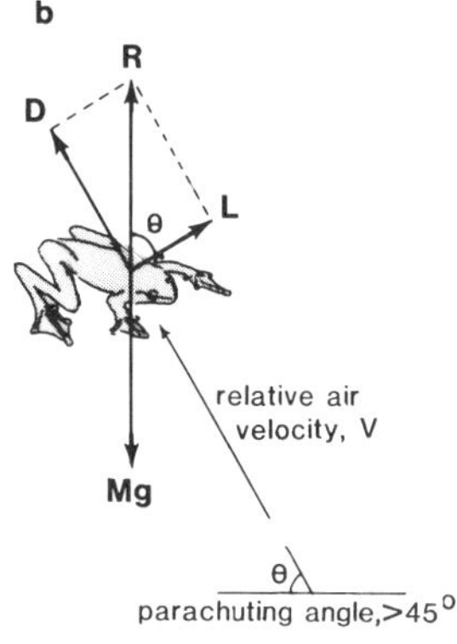

PARACHUTING:

L/D<1

Figure 8-4 Aerodynamics of (a) gliding flight and (b) parachuting. In parachuting the force and speed relationships are as for gliding, but because the area supporting weight is so small, lift becomes small, and the overall L/D ratio is lower, resulting in a correspondingly steeper glide path.

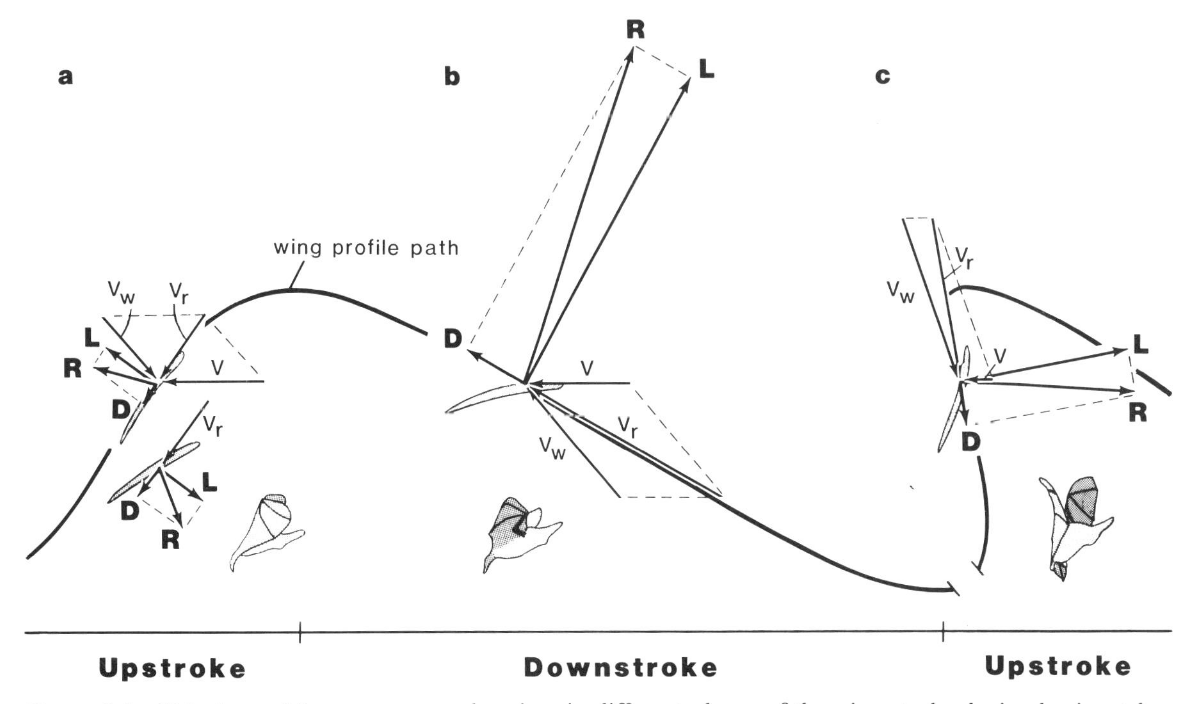

Figure 8-5 Velocity and force systems at the wings in different phases of the wing stroke during horizontal flapping flight and hovering in a long-eared bat. Upstroke and downstroke in forward flight at speed V are shown in a and b, respectively, and upstroke in very slow flight and hovering (low V) is shown in c.

meets the lower or upper surface of the wing. In long-winged birds, such as falcons, the upstroke can even provide high positive lift (J. M. V. Rayner, personal communication). Part of the upstroke can provide thrust, especially in hovering and slow flight (Fig. 8-5c). For a net lift force to be produced on the upstroke, the wings must be moved backward relative to the still air (Fig. 8-5c). In hovering the thrust forces are counteracted by drag forces produced during the downstroke, so that the net horizontal force is zero. Analysis of the forces generated in the slow flight of a bat (*Plecotus auritus,* 9 g) shows that lift is provided by the downstroke and thrust primarily by the upstroke (U. M. Norberg, 1976a).

Hovering flight can be understood by combining the theory of airscrews (momentum theory) with blade element theory. The momentum theory is based on the motion imparted to the air during hovering. The forces acting on the wings are those necessary to generate that motion. The wings sweep out a disk area S_d (see Fig. 8-7c) through which a downward airflow is sucked. Although each wing usually does not sweep out a full half-circle, the wing disk area is taken to be the area of the circle having the wing span as its diameter (since air between the swept sectors is also sucked in), so $S_d = \pi b^2/4$, where b equals wingspan.

The velocity with which the air passes through the wing disk is termed the induced velocity V_{ind}. Air reaches this velocity as a result of being accelerated downward into the area of reduced pressure above the wing disk. The cross-sectional arc of the downward flowing air diminishes below the flyer, so a similar pressure gradient occurs also below the animal and the air accelerates further to a velocity of $2V_{ind}$ far below (Fig. 8-7c). The forces on the wings, which must equal the weight of the hovering animal, are given (by Newton's laws) as the rate at which downward momentum is imparted to the air. This rate of change of momentum is the product of the air velocity below the animal and the mass flow f_m through the wing disk, and is

$$2V_{ind}f_m = 2V_{ind}(\rho S_d V_{ind}) = 2\rho S_d V_{ind}^2 = Mg, \quad (8\text{-}4)$$

so that $V_{ind} = \left(\frac{Mg}{2\rho S_d}\right)^{1/2}$.

Thus, the induced velocity is proportional to the square root of the disk loading Mg/S_d. The power required to maintain the flow through the wing disk is termed the induced power (see below) and is $P_{ind} = MgV_{ind}$ (compare with Fig. 8-2).

Power components Flight requires power for various purposes. We recognize four kinds: parasite power (P_{par}), or work against form and frictional drag of the body; profile power (P_{pro}), or work against form and frictional drag of the wings; induced power (P_{ind}), or work needed to generate lift and thrust; and inertial power (P_{inert}), or work needed to accelerate the wings at each stroke.

These power drains are associated with mechanical efficiency, which is the ratio of mechanical power output to metabolic power input. Its value is calculated to be about 0.19 to 0.28 for various birds and bats. In addition to these components comes the basal metabolism P_b, which represents the cost of internal body functions such as circulation, respiration, and so on. The principal power drains are the parasite, profile, and induced powers.

To minimize parasite power the body should be streamlined. This is important for fast-flying animals, since parasite power rises with the third power of the flight speed (Table 8-2). Profile power is considered by Pennycuick (1975) to be independent of speed, but to increase with increasing speed by Rayner (1979b,c; Table 8-2). According to Rayner's theory (but not Pennycuick's) long wings can be responsible for a slight increase in profile power. Induced power is the main power in hovering and slow flight. To minimize this power, hovering and slow flying animals should have low wing disk loading (Mg/S_d) and large wingspan (Pennycuick, 1968), as well as the capacity for a large amplitude and a short period of wingstroke (Rayner, 1979c). Inertial power is proportional to the moment of inertia of the wing (mass of wing times the square of its radius from the center of rotation) and to the angular acceleration of the wing. Most of the inertial power is considered to be converted into useful aerodynamic power and is, therefore, usually neglected (although it may be of some importance in hovering and slow flight).

The power needed to fly is then approximately

$$P = \frac{1}{\eta}(P_{par} + P_{pro} + P_{ind}) + P_b, \quad (8\text{-}5)$$

where η is the mechanical efficiency.

Pennycuick (1969) developed a theory for estimating the power required by a flying animal. It is based on classical aerodynamic theory, which includes blade element theory combined with the induced fluid velocities predicted for a momentum jet (Table 8-2). Tucker (1973) and Pennycuick (1975) have elaborated the theory further. Most existing studies (for example, Tucker and Parrott, 1970; Tucker, 1973; Greenewalt, 1975; U. M. Norberg, 1976a,b) use this theory, which assumes that steady-state or quasi-steady-state aerodynamics prevails.

On the basis of morphologic and kinematic data, Weis-Fogh (1972, 1973) derived analytical expressions for the average lift coefficient, the aerodynamic power, the moment of inertia of the wings, and the dynamic efficiency in animals that perform symmetrical hovering (that is, the wing tip describing a more or less horizontal figure eight, and the morphological upstroke contributing lift while the wing's upper surface faces downward), such as hummingbirds and various insects. U. M. Norberg (1976a,b) devised equations based on Pennycuick and Weis-Fogh, which, together with kinematic and morphologic data, allow the exact calculation of the lift and drag coefficients (as averaged over the whole wing and entire wing stroke) for forward flight and asymmetrical hovering (that is, hovering with stroke plane usually tilted and wings flexed during an upstroke, which provides little or no useful force). Power consumption can then be calculated.

But any type of flapping flight also involves nonsteady periods, particularly at the reversal points where active pronation and supination

Table 8-2 Equations and proportionalities for various power components according to Pennycuick's (1969, 1972, 1975) and Rayner's (1979c) models.

Power types	Pennycuick	Rayner
Induced power	Momentum theory	Vortex theory
Hovering flight	$P_{ind} = \frac{k(Mg)^{3/2}}{(2\,\rho S_d)^{1/2}} \propto b^{-1}$, where $k \sim 1.2$.	$P_{ind}(\textit{symmetr. hov.}) = \frac{(Mg)^{3/2}}{R'(2\,\rho S_d)^{1/2}} \propto b^{-1}$, where $R' = \left(\frac{0.923\phi}{n}\right)^{1/2}$. $P_{ind}(\textit{asymmetr. hov.}) = \left(\frac{0.95}{R'} + \frac{1.2f}{R'^5}\right)\frac{(Mg)^{3/2}}{(2\rho S_d)^{1/2}} \propto b^{-1}$, where $R' = \left(\frac{0.808\phi}{n}\right)^{1/2}$.
Forward flight	$P_{ind} = \frac{k(Mg)^2}{2\,\rho S_d V} \propto b^{-2}V^{-1}$, where $k \sim 1.2$.	$P_{ind} \propto b^{-3/2}V^{-3/2}T^{-1/2}$.
Profile power	Blade element theory	Blade element theory
Hovering flight	$P_{pro} = \frac{0.877Xk^{3/4}(Mg)^{3/2}A^{1/4}}{\rho^{1/2}S_d^{3/4}} \propto b^{-3/2}$, where $X \sim 2$ at medium speeds, $k \sim 1.2$, and	$P_{pro} \propto b^5T^{-3}$.
Forward flight	$A = (2.85 \times 10^{-3})M^{2/3}$	$P_{pro} \propto b^2V^3$.
Parasite power	$P_{par} = \frac{1}{2}\rho AV^3$.	$P_{par} = \frac{1}{2}\rho AV^3$.

Notation: Mg = body weight; M = body mass; A = the equivalent flat plate area of the body; S_d = the wing disk area ($\pi b^2/4$); b = wingspan; ρ = air density; k and X are constants; T = stroke period; V = speed; R' is a nondimensional radius of the vortex ring; Φ is the angular position of the wing during the downstroke; n = stroke frequency; and f is a "feathering parameter" (which is, for instance, 0.01 for a hummingbird, 0.027 for a bat, and 0.175 for a wren).

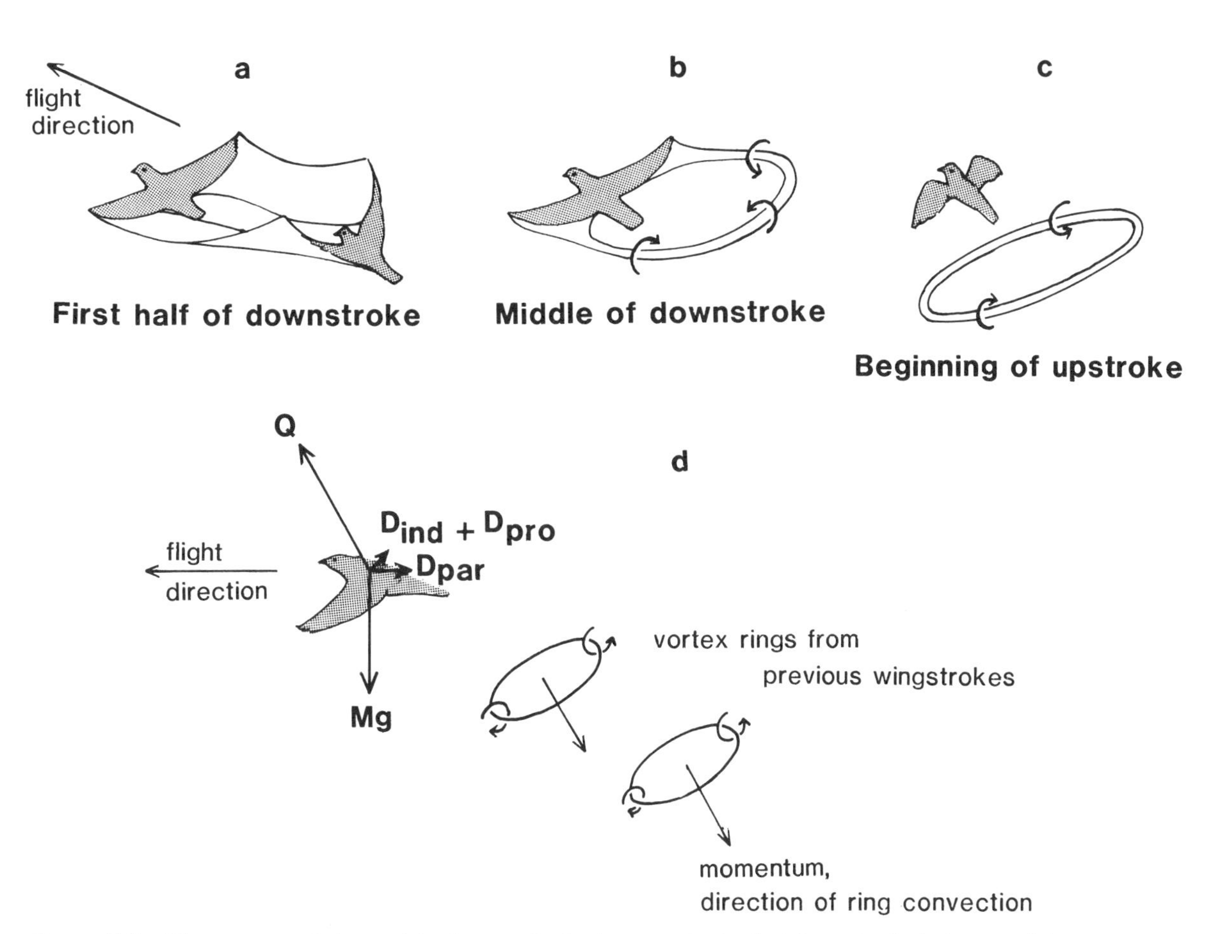

Figure 8-6 The pattern of air vorticity in one single downstroke (a-c). A hypothetical picture of the vortex sheet formed during the first half of the downstroke is shown in a. A vortex ring is then formed (b,c), which convects downward. The forces acting on the bird must be in equilibrium in steady horizontal flight (d), and are as follows: Mg = weight; D_{ind} = induced drag; D_{pro} = profile drag; D_{par} = parasite drag; and Q = reaction of vortex ring momentum. (After Rayner, 1979c.)

occur. Nonsteady effects are especially important when stroke amplitude and rate of twisting are large, which they are at low speeds and in asymmetrical hovering. The previous theories, therefore, give approximate values that tend to err on the low side. For flight modes where nonsteady aerodynamic conditions are important, all models treated so far underestimate the induced power. But the momentum jet approach is still valuable because it gives a minimum limiting value for induced power. Induced flow fluctuates because of the nature of the wing stroke. Induced speed must, therefore, take higher values during some phases of the stroke to compensate for the lower speed when the wing stroke is reversed, and more power is required than with a constant jet.

An alternative, but complementary, view to the combined blade element theory and momentum theory resulted from consideration of how an airfoil influences fluid flow in order to generate lift (see Fig. 8-1). The lift and induced drag on an airfoil result from the interaction between a vortex on the airfoil and the wing's movement. Ellington (1978) and Rayner (1979a,b,c), largely independently, took this new approach to the theory of animal flight based on an idealized pattern of induced air flow behind the flapping wings. The resulting vortex theory was applied by

Ellington to insect flight and by Rayner to bird flight. It is derived from the fact that the mechanism by which both lift and thrust are generated results from trailing edge vortices, which are shed behind an animal's wings. The stronger the vortex, the more lift is generated. Further, the vortex becomes stronger as the angle of attack increases (up to stall). The values of the lift and drag coefficients need not be known when calculating the forces, and the theory is not bound to the constraints of steady-state (or quasi-steady-state) aerodynamics.

Rayner's model for calculating power according to the vortex theory is a combination of that of Pennycuick (1975) with a three-dimensional vortex theory used for the calculation of induced power. The vortex interpretation of air flow is undoubtedly realistic, and it has gained empirical support from experimental flow visualization for flying finches (Kokshaysky, 1979). A disadvantage is that the calculations are extremely involved and tedious.

The essentials of the vortex theory can, however, be easily summarized. The wing as it travels leaves behind it a vortex sheet, which is formed during the first half of the (lift-producing) downstroke (Fig. 8-6a). During the upstroke the wings are more or less flexed, have no bound vorticity, and contribute little or no lift (Fig. 8-6c). The flow is unsteady because of the wing's oscillations. The sheet formed by the downstroke of both wings rolls up into one loop, or torus, of concentrated vorticity. As the animal continues forward, the successive downstrokes produce a stack of vortex rings behind the body, each inclined to the direction of flight and moving backward and downward. The rings are elliptical, elongated in the direction of flight (Fig. 8-6b). The forces acting on the bird are shown in Figure 8-6d. The parasite drag of the body, the profile drag of the wings, and the animal's weight must be balanced by the reaction of the momentum of the vortex rings. The parasite power and profile power required are the consequence, respectively, of the rate at which work is done against drag of body and wings. The energy of the vortex shed behind the wing tips (a necessary result of the pressure difference on the two surfaces of the wing) is responsible for the induced drag.

In hovering the wake is composed of a stack of horizontal, coaxial, circular vortex rings. Figure 8-7 shows the formation of the wake of vortex rings in (a) symmetrical and (b) asymmetrical hovering. In both cases the momentum of the wake is vertical. Circulation is determined by the animal's weight and the time for which a single ring must provide lift. The main power drain in hovering is the induced power. For symmetrical hovering the momentum theory may underestimate the induced power by 10% to 15%, and for asymmetrical hovering, by as much as 50% (Rayner, 1979a). In hovering, non-steady-state aerodynamics certainly prevails, particularly when it is asymmetrical. Insofar as the upstroke generates useful forces, which it does in some bats at least (U. M. Norberg, 1976a), the vortex ring model as it now stands (Rayner, 1979a,b,c) is inadequate because it assumes that no vorticity is shed by the upstroke.

Figure 8-8 shows power curves for a pigeon calculated with conventional aerodynamic theory (a) and with vortex theory (for induced power only) combined with profile power calculated differently (b). The sum curve of the power components typically has a shallow U-shape in both cases. The maximum range speed V_{mr} (speed for maximum distance that can be flown on a given amount of energy), at which the power to speed ratio reaches its minimum, can be found by drawing a tangent to the curve from the origin. This speed is higher than the minimum power speed V_{mp}. Table 8-1 shows how the power components vary with wingspan, speed, and stroke frequency according to Pennycuick's and Rayner's models.

Because these theories synthesize principles from basic mechanics and aerodynamics, the final models not only give predictions about power consumption but also provide insight and understanding of the mechanism of animal flight. Pennycuick's equations are admirably simple; using only total mass, wingspan, and flight speed they generate predictions on power consumption that are correct to within 10% of measured values (Tucker, 1973).

Scaling Because scaling is treated in Chapter 2, only the scaling effects that are important for flight speeds and flight powers are discussed here.

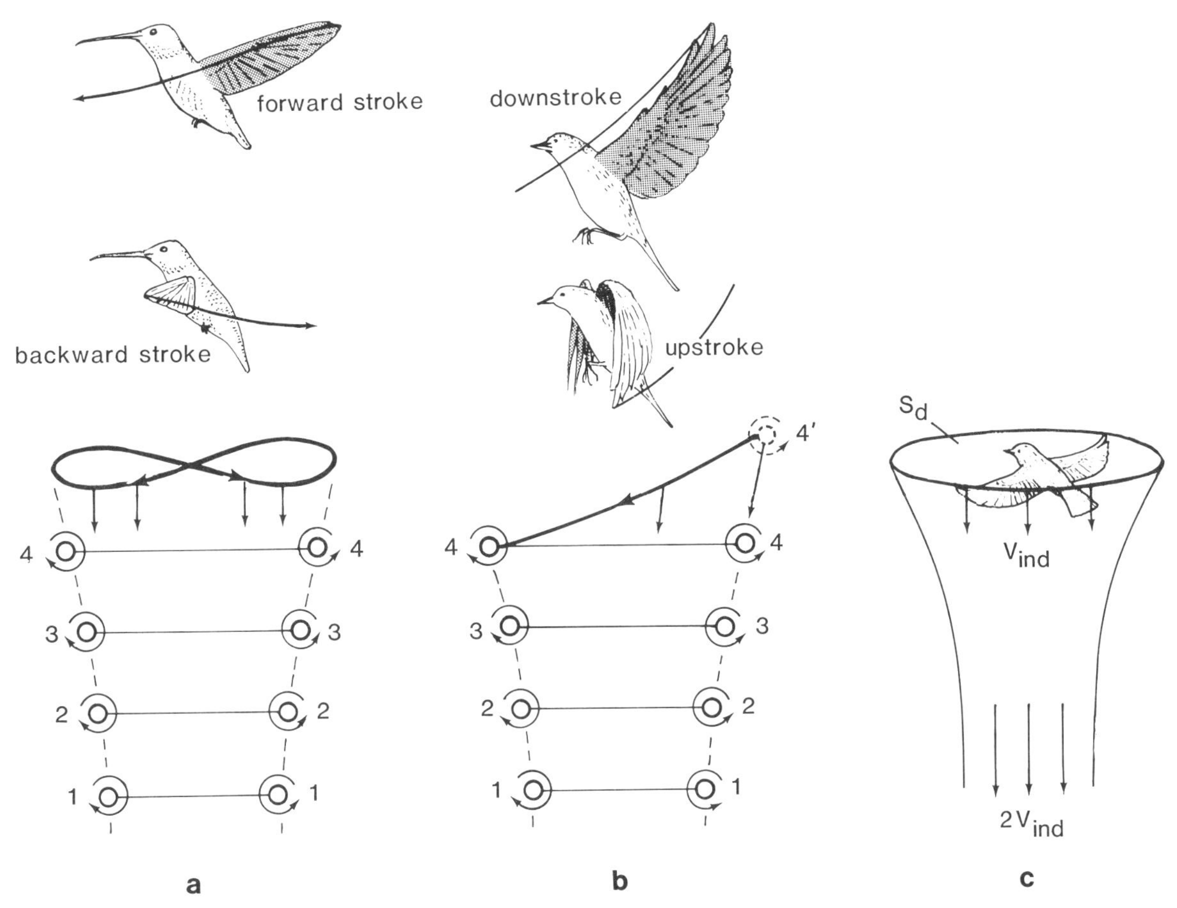

Figure 8-7 Formation of vortex ring wake in four wingstrokes (1-4) in symmetrical hovering (a) and asymmetrical hovering (b). By the time the wings have reached the bottom of the stroke, the first-formed part (4′ in b) of the vortex ring has convected down to the same level; the complete rings, therefore, convect downward in a horizontal orientation in spite of the inclination of the stroke plane. A continuous momentum jet past a hovering animal is shown in c. Air is sucked down from rest far above the animal; it passes through the wing disk at the induced velocity V_{ind} and reaches $2V_{ind}$ farther down in the most contracted zone. The momentum jet gives the lower limiting value of induced power. (a and b after Rayner, 1979c, and c after Pennycuick, 1969, 1975.)

For geometrically similar animals, volume varies with the cube of length and area with the square of length. Then, for any characteristic speed and length,

$$V \propto \left(\frac{Mg}{S}\right)^{1/2} \propto \left(\frac{l^3}{l^2}\right)^{1/2} = l^{1/2} \propto M^{1/6=0.17} \qquad (8\text{-}6)$$

for geometrically similar flying animals. Thus, if a bird is geometrically similar to another bird but has a times the wingspan, then it should fly $a^{1/2}$ times as fast. Similarly, if the bird weighs d times as much, it should fly $d^{1/6}$ as fast. The speed contrasted can be, for instance, V_{mp} or V_{mr}.

The largest flight muscle, pectoralis major, constitutes a roughly constant proportion (about 17%) of the total mass of a bird regardless of size (Greenewalt, 1962). Therefore, the power available increases more slowly with total mass than does the power needed for active flight. This relationship alone may set the upper size limit of flying birds.

Insofar as geometrical similarity prevails, the

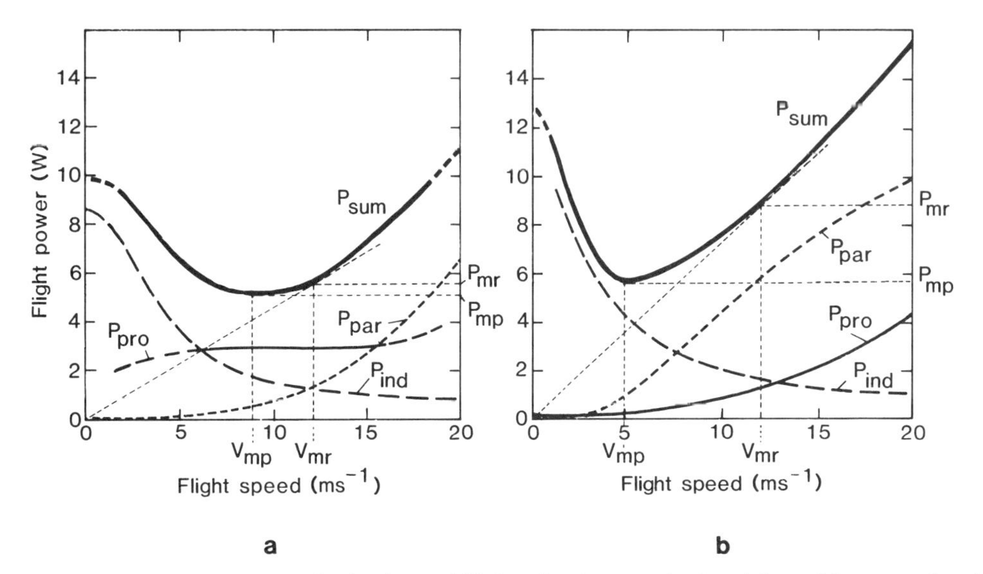

Figure 8-8 Power curves for horizontal flight of a pigeon calculated from (a) conventional aerodynamic theory according to Pennycuick (1968, 1975) for a 400-g bird, and from (b) a combination of vortex theory for induced power only according to Rayner (1979c) for a 333-g bird, and profile power calculated differently from Pennycuick. V_{mp} = minimum power speed; V_{mr} = maximum range speed; P_{mp} = power required at V_{mp}; P_{mr} = power required at V_{mr}; P_{ind} = induced power; P_{par} = parasite power; P_{pro} = profile power; and P_{sum} = sum of the power components P_{ind}, P_{par}, P_{pro}.

muscle power P_r required to fly can be represented as the product of the average, effective thrust and the forward speed. That is,

$$P_r = T'V$$
$$= Mg\frac{D'}{L'}V \propto Mg \times Mg^0 \times Mg^{1/6} \propto M^{7/6=1.17} \quad (8\text{-}7)$$

(compare with Fig. 8-2). Thus, if a bird weighs twice as much as another, geometrically similar bird, it will require $2^{1.17} = 2.25$ times as much power to fly at any characteristic speed under corresponding conditions.

The power available P_a from the flight muscles may be represented as the product of the mass m of the flight muscle and the power available from each gram of muscle. The latter is the specific work $\bar{Q}$ (work done in one contraction) times the flapping frequency f. Thus,

$$P_a = m\bar{Q}f \propto l^3 \times l^0 \times l^{-1} = l^2 \propto M^{2/3=0.67} \quad (8\text{-}8)$$

for geometrically similar birds (Pennycuick, 1969). If a bird weighs twice as much as another bird, it will need 2.25 times as much power to fly, but it will have only 1.59 times as much power available from its muscles. The power margin thus decreases with increasing size. The two power curves meet at the maximum practicable mass. This mass would appear, on empirical grounds, to be about 12 kg for a bird with its crop empty (Pennycuick, 1969), which is about the weight of the largest birds that use flapping flight. Examples are the Kori bustard *Ardeotis kori*, the white pelican *Pelecanus onocrotalus*, the mute swan *Cygnus olor*, and the California condor *Gymnogyps californianus*.

Bramwell and Whitfield (1974) stated that the pterosaur *Pteranodon ingens* had a probable mass of 16.6 kg and a wingspan of 6.95 m. Lawson (1975) suggested a wingspan of 15 m for the largest known pterosaur, *Quetzalcoatlus northropi*, but McMasters (1976) and Langston (1981) favor Lawson's alternative estimate of 11–12 m. Langston estimated its mass as 86 kg. These creatures must have flown mostly by soaring and gliding. The same would apply to the huge pleistocene *Teratornis*, with an estimated mass of 20 kg. The power available versus the power needed for takeoff and active flight must have posed immense problems for these incredibly large flyers.

Not all birds and bats are geometrically similar. However, deviations from geometrical similarity are small for most groups, and, therefore, the relationships of speed and power to mass shown in Eqs. (8-6) to (8-8) may be regarded as good approximations.

Greenewalt (1975) and U. M. Norberg (1981a) calculated the relationships between body mass and various wing characters in several groups of birds and bats, respectively. Wing loading (Fig. 8-9), for instance, actually increases less in birds than would be predicted according to geometric similarity. For passeriforms $Mg/S \propto M^{0.22}$, and for shorebirds and ducks $Mg/S \propto M^{0.29}$. The latter does approach expectation according to geometric similarity, where $Mg/S \propto M^{0.33}$. Hummingbirds have a constant wing loading that is independent of body mass. Frugivorous bats (mega- as well as microbats) follow geometric similarity, whereas in insectivores wing loading is proportionately larger in larger ones than in small ($Mg/S \propto M^{0.4}$). The wing loadings of animal gliders do not increase with size (Rayner, 1981). This means that small gliders have proportionately small flight surfaces and large animals large ones. As a consequence, they should have similar gliding speeds.

Rayner (1979c) calculated the relations of power and flight speed to body mass according to the vortex ring model, using empirical data from Greenewalt (1962). He found that minimum power speed and minimum power scale as $V_{mp} \propto M^{0.13}$ and $P_r \propto M^{1.05}$ for passeriform birds, as $V_{mp} \propto M^{0.10}$ and $P_r \propto M^{1.07}$ for shorebirds, and as

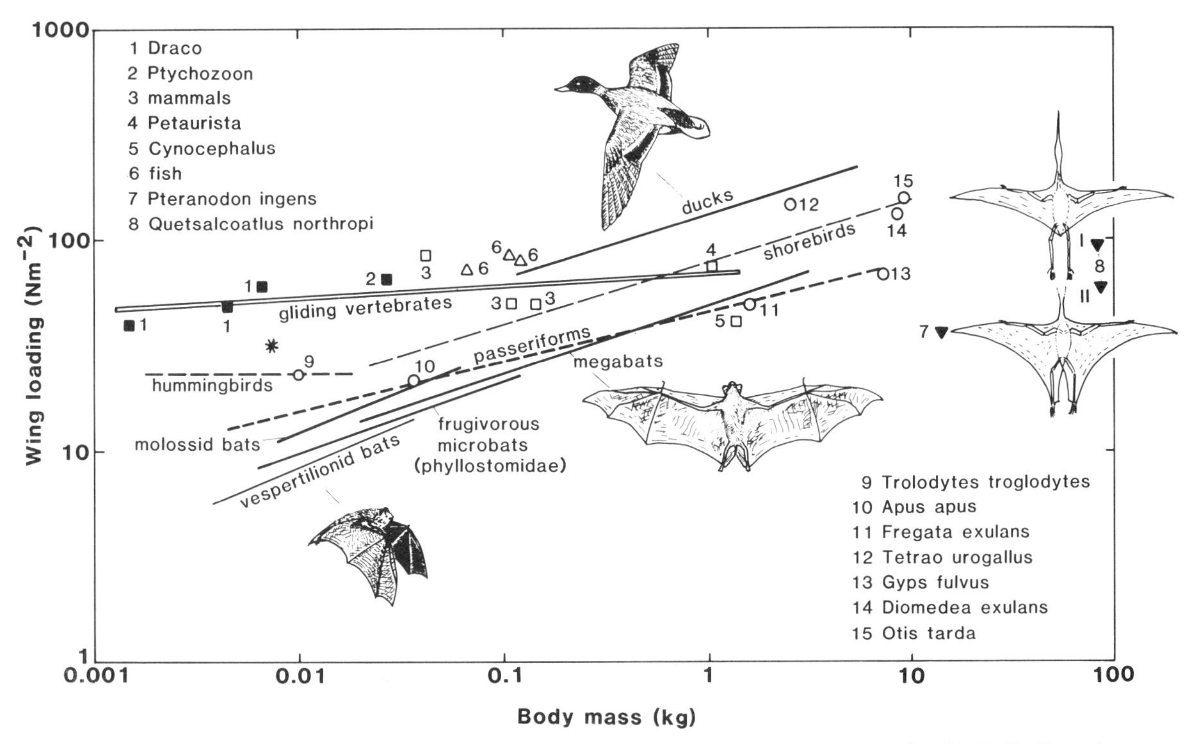

Figure 8-9 Wing loading versus body mass of some gliding vertebrates: fish (6), reptiles (1, 2, 7, 8), and mammals (3, 4, 5), all from Rayner (1981); birds (9–15), recalculated from Greenewalt (1975); bats, from U. M. Norberg (1981a); and pterosaurs (7, 8). The wing area of the pterosaur *Quetzalcoatlus* was calculated from a drawing in Langston (1981), where the wing membrane ends at the tip of the tail (resulting in the higher wing loading). The alternative estimate is based on a modified version of the same drawing *(lower inset figure)* where I let the membrane reach back to the feet as in *Pteranodon,* as reconstructed by Bramwell and Whitfield (1974), from whom the *Pteranodon* data are also taken.

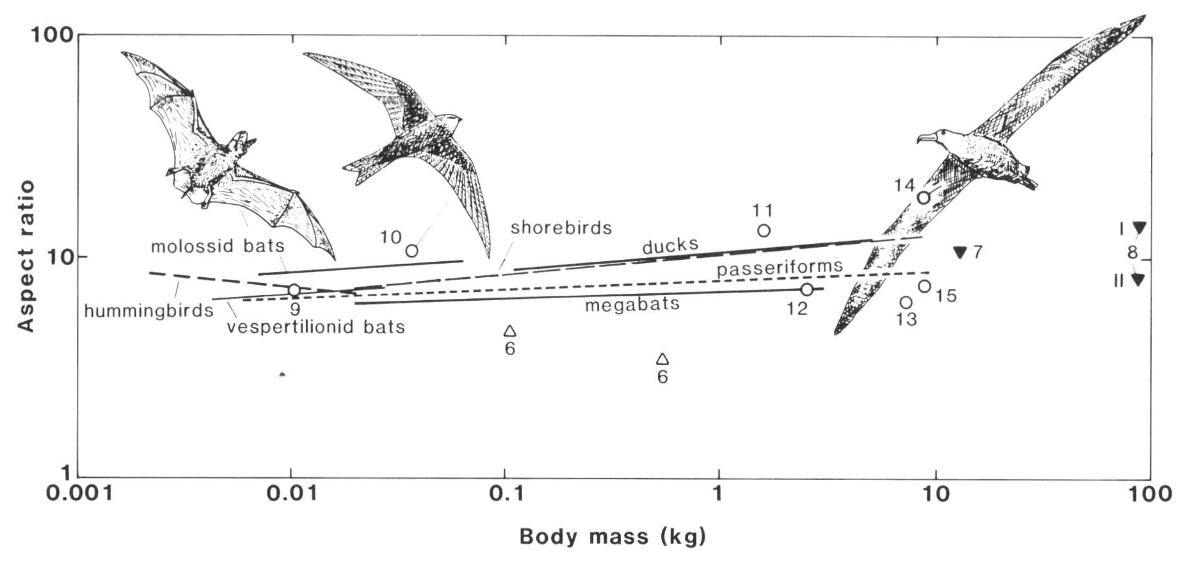

Figure 8-10 Aspect ratio versus body mass of the same species as in Fig. 8-9 (with the same sources and symbols).

$V_{mp} \propto M^{0.22}$ and $P_r \propto M^{1.14}$ for ducks. The deviations from geometric similarity (for which $V_{mp} \propto M^{0.17}$ and $P_r \propto M^{1.17}$) are small, but we see that the larger the birds, the proportionately lower is the flight speed (ducks excepted) and thus also the flight power. Although the speed of large ducks is proportionately greater than for smaller ones, their flight power is somewhat lower. Ducks in general have a higher optimum velocity than other birds, which follows from their high wing loadings (Fig. 8-9).

One might expect that increase of size would have been accompanied by such departures from geometric similarity as would reduce the minimum specific power (power divided by mass) by increasing L/D. The way to increase L/D as body size increases is to increase wingspan more rapidly than the square root of wing area; that is, by increasing the aspect ratio b^2/S (Lighthill, 1977). This is because induced drag and power both decrease with larger span. For geometrically similar flying animals aspect ratio should not vary with size, but in fact, departures from constant aspect ratio are statistically significant (though very small) in birds (Greenewalt, 1975). In shorebirds and ducks aspect ratio varies with body mass raised to about 0.1 (Fig. 8-10), as is the case for molossid bats. For other bat groups and passeriform birds departures from geometric similarity are less (Greenewalt, 1975; U. M. Norberg, 1981a).

Cost of transport C is the aerodynamic work done by a bird in transporting unit weight over unit distance. Rayner (1979c) found, using his vortex ring model, that for birds C scales with body mass as $C = 0.212M^{-0.07}$, which is an extremely small correlation. Oehme, Dathe, and Kitzler (1977) did not find any correlation between C and M. Cost of transport equals $(P/Mg)V$, which is the inverse of the lift to drag ratio (see Eq. 8-7). Body dimensions thus differ from geometric scales in such a way that larger birds have a lower C.

Dimensional scaling is a valuable tool. It can be used to estimate unknown parameters when only a few values are known, and it also contributes to the understanding of the relations among structure, function, and various constraints to increase in size.

Gliding and Parachuting

A gliding or parachuting animal (including those capable of powered flight) cannot do so more slowly than stalling speed. This speed can be reduced by increasing the wing area, improving the airfoil planform and profile, or by delaying stall by other structural means (such as wing slots and flaps; see below). By flexing its wings (making the wing area smaller), a flying animal can increase its gliding speed. The gliding angle (Fig. 8-4) depends upon the geometry of the wing. Because the induced drag of the wings decreases with increased span (see Table 8-1), wings having high aspect ratio have high L/D ratios, and thus allow shallow glide angles. Best glide ratios (which equal L/D ratios) range from 10:1 to 15:1 for birds of prey and vultures and may reach 24:1 in albatrosses (Pennycuick, 1972a). Bramwell and Whitfield (1974) estimated the glide ratio in the pterosaur *Pteranodon ingens* to be 19:1. Measured maximum lift coefficients are 1.5 to 1.6 in gliding birds and 1.5 in a gliding bat (Pennycuick, 1971c, 1972a).

Gliding animals have to be capable of steady, controlled flight. If the equilibrium of forces in steady flight is disturbed, moments are set up to restore it. The earliest forms of birds *(Archaeopteryx)* and pterosaurs *(Rhamphorhynchus)* had long tails, which probably served for control and stability. Longitudinal stability (pitch) is closely related to the control of speed. It can be achieved in several ways, for instance, by fore-and-aft movements of the wings, by deflecting the tail and rear part of the wings (bats), and by deflecting the hand wings downward (birds and bats; see Pennycuick, 1972a). Stability about the long axis (roll) is achieved by increasing the dihedral (V-attitude) and sweep-back of the wings or by twisting and flexing the wings to adjust their lift. Stability about the vertical axis (yaw) is controlled by the tail or, again, by twisting and flexing the wings to change the drag coefficient.

Very few animals are pure parachuters. Most are gliders, which travel usually from tree to tree and can voluntarily steepen their angle of descent. Escape from predation may have furthered the evolution of parachuting or gliding abilities. However, other factors, such as optimal foraging over some vertical zone in trees, may have played important roles. Even for animals with good ability of powered flight it takes less energy to climb and hop upward in a tree and fly downward to the next one than to do the reverse (R. Å. Norberg, 1981). Parachuting or gliding animals are represented in five vertebrate classes; they have evolved independently from different groups that were originally arboreal (except for the "flying" fishes).

Marine "flying" fish, which belong to the families Exocoetidae and Hemirhamphidae, occur in the tropical oceans. Examples are the two-winged *Exocoetus* and four-winged *Cypsilurus.* Flying fish have large pectoral fins, a thin body, and an enlarged lower lobe of the caudal fin. While totally submerged, the fish accelerates using the tail. It then breaks the surface, extends the pectoral fins to support the body in the air, and, while escaping water drag from the body, accelerates further by rapidly beating the lower lobe of the tail, which remains in the water. The fish then ascends in a glide on outstretched fins. Escape from predation has usually been considered to be the reason for the evolution of flight in marine fish.

"Flying" frogs, which belong to the families Rhacophoridae and Hylidae, occur in Southeast Asia, Australia, and South and Central America. Many species, including *Phrynohyas venulosa,* are only able to parachute. Their airfoil surface is enlarged by spreading the webbed feet, and many are able to flatten their bodies so that the entire animal contributes to the generation of lift. Some species have a membranous flange along the hind edge of each limb. An example is *Hyla miliaria,* which can glide at an angle of about 18° to the horizontal (Duellman, 1970). *Rhacophorus reinwardtii* is another good glider. Scott and Starrett (1974) observed the glide angles of various flying frogs, which in most cases were as shallow as 30°.

Among reptiles many pterosaurs were undoubtedly good gliders. Even some other prehistoric reptiles, such as *Icarosaurus* and *Daedalosaurus,* were able to glide. These latter had a flight surface formed from elongated ribs, as in the recent genus *Draco* (Agamidae, with about 20 species). The gliding surface in *Draco* is formed by a thin membrane supported by five to seven ribs.

Ptychozoon, the "flying" geckos, have broad flaps of skin on either side of the body, which are spread out by the limbs during flight. They also have small flaps on either side of the neck and tail, webbed feet, and flattened body and tail. Their angles of descent are 45° or steeper (Russel, 1979), which makes them parachutists.

A colubrid snake of the genus *Chrysopelea* (from

Borneo) can draw in the ventral surface of its body to form a deep concavity. During gliding the body is held rigid in coils that form a triangle-shaped airfoil. The tail is used for control of the glide, the angle of descent being about 30° (Rayner, 1981). The transverse coils of the body are close to one another, and Rayner suggested that in this configuration the snake operates as some kind of "slotted delta wing."

Many birds are very good gliders. Those with low wing loadings (large relative wing areas) can glide slowly with low sinking speed, since $V \propto (Mg/S)^{1/2}$. Most of these birds possess separated primary feathers. Rayner claimed (1981) that the probable function of the separated feathers during gliding "is to spread the vortex wake behind the wing, thereby reducing the glide angle with little change in the glide speed."

In gliding birds the feet, which produce drag but no lift, can be used as airbrakes to steepen the gliding angle for making controlled descents. The webbed feet of waterbirds make particularly effective airbrakes. Even some vultures that use their feet as airbrakes have a small web between the middle and outer toes (Pennycuick, 1971a).

A fold of skin stretched between the fore and hind legs has evolved independently in three orders of mammals for use as a gliding wing: Marsupialia (three genera of flying phalangers), Dermoptera (two species of *Cynocephalus,* or colugos, or "flying lemurs"), and Rodentia, including about 12 genera of sciurids and 3 genera of anomalurids. All are arboreal and most are tropical forest species, although the flying squirrels are distributed throughout the Holarctic region. Gliding mammals have low aspect ratios (1 to 1.5), but their wing loadings are similar to birds' (Rayner, 1981; Fig. 8-9). Their size range (10 to 1500 g) is similar to that of bats.

Cynocephalus is the only species for which no part of the body extends beyond the membrane —even the digits are included. In all gliding mammals the flight membrane can be controlled by the limbs. Gliding rodents have a cartilaginous spine that extends into the airfoil, from the wrist in sciurids and from the elbow in anomalurids. It stretches the membrane and somewhat increases the span. The long tail of gliding mammals and some reptiles is used for stability and flight control.

Measurements on the flight of the gliding marsupial *Petaurus breviceps* show excellent control and gliding angles of 11° to 27° (Nachtigall, 1979). Bats, on the other hand, are not able to vary wing area to control gliding, since they are less able than birds to flex their wings during flight while maintaining good aerodynamic performance. Some bats *can* glide, but they seldom do.

Wing Adaptations for Powered Flight

Flight is very energy-consuming even though the cost of transport over unit distance is fairly cheap. Powered flight, therefore, requires a high degree of morphological adaptation. For instance, the wings have to be long enough to reduce the induced power, which is very high at low speeds. They have to be resistant to bending, but at the same time light to keep inertial forces within reasonable limits. The muscular system must be modified to be able to transmit a great deal of power to the wings. It is necessary for the flight muscles to have large areas of origin. Animals having powered flight also must have low total weight. Three groups of vertebrates have evolved a wing system capable of powered flight: the pterosaurs, birds, and bats.

Lift-Enhancing Characteristics The wings of birds and bats are movable in various ways that permit the animal to change the geometry and aerodynamic characteristics of the wing to control motion or improve performance in some desired manner. Bats have the capability to vary the chordwise *camber* (anteroposterior curvature) of the wing by lowering the thumb and legs. In this respect they highly surpass birds. Bat wings, especially those of low aspect ratio, have high chordwise camber (Fig. 8-11), which is efficient at low-speed flight. Broad-winged bats and several slow-flying microchiropteran species usually have relatively broader membrane parts anterior to the forearm and third digit than have long- and narrow-winged bats.

Figure 8-11 *Rousettus aegyptiacus* (Megachiroptera) in flight. Note the angled-down leading edge flap formed by the parts of the membrane anterior to the arm and third digit. (From U. M. Norberg, 1972.)

For a simple nonpermeable wing the maximum C_L obtainable is between 1 and 2. When a critical α (stalling angle) is exceeded, the airstream separates from the wing with a sudden fall of C_L and increase of C_D (Figs. 8-3a and 8-12a). But the stalling can be delayed and C_L increased up to a value of about 3. In birds this can be accomplished by wing slots (the separation of the wing-tip primary feathers), which delay stalling in two ways. First, they permit through-wing suction, and hence prevent flow separation at the upper surface of the wing parts behind the slots. The alula and the free ends of the hand remiges (primary feathers) form such slots. Bats lack similar slots.

A second effect of separated primaries is that the feathers can then act as individual airfoils, each of which produces lift and each of which can

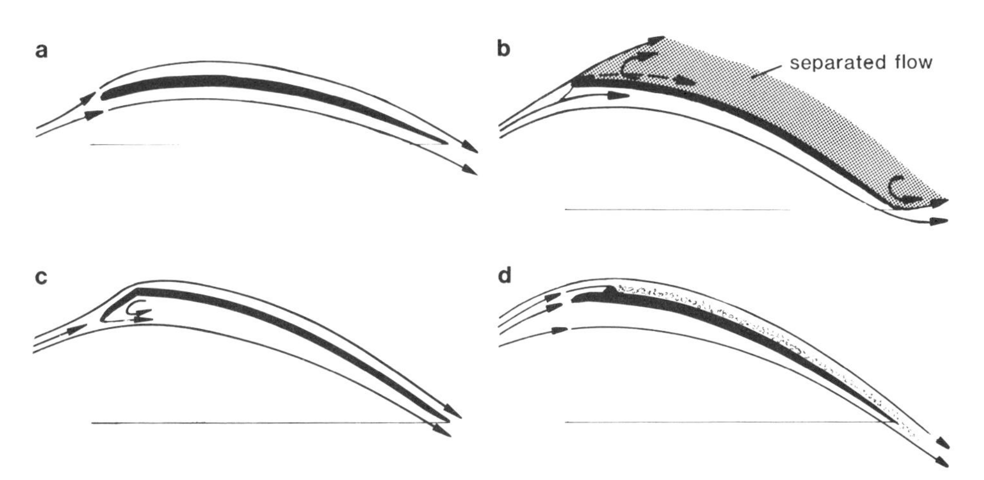

Figure 8-12 a: Schematic view of the laminar flow over a wing held below the stalling angle. b: Airflow separation destroying the underpressure at the upper surface of the wing at an angle of attack larger than the stalling angle. c: Leading edge flap keeping the airflow attached to the surface at the same angle of attack as in b. d: A protruding ridge, a turbulence generator, causing a transition of the boundary layer from laminar to turbulent flow, thus keeping the flow attached.

be twisted individually in the leading edge-down direction under aerodynamic load. As a result of the increasingly larger flapping velocity from base to tip of the wing, the air meets the wing increasingly more from below (thus increasing α) toward the tip in the downstroke. The separated primaries can be twisted more than could the entire wing tip if the feathers were not separated, and hence higher local angles of attack are allowed for the inner parts of the wing before the outer parts reach stalling angles. As a result of separated primaries, the average lift coefficient of the entire wing may be increased.

A leading edge-down pitch moment of a primary feather results if its chordwise aerodynamic center of pressure is located behind the pitch (torsion) axis. This is achieved by the asymmetrical location of the rachis near the leading edge of the primaries, and the primaries nearest to the leading edge are the most asymmetrical.

A third way to delay stalling is by providing leading edge flaps, which have been developed for airplane wings as high-lift devices. These flaps act to keep the flow laminar over the wing at higher angles of attack, thus permitting higher lift coefficients without separation (Fig. 8-12b), especially for wings having thin sections and a sharp leading edge (Abbott and von Doenhoff, 1949). In bats the patagia (flight membranes) anterior to the arm and third digit together may function as a leading edge flap (with no slot behind) when lowered by the thumb and the second digit (Hankin, 1913; U. M. Norberg, 1972; Fig. 8-11). The effectiveness of leading edge flaps increases as the radius of the leading edge decreases (that is, as the leading edge becomes sharp). This radius is extremely small in bats.

A fourth way to delay stall is to make the boundary layer of the wing turbulent. A turbulent boundary layer is able to remain attached to the surface at higher angles of attack than can a laminar one and hence permits a larger C_L. For bats, projection of the digits and arm above the contour of the wing and concentration of hairs on the skin at the arm, making the upper wing surface near the leading edge rough, function as turbulence generators (Fig. 8-12c).

In flying birds turbulence generates some aerodynamic noise. This can be disadvantageous to birds of prey, especially to those that hunt by ear and whose prey have acute hearing. Owls have achieved a silent flight by three features in feather structure, which were first described by Graham (1934). First, a stiff serration of the front margin of the leading edge feathers of the hand wing act to keep the flow attached near the leading edge, even at large angles of attack, hence decreasing noise as well as increasing lift. Second, soft, fringed feather margins at the trailing edge of the wing probably suppress trailing edge noise. Third, the soft downy covering of the upper surface of the wings makes the feathers slide soundlessly on one another.

In birds the main function of the tail may not be longitudinal control, although it is undoubtedly used for this, especially in rapid maneuvers. The principal function appears instead to be analogous to that of trailing edge flaps on airplane wings (Pennycuick, 1972a). At very low speeds, especially at takeoff and landing, the tail is typically spread and depressed, causing an increase of the supporting area, and thus in lift. Further, the action of the tail helps to suck air downward over the bases of the wings, thus increasing the maximum C_L of the wing itself. Birds with long forked tails, such as some swifts, swallows, terns, and frigate birds, use the spread tail as a long flap that is preceded by a slot posterior to the wing. Such birds are specialized for very slow flight and maneuverability (although their cheap flight enables them also to fly fast). This has been discussed by Pennycuick (1972a).

In microbats the tail membrane, which is connected to the wing membrane via the legs, provides a supplementary wing area. The inclusion of the hind legs in the wing and tail membranes adds greatly to the control of wing camber and wing twisting (Fig. 8-13). Those species with a large tail membrane, such as megadermatids and nycterids, are specialized for hovering and slow maneuverable flight.

Energy-Saving Elastic Systems One uncertainty is to what extent inertial power is converted into useful aerodynamic power in flapping flight. Work has to be done to accelerate the wing at the beginning of the downstroke, but at the lower reversal point in fast forward flight the kinetic

Figure 8-13 Long-eared bat *Plecotus auritus* in slow forward flight, with wings in the later part of the upstroke (all three views photographed on the same flash). The feet are raised (but the tail tip is lagging behind), which reduces the camber at the base of the wings. The middle part of the wings is still cambered, but the wing tips are momentarily inverted, producing a large thrust force but negative lift. (From U. M. Norberg, 1976a.)

energy of the wing can be transferred to the air, giving lift. In very slow flight and hovering this transfer of energy is not as easily achieved, since airspeed at the wing is then probably too low at the turning points. Hence, the loss of inertial power should be of importance in hovering flight, unless kinetic energy can be removed and stored by some other means.

Insects have a highly elastic, energy-saving mechanism at the hinge of the wing. The major benefit derives from the sclerotized cuticle of the thorax and from the rubberlike ligaments whose main component is the protein resilin (Weis-Fogh, 1965). Many insects would indeed not be able to produce the power necessary to fly if they lacked this elastic system. Vertebrates lack a similarly effective system, but possess some structures that are possibly energy-saving. However, inertial power is much higher in insects than in birds and bats, so there is more need to save in the former.

The wing membrane of bats (and probably that of pterosaurs) is a highly elastic structure that contains the protein elastin, and in most bats the tips of the third to fifth digits are cartilaginous and flexible. These features might be of some importance for absorption of kinetic energy and release of elastic energy at the top and bottom of the stroke.

In birds there are some possibilities of energy storage. Schaefer (1975) suggested that the clavicle may store and release energy at the top and bottom of the wing stroke. Although the clavicles are relatively slender, they may, with muscular support, be strong enough to function in the suggested manner.

Pennycuick and Lock (1976) described a mechanism whereby primary feathers might increase the efficiency of transfer of the wing's kinetic energy to the air toward the end of the downstroke when the primary feathers unbend (after being bent by aerodynamic loads in the beginning of the stroke). The authors proposed that the mechanism exists, but that it cannot transfer all the wing's kinetic energy to the air in hovering; only at some forward speed might the mechanism become fully effective.

Skeletal and Membranous Systems The force of the airstream subjects the wings of flying animals to great strains during flight. The wings of all flyers, whether feathered or membranous, are supported by a framework of skeletal elements that is controlled by muscles. Special arrangements in the wing reduce the demand for powerful muscles and for skeletal elements having large cross-sectional areas. This reduces the mass of the wings, and consequently their inertial loads, and also reduces the total mass of the animal.

Pterosaurs and most birds have (or had) pneumatic bones to reduce weight, whereas bats do not. The skeletal elements of bats, especially in the wings, are instead very slender, which contributes to low weight.

The leading edges of the wings are subject to especially great strains during flight. The wing skeleton in pterosaurs, birds, and bats all have stay systems in the arm, and birds and bats also have them in the hand wings. These systems provide good support for the anterior parts of the wings (Fig. 8-14). Because the elbow joint is more elevated than the shoulder and wrist joints when the wing is outstretched laterally (in birds and bats, probably also in pterosaurs) and because the muscle along the leading edge of the arm wing tightens the patagium anterior to the arm, the chordwise profile of the arm wing becomes very convex, which promotes lift production. The muscle and the membrane of the leading edge prevent the angle between the humerus and the ulna and radius from opening excessively.

In birds the ulna and radius form a slightly convex unit with a relatively long distance between them, an arrangement that greatly increases resistance to bending forces in the plane in which they lie. Further, the radius and ulna move in a fashion similar to a pair of "drawing parallels," causing automatic extension and flexion of the manus with the elbow. This mechanism, investigated by Fisher (1957), largely relieves the wings from large muscles to move the manus (see illustration in Hildebrand, 1982). Bramwell and Whitfield (1974) found a similar arrangement in the pterosaur *Pteranodon.* In bats the ulna is rudimentary. Extension and flexion of the bat manus are caused by extensor and flexor muscles of the arm and manus.

During flight aerodynamic forces cause (caused) a curving of, and hence tension in, the wing membrane of bats (and pterosaurs), which then pulls at the lines of attachment, that is, the leg, arm, and digits. The tension occurs in the tangent planes of the membrane at these lines. Although the resultant force of the wing is directed nearly normal to the wing chord, it is *this* force that is transformed into tension in the membrane. The strain is especially great on those skeletal elements that constitute the leading edge of the wing (particularly of the hand wing), which stretches out the membrane and leads the wing movements.

In the pterosaurs the enormously elongated fourth digit, with its four long phalanges, formed the leading edge of, and the entire support for, the membrane of the hand wing (together with the much reduced first to third metacarpals, which were more or less fused to the fourth metacarpal). The fourth digit alone had to resist the bending forces caused by air resistance and by the pull of the membrane of the hand wing. In accordance with the strong bending forces on the leading edge, the fourth digit was remarkably thick.

In bats rigidity of the leading edge distal to the thumb is obtained by a special arrangement of the second and third digits (U. M. Norberg, 1969). Determining factors are, first, a ligamentous connection between the second and third digits, and second, the fact that the third metacarpophalangeal joint is angled somewhat less than 180° in the plane of the membrane anterior to the third digit (Fig. 8-14c). During the wing stroke the second digit and the leading edge of the third digit are subjected to bending forces. The ligament connecting the second and third digits can take up only tension, whereas the metacarpal and first phalanx of the third digit are subjected mainly to longitudinal compression. Because of its convexity, this arrangement constitutes a rigid unit — in the plane of the membrane — between the thumb and the joint between the first and second phalanges of the third digit. This unit is especially wide in megabats (Fig. 8-11). The second and third digits in bats are together analogous to the fourth digit in pterosaurs. Because of the mechanically superior design of the leading edge of the hand wing in bats, the digits can be considerably thinner and lighter than in pterosaurs.

In birds the proximal part of the leading edge of the hand wing owes its rigidity to the first, second, and third digits — in particular to the carpometacarpus and the phalanges of the second digit. The distal part of the leading edge is formed by the anteriormost primary (or primaries when they are graded; Fig. 8-14b). The two long metacarpals form a slightly convex unit that encloses an intermetacarpal space resembling the one in the forearm between the ulna and radius. This arrangement increases resistance to bending forces caused by the airstream, which are

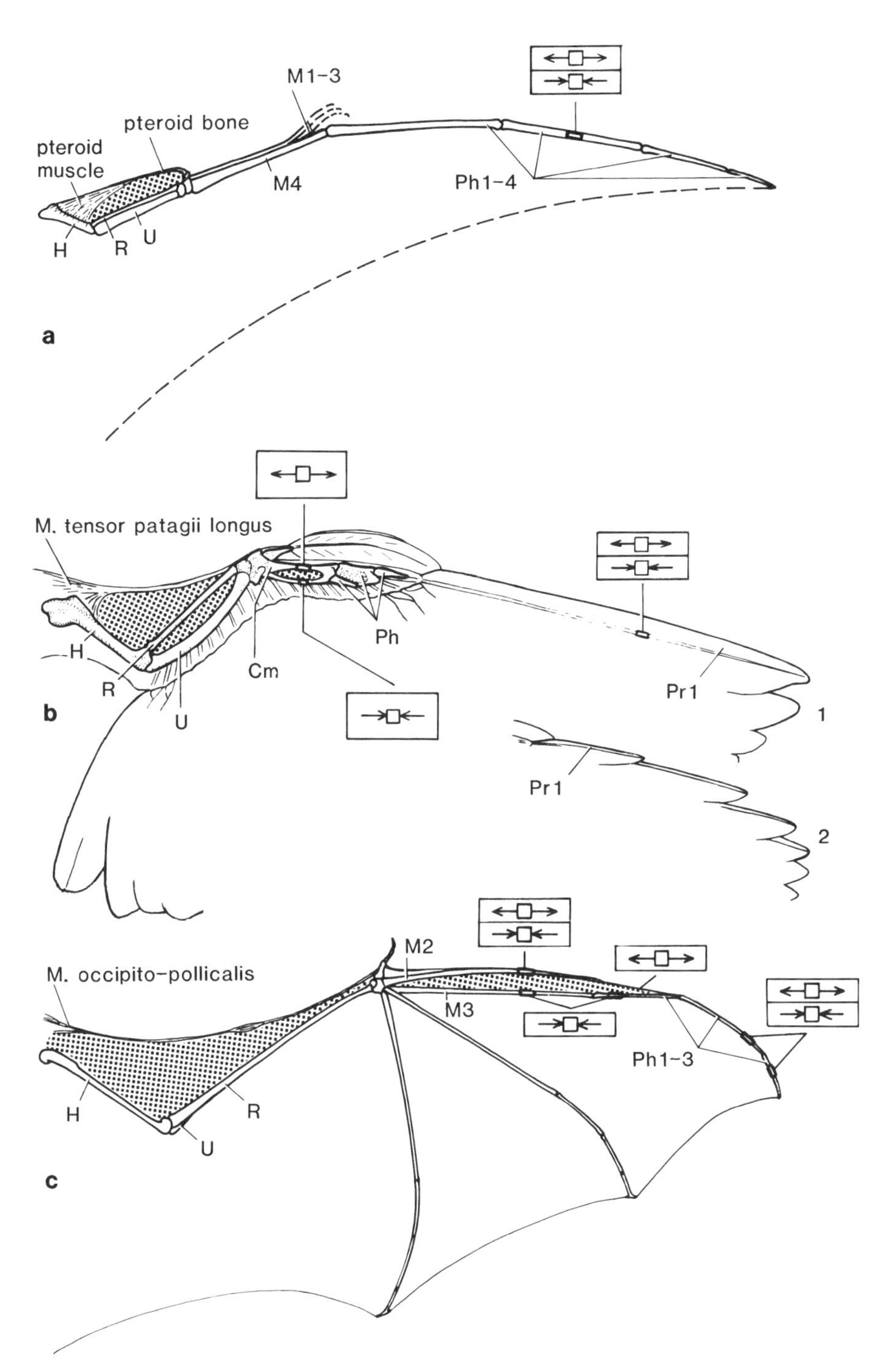

Figure 8-14 Wings of (a) a pterosaur *(Pteranodon)*; (b) birds, including a pigeon *Columba livia* (1) and a magpie *Pica pica* (2); and (c) the long-eared bat *Plecotus auritus.* The stippled areas show the stay systems formed by the skeletons and tendons. c shows a simplified mechanical analysis of an arrangement of the second and third digits of the bat wing, making the leading edge of the hand wing rigid in the membrane plane. *H* = humerus; *R* = radius; *U* = ulna; *M* = metacarpal; *Cm* = carpometacarpal; *Ph* = phalanx; *Pr* = primary. (Modified from U. M. Norberg, 1970.)

in the plane in which the fused metacarpals lie. The phalanges of the second digit are thick and dorsoventrally flattened. The first primary attaches to the tip of the last of these phalanges. In many bird species the first primary is shorter than the next primary. The bending forces are then spread over more than one primary. The air pressure on the feathers behind those of the leading edge cannot be transformed into backward pull on the leading edge of the hand wing, as in pterosaurs and bats, because the feathers take up their own bending forces more or less independently.

The digits of bats are shaped in such a way as to have their greatest cross-sectional diameters in those planes where the bending forces are largest, thereby keeping their mass low while still maintaining rigidity in the important directions. The less the curvature of the membrane, that is, the tauter the membrane, the larger the tensional forces at the lines of attachment. Since the wing is divided into several parts by the digits, each patagium can be regarded as a unit for the acting forces. In equilibrium the tensions transmitted to digits four and five from the membranes are equal on each side (unless lateral muscle forces are involved). Therefore, these digits are subjected mostly to dorsoventral bending. Accordingly, their phalanges are laterally compressed.

Figures 8-15a and c show the principal forces in a bat and pterosaur wing, respectively. The membrane of the bat is trisected by two bones. If the horizontal projections of each entire membrane and the angle β (see figure) are equal in the trisected and intact membranes, then the force per unit of length of the periphery of one segment of the trisected membrane becomes a third of the corresponding force of the intact membrane. This is because the area is reduced to a third, and because F is proportional to the radius R of the membrane (Fig. 8-15c).

The distal part of the wing's leading edge is exposed to tensional forces transformed only from the adjacent patagium, whereas the fourth and fifth digits act as compression members that alter the direction of the tensional forces (Pennycuick, 1971c; Figure 8-15b). In consequence, the distal part of the wing's leading edge is relieved of large tensional forces and the need of a powerful extensor muscle. If the wing membrane were outstretched only by one digit, as in pterosaurs, the digit forming the leading edge, as well as the hind limb, would have to resist tensional forces transformed from aerodynamic pressure on the entire wing membrane posterior to the wing skeleton, and the tension would be much larger on the outer part of the wing's leading edge than on the inner part.

In addition to the digits, the elastin fibers determine the directions of the tension in the membrane. By an appropriate structure of the membrane of pterosaurs, as for instance by the occurrence of elastic strands along the path indicated by dotted lines in Figure 8-15d, the tensional forces might have been deflected, relieving the distal part of the leading edge from large forces.

Muscle System In birds and bats (as well as pterosaurs), the pectoralis muscle is the main depressor of the wing, and it usually acts also as a pronator (nose-down rotator) of the wing. It is highly developed, and in birds and bats it varies from 10% to 20% of the animal's mass. The larger value is found in animals with high wing loadings, which need to be compensated for by high flight speeds and/or high flapping frequencies.

Figure 8-16 shows the direction of pull of the main muscles acting during the wing stroke in pterosaurs, birds, and bats. In pterosaurs the pectoralis and coracobrachialis muscles may have been the depressors of the wing. Three muscles may have been concerned with raising the wing, namely the subcoracoscapularis, deltoideus, and latissimus dorsi muscles (Bramwell and Whitfield, 1974; Fig. 8-16a).

In birds the downstroke is performed mainly by the pectoralis major and anterior coracobrachialis muscles. The pectoralis is divided into two parts, major and minor, which have different functions. Both originate on the sternum, and they are adjacent, but whereas the pectoralis major passes directly to the humerus and inserts anteriorly on the deltoid crest, the pectoralis minor (supracoracoideus) passes through the foramen triosseum, between the proximal ends of the scapula, coracoid, and clavicle, and inserts posterodorsally on the humerus (Fig. 8-16b). It

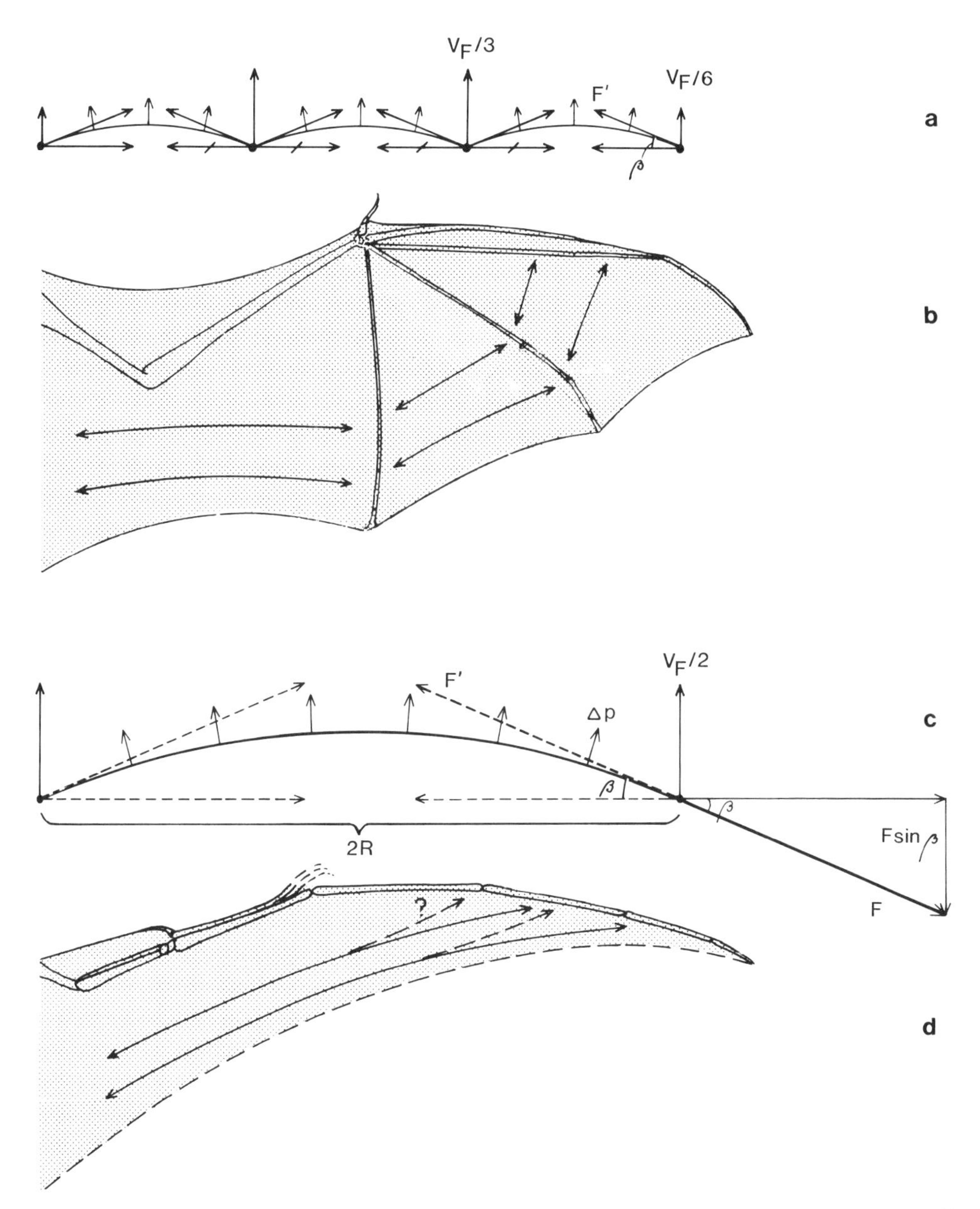

Figure 8-15 Mechanical principles of the membrane-skeletal arrangement in the wings of bats (a,b) and pterosaurs (c,d). a: Force diagram of a cross section of a schematic wing (bat type) extended by the body and three digits forming equal interjacent areas. The distal digit *(far right)* is exposed to tensional forces transformed from forces from one section only. The lateral components of tensional forces on each side of the middle digits cancel out. b: The digits act to alter the direction of tensional forces, relieving the wing tip from large forces (Pennycuick, 1971). c: Force diagram of a cross section of a schematic wing (pterosaur type) extended by the body and one digit. The digit *(far right)* has to take up bending forces transformed from the entire wing membrane (no canceling of lateral forces at interjacent digits as in bats). d: Arrows indicate the assumed direction of tensional forces in a pterosaur wing. (Modified from U. M. Norberg, 1972.)

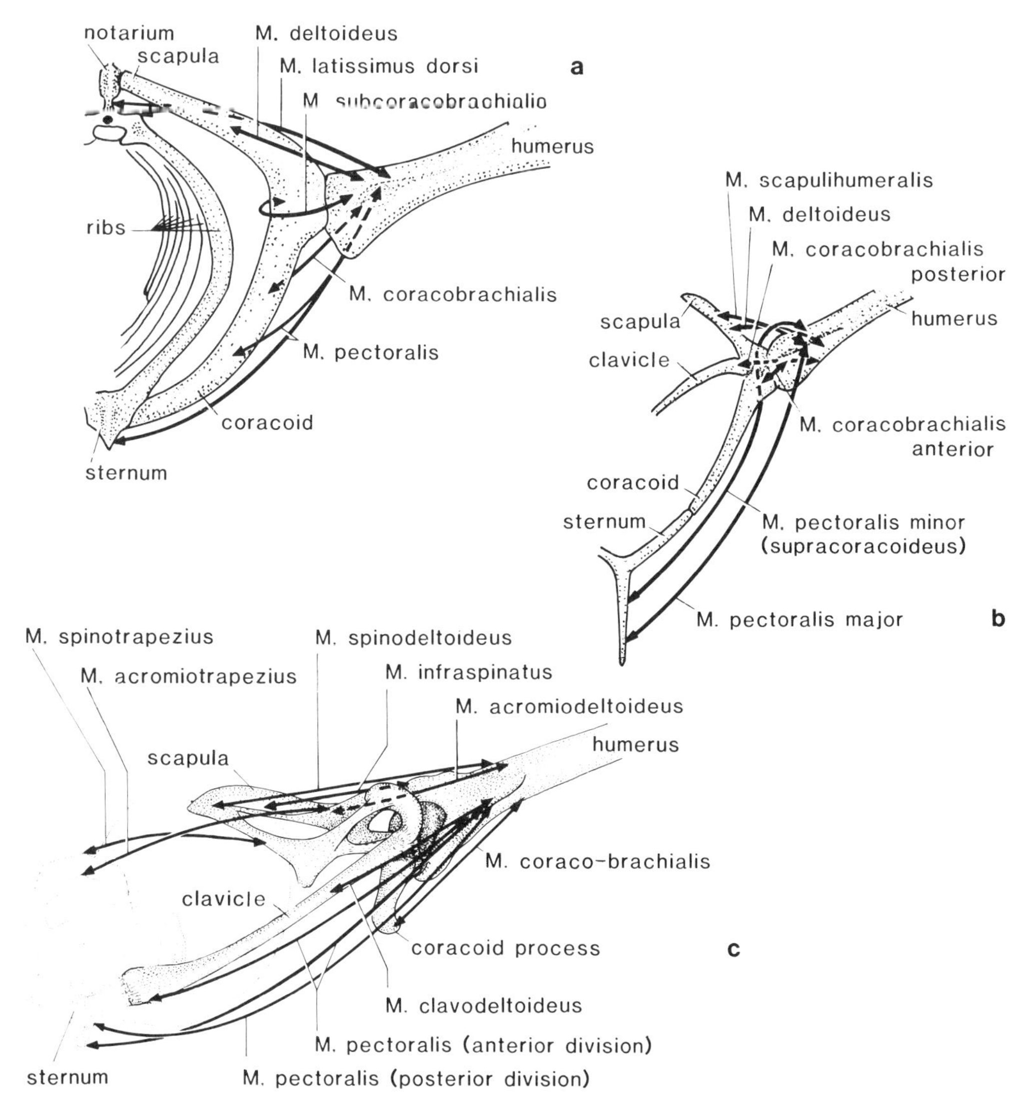

Figure 8-16 Body cross section at the shoulder joint showing the direction of pull of the main muscles acting during the wingstroke in (a) pterosaurs, (b) birds, and (c) bats. (c is modified from U. M. Norberg, 1970.)

thereby exerts an upward pull on the humerus. Additionally, the deltoideus and the posterior coracobrachialis muscles act as wing elevators. The latissimus dorsi and scapulihumeralis muscles pull the wing backward and are thus also involved in the upstroke. The flight muscles of birds have been investigated by, for instance, Herzog (1968).

In bats the downstroke is caused mainly by the pectoralis, subscapularis, and clavodeltoideus muscles (Fig. 8-16c). The upstroke is controlled mainly by the spinodeltoideus and acromiodeltoideus muscles. In birds and bats (and pterosaurs) the latissimus dorsi is the main supinator (nose-up rotator) of the wing.

Greenewalt (1975) plotted the mass of the large pectoral muscle against body mass for three bird groups: "passeriform," "shorebird," and "duck" (each group containing more than the named birds). He found that values for passeriforms are lower than those for the other two groups. The ratio between the mass of the large pectoral muscle and wing mass shows significant differences among the groups, being about 0.7 in passeriforms, 1.2 in shorebirds, and 1.7 in ducks. Ducks have much higher wing loading and hence higher flight speed and requirements for power than do the other birds. Shorebirds are intermediate in these respects (Fig. 8-9). If we assume constant power output per unit mass of muscle tissue, total potential power increases consistently from passeriforms through shorebirds to ducks in response to the increasing demands for aerodynamic power.

In slow flight and hovering the upstroke gives some lift and/or thrust, whereas in fast flight it probably is more or less passive. Therefore, the pectoralis minor is most developed in hovering and slow-flying birds. Greenewalt found that the pectoralis major of his passeriform group has about twelve times the mass of pectoralis minor. Hummingbirds are exceptional for the degree to which aerodynamic power is produced also in the upstroke, and their pectoralis major has only twice the mass of the pectoralis minor (Greenewalt, 1962).

Birds have a large sternum with a well-developed sternal ridge, which provides much surface for the origins of the pectoral muscles. Pterosaurs also had a large sternum with a ridge, which, however, was not as large as in most birds. Some bats have a well-developed sternal ridge, whereas others have scarcely any. However, in the latter it is common for the anteriormost part of the manubrium of the sternum to have a ventrally projecting tubercle from which a median ligamentous sheet passes ventrally. Together with the sternum, this sheet forms an attaching surface, on either side, for the pectoralis muscle (U. M. Norberg, 1970). This ligamentous sheet and the sternal ridge are together analogous to the sternal ridge in birds and pterosaurs. Since these sheets do not fossilize, pterosaurs might have had similar ligamentous sheets along the ridge, and hence better developed flight than generally thought.

Flight Types and Wing Form

Flight style and optimal flight speed vary among flying species and depend on habitat structure, choice of food, foraging behavior, density of food, and so on. Natural selection apparently works toward a wing shape that minimizes the power required to fly in the manner and at the speed that are optimal for the animal.

Hovering Hovering is the most energy-demanding type of flight; it puts large demands on the strength and rigidity of the wing skeleton. Various small birds and bats hover. Examples are the hummingbirds (Trochilidae), sunbirds (Nectariniidae), goldcrests *(Regulus),* nectarivorous bats (Phyllostomidae), and long-eared bats *(Plecotus).* Hummingbirds, whose sizes range from 2 to 20 g, are the most excellent hoverers among birds. Their kinematics is more similar to those of insects than to other birds. Hummingbirds hover with wings fully extended during the entire wing stroke, the wing tip describing a horizontal figure eight (symmetrical hovering; Fig. 8-6a). During the backstroke the bird turns the anatomical upper side of the wing downward, thus establishing a positive angle of attack in the new direction. Hence, the wings create lift during the entire stroke. The horizontal components of the thrusts, which have opposing directions over the two half strokes, cancel in stationary hovering. The wingbeat rate is 60 to 70 Hz for the smallest hummingbirds and about 12 for the largest (Greenewalt, 1975). The strategy an animal should "choose" to hover most efficiently is to reduce stroke frequency and increase stroke amplitude unless an unfavorable increase in profile power results (Rayner, 1979c).

Since the flapping frequency, and hence the angular acceleration, are large in hovering flight, the inertial loads on the wing skeleton are particularly high. Therefore, the supporting structures of the arm wing must be strong, and hence relatively heavy. Being heavy, they should have the center of mass of the wing located as near to the wing base as possible, where the acceleration of the wing is lowest. In birds this can be achieved by having a short arm wing in relation to total wing length (U. M. Norberg, 1979). In hummingbirds the wrist and elbow are entirely stiff (the wing cannot flex), and the bones of the arm wing are robust and short. They are angled relatively sharply at the elbow joint, which contributes to making the arm wing extremely short. Hence, although the wing skeleton is a rigid structure, its moment of inertia is low because of its shortness.

The hand wing is compensatorily long, making the total wing length long for the size of the bird. This reduces the induced power, which is the main power drain in hovering. The hand wing consists mainly of the primary feathers and has low mass, even though long.

Unlike hummingbirds, other hovering birds are not as well adapted for hovering. They are not able to perform the rotation at the shoulder joint that enables hummingbirds to turn the wing's upper surface down during the morphological upstroke, thus producing lift also during the upstroke. Birds that hover asymmetrically beat their wings forward and downward, and then flex them to various degrees during the upstroke (Fig. 8-6b), which contributes only small lift forces. Some larger birds, such as pigeons, can hover for a few strokes. Thus they have sufficient aerodynamic capacity and structural strength, but insufficient power for prolonged hovering.

Bats hover in almost the same manner as birds with asymmetrical hovering. The stroke plane is more nearly horizontal than in forward flight. It is inclined about 30° to the horizontal in the long-eared bat *(Plecotus auritus)* in hovering, contrasted to as much as 58° in slow forward flight (U. M. Norberg, 1976a,b). As compared with other bats, hovering species (size range 6 to 10 g) usually have relatively large wing areas and hence low wing loading. They usually are slow flyers and are highly maneuverable.

Slow Flight and Maneuverability Birds and bats that are capable of slow and maneuverable flight usually have wings with low aspect ratio, large wing area, and low wing loading. These animals are often small or medium-sized. Birds in this group often have a large alula and the ability to separate the primary feathers, which increases lift at slow speed. Camber is usually pronounced, especially in the bats. Those species flying among bushes and trees (or cacti) have limited wing span and hence high wing and disk loading Examples are gallinaceous birds, the wren (*Troglodytes;* Fig. 8-9), and Darwin's finches. These species have rounded wings, and make short, not agile, flights. Gallinaceous birds, as well as some other large birds with similar wing form, are adapted to take off vertically (for instance, to escape predators) in dense vegetation.

Some other species are able to fly slowly, but are not precluded from flying fast. Birds in this category are swallows and kites. They have long wings and large wing area (and hence low wing and disk loadings), pointed wing tips, and no slots. They usually have moderate aspect ratios.

Weis-Fogh (1973) described two unusual, nonsteady-state mechanisms used by some insects and birds (and possibly bats) to create lift in slow flight and hovering: the clap-and-fling and the flip. The former begins with the wings being brought together at the top of the morphological upstroke (clap). They are then moved apart (flung open) and the air is abruptly sucked into the gap between them, so that a circulation is set up almost instantaneously, thus giving some lift. Hence, the lift-delay that is usual at the beginning of a stroke (Wagner effect) is avoided. The clap-and-fling can be seen in pigeons and other birds at takeoff and landing. In the flip, which is a supination at the beginning of the upstroke and pronation at the beginning of the downstroke, the wings are rapidly twisted, thus setting up an appropriate circulation around them. These two mechanisms create circulation prior to, and independently of, the translation of the wing through the air. During slow forward flight and hovering, the nonsteady phases occupy a significant part of the wing-stroke cycle.

Finally, in slow flight and hovering the wings of birds and bats can be moved backward relative to the still air during the upstroke (Figs. 8-5c and 8-13, *bottom*). This backward flick may produce a large thrust.

Fast Flight Wings having high aspect ratio, tapered tips, and high wing loading but low disk loading are characteristic of fast-flying species, such as ducks, shorebirds, and freetailed bats (Molossidae; Figs. 8-9 and 8-10). Their wings are unslotted (as are those of all bats), only slightly cambered, and thin. They have rapid wing beats and small stroke amplitudes, except at takeoff and landing when speed is low. They have streamlined, thin bodies to reduce parasite power, which is high at fast speeds. Large water birds adapted to fly fast, such as divers and swans, must skitter at takeoff to gain adequate air velocity and lift. During this run across the surface of the water, the birds probably also gain some lift by the aerodynamic ground effect. Ground effect is an increase in lift and a decrease in drag of an airfoil that is moving close to the ground.

Intermittent Flight Some bird species use intermittent flight, which can be of two forms, undulating and bounding (Rayner, 1977, 1979c). Undulating flight alternates between flapping and gliding and is used by many large birds that cannot soar efficiently. Bounding flight consists of periods of a few wing strokes alternating with periods with the wings folded against the body. It is used by many small passerines having large profile drag, low disk loading, and high stroke frequency (such as flycatchers), for which flight power is generally high (Rayner, 1977). The low aspect ratio and high parasite drag (giving a low average L/D ratio) that are common among passerines rule out gliding and undulating flight, leaving bounding flight and continuous flapping as the alternatives. Many small birds, but no bats, use bounding flight. Rayner points out that, in addition to the aerodynamic benefit, there may also be physiological advantages from bounding flight.

Soaring Soaring costs a minimum of energy as compared with other types of flight. During soaring, energy is extracted from natural winds and is converted to potential (height gain) or kinetic (speed gain) energy. Because of this use of air movements, both vertical and horizontal, the only additional energy needed is for corrective maneuvers and for holding the wings down in the horizontal position. Many large birds use soaring when searching for food and during migration.

In slope soaring the bird flies in a region of rising air caused by upward deflection of the wind (slope lift) over a slope, forest edge, water wave, and so forth. If the vertical component of the air velocity exceeds the sinking speed of the gliding bird, the bird is carried up. Slope soaring along ocean waves is frequently used by large petrels and albatrosses (Procellariiformes; Pennycuick, 1982).

Thermal soaring occurs in thermals, which are rising volumes of warm air caused by the thermal instability in the atmosphere. The unstable vertical distribution of air of different temperatures, which results from differential heating from the ground or cooling from above, results in turn in the moving thermals. Thermals vary in form and structure but can be classified into two main types, the dust devil (columnar type) and the bubble or vortex ring (doughnut type), which are described in Pennycuick (1972b, 1975).

Some birds, such as vultures, eagles, buzzards, and storks, use thermals for cross-country soaring. They climb in a thermal to some substantial height, glide off in the desired direction, losing height as they go, and then climb in a new thermal. Birds that use thermal soaring as their main means of locomotion typically have short soaring wings, which have a very low aspect ratio (≤ 7; Pennycuick, 1971b), large wing area, and hence low wing loading (Figs. 8-9 and 8-10).

Soaring birds with long soaring wings have narrow wings and an aspect ratio greater than 10. Examples are albatrosses, frigate birds, and gulls. The long soaring wing has no slots, low camber, and very low induced power. The gliding angle is small. Albatrosses have high wing loading and very high aspect ratio (13 to 15; Pennycuick, 1982; Figs. 8-9 and 8-10), and are believed to use dynamic soaring (see below).

Birds with low wing loading can glide slowly, with low sinking speed, and are good at exploiting weak and narrow thermals. Birds with high wing loading can glide fast without excessive steepening of the gliding angle, but with a larger turn radius (the radius of a banked turn is proportional to wing loading, $r \propto Mg/S$). The short soaring wing and the long soaring wing thus are adapted to different requirements.

Dynamic soaring is not dependent on vertical air movements, but instead on variations in horizontal wind speed. Dynamic soaring might be possible in random turbulence but it is more likely in the wind shear that occurs over a flat surface, such as the ocean. Near the surface the wind speed is slowed by friction, so a wind gradient is formed.

The classical interpretation of dynamic soaring technique is described in, for instance, Lighthill (1975) and Pennycuick (1982), and can be briefly

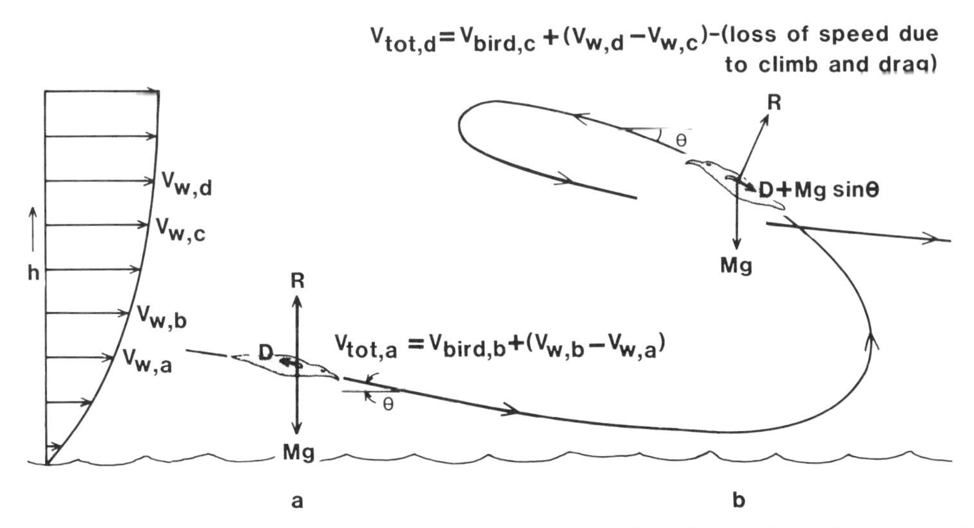

Figure 8-17 Simplified analysis of the dynamic soaring of an albatross in a wind shear. The wind speed increases with height h above the surface. The bird makes use of the wind gradient (change of wind speed with height), which is steep near the surface and decreases further up. $V_{w,a\text{-}d}$ are the wind speeds at heights a-d, and $V_{bird,a\text{-}d}$ the gliding speeds of the albatross relative to the surface at heights denoted a-d. V_{tot} is the bird's speed relative to the surface at given heights.

summarized. When the bird glides downward in a downwind direction (Fig. 8-17a), it gains airspeed and thus kinetic energy. At sea level it soars in slope lift along the windward face of a wave and also makes use of some of the kinetic energy for maneuver. When it has used up most of this energy, the bird climbs into the wind (Fig. 8-17b) and, although it tends to slow down (relative to the water) because it is expending work against gravity, it gains airspeed (and thus kinetic energy) as it climbs into the wind because the wind is blowing progressively faster the higher the bird rises. Thus it converts kinetic energy to potential energy. When the wind gradient becomes too weak to allow any further climb, the bird turns downwind again and can use the gained energy for maneuvering. Using only these techniques, albatrosses would be able to zigzag over the ocean with only infrequent flapping of the wings.

Pennycuick (1982) found, however, that the main soaring method of albatrosses and large petrels appeared to be slope soaring along waves, although windward "pull-ups" in the dynamic soaring technique were also seen in large and medium-sized procellariforms. He concluded that most of the energy for the pull-up must come from the kinetic energy gained by accelerating gliding along a wave in slope lift, since the wind gradient would not be strong enough for the bird to sustain airspeed in a windward climb.

Pennycuick (1982) found that the albatrosses *Diomedea* and *Phoebetria* have a lock on the humerus that works with a tendonous sheet that is in parallel with the pectoralis muscle. The lock restrains the wing from elevating above the horizontal when the humerus is fully protracted, although the arm can be raised when retracted. He also found that giant petrels *(Macronectes)* are similar functionally, but that their anatomy is a little different. This arrangement reduces the energy cost of gliding flight. It was not found in small procellariform species. A functionally similar system occurred also in the large pterosaur *Pteranodon.* As the humerus moved upward there was one position in which the rigid articular head of the humerus locked in the glenoid cavity (Bramwell and Whitfield, 1974). This was due to the peculiar shape of the cavity, which had its upper and lower surfaces set at an angle to each other. The humerus was calculated by the authors to be directed upward at 20° and swept backward at 19° in the locked position, which can be assumed to have been the natural position for gliding.

Bats cannot use soaring flight because convective air currents are absent at night, but some bats can slope soar. The largest pterosaurs were certainly good gliders and soarers. McMasters (1976) pointed out that the wings of large pterosaurs appear to be direct natural counterparts of hang gliders with high aspect ratio and cylindrical camber. The largest hang glider of this type has a span of 11.8 m and a load mass of 100 to 120 kg, which probably is very similar to the largest known pterosaur.

Migration

Many birds of the north temperate zone, especially insectivorous birds, migrate south during autumn when food becomes scarce. Such a journey can span 6000 km and take several weeks. Most European bats hibernate in the winter, but some species, in Europe as well as North America, are known to make regular migrations of a few hundred kilometers and sometimes more.

The distance a bird or bat can fly nonstop depends on its effective lift to drag ratio L'/D', which is the ratio of the weight to some average horizontal force needed to propel the animal along, or, more exactly, the ratio (weight × speed)/(mechanical power for flight) (Pennycuick, 1972a, 1975). The mechanical power required to fly in steady horizontal flight at speed V is

$$P = T'V = Mg\frac{D'}{L'}V,$$

where T', D', and L' are the effective thrust, drag, and lift forces, respectively, as averaged over a number of wing-beat cycles (see Eq. 8-7).

In long migratory flights the situation is more complicated because the animal has to carry extra fuel (that is, weight). Before migration some passerines store up to half their lean body mass in fat (Odum, Connell, and Stoddard, 1961). As this fuel is used up, the animal's mass progressively decreases, and so does the power required to fly and the rate of using fuel. The actual distance Y flown is given by Pennycuick (1969, 1975) as

$$Y = \frac{e\eta}{g}(L'/D')\ln(M_1/M_2), \tag{8-9}$$

where e is the energy released on oxidizing unit mass of fuel, η is the mechanical efficiency, g is the acceleration owing to gravity, M_1 is the body mass at the beginning of the flight, and M_2 the mass at the end. The effective lift-to-drag ratio is a maximum at the maximum range speed V_{mr}, where the distance traveled per unit work is a maximum, and the power required is P_{mr}. From Eq. (8-7), it follows that

$$\left(\frac{L'}{D'}\right)_{max} = \frac{MgV_{mr}}{P_{mr}}. \tag{8-10}$$

Pennycuick (1969) calculated the P_{mr} required to fly at V_{mr} to be proportional to $b^{-1.5}$. Hence, migratory birds should have a long wingspan to be able to cover as long a range as possible on a given amount of fuel.

Some birds, such as geese, fly in formation flight during migration. Theoretical aerodynamic studies of horizontal V configurations have shown that current interaction decreases the induced drag of the birds (Lissaman and Schollenberger, 1970). The vortex sheet left behind a flying bird rolls up into a loop of concentrated vorticity. The adjacent bird in the flock is considered to make use of this upwash, thus decreasing the net induced drag of its wings. It is uncertain whether this interpretation is reconcilable with the vortex theory of flight.

Using photographic data, Nachtigall (1970) demonstrated a phase relationship between wing beats of neighboring birds in flocks of geese flying in V formation, which would be expected in view of the vortex theory. Gould (1972), however, found no such relationship in Canada geese. Heppner (1974) suggested that the V, or echelon, formation might be a response to the necessity for maintaining a high-resolution visual image of neighboring birds.

Most migrating birds do not, however, travel in regular formation, but rather in large clusters. Higdon and Corrsin (1978) used classical aerodynamic theory to estimate the change in total induced drag for both two- and three-dimensional lattices of birds as compared with that for the same number of birds flying individually. They found a decrease in total drag when the flock extends farther laterally than vertically, but an increase in drag in a high, narrow flock. The authors suggest that further research on this topic should include photographic data on the shapes and internal structures of migrating flocks, as well as information on the relationships of the flapping phases among the birds.

Chapter 9

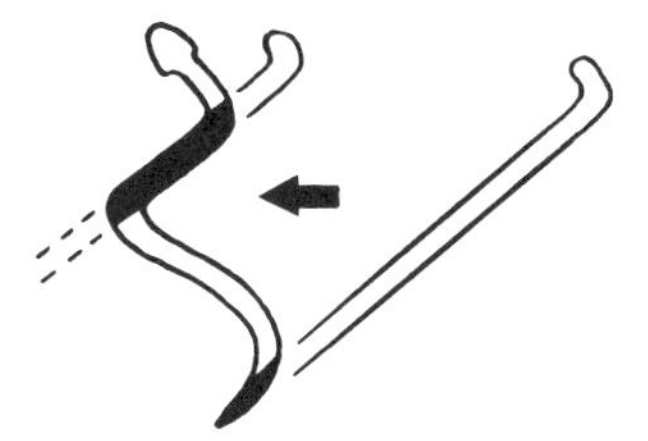

Terrestrial Locomotion without Appendages

James L. Edwards

Although limbs are excellent structures for propelling an animal quickly across a terrestrial habitat, there are situations in which limbs are a hindrance. In movement through dense undergrowth or small crevices, both above and below ground (including those constructed by the animal), limbs are not able to swing freely. Indeed, many limbed lizards and salamanders fold their legs against the sides of the body in such situations and progress by means of the axial musculature, thus adopting a kind of behavioral limblessness. Limbless locomotion is also quite efficient. Snakes crawling on a treadmill have a cost of transport only about half as large as that of comparably sized limbed reptiles (Chodrow and Taylor, 1973). As with undulatory swimmers (see Chapter 7), a partial explanation of this efficiency lies in the absence of the recovery stroke that limbed animals need to return the limb to the propulsive position.

Several different lineages of tetrapods have reduced their limbs or lost them entirely, forcing a return to a reliance on axial musculature to power their propulsion. Such reduction or loss of limbs is always associated with elongation of the body (Gans, 1975). Locomotion in restricted spaces could alternatively have been facilitated by a simple reduction in overall size. The advantage of elongation is that it allows crevice traversal while at the same time maintaining the total volume of the body cavity. Elongation may also increase the efficiency of axial locomotion by providing both more axial musculature and a longer, more flexible vertebral column through which that musculature can act.

Although the above scenario may well describe the initial adaptive pressures leading to loss of limbs, such loss has not restricted limbless vertebrates to crevice-defined habitats. For example, the snakes, perhaps the quintessential limbless group, now occupy nearly all habitats available to vertebrates, including aquatic, arboreal, and even gliding niches.

Survey of Limbless Tetrapods

Among the living amphibians one order, the caecilians or Apoda, is characterized by the complete lack of limbs, with no trace of either pectoral or pelvic girdles. Although some caecilians are aquatic, the majority are fossorial and use lateral undulatory and modified concertina locomotion (as defined below). Caecilians typically have 90 to 120 vertebrae; more than 250 are present in some very elongate species. Most of the body is divided into annulate segments. The skull is robust and compact and is adapted for active burrowing (Wake and Hanken, 1982).

Although the salamanders (order Caudata) contain no completely limbless forms, the elongate Amphiumidae have very reduced limbs and the Sirenidae lack pelvic limbs. Both these families are primarily aquatic, however. Facultative terrestrial limblessness is commonly seen in fully limbed salamanders when they are harrassed; in such situations the animals fold their limbs against the side of the body and progress by a relatively undirected form of fast lateral undulation.

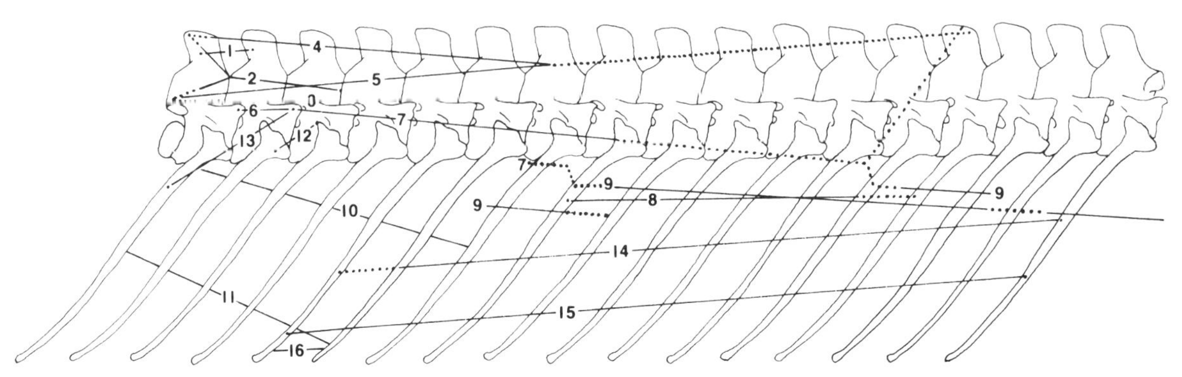

Figure 9-1 Schematic representation of the midtrunk axial musculature of a representative snake *(Python)*, with anterior to the right. Muscular tissue indicated by solid lines, tendinous connections by dotted lines. Note the very large number of segments crossed by several of the muscles. Muscle names: 1 = interneuralis; 2 = multifidus; 3 = interarticularis superior; 4 = spinalis; 5 = semispinalis; 6 = interarcualis inferior; 7 = longissimus dorsi; 8 = supracostalis dorsalis; 9 = iliocostalis; 10 = obliquus internus dorsalis; 11 = obliquus internus ventralis; 12 = tuberculocostalis; 13 = levator costae; 14 = supracostalis lateralis superior; 15 = supracostalis lateralis inferior; 16 = intercostalis ventralis. (Modified from Gasc, 1974.)

Two groups of fossil amphibians in the subclass Lepospondyli contained fully limbless forms. Aistopods (order Aistopoda) had up to 200 vertebrae and were rather snakelike in overall appearance. Apparently they were terrestrial animals that inhabited shallow crevices or leaf litter. Some nectrideans (order Nectridea) were elongate, especially in the tail region, and had reduced limbs or none at all. The presence of fan-shaped neural and hemal spines on the tail vertebrae and the nature of the sediments in which they are found suggest that these animals were pond dwellers. Finally, a third order, Microsauria, had several genera that were elongate with somewhat reduced limbs. Based on cranial characters, Carroll and Currie (1975) have suggested that microsaurs were the ancestors of the caecilians, even though their postcranial skeletons were very different from those of caecilians (Wake and Hanken, 1982).

Among the reptiles only members of the order Squamata have become limbless. At least ten different lineages of lizards (suborder Sauria) have exhibited tendencies toward elongation and limblessness, with several genera of skinks, some anguids, one family of geckoes (the Pygopodidae), some gerrhosaurids, and the dibamids and anniellids all containing species lacking one or both pairs of limbs. Lizards are known to use lateral undulatory and concertina locomotion (Gans, 1975).

All amphisbaenians (suborder Amphisbaenia) except the genus *Bipes* lack both pairs of limbs, and the latter genus lacks the pelvic pair. Amphisbaenians are fossorial and have elongate, annulate bodies somewhat reminiscent of caecilians. They use their modified heads or, in the case of *Bipes,* their robust forelimbs to excavate tunnels. Lateral undulatory, rectilinear, and concertina locomotion have been reported in amphisbaenians (Gans, 1975).

The snakes (suborder Ophidia) are the most diverse group of limbless tetrapods, with only small, clawlike rudiments of hind limbs in some boids. Many snakes are extremely elongate, with more than 500 vertebrae reported for some extinct species. The axial musculature of snakes is extremely complex (see Fig. 9-1), with the origin and insertion ends of a single muscle of the spinalis-semispinalis group (muscles 4 and 5 in Fig. 9-1) being separated by up to 45 vertebrae (Jayne, 1982). Although snakes today occupy a wide variety of niches, including burrowing, arboreal, and aquatic habitats, they appear to be derived from burrowing ancestors that had nearly lost their eyes. Snakes use all the locomotor modes found in other limbless tetrapods, as well as one mode (sidewinding) restricted to snakes.

Modes of Limbless Locomotion

Limbless locomotion is usually divided into four distinct types: lateral undulation, concertina, sidewinding, and rectilinear. The first three modes are powered by waves of contraction that travel from anterior to posterior along the axial musculature of alternating sides of the trunk. The horizontal traveling waves of muscular activity are derived from the myomeric swimming mode of fishes (see Chapter 7), with the propulsive forces now being applied to the terrestrial substrate. In fact, limbless locomotion has evolved only in those tetrapod groups that retain a large amount of lateral axial movement even in their normal limbed locomotion—the nonanuran amphibians and the squamate reptiles. Frogs (with their short vertebral columns and heavy reliance on hind-leg extension), birds (with stiff, fused vertebral columns), and mammals (with primary reliance on vertical movements of the axial skeleton) have never produced limbless forms.

In rectilinear locomotion, although waves of contraction pass down the body, these waves are confined to the cutaneous and costocutaneous muscles. In addition, the muscular activity is synchronous, with muscles on the right and left sides of the body acting together. Such propulsion presumably requires a more specialized type of motor control and has evolved only in snakes and amphisbaenians.

Although the limbless locomotor modes will be discussed separately, it should be recognized that a single animal is usually capable of using more than one mode, even at a single time. For example, Gasc (1974) described an African sidewinder *(Cerastes cerastes)* moving with concertina in the posterior portion of the body and beginning sidewinding in the anterior portion; and as soon as a part of the body touched a wall, that portion of the body began to use rectilinear locomotion.

Lateral Undulation Lateral undulation is the most common mode of limbless locomotion and is known to be used by all limbless forms and in most terrestrial habitats (on the surface, below ground, in arboreal situations). It requires projections from the substrate to act as pegs or pivotal points to provide the reaction forces necessary to push the animal forward.

Lateral undulation is best understood by a comparison with the undulatory swimming of fishes (see Fig. 7-2): the propulsion for forward movement is derived by nearly identical means on water and on land. In both situations muscular contraction is used to produce a bend, which passes posteriorly along the length of the body. In water the bend is continuously pushing obliquely backward against the surrounding water, which responds with an equal but opposite reaction force (Fig. 7-2b); therefore, all parts of the body involved in forming the wave produce forward propulsion. In contrast, in the terrestrial situation only those parts of the body in contact with pivotal points provide propulsion.

Another major difference is that in water the traveling waves of muscular activity progress down the body at a speed greater than that of forward propulsion of the animal. As a result, the waves also travel with respect to the substrate. On land the lateral undulatory waves pass posteriorly along the body at the same speed as the animal moves forward; thus the wave stands with respect to the ground and the body follows the same course along its whole length.

The differences between the sinusoidal swimming of fish and lateral undulation of limbless tetrapods are relatively minor and are due primarily to the differences in substrates. Elongate fishes such as anguillid eels move by lateral undulation when on land, and limbless tetrapods swim with the fish mode (Gans, 1962).

Lateral undulation is also used by many sand swimmers, which move through loose sand as if it were a fluid medium. Since the "pegs" against which a sand swimmer pushes are located all around its body, the animal does not need to form a standing wave with respect to its surroundings. Instead, it passes traveling waves down its body at a faster rate than its forward progression, just as a fish does in water (Gans, 1974). Sand swimming is seen in some fully limbed lizards (such as *Uma*), several skinks with reduced or absent limbs, and some snakes.

Lateral undulation is commonly used in arboreal situations. Here the animal must derive an upward as well as forward component from the pivot points. Arboreal snakes often have extremely long axial muscles, presumably to hold up sections of the body as the snake stretches from branch to branch.

Figure 9-2 indicates the forces involved in lateral undulation on a terrestrial substrate. *F*1 is the force directed obliquely backward on the peg by a portion of the body. The peg responds with an equal but opposite reaction force *R*1, which can be factored into a longitudinal component *L*1, which drives the body forward, and a horizontal component *H*1, which is wasted energy as far as propulsion is concerned.

Since the body is sliding past the peg, at that point it will produce sliding friction, which must be overcome. The sliding friction *f* acting along the surface of the body can be visualized as deflecting the reaction force posteriorly (*R*2), thus decreasing the propulsive component *L*2 and increasing the amount of energy lost to horizontal movement *H*2. Sliding friction is equal to the coefficient of friction (determined in large part by the roughness of the peg and the body segment touching it) times the force exerted against the peg. As the ventral side of the animal glides over the substrate, it also produces sliding friction, which increases with the mass of the organism.

Certain features of the integument of snakes appear to reduce frictional forces. The smooth ventral scales and the imbricating (overlapping) nature of the scales reduce ventral sliding friction. The regular cycles of shedding of the skin may also aid in reducing friction by periodically renewing the unroughened outer layer of the scales. The microornamentation of the body scales of uropeltid snakes inhibits wetting of the scales (Gans and Baic, 1977); wetting would cause the adhesion of soil particles, leading to increased friction with the walls of the tunnels of these burrowing animals.

Gray and Lissmann (1950) measured the propulsive forces exerted by a snake when traveling by lateral undulation through a metal channel (Fig. 9-3). The channel was fitted with small steel strips, the deflection of each strip measuring the force acting normally on the channel at that point. When moving from left to right through

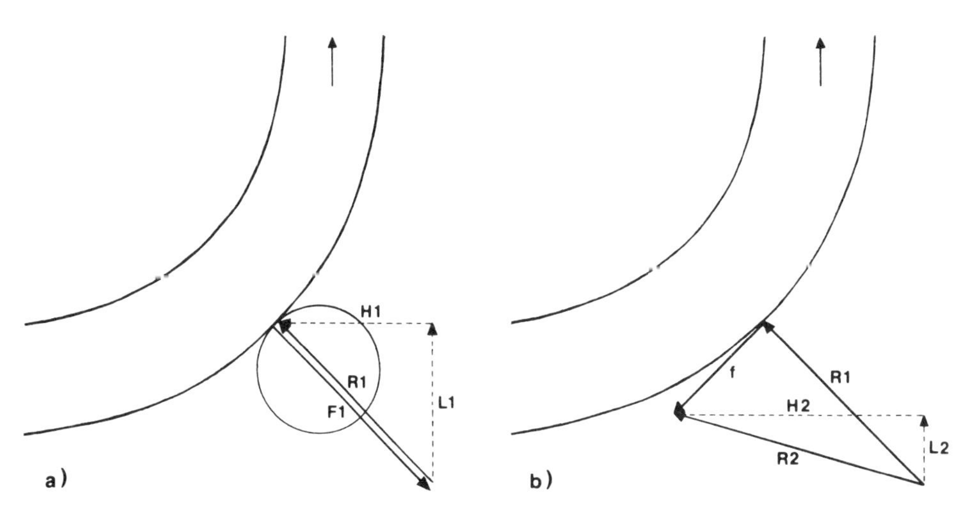

Figure 9-2 The forces acting on a portion of the body of a limbless tetrapod in contact with a pivot point during lateral undulation. a: First approximation, ignoring sliding friction. *F*1 = force exerted by body on peg; *R*1 = reaction force exerted by peg on body; *L*1 = longitudinal component of reaction force; *H*1 = horizontal component of reaction force. b: The effect of sliding friction. *f* = sliding friction; *R*2 = reaction force reduced by friction and factored into its longitudinal (*L*2) and horizontal (*H*2) components.

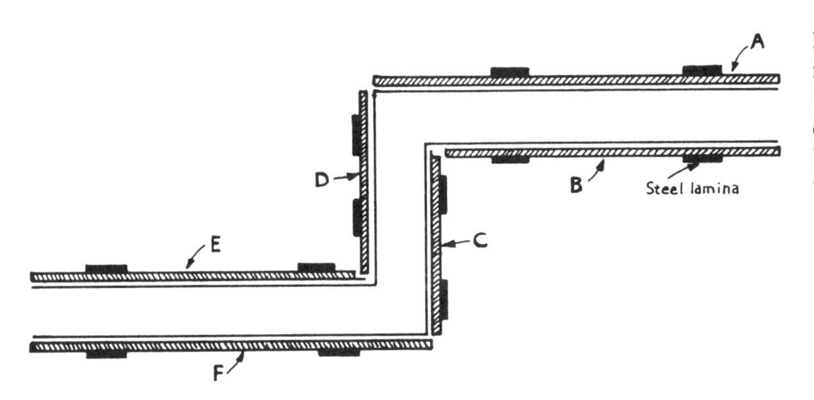

Figure 9-3 Channel (as viewed from above) used by Gray and Lissmann (1950) to measure forces exerted by a snake moving by lateral undulation. Letters identify the walls of the channel.

the channel, the snake was found to exert forces only on the bottom and left walls (at *B*, *D*, and *F* in Fig. 9-3).

The frictional forces encountered by snakes were determined by Gray and Lissmann in the following way. The forces needed to tow the bodies of dead snakes across various substrates were measured. Since the towing forces must be equal to the ventral friction and since ventral friction is equal to the coefficient of ventral friction times the mass of the body, the coefficient of ventral friction was easily determined. For surfaces of dry wood, metal, or glass the coefficients of ventral friction ranged from 0.2 to 0.4.

The coefficient of lateral friction was more difficult to estimate. Gray and Lissmann constructed a channel with a movable section (Fig. 9-4); the movable section was mounted on a balance capable of measuring forces both normal and tangential to the direction of the snake's travel. The force acting normal to the channel (R in Fig. 9-4) could be measured directly from the balance, as could the total tangential force, comprising both lateral and ventral friction (μR and Fv, respectively). Ventral frictional forces were again measured by towing a dead snake of the same mass across the movable channel section. The lateral frictional force for the section was then equal to the total tangential force minus the ventral frictional force. Finally, the coefficient of lateral friction was determined by dividing the lateral friction force μR by the normal force R. Coefficients of lateral friction for Gray and Lissmann's apparatus were in the range of 0.35 to 0.4, nearly identical to the coefficients of ventral friction.

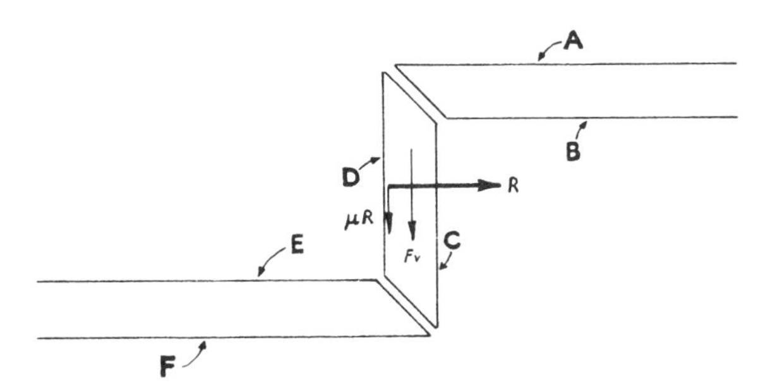

Figure 9-4 Channel with movable section for measuring forces normal and tangential to the channel. Normal forces R provide forward propulsion. Frictional forces include lateral friction μR and ventral friction Fv. Other letters identify the walls of the channel. (Modified from Gray and Lissmann, 1950.)

For a snake moving at a constant velocity, the total propulsive forces should equal the total frictional forces. The propulsive forces at each pivotal point can be measured directly by the deflection of the flexible steel strips of the apparatus in Figure 9-3, and the frictional forces can be calculated as shown in Figure 9-4. For a grass snake *(Natrix natrix)* weighing 80 g, Gray and Lissmann measured a propulsive force of 68 dynes when the animal moved through a zigzag channel with five rectangular bends. Using a value of 0.4 for both lateral and ventral coefficients of friction, the calculated frictional value was 59 dynes. Given the many approximations used in arriving at the calculated figure, the disagreement between the calculated and measured figures is not surprising.

In order to measure the forces exerted on single pegs by snakes using lateral undulation, Gray and Lissmann (1950) used hanging pendulums as pivot points; the deflection of the pendulums indicated both the direction and magnitude of the forces exerted against them. Snakes moved past a single peg by pressing directly posteriorly on it (Fig. 9-5), with all the reaction force contributing to forward propulsion. As the number of pegs was increased, the amount of energy lost to lateral forces increased dramatically; the forward component increased much more slowly. This suggests that for any given substrate and spacing of pegs, there may be an optimum number of pivot points that a particular size snake could contact to maximize forward movement per amount of energetic output. Gans (1962) suggested that at least two pivot points are necessary at any one time for lateral undulation, and later he stated (1970) that at least three are needed for sustained forward motion. To test the latter statement Bennet et al. (1974) measured maximum speed as snakes of three different species were chased across boards studded with pegs arranged in various patterns. In agreement with Gans's speculation, the speeds attained when the animals could contact fewer than three pegs at one time were much lower than those attained when more than three pegs were touched. This also explains why longer snakes are better lateral undulators than short ones (Gasc, 1974): at any moment longer animals have a large array of pivot points from which to choose and thus should be able to contact three or more points more often than their shorter relatives can.

Gray and Lissmann (1950) theorized on the probable sites of muscular activity used by a snake to produce lateral bending. They reasoned that if a flexible rod is pushed through a sinuous channel, the rod must shorten at two points with respect to each bend of the channel: on the concave side going into each curve and on the convex side leaving the curve (Fig. 9-6). Lateral undulation is like moving through a tube in that each point of the body follows in the same path as that directly before it; therefore, they felt that active contraction must occur in the muscles at analogous points during lateral undulation. They also proposed a mechanical model consisting of rigid elements connected by ball and socket joints and moved by short muscles; they then showed that contractions of the muscles at the suggested points (concave side behind a curve, convex in front of it) could

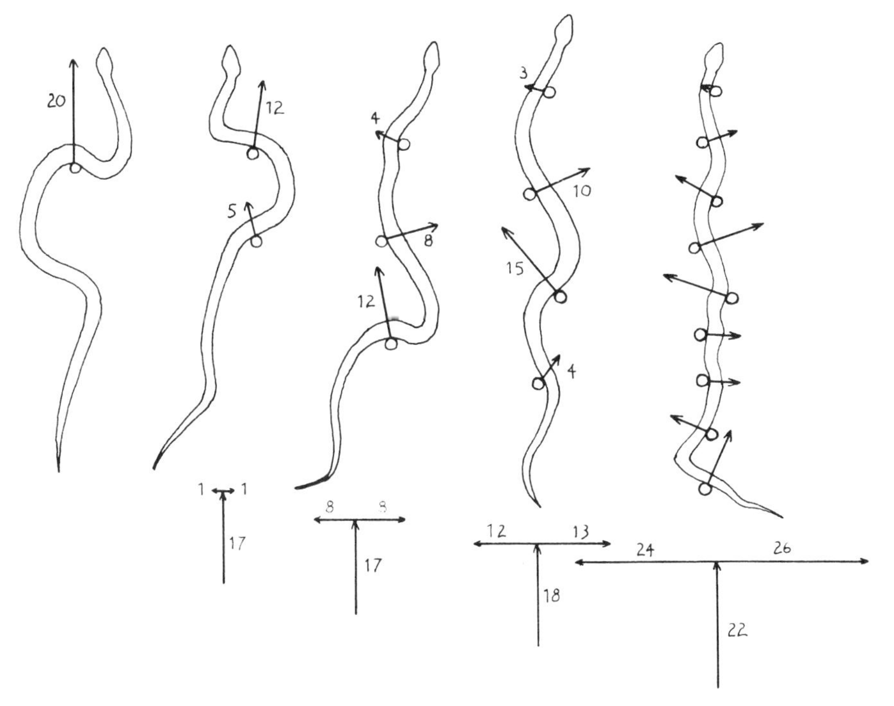

Figure 9-5 Force data from snakes using variable numbers of pegs for lateral undulation. Vectors below figures indicate summary forces (reported in grams as in the original) in the directions indicated. (Modified from Gray and Lissmann, 1950.)

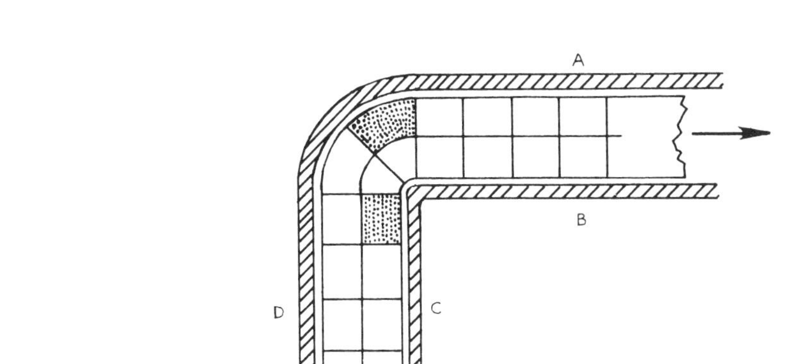

Figure 9-6 Theoretical changes in segment length required by a snake locomoting through a sinuous tube in the direction of the arrow. Segments must decrease in length (presumably through muscular contraction) at sites indicated by stippling: on the concave side going into a curve and on the convex side coming out of a curve. (Modified from Gray and Lissmann, 1950.)

produce the same distribution of forces against the walls of the channel as they had actually measured in the real situation.

Unfortunately, the model of Gray and Lissmann is not realistic. Rather than being short, simple elements extending directly from one vertebra to the next, the axial musculature of snakes is complex (Fig. 9-1), with some muscles extending over large numbers of vertebrae by means of long tendons. In addition, tendinous connections exist between many of the muscle groups. For example, in many snakes the spinalis (muscle 4 in Fig. 9-1), semispinalis (muscle 5), longissimus dorsi (7), and iliocostalis (9) are interconnected by a complex series of tendons and presumably act as a coherent unit. In such situations the insertions of the anterior tendons of the complex may lie more than 40 vertebrae anterior to the origin of the longissimus dorsi (Gasc, 1974, 1981; Jayne, 1982). Because much of the actual muscular tissue is located toward the posterior end of the complex and because of interactions between adjacent muscular groups, it is difficult to predict the resultant effect of contraction of any of these muscles. However, it appears likely that the site of action is usually located far anterior to the contractile tissue, throwing much doubt on Gray and Lissmann's model.

Gasc (1974) argued that the longissimus (muscle 7 in Fig. 9-1), aided by the iliocostalis (9), is the fundamental unit for lateral bending in snakes. However, owing to its length and relatively low ratio of muscular to tendinous tissue, the longissimus dorsi might have some difficulty in actually initiating lateral flexion. Gasc therefore suggested that the levator costae (muscle 13) serves to initiate flexion. The levator costae is a relatively short, robust muscle running from the prezygapophyseal process of a vertebra to the proximal region of a rib located one or two segments posteriorly. Gasc is thus proposing that the rib acts as the origin of the levator costae, at least during the initiation of lateral flexion.

I see two difficulties with Gasc's hypothesis. First, the line of action of the levator costae is oriented posteroventrally, and it would therefore cause torsion as well as lateral bending of the vertebrae. Snake vertebrae are replete with accessory processes that combat torsion, so that the ventral component of the muscle's action would be wasted energy. Second, the levator costae is the most obvious muscle to stabilize the rib against the posterior pull of the iliocostalis during lateral bending; the iliocostalis and levator costae attach at nearly the same point along the rib and their lines of action are in large part antagonistic

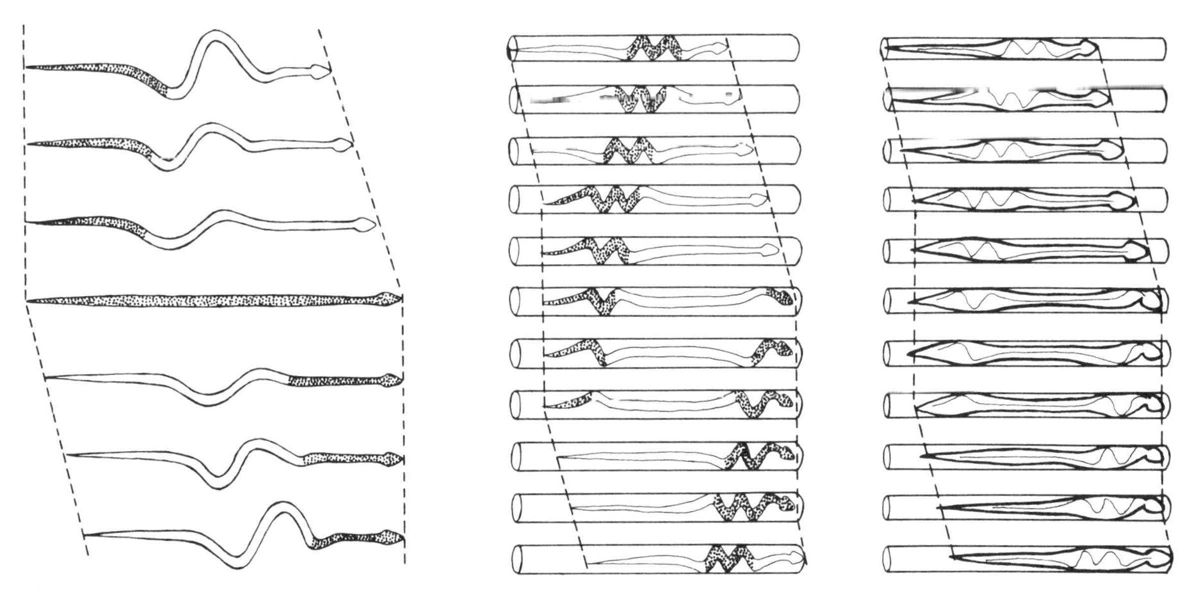

Figure 9-7 Three types of concertina locomotion: surface *(left)*, subsurface *(middle)*, and burrowing by means of internal bending of the axial skeleton *(right)*. Stippled areas denote portions of body in static contact with the substrate.

(if the levator costae is considered to have its nominal origin on the vertebra). Perhaps the interarcualis inferior (muscle 6 in Fig. 9-1) or the multifidus (2), muscles whose line of action is almost wholly in the anteroposterior plane, is the initiator of lateral bending, or at least aids the levator costae in this action.

As the above discussion indicates, our understanding of the muscular activity underlying lateral undulation in snakes is at a very primitive level. Functional speculations that are derived solely from lines of action on diagrams or from manipulation of dissections must be considered only tentative hypotheses. Data from living organisms, such as electromyographic recordings coupled with filmed motion analysis, must be gathered before the controversy over the generation of lateral bending in snakes can be resolved.

Concertina Locomotion Lateral undulation is impossible on surfaces where pegs are very small or very far apart or in straight-sided tunnels where the animal cannot push backward. In such situations limbless tetrapods move by concertina locomotion. Concertina propulsors throw a large part of the body into S-shaped curves and place these curves in contact with the substrate, thus providing a stable region with a large reservoir of static friction. More anterior portions of the body are extended forward from the stable region, while more posterior portions are pulled forward to be added into the stable region. The resulting movement resembles the folding and unfolding of the bellows of an accordion or concertina (Fig. 9-7).

Any part of an animal in static contact with the substrate will provide a stable base against which the rest of the body can act. The force that can be exerted against the substrate without causing sliding is determined by the mass in contact with the substrate times the coefficient of friction. By folding a large portion of its body into an S curve and keeping it in static contact, a limbless animal thus maximizes its static reservoir. Forward propulsion is then achieved by removing the anteriormost part of the folded portion from contact with the substrate and unfolding it presumably by contraction of the muscles on the convex side of each curve. However, removal of the anterior part of the body from static contact reduces the static friction reservoir, which must be replenished by folding more posterior portions and adding them to the static S curve. Thus the portion of the body in contact with the substrate moves posteriorly. Eventually the anteriormost body portions must again be folded and brought into contact with the substrate to initiate a new region of static contact.

This type of surface concertina locomotion is used by all limbless tetrapods on flat substrates on which projections are either too small or too sparse to act as good pivotal points. It is also used by snakes in traversing thin structures, such as branches and even telephone wires, where pegs are not available for lateral undulation. The snake loops its body transversely several times over the branch or wire to provide the requisite region of stable contact. Surface concertina locomotion is occasionally used by limbed animals as well. On slippery substrates a California slender salamander *(Batrachoseps attenuatus)* will fold its tail into S curves, apply the curves to the substrate, and then use the tail in typical concertina mode to aid the limbs in forward progression.

A slightly different kind of concertina locomotion is used by many snakes to traverse straight-sided tunnels. Again the body is thrown into S curves, but the animal increases its static friction reservoir by pressing the sides of the body against the tunnel walls (Fig. 9-7). A variant of this type of locomotion is even used by some snakes for climbing; the irregular bark of a tree trunk provides the points against which the body can push.

A third type of concertina locomotion depends on specializations that allow certain vertebrates to bend the vertebral column separately from the outer parts of the body. This type is used in burrows and crevices only slightly larger in diameter than the body of the animal. By bending a portion of the column into S curves, the animal expands the body wall, bringing it into contact with the walls of the burrow or crevice and thus providing a point of stable support. This type of concertina locomotion is seen in caecilians (Gaymer, 1971), amphisbaenians and uropeltid snakes (Gans, 1973), and possibly in typhlopid snakes (Gasc, 1974).

The expansion is achieved by different means in different groups. In caecilians the vertebrae, ribs, and immediately adjacent axial muscles are separated from the viscera and more lateral muscles of the body wall by connective tissue sheaths. Movement of the axial core then expands the rest of the body. In amphisbaenians and uropeltid snakes the vertebral column, axial muscles, and viscera all move together within the rather loose skin; thus only a thin envelope of skin and muscle is not involved in producing the body expansion. These reptiles use body expansion not only for locomotion, but also for tunnel expansion and as a reservoir of static friction from which to launch penetration movements. Uropeltids are able to expand the anterior part of the body to more than twice the normal diameter; the anterior body musculature has several ultrastructural and enzymatic specializations to cope with this activity (Gans et al., 1978).

Although concertina locomotion allows propulsion in situations where lateral undulation is impossible, this form of movement has several liabilities. Since it requires the alternate starting and stopping of each body segment, momentum is not conserved, and the locomotion is generally quite slow. It is also difficult for short, stout-bodied snakes to achieve tight enough S curves to make concertina movement a viable locomotor mode.

Sidewinding Sidewinding is known only among snakes. It is typically used on yielding substrates, such as sand and mud, but is also commonly found interspersed with concertina and lateral undulatory locomotion on mixed or smooth surfaces.

As in lateral undulation and concertina locomotion, waves of contraction pass down alternate sides of the body of a snake moving by sidewinding, and as in concertina, portions of the body are in static contact with the substrate. However, sidewinding is much quicker than concertina locomotion and appears to be very efficient; Cowles (1956) reported movement of more than 1.5 km in part of a single night's travel by a sidewinder *(Crotalus cerastes).*

On sandy substrates a sidewinding snake produces a highly distinctive trackway (Fig. 9-8) consisting of a series of J-shaped elements. A sidewinding movement is initiated by the snake's lifting its anterior portion from the substrate. The head and "neck" are then placed on the ground with the head pointed in the direction of

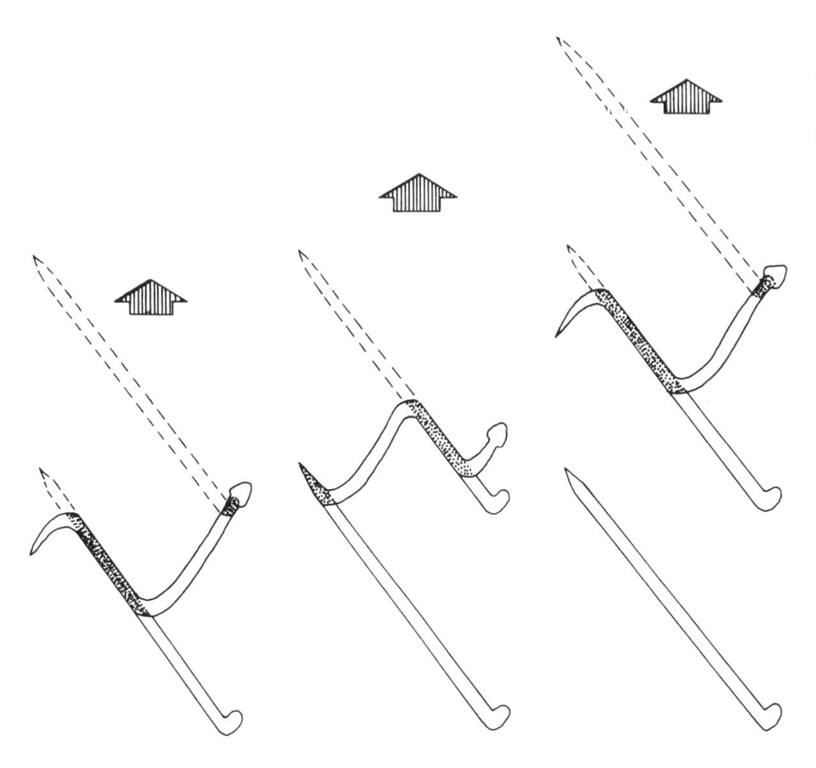

Figure 9-8 Successive moments during sidewinding of a snake moving in the direction of the arrows. Shaded portions of the body are in contact with the substrate; open portions are lifted from the substrate.

travel. The "neck" is then bent tightly to one side, producing the hook of each J-shaped track. The snake continues to lift more posterior portions of the body from the preceding track and place them in a straight line behind the "neck," thus lengthening the stem of the J. A new track is generally initiated before the snake has completed the old one. Thus at any moment a sidewinding snake usually has two portions of the body in contact with the substrate and two portions free of it. Also note that at any time the caudalmost parts of the body in any track are advanced in the direction of travel.

As in concertina locomotion, it is the mass of a sidewinding snake in contact with the substrate that provides the reservoir of static friction against which the moving portions of the snake can push. On yielding substrates the snake's body often amasses a small pile of particles on the side away from the direction of travel. This "windrow" helps to inhibit backward sliding. The portion of the body in contact with the substrate is often twisted around the long axis of the snake, maximizing the area in contact with the windrow (Gans, 1974). The sidewinders of the African and American deserts *(Cerastes, Eristicophis,* and *Crotalus cerastes)* also have vertical keels on the scales at the sides of the body that may act to reduce any sliding tendencies in the direction of travel. Furthermore, sliding friction is minimized because the moving portions of the body are lifted out of contact with the substrate.

The height that a sidewinder lifts the moving parts of its body varies according to the coefficient of friction of the substrate (Gans and Mendelssohn, 1972). On smooth substrates with low coefficients of friction a snake often drags its body between tracks, whereas high-friction substrates elicit high loops in the body between tracks. The loops must be lifted high enough to avoid contacting any irregularities in the substrate; sidewinding snakes periodically use the ventral surface of the body to test the substrate for irregularities.

Sidewinding is often characterized as an adaptation for overcoming the unstable pivot points provided by a shifting substrate. Cowles (1956) suggested that in desert environments sidewinding may also allow movement across hot substrates by minimizing the amount of time any portion of the body is in contact with the ground. (Unfortunately, no controlled experiments have been performed with substrates of various temperatures to test Cowles's hypothesis.) However, sidewinding is not restricted to desert species and nearly every snake will use it under the right conditions. For example, I have observed that neonatal Florida water snakes *Nerodia fasciata)* use sidewinding when placed on fine sand and resort to a mixture of sidewinding, concertina, and lateral undulation when placed on a smooth linoleum floor.

There is some controversy regarding the evolution of sidewinding. Gray (1946) suggested that sidewinding results when normal lateral undulation is unopposed by sufficient external forces. He showed that the wave form of the body at any point during sidewinding is identical to the form that the body would have assumed at a comparable point in lateral undulation. He reasoned that the key to sidewinding lies in the ability of the snake to transfer its weight to the segments in contact with the substrate, which in turn depends on the action of the dorsoventral musculature.

The similarity of sidewinding to lateral undulation was also stressed by Cowles (1956) and Brain (1960). Brain likened sidewinding to lateral undulation with every other curve of the body raised from the ground (Fig. 9-9). Because of the lifted segments, the lateral forces produced at each pivot point are unbalanced and the snake moves laterally as well as forward. Cowles noted that when entering water, a sidewinding snake begins lateral swimming without any perceptible change in pattern; the same imperceptible change, this time from swimming to sidewinding, occurs when a swimming snake comes onto land. Gans and Mendelssohn (1972), on the other hand, were impressed with the similarities of sidewinding to concertina locomotion. They noted the use of static friction in both these locomotor types and suggested that sidewinding evolved from concertina locomotion.

An unanswered question is why sidewinding is restricted to snakes. Gray's (1946) analysis seems to suggest that nonophidian limbless tetrapods are unable to sidewind because they lack either the requisite dorsoventral musculature or a suitable neural control mechanism.

Rectilinear Locomotion In contrast to all other modes of limbless propulsion, rectilinear locomotion is not produced by lateral bending of alternating sides of the body. Instead, the body is held in a straight line—hence the name—and waves of contraction pass down the costocutaneous musculature of both sides of the body simultaneously. As a result of these contractions the ventral body wall is alternately bunched and stretched, a process that early investigators ascribed to anteroposterior movement of the underlying ribs. Bogert (1947) performed the simple experiment of cutting a small window in the skin of a living snake and observed that the ribs do not move at all during rectilinear locomotion.

The most thorough study of rectilinear locomotion is that of Lissmann (1950). He painted small dots on various portions of the body of a boa constrictor and filmed the animal moving by rectilinear locomotion. He found that the dorsal portion of the body moved forward at a steady velocity, while the ventral scutes followed a stepwise progression (Fig. 9-10). At some times the scutes were rapidly accelerated forward, while at other times they stood still or even slightly lost ground.

The only snake muscles that seem to be correctly situated to produce rectilinear movement are the costocutaneous muscles, running from the ribs to the skin. Snakes have two sets of these muscles (Fig. 9-11). The costocutaneous inferior runs from the ventral tip of a rib to attach to the skin a few segments anteriorly; its line of action is almost wholly in the anteroposterior plane. In snakes that are good at rectilinear locomotion, this muscle is robust. The weaker costocutaneous superior muscle originates on the middle of a rib and inserts on the skin a few segments posteriorly; its line of action thus has a distinct dorsal component.

Lissmann suggested that a sequence of rectilinear locomotion begins with the contraction of the costocutaneous inferior muscles in one region of the body. The resultant pull on the ventral scutes fixes them to the ground and may even dig the posterior edges of the scutes into the substrate, thus providing a pivot point against which other muscles can act. As the inferior costocutaneous muscles immediately posterior to this point con-

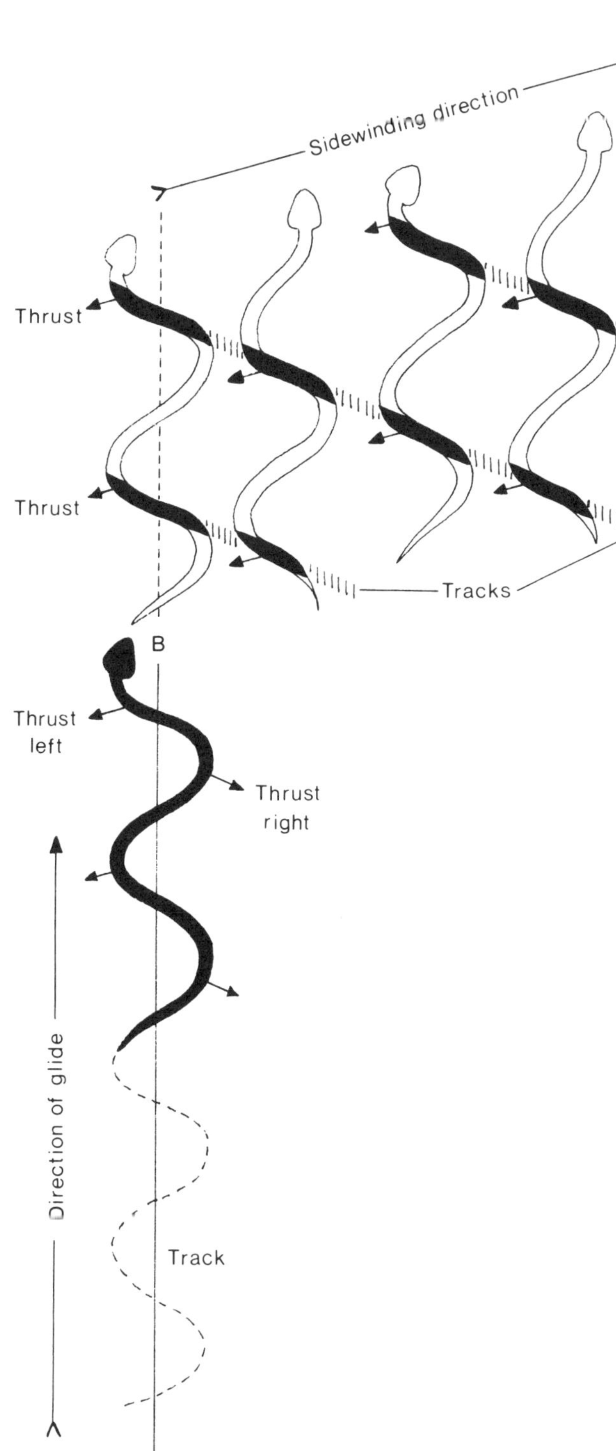

Figure 9-9 Diagram modified from Brain (1960) correlating lateral undulation and sidewinding. Portions of body in contact with substrate are shaded. Snake at bottom is using lateral undulation with entire body in contact with substrate; lateral forces from right and left sides balance each other and the snake moves straight ahead. In sidewinding *(above)*, snake lifts those portions that would have provided thrust on the right side of the body; as a consequence, lateral forces are unbalanced and the snake drifts to the right.

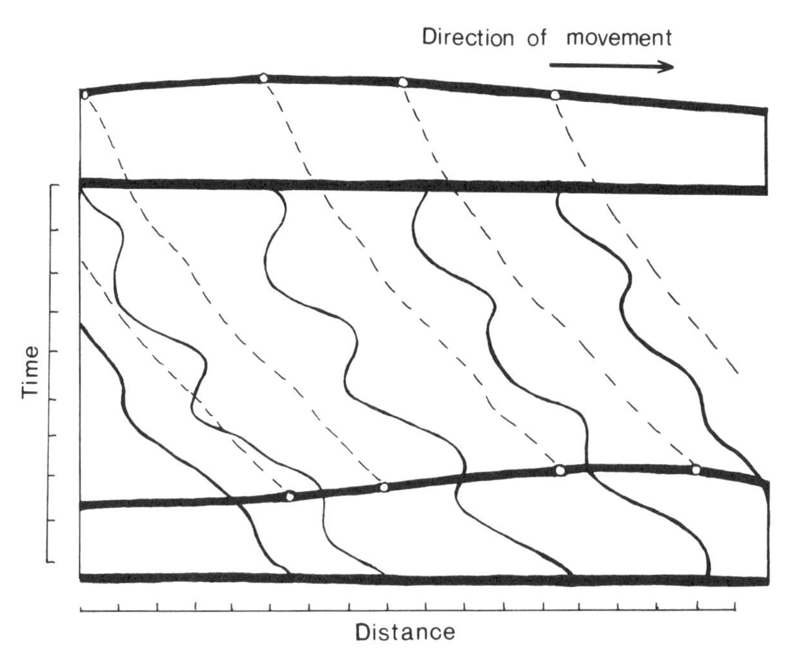

Figure 9-10 Tracings of a portion of the body of a snake using rectilinear locomotion. Broken lines show successive positions of points on the dorsal surface of the animal; solid lines connect ventral points. Note that the dorsal side of the snake moves continuously forward, whereas the ventral side progresses in jerks.

tract, they pull the vertebral column forward and at the same time add their affected scutes into the pivot point. Simultaneously, the superior costocutaneous muscles anterior to the pivot point contract to lift scutes off the substrate and out of the pivot point. As a result, each pivot point moves posteriorly. Several pivot points are generally present along the body of a snake using rectilinear locomotion.

In addition to their antagonistic actions on the ventral scutes, the two sets of costocutaneous muscles also affect the rest of the body very differently. The sequential contractions of the inferior costocutaneous muscles impart a continuous momentum to the vertebral column, leading to the steady forward movement of the dorsal side of the body noted by Lissmann. The ventral part of each body segment is alternately fixed to the substrate and kept in place by the inferior muscles, then lifted out of contact with the substrate by the superior costocutaneous and quickly accelerated forward to catch up with the dorsal side of

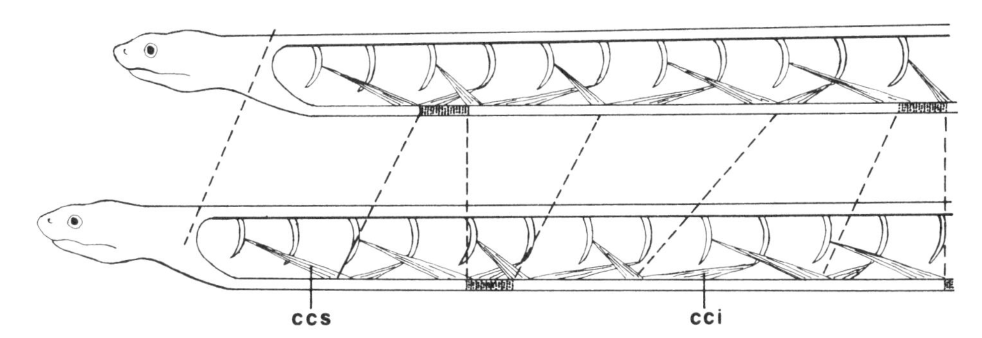

Figure 9-11 Lissmann's model (1950) for the production of rectilinear locomotion by costocutaneous muscles. Contraction of costocutaneous inferior muscles *(cci)* propels vertebral column forward and fixes ventral scutes into pivot points *(stippled regions)*. Costocutaneous superior muscles *(ccs)* lift scutes and accelerate the ventral side of the body forward to catch up with the vertebral column. (Modified from Engelmann and Obst, 1981.)

the body. The large inferior costocutaneous muscles thus provide the propulsive action for the majority of the body, whereas the slender superior muscles are only required to lift and transport the scutes.

Snakes are able to move only forward with rectilinear locomotion. Amphisbaenians, however, can move either forward or backward. In addition to the two sets of costocutaneous muscles, amphisbaenians possess another set of muscles connecting the dorsal portion of each rib to the skin anterior to the rib. It is apparently this third set of muscles, coupled with the loose, annulate skin, that allows posterior rectilinear locomotion in amphisbaenians (Gans, 1974).

Rectilinear locomotion is used on the ground, on tree limbs, and in burrows. Although, like concertina movement, it relies on areas of static friction, rectilinear locomotion has some distinct advantages. Momentum is maintained in the dorsal side of the body, thus making rectilinear movement a relatively efficient, although slow, locomotor mode. The static reservoir is divided among a few small areas of the body, so that the mass that must be alternately fixed and then moved at each static site is smaller than in concertina locomotion. Rectilinear propulsion is well suited for stout-bodied snakes, such as many boids and some vipers, that are not adept at concertina locomotion. Rectilinear locomotion is also excellent for the stealthy stalking of prey, both because of its slow speed and because it does not require lateral movements of the head or body.

Conclusions

Although the major modes of limbless locomotion are well defined, our knowledge of their use in nonlaboratory situations is still meager. Particularly lacking are studies describing the transitions between modes, as well as those situations in which more than one mode is in use at a single time. Comparative studies that correlate body form, locomotor mode, and substrate characteristics, both within and between species, are also needed. Cost of transport measurements should be gathered for all locomotor modes to test the suggestions made above about the relative efficiencies of each of the limbless types.

Aside from cinematography, the modern techniques of functional morphology have hardly been applied to limbless locomotion. In particular, electromyography should be used to determine which muscles are active, and in what sequence, in producing the lateral curves of lateral undulation, concertina propulsion, and sidewinding, as well as to test Lissmann's model for the production of rectilinear locomotion.

Chapter 10

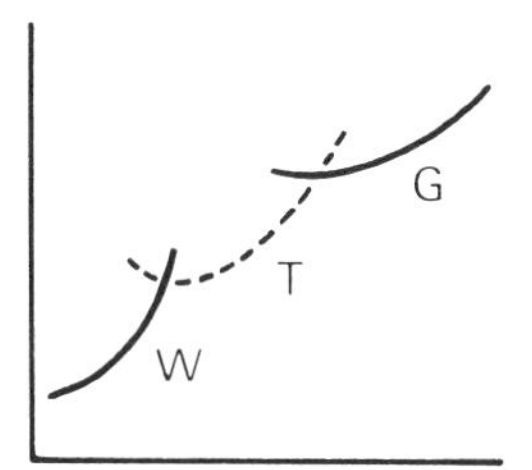

Energetics and Locomotion

Albert F. Bennett

Locomotion is energetically expensive. Moving an animal's mass through space requires large increments in metabolic energy input, and these costs rise rapidly with increasing speed. Only thermoregulation by birds and mammals in very cold environments can cause a similar rise in acute energy expenditure. Locomotion may thus have a major impact on the energy budgets of animals, particularly animals that forage actively and continuously and animals that migrate.

Locomotory Costs and Metabolic Energetics

Whether a large or small proportion of the daily energy budget is spent on activity, the expense of locomotion greatly affects the behavior of all vertebrates. High or even moderate levels of movement may tax energy delivery systems to their limits and result in fatigue or inadequate performance. We can appreciate these relationships qualitatively in considering our own responses to activity. A steady walk of about 6 km/h can be undertaken by most people without noticeable difficulty. Doubling that rate to a run (12 km/h) requires considerable exertion with greatly increased rates of ventilation and circulation. This speed can be sustained only with difficulty. Increasing the rate again to a sprint of about 20 km/h exceeds our limits of sustainable activity and results in very rapid exhaustion. In many vertebrates the speeds at which pronounced exertion and exhaustion occur are considerably less than those of humans. These limits vary greatly among different vertebrate groups and also vary with such factors as body size and temperature. The failure of locomotory performance, either in maximal speed or stamina, can result in a spectrum of consequences up to and including death. Such factors as pursuit or escape from predation, territorial defense, and competition for a mate and courtship may require high or sustained levels of performance at the limits that an animal can undertake.

The energy demands even of resting vertebrates can be substantial. Protein synthesis, ion transport, and circulation of body fluids, along with other processes, use energy continuously. Minimal rates of energy use are designated standard metabolic rates or, for endotherms, basal metabolic rates. These depend on many factors, principally body size and temperature. In vertebrates this maintenance metabolism is fueled aerobically, that is, by complete combustion of foodstuffs with oxygen to carbon dioxide and water. These metabolic rates are generally determined by measurement of the rate of oxygen consumption ($\dot{V}_{O_2}$). During physical activity the contraction of skeletal muscle requires additional energy, and respiratory and cardiovascular systems must increase the rates of oxygen delivery to the metabolizing tissue. In turn, the increased activity of these systems themselves requires increased oxygen delivery and energy use.

The capacities of oxygen delivery in vertebrate systems are not enormously expansible. Maximal rates of oxygen consumption ($\dot{V}_{O_2max}$) exceed standard levels on the average by approximately 10-fold, varying in range in different animals from 4-fold to about 25-fold (Brett, 1972;

Lechner, 1978; Bennett, 1978). The functional basis of these limitations is not completely known. However, the use of a cardiovascular system to transport oxygen from the external environment to the tissues may be an important limiting factor. Insects, for instance, which have tracheal respiratory systems, may in some cases increase $\dot{V}_{O_2}$ 100- or 200-fold above resting levels. For whatever reason, in vertebrates (particularly in ectothermic vertebrates) aerobic scope, which is the difference between $\dot{V}_{O_2max}$ and $\dot{V}_{O_2standard}$, may be rather low.

If energy demand exceeds aerobic energy supply, anaerobic metabolism is utilized. Over the period of a few seconds, muscle can sustain contractile function by the catabolism of endogenous stores of adenosine triphosphate and creatine phosphate. However, for longer periods, anaerobic metabolism in vertebrates primarily involves the production of lactic acid by carbohydrate catabolism. As these reactions do not depend on cardiovascular transport, they can be activated within a muscle and produce large quantities of available energy in a short time. The efficiency of production of energy per unit substrate traversed is considerably less than for aerobic metabolism, but energy conservation may not be foremost in the priorities of a fleeing animal. The behavioral repertoire and performance capacities of vertebrates, particularly those of ectothermic vertebrates, are considerably expanded by the use of anaerobic metabolism. However, activation of extensive anaerobiosis generally results in fatigue and decreased performance capacity. The physiological causes of these relationships are not well understood. For locomotory performance, however, the implications are clear.

Anaerobic metabolism may be used during the initial stages of activity, before oxygen consumption has risen to a new elevated level, or for short (approximately 1 min) bursts of intense activity. Activity of longer duration must be undertaken within aerobic limits or fatigue may result. The aerobic metabolic rate must lie within the boundaries set by maximal and standard levels of oxygen consumption. Maximal levels of oxygen consumption thus set limits on the sustained activities and behaviors of animals. Most of the locomotion and locomotory energetics discussed in this chapter are carried out within those limits.

Metabolic energy expenditure may be regarded as the power input into locomotion. The power output, the rate of work actually performed by the locomoting animal, may be considerably less. Analysis of power output requires a detailed examination of the biomechanical performance of an animal and has not been attempted extensively. The ratio of power output to power input, the mechanical efficiency, has been commonly assumed to be about 20% to 25% for vertebrate skeletal muscle. Similar maximal values for mechanical efficiency have been found for swimming fish (Webb, 1975, 1977). However, recent determinations on terrestrial animals (Taylor, 1980; Taylor and Heglund, 1982; Heglund et al., 1982) have found that peak mechanical efficiency is size-dependent and varies from less than 10% in quail to approximately 70% in humans. The higher efficiencies are thought to be due primarily to storage and release of elastic energy during individual limb cycles. More information on locomotory efficiencies would be very helpful in understanding the return that animals receive for their metabolic energy expenditure. (The storage of elastic energy is discussed in Chapters 2, 3, and 4.)

The analysis of locomotory energetics requires the determination of $\dot{V}_{O_2}$ of an animal moving at a steady rate. These measurements can be made on animals walking on a treadmill, flying in a wind tunnel, or swimming against a current. An effective method for measuring $\dot{V}_{O_2}$ in aerial breathers entails fitting an animal with a light-weight, clear plastic mask (see Fig. 10-1). A pump draws room air into and through this mask. The animal consumes a portion of the oxygen passing in the air stream, lowering the oxygen content of the air in the flow stream below that of the room air. The animal's rate of oxygen consumption can be calculated from this differential in oxygen concentration and the rate of air flow through the mask. Speed can be set by regulating the rate of tread movement for terrestrial walkers or wind flow for flyers. Considerable perseverence and ingenuity are sometimes required to elicit cooperation from the subjects, but animals will often attempt

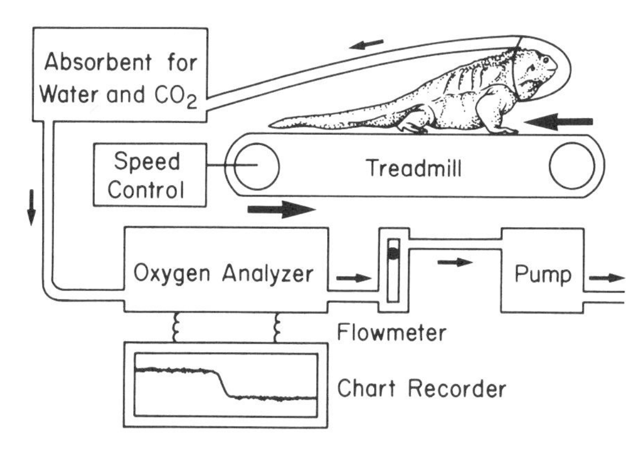

Figure 10-1 Measurement of oxygen consumption of an animal walking on a treadmill.

to match speed and maintain a constant station during the experiment. Data have been collected on a surprisingly large and diverse group of terrestrial animals. The logistical operations, including design concerns for laminar flow and drag, are more difficult for flying animals, and relatively fewer observations have been made on them.

Many fish will also maintain station against a current. If the oxygen content of the water upstream and downstream from the fish and the rate of the current flow are measured, $\dot{V}_{O_2}$ can be calculated in a manner similar to that for aerial breathers.

Each locomotory mode, walking or running, flying, and swimming, has its own relationship between energy use and speed. The patterns are quite distinct and are discussed separately below, after which the relative costs of these different locomotory modes are compared.

Terrestrial Locomotion

As terrestrial animals walk and run, $\dot{V}_{O_2}$ increases approximately linearly with speed (Taylor et al., 1970; Taylor, 1977; Taylor et al., 1982) until $\dot{V}_{O_2max}$ is attained. This relationship between $\dot{V}_{O_2}$ and walking speed is shown in Figure 10-2A for a lizard, the gila monster. Also shown (Fig. 10-2B) are a variety of terms that have been used to describe locomotory energetics of terrestrial animals (Tucker, 1970; Schmidt-Nielsen, 1972). The speed at which $\dot{V}_{O_2max}$ is attained is the maximal aerobic speed, which is approximately 0.6 km/h in this lizard at 25°C. Speeds below this can be sustained for long periods because they are dependent only on aerobic metabolism. Walking at speeds greater than the maximal aerobic speed elicits the same aerobic power input, since it is already maximal. The additional power input required is provided anaerobically. These speeds cannot be sustained without resulting in fatigue.

If the values for $\dot{V}_{O_2}$ are extrapolated to zero speed, they intersect the ordinate at levels considerably in excess of those measured for resting animals. This value of $\dot{V}_{O_2}$ is called (accurately but somewhat inelegantly) the *y*-intercept. It ranges from 1.3 to 2.9 times resting metabolic rate and generally averages approximately 1.7 (Paladino and King, 1979). The functional basis of this increment is not well understood. It is often attributed to a "postural" component of locomotion,

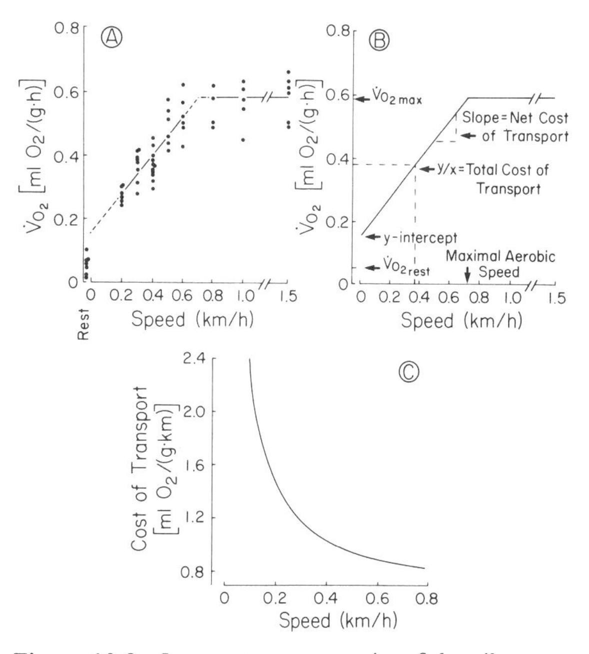

Figure 10-2 Locomotory energetics of the gila monster *(Heloderma suspectum)*. A: Oxygen consumption at different walking speeds; B: locomotory terminology; C: total cost of transport at different speeds. (Data from John-Alder, Lowe, and Bennett, 1983.)

a cost associated with maintaining an upright stance. Whatever its cause, it greatly increases energy expenditure even during very slow locomotion. Its presence partly accounts for the high cost of locomotion in general and the narrow range of sustainable speed of animals with low aerobic scopes.

The slope of the relation of $\dot{V}_{O_2}$ to speed between zero and maximal aerobic speed is defined as the net cost of transport. This quotient is expressed as volume of oxygen consumed (aerobic power input) per distance traveled, that is, volume of O_2 per unit time divided by distance per unit time, generally as ml or l of O_2 per km. It represents the amount of fuel that must be expended to traverse a given distance and is independent of time and speed. Because of this independence, it is widely used as the standard of comparison among animals of different taxonomic groups, body sizes, or body temperatures.

It should be evident, however, that the real locomotory costs to animals are quite different from these values, which specifically exclude the portion of energy expenditure associated with maintenance and posture. The entire aerobic power input to locomotion is taken into account by the total cost of transport, the quotient of $\dot{V}_{O_2}$ at any speed (between zero and maximal aerobic speed) and that speed. Although this value also has units of volume of O_2 consumed per distance traveled, it is highly speed-dependent. This dependence is shown in Figure 10-2C for total cost of transport in the gila monster. As the animal walks faster, the total amount of energy expended while traversing a given distance declines. As an animal spends less time crossing a given distance at a greater speed, maintenance and postural processes do not have to be sustained as long and consequently require less energy expenditure. Although maintenance energy costs continue whether the animal is locomoting or stationary, the postural costs do not. Therefore, the most economical transport is attained when an animal walks at close to its maximal aerobic speed.

The linear increment of aerobic power input with increasing speed is perhaps unexpected, as power inputs in mechanical vehicles must often increase exponentially as speed increases. A partial explanation for this linearity may be found in

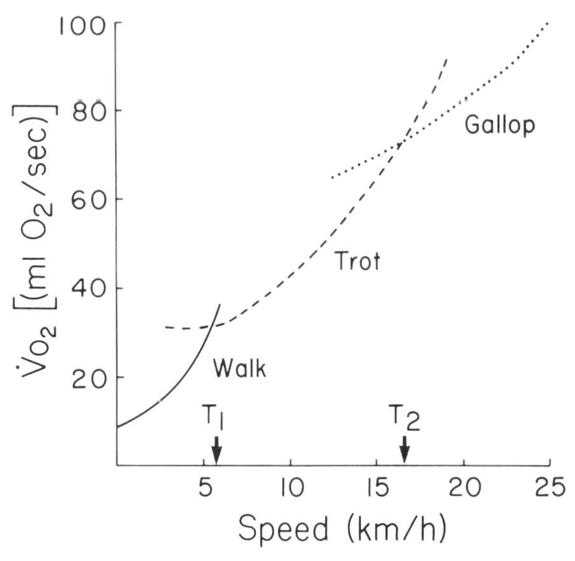

Figure 10-3 Oxygen consumption of a horse trained to continue different gaits beyond their normal range of speed. The horse normally changes from a walk to a trot at T_1 and from a trot to a gallop at T_2. (Data from Hoyt and Taylor, 1981.)

gait transitions at different speeds (Hoyt and Taylor, 1981). As speed increases, quadrupeds often change gait from a walk to a trot to a gallop. Oxygen consumption in a horse trained to continue these gaits beyond their normal range of speeds is shown in Figure 10-3. Within each gait, $\dot{V}_{O_2}$ increases exponentially with speed, but the horse naturally selects the gait that has the lowest energy expenditure at any given speed. The minimal cost of transport is the same in each gait. A linear approximation is an excellent description of the $\dot{V}_{O_2}$ maintained over the entire range of speeds.

These relationships have proved highly useful in analytical and comparative studies of locomotory energetics. It is important to keep in mind, however, that they represent determinations based on steady rates of $\dot{V}_{O_2}$ attained after several minutes of activity. When an animal is initially active, power output may increase almost immediately as a step function. Aerobic power input lags significantly for 1 – 2 min before a new level is attained. Anaerobic metabolism provides the differential energy requirement during this period. Often, $\dot{V}_{O_2}$ will overshoot before settling at an ultimate plateau level. After activity has ceased, oxygen consumption almost always remains elevated above resting levels for some time. This excess $\dot{V}_{O_2}$ is termed the oxygen debt. Thus, the metabolic patterns and energetics of animals initiating activity, changing speeds, or recovering from activity are far more complex than the above relationships would suggest. These steady-state models attempt only to approximate natural activity and energy expenditures.

Quadrupedal Locomotion: Mammals and Lizards Endothermic (mammals and birds) and ectothermic (reptiles, amphibians, and fish) vertebrates have very different capacities for metabolic power input and consequently differ greatly in their locomotory capacities. These factors can readily be seen in comparisons of the locomotory energetics of lizards and mammals. Such direct comparisons can easily be undertaken in animals of similar body size and temperature. Mammals have considerably greater levels of resting $\dot{V}_{O_2}$, exceeding those of resting ectotherms of equal size and temperature by 6- to 10-fold (Benedict, 1932; Dawson and Bartholomew, 1956; Bennett, 1978, 1980a). This differential in metabolic rest and heat production is the basis of endothermy, and it represents a major energy commitment to this metabolic mode. Maximal oxygen consumption, too, is substantially greater in mammals, also exceeding $\dot{V}_{O_{2}max}$ values of ectotherms by approximately 10-fold. Thus, oxygen processing abilities are nearly an order of magnitude greater in endotherms.

Mammals, with their greater aerobic scopes, are able to sustain substantially greater levels of metabolic power input and achieve greater maximal aerobic speeds (Fig. 10-4). Energy expenditure at rest or while walking at low speeds is substantially less in a lizard than in a mammal. The net cost of transport is very similar in both animals. Increments in speed thus require similar increments in energy input, indicating that the differences in limb configuration among these animals have no substantial energetic consequences (Bakker, 1972). However, a lizard of approximately 1 kg in mass reaches its aerobic limits at rather low speeds of 1 km/h or less. Maximal levels of aerobic energy expenditure are achieved

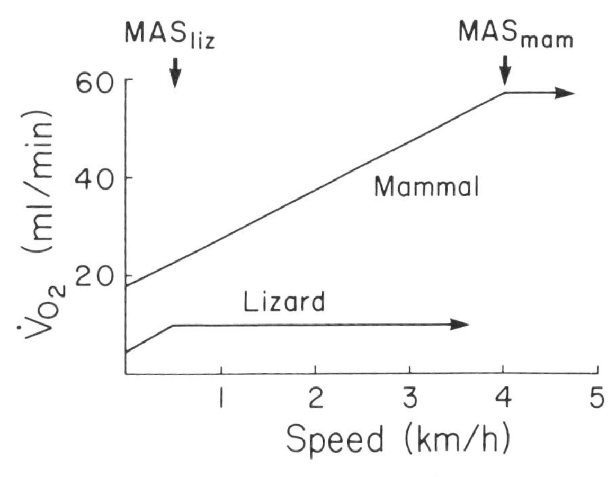

Figure 10-4 Oxygen consumption of a 1-kg mammal and a 1-kg lizard (35° C) walking on a treadmill. *MAS* = maximal aerobic speed. (Redrawn from Bennett and Rubin, 1979.)

by a mammal of similar size at 4-5 km/h. There is thus a substantial range of locomotory activities that is sustainable by endothermic vertebrates but cannot be maintained by ectotherms.

Maximal burst speeds, however, are similar in both mammals and lizards. Since these are fueled primarily by anaerobic metabolism, the greater aerobic power inputs of mammals do not greatly affect this aspect of locomotion. Ectotherms may thus be equally fast and may escape from endotherms during brief periods of rapid pursuit.

Size and Locomotory Energetics One might reasonably expect that the energy cost of locomotion should increase directly as body mass increases. That is, a doubling of body size might be expected to double locomotory costs. In fact, energy expenditure does not increase in proportion to mass, but rather as $mass^{0.7}$. Allometric equations have been calculated summarizing the net cost of transport for many species of mammals (Taylor et al., 1982) and lizards (Gleeson, 1979):

$$\dot{V}_{O_2} = 0.53\ m^{0.68} \qquad \text{(mammals)}$$

$$\dot{V}_{O_2} = 0.67\ m^{0.75} \qquad \text{(lizards)}$$

where $\dot{V}_{O_2}$ is l O_2/km and m is mass in kg. As there is considerable scatter about these regressions, they are not significantly different and net costs of transport of equal-sized mammals and lizards

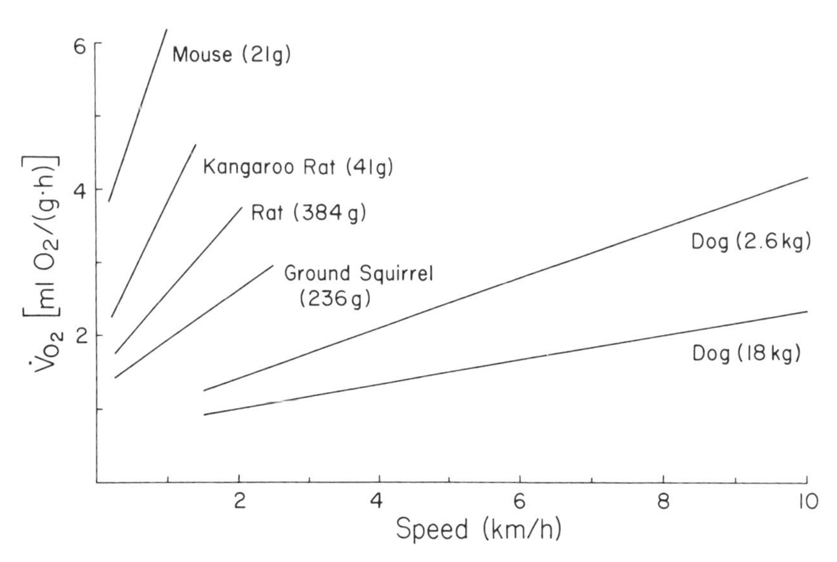

Figure 10-5 Mass-specific rates of oxygen consumption of different-sized mammals running on a treadmill. (Redrawn from Taylor, Schmidt-Nielsen, and Raab, 1970.)

are not different. Total locomotory costs may be estimated by adding the *y*-intercept value to the product of speed and net cost of transport. For mammals, this relationship is

$$\dot{V}_{O_2} = 1.09\ m^{0.69} + 0.53\ m^{0.68}S,$$

where $\dot{V}_{O_2}$ is l O_2/h, *m* is mass in kg, and *S* is speed in km/h (Taylor et al., 1982).

Net costs of transport have been found to be very similar among several groups of terrestrial vertebrates. Transport costs are, for instance, indistinguishable between quadrupedal and bipedal walkers (Fedak and Seeherman, 1979; Paladino and King, 1979), contrary to an earlier report based on fewer observations (Fedak, Pinshow, and Schmidt-Nielsen, 1974). Even within single species of primates, quadrupedal and bipedal gaits have equal transport costs (Taylor and Rowntree, 1973). The relationship of oxygen consumption, speed, and transport costs in most hopping mammals are also adequately described by these equations (Thompson et al., 1980). The evolution of bipedality or bipedal hopping was probably not, therefore, based on energetic grounds. One conspicuous exception to these general patterns of energetics of both bipedal and quadrupedal locomotion is the red kangaroo. It maintains a constant oxygen consumption over a broad range of hopping speeds while altering stride length (Dawson and Taylor, 1973). Elastic storage in the conspicuous tendons of the hind limbs may be involved in this unique pattern of locomotory energetics (see also Chapter 3).

If these relations are expressed on a mass-specific basis, it is clear that the cost of moving a unit of mass is less in larger organisms and is approximately proportional to body mass$^{-0.3}$. Thus, at equal speeds a mouse or rat must utilize more energy and consume more oxygen per gram body mass than does a dog or human (Fig. 10-5). The functional basis of this allometry is not understood completely, but it may reflect differences in the inherent contractile properties of muscle tissue between large and small animals (Taylor et al., 1980).

These regression equations may be used to predict transport costs of a previously uninvestigated animal or to compare differences in locomotory modes. For instance, the net cost of transport of a garter snake is only one third that predicted by the equation for lizards (Chowdrow and Taylor, 1973), indicating that crawling may be considerably less expensive than walking. These equations are powerful in theoretical or generalized analyses of locomotory or energetic costs. However, there is real variability among species that is obscured by these regressions, and predictions should be made with a note of caution for any particular species.

Carrying external loads, such as food or an infant, also increases locomotory costs. Oxygen consumption rises in direct proportion to the percentage mass increment (Taylor et al., 1980). If a load equivalent to 10% of its mass is placed on a walking animal, its oxygen consumption will rise 10%. This relationship has interesting allometric implications. A smaller animal expends more energy per unit mass walking at a given speed than does a larger animal. If a load representing an equal percentage of body mass is placed on each, the amount of energy input per unit mass must rise a greater amount in the smaller animal to sustain the load. A smaller animal must consequently expend more energy to produce an equivalent force at a given speed than does a larger animal.

Body Temperature and Locomotion Most metabolic processes are directly temperature-dependent. As body temperature changes, $\dot{V}_{O_2rest}$ and $\dot{V}_{O_2max}$ are markedly affected and are substantially lower at low body temperatures. Changes in these aerobic metabolic factors may be expected to produce changes in locomotory capacities and patterns. Endotherms generally avoid these fluctuations by maintaining constant body temperature, although there is much more variability in mammalian muscle temperature than is generally appreciated. Body temperature changes may, however, be pronounced in ectotherms. Relatively few analyses of the effects of temperature on locomotory capacities and energetics have been undertaken.

The contrasting advantages and disadvantages of different body temperatures for locomotory performance are shown in the desert iguana (Fig. 10-6; John-Alder and Bennett, 1981). Low body temperatures permit energy conservation and economical locomotion: resting or walking at any given speed costs less at the lower temperature. However, the range of aerobically sustainable behaviors is relatively low because of low $\dot{V}_{O_2max}$. At 40° C, lizards can walk three times faster without tiring than at 25° C. Thus, the higher body temperature permits a greater range of sustainable behaviors while requiring a greater energy input. These increased costs of the higher temperature can be somewhat offset by the expanded behavioral capacity. The minimal total cost of transport, that is, the least amount of energy required to traverse a given distance, is equal at both temperatures (Fig. 10-6B). Thus, an animal with a high body temperature may reduce its transport cost by traveling at a faster speed. Behavioral thermoregulation permits an ectotherm to take advantage of both aspects of the thermal dependence of locomotory energetics: energy conservation at low temperatures and expanded locomotory capacities at high temperatures.

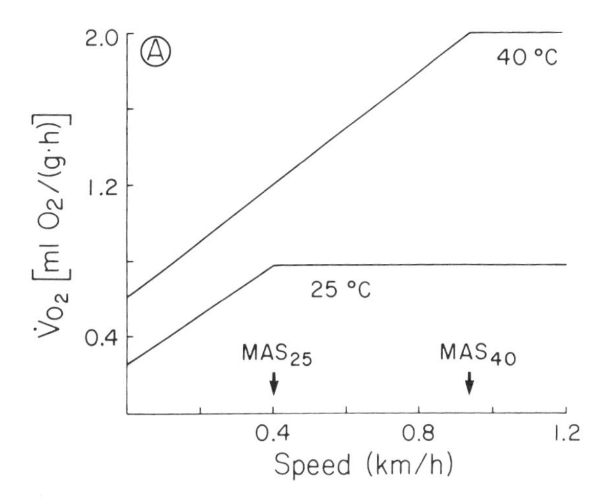

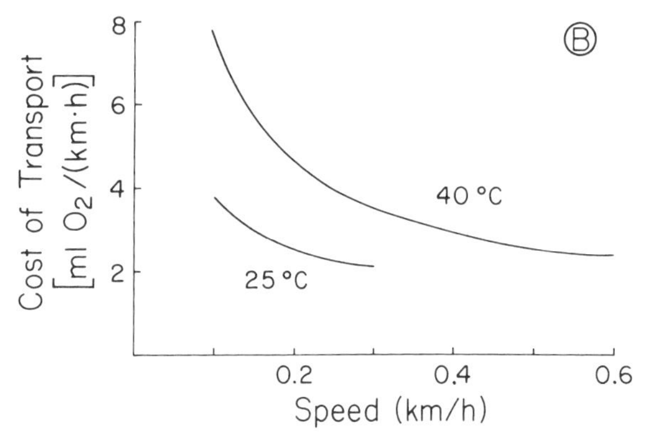

Figure 10-6 Locomotory energetics of a lizard *(Dipsosaurus dorsalis)* on a treadmill at two body temperatures. A: Oxygen consumption; B: total cost of transport. (Data from John-Alder and Bennett, 1981.)

In contrast to the thermal dependence of oxygen consumption and transport, anaerobic metabolism in ectothermic vertebrates can be relatively independent of body temperature (Bennett, 1978). The rate of lactic acid formation during intense activity in small lizards is nearly constant between 20° and 40° C (Bennett and Licht, 1972). Burst speeds show a similar lack of thermal dependence in these animals (Bennett, 1980b). Anaerobic metabolism provides a capacity for rapid activity and escape over short distances over a considerable range of body temperatures. Locomotory capacity in ectotherms is thus not completely tied to the strict thermal dependence of aerobic metabolism.

Flight

Flight requires high levels of power input. Although brief bursts of flight might be fueled anaerobically, sustained flight of more than a minute's duration must be supported aerobically. Among the vertebrates only birds and mammals possess levels of oxygen consumption capable of sustaining flapping flight. Among the mammals, only bats, which have particularly high levels of $\dot{V}_{O_2max}$, fly. Although it requires high energy expenditure, flight is a relatively economical way of covering long distances, particularly during migration.

Several models have been advanced to predict the requisite power input for avian flight (for example, Tucker, 1973; Pennycuick, 1975; Greenewalt, 1975; Rayner, 1979; see also Chapter 8). Although these differ in several details, they all predict a U-shaped relationship between power input and speed (see Fig. 10-7A). That is, power requirements for hovering flight (stationary flight with no forward velocity) are predicted to be high. As speed increases, costs decrease to a minimal value at velocity V_0, the minimum power speed. At greater speeds, power input increases again. Thus, there is a predicted minimal cost associated with flight, for if a flier attempted to minimize its rate of energy expenditure, it would fly at V_0. This speed is not, however, the most economical in terms of energy expended per distance traveled (that is, the minimal cost of transport). This is obtained at the V_1, the maximum range speed, at which a line from the origin is tangent to the power curve. Distance traveled per unit fuel would be minimal at V_1, and a flier traveling long distances might be expected to fly at this speed.

The predicted speed dependence of locomotory costs in fliers is thus fundamentally different from that of terrestrial animals. In the latter minimal rates of energy expenditure are obtained while standing still, and minimal total cost of transport occurs at maximal aerobic speed. In fliers flight is least expensive and minimal transport costs occur at intermediate speeds.

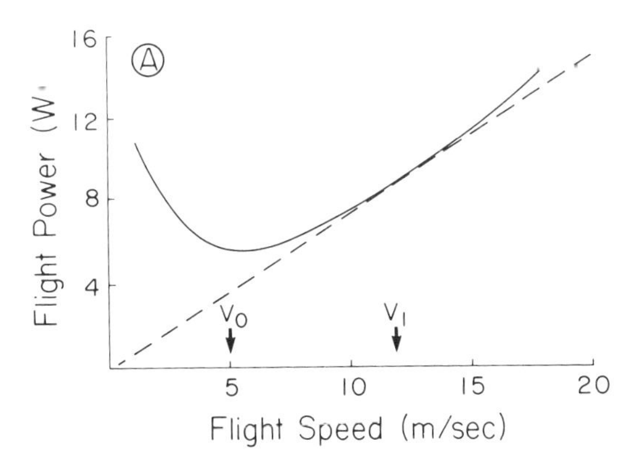

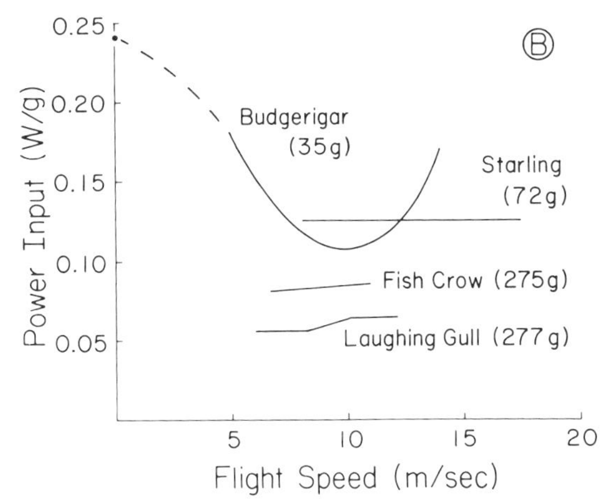

Figure 10-7 Flight energetics of birds. A: Theoretical predicted power input required for flight by a 330-g pigeon. V_0 = minimum power speed; V_1 = maximum range speed. (Redrawn from Rayner, 1979.) B: Observed power requirements for four species of birds. (Redrawn from Bernstein, Thomas, and Schmidt-Nielsen, 1973.)

Relatively few measurements of oxygen consumption as a function of speed have actually been made on flying birds (budgerigar and laughing gull, Tucker, 1968, 1972; fish crow, Bernstein, Thomas, and Schmidt-Nielsen, 1973; starling, Torre-Bueno and Larochelle, 1978; see Fig. 10-7B). A distinct U-shaped relationship between oxygen consumption and flight speed is seen only for the budgerigar. Oxygen consumption increases slightly with increasing speed in the laughing gull, but it is independent of speed in the fish crow and starling. The data do not correspond well to model predictions. Many variables are built into such models, and flying birds alter these variables so as to minimize energy expenditure. Moreover, the efficiency of energy generation may be changing as a function of speed. At any rate, the predictions concerning a minimal cost of flying at a specific speed (V_0) and an intermediate speed (V_1) with a minimal total cost of transport are not found in any bird examined besides the budgerigar. Total cost of transport is minimal at the fastest speeds flown for the other species. Further observations on the actual cost of avian flight and the reconciliation of the models with empirical data are required before we can understand avian flight energetics.

These data and models refer to the cost of level, continuous flapping flight, but other patterns of flight are also commonly seen in some birds. Gliding flight on air currents with fixed and extended wings can be used by birds with large wingspans. The cost of gliding as estimated for vultures (Pennycuick, 1972) and measured for gulls (Baudinette and Schmidt-Nielsen, 1974) is less than one third that of flapping flight. Gliding approximately doubles energy expenditure above basal metabolic levels, whereas flapping flight requires a substantially greater increment. It has also been proposed (Tucker, 1968) that undulating flight, which is seen in many small birds, may be more economical than level flight (Rayner, 1977).

Determination of the energetics of bat flight has received relatively less experimental attention. Maximal oxygen consumption, net cost of transport, and total cost of transport are similar in birds and bats, indicating a convergence of locomotory energetics and physiological capacity in these groups (Thomas and Suthers, 1972; Thomas, 1975). Oxygen consumption is relatively independent over a range of flight speeds in two species of bats examined (Thomas, 1975).

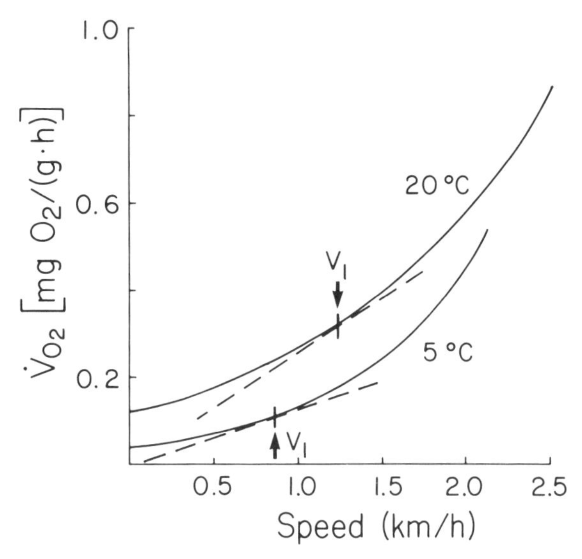

Figure 10-8 Oxygen consumption of a 50-g salmon acclimated to two temperatures. (Data from Brett, 1964.) V_1 = speed at which the normal cost of transport is attained at each temperature.

Swimming

Energy expenditure during swimming has been measured for several species of teleost fish, and its speed dependence is different than that of either flying animals or terrestrial walkers. Oxygen consumption increases exponentially with increasing speed (Fig. 10-8) and is generally best described by the relation $Y = ae^{bx}$, where Y is $\dot{V}_{O_2}$, x is swimming speed, and a and b are constants (Brett, 1964). More specifically, $\dot{V}_{O_2}$ often increases approximately as the square of swimming speed (Fry, 1957). Standard metabolic rate is often measured in fish by extrapolating $\dot{V}_{O_2}$ to zero swimming speed, as it is operationally difficult to measure this rate directly (Fry and Hochachka, 1970). Under these circumstances, there would be by definition no "postural" costs associated with swimming locomotion. In some fish standard metabolic rate is less than that observed by extrapolation to zero speed (Puede and Holliday, 1980; Duthie, 1982). The basis of this cost, designated as "postural" by analogy with terrestrial animals, is unknown. Oxygen consumption increases with increasing speed up to the critical swimming speed, the equivalent of maximal aerobic speed in terrestrial animals. This is the greatest speed at which activity can be sustained and is used to determine $\dot{V}_{O_2max}$. Very active fish, such as salmon *(Oncorhynchus)*, have greater levels of $\dot{V}_{O_2max}$ and can attain greater critical swimming speeds than more sluggish species such as goldfish *(Carassius)* (Fry, 1947; Brett, 1964). Swimming at levels in excess of critical speeds, fish activate anaerobic metabolism and very rapidly accumulate high levels of lactic acid. Exhaustion ensues rapidly and recovery from burst activity may be protracted, particularly at low body temperatures.

The minimal cost of transport in fish, as in birds, is attained at some intermediate locomotory speed V_1 (Fig. 10-8), determined by a line through the origin and tangent to the curve relating oxygen consumption to speed. This is the most economical speed for traversing a given distance. From the limited data available the minimal cost of transport does not appear to vary greatly with structure or swimming mode among fish of different families: anguillids, sparids, and salmonids, for instance, have similar locomotory costs (Brett, 1964; Rao, 1968; Wohlschlag, Cameron, and Cech, 1968; Schmidt-Nielsen, 1972). Difference in body form and locomotory mode may, however, influence other aspects of locomotory performance, such as burst speed.

Body size greatly influences locomotor capacities of fish. Larger fish can sustain substantially greater critical swimming speeds than can smaller fish (Brett, 1965; Webb, 1975; Hoar and Randall, 1978; Wardle, 1977; Weihs, 1977). For example, at 15° C a 3-g salmon has a critical swimming speed of 1.9 km/h, whereas that of a 1400-g fish is 6.4 km/h (Brett and Glass, 1973). Speeds measured in body lengths decrease with increasing size: in the former examples, the speed of the smaller fish is 6.7 body lengths/sec and that of the larger is 3.3 body lengths/sec. The minimum cost of transport per unit mass in salmon decreases as mass$^{-0.24}$ (Brett, 1965; Schmidt-Nielsen, 1972). As in terrestrial animals, the speed dependence of mass-specific metabolic rate is less in larger fish.

As in locomotion in terrestrial ectotherms, temperature exerts a controlling influence on energy expenditure and performance capacity. For experiments on thermal influences fish are previously acclimated to all experimental temperatures. Consequently, the resulting patterns are not strictly comparable to those determined during the acute temperature exposure of terrestrial ectotherms. As in the latter group, activity at any given speed is more expensive at a higher temperature (Fig. 10-8). For example, $\dot{V}_{O_2}$ of a salmon swimming at 1 km/h is approximately twice as high at 20° C as at 5° C (Brett, 1964). Critical swimming speed also increases at higher temperatures, but this increment is relatively small in comparison with that in terrestrial ectotherms. In salmon critical speed increases from 2.1 km/h at 5° C to only 2.5 km/h at 20° C (Fig. 10-8; Brett, 1964). For several species of fish examined, $\dot{V}_{O_2max}$ and maximal critical swimming speeds are attained at intermediate acclimation temperatures and not at maximal temperatures (15° C in salmon, Brett, 1964; 20° C in goldfish, Fry and Hochachka, 1970; 30° C in bass, Beamish, 1970). Critical swimming speeds and $\dot{V}_{O_2max}$ remain relatively constant as temperature is increased above these levels. These plateaus in performance and $\dot{V}_{O_2max}$ may be the result of a limitation on environmental oxygen availability in some species (Brett, 1964).

The energetics of fish locomotion has received considerable theoretical attention (Weihs, 1973a,b, 1974, 1977; Webb, 1975, 1977; Wardle, 1975, 1977; Alexander, 1977; Wardle and Videler, 1980; Videler and Weihs, 1982). Several of these authors have proposed models for power input necessary at different speeds. Some have suggested that schooling (Weihs, 1973a) or burst-and-coast swimming behavior (Weihs, 1974; Videler and Weihs, 1982) can substantially reduce energy expenditure over solitary or steady-rate swimming. These models now require empirical testing.

Energetic measurements have also been made on locomotory costs of several nonpiscene swimmers (humans, Andersen, 1960; ducks, Prange and Schmidt-Nielsen, 1970; sea turtles, Prange, 1976; marine iguana, Gleeson, 1979; Vleck, Bartholomew, and Gleeson, 1981). In all cases energy expenditure during swimming is substantially in excess of the costs measured or estimated for fish of similar size. The cost of swimming in ducks and humans is 20 to 30 times higher than levels predicted for fish (Schmidt-Nielsen, 1972). These values do reflect also the greater metabolic costs associated with endothermy in these species, but even in ectothermic turtles and lizards, metabolic costs are 2 to 3 times those anticipated for fish. Surface swimming involves considerable wake formation, which should increase power input requirements, but this undoubtedly does not account for all the energetic differential. It is clear that the other vertebrates measured have not been able to reduce swimming costs to those characteristics of fish. Presumably, marine mammals such as cetaceans have substantially lower locomotory costs during swimming (Schmidt-Nielsen, 1972), but these have not been examined.

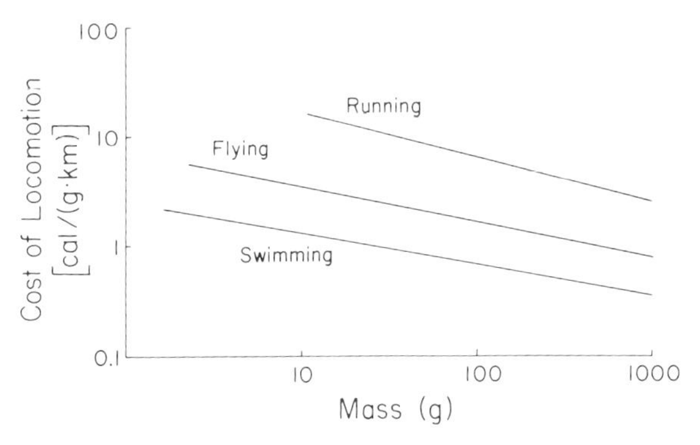

Figure 10-9 Cost of locomotion of running, flying, and swimming animals. (Redrawn from Schmidt-Nielsen, 1972.)

Locomotory Costs and Evolution of Energetic Patterns

A comparison of the energy costs of different locomotory modes was undertaken by Tucker (1970) and Schmidt-Nielsen (1972), revealing striking differences in the levels of energy expenditure associated with walking, flying, and swimming (Fig. 10-9). (See Chapter 6 for the energetics of fossorial vertebrates.) Terrestrial locomotion is by far the most expensive way of getting around. Much of the energetic input is expended in raising the center of mass and accelerating and decelerating the limbs through successive limb cycles. Flight is a less expensive way of traversing a given distance, despite the relatively large amounts of energy input required. Swimming by fishes is the least expensive of the three locomotory modes, requiring only about one half to one third the input required for flight and only one eighth that for walking. In view of these large differentials it is not surprising that most animals that undertake long distance movements are swimmers or flyers.

Locomotory energetics undoubtedly played a significant role in shaping the evolution of vertebrate metabolic and behavioral patterns. Levels of standard and maximal oxygen consumption are similar throughout the ectothermic vertebrates and most invertebrate groups (Hemmingsen, 1960; Bennett, 1978), suggesting a similar pattern of energy use and capacity for oxygen transport in ancestral forms. Likewise, anaerobic capacities of the vertebrates have been well developed since their very early evolution (Ruben and Bennett, 1979). Since ancestral vertebrate systems evolved in aquatic environments, costs of transport must have been relatively low for ancestral fish, as they are for modern fish. The low cost of swimming permits high cruising speeds and a great range of aerobically sustainable behavior, even though $\dot{V}_{O_2max}$ is relatively low in fish in comparison with modern birds and mammals. A salmon of 1-2 kg, for instance, can swim at nearly 5 km/h for long periods (Brett and Glass, 1973). This speed is fully equivalent to the maximal aerobic speeds of terrestrial mammals of equal size, animals that possess substantially greater aerobic power inputs. The well-developed anaerobic capacities of modern fish increase capacities above sustainable levels of performance. Burst speeds are approximately four times critical swimming speeds in modern fish (Brett, 1965; Wardle, 1975). Thus, in extant fish anaerobic metabolism plays a supplementary role for escape or pursuit on top of a substantial behavioral repertoire that is supported aerobically. These energetic relationships were probably similar in ancestral fish.

These metabolic patterns and capacities would then have characterized the vertebrate groups that colonized the land. However, their range of sustainable levels of activity could have been severely constrained because of the greater transport costs associated with terrestrial locomotion. Similar aerobic scopes could no longer provide previous levels of activity. This condition may be exemplified by modern lizards, which have maximal aerobic speeds of only 0.5 – 1 km/h in animals of 1 – 2 kg and even lower capacities at low body temperatures. Anaerobic capacities would still have been well developed and would have necessarily provided a much greater margin of behavioral capacity than previously. Burst speeds in lizards this size may be 15–20 km/h, increasing performance levels by 15- to 40-fold. Activation of this anaerobic metabolism results, however, in rapid exhaustion and loss of performance capacity.

With the evolution of higher rates of metabolism associated with endothermy, aerobic capacities greatly expanded. These permitted high levels of sustainable behavior during terrestrial locomotion in spite of the high cost of transport. Additionally, sufficient aerobic power was then available to permit flapping flight, with its advantages of low locomotory costs, once a certain level of aerobic power input was attained. In endotherms anaerobic metabolism was again relegated to use only during the initial stages of activity or under very substantial exertion, and it provided support for a relatively smaller proportion of maximal behavioral performance. In mammals, for instance, burst speeds are only 2 to 3 times maximal aerobic speeds. The evolution of endothermy thus substantially altered patterns of locomotion and metabolic support among the evolving groups. The expanded capacities for sustainable behavior were undoubtedly significant selective factors in promoting development of greater aerobic levels (Bennett and Ruben, 1979).

Chapter 11

Ventilation

Karel F. Liem

Ventilation is the process either of pumping a current of water over gills or of moving air over the respiratory surfaces of lungs. It is brought about by muscle contractions and by elastic recoil phenomena in the skeletal framework surrounding the respiratory organs, and it may include the operation of valves at the entrance and exit of the respiratory system. This chapter is an examination of the mechanism of ventilation in selected vertebrates.

The Water Pump of Teleost Fishes

The respiratory pump in teleost fishes consists of a double pump in which the synchronous expansions and contractions of the buccal and opercular (gill) cavities drive a nearly continuous and unidirectional flow of water over the gills (Fig. 11-1). The continuous flow is made possible by the maintenance of a differential pressure between the buccal and opercular cavities (Ballintijn, 1969a,b,c). Although as many as four phases have been recognized, it is agreed that two major phases in the respiratory cycle can be correlated with distinct pressure and electromyographic events: the force pump phase and the suction pump phase.

During the buccal force or pressure pump phase the mouth is closed by means of the oral valves and a high pressure is created within the buccal cavity (Fig. 11-1). Synchronously, the pressure within the opercular cavity is also elevated and the opercular valve is opened. Because the pressure generated in the buccal cavity is higher than that in the opercular cavity, water flows from the buccal cavity across the gills into the opercular cavity and thence to the exterior via the opened opercular valve.

During the opercular or suction pump phase the mouth is opened and a low pressure is created within the expanding buccal cavity, drawing water from the exterior into the buccal cavity. Synchronously, the pressure in the expanding opercular cavity with its closed valve is reduced even further than that in the buccal cavity. Consequently, water continues to flow from the buccal cavity across the gills into the opercular cavity.

According to this general model, the differential pressure between the buccal and opercular cavities is induced by the resistance created by the sieve formed by gill filaments on the four gill arches. The pores of the gill sieve can be adjusted by changing the angles of the gill filaments using adductor and abductor muscles. In this way the gill resistance, and thus the differential pressure, can be regulated. Measurements have shown that the water flow across the gills is directly proportional to the differences in pressures in the buccal and opercular cavities. This model of the aquatic respiratory pump applies to the vast majority of the Teleostei and to *Amia*.

Functional-Morphological Basis In spite of the great morphological diversity of the more than 20,000 species of teleosts, there is a common structural and functional design in their skulls. All teleosts possess a kinetic skull, which is conducive to the development of interconnected cavities that can expand and contract through the

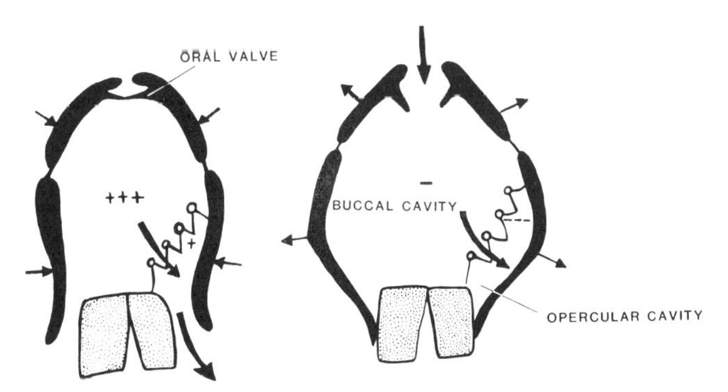

Figure 11-1 The position of the gill arches and gill filaments of one side, shown in horizontal section, during the buccal force pump phase *(left)* and opercular suction pump phase *(right)*. Directions of movement of the walls of the respiratory cavities are shown by thin arrows. Pressures in the buccal and opercular cavities are indicated relative to the outside water. Thick arrows indicate water flow.

action of multiple muscles acting upon complex linkage systems of skeletal units. For a description of the major skeletal components and muscles in a teleost head, see Chapter 12, Figure 12-2. In the aquatic medium, such a kinetic design with multiple muscles is well adapted for creating controlled water currents during respiration, coughing, and feeding.

Muscle Coordination The kissing gourami *Helostoma temmincki* serves as an example for the general pattern of muscle coordination in teleosts during aquatic respiration. *Helostoma* is a bimodal breather, relying on both aquatic and aerial respiration. It retains the primitive aquatic respiratory pump found in most teleosts, and it also possesses all the specializations of an air breather in the transitional stages from aquatic to aerial respiratory adaptations.

During the buccal force pump phase, pars A_3 of the adductor mandibulae is the first muscle to contract, adducting the jaws to close the mouth (Fig. 11-2). It is immediately followed by activity in the adductor arcus palatini and geniohyoideus muscles, which, respectively, adduct the suspensorium or side walls and raise the floor of the buccal cavity, thereby decreasing its volume and increasing its pressure. As the pressure in the buccal cavity builds up, water is forced through the gills into the opercular cavities, which increases the pressure in the latter. At the end of the buccal force pump phase, the adductor operculi contracts and water is forced out from the opercular cavities through the open opercular valve.

The onset of the opercular suction pump phase is indicated by activity of the levator operculi muscle, immediately followed by contraction of the levator arcus palatini. As explained in Chapter 12, contraction of the levator operculi results in the lowering of the mandible. Once the mouth opens, the contraction of the levator arcus palatini expands the buccal cavity by abducting the side walls of the buccal cavity. At the same time the dilatator operculi muscle also contracts to expand greatly the opercular cavity, thereby lowering the pressure to such an extent that water is drawn through the gills from the buccal cavity (Fig. 11-2). At the end of the opercular suction pump phase the pressures in the buccal and opercular cavities rapidly equalize as a preparation for the start of the buccal pressure pump phase of the next cycle. Adduction during this short preparatory phase is thought to occur under influence of elastic recoil forces, since the muscles are inactive.

General and Comparative Aspects Gill ventilation in most teleosts conforms to the pattern described for *Helostoma temmincki:* it is brought about by action of a pressure pump in front of the gills and of suction pumps behind them. The lateral expansions and contractions of the buccal and opercular cavities are synchronous, whereas the lowering of the floor of the buccal cavity may be synchronous with or may precede the lateral expansion. Fishes react to fluctuations in water temperature by changing the frequency and the amplitude of the ventilatory movements without fundamental change in the motor coordination (Hughes and Roberts, 1970; Heath, 1973; Elshoud, 1978). However, during the high-intensity respiration of most fishes studied, the movements of the floor of the buccal cavity are maximized by activity in the sternohyoideus muscle, which is inactive during lower intensity breathing.

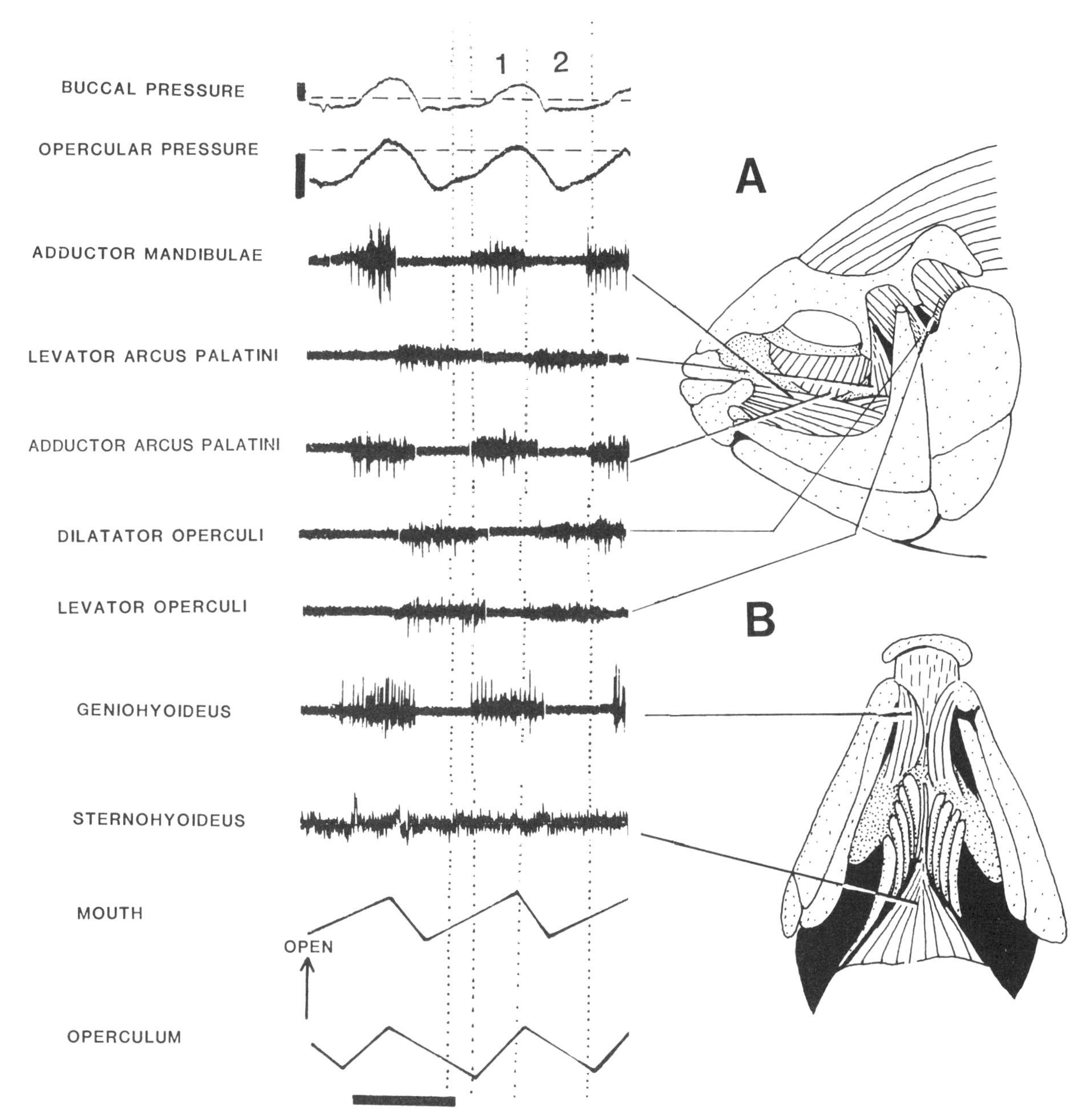

Figure 11-2 Electromyograms, together with the movement of the lower jaw and operculum, and pressures in the buccal and opercular cavities in *Helostoma temmincki.* Lateral (A) and ventral (B) aspects of the cephalic muscles. 1 = buccal force pump phase; 2 = opercular suction pump phase.

In different teleosts either the buccal or the opercular pump may become dominant, according to the development of the branchiostegal and opercular apparatus. Studies using electromyography and pressure transducers have revealed that bottom-living fish usually have well-developed opercular suction pumps and low-frequency pumping rhythms, whereas active pelagic fish have shorter rhythms and dominant buccal pressure pumps combined with "ram-jet" ventilation

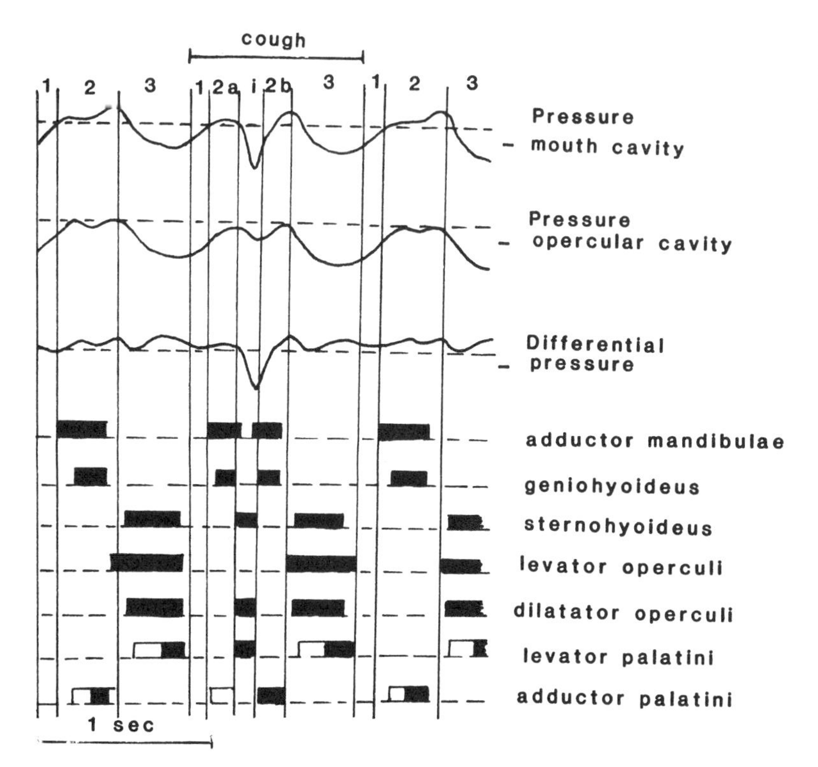

Figure 11-3 Time relationships of respiratory muscle activity, movements of lower jaw and operculum, and the pressures in buccal and opercular cavities during the cough in a carp. (From Ballintijn, 1969b.)

(Hughes and Morgan, 1973). Ram-jet ventilation makes use of the forward motion of the fish to force water backward over the gills (Muir and Buckley, 1967). In these fishes the ventilation volume can be regulated by the size of the mouth and/or the opercular apertures. Certain fish (such as mackerel and tunas) rely entirely on ram-jet ventilation.

Comparative studies (Teichmann, 1959; Hughes and Ballintijn, 1965; Grigg, 1970) have revealed that the general pattern of ventilation is similar in elasmobranch and teleost fishes. Although detailed time relations of the buccal and opercular pumps differ among species, recorded pressures on either side of the gills show a constant pressure differential, suggesting a continuous unidirectional flow (Hughes and Morgan, 1973).

Coughing Coughing appears to be a normal part of the respiratory rhythm of many fishes (Ballintijn, 1969b; Hughes, 1975; Osse, 1969), but it increases in frequency when the fish is put into water containing suspended solids or chemicals or when the fish enters a new environment. A cough is characterized by a reversal in the direction of flow of water through the gills. The buccal pressure first rises and then suddenly reverses to a steep drop (Fig. 11-3). As a result the buccal pressure has become much lower than the pressure in the opercular cavities and water is drawn through the gills from the opercular cavities into the buccal cavity. It is hypothesized that in this way debris and other particles are cleared from the gill sieve.

The basic muscular coordination underlying the coughing movements is somewhat similar in the various teleosts studied (Ballintijn, 1969b; Osse, 1969). Typically, the adductor mandibulae and the adductor arcus palatini muscles initiate the rise in pressure in the buccal cavity (Fig. 11-3). Next, contractions of the levator arcus palatini and sternohyoideus muscles create the sudden and extensive increase in volume of the buccal cavity by abducting its side walls and lowering its floor while the mouth remains shut because of activity of the adductor mandibulae muscles. This lowers the pressure in the buccal cavity. This phase is immediately followed by activity in the dilatator operculi and levator operculi muscles, which reverses the flow of water by creating an open communication between the opercular cavity, the surrounding water, and the buccal cavity with an opened mouth. The next two stages of muscle coordination resemble the opercular suction and buccal pressure pump phase of the respiratory cycle: a second burst of activity in the levator arcus palatini, sternohyoideus, and dilatator operculi muscles, followed by a second burst of activity in the adductor mandibulae and adductor arcus palatini muscles (Fig. 11-3).

Electromyographic and behavioral observations disclose a wide scale of intensities in coughing. Some coughing intensities are associated with the expulsion of particles out of the mouth, whereas others probably reflect displacement activities elicited by sensory stimuli received from the environment. Much more research must be done before the adaptive significance of these phenomena can be explained.

Branchial Air Breathing in Teleosts

In *Helostoma,* as in all Anabantoidei and some other advanced acanthopterygian, air-breathing fishes, the principal site for gas exchange with air is the epithelium covering the highly modified, folded first epibranchial bone (Fig. 11-4; Liem, 1963, 1980; Peters, 1978; Lauder and Liem, 1983), which is contained in a specialized subdivided suprabranchial chamber above the gills. This chamber communicates with the buccopharyngeal cavity by means of a pharyngeal opening and with the opercular cavity via both a branchial aperture between the first and second gill arches and an opercular opening between the first gill arch and the operculum (Fig. 11-4). Ventilation of the suprabranchial chamber takes place via these three openings. Many advanced air-breathing teleosts can ventilate their suprabranchial chambers in two ways, termed triphasic and quadruphasic.

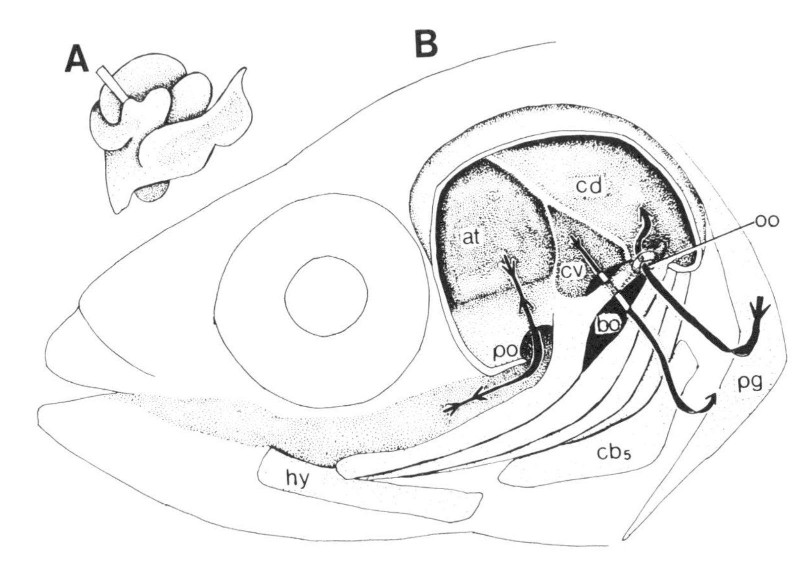

Figure 11-4 A: Lateral view of the expanded and folded first epibranchial bone serving as an air-breathing organ in *Helostoma temmincki.* B: Lateral view of the suprabranchial chamber of an anabantoid fish opened by incision through the wall. Gills and first epibranchial bone have been removed. Arrows indicate pathways for air and water. The structure over the opercular opening *(oo)* is a thickening of the operculum, shown here as a C-shaped, sausage-like bulge. This bulge can be pressed tautly against the muscular process of the first epibranchial on which it lies, thus closing the opening. *at* = atrium; *bo* = branchial opening; cb_5 = fifth ceratobranchial; *cd* = caudodorsal compartment; *cv* = caudoventral compartment; *hy* = hyoid; *oo* = opercular opening; *pg* = pectoral girdle; *po* = pharyngeal opening in floor of the suprabranchial cavity. (From Liem, 1980.)

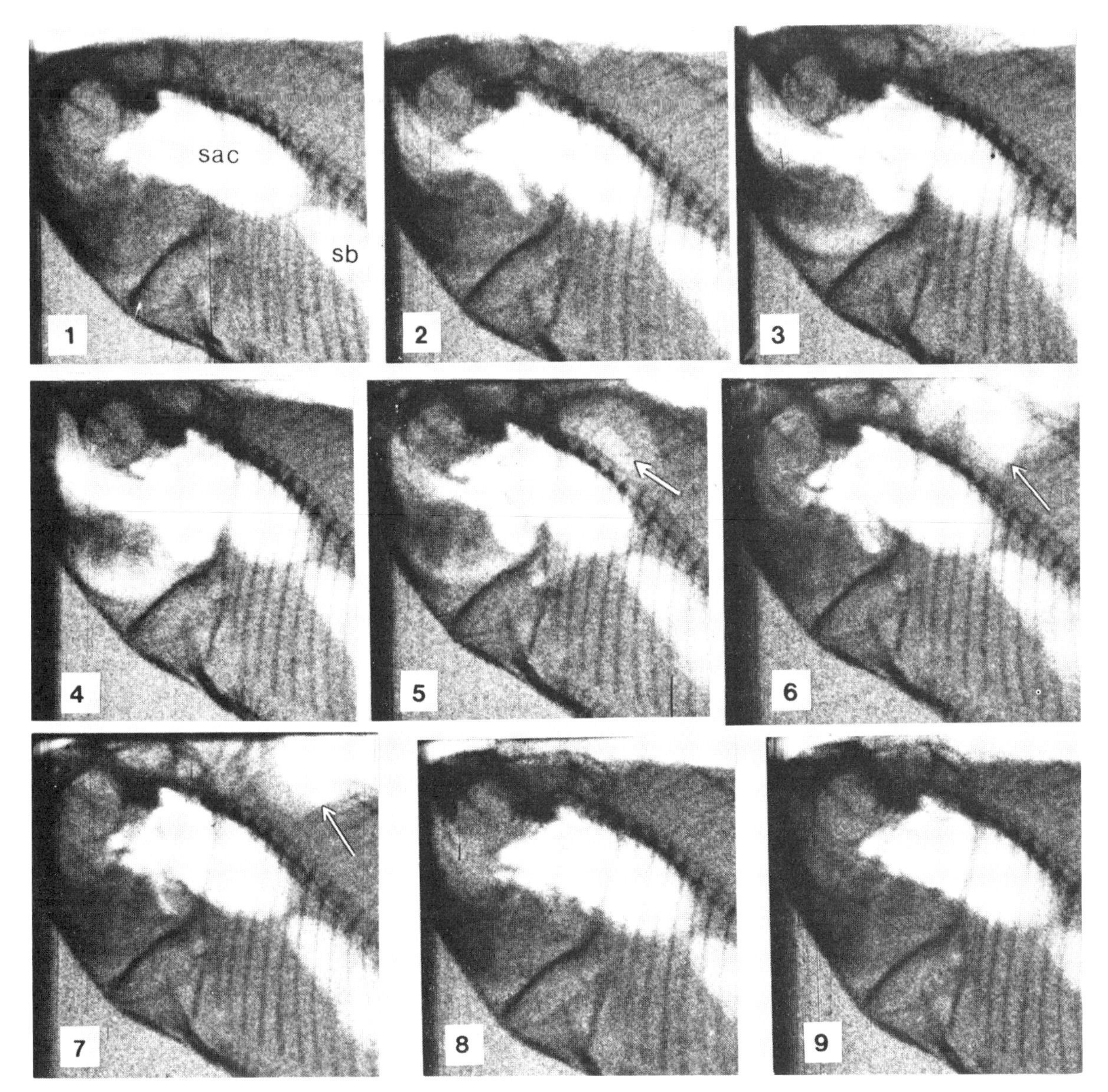

Figure 11-5 Prints from an X-ray film taken at 150 frames/sec of *Helostoma temmincki* during triphasic air ventilation. Note that the suprabranchial air chamber *(sac)* maintains its volume and shape throughout the cycle. The arrows point to air bubbles that have escaped from under the operculum; *sb* = swim bladder. (From Liem, 1980.)

The triphasic pattern of air ventilation proceeds by creating a draft of fresh air from the buccal cavity through the branchial opening into the suprabranchial chamber, and then out of the chamber into the opercular cavity via the branchial and opercular openings into the opercular cavity. Because the operculum is abducted, the air escapes from under the gill cover. During ventilation the suprabranchial chamber remains filled with air. In the preparatory phase the fish rises to the surface (Fig. 11-5), squeezes the water out of the buccopharyngeal cavity, and thus raises the buccopharyngeal pressure by action of the geniohyoideus and adductor arcus palatini mus-

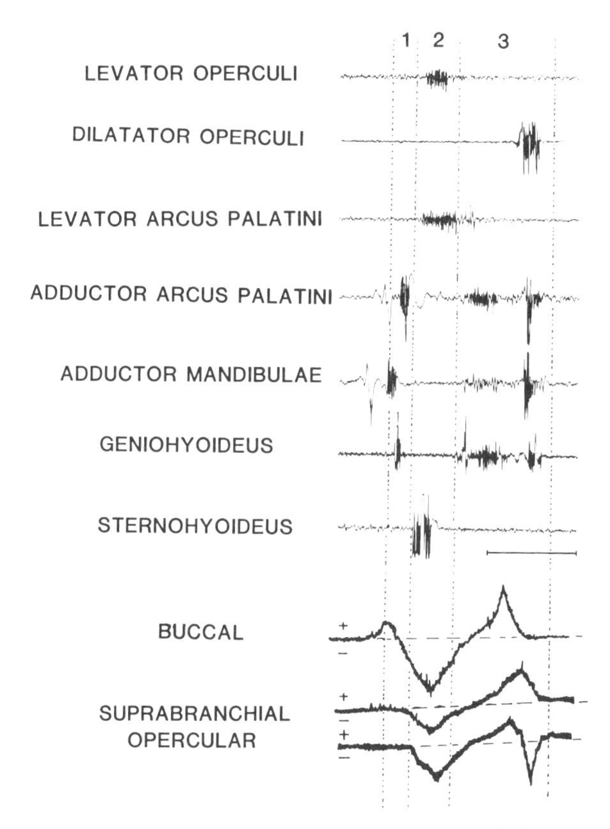

Figure 11-6 Electromyograms together with the suprabranchial, opercular, and buccal pressures of *Helostoma temmincki* during triphasic air ventilation. Phases: 1 = preparatory; 2 = expansive; 3 = compressive.

cles (Fig. 11-6). Once at the surface it opens its mouth and expands the buccopharyngeal cavity by means of contractions of the levator arcus palatini and sternohyoideus muscles, thereby creating a sudden drop in pressure. Air flows into the buccopharyngeal cavity and the mouth is closed. The compressive phase then begins by prolonged actions of the geniohyoideus and adductor arcus palatini muscles, which impart significant compression in order to move the entrained gas into the suprabranchial chamber. Because the operculum is abducted by action of the dilatator operculi, a draft is created. The tidal volume is determined by the size of the buccopharyngeal cavity and comprises about 60% of the total volume of the suprabranchial cavity.

The quadruphasic pattern of ventilation closely resembles coughing. As in coughing, four phases can be distinguished (Figs. 11-7, 11-8). Ventilation begins with a compressive preparatory phase by action of the adductor arcus palatini and adductor mandibulae muscles. It is followed by a reversal phase that begins with a sudden drop in buccopharyngeal pressure created by high-amplitude firings of the levator arcus palatini and sternohyoideus muscles, while action of the dilatator operculi opens the operculum. As the operculum opens, water enters into the opercular cavity and is drawn through the opercular and branchial openings into the suprabranchial cavity, thus flushing the gas bubble into the buccopharyngeal cavity via the branchial opening and thence out of the opened mouth. Thus the suprabranchial chamber becomes filled with water. The reversal phase is followed by the expansive phase, the main feature of which is the opening of the mouth and expansion of the buccopharyngeal cavity, which becomes filled with fresh air. The ventilatory cycle is then completed by the compressive phase, during which the mouth is closed by a second burst of the adductor mandibulae. The trapped air in the buccopharyngeal cavity is then forced into the suprabranchial chamber via the branchial opening by action of the adductor arcus palatini and geniohyoideus muscles, which must impart a significant compression in order to flush out the water from the suprabranchial chamber and replace it with air. During quadruphasic ventilation the residual volume is eliminated, resulting in a 100% tidal volume.

Origin and Evolutionary Significance Electromyographic and pressure profiles of air ventilation in advanced teleosts have shed light on the physiological origin of the mechanisms. The profiles of the triphasic mode of air ventilation are only slight modifications of the patterns so characteristic during prey capture. In the triphasic pattern of air ventilation the basic function of prey capture is put to a dramatically different use by simply substituting an air bubble for the prey item. Neuromuscular patterns are identical during air ventilation and prey capture, except that activity of the dilatator operculi muscle is delayed to the compressive phase of ventilation, whereas during prey capture the muscle is active earlier

Figure 11-7 Tracings from an X-ray film taken at 150 frames/sec of *Helostoma temmincki* during quadruphasic ventilation. Note that the suprabran chial chamber *(sac)* is emptied first and air invariably escapes from the mouth. Filling takes place in exactly the opposite sequence. Phases: 1 = preparatory; 2–5 = reversal; 6 = expansive; 7, 8 = compressive. *sac* = suprabranchial air chamber; *sb* = swim bladder. (From Liem, 1980.)

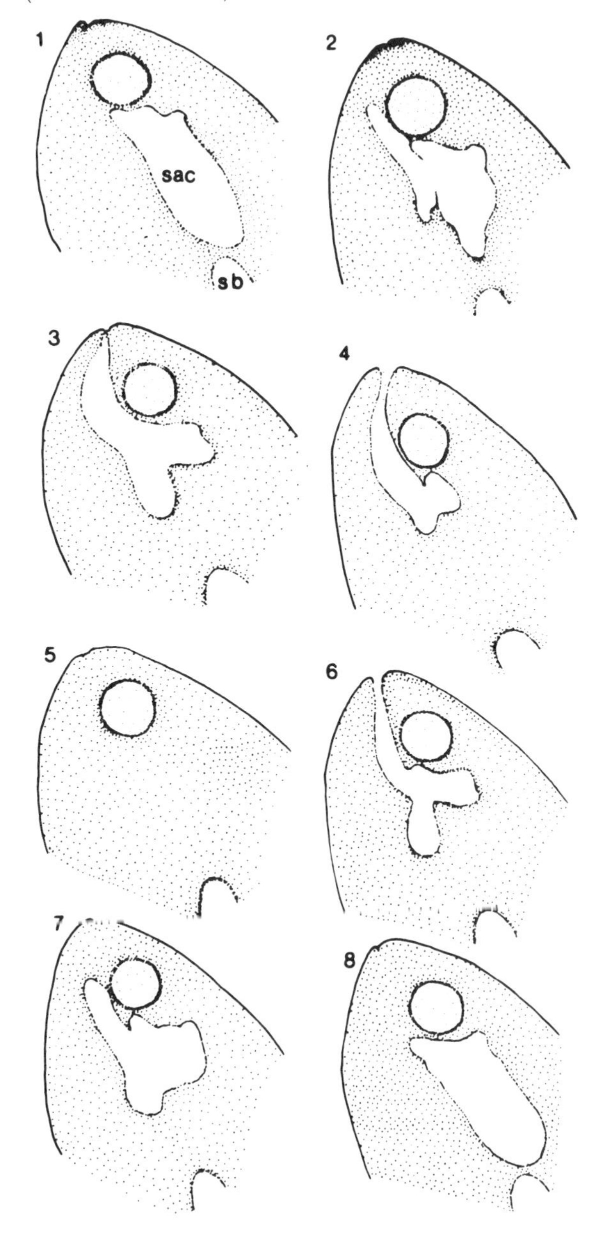

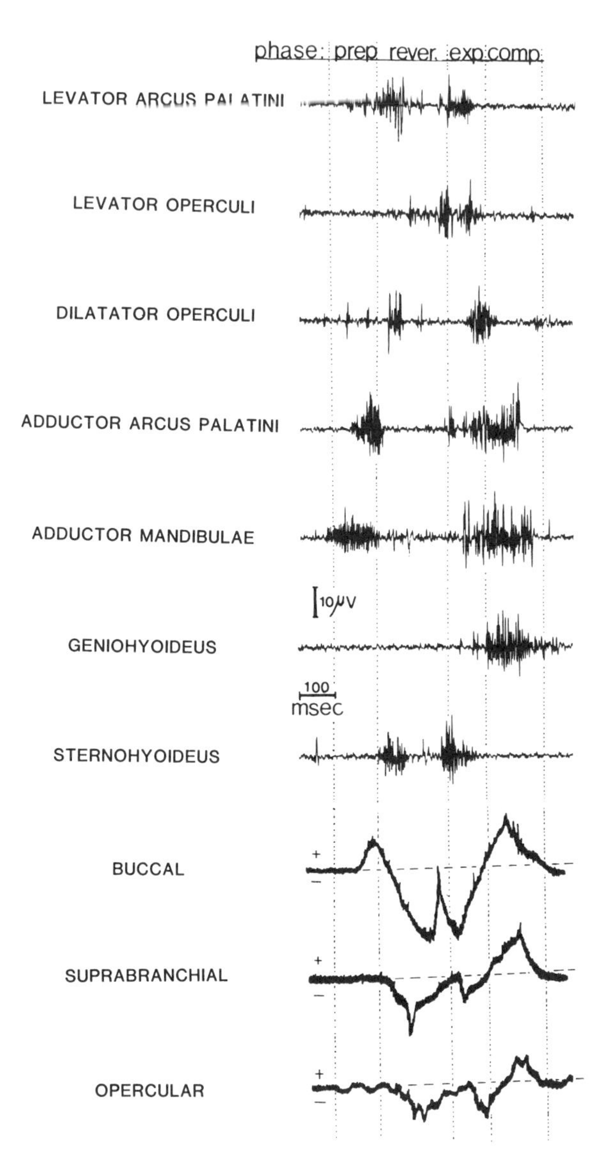

Figure 11-8 Electromyograms together with the pressures of the buccal, opercular, and suprabranchial chambers of *Helostoma temmincki* during quadruphasic air ventilation.

during the expansive phase. Likewise, the quadruphasic pattern of air ventilation is identical to the pattern of muscle actions and pressure changes during the cough. Thus the quadruphasic cough pattern, with its original gill-clearing functions, assumes a totally new function in air-breathing advanced teleosts. In the first half of the cycle it is the air bubble, rather than debris on the gills, that is being flushed out; and in the second half of the cycle a new air bubble, rather than water, is sucked in. Thus, what appear to be drastic changes in functions, that is, two modes of ventilation, are actually the assumption of new functions by an existing functional complex without interference with the original functions, prey capture and coughing (Gans, 1970). The same pattern generators of the original function are retained.

The dual strategy of air ventilation in advanced teleosts gives the fish a built-in bimodal versatility. Because quadruphasic ventilation uses water to move the air bubble, it can only be effective in the aquatic medium: a fish that relies solely on quadruphasic ventilation will not survive out of water. Triphasic ventilation, even though it is hampered by a large residual volume, is quite effective out of water. Bimodal fishes can switch from one mode to the other depending on the medium they are in. It is therefore not surprising that the two modes of ventilation are widely distributed among air-breathing advanced teleosts.

Ventilation in Fish Larvae

Data on respiration and ventilation in fish larvae are scarce, probably because the process is greatly influenced by temperature, salinity, time of day, and especially by change in activity and metabolism (Holliday, Blaxter, and Lasker, 1964). Thus an analysis of fish larvae would not be valid in any but lightly anaesthetized specimens under carefully controlled conditions. Larvae of different taxa cannot be compared because the experiments used by the various investigators differ significantly in design.

Fish larvae, which lack gills, breathe with their integument. Most teleost species live in well-oxygenated, continuously flowing waters and do not require special ventilatory mechanisms or other adaptations. However, fish larvae inhabiting stagnant and deoxygenated waters have evolved specialized proliferations of capillary beds, specialized larval respiratory appendages, and specialized ventilatory mechanisms. Thus the circumtropical synbranchiform fishes that inhabit swamps, rice fields, and permanent or temporary ponds have larvae with large muscular and vascularized pectoral fins that function both as external gills and as ventilators (Fig. 11-9). Dye tracer experiments show that the active movements of the pectoral fins propel water from a well-circumscribed area anterodorsal to the head posteriorly along the length of the entire larva and its yolk sac (Liem, 1981). This ventilating mechanism, which draws the layer of water ventrally, enables the buoyant larva to exploit the thin surface layer of water in which diffusion provides sufficient oxygen in an otherwise depleted water column. Dye tracers placed in various other positions in front of the larva do not get caught in the respiratory current generated by the pectoral fins.

The concentration of oxygen in the water has an important influence on the nature and frequency of pectoral fin movement. When a larva is prevented from approaching the surface layer of the water it invariably dies within 50 minutes. Larvae of the Australian lungfish *Neoceratodus* (Whiting and Bone, 1980) and some Amphibian larvae may possess analogous adaptations. Their body surface is covered with cilia arranged in a pattern conducive to the generation of a posteriorly directed flow of water. Because the principal flow of blood runs countercurrent to the water stream in synbranchiform as well as in lungfish larvae, the ventilatory pattern has been hypothesized to be a countercurrent mechanism (Liem, 1981). However, alternate explanations exist (F. L. Powell, personal communication). Because the oxygen in the ventilating water is likely to be high all along the respiratory exchange surface, the ventilatory pattern in synbranchiform fish larvae may correspond to a mixed pool model of gas exchange as is seen in nonavian vertebrate lungs (Piiper and Scheid, 1973). Another interpretation is that because the capillaries do not traverse the entire length of the exchange surface (Fig. 11-9) the cross-current model as described for the avian lung (Piiper and Scheid, 1973) is more appropriate than the proposed countercurrent mechanism. It is clear from these controversies that ventilation and gas exchange in variously adapted fish larvae are very promising research areas that may uncover a family of new and exciting questions, problems, and issues relating to the structure, function, ontogeny, evolution, and be-

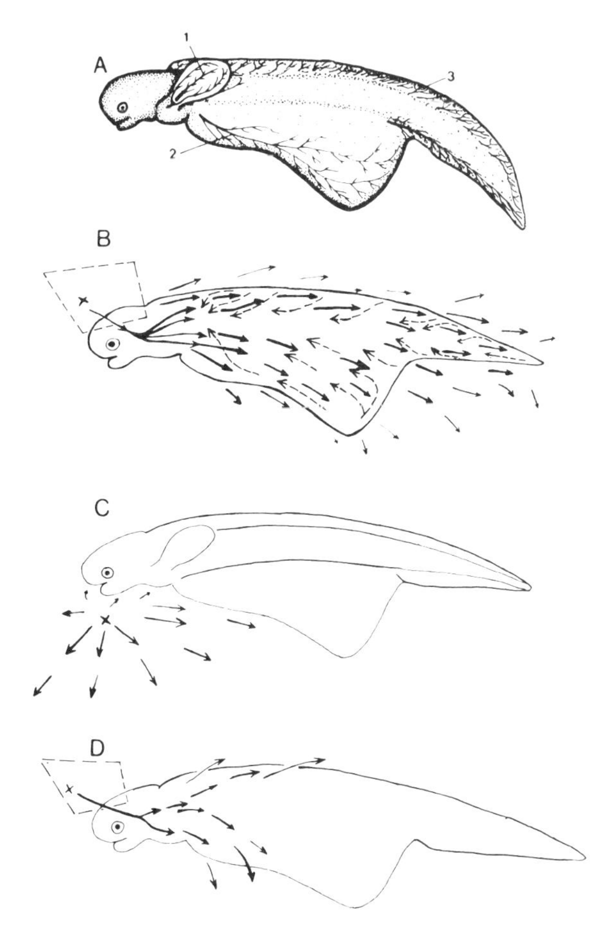

Figure 11-9 A: Larva of *Monopterus albus* four days after hatching, with (1) large vascular pectoral fins, (2) well-developed respiratory capillary networks on the yolk sac, and (3) vascular unpaired (median) fins. B: Water currents generated by pectoral fins (not shown) as indicated by solid arrows; *x* indicates the placement of the dye (methylene blue); dashed arrows show the principal directions of blood flow. This pattern occurs when air equilibration of the water drops below 40 percent. C: Fate of tracer dye when placed at *x*, outside the sphere of the pectoral fins. D: Water currents generated by the pectoral fins as indicated by dye movements when air equilibration is above 40%; the major current takes a ventral direction after passing the hepatointestinal capillary network, whereas the dorsal current is only weakly developed so that the tracer dye is diffused. (From Liem, 1981, courtesy of *Science.*)

havioral ecology of piscine respiratory mechanisms.

The Bimodal Breathing of Dipnoi

In the bimodally breathing lungfishes *(Neoceratodus, Lepidosiren,* and *Protopterus)* aquatic ventilation proceeds, as in teleosts, by interacting buccal and opercular pressure changes (McMahon, 1969). The buccal force pump is initiated by contraction of the adductor mandibulae muscles, which close the mouth. Next the constrictor hyoideus, rectus communis, and geniothoracicus muscles pull the hyoid and pectoral girdle up and forward to raise the floor of the mouth and drive the water over the gills and out through the opercular openings. Finally, the adductor mandibulae become inactive and the unopposed actions of the ventral longitudinal muscles (rectus communis, constrictor hyoideus, and geniothoracicus) open the mouth and pull the hyoid and pectoral girdle back and down, lowering the floor of the mouth and thereby drawing water in.

Aerial ventilation in lungfishes proceeds by the action of a well-developed pulse pump. As the fish rises toward the surface, it closes its mouth by contracting the adductor mandibulae muscle complex and forces water through the gills by action of the buccal compressors (the geniothoracicus, rectus communis, and intermandibularis muscles; Fig. 11-10). The operculum then closes, the snout extends through the water's surface, and the mouth opens by unopposed action of the constrictor hyoideus muscle, which also causes a distention of the buccopharyngeal cavity. At this stage the glottis opens and gas passes from the lung into the buccopharynx and leaves the mouth at the completion of expiration. Closure of the mouth by action of the adductor mandibulae muscle traps a fresh air bubble within the distended buccopharyngeal cavity. Renewed action of the buccal and pharyngeal compressors forces the new air into the lung to complete inspiration. Often this inflow pulse is repeated in each respiratory event.

The pulse pump is clearly derived from the aquatic pump. The pressure and electromyographic patterns of the inspiration phase of the

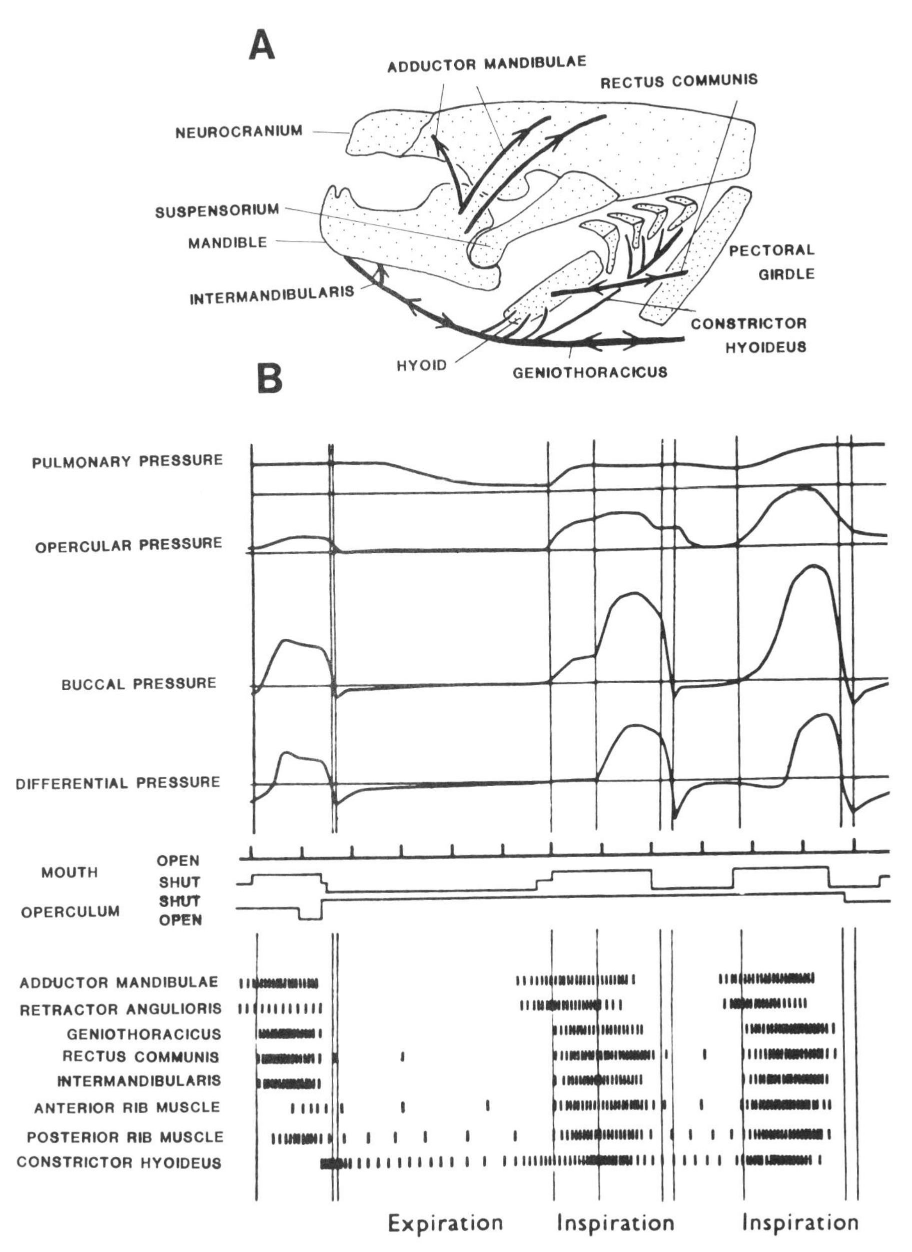

Figure 11-10 A: The major head and branchial structures on the left side of the lungfish *Protopterus.* B: Combined data from pressure records and electromyograms showing the interrelation of muscular action and pressure development during the aerial respiratory cycle. (From McMahon, 1969.)

pulse pump are virtually identical to those of the buccal force pump phase of the aquatic pump (Figs. 11-2, 11-10). The expiration phase of the pulse pump is dominated by the constrictor hyoideus muscle to the same extent as the suction pump is of the aquatic pump (Fig. 11-10). It has been argued (Gans, 1970) that in pulse pumping the volume of air that may be moved during a single pulse is a function of the pressure differential between lung and pharynx and of the volumetric change possible in the buccal cavity. This places a significant limitation on the ratio of tidal volume to lung volume. Furthermore, the pressure in the buccal cavity must be much higher than that in the lung, where the pressure is always higher than atmospheric. Because pulse pumping is ineffective for driving air through narrow tubes, pulse-pumping vertebrates tend to have neither tracheae nor long bronchi. Finally, the functional design of the buccal cavity in an air-breathing fish is likely to be the result of a compromise between the possibly conflicting demands of respiration and feeding, rather than an optimization for effective ventilation.

Ventilation of Actinopterygian Respiratory Gas Bladders

Many actinopterygian fishes possess multifunctional gas bladders, which can function to regulate buoyancy, serve as resonators in hearing (among Cypriniformes, see Chapter 15), and may become lunglike to function as principal or accessory breathing organs. Respiratory gas bladders have evolved independently in several actinopterygian lineages (Lauder and Liem, 1983): gars (Ginglymodi), *Amia* (Halecomorphi), *Arapaima, Notopterus,* and *Pantodon* (Osteoglossomorpha), *Megalops* (Elopomorpha), *Gymnotus* (Ostariophysi; Liem, Eclancher, and Fink, 1984), and *Erythrinus* and *Hoplerythrinus* (Characoidei; Kramer, 1978).

Cineradiographic and electromyographic data, together with evidence obtained with pressure transducers, indicate that all actinopterygian fishes with respiratory gas bladders use the pulse pump to move air out of and into the bladder. In various lineages the pulse pump becomes specialized in different ways. Ventilation in the gar *Lepisosteus* exhibits all the characteristic features of a representative actinopterygian pulse pump (Rahn et al., 1971). Exhalation begins with the gar approaching the surface at an angle (Fig. 11-11) while compressing its buccopharyngeal cavity by action of the adductor mandibulae and intermandibularis muscles. This is followed by a relaxation of all the muscles and an opening of the glottis. Because the pressure within the gas bladder is higher than in the buccal cavity, gas passes passively by hydrostatic pressure from the bladder to the mouth cavity. Once the buccal cavity is filled with gas, the buccal floor is raised by activity of the adductor mandibulae, intermandibularis, and interhyoideus muscles, forcing the air to escape from under the operculum. By this time the snout of the fish has broken the surface and exhalation is completed.

Inhalation begins with the snout out of water. The mouth opens and the floor of the buccal cavity is lowered by unopposed activity of the sternohyoideus muscle. In this way the buccopharyngeal cavity is filled with air. Action of the adductor mandibulae closes the jaws, and the gar sinks below the surface with the distended, air-filled buccopharyngeal cavity. Action of the adductor mandibulae, intermandibularis, and interhyoideus muscles raises the buccopharyngeal floor, thus pumping the air into the gas bladder.

The ventilatory event lasts about 0.5 sec and consists of a partial exhalation, leaving a very large residual volume, followed immediately by an inhalation. The tidal volume is dictated by the size of the buccal cavity and averages about 40% of the total volume of the gas bladder. Initially, water is squeezed out by action of the buccopharyngeal compressors. The expiratory gas is then forced from the buccal cavity through the opercular slit. All air inflow occurs through the mouth, with the buccal force pump driving the fresh air into the gas bladder. Because the pressure within the gas bladder is higher than atmospheric, the buccal force pump must impart significant compression in order to move the entrained gas. The greater the pressure within the gas bladder, the greater the fraction of the energy in a given pulse that serves compression rather than flow. Thus the ventilatory mechanism of *Lepisosteus* clearly exhibits the constraints and limitations of pulse pumping.

Ventilation in Amphibian Tadpoles

All free-living anuran larvae use an essentially similar buccal pump, despite differences in the morphology of the elements, to create a unidirectional, continuous flow of water across the gills, as

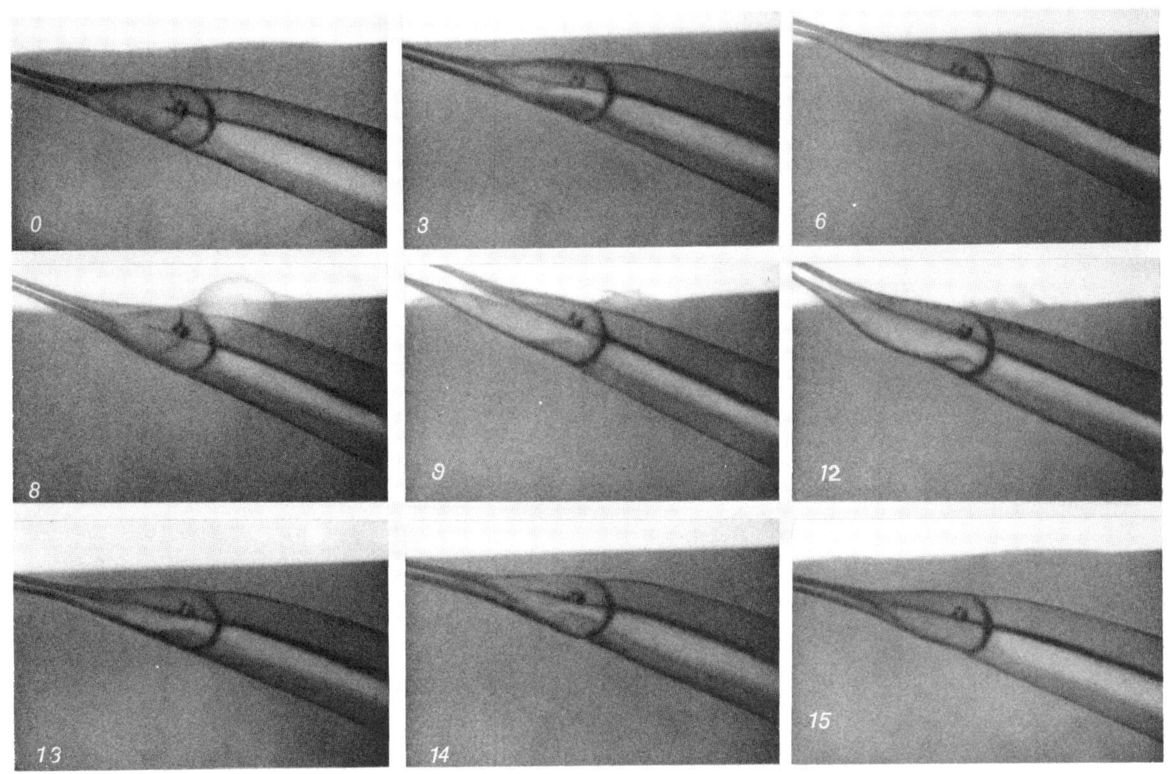

Figure 11-11 Individual frames of an X-ray film taken at 100 frames/sec of the garfish *Lepisosteus oculatus* during air ventilation. 0: Fish approaches the surface. 3: Buccal region distends just before the tip of the snout reaches the surface. 6: Snout breaks the surface with the mouth closed and the expanding buccal cavity fills with gas from the lung, which decreases its volume. 8: With the mouth closed and the snout out of the water, the buccal floor is raised with coincident expulsion of large bubbles from the posterior opercular edge. 9: With some of the head and the angle of the mouth protruding out of the water, the buccal floor drops very sharply and the mouth opens. The buccal cavity is completely filled with fresh air. 12: The mouth is closed, the distended buccal cavity filled with air, and the fish begins to sink. 13, 14: As the fish sinks below the surface, the buccal floor is raised, the volume of air in the buccopharynx is drastically reduced, and the lung can be seen to expand. 15: All air has been forced into the lung through the glottis by the buccal force pump and the fish continues to sink.

well as to generate food-bearing currents to the branchial food traps and larval gill filters (Wassersug and Hoff, 1979).

Basic Mechanism The major components of the aquatic pump are the buccal cavity, which is separated from the mouth by an oral valve, the pharyngeal cavity, which is separated from the buccal cavity by a ventral velum, and the gill cavity containing the gill curtain (Fig. 11-12). The chambers are surrounded by several cartilages, as depicted in Figure 11-13. The "piston" for the buccal pump consists of paired ceratohyal cartilages underlying the floor of the buccal cavity. Each articulates with the ventral surface of the palatoquadrate. These articulations act as fulcrums about which the ceratohyals rotate when the large depressor muscles, the orbitohyoidei, contract. Contraction of the orbitohyoidei elevates the lateral margins, and consequently depresses the medial margins, of the ceratohyal. This action expands the buccal cavity, creating a low pressure in the mouth cavity. The medial margins of the ceratohyals are returned to their

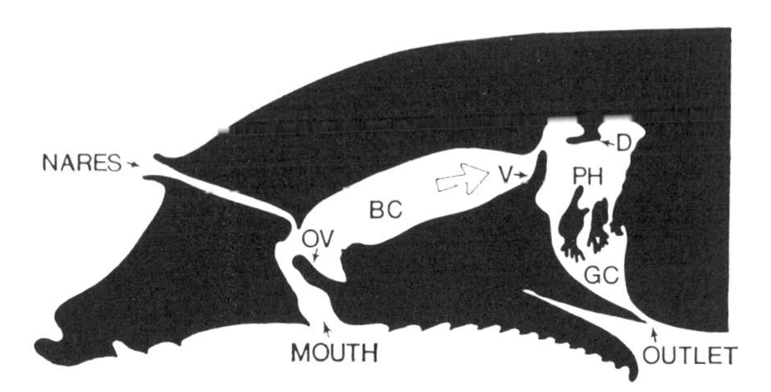

Figure 11-12 Composite diagram of sagittal and parasagittal sections through the head of an *Ascaphus* tadpole with its sucker disengaged. The valves are shown as oriented at the beginning of the expiration phase of the ventilatory cycle. The large arrow indicates the beginning of flow of buccal water over the ventral velum and into the pharynx. *BC* = buccal cavity; *D* = dorsal velum; *GC* = gill cavity; *OV* = oral valve; *PH* = pharynx; *V* = ventral velum. (From Gradwell, 1971.)

original elevated position by contraction of a single, transversely oriented, interhyoideus muscle, which is attached to the ventral surfaces of the ceratohyals (Wassersug and Hoff, 1979).

Pressure transducers show that ventilation proceeds by alternate action of the buccal and pharyngeal force pumps, which produce a continuous branchial water flow (Gradwell, 1971). Inhalation is accomplished by depression of the buccal floor by contraction of the orbitohoidei muscles. In this way a fall in the buccal pressure relative to the ambient pressure is created. The fall in buccal pressure closes the ventral velum, occluding the buccal from the pharyngeal cavity. Water is drawn into the buccal cavity through the nares and mouth. Toward the end of inhalation pharyngeal constriction occurs, causing a rise in pharyngeal pressure (Fig. 11-14), which keeps the ventral velum shut. Therefore, water passes from the pharyngeal cavity through the gill curtain into the gill cavity. Exhalation begins with elevation of the buccal floor by action of the interhyoideus muscle. Compression of the buccal cavity raises buccal pressure above ambient pressure, thereby closing the valvular internal nares and the oral valve. At this stage the ventral velum opens. No reflux occurs at the nares or mouth, and water is deflected from the buccal cavity against the dorsal velum. Hence the water is further deflected until it leaves the pharyngeal cavity.

Model of the Buccal Pump Wassersug and Hoff (1979) have described a general model of the buccal pump of tadpoles, which sacrifices realism

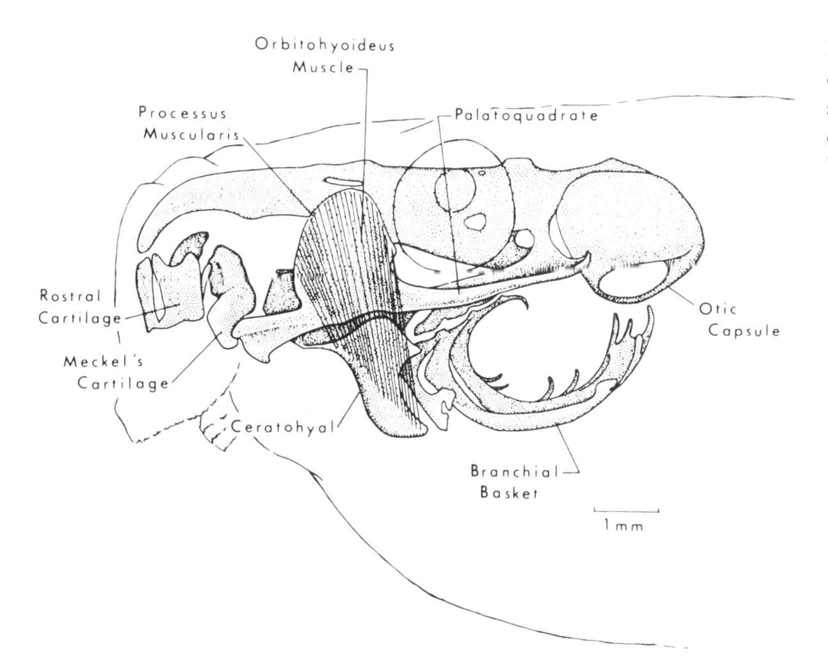

Figure 11-13 Lateral view of the chondrocranium of a tadpole showing major structures and the orbitohyoideus muscle. (From Wassersug and Hoff, 1979.)

to gain generality and precision. The model is represented in Figure 11-15. It represents half the buccal pump as seen in anterior view. The floor of the buccal cavity, supported by the ceratohyal and hypobranchial plate, is represented as a triangle, shown in two positions, with base length *A* and altitude *C*. The depressor musculature (orbitohyoidei), of which the longest fibers are labeled *H*, rotates the ceratohyals through an angle. This causes the floor of the mouth to drop a distance *G*. This action displaces a volume of water shaped like an inverted, truncated, right triangular pyramid. There are various ways of calculating this volume on the basis of several parameters of the model (Wassersug and Hoff, 1979). The model can be used to predict buccal volume for any specified amount of contraction in the depressor musculature *F*. A low value of *F* indicates a large buccal volume. For any value of *F* the angle α can be calculated, giving an estimate of how much the buccal floor is depressed with each stroke of the buccal pump. A lever-arm ratio, taken as the projected width of the lateral portion of the ceratohyal (*B-C*) over half the total width of the buccal floor (*B*), can also be calculated. This gives a good measure of the mechanical advantage of the musculature that deflects the buccal floor down from its resting position. This model has proved to be sufficiently accurate to predict how changes in several elements of the buccal pump can affect the functional and adaptive features of both respiration and feeding in tadpoles.

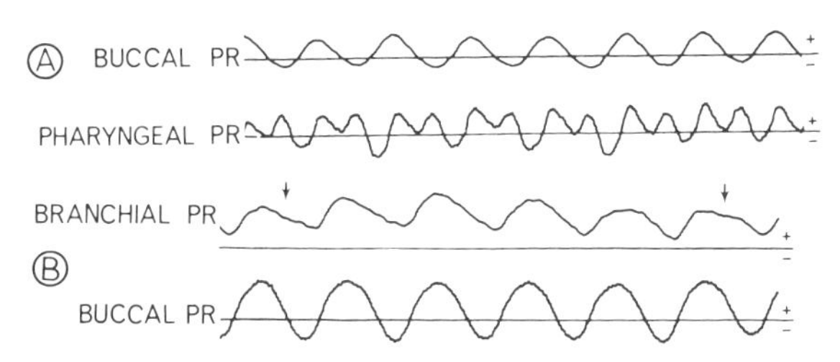

Figure 11-14 A: Pressures recorded simultaneously from the buccal cavity and pharynx of a lightly anesthetized tadpole. In each ventilatory cycle the first peak in pharyngeal pressure is the transmitted effect of the buccal force pump. The second peak is caused by constriction of the pharynx during its occlusion from the buccal cavity by the ventral velum. B: More fully anesthetized tadpole. Branchial pressures remain positive throughout each irrigation cycle, ensuring a continuous branchial outflow. Calibrations: 1 cm water; 1 sec. (From Gradwell, 1971.)

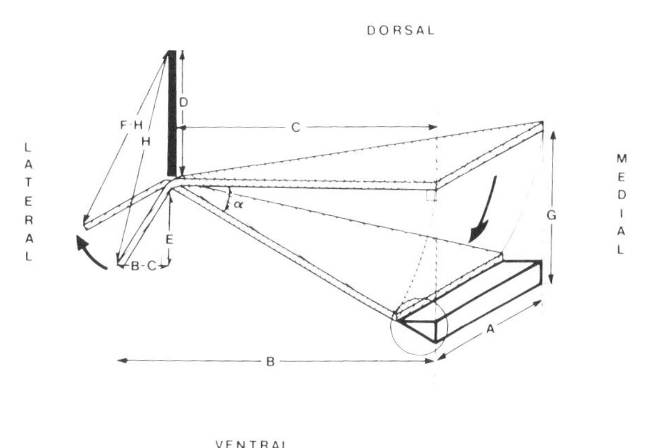

Figure 11-15 Generalized mechanical model of the buccal pump of tadpoles as formulated by Wassersug and Hoff (1979) and seen in an anterior oblique view. The triangle of height *C* and base *A* represents the functional area of the buccal floor. B is width of buccal floor, the black bar *D* represents the processus muscularis of the palatoquadrate, and *H* represents the depressor musculature of the buccal pump in a relaxed state. *G* is the distance the floor is depressed and α is the angle through which the ceratohyal rotates when *H* contracts to length $F \cdot H$. The model treats the buccal volume on each side as an inverted, truncated, right triangular pyramid. The darkly outlined wedge is the truncated portion of the pyramid. (After Wassersug and Hoff, 1979.)

Ventilation in Frogs

As in the lungfish, ventilation in most frogs depends on the activity of a buccal force pump, which determines pulmonary pressure. The elevated pulmonary level facilitates the expulsion of gas during ventilation of the lung (de Jongh and Gans, 1969).

The anatomical basis of the buccal force pump of frogs is the corpus of the hyobranchial apparatus under the tongue and the muscular floor of the mouth, which is made up by the thick intermandibularis muscle running transversely between the mandibular rami. Two pairs of longitudinal muscles, the anterior geniohyoidei and

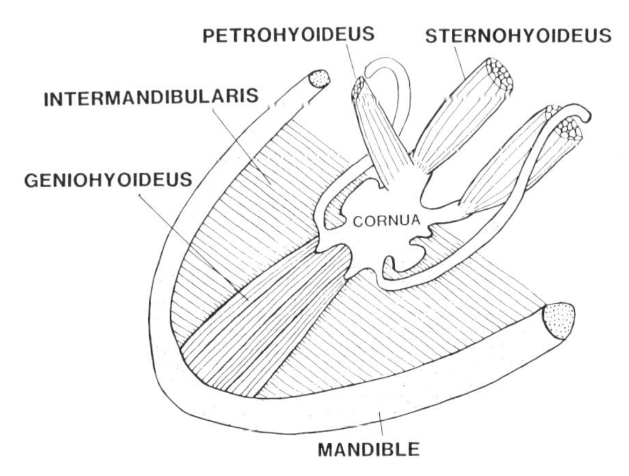

Figure 11-16 Major muscles of the buccal floor of the frog seen obliquely and dorsally.

posterior sternohyoidei, complete the principal muscular complex of the buccal force pump (Fig. 11-16). The petrohyoid muscles run between the hyoid and otic capsule. The sternohyoideus pulls the corpus posteriorly and ventrally, enlarging the buccal cavity, whereas the intermandibularis, geniohyoidei, and petrohyoidei raise the floor of the buccal cavity reducing the volume.

Ventilation in most frogs is divided into five phases (Fig. 11-17). In the first phase contraction of the sternohyoideus lowers the buccal floor. The external nares are open and the glottis closed, so fresh air flows into the posteroventral portion of the buccopharyngeal cavity. In the second phase the glottis opens and pulmonary gas enters the buccal cavity as a result of elastic forces in the lung. The pressures in lung and buccal cavity are about equal during the ill-defined third phase, in which behavioral variability has frequently been observed. In the fourth phase the nares are closed and contraction of the muscles of the buccal floor rapidly raises first the buccal pressure and then the lung pressure. At the end of the fourth phase the nares open and the glottis closes. During the fifth phase the buccal pressure drops sharply below that of the lung as the buccal floor is depressed by gravitational and elastic forces.

The mechanism of ventilation can be modeled as a pump with a twin (buccal and pharyngeal) piston and twin (narial and glottal) valve, which inflates an elastic sac (de Jongh and Gans, 1969). Even though there is some basic resemblance to the pulse pump of lungfish, the frog's buccal force pump is much more complex. Significant mixing is avoided because the exhaled gas from the lung escapes rapidly in a forward direction. The resultant jet stream–like expiration is confined to the upper half of the buccopharyngeal airway and effectively bypasses the posterior portion of the buccal cavity where the fresh air, ready for the next inhalation, is temporarily stored. After inhalation a period follows when the glottis remains shut, but oscillatory pumping movements force air in and out of the open nares. These oscillations have been thought of as part of an olfaction, or "sniffing," process, but it can also serve to flush out residues of expired air in the buccal cavity as a preparation for the next ventilatory cycle.

Ventilation in Reptiles

All air-breathing fishes and amphibians use pulse pumping to ventilate their air-breathing organs (Fig. 11-18A). Within the body plan of anamniotes and in the aquatic medium, pulse pumping is clearly a multifunctional strategy: it allows the maintenance of gill functions, serves bimodal (aquatic and aereal) breathing, and furnishes the proper mechanism for bubble, foam, or other types of nest building. When lower vertebrates became terrestrial, a new mode of ventilation evolved: the aspiration pump. All reptiles have aspiration breathing, although each group uses a different mechanism. In an aspiration pump air follows, rather than induces, the changing shape of the lung; that is, airflow is induced by mechanical deformation of the pulmonary walls (Fig. 11-18B). With the separation of the driving muscle from the pulmonary walls, the response becomes a function of the materials connecting the pulmonary wall and the transverse, intercostal, or other muscles generating the power (Gans, 1970). An advantage of the aspiration pump is that the lungs may be filled in a single stroke, independent of the size of the buccal cavity. Feeding and respiratory couplings, which constrain both those systems, become decoupled. Once the constraint was released, each system

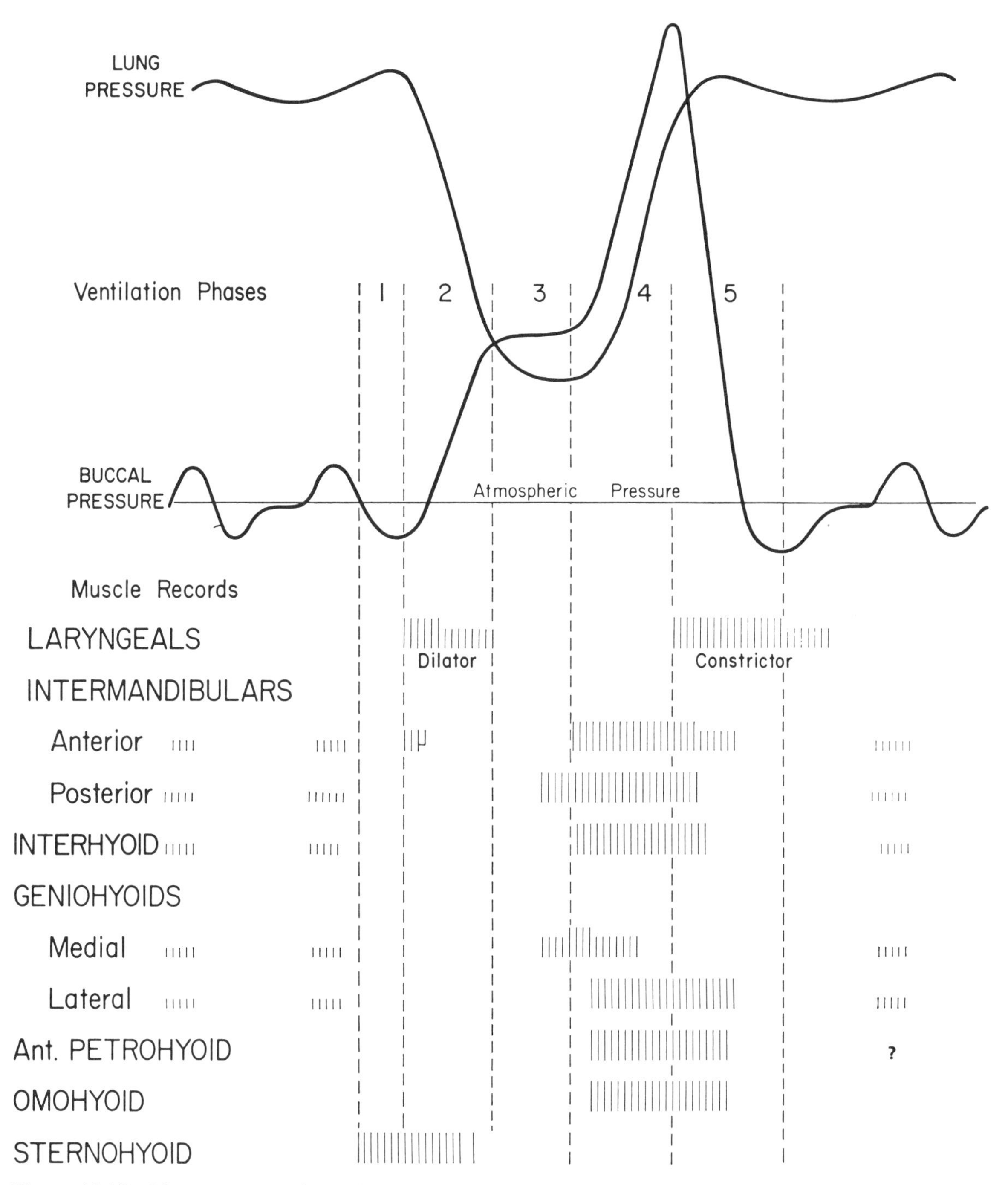

Figure 11-17 Diagram summarizing the time relationships among lung pressure, buccal pressure, and the electromyograms of the respiratory muscles in the frog. The magnitudes of the firings as indicated by the vertical hatching are rough approximations. (From de Jongh and Gans, 1969.)

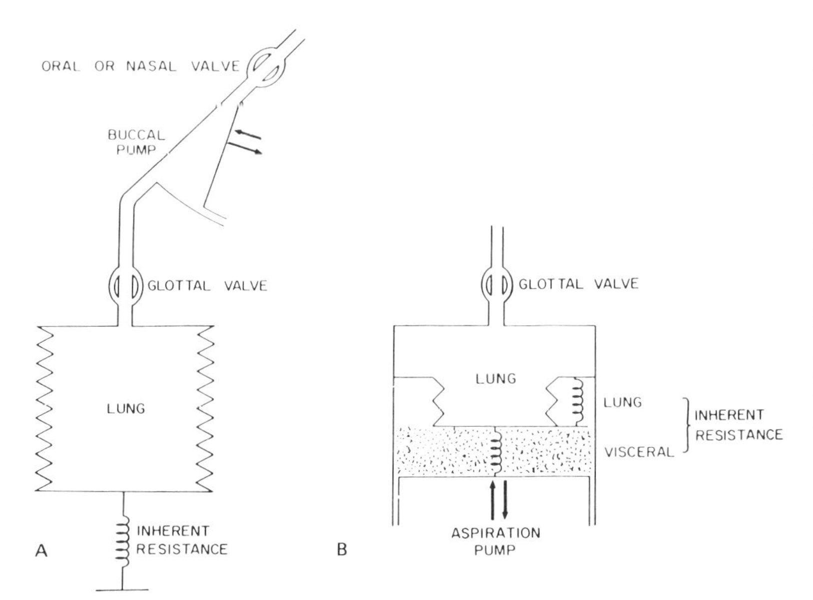

Figure 11-18 Contrast between a pulse-pumping (A) and an aspiration-pumping (B) mechanism for filling the lungs. Two valves are required in pulse pumping and the distention against inherent resistance is here transmitted by the compressed gas. In contrast, the aspiration pump may or may not have a valve and airflow is induced by mechanical deformation of the pulmonary walls. (Gans, 1970, reprinted with permission from *Forma et Functio,* Pergamon Press, Ltd.)

could develop independently into a great diversity of morphologies representing multiple solutions to environmental problems.

Basic Mechanism The ventilatory mechanism of freely moving *Caiman crocodilus* as studied by cineradiography and electromyography by Gans and Clark (1976) is discussed here as representative of reptiles even though each group has a differently modified aspiration pump. The anatomical components of the aspiration pump are two separate channels running from the valvular external nares, located on the snout past the olfactory chamber, to the internal nares on the roof of the posterior pharynx. The glottis is positioned immediately ventral to the internal nares. From here the trachea passes posteriorly, connecting via the bronchi to the extensive lungs within the rib cage.

Three major groups of muscles play a key role in driving the aspiration pump: (1) various intercostal groupings that can increase and decrease the volume of the pleural cavity by adjusting the positions of the ribs; (2) various abdominal muscles that can force the liver anteriorly, thus reducing the volume of the thoracic cavity, which includes the lungs in their pleural cavity; and (3) the diaphragmatic muscles that attach the liver to the ilia and epipubic elements of the pelvic girdle and can pull the liver posteriorly, thus increasing the pleural volume.

Ventilation is diphasic, involving exhalation followed by inhalation (Fig. 11-19). Exhalation is the result of the action of the transverse abdominal muscles, which change the volume of that portion of the visceral cavity posterior to the liver, thereby shifting the liver anteriorly. As a result the pleural volume is reduced, the pulmonary pressure increases, and the glottis opens. Inhalation is caused primarily by contraction of the diaphragmatic muscle, which pulls the liver posteriorly. The shift of the liver increases pleural volume and is coincident with the drop of pulmonary pressure below atmospheric pressure. It is thought that the intercostal muscles facilitate airflow by shifting the position of the ribs. Their activity can be interpreted as a fixation of the flexible rib cage so that it resists aspirating and compressing actions of the hepatic piston. The pattern of muscle actions during ventilation changes as the body is immersed. Exhalation tends to become more passive and inhalation requires increased muscular effort (Fig. 11-19).

Comparative Aspects Each reptilian breath begins with an exhalation induced by an increase in intrapulmonary pressure coincident with, or immediately followed by, an opening of the

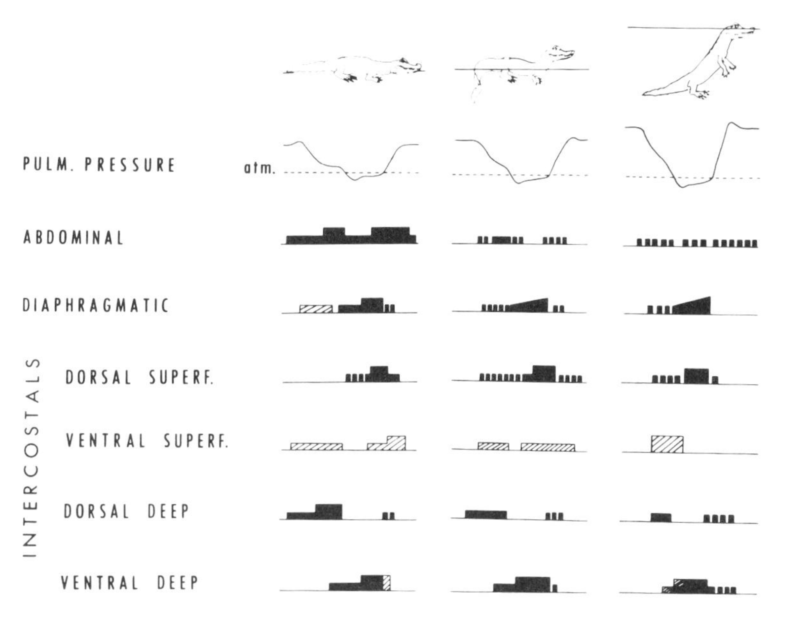

Figure 11-19 Activity patterns of various muscle groups at different depths of immersion in *Caiman*. (From Gans and Clark, 1976.)

glottis. The outflowing gas causes the intrapulmonary pressure to drop. During inhalation various muscular mechanisms enlarge the pleural cavity, dropping the intrapulmonary pressure and inducing air to flow in until the glottis closes. In reptiles the movement and pressure profiles during a single breath are normally triphasic (rise-fall-rise), but the flow is always biphasic (outflow-inflow). Even though the pressure, movement, and airflow patterns in reptiles show general similarities, the underlying muscular mechanisms are different in the various lineages.

Although the mechanics of lizard respiration are not yet completely known, experimental analyses have shown that portions of the intercostal muscles fire during exhalation, whereas other portions are active during inhalation. The complex musculature of the thoracic and abdominal wall suggests that multiple contraction patterns are likely. Study is needed to learn how the complex muscle contractions act directly and indirectly via the viscera to bring about ventilation in lizards. In snakes compression and inflation of the lung occur by regionally controlled shifts in the volume of the visceral cavity, either by movements of the ribs or by an elevation of the midventral wall by action of the transverse abdominal muscles (Rosenberg, 1973; Clark, Gans, and Rosenberg, 1978). In turtles the lungs lie in the dorsalmost position beneath the carapace (Fig. 11-20). Ventilation results from changes in visceral volume brought about by inward and outward movements of the limbs and girdles (Gaunt and Gans, 1969). As the entire visceral mass lies in a single cavity, any shift of its volume by changing positions of the limb girdles or muscular septa causes a change in pulmonary pressure. Furthermore, relative filling of the bladder with water will affect the gas content of the lung and with this, the buoyancy. During exhalation the trans-

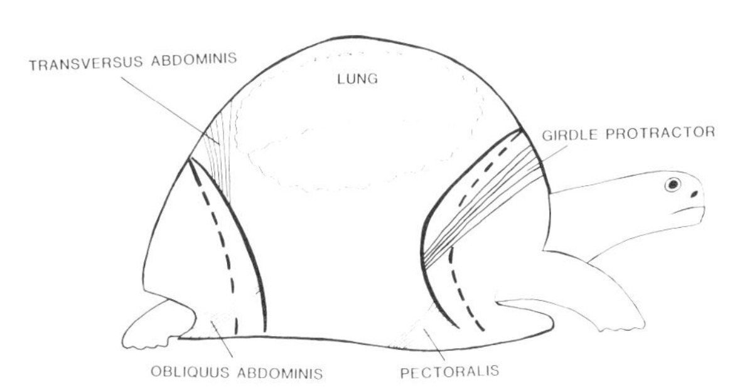

Figure 11-20 The mechanism of lung ventilation in a turtle as deduced from pressure recordings and electromyography. Action of transversus abdominis and pectoralis muscles decreases pressure in the abdominal cavity and therefore increases intrapulmonary pressure. The obliquus abdominis and serratus are active during inhalation.

versus abdominis and pectoralis muscles are active, causing an inward movement of the pelvic and pectoral girdles, respectively. The movements increase the visceral, and thus the intrapulmonary, pressure. Inhalation is brought about by action of the obliquus abdominis and serratus muscles, which move the girdles outward (Gaunt and Gans, 1969). As a result the visceral volume is increased and the intrapulmonary pressure drops below atmospheric pressure. Thus pulmonary pressure in turtles is under voluntary control and depends on the position of the limbs, external influences, and the state of the limb muscles.

Ventilation in Birds

The structure of avian lungs differs fundamentally from that of all other vertebrates. There are no dead-end alveolar ducts and alveoli. The paleopulmonic and neopulmonic parabronchi of the lungs are relatively rigid and maintain a virtually constant volume (Fig. 11-21). Consequently, the mode of ventilation deviates drastically from that of mammals and reptiles.

Morphology and Airflow The gross anatomy of the avian respiratory system is schematically represented in the Figure 11-21. The trachea divides into two bronchi, of which the intrapulmonary portions are called mesobronchi. Each mesobronchus gives rise to a more proximal group of secondary bronchi, the ventrobronchi, and to a more distal group of secondary bronchi, the dorsobronchi and the laterobronchi, before terminating in the abdominal and postthoracic air sacs.

Because the lungs (paleo- and neopulmonic parabronchi) are rigid, the only changes in volume during the ventilatory cycle occur in the voluminous air sacs (Duncker, 1971, 1972). The cranial air sacs (cervical, interclavicular, and prethoracic) are connected to ventrobronchi, whereas the caudal air sacs (abdominal and postthoracic) communicate with the mesobronchus. Since there are no structural valves in the bronchial system, the pattern of airflow through this rather complex system, which seems to offer alternative pathways, has been a challenging problem. Bretz and Schmidt-Nielsen (1971) offer an excellent historical perspective.

The use of thermistor probes into the primary and secondary bronchi of ducks (Bretz and Schmidt-Nielsen, 1971) and flow meters in the dorsobronchi of geese (Brackenbury, 1971) showed beyond doubt that the parabronchial gas flow is from the dorsobronchi via the parabronchi to the ventrobronchi during both inhalation and exhalation. Expiration is always active, but during quiet breathing inspiration is a passive process. Electromyographic analysis (Fedde, Burger, and Kitchell, 1961) has shown significant overlaps in the timing of activity of different muscle groups. The movements produced by the action of the various costal and abdominal groups are transmitted either directly or indirectly to the air sacs. Pressure recordings from the air sacs have shown that during easy breathing (eupnea) the pressure waveforms are characterized by an exponentially decaying expiratory phase of longer duration than the inspiratory phase. These waveforms indicate a constant muscular effort during expiration. Differences in pressure between air sacs exist because of differences in resistance of pathways between the outside air and the individual sacs. Passage of air through the parabronchi during inhalation is caused by the aspiratory action of the anterior air sacs, and during exhalation the posterior sacs force air forward through the parabronchi (Fig. 11-21).

Comparative Aspects This description of the ventilatory cycle probably applies to most birds. In primitive birds (such as penguins) the neopulmo or neopulmonic parabronchi are absent. In more advanced birds (such as fowllike birds and songbirds) the neopulmo reaches its highest development, although it never exceeds 20% of the total mass of the lung tissue (Duncker, 1971). Airflow in the neopulmonic parabronchi reverses during inhalation and exhalation (Fig. 11-21). The efficiency of ventilation in the avian lung is based on the unidirectional airflow through the paleopulmonic parabronchi, which are open ended. Thus the air content in the air capillaries is renewed twice within one ventilatory cycle: air flows from the posterior sacs through the para-

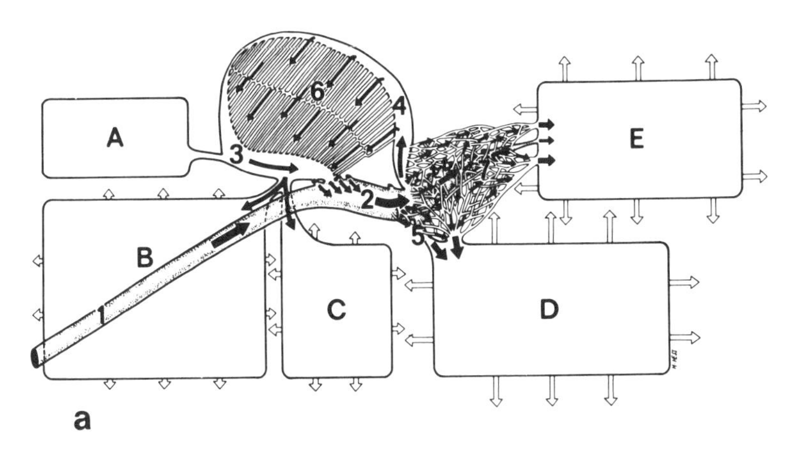

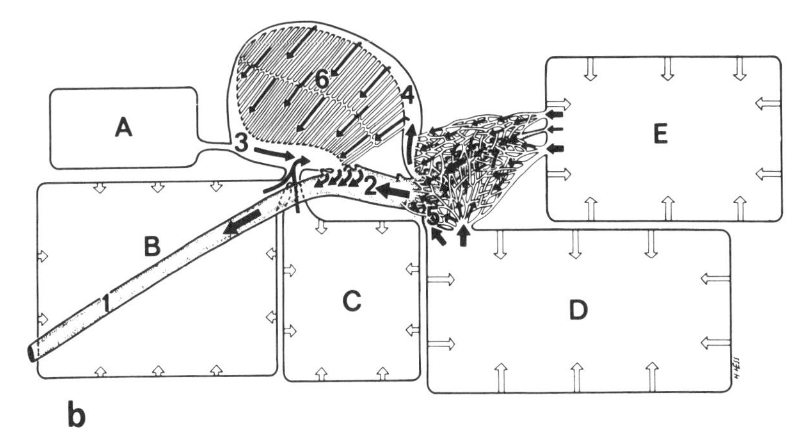

Figure 11-21 Airflow through paleopulmonic (6) and neopulmonic parabronchi (7) during inhalation (a) and exhalation (b) in birds. *A, B, C, D, E,* respectively, are the cervical, interclavicular, prethoracic, postthoracic, and abdominal air sacs. 2 = mesobronchus; 3 = ventrobronchi; 4 = dorsobronchi; 5 = laterobronchi. (From Duncker, 1971.)

bronchi in one direction during both inhalation and exhalation. Through a bellows action of the costal musculature on the extensive air sacs, the airflow in birds can reach very high values. This avian system offers a unique structural advantage in that residual air volumes of the gas exchange portion of the lung can be kept very low. In addition to the usefulness of air sacs in increasing the total air capacity and facilitating the exchange in flow through tubes, they help to reduce the density and to regulate the body temperature. The avian ventilatory mechanism is not only anatomically linked with the locomotory apparatus for flight, but flying birds have achieved phase-locked locomotor and ventilatory cycles. The locomotor-respiratory coupling may serve as an important adaptation so that flying birds can effectively ventilate their lungs while in motion and sustain aerobic metabolism.

Ventilation in Mammals

There are surprising gaps in our knowledge of the aspiration pump of mammals. It is generally agreed that the powerful muscular diaphragm that separates the thoracic and abdominal cavities provides the structural basis for a very efficient aspiration pump. Although the intercostal muscles are also hypothesized to play an important role, their precise functions and actions have not been determined experimentally, and all statements concerning their role in the ventilatory cycles should be considered speculations (De Troyer, Kelly, and Zin, 1983). Owing to the complexity of the rib movements, the mechanical actions of these muscles have defied analysis. The mechanical link between the rib cage, intercostals, the diaphragm, and the abdomen remains poorly understood and is still a subject of active research.

Muscular Basis Several muscles have been associated with respiratory functions. Characteristic of mammals is the muscular diaphragm, which is composed of three parts: (1) the crural or vertebral, which arises as the crura from the lumbar vertebrae and arcuate ligament; (2) the costal, which arises from the costal margin; and (3) the sternal, which arises from the xiphoid process. All fibers converge on the central tendon. The external intercostals extend from the tubercles of the ribs to the costochondral junction, and their fibers slope obliquely downward and forward from one rib to the next. The deeper internal intercostals extend from the ventral end of the intercostal space to the angles of the ribs dorsally, where they become continuous with the dorsal intercostal membrane. The fibers slope obliquely downward and backward from one rib to the next.

In *Homo* the abdominal muscle complex consists of several major muscles. The external oblique originates from the outer surfaces of the last eight ribs, passing downward to the iliac crest and rectus sheath. The internal oblique arises from the lumbar fascia, the iliac crest passing forward to the last three ribs and the rectus sheath. The transversus abdominis originates from the costal cartilages of the more posterior ribs, the lumbar fascia, and the iliac crest and passes forward to the rectus sheath. The rectus abdominis arises from the pubic crest and symphysis and passes anteriorly in a sheath formed by the aponeurosis of the oblique and transversus muscles to the fifth, sixth, and seventh costal cartilages. The scaleni arise from the transverse processes of the last five cervical vertebrae and pass to the anterior surfaces of the first and second ribs. The sternomastoid arises from the manubrium sterni and the medial part of the clavicle to insert on the mastoid process and the occipital bone.

Respiratory function has often been attributed to such muscles as transversus thoracis, levatores costarum, trapezius, pectoralis major and minor, latissimus dorsi, serratus anterior and posterior, quadratus lumborum, and sacrospinalis. However, electromyographic analysis has failed to identify respiratory function for any of these muscles (Campbell, 1958).

During quiet ventilation only the inspiratory muscles show much activity. The diaphragm is active, whereas the external intercostals and scaleni are only sporadically active. However, the mechanical action of individual intercostal muscles has not yet been definitely established (Rankin and Dempsey, 1967; De Troyer, Kelly, and Zin, 1983).

During high-intensity ventilation the sternomastoids and extensors of the vertebral column are recruited and become active toward the end of inspiration. All abdominal and intercostal muscles begin to contract with increasing vigor toward the end of expiration. The muscles of the abdominal wall are probably the most important muscles of forced expiration. There is no doubt that the diaphragm is the major muscle to contract during inhalation. It uses the intraabdominal pressure and viscera as its fulcrum. Its contraction leads to an expansion of the base of the thorax by moving the ribs outward and up or anteriorly. During forced ventilation the sternomastoid acts on the sternum in such a way as to increase the dorsoventral diameter of the thorax, thereby causing an even greater inhalation.

General Aspects of the Mechanics During shallow ventilation there is a balanced antagonism between the muscles of inspiration (diaphragm, intercostals, and scaleni) and the passive elastic forces, with expiratory muscles only exceptionally and coincidentally involved. Thus inspiration is produced by contraction of the main inspiratory muscle, the diaphragm, and expiration is produced by elastic recoil of the lungs (Otis, Fenn, and Rahn, 1950). Even during forced ventilation, the expiratory muscles (the abdominal muscle complex) become active only in the later phase of expiration. Apparently, the potential energy created by the inspiratory muscles is stored in the elastic tissues of the lung and thorax, and it is sufficient to supply most of the necessary force for expiration when the lung and the thorax recoil. Electromyography and pressure recordings have verified that an increase in the force of aspiration leads to increased distention of the lungs and greater elastic recoil. During forced ventilation a progressively greater number of principal and accessory inspiratory as well as expiratory muscles are recruited.

Current models of the ventilatory apparatus of mammals take into consideration the mechanical linkage of the diaphragm, the intercostals, the sternomastoid, the rib cage, and the abdomen. The major component of the proposed link is that the muscles of inspiration, the pressure generators, operate on the rib cage as though they were arranged in series. In such a model the diaphragm is thought to lift and expand the rib cage only to the extent that abdominal pressure increases. However, results of recent research require a slight modification of the model. The diaphragm of the dog consists of two muscles, contraction of which has a different effect on the rib cage (De Troyer et al., 1981). The costal part of the muscle has a direct inspiratory action on the lower rib cage, even without the aid of increase in abdominal pressure. The crural part seems to have an expiratory effect, which is, however, balanced by the rise in abdominal pressure. Thus, contraction of the costal part expands the rib cage both through a direct effect and through the increase in abdominal pressure. In the absence of abdominal pressure the crural part has an expiratory effect, which can only be balanced by a rising pressure in the abdomen.

Locomotor-Respiratory Coupling Although the current model of mammalian mechanics has been formulated without consideration of the potential linkages between the locomotor and respiratory functions, it is clear that locomotor and respiratory functions both rely on cyclic movements in shared components: ribs, sternum, abdominal muscles, vertebral flexors and extensors, and other muscles associated with the thorax. Visceral motion during running could influence ventilation by altering intraabdominal and intrathoracic pressures. Bramble and Carrier (1983) have shown that mammalian locomotion and respiration are closely coupled during running.

Simultaneous recordings of gait and lung ventilation in hares, horses, and humans disclose a consistent locomotor-respiratory coupling. Inhalation and exhalation during the canter of the horse are represented in Figure 11-22 by large, distinct bursts of sound at the nostrils. Inhalation is coincident with the lifting of the lead forelimb. At this point the horse leaves the ground and enters the floating or gathered suspension phase (position 4 in the figure). Air continues to flow into the lungs during the initial or on-loading portion of the thoracic cycle, but the flow decreases rapidly or stops altogether as compressive loading of the rib cage escalates. Exhalation begins at a high rate near the point at which the thoracic complex experiences its peak load.

In the gallop ventilation remains coupled to gait, with some adjustments. Inhalation is now confined to the off-loaded interval when the horse is either entirely off the ground or is supported by the hind limbs only. Conversely, exhalation begins as the first forelimb touches the ground and thoracic loading begins.

It is proposed that locomotor-respiratory coupling may be a general requirement for sustained aerobic activity among endothermic vertebrates. It is not yet known, however, whether gait and respiration are phase-locked in marsupials or in very small mammals with jerky, irregular motions — or in others at very low rates of travel. Because of the locomotor-respiratory coupling, respiration may in many instances affect or even dictate locomotor pattern in mammals (Bramble and Carrier, 1983).

Future Areas of Research

One important task remaining in the study of vertebrate ventilation is to unravel the basic mechanism of the mammalian ventilatory cycle by the use of electromyography and pressure recordings synchronized with cineradiography. Once this is better understood, one can determine the anatomical and functional couplings between the locomotor and respiratory apparatus with greater precision. A comparative analysis can then evaluate the nature and variability of coupling patterns in different mammalian lineages and assess the effects of relaxation or elimination of the coupling or constraints on adaptive radiation of the locomotor apparatus.

It is not certain whether the locomotor-respira-

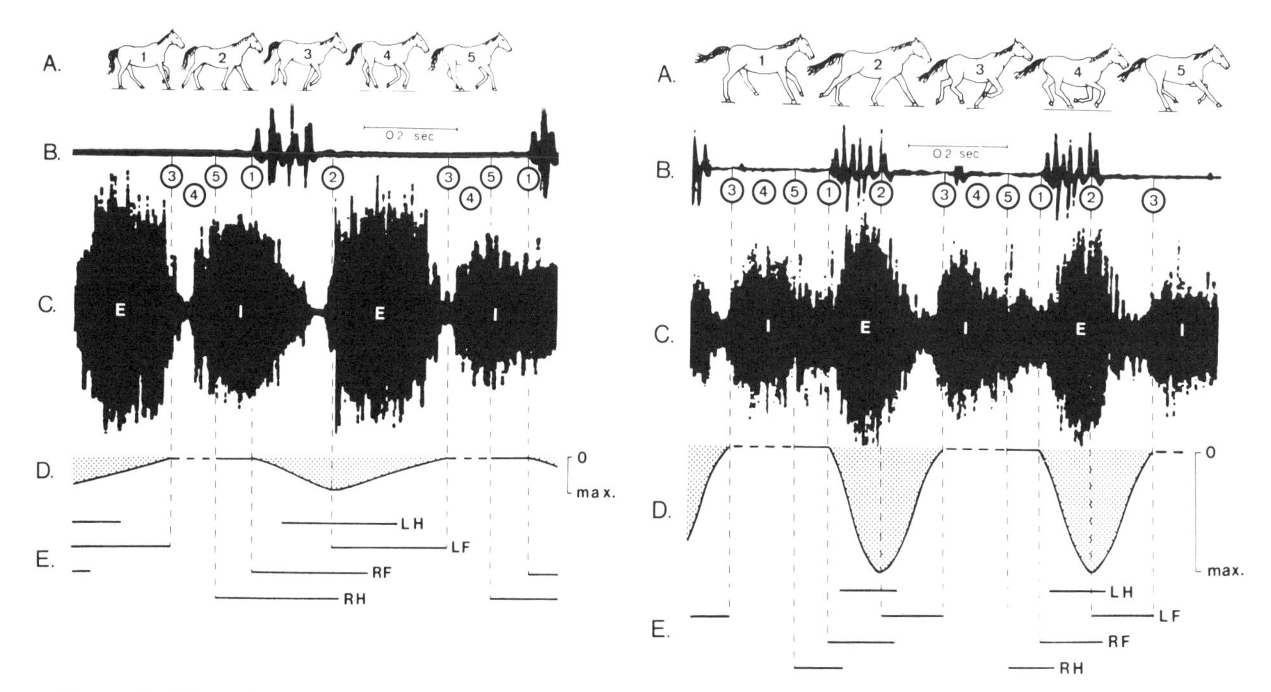

Figure 11-22 Relationship of locomotor and ventilatory cycles in the same horse at a canter *(left)* and gallop *(right)*. A: Tracings of body position at five selected points in locomotor cycle. B, C: Oscilloscope tracings of footfall and breathing. D: Hypothetical loading profile for thoracic complex. E: Standard support diagram for the gait. Loading increases in the gallop as the animal spends a large fraction of each stride off the ground. *E* = exhalation; *I* = inhalation; *LF, RF* = left and right forelimbs; *LH, RH* = left and right hind limbs. (From Bramble and Carrier, 1983.)

tory coupling is characteristic only of endotherms (Bramble and Carrier, 1983). In fact, there is preliminary evidence that some teleosts and turtles also have locomotor-respiratory coupling. The linkage may have a far wider distribution among vertebrates for anatomical and/or functional reasons.

In fishes study is needed to test for the relative influences of countercurrent (Liem, 1981), mixed-pool, and cross-current (Piiper and Scheid, 1973) exchange in larval respiration. The functional implications are important not only for a more accurate understanding of respiratory physiology, but also for life-history and ecological theory. Currently, the water flow dynamics in the respiratory tract of fishes are still poorly understood (Holeton and Jones, 1975; Lauder, 1984). Simultaneous measurements of water velocity in the buccal chamber and of buccal and opercular hydrostatic pressure of carp have revealed much higher velocities than predicted by current models. The high flow velocities mean that the kinetic energy of flow makes a substantial contribution to the total fluid energy in fishes. Thus there is potential for considerable error in estimates of the energetics of fish ventilation based solely on pressure measurement.

Because various fishes have independently evolved air ventilatory mechanisms using either gas bladders or accessory air-breathing organs, a comparative functional morphological analysis may reveal how feeding-respiratory couplings and developmental factors may canalize the morphological and functional specializations in phylogenetically unrelated lineages.

The level of muscular activity during ventilation varies drastically in all vertebrates depending upon behavioral and general environmental factors. An in-depth experimental analysis of elastic recoil mechanisms is much needed to understand the energetics and efficiency of ventilation. A general pattern seems to emerge indicating that during quiet respiration exhalation takes place by active muscular activity, whereas inhalation occurs by elastic recoil in fishes (Liem and Osse, 1975), some reptiles (Gaunt and Gans, 1969), and mammals. However, more research is needed to establish the generality of passive inhalation by elastic recoil in vertebrates.

Finally, the reptiles offer unique opportunities to understand the process by which the aspiration pump has evolved and how its remarkable diversity can be related to the invasion and opportunistic exploitation of the new terrestrial habitats.

Chapter 12

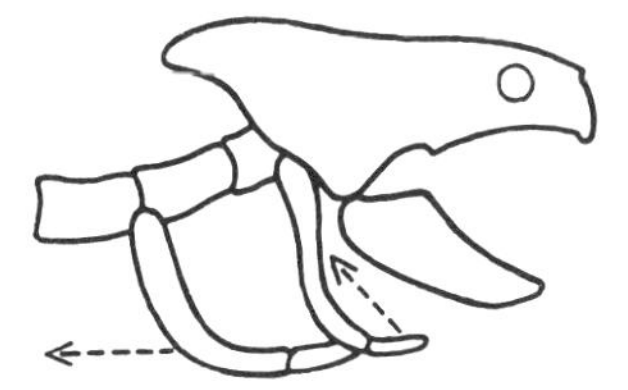

Aquatic Feeding in Lower Vertebrates

George V. Lauder

Aquatic feeding is the primitive mode of prey capture in vertebrates and is therefore of special interest in the study of vertebrate evolution and trophic biology. This fact has an especially important implication for the analysis of evolutionary patterns of structure and function in the feeding mechanism: the diversity of terrestrial methods of prey capture has been derived from a musculoskeletal apparatus and neural control system that function in a medium 900 times as dense as air and 80 times as viscous. Accordingly, the hydrodynamic properties of water have played a fundamental role in shaping the basic mechanism of aquatic feeding and may have determined the nature of the feeding mechanism in a number of primitive tetrapods. Many underlying similarities within the diversity of aquatic feeding modes appear to result from hydrodynamic problems associated with the removal of small particles from a dense and viscous medium.

All the major groups of vertebrates contain species that feed aquatically. This chapter examines the two feeding processes that are used by more than 99% of all aquatic vertebrates, suspension feeding and suction feeding, with emphasis on the mechanics of feeding in selected lower vertebrates. Other taxa and modes of feeding will not be discussed here, specifically elasmobranchs (Springer, 1961; Moss, 1972, 1979); the Galapagos iguana *Amblyrhynchus;* adult lampreys and hagfishes (Reynolds, 1931; Gradwell, 1972; Hardisty, 1979); crocodiles and gavials; birds such as the penguin, loon, skimmer *(Rynchops),* and flamingo (Jenkin, 1957; Zweers, 1974); cetaceans, sirenians, and pinneped carnivores (Gordon, 1984).

The study of aquatic feeding in vertebrates has undergone a renaissance in recent years with the introduction of modern experimental techniques for the study of functional morphology. Many early studies on feeding mechanisms used manipulations of preserved specimens to deduce the sequence of movements occurring during feeding or employed macroscopic observations of filtering structures and surfaces. Within the last 20 years high-speed cinematography has allowed the accurate determination of bone movement during feeding and, coupled with simultaneous electromyographic recording of muscle activity, has permitted the quantitative study of musculoskeletal mechanics beyond what can be achieved by merely studying muscle origins, insertions, and lines of action. Other techniques that have vastly improved the description of feeding mechanics include strain gauges, which permit the direct recording of bone deformation patterns (Lanyon, 1976; Lauder and Lanyon, 1980), pressure transducers (Lauder, 1980a), and accurate particle density and size counters for the examination of the dynamics of suspension feeding (Wassersug, 1972). These new experimental approaches, coupled with theoretical analyses of feeding methods (Rubenstein and Koehl, 1977), have resulted in a more comprehensive understanding of the functions of head bones, muscles, and filtering systems than was previously possible. Many of the misconceptions and assumptions of earlier nonexperimental investigations have been revised.

Principles of Aquatic Prey Capture

The high density and viscosity of water relative to air has four important consequences for suspension and suction feeding in the aquatic environment. First, movement of the predator in approaching the prey affects the position of the prey because of the lateral deflection of flow streamlines anterior to the predator's head. Prey tend to be deflected to the side unless the flow streamlines are reoriented by expansion of the mouth cavity. Second, constraints are placed on the design of filters that remove small particles from the fluid. Small pore sizes in sieve filters greatly increase the drag experienced by predators during locomotion and increase the muscular effort and hence the use of available oxygen during feeding. This problem can be partially overcome by having a sieve with an adjustable pore size, as seen in the gill rakers of filter-feeding fishes. Third, the gravitational deposition of particles on a filter is regulated by the relative densities of the particles and of the fluid environment. Finally, inertial factors, such as kinetic energy at high flow velocities and the momentum of the flow (factors that can usually be neglected when analyzing feeding in air), must be taken into account.

Suspension feeding in vertebrates differs from suction feeding in that the selection of prey to be swallowed occurs primarily at the filter (gills, mucus net, and so on), usually after the prey has entered the mouth. In most cases the suspended particles are filtered out primarily by size, shape, and density, with little regard for the food value of the particles (Jorgensen, 1966). The filtered particles may then be sorted into those to be swallowed and those to be rejected. In suction feeding prey selection is usually performed prior to ingestion and the determination of food value is made before particle capture.

Rubenstein and Koehl (1977) have recently summarized the six possible mechanisms of suspension feeding available to aquatic organisms. The mechanism often assumed to be the most common is sieving, in which all particles captured are larger than the pores in the filter, and no items smaller than the pores are retained (Fig. 12-1A). This may actually be a rather rare mechanism of prey capture. Direct interception occurs when particles in the flow contact and stick to a filter surface (Fig. 12-1B). Inertial impaction on a filter results from particle inertia as flow is diverted through the open areas of the filter (Fig. 12-1C). Particles cross streamlines and adhere to the filter surface. Gravitational deposition (Fig. 12-1D) occurs as denser items settle out onto the filter surface. Motile particle deposition on a filter results from active movement of particles (for instance, locomotion by planktonic copepods) or other external influences that give potential food particles sufficient energy to deviate from the streamlines (Fig. 12-1E). Finally, electrostatic attraction between particles in the water and the filter may result in adherence to the filter surface. This mechanism may be quite important when mucus secretions cover the filter. LaBarbera (1978) has shown in invertebrates that both positively and negatively charged particles are preferentially attracted to a mucus filter over those neutrally charged. Even irregularly shaped filters may be very effective at removing particles from fluids when nonsieving mechanisms dominate.

Rubenstein and Koehl (1977) noted that as a consequence of these six possible filtration methods, the diet of suspension feeders may be varied significantly by several different mechanisms. First, the pore size of the filter may be actively changed. Second, the rate of filter cleaning governs the size of food captured because the accumulation of food on a filter surface greatly affects the size of particles removed from the water. Finally, increasing the diameter of fibers in a fibrous netlike filter modifies the size distribution of captured particles primarily by increasing the proportion of particles trapped by direct interception.

In sharp contrast to suspension feeding, which may be modeled by steady-state hydrodynamic equations, the salient feature of suction feeding is its nonsteady, dynamic nature. For example, prey capture by the gar *Lepisosteus* may take place in 0.025 sec, and anglerfish can completely engulf their prey in 0.015 sec (Grobecker and Pietsch, 1979). Flow velocities in the mouth cavity of

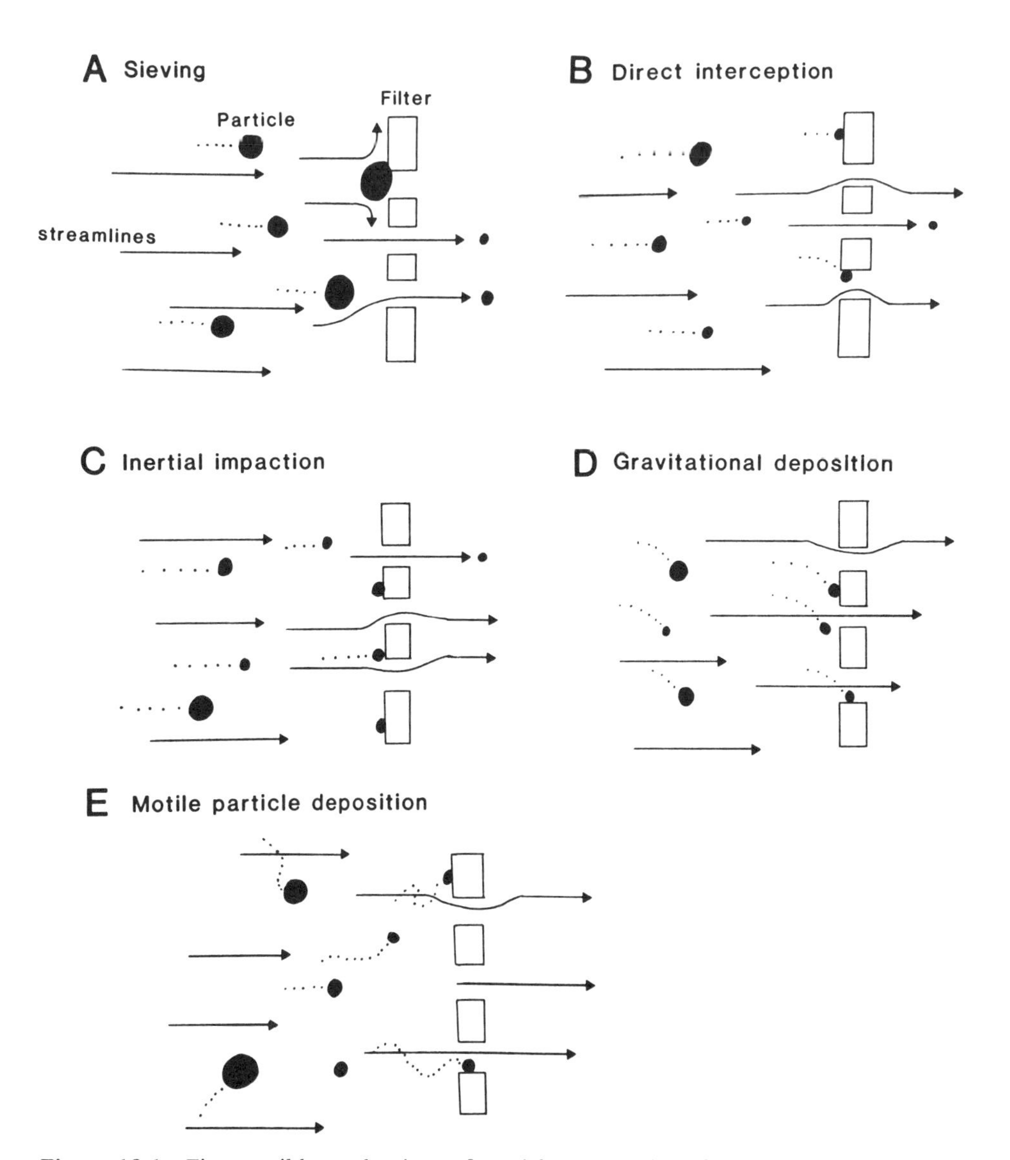

Figure 12-1 Five possible mechanisms of particle capture by a filter. Sieving appears to be the system used by ray-finned fishes, whereas direct interception and inertial impaction are used by most other suspension-feeding vertebrates. A sixth mechanism of particle capture, electrostatic attraction, may be important if mucus has a different electrical charge from the particles. Note that different predictions about the size distribution of particles on each side of the filter follow from the different mechanisms of particle capture. Dotted line represents particle path. (Modified from Rubenstein and Koehl, 1977.)

fishes often go from 0 to 12 m/sec in 0.03 sec (Osse and Muller, 1980). It is thus inappropriate to apply hydrodynamic models such as the Poiseuille relation and Bernoulli equation, which assume steady flows, to the analysis of suction feeding (Lauder, 1980b). Furthermore, the walls of the mouth cavity are compliant, and we have no information to suggest that pressure-to-velocity relationships are linear for nonsteady flows.

As an example of a steady-state model, consider an application of the Bernoulli relation to the mouth cavity of a teleost fish. The equation

$$\frac{P_1}{\rho g} + \frac{\frac{1}{2}V^2}{g} = \frac{P_0}{\rho g},$$

where P = pressure, V = flow velocity, ρ = fluid density, and g = gravitational acceleration, has been applied to aquatic feeding in fishes (Osse, 1969; Pietsch, 1978). Since P_0 is the pressure in the surrounding water and a very rough estimate of flow velocity can be made by measuring changes in buccal volume in preserved fishes, the pressures generated in the mouth cavity can be estimated. However, one consequence of the fact that the dynamic suction-feeding mechanism of aquatic vertebrates violates the assumptions of the Bernoulli equation is that such predictions of change in buccal pressure are inaccurate by a factor of 10 or more; also, they give no indication of the pattern of pressure change in the mouth cavity.

Suction feeding (also called inertial suction feeding or gape-and-suck feeding) takes its name from the rapid expansion of the mouth cavity that causes a negative (suction) pressure relative to the surrounding water. This causes a flow that carries prey into the mouth. The jaws then close on or behind the prey while the water exits posteriorly between the gill arches (in fishes and aquatic urodeles). The inertia of the prey determines the flow velocity necessary to capture it (Gans, 1969). The greater the inertia, the higher the volume flow into the mouth cavity must be to draw the prey within reach of the jaws.

The dynamic nature of suction feeding has important implications for the understanding of the relationship between form and function in aquatic vertebrates that contrast with those extrapolated from steady-state situations. In dynamic events pressures within a single cavity need not be uniform. It is thus possible that pressures measured at different locations in the buccal cavity might be very different. Second, inertial effects (Streeter and Wylie, 1979), such as the momentum of water entering the mouth, may largely determine the dynamic characteristics of suction feeding. The role of inertial effects in the suction feeding of teleost fish is considered below.

Suspension Feeding

Suspension feeding was probably the primitive method of prey capture for both chordates and gnathostomes (Denison, 1961; Moy-Thomas and Miles, 1971; Mallatt, 1981). Many of the earliest groups of fossil agnathans such as the osteostracans and heterostracans are believed to have filtered out detritus and suspended microorganisms with a pharyngeal mucus filter. Suspended material was probably captured by a combination of forward body motion, ciliary tracts in the pharynx, active expansion of parts of the head, especially of the floor of the mouth (Denison, 1961), and an active muscular velum inside the mouth opening. Active feeding on bottom sediments may have been common.

An important aspect of the suspension feeding of early chordates remains common in extant filter feeders: food is brought into contact with the filter by modifications of the respiratory pump. Respiratory movements in chordates result in a relatively steady flow of water into the mouth and out over the gills (Hughes and Shelton, 1958; Randall, 1971). This flow is used to transport suspended particles, and early agnathans may have used exaggerated movements of the respiratory pump to scoop up bottom sediments.

Larval Lamprey Larval lampreys (ammocoetes) have long been known to trap suspended algae and detritus on mucus within the pharynx, but it is only recently that the details of the trapping mechanism and feeding dynamics have been elucidated (Mallatt, 1979, 1981; Moore and Mallatt, 1980). Ammocoetes trap suspended particles by the mechanism of direct interception (Fig. 12-1B), not by sieving. In contrast to several previous hypotheses, Mallatt (1979) has shown that mucus is secreted in the lateral parabranchial chambers and on the lateral surface of the gill filaments. Mucus strands from the seven lateral chambers on each side move medially to join a central mucus food cord, which is propelled posteriorly into the esophagus by a central dorsal band of cilia. Food impacts directly onto mucus strands, lateral surfaces of gill filaments, and on the walls of the parabranchial chambers, but seldom on gill lamellae. The ventral endostyle adds digestive enzymes to the food cord and does not produce mucus.

Ammocoetes are partially selective feeders in

that they do not consume algae and detritus in the proportions present in the surrounding water. Selection is not based on particle size, although size does in part determine where particles are trapped in the pharynx. Ammocoetes are capable of capturing food in the size range of 5 to 340 μm with high efficiency (about 76%). The rate of water flow through the pharynx is slow (on the order of 40 ml/g/h) as compared with other suspension feeders, although ammocoetes gradually increase flow rates with increasing food concentration up to a critical concentration of 100 to 330 mg food/l (Mallatt, 1982).

Actinopterygian Fishes Surprisingly little research has been done on the mechanics of filter feeding in ray-finned fishes. Their mechanisms of filtration differ from the primitive direct interception of prey by mucus strands and thus provide examples of the other filter mechanisms listed earlier. Most discussions of the mechanics of particle filtration are highly speculative.

Filter feeding is accomplished with long gill rakers that extend anteriorly into the pharynx from the epibranchial and ceratobranchial gill arch elements. Mucus-producing cells are present on the gill arches at the base of the gill rakers and serve to coat the arches, and perhaps also the rakers, with mucus (Weisel, 1973). Experimental evidence suggests, however, that this mucus does not contribute significantly to particle entrapment by direct interception because the smallest particles filtered are generally much larger than the minimum mesh size — a circumstance consistent with a sieving filtration mechanism (Durbin and Durbin, 1975; Rosen and Hales, 1981). Particles enter the pharynx through the mouth, which is usually held widely open as the feeding fish swims through the water column. The particles then make contact with the gill rakers as the flow streamlines diverge laterally to pass over the gills. Filtration is primarily by sieving, and clogging of the filter may be prevented by water pressure that forces trapped particles along the gill rakers to the gill arch where they could become attached to mucus strands. There is no information on how the food-laden mucus is collected and ultimately swallowed. Perhaps the intermittent "coughing" movements observed during filter feeding serve to agglutinate separate mucus strands from the gill arches into a larger mass, which is then swallowed.

Rosen and Hales (1981) have studied filter feeding in the paddlefish *Polyodon spathula* and have compared the size of spacings between gill rakers with the size of prey captured. *Polyodon* is an indiscriminate filter feeder and swallows all filtered particles including large amounts of detritus. The mean size of "pores" between the gill rakers is 0.07 mm, whereas the smallest prey filtered from the water averages about 0.12 mm in size. Very small prey items are not captured, even though available in the water column, and the smallest prey swallowed are generally twice as large as the filter pore size. Menhaden *Brevoortia tyrannus,* one of the most common filter-feeding fishes, can filter particles of 0.015 mm in size, although capture efficiency is then low (25%) as compared with that for larger particles (0.4 mm), where capture efficiencies of 60% to 80% are measured. Menhaden can filter an enormous quantity of water: each fish can filter 20 l/min at the mean swimming speed of 40 cm/sec (Durbin and Durbin, 1975). The mean size of "pores" in the gill raker filter of menhaden has been estimated at 0.08 mm, although the range is 0.01–0.10 mm.

Two mechanisms have been proposed for the selection of particles at the filter: active changes in size of filter pores by movements of the gill rakers or of the hyoid and gill arches, and changes in body velocity during filter feeding. The latter mechanism has been documented in the menhaden, which changes filtering velocity according to the size and availability of prey by adjusting the speed of locomotion. Changes in pore size may result from movements of the gill rakers by intrinsic muscles of the gill arches (as in *Polyodon*), but few fishes possess musculature attaching to the gill rakers. In menhaden the spacing between rakers or adjacent arches may be varied by relative movement of the arches as a result of dorsal and ventral branchial musculature (especially the transversi ventrales and dorsales muscles). The spacing between rakers on the same arch can be adjusted by dorsoventral movement of the arch that increases the angle between the epibranchial and ceratobranchial elements, thus fanning out the gill rakers anteriorly. Both of these methods may result in a "dead space" between the gill rakers of adjacent arches through which water can flow without contacting the filter surface, thereby reducing both the drag and the efficiency of particle capture.

Most filter-feeding fishes are facultative feeders and can switch to picking individual prey items out of the water by inertial suction (Leong and O'Connell, 1969; Rosen and Hales, 1981). The menhaden is an exception, being an obligate filter feeder (Durbin and Durbin, 1975).

Anuran Larvae Anuran larvae, or tadpoles, are remarkable for their ability to filter extremely small particles from the water. The complexity of their suspension feeding apparatus provides an interesting comparison to the feeding of ammocoetes and ray-finned fishes. Particles enter the mouth cavity with the respiratory flow created by the buccal pump. The buccal pump consists of paired ceratohyal cartilages in the buccal floor that are moved dorsoventrally by the orbitohyoideus muscle extending between the palatoquadrate and the ceratohyal on each side (de Jongh, 1968; Wassersug and Hoff, 1979). Large particles may be rejected by the labial papillae; moderately sized particles pass directly back to the esophagus. Smaller suspended material is carried laterally with the flow of water to be trapped either on the gill filters or on mucus of the branchial food traps. Gradwell (1975) described cilia-driven food cords passing medially and caudally into the esophagus. The removal of food particles by the branchial food traps and gill filters has been proposed to occur by three methods: inertial impaction, direct interception, and possibly by electrostatic attraction between particles and mucus (Wassersug, 1980). The gill filters do not contain mucus-secreting cells; mucus is secreted initially on the ventral velum and branchial food traps in the pharynx before moving posteriorly to the esophagus.

The gill filters of tadpoles can have pore sizes of 5 μ or less and, in contrast to the fish species discussed above, can filter suspended particles down to 0.13 μ (Wassersug, 1972), a size considerably smaller than the pore size. This fact suggests that sieving is not an important mechanism of ultraplanktonic food capture and that direct interception of food particles by mucus strands may be dominant. Tadpoles are not all obligate filter feeders; many create suspensions of food particles by biting off food from the bottom and drawing this suspension into the pharynx where filtering and trapping of the food pieces takes place. The feeding dynamics of tadpoles differ from those of fishes in that swimming velocity is not adjusted to change flow over the filter: either changes in the rate of buccal pumping or changes in stroke volume are used to vary filtration (Seale and Wassersug, 1979). Although no evidence has been obtained for clogging of the sieve filters of fishes, the ingestion rate declines and food may be ejected from the mouth when the concentration of food surpasses about $1.5 \times 10^3\ \mu m^3/ml$. Both fishes and tadpoles exhibit a threshold in food concentration below which filter feeding does not occur.

Common to all three groups of filter feeders discussed is an intimate functional association between prey capture and relatively steady-state, low-velocity respiratory mechanisms and flows. By contrast, although the same basic muscular and skeletal elements are used for both respiration and suction feeding, the mechanisms underlying these two processes are significantly different. An understanding of suction feeding requires analysis in terms of rapidly varying kinematic parameters and dynamic flow.

Suction Feeding

Prey capture by suction feeding is primitive for the Osteichthyes and is common in all the major lineages of lower vertebrates, including ray-finned fishes (Actinopterygii), coelacanths (Lauder, 1980c), and lungfishes. Suction-feeding mechanisms are thus fundamental to considerations of the origin and evolution of tetrapod feeding mechanisms discussed in Chapters 13 and 14.

Actinopterygian Fishes The most comprehensive and detailed recent research on suction feeding has been done on ray-finned fishes, which allows us to consider the mechanics and hydrodynamics of feeding in this group in detail. Four

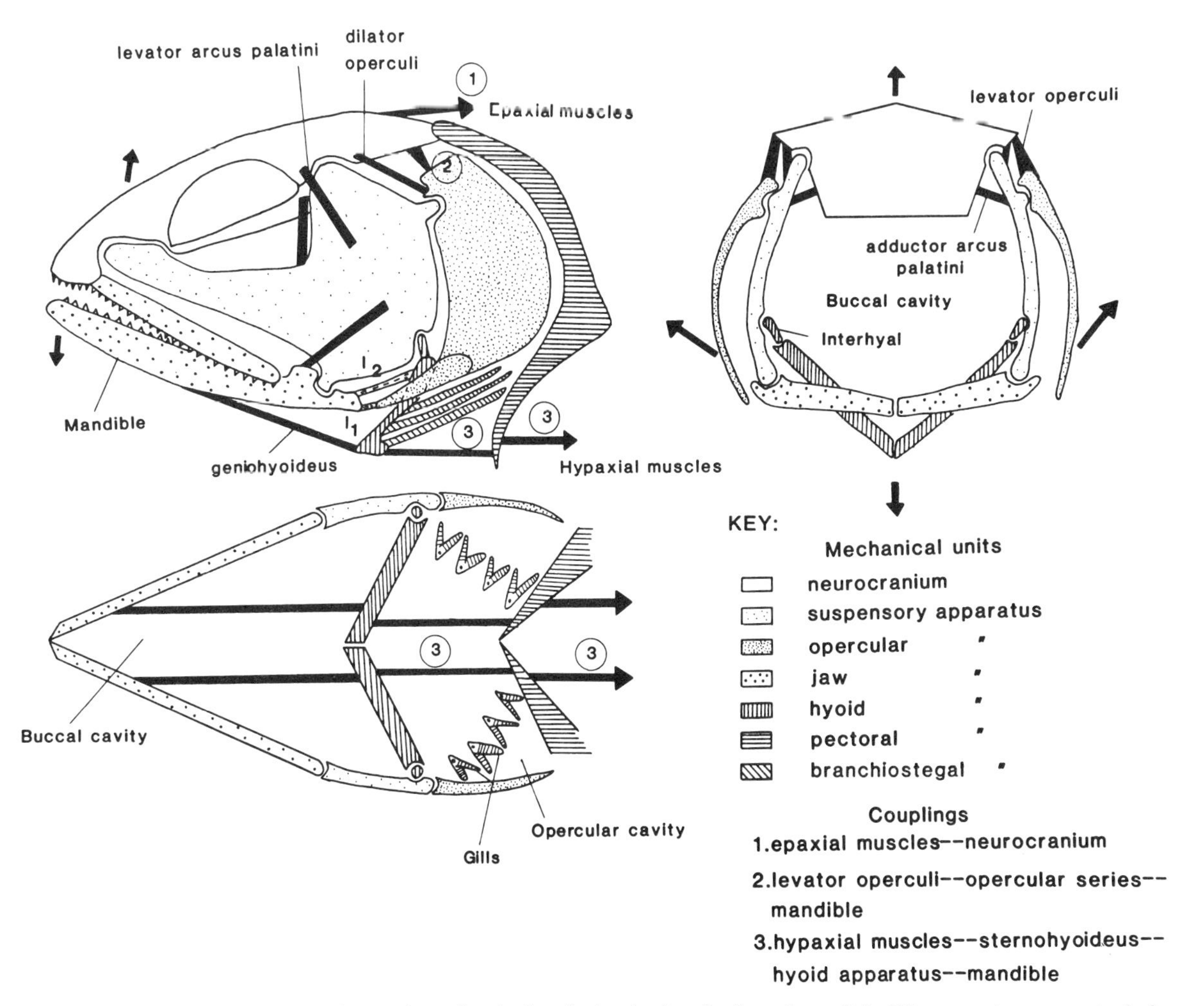

Figure 12-2 The major muscles and mechanical units in the head of a teleost fish. Three major musculoskeletal couplings are involved in the expansive phase of prey capture: (1) the epaxial muscles coupling, which causes cranial elevation; (2) the levator operculi coupling, which mediates mandibular depression via the opercular apparatus and interoperculomandibular ligament (ligament l_1); and (3) the hyoid coupling, which also governs mandibular depression via the mandibulohyoid ligament (l_2).

phases of prey capture may be defined by the timing of movements of cranial bones: a preparatory phase, an expansive phase, a compressive phase, and a recovery phase.

The preparatory phase (which has been found only in advanced percomorph teleosts) occurs as the prey item is being approached and before the mouth begins to open. The key feature of this phase is that the volume of the buccal cavity is decreased (Fig. 12-2), thus reducing the volume of water inside the mouth prior to rapid expansion as the mouth opens (Liem, 1978). Compression of the mouth cavity occurs by the action of the adductor mandibulae, adductor arcus palatini, adductor operculi, and geniohyoideus muscles (Figs. 12-2, 12-3A). Buccal volume is thus reduced mainly by lateral compression of the head as the suspensory apparatus on each side moves medially and by elevation of the floor of the buccal cavity as the hyoid is protracted. Electrical activity in the adductor mandibulae stabilizes the lower jaw so that hyoid protraction by the geniohyoideus can occur.

The expansive phase of the strike extends from

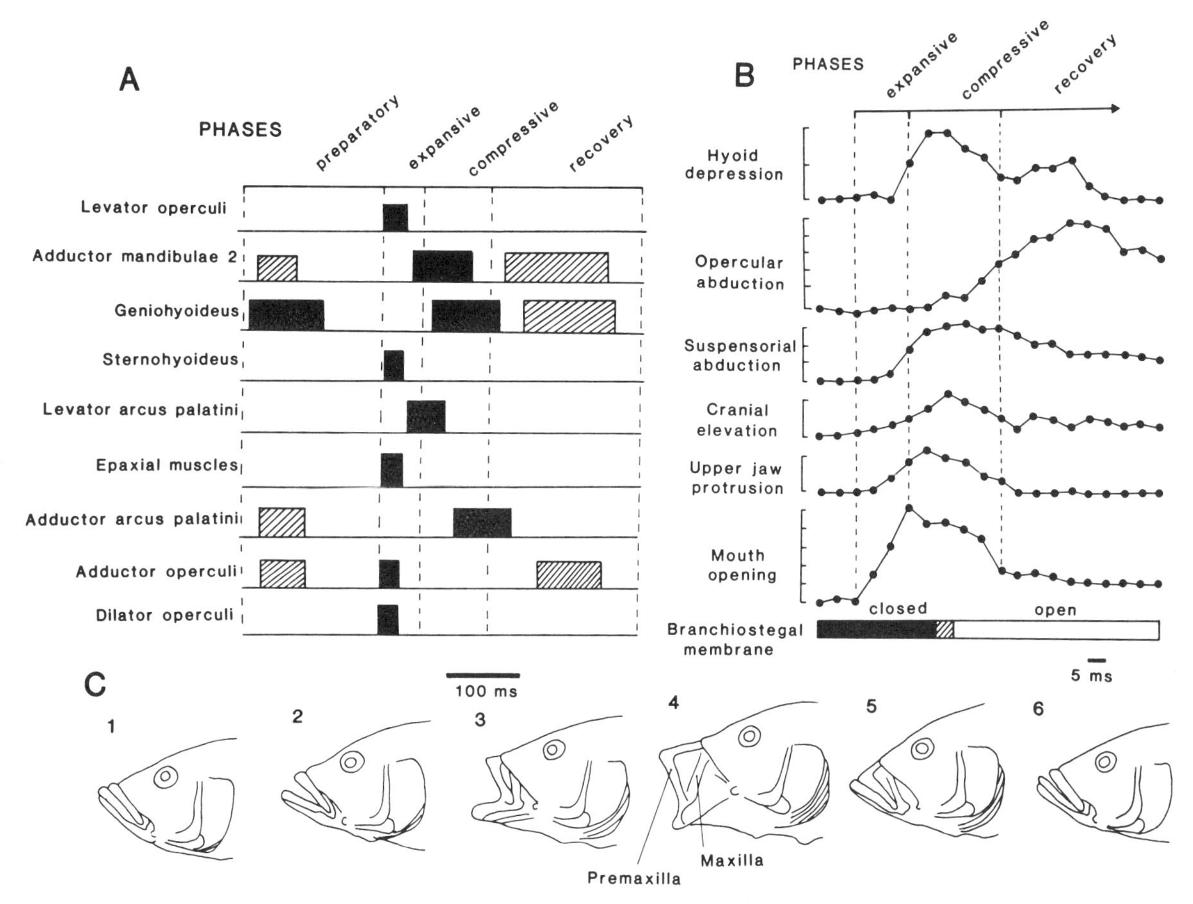

Figure 12-3 A: Periods of electrical activity in cranial muscles of an advanced teleost fish, *Lepomis,* during feeding. Note that activity of adductor and abductor muscles overlaps the boundaries of the expansive and compressive phases. Black bars indicate the duration of electrical activity common to all strikes; hatched bars indicate occasional activity. B: Kinematic pattern of head movements in *Lepomis* during capture of a goldfish. Note the delay in opercular abduction until the compressive phase, as well as the relative duration of the phases. C: Sequence of jaw movements in a cichlid fish, *Serranochromis,* showing protrusion and retraction of the upper jaw (modified from Liem, 1978).

the start of mouth opening to peak gape. Three musculoskeletal linkage systems function to increase the gape. (1) The epaxial muscles cause a dorsal rotation of the cranium on the vertebral column, thus expanding the roof of the mouth cavity. (2) The mandible is depressed ventrally by the action of the levator operculi muscle, which rotates the gill cover posterodorsally (Liem, 1970). This dorsal rotation is transmitted to the retroarticular process of the mandible by a ligament between the opercular series and the mandible (see Fig. 12-2, coupling 2; Fig. 12-4, *LOP*). (3) The mandible is also depressed ventrally by a second musculoskeletal coupling involving the hyoid apparatus. Contraction of the sternohyoideus muscle causes an initial posterior retraction of the hyoid, and this movement produces depression of the mandible via a ligament from the hyoid to the opercular series or mandible (see Fig. 12-2, coupling 3; Fig. 12-4, *SH*). Shortly after the mouth begins to open, lateral expansion of the buccal cavity occurs as a result of lateral movement of the palatoquadrate, which is mediated by the levator arcus palatini (Figs. 12-2, 12-3), and the hyoid apparatus swings ventrally, greatly expanding mouth cavity volume. It is during this

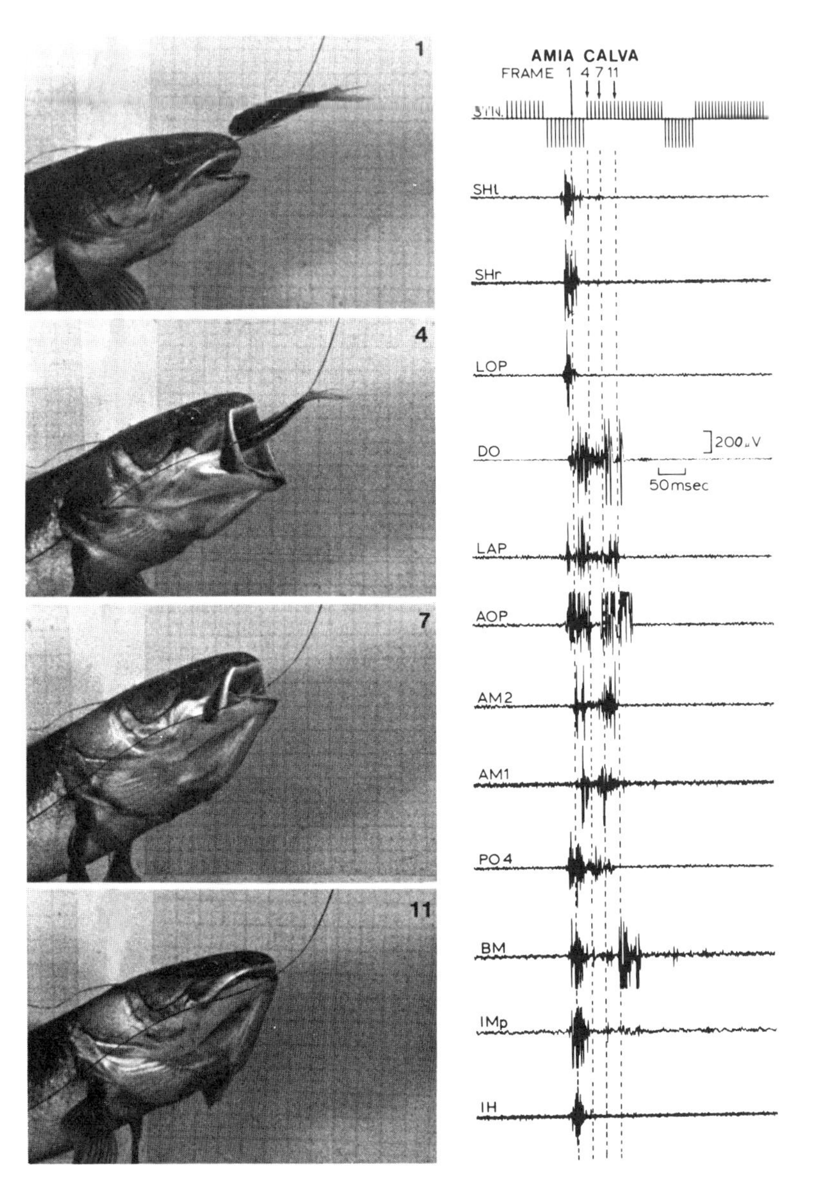

Figure 12-4 Prey capture in the bowfin *Amia calva*. The frames *(left)* are from a high-speed (200 frames/sec) film that is synchronized with electromyographic recordings of cranial muscles *(right)*. The recordings are a summary of 45 feeding events. The wire leading from the head muscles to the recording apparatus can be seen in the photographs. Note that the maxilla swings anteriorly to produce a nearly circular mouth opening at peak gape (frame 4) as the prey enters the mouth. Both the levator operculi and sternohyoideus muscles are active at the start of the expansive phase and activate couplings 2 and 3 (Fig. 12-2) to cause mouth opening. Muscles: *SHl*, *SHr* = left and right sternohyoideus muscles; *LOP* = levator operculi; *DO* = dilator operculi; *LAP* = levator arcus palatini; *AOP* = adductor operculi; *AM2*, *AM1*, and *PO4* = divisions of the adductor mandibulae; *BM* = branchiomandibularis; *Imp* = intermandibularis posterior; *Ih* = interhyoideus. (From Lauder, 1980d.)

time that protrusion of the premaxillae toward the prey occurs in advanced teleost fishes (Fig. 12-3C, frame 3). The volume of the opercular cavity remains relatively constant throughout the expansive phase. Abduction of the gill cover usually begins after peak gape has been reached (Figs. 12-2; 12-3B), and the branchiostegal membrane and rays then tightly seal the posteroventral opening of the opercular cavity (Fig. 12-3B). The expansive phase is usually shorter than either the preparatory or compressive phases. In some fishes, expansion of the mouth cavity is extremely rapid: in *Lepisosteus* peak gape is reached within 15 msec of the start of mouth opening (Lauder, 1980d), and Grobecker and Pietsch (1979) have described an expansive phase lasting only 5 msec in antennariid anglerfishes. In *Amia calva* the expansive phase lasts 20–40 msec (Fig. 12-4).

The compressive phase (the time from peak gape to closure of the jaws) lasts, on average, twice as long as the expansive phase (Figs. 12-3B, mouth opening; Fig. 12-4). The compressive phase is characterized by activity in the adductor mandibulae (actually beginning during the end of the expansive phase), which initiates closure of the jaws; adduction of the suspensory apparatus by the adductor arcus palatini; protraction of the hyoid by the geniohyoideus; and return of the cranium nearly to its initial resting position (Figs. 12-2; 12-3A,B; 12-4). A feature of the compressive phase kinematic pattern that has significant implications for the hydrodynamic model of feeding discussed below is the timing of opercular abduction and opening of the branchiostegal valve. In nearly all ray-finned fishes studied, lateral movement of the operculum begins only at or near peak gape (Fig. 12-3B), and in some cases the mouth is nearly closed before the gill cover has undergone significant lateral excursion (Liem, 1970; Lauder, 1979; 1980a,d). In addition, opening of the opercular and branchiostegal valves occurs during the compressive phase, indicating that water sucked into the buccal cavity during the expansive phase has only just begun to exit to the outside after passing through the gills (Fig. 12-3B). In some cases water does not begin to emerge from the opercular cavity until the jaws have closed completely. Retraction of the protruded premaxillae also occurs during the compressive phase (Fig. 12-3C, frames 4,5,6), and the maxilla, which has rotated anteriorly about its dorsal palatal articulation, returns to its position during the preparatory phase. The lower jaw is often adducted against partially protruded premaxillae, but the relative timing of lower jaw adduction and premaxillary retraction varies with the type of prey and the velocity of jaw movement.

The recovery phase is defined as the time from the end of the compressive phase (zero gape) to return of the skull bones to their initial position before the preparatory phase. This phase may last 0.5 sec or longer if a large prey has been captured. Usually the gill cover is the last bone to return to the resting position (Fig. 12-3B), although hyoid depression, cranial elevation, and suspensory abduction all are present at the time the jaws close and are thus major movements also occurring during the recovery phase.

The pattern of bone movement during prey capture exhibits an anterior-to-posterior sequence of peak excursions. Thus, mouth opening peaks before hyoid depression, which peaks before suspensorial abduction and opercular abduction (Fig. 12-3B). This pattern is remarkably consistent in all suction-feeding ray-finned fishes studied, and it is also found in the aquatic turtles and salamanders discussed below.

Comparative analyses of suction feeding in lower vertebrates have revealed that the hyoid apparatus produces the main change of volume in the mouth cavity during the expansive phase (and thus contributes the most to the creation of negative pressures), and that movement of the hyoid is similar in all major groups of lower vertebrates. In ray-finned fishes, coelacanths, lungfishes, and probably also acanthodians, the hyoid bar (ceratohyal, epihyal, and hypohyals) on each side articulates with the palatoquadrate by a small rodlike interhyal (Fig. 12-2). The primitive living members of each lineage also possess a strong ligament between the ceratohyal or epihyal and the posteroventral aspect of the mandible (ligament l_2 in Fig. 12-2). It thus appears that coupling 3, discussed above as one of the two mechanisms in teleost fishes that mediate mandibular depression during the expansive phase, is also the primitive system for abducting the lower jaw in teleostome vertebrates.

Mandibular depression via the hyoid apparatus involves both dorsoventral and anteroposterior movements of the hyoid, and these reach a maximum at different times (Fig. 12-5). Anteroposterior excursion peaks in the expansive phase shortly after the mouth begins to open and the initial posterior component of hyoid movement is primarily responsible for rapid depression of the lower jaw. Dorsoventral excursion of the hyoid peaks during the compressive phase and produces a large increase in buccal volume, which results in high flow velocities into the mouth. In lungfishes (Fig. 12-6), as in coelacanths, primitive ray-finned fishes, and probably acanthodians, the hyoid coupling is the only mechanism mediating mandibular depression. This pattern of hyoid movement is basic to lower vertebrates (Fig. 12-5) and was a fundamental feature of the early tetrapod feeding mechanism.

With this background on the pattern of muscle activity and bone movement during suction feed-

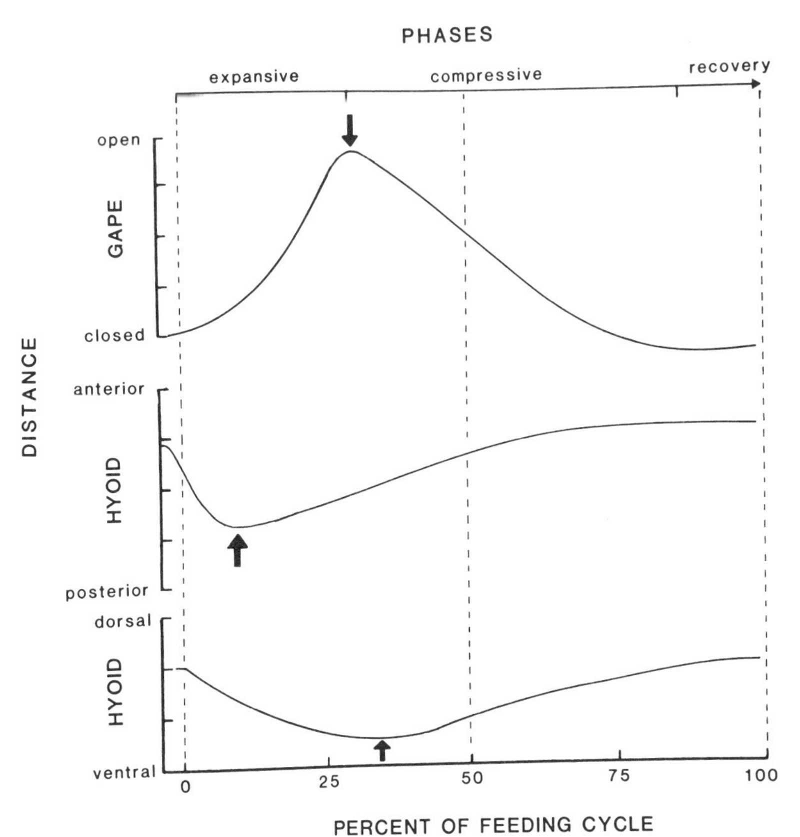

Figure 12-5 The relationship between hyoid movement and gape in primitive gnathostomes (lungfishes, coelacanths, and ray finned fishes). Arrows indicate peak excursion. Note that anteroposterior excursion of the hyoid peaks during the expansive phase, whereas its dorsoventral movement peaks during the compressive phase. The initial posterior movement of the hyoid apparatus is related to mandibular depression via coupling 3 (Fig. 12-2).

ing, the pattern of pressure change in the mouth cavity can be examined and a general hydrodynamic model for suction feeding proposed. One experimental approach used to measure pressure in the buccal and opercular cavities during feeding is illustrated in Figure 12-7A. Cannulae are chronically implanted in the buccal and opercular cavities and attached to two transducers located above the aquarium. The cannulae and the pressure transducers are filled with a viscous fluid that allows pressure changes in the mouth cavity to be transmitted to the transducer diaphragm. High-speed FM tape records of pressure changes are played back at a greatly reduced speed through a chart recorder for analysis (Fig. 12-7B). Many technical problems are associated with measuring pressures with fluid-filled cannulae (summarized in Lauder, 1980b), the most serious of which is the low-frequency response (about 100 Hz) of the transducer. Newer techniques avoid these difficulties (Lauder, 1983a).

The pressure patterns recorded during suction feeding may be correlated with the four kinematic phases defined above by synchronizing a high-speed film with the pressure traces. When a preparatory phase is occurring, a sharp positive pressure pulse, caused primarily by hyoid protraction, is recorded in the buccal cavity (Fig. 12-7B, buccal pressure). During the start of the expansive phase, pressure in the buccal cavity drops precipitously, whereas pressure in the opercular cavity usually increases slightly. In a rapid and high-speed act of suction the pressure differential across the gills may reach 400 cm H_2O. At about the middle of the expansive phase, pressure in the opercular cavity begins to drop. It reaches a low value of -150 cm H_2O, which is considerably less than the -650 cm H_2O recorded in the buccal cavity (Fig. 12-7B). Buccal pressure reaches a minimum prior to opercular cavity pressure, and maximum negative values for both are reached in the compressive phase or at the end of the expansive phase. Toward the end of the compressive phase, buccal pressure rapidly returns to ambient, whereas opercular pressure remains negative (Fig. 12-7B: t_4-t_5), thus reversing the pressure differential from that of the beginning of the expansive phase. At the end of the compressive phase and the start of the recovery phase there occur a positive opercular pressure and both a short positive and a final negative pulse in the buccal pressure trace (Fig. 12-7B).

These patterns of pressure change may be used to formulate a general model of fluid flow through the mouth cavity during high-speed suction feeding (Fig. 12-7C). At the start of the expansive phase the rapid reduction in buccal pressure causes a flow into the buccal cavity from in front of the mouth and perhaps also from the opercular cavity posteriorly. As negative buccal and opercular pressures decrease, the dominant anteroposterior flow pattern is established (Fig. 12-7C, t_5). A brief flow reversal may occur during the initial stages of high-speed suction feeding. Adduction of the operculum and branchiostegal rays forces water anteriorly between the gill arches as the suspensorium is adducted. The sharp decrease of buccal pressure then begins as the sides of the head are adducted, thus compressing the gill arches and effectively isolating the opercular cavity from the buccal cavity. No flow passes over the gill arches in the anteroposterior direction until after suspensorial abduction has begun, which comes shortly before peak gape (Fig. 12-3B). The mouth then closes and water flows out between the operculum and pectoral girdle. Rapid closure of the mouth while fluid is rapidly moving posteriorly through the mouth cavity has an important hydrodynamic conse-

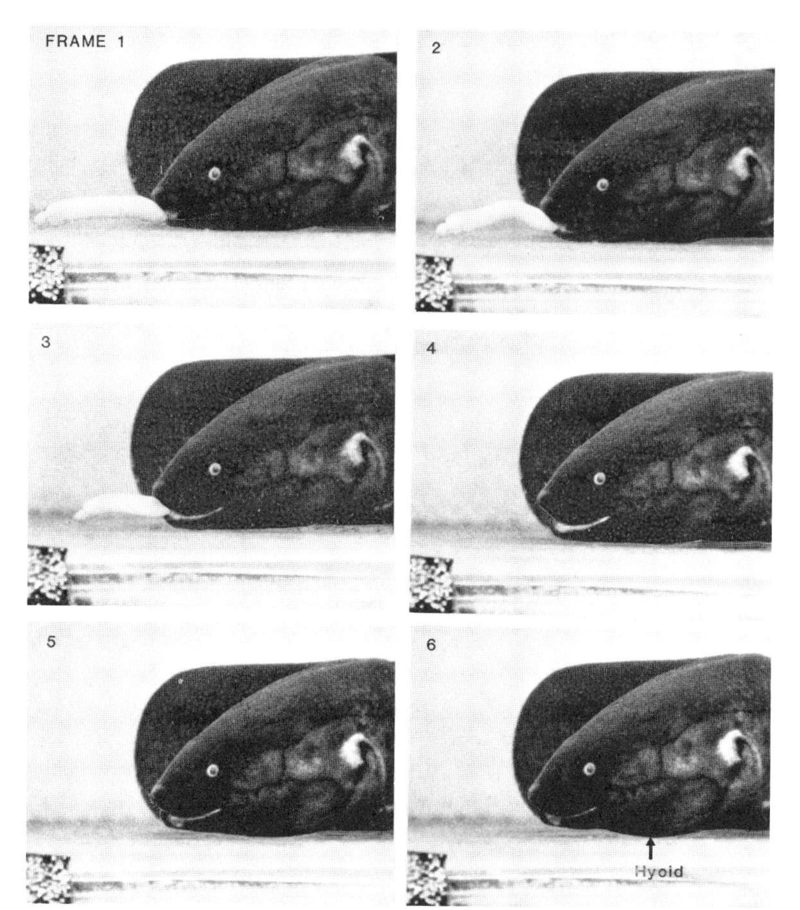

Figure 12-6 Prey capture by the South American lungfish *Lepidosiren paradoxa. Lepidosiren* feeds by closely approaching the prey, using a relatively narrow gape (with a consequent high velocity of flow) and rapid and extreme hyoid depression *(arrow)*. The film was taken at 200 frames/sec; the time between sequential frames is 0.05 sec. The duration of the expansive phase in this sequence is 20 msec, and the total feeding sequence takes 35 msec.

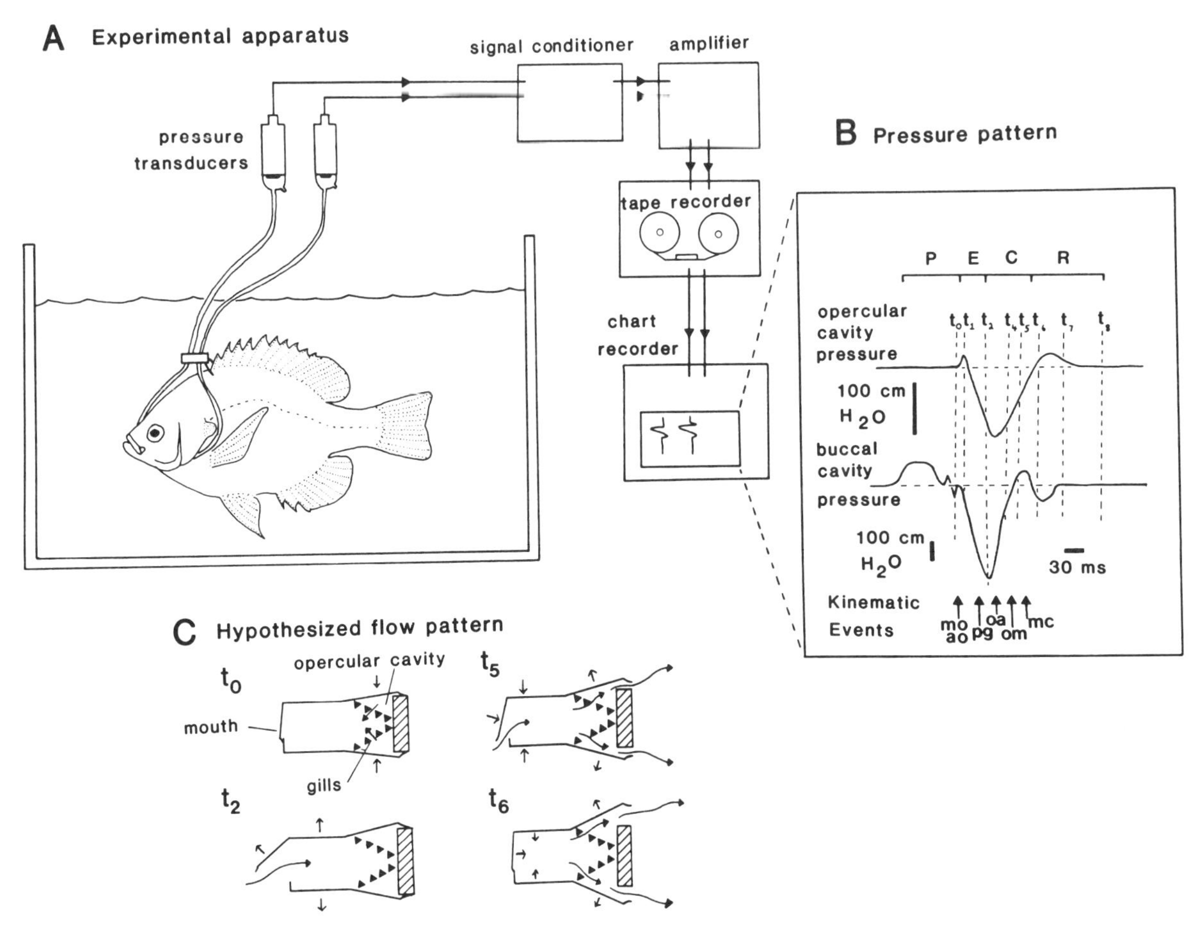

Figure 12-7 A: Experimental apparatus for recording pressures in the mouth cavity of fishes. Cannulae are implanted in the buccal and opercular cavities and are attached to transducers located above the aquarium. B: Representative traces from the buccal and opercular cavities of a bluegill sunfish *Lepomis macrochirus* feeding on a goldfish. The phases are: P = preparatory; E = expansive; C = compressive; and R = recovery. The preparatory phase is not found in all feeding events. Kinematic events are: *mo* = mouth opening; *ao* = opercular adduction; *pg* = peak gape; *oa* = opercular abduction; *om* = branchiostegal valve opens; *mc* = mouth closure. C: Proposed fluid flow through the mouth cavity during high-speed suction feeding over the period t_0 to t_6. Large arrows indicate directions of flow; small arrows indicate movements of bones.

quence that is responsible for the second negative pressure pulse in the buccal cavity (Fig. 12-7B). This second pressure reduction is due to the "water hammer effect"—rapid closing of a valve in a pipeline having fluid flowing through it results in a sharp pressure decrease on the downstream side of the valve. This is exactly what happens when the mouth rapidly closes (in 15 msec) during feeding; a sharp pressure reduction occurs at the level of the vomer and anterior end of the parasphenoid. During feeding on slowly moving prey, when the velocity of mandibular adduction is low, the second negative pressure phase is absent.

The key concept to emerge from the analysis of pressure records of suction feeding in ray-finned fishes is the role of the gills and gill arches as a resistant element in the mouth cavity: the branchial apparatus functionally segregates the mouth cavity into two distinct subsections that exhibit different patterns of volume, flow, and pressure change during feeding. The branchial apparatus thus plays a crucial role in suction feeding by decoupling events occurring in the opercular cavity from those in the buccal cavity. The operculum, often cited as contributing to negative pressure in the mouth cavity, appears to be relatively unimportant in directly generating suction pressure. Instead, it regulates opening of the mouth and serves to reduce the inflow of water at the back of the mouth cavity during the expansive phase.

One final hydrodynamic aspect of aquatic feeding in ray-finned fishes, the role of mouth shape, merits some consideration. All fishes that feed by high-speed suction exhibit some mechanism for occluding the lateral area between the upper and lower jaws. This results in a nearly circular gape (see Fig. 12-4), which orients flow streamlines in a more anteroposterior direction than if the corners of the mouth were open (Alexander, 1967; Lauder, 1979). Consequently, water is drawn in from in front of the head, and prey may be captured from a greater distance. In *Polypterus* the jaw margin is occluded by a thickened and expanded lateral fold of skin; in primitive teleosts the maxilla swings anteriorly to occlude the corner of the mouth (Fig. 12-4); whereas in more advanced teleosts the premaxilla serves this function.

Protrusion of the upper jaw, or movement of the premaxilla and maxilla (relative to the neurocranium) toward the prey, has evolved independently in a number of teleost lineages. Protrusion is an especially important aspect of suction feeding in many advanced teleost fishes (Fig. 12-3C), yet the protrusion mechanisms remain poorly understood. Numerous hypotheses have been proposed to explain the "advantage" of protrusible jaws, but each suffers from many counterexamples (protrusion mechanisms are reviewed in Lauder and Liem, 1981). At least four separate mechanisms have been proposed, but although some kinematic and electromyographic data are consistent with each, no mechanism has been subjected to a controlled experimental analysis. In short, jaw protrusion has been the subject of considerable speculation, yet further work is badly needed, both on the diversity of protrusion mechanisms in teleosts and on the detailed mechanics of musculoskeletal couplings (see also Fig. 18-2 in regard to protrusion mechanisms).

Following the initial strike, buccal and pharyngeal manipulation of the prey may occur prior to swallowing. In most teleosts the prey is simply maneuvered into a position for transport to the esophagus and little "chewing" or maceration occurs; prey are swallowed whole. Manipulatory mechanisms involve both the oral jaw apparatus (mandibular arch) and the hyoid arch and pharyngeal jaws. The pharyngeal jaw apparatus of most teleost fishes consists of paired fifth ceratobranchials, which usually bear teeth, and toothed dermal plates associated with the posterior epibranchial and pharyngobranchial bones (Nelson, 1969). These structures form the lower and upper pharyngeal jaws, respectively. Prey are swallowed by a regular, rhythmic simultaneous protraction and retraction of the pharyngeal jaws (Lauder, 1983b). Both the upper and lower jaws move anteriorly to grip the prey and then move posteriorly, pulling it into the esophagus. This cycle is then repeated until the entire prey item has entered the esophagus. This description probably applies to many teleostean fishes, but experimental analyses of more primitive ray-finned fishes are lacking.

Several clades of teleosts have modified the primitive pharyngeal transport mechanism, in which prey are swallowed whole, to perform masticatory and crushing functions. The most prominent lineage is the Pharyngognathi (Liem and Greenwood, 1981), which includes the cichlids, wrasses, parrotfishes, and damselfishes. In these fishes the two lower pharyngeal jaws are firmly attached in the midline and form a single unit. The upper pharyngeal jaws are hypertrophied and articulate with the base of the skull. Prey, or food-encrusted shells and rocks, are crushed by strong adduction of the pharyngeal jaws and may be subjected to extensive manipulation or shredding prior to swallowing (Liem, 1978).

Salamanders Prey capture by suction feeding is found in a variety of salamanders from different lineages. Presumably, aquatic feeding is primitive

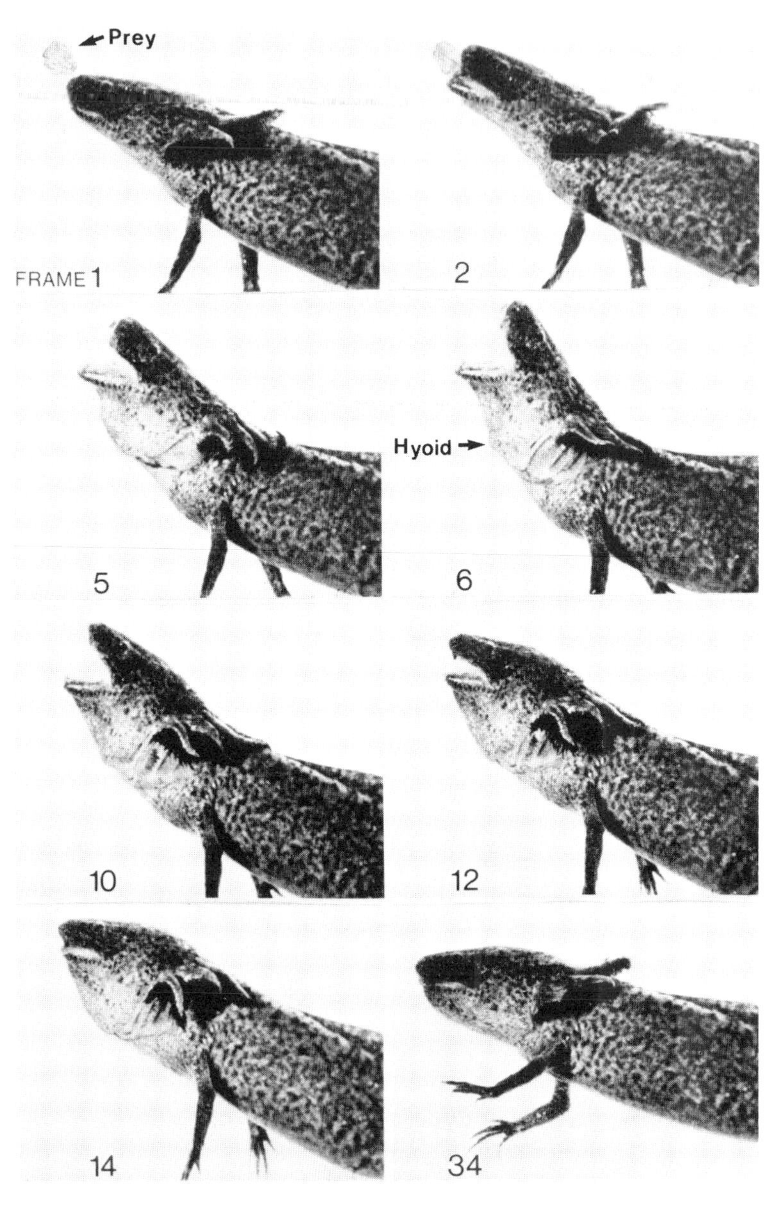

Figure 12-8 Frames from a high-speed film (200 frames/sec) of aquatic prey capture by *Ambystoma mexicanum.* The prey is a small piece of earthworm dropped down from above the head. Note the rapid expansive phase (frames 1–6), the extreme hyoid depression, and the dorsoventral extension of the gill arches.

for urodeles, and many living salamanders are either totally aquatic *(Pachytriton)*, feed aquatically during larval periods *(Ambystoma)*, or return to the water to feed during the breeding season *(Taricha, Notophthalamus)* (Ozeti and Wake, 1969).

The dynamics of aquatic feeding in salamanders has not been examined in detail since the early work by Matthes (1934), but high-speed films of feeding (Fig. 12-8) and an analysis of the major muscles and mechanical units of the head (Fig. 12-9) strongly suggest a close similarity to ray-finned fishes (Shaffer and Lauder, in press,a,b). For the purposes of comparison the suction-feeding act in aquatic salamanders may be divided into the four phases used to describe ray-finned fishes above. A preparatory phase is difficult to demonstrate by kinematic data alone and thus the high-speed film frames shown in Figure 12-8 provide no evidence for a preparatory

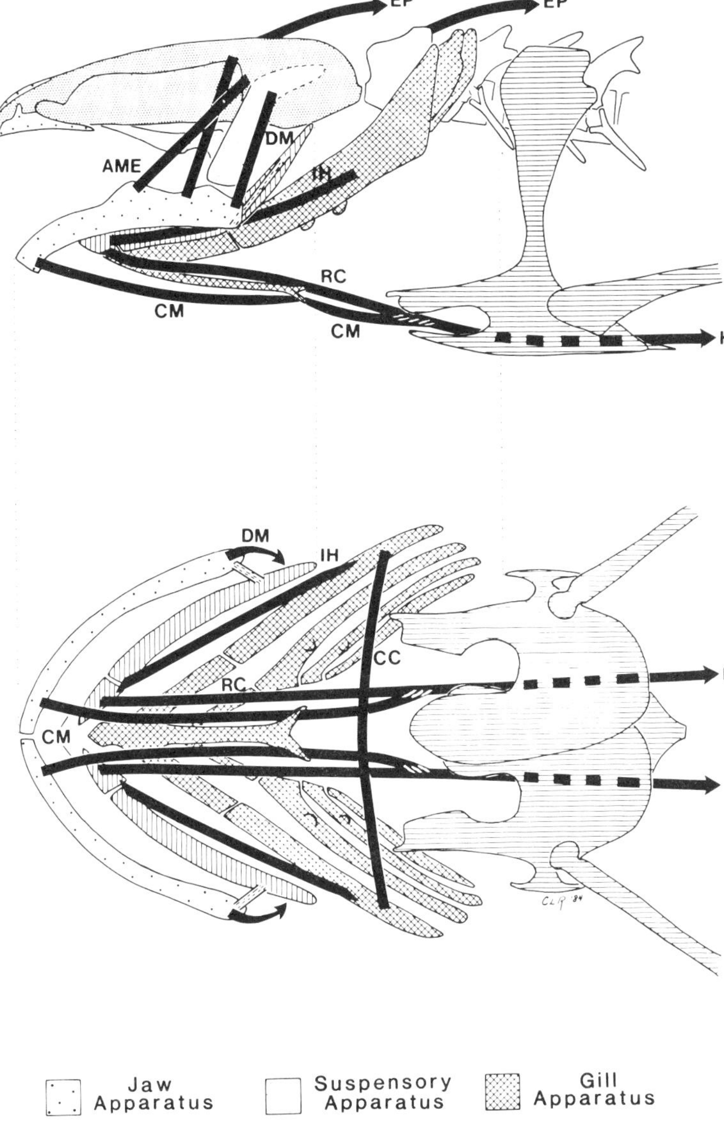

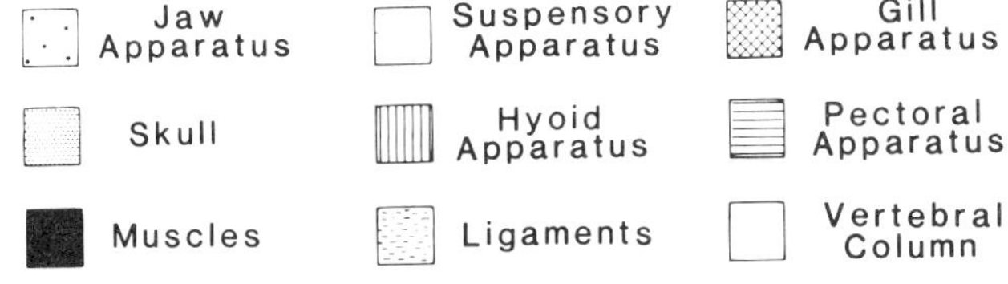

Figure 12-9 The major muscles and mechanical units in the head of an aquatic salamander such as *Ambystoma mexicanum.* The major muscles involved in the rapid expansive phase are the epaxialis *(EP)*, rectus cervicis *(RC)*, hypaxialis *(HY)*, and coracomandibularis *(CM)*. Note the ligament between the hyoid apparatus and the lower jaw: this ligament is important in transferring the posteroventral movement of the hyoid to the mandible and aids mandibular depression. Muscles active during the compressive phase include the adductor mandibulae *(AME)* and constrictor colli *(CC)*. Other muscles are the depressor mandibulae *(DM)* and interhyoideus *(IH)*.

reduction in buccal volume. However, a decrease in buccal volume could easily be achieved by synchronous activity of the adductor mandibulae and coracomandibularis (Fig. 12-9) prior to mouth opening (for instance, in Fig. 12-8, frame 1). Electromyographic evidence will be needed to demonstrate a preparatory phase. The expansive phase (Fig. 12-8, frames 1 – 6) lasts only 25 msec, as maximum gape is achieved very rapidly. Hyoid depression begins early in the expansive phase as the result of electrical activity in the rectus cervicis and its intrabranchial muscle slips (Fig. 12-9). As in ray-finned fishes the pectoral girdle moves posteriorly during the expansive phase as a result of strong activity in the hypaxial body musculature that attaches to it. The forelimbs also move posteriorly as the girdle is retracted.

Mandibular depression is the result of activity in the depressor mandibulae, but based on the kinematic pattern illustrated by Figure 12-8, the ventral body and throat muscles and the hyoid and branchial apparatus appear to play an equally important role. Posterior movement of the pectoral girdle and hyoid resulting from contraction of the hypaxial and rectus cervicis muscles could be transmitted to the mandible either through the geniohyoideus muscles, which may be synchronously active, or via ligamentous connections between the ceratohyal and mandible. Elevation of the neurocranium by the epaxial musculature is also a prominent feature of the expansive phase.

The compressive phase is slightly longer in duration than the expansive phase, lasting about 35 msec. Hyoid depression reaches its peak during the compressive phase, and by the end of this phase the ceratohyal cartilages have reached a nearly vertical position, which causes a small prominence in the throat region beneath the eye (Fig. 12-8, frame 10). The adductor mandibulae appears to be primarily responsible for adduction of the lower jaw. During the recovery phase the hyoid apparatus returns to its resting position, presumably as a result of activity in the coracomandibularis or geniohyoideus, and the pectoral girdle and forelimbs are protracted.

One point of great similarity in aquatic prey capture between fishes and *Ambystoma mexicanum* is that flow does not begin to emerge from between the gill arches until the very end of the expansive phase. In the salamander maximum dilation of the skin covering the gill bars laterally is not achieved until the end of the compressive phase (compare with the delay in opercular abduction shown in Fig. 12-3B).

Perhaps the salient feature that distinguishes the dynamics of aquatic feeding in salamanders from that of fishes is the lack of a well-defined opercular cavity bounded laterally by bony elements. This morphological difference may have relatively little effect on suction dynamics, however, if, as in fishes, the branchial apparatus constitutes an important barrier to flow. If this is true, then the gills and gill arches of aquatic salamanders are analogous to the operculum of teleosts in preventing fluid inflow from the back of the buccopharyngeal cavity during the expansive phase and in delaying outflow until the compressive phase. Alternating gill rakers on adjacent arches, may, as in fishes, function to create a resistance to flow until the expansive phase is over.

Ozeti and Wake (1969) identified several anatomical features of aquatic or partially aquatic salamanders that seem to be related to specialization for suction feeding. A reduction in the size of the tongue and tongue pad is common, as is a relatively robust hyoid and branchial skeleton. Figure 12-8 demonstrates that the pattern of hyoid and branchial movement in at least some aquatic salamanders very closely resembles that of fishes, especially in the sharp posteroventral hyoid rotation near the start of the expansive phase. In an interesting parallel with fishes, salamanders that return to the water to breed develop labial "lobes" (Ozeti and Wake, 1969), which serve the same function as the various structures of actinopterygians that prevent water inflow lateral to the jaws.

Primitive salamanders exhibit many features of the cranial musculature that are common to other lower vertebrates, and the pattern of jaw movement during aquatic feeding is similar to that seen in fishes. Thus, despite several major differences in skull morphology, the primitive suction-feeding mechanism is clearly comparable with that of fishes and exhibits many of the jawbone movements that are primitive for teleostomes.

Turtles Many turtles feed aquatically, and several possess a high-speed suction-feeding mechanism capable of capturing prey in only 30–50 msec (Shafland, 1968; D. M. Bramble, personal communication). The snapping turtle *Chelydra* and the mata-mata *(Chelus)* are two of the best known turtles that feed by suction, but aquatic feeding has also evolved in several other lineages. From the preliminary data available in the Shafland and Bramble papers (see Fig. 12-10), suction feeding in turtles seems to show many similarities to that of aquatic salamanders and fishes, although, to be sure, there are also key differences. A preparatory phase has been ob-

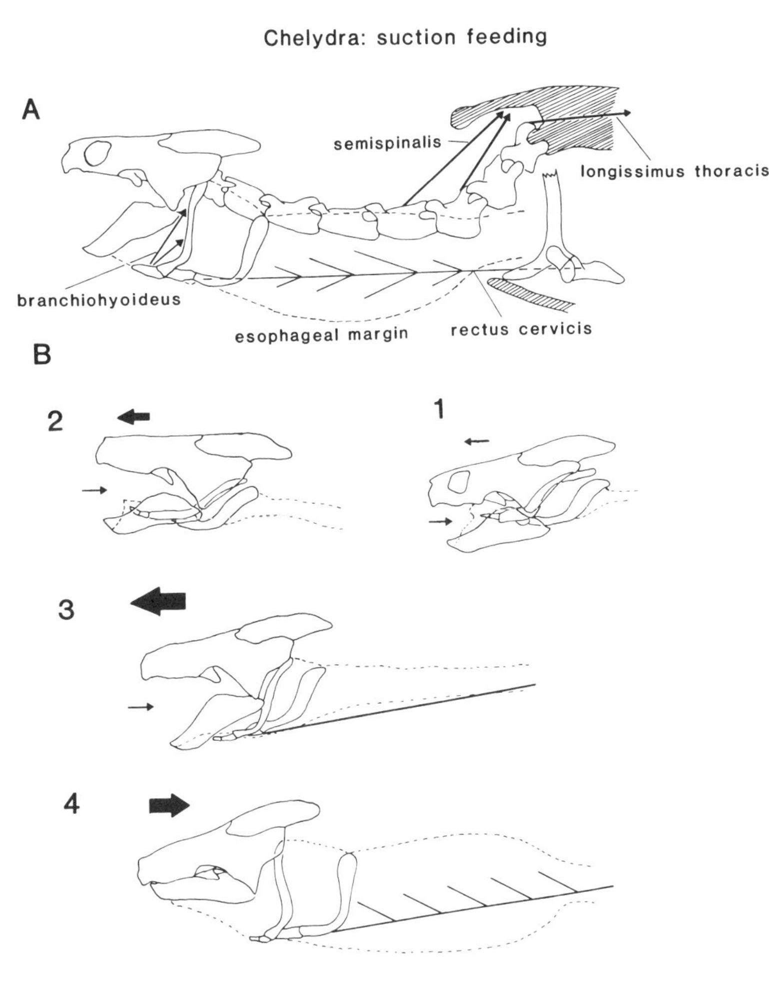

Figure 12-10 A: The major muscles and mechanical units involved during aquatic prey capture in turtles. B: Major movements occurring during prey capture in *Chelydra.* Large arrows indicate movement of the head, which is rapidly thrust at the prey during the expansive phase and part of the compressive phase; small arrows indicate fluid flow into the mouth. Stage 1: preparatory phase. Bramble (1982) has shown that this phase differs from that in teleost fishes in that the mandible is depressed while the hyoid is protracted, and water is forced anteriorly out of the esophagus. Stages 2 and 3: beginning and end of the expansive phase. Stage 4: end of the compressive phase. Head retraction is occurring at this time and water flow continues posteriorly within the esophageal bag. (Figure modified from Bramble, 1982.)

served by Bramble in which the mandible is partially depressed while the hyoid and tongue are protracted (presumably by the branchiohyoideus muscles, Fig. 12-10A), thus reducing buccal volume (Fig. 12-10B, drawing 1). Esophageal compression also occurs at this time to reduce intraoral water volume. During the expansive phase (Fig. 12-10B, drawing 2,3) the head and neck are rapidly accelerated toward the prey, and water begins to flow into the buccal cavity as depression of the hyoid begins. Mandibular depression may be partially due to activity in the rectus cervicis and geniohyoideus muscles, and the epaxial musculature is active to elevate the cranium. The expansive phase may be extremely rapid (50 msec or less). During the compressive phase jaw closure occurs as a result of activity in the adductor mandibulae, and the head and neck begin to move posteriorly (Fig. 12-10B, drawing 4). The hyoid reaches maximum posteroventral rotation during this phase, and expansion of the esophagus occurs as a result of tendinous insertions from the rectis cervicis (D. M. Bramble, personal communication). Esophageal expansion reaches a maximum in the recovery phase (Shafland, 1968), although eventually the hyoid and esophagus return to their resting positions. Turtles thus show the two basic attributes of suction feeding mentioned earlier: posteroventral rotation of the hyoid apparatus, which makes a major contribution to expansion of the buccal cavity; and an anteroposterior sequence in peak excursion of successive movements of head elements. In turtles this sequence is mouth opening, hyoid depression, and esophageal expansion. Esophageal expansion, which starts late in the compressive phase and reaches a peak in the recovery phase, appears to be analogous to expansion of the opercular cavity in fishes, serving to maintain unidirectional flow at the mouth opening until the jaws have closed on the prey.

One obvious difference between aquatic feeding in turtles and adult salamanders and feeding frogs on the one hand and most fishes on the other is the presence of bidirectional flow in most of the former. Water entering the mouth during the expansive phase must ultimately exit by the same path it entered, and this reverse flow must be timed so as not to greatly reduce the intake velocity and jeopardize the chances of prey capture.

The principles of the dynamics of aquatic feeding discussed earlier permit several predictions concerning comparative pressure and flow patterns in unidirectional and bidirectional feeding systems. First, because of the nonsteady flow pattern during suction and the dominance of inertial effects, almost no difference in either flow or pressure patterns will be observed during the expansive and early compressive phases. Second, flow velocities will peak in the expansive phase and as the jaws close, and a second negative pressure will be recorded just inside the jaws owing to the water hammer effect. Third, in bidirectional flow systems pharyngeal pressure will show either a single prolonged positive pressure pulse or a biphasic pattern with a negative phase occurring during the flow reversal.

The major difference between unidirectional and bidirectional flow systems concerns the patterns of pharyngeal and opercular pressure that occur late in the feeding cycle. The highly nonsteady flow dynamics of suction feeding ensures that the inability of water to escape posteriorly will have little effect, if any, on flow and pressure patterns in the expansive phase and most of the compressive phase.

Conclusions and Summary

A remarkably consistent pattern of head movements is found in aquatic vertebrates that feed by inertial suction. The hyoid apparatus in particular appears to represent a fundamental element in aquatic feeding systems, at least since the appearance of the first teleostomes (the acanthodians) in the Silurian. In forms as diverse as sturgeons, perch, coelacanths, lungfishes, turtles, and salamanders, the hyoid apparatus retains its dominant role in producing large and rapid volume changes of the oral cavity by means of posteroventral rotation.

A second consistent aspect of suction feeding is the anteroposterior sequence of cranial expansion and the importance of this sequence in mediating nonsteady flow regimes during suction. The importance of distinguishing between nonsteady and steady flow regimes cannot be overemphasized, because the hydrodynamic models, fluid properties, and assumptions underlying cause-effect relationships between kinematics and pressures are radically different for the two flow patterns (Webb, 1978; Streeter and Wylie, 1979; Osse and Muller, 1980). Steady-state models (for instance, the Poiseuille relation) result in gross errors when applied to dynamic flows.

The dynamics of prey capture in fishes has been the subject of many investigations in the last decade, and the use of techniques such as high-speed cinematography, pressure transducers, and electromyography has resulted not only in a greater understanding of trophic mechanics in fishes, but also in an emerging set of general principles and concepts for aquatic prey capture. But in order to achieve a comprehensive understanding of the process, data comparable to that on fishes must be acquired for turtles, salamanders, and anurans. Comparative information will allow the testing of models of pressure change and fluid flow and will permit the assessment of differences between bidirectional and unidirectional feeding systems. It is important to realize that in any comparative study of functional morphology, the techniques applied to examine structure and function have a significant influence on the type of questions asked and the level of detail in which the system is analyzed. Unless pressure transducers and electromyography are used in the study of aquatic feeding in vertebrates other than fishes, it will be difficult to develop general concepts of aquatic prey capture. New techniques not only allow greater descriptive precision, but also open up new questions and problems and expose inconsistencies in previous accounts. In fishes, for example, prior to the in vivo use of pressure transducers, steady-state models were the descriptive paradigm for suction feeding. It is now clear that this approach is inadequate and that descriptions of aquatic feeding will need to be framed in a dynamic context.

Research on the dynamics of filter feeding in fishes, by contrast, lags far behind studies on other vertebrates. The feeding dynamics and mechanics of particle capture are virtually unknown despite the importance of filter-feeding fishes in aquatic food chains. It remains to be seen if the properties of filter-feeding systems are as consistent as those of suction feeding appear to be. The existence of multiple theoretical mechanisms for entrapping particles, as well as the independent evolution of filter-feeding systems in several vertebrate lineages, suggests, however, that a considerable diversity of approaches to filter feeding may occur. The extent to which this diversity is a consequence of the variation available in systems with steady-state flow regimes (in contrast to the apparently rigid constraints imposed by the use of nonsteady dynamic flows) will be of considerable interest and significance in determining the role of extrinsic, environmental factors in governing the design of vertebrate structure. Relatively steady, low-velocity flows may permit a wider range of functional and structural solutions to the problem of capturing prey than high-speed flows having nonsteady dynamic characteristics. The constraints imposed on jaw function by the use of high-velocity flows may be responsible for the remarkable uniformity in kinematic patterns observed during suction feeding in lower vertebrates.

Chapter 13

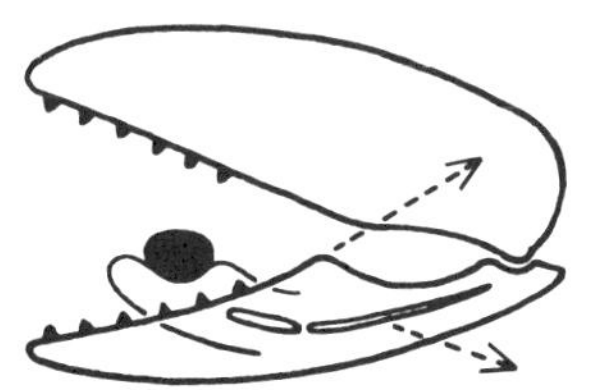

Feeding Mechanisms of Lower Tetrapods

Dennis M. Bramble
David B. Wake

Functional and experimental investigations conducted during the past two decades have vastly extended our knowledge of the feeding mechanisms of fish and mammals. Despite some exceptional work, however, comparable advances have not characterized study of the trophic complexes of amphibians, reptiles, and birds. The result is an information gap that necessarily impairs appreciation of evolutionary patterns and processes in the history of vertebrate feeding. Our primary goal here is to provide a review and synthesis of current knowledge of the feeding mechanisms of nonmammalian tetrapods, with particular attention to the biophysical constraints associated with terrestrial as opposed to aquatic feeding.

Influences of the Feeding Medium

The structural and functional organization of the tetrapod feeding complex reflects modifications associated historically with the crossing of the air-water interface. The biophysical constraints associated with feeding in air and water influence the basic structure and performance of the tetrapod feeding mechanism. A more detailed treatment of aquatic feeding is given in Chapter 12.

Constraints on Food Transport The processes by which food is taken into the mouth, conveyed through it, and swallowed are greatly influenced by differences in the density and viscosity of air and water. Substantially different morphological designs of the feeding apparatus are required because of the different dynamic behavior of food particles when in water or air. The net gravitational load acting on most food objects in water is small compared with frictional or drag forces; the opposite is true in air. Thus, the relative influence of gravitational and drag forces appears to be the central determinant of medium-dependent form and function in the vertebrate feeding complex.

Suction is a nearly universal component of feeding in aquatic vertebrates, but it normally has no role in the ingestive process of terrestrial tetrapods except in drinking. The large, rapidly produced negative pressures within the oropharyngeal chambers of fish and aquatic amphibians (see Chapter 12), as well as turtles (Shafland, 1968; Bramble, 1973), serve to accelerate water —and any entrained food particles—into the mouth. The particles move with the medium if drag forces acting on their surface exceed the static inertia or any resistive forces (for example, propulsive) they may generate. Once food is within the mouth its further transport continues to involve fluid flow, although at lower velocities and intraoral pressures than are associated with initial ingestion. The use of controlled water flow to manipulate food within the mouth has been observed in fish (Alexander, 1969), salamanders (Regal, 1966; Severtsov, 1966), and turtles (Gans, 1969).

The mechanism of swallowing (deglutition) in aquatic tetrapods probably depends on the physical properties of water. Several species of aquatic turtles (such as *Chelydra, Chrysemys,* some *Clemmys*) readily seize food on land, but they are unable to continue the feeding process unless permitted to submerge the head in water. In vivo pressure recordings of the snapping turtle *Chelydra serpentina* indicate that swallowing is associated with characteristic profiles of intraoral and pharyngeal fluid pressure (Bramble, unpublished data).

Suction can rarely be exploited in air because the drag forces are too small to overcome the combined static inertial and gravitational forces operating on any but very small food objects. Consequently, in land-dwelling vertebrates coordinated movements of the tongue and head replace fluid transport in the feeding process.

The essential difference between aquatic and terrestrial feeding, therefore, rests with the mechanism(s) of food transport. It is useful to separate transport into three phases (see also Chapter 14): ingestion, transport, and deglutition. Ingestion is the passage of food into the jaws or mouth cavity. Except in frogs, some salamanders, and turtles, ingestion rarely involves the use of the forelimbs in lower tetrapods. Intraoral transport is the movement of food through the oropharyngeal chamber for mechanical reduction and preparation for swallowing. Finally, deglutition is the transfer of the food mass (bolus) from the pharynx to the esophagus.

Ingestion and intraoral transport may be accomplished by three distinct biomechanical modes. Inertial feeding relies on rapid displacement of the head and jaws relative to the food item (Gans, 1961a, 1969), thus exploiting the inertia (static or dynamic) of the food object to shift it rearward through the oral cavity. The two remaining modes are suction feeding (aquatic only) and lingual feeding (terrestrial only), both dependent on similar actions of the hyobranchial skeleton and related musculature. Each may be combined with inertial feeding but does not require this coupling to transport food effectively.

The tongue is the principal agent of food transport in lingual feeding, and its use is subject to several biophysical constraints. To accelerate a food particle the tongue must simultaneously oppose both its gravitational (weight) and inertial resistance. An upward muscular force from the tongue opposes the weight of the food object, but its inertial resistance must be overcome by composite bonding forces acting between the food and lingual surfaces. Thus, the mass of the food item must be scaled to the size and strength of the tongue, and resistive forces acting on the free surface of the food particle must be small compared with those joining the tongue and food. The physical bond between tongue and food arises from two principal sources: wet adhesion and interlocking. Wet adhesion occurs when a thin layer of bonding fluid is placed between two solid surfaces. There are two kinds of wet adhesion, stefan and capillary (Emerson and Diehl, 1980). In the former the strength of the bond is proportional to the square of the surface area over which adhesion takes place and to the viscosity of the adhesive fluid. The force of capillary adhesion derives from the surface tension generated at the air-fluid interface and is directly proportional to the surface area. Capillary adhesion is relatively great for materials having high surface energy and surfaces that are easily wetted (Hilyard and Biggin, 1977). Since the relative importance of stefan and capillary adhesion is currently unknown, we use adhesive bonding as an inclusive term.

The tongue may also gain traction on food by interlocking, or physically engaging irregularities on the surface of the food. The strength of interlocking depends on the magnitude of the perpendicular force pushing the surfaces together as well as on the texture (roughness) of the two surfaces (Emerson and Diehl, 1980). Interlocking is therefore likely to be more important when the tongue manipulates rougher, heavier items. Nonetheless, substantial interlocking forces may be achieved with lighter food objects provided they are pressed against the roof of the mouth by the tongue.

Finally, if the lubricated tongue is first flattened against food having a smooth surface and low porosity and then pulled away in such a manner as to enlarge vacuities or spaces between surface projections without breaking the surrounding vacuum seal, purchase on the food would be established by lingual suction. Although there is so far no direct evidence of true lingual suction in any vertebrate, it is suspected in both toads (Gans and Gorniak, 1982b) and chameleons (Schwenk, 1983).

Morphological Correlates Comparison of aquatic and terrestrial turtles indicates that the

Table 13-1 Comparison of selected morphological features in hypothetical vertebrates specialized for aquatic (suction) and terrestrial (lingual) feeding.

Feature	Aquatic feeding	Terrestrial feeding
Palate	Flat, smooth	Vaulted, with relief
Internal nares	Small, anterior	Larger, functionally posterior
Hyoid skeleton	Large, relatively rigid, well ossified	Moderate to small, flexible, less well ossified
Hyoid musculature	Intrinsic and (especially) extrinsic muscles massive	Intrinsic (usually) and extrinsic muscles more modest in development
Tongue	Relatively small, with simple surface topography	Relatively large with complex (pappilose) topography
Tongue musculature	Intrinsic (especially) and extrinsic muscles poorly developed	Complex, well-formed intrinsic and extrinsic muscles
Oral epithelium	Smooth, noncornified, few mucous glands	Often punctate, ridged; locally cornified; abundant mucous glands
Gape	Relatively short (restricted laterally)	Relatively long (not restricted)

same stereotyped sequence of kinematic (and probably motor) events can effect food transport in both air and water, but by different mechanical means (Bramble, 1973, unpublished data). This suggests that the major functional and biomechanical distinctions between aquatic and terrestrial feeding stem mainly from morphological differences reflecting adaptation to the physical properties of the two fluids. The strongest empirical support for this notion comes from studies on salamanders (Özeti and Wake, 1969) and turtles (Bramble, 1973, unpublished data; Winokur, 1974), two primitive tetrapod assemblages having both obligate aquatic and terrestrial species as well as intermediate forms. Based on data from these two groups and on theoretical considerations, the more important structural differences in the feeding apparatus of aquatic and terrestrial tetrapods are contrasted in table 13-1. The comparison is between extreme forms in order to emphasize the consequences of fluid properties on morphological design. The aquatic tetrapod is an obligate feeder in water and relies heavily on suction for ingestion and intraoral transport, whereas the terrestrial form uses the tongue extensively in the same feeding activities.

The central distinction in structure between aquatic and terrestrial feeders lies in the organization of the hyolingual complex. The morphology of the hyoid and tongue of lower tetrapods has been studied for systematic purposes (Cope, 1900; Camp, 1923; McDowell and Bogert, 1954; Poglayen-Neuwall, 1954; Regal, 1966; Tanner and Avery, 1982), but investigations that relate structure and function to feeding behavior are rare. Even where this has been attempted (for example, Özeti and Wake, 1969; McDowell, 1972; Regal and Gans, 1976; Lombard and Wake, 1977; Smith, 1984), there are few quantitative or experimental data on hyolingual function in relation to natural diet and feeding strategy.

Within the hyolingual complex there is an inverse relationship between the relative development of the hyoid apparatus and the tongue (Fig. 13-1). The hyoid skeleton tends to be a massive,

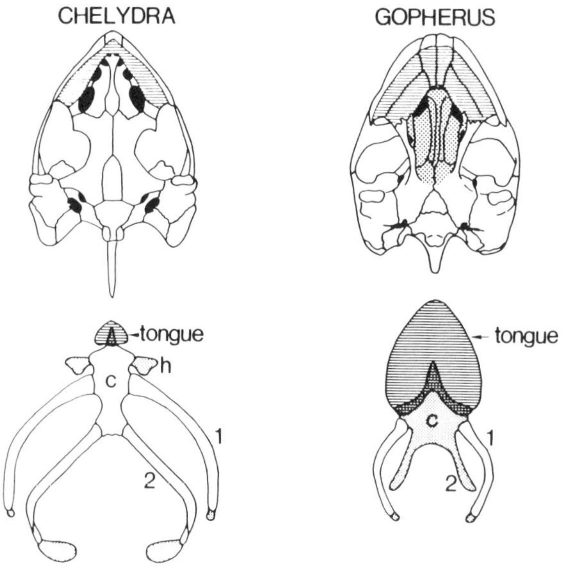

Figure 13-1 Skulls (ventral views) and associated hyolingual complexes (dorsal views) of aquatic *(Chelydra)* and terrestrial *(Gopherus)* turtles. *Chelydra* is a carnivorous inertial-suction feeder, whereas *Gopherus* is a herbivore that uses lingual-inertial and lingual feeding techniques. In *Chelydra* the hyoid skeleton is massive and the tongue is small; the opposite is true of *Gopherus.* The corpus of the hyoid *(c)* and the reduced second pair of branchial horns (2) are cartilaginous (stippled) in the land tortoise. The cranium of *Gopherus* has broad triturating surfaces *(hatched)* and a strongly vaulted palate *(stippled).* The triturating surfaces are restricted in the snapping turtle and the palate is broad and flat. *c* = hyoid corpus; h = cornu; 1, 2 = first and second branchial horns (ceratobranchials).

well-ossified, and fairly rigid frame in aquatic species. Its paired horns are hinged to the median elements in such a way that its range of motion is restricted. The intrinsic and extrinsic hyoid musculature is usually strong. Both the development of these muscles and the size and rigidity of the hyoid skeleton are functionally associated with the large mechanical forces required to accelerate water rapidly.

The tongue of aquatic tetrapods is typically small, smooth, and relatively immobile, and it has poorly developed intrinsic lingual musculature (Fig. 13-2). Three factors account for this. First, effective lingual transport and manipulation of food items is difficult because the composite forces (especially adhesion) that normally bond food and tongue surfaces in air are generally absent in water. Second, the tongue occupies space within the oral cavity that could otherwise be utilized for volumetric expansion and suction. Finally, the tongue may constitute an obstruction to smooth, high-velocity fluid flow within the oropharyngeal chamber. This could induce turbulence or even local cavitation, both of which would increase the force and energy requirements of suction feeding.

Extensive use of the tongue in food manipulation has been documented in turtles (Bramble, 1973, unpublished data), *Sphenodon* (Gorniak, Rosenberg, and Gans, 1982), several lizards (Frazzetta, 1962; Throckmorton, 1976, 1980; Smith, 1984), and birds (Hutchinson and Taylor, 1962; White, 1968; Homberger, 1980a,b; McLelland, 1980; Zweers, 1982a). The organization of the hyolingual complex in terrestrial tetrapods reflects the dependence on lingual transport of food (Fig. 13-1). The hyoid apparatus is more lightly built and typically more mobile than in aquatic feeders. Mobility is conferred by thinness and flexibility of the individual skeletal elements and by greater range of motion allowed by the joints between these elements. Although the hyoid apparatus of terrestrial vertebrates may effect substantial enlargement of the pharynx, particularly in lizards that specialize in large or bulky prey (Smith, 1979, 1982, 1984), this action lacks the speed and strength that characterize hyoid expansion in aquatic suction feeders.

The tongue of terrestrial species also tends to be larger, more muscular, and to have a more complex surface than that of aquatic forms (Fig. 13-2). Greater strength is required to counter the gravitational loads imposed by food objects in air; larger size promotes increased surface contact with the food, which enhances adhesive bonding. Slow-motion cinematographic records show that adhesive bonding may permit relatively light items to be picked up and ingested through tongue movements alone in salamanders, frogs, tortoises, some lizards, and *Sphenodon.* This is most dramatically seen in chameleons and those

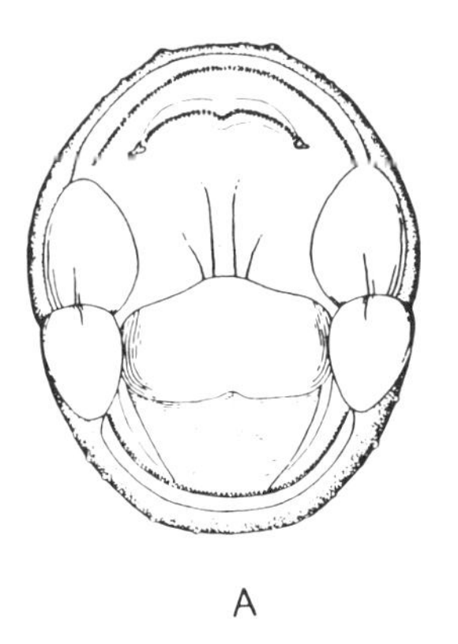

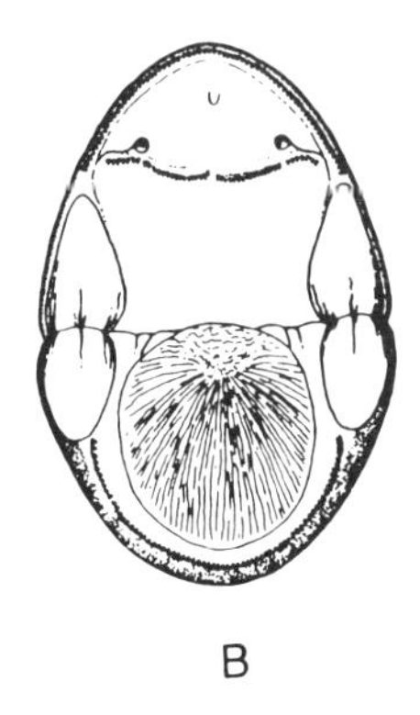

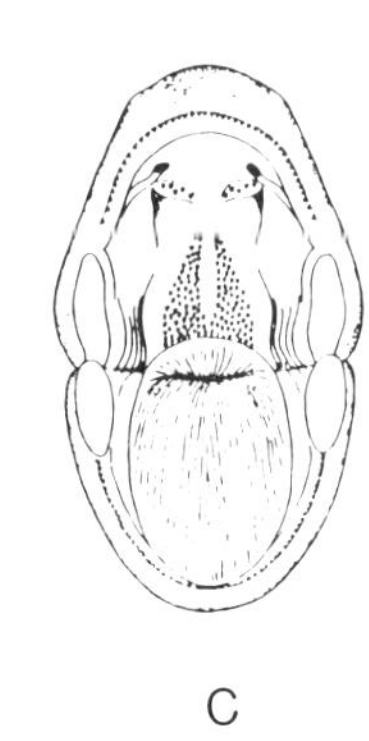

Figure 13-2 Oral morphology of aquatic and terrestrial salamanders: A *(Cryptobranchus alleganiensis)* is fully aquatic; B *(Ambystoma maculatum)* is a primitive terrestrial species retaining aquatic reproduction; C *(Plethodon cinereus)* is fully terrestrial. Note relatively small, smooth tongue of the aquatic species compared with the larger, textured tongues of the terrestrial salamanders. The palate of *Plethodon* is slightly vaulted and possesses numerous teeth posteriorly. This contrasts with the wider, flatter, and smoother palatal surfaces of *Cryptobranchus* and *Ambystoma.* (From Bishop, 1941.)

salamanders that use ballistic tongue projection (Gans, 1967; Lombard and Wake, 1976, 1977).

Interlocking forces between tongue and food are strongly influenced by surface texture. The surface of the tongue in terrestrial tetrapods is made rougher by the presence of lingual papillae. These surface projections are well developed in terrestrial turtles (Winokur, 1974; Tanner and Avery, 1982), many lizards (McDowell, 1972; Tanner and Avery, 1982; Schwenk, 1984), and *Sphenodon* (Gabe and St. Girons, 1964), but are poorly developed or absent in most aquatic turtles and crocodilians. They are generally soft and filamentous in land tortoises and herbivorous lizards, but may be short and cornified in omnivorous and carnivorous lizards (McDowell, 1972) and birds (Ziswiler and Farner, 1972). A recent survey, however, reveals no consistent relationship between tongue surface morphology and dietary habits in lizards (Schwenk, 1984). In lizards papillae are best developed on the "hind-tongue." The "foretongue" is to varying degrees specialized for the transfer of chemical stimuli to Jacobson's organ (Schwenk, 1984). It is presumably less involved in food transport than is the hind-tongue, and in some lizards the foretongue may be invaginated into the hind-tongue while feeding (McDowell, 1972). Nonetheless, use of the anterior region of the tongue to capture prey appears to be primitive for lizards (Schwenk, unpublished data). In lizards presumably specialized for feeding on large prey (for instance, varanids) and in snakes, only the foretongue is retained (McDowell, 1972). It is highly modified for chemical transfer, has little or no surface texture, and is not used for intraoral transport of food (Smith, 1984). Surface folds and projections also occur on the tongue of terrestrial salamanders and frogs (Regal, 1966; Larson and Guthrie, 1976; Regal and Gans, 1976; Lombard and Wake, 1977), but they are generally larger, noncornified, and fewer in number than those of reptiles. A significant exception is the large, relatively complex tongue of the specialized ant- and termite-eating frog *Rhinophrynus* (Trueb and Gans, 1983).

The ability to shape the tongue surface to fit that of the food improves the capacity for intraoral transport by increasing the traction of the tongue through both adhesion and interlocking. The control of tongue shape is largely dependent on the development and complexity of its intrinsic and extrinsic musculature, but it may also involve hydraulic forces (Homberger, 1982; Kier and Smith, 1983). The tongue musculature varies greatly among terrestrial amphibians and reptiles (Horton, 1982; Lombard and Wake, 1977; Regal and Gans, 1976; Sondhi, 1958; Tanner and Avery, 1982), but apart from that associated with special tongue-projecting mechanisms (see below), little is known about the manner in which these muscles actually govern lingual movement. However, Smith (1984) has outlined several potential mechanisms for tongue movement in lizards and has provided experimental data on tongue use in *Ctenosaura* and *Tupinambis.* Her evidence, including electromyograph recordings from the intrinsic lingual musculature, suggests that the highly protrusible tongue of *Tupinambis* may function, in part, as a muscular hydrostat. The more generalized tongue of *Ctenosaura* appears not to operate in this manner.

The capacity to shape the tongue to the food appears to be especially well developed in herbivorous and insectivorous lizards, tortoises, and *Sphenodon,* wherein the tongue tends to be large, muscular, and heavily papillated. The tongues of generalized terrestrial salamanders, such as *Ambystoma,* also reshape considerably during prey capture (Larson and Guthrie, 1976). Here, however, changes of shape are effected primarily by movements of the associated hyoid skeleton and perhaps by hydraulic pressure rather than by the direct actions of the intrinsic musculature. The tongue of birds usually lacks well-developed intrinsic musculature (Ziswiler and Farner, 1972), a fact that would appear to preclude precise shaping to food items. Tongue movements are largely attributable to the very mobile hyoid apparatus upon which the tongue is mounted. Nonetheless, relatively large, mobile tongues are found in fruit- and seed-eating birds, especially those that peel or husk food objects before swallowing. Parrots are exceptional in that they have a complex set of intrinsic tongue muscles that are used extensively in food handling (Mudge, 1903; Homberger, 1980a,b, in press). In parrots, at least, shaping of the tongue may also involve special hydraulic mechanisms (Homberger, 1982, in press).

Medium-dependent feeding mechanics are also reflected in the structure of the palate (Figs. 13-1, 13-2). Tetrapods specialized for aquatic suction feeding generally have a flat, smooth palate without major topographic relief. The internal choanae are small, forwardly positioned, and generally beveled into the palatal surface. Such palates maximize the potential for volumetric change within the oral chamber and promote smooth fluid flow. This simple morphology is retained in virtually all amphibians, including primitive terrestrial species (see, for example, Fig. 13-2B). Land-dwelling amniotes, in contrast, typically possess a vaulted palate into which a pair of prominent choanal grooves or channels are recessed (see, for example, *Gopherus,* Fig. 13-1). The two channels, often separated by a midline septum, normally extend rearward to terminate close to the glottal aperture above the root of the tongue. This arrangement appears to facilitate lung ventilation by providing an airway that bipasses (dorsally) the oral cavity. It is necessary because the tongue of most amniotes nearly fills the oral cavity when the mouth is closed and the tongue is in a resting position.

A Model Generalized Feeding Cycle

Because the feeding cycles of such phyletically diverse groups as mammals, reptiles, and amphibians appear to incorporate very similar trains of stereotyped kinematic and motor events, involving homologous sets of muscles, they have probably derived from a common ancestral feeding mechanism utilizing much the same motor pattern. If so, a model defining the properties of such a primitive mechanism should have heuristic value for future efforts to understand the evolutionary processes that have led to the array of feeding mechanisms observed among tetrapods, living and extinct.

The model developed here, although hypothetical, is based largely on a synthesis of available experimental data (kinematic, electromyographic, and so on), chiefly from generalized salamanders, turtles, and lizards, together with consideration of the biophysical constraints noted above. The model is restricted to the rhythmic "chewing" or intraoral transport cycles, which appear to exist in nearly all modern tetrapods. Excluded, therefore, are ingestive feeding behaviors, which, however, are sometimes rhythmic and frequently bear strong kinematic resemblance to the transport cycle.

The principal functions of the generalized feeding cycle are ingestion, mechanical reduction (if any), and transport of the food to the esophagus. Mechanically, the mechanism can be likened to a ratchet. In successive strokes the ratchet ad-

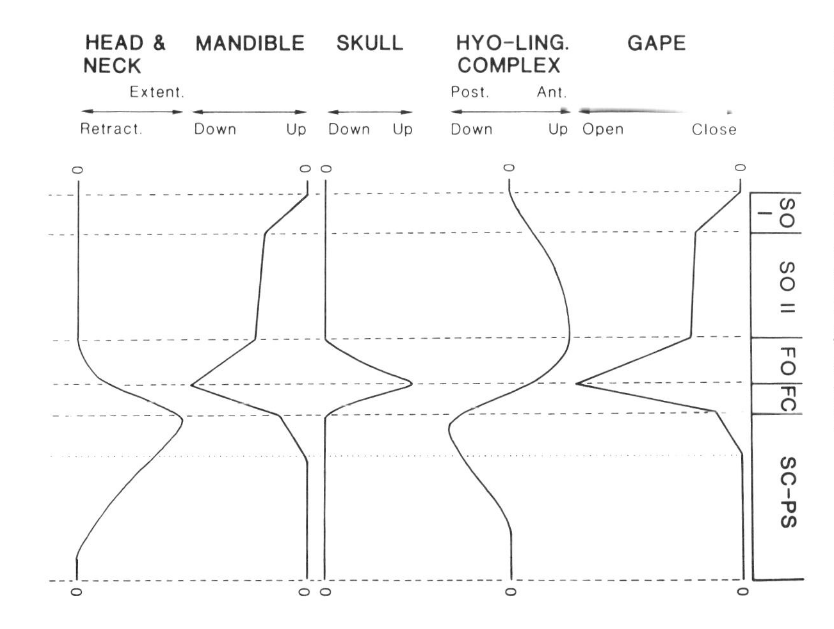

Figure 13-3 Composite kinematic profile of a theoretical model feeding cycle as would be expected in a primitive, generalized tetrapod. Major kinematic stages of the cycle are slow open I (*SO*-I), slow open II (*SO*-II), fast open *(FO)*, fast close *(FC)*, and slow close–power stroke *(SC-PS)*. The dotted line indicates the boundary between the *SC* and *PS* intervals of the final stage. For each profile the zero point is the baseline or resting position.

vances, engages the food, and then carries it posteriorly through the mouth and pharynx. This basic mechanism may be assisted by properly timed inertial feeding movements of the head.

Kinematics and Mechanics The cyclic movements of the generalized feeding program are organized into the same four kinematic stages that characterize the masticatory cycle of mammals (Hiiemae, 1978; Chapter 14). The stages are: slow open (SO), fast open (FO), fast close (FC), and slow close–power stroke (SC-PS). As in mammals these stages are identified on the basis of characteristic changes in the gape profile. Additionally, each is attended by integrated movements of the cranium, mandible, and hyolingual complex. Figure 13-3 shows the kinematic profile of this model feeding cycle, Figure 13-4 predicts some of the associated muscle activity patterns, and Figure 13-5 illustrates the mechanics of food transport in the intact animal.

Slow open is regarded in most analyses of the tetrapod feeding program as a single kinematic stage. However, it is often possible to recognize two distinct episodes within the stage (SO-I and SO-II), which we believe may have actual as well as theoretical importance.

The model feeding cycle commences with SO-I, during which the mandible is depressed to a comparatively low (usually less than 30% of maximum) gape angle. The tongue and hyoid apparatus are simultaneously shifted in an anterodorsal direction (Fig. 13-5B). This action serves to set (or reset) the lingual ratchet prior to the initiation of food transport. While the tongue is advancing beneath the food, it is important that the food not be carried with it. This is usually prevented by having the food item pinned between the jaws or marginal teeth. Because the oral cavity usually tapers anteriorly, larger food items may also be prevented from shifting forward by pressure against the marginal teeth or the margins of the oral vault. However, for smaller items, or those already well within the mouth, the food must be fixed against the palate. Palatal teeth, when present, help to fix the food, but otherwise traction between the food and palatal epithelium must suffice to maintain the position of the food.

Positional stability of the food during SO-I necessitates minimizing the sliding friction between the tongue and food. This is particularly important since the tongue must also apply an upward static force on the food to help fix it to the palate. Sliding friction is reduced by streamlining of the lingual surface through passive rearward folding of its papillae and by the expression of lubricating secretions onto the tongue's surface.

Depression of the mandible in the SO-I stage is

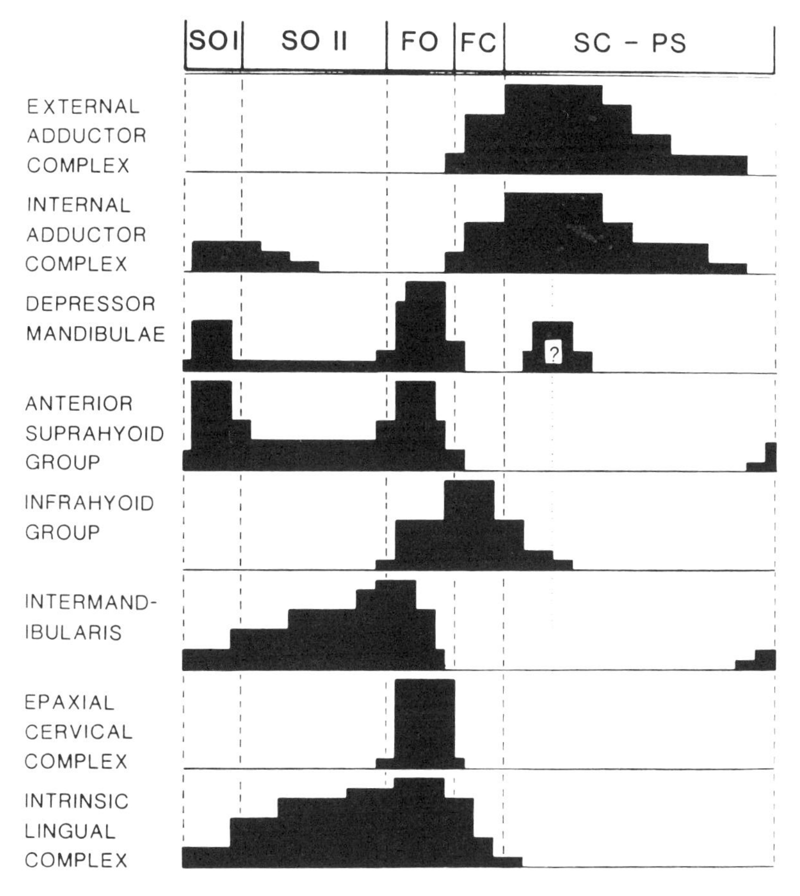

Figure 13-4 The expected electromyographic activity patterns of eight selected muscle units during the model feeding cycle. Width between dotted lines indicates timing of stages. External adductor complex = chiefly M. adductor mandibulae; internal adductor complex = chiefly M. pterygoideus; anterior suprahyoid group = Mm. geniohyoideus and genioglossus; infrahyoid group = chiefly M. rectus cervicis; epaxial cervical complex = muscles linking neck with occiput.

associated with moderate contraction of the depressor mandibulae and at least the posterior portion of the pterygoideus complex (Fig. 13-4). The latter contributes to the stability of the jaw joint and may also serve to protract the mandible. Anterodorsal displacement of the hyolingual unit derives from strong activity in the anterior suprahyoid group (especially the geniohyoideus and genioglossus) as well as in the intermandibularis. The intermandibularis contributes significantly to the trajectory of the tongue and hyoid by tensing and lifting the floor of the mouth, thereby forming a platform upon which the hyolingual unit slides. The intrinsic tongue muscles are expected to be only moderately active at this stage of the feeding cycle.

The duration of the SO-II stage is typically longer than that of SO-I. It is recognized by a distinct decline in the rate of change of gape (Fig. 13-3). Thus, the gape angle may increase little over that attained in SO-I—or may even decrease slightly, as is sometimes observed in mammals (Kay and Hiiemae, 1974; Crompton et al., 1977). The tongue and hyoid continue to move forward and upward, but their anterior velocity gradually declines to zero. By the end of the SO-II interval the hyoid skeleton normally has reached the anterior limit of its cyclic motion. This is the condition observed in the transport cycles of mammals (Crompton et al., 1975; Hiiemae, Thexton, and Crompton, 1979), lizards (Smith, 1984), and tortoises (Bramble, unpublished data).

The mechanical function of SO-II is that of fitting the tongue to the food (Fig. 13-5C). To do this, the tongue is forced upward against the food item and its surface configuration is adjusted so as to maximize bonding. In theory this should require reshaping the tongue to increase surface contact and to reorient its surface projections so

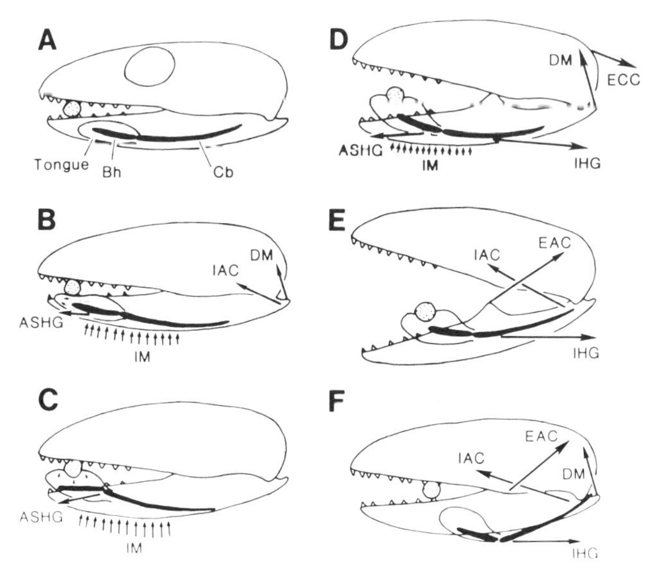

Figure 13-5 The mechanism of intraoral transport in the model generalized tetrapod. A: System at rest with food object *(stippled)* in jaws. B: Slow open I, hyolingual unit advancing beneath food. C: Slow open II, tongue being fitted to food. D: Middle of fast open with tongue cradling food and cranium and mandible accelerating in opposite directions. E: Early in fast close with cranium descending, mandible elevating, and tongue and hyoid rapidly accelerating rearward. F: Slow close stage of slow close–power stroke, showing food fixed by teeth and tongue and hyoid at maximally retracted position. *ASHG* = anterior suprahyoid group; *Bh* = basihyal; *Cb* = ceratobranchial; *DM* = depressor mandibulae; *EAC* = external adductor complex; *ECC* = epaxial cervical complex; *IAC* = internal adductor complex; *IHG* = infrahyoid group; *IM* = intermandibularis; arrows indicate direction of muscle force.

as to better engage irregularities on the underside of the food. That slips of the intrinsic tongue musculature extend to individual papillae in the tongues of some reptiles (such as turtles and lizards; Winokur, 1974, unpublished data; Schwenk, 1984) clearly suggests the possibility of rather precise control of surface topography. Slow-motion films of tortoises reveal that considerable changes in tongue shape do indeed accompany SO-II (Bramble, unpublished data). In any case, appropriate physical engagement between the ratchet and the food surface seems prerequisite to the effective acceleration and transport of the food during the subsequent FO and FC stages.

Low-level activity in the depressor mandibulae during SO-II correlates with the relatively static position of the lower jaw (Figs. 13-3, 13-4). Conversely, activity in the intermandibularis and intrinsic tongue muscles increases as the tongue is actively applied to the food. Strong contraction of the suprahyoid musculature continues throughout this stage.

The FO stage involves the sudden and rapid opening of the mouth to maximum gape (Fig. 13-3). The method of achieving this is established by the degree of gape required. For minimal to moderate gapes, only the mandible is depressed. If the mouth must be opened more widely, much of the gape is attained by upward rotation of the skull about the craniocervical joint. Cranial elevation (when present) distinguishes FO from SO, as do the initiation of a forwardly directed inertial thrust of the head and neck (and sometimes the trunk as well) and the very rapid posteroventral retraction of the hyolingual unit. Again, however, the extent of the inertial maneuver depends upon relative food size and mass. Larger, heavier objects elicit a thrust, whereas smaller, lighter items may be transported solely by the tongue.

Fast open marks the start of active food transport. The tongue cradles and supports the food while beginning to accelerate it backwards (Fig. 13-5D). Considerable change in tongue shape may occur. Frequently its lateral margins are raised while the central region is depressed, a maneuver that helps the tongue to "grasp" the food item. Such changes in lingual shape are evident in tortoises (Bramble, ms.) and generalized salamanders (Larson and Guthrie, 1976). As the tongue carries the food rearward, the gape is rapidly increased so as not to impede its progress. If most of the gape increase is effected by cranial elevation, the mandible and associated hyolingual unit remain relatively steady as the mouth is opened. The upward swing of the skull also serves to break the bond between the roof of the mouth and food at the onset of the transport phase. When incorporated into the FO stage, the inertial thrust can contribute significantly to the transport process by causing the cranium to accelerate over the food as the food itself is moving rearward.

Rapid opening of the mouth during FO is reflected in strong contraction of the depressor mandibulae as well as the dorsal neck musculature (Fig. 13-4). Retraction of the tongue and hyoid stems chiefly from a strong rearward pull on the hyoid skeleton by muscles of the infrahyoid group, chiefly the rectus cervicis. This muscle may also contribute to depression of the lower jaw by virtue of its linkage to the symphysis via the anterior suprahyoid muscles (see Fig. 13-5D). This function is suggested by experimental evidence in a broad spectrum of vertebrates including tortoises (Bramble, unpublished data), lizards (Smith, 1984), aquatic salamanders, and bony fish (Lauder, in press). The same mechanism of mandibular depression has been demonstrated in mammals, where it involves the homologous infrahyoid muscles, the sternohyoideus (Crompton et al., 1975, 1977; Hiiemae, Thexton, and Crompton, 1979).

The FC stage results in the rapid closing of the jaws to engage the food. The mandible is elevated and the cranium lowered (if it has been raised) (Fig. 13-4). The hyolingual unit continues its rearward movement and by the end of the stage has nearly reached the most posterior and ventral point of its orbit. By the end of FC the inertial thrust of the head has also been completed.

Closure of the jaws on a large food object prevents its immediate further transport (Fig. 13-5E,F). Smaller items may continue to progress posteriorly until stopped by contact with the palate or until the tongue and hyoid cease moving. In reducing gape, actions of the skull and mandible are essentially the reverse of those seen in FO. Retraction of the hyoid skeleton by the infrahyoid musculature spreads this framework, thus expanding the posterior oral and pharyngeal chambers to accommodate the incoming food object. The skull is accelerated downward by the ventral neck musculature. Strong contraction of the external adductor and pterygoideus complexes signals the beginning of FC, as does the cessation of activity in the depressor mandibulae (Fig. 13-4).

The onset of the SC-PS stage of the generalized feeding program is associated with a sharp decrease in the rate of gape reduction (Fig. 13-3), which is the consequence of the resistance encountered by the jaws as they meet the food. The abruptness of the change in the gape profile and, hence, of the onset of the SC-PS stage, is strongly influenced by the physical properties of the food. As the jaws engage the food, the hyolingual unit begins to rebound anterodorsally towards its resting position.

Primitively, the PS portion of the SC-PS stage consists of simple static adduction of the jaws against the food object, with little or no change in gape angle. The intensity and duration of this interval depend on the species under consideration and the nature of the food itself. In some reptiles (living and extinct) the power stroke includes a well-defined slicing and grinding action in which there is horizontal translation of the adducted mandible relative to the cranium. In tortoises (Bramble, 1974) and dicynodont therapsids (Crompton and Hotton, 1967) the lower jaw is retracted, whereas in *Sphenodon* it is drawn forward (Rosenberg and Gans, 1977; Gorniak, Rosenberg, and Gans, 1982).

Activity in the adductor musculature rises sharply as the jaws meet the resistance of the food at the beginning of SC (Fig. 13-4). The peak of their contraction is reached early in PS and then declines. There may then be renewed activity in the depressor musculature. Electromyograph recordings have shown that the depressor mandibulae is sometimes active during jaw closure in lizards (Throckmorton, 1980), but apparently not in the primitive lepidosaur *Sphenodon* (Gorniak, Rosenberg, and Gans, 1982). There are also indications of depressor activity at this stage of the cycle in land tortoises (Bramble, unpublished data). Contraction of the reptilian depressor muscle during PS may be required to help stabilize the jaw joint in at least some instances (Bramble, 1978). Contraction of the infrahyoid musculature ceases during SC, permitting the hyoid and tongue to begin their return to the resting position. If the cycle is to be followed immediately by another, activity may resume within the suprahyoid group prior to the termination of the power stroke. This stage also allows the head to return to its original position following the inertial thrust.

Implications of the Model We envision a central role for the tongue in food transport: it is lingual transport that distinguishes the typical feeding of terrestrial from aquatic vertebrates. It is likely, as Severtsov has argued (1964, 1971), that the development of a mobile tongue was the key problem to be surmounted in the complete transition of early amphibians to land life. Indeed, it appears that lingually mediated intraoral transport continues to be the central organizing factor of the complex and highly evolved masticatory program of mammals (Hiiemae, Thexton, and Crompton, 1979). Assuming that the model approximates the feeding program utilized by early tetrapods, new and potentially useful insights, predictions, and tests arise concerning the evolution of the tetrapod feeding complex.

One of these concerns the significance of SO. This stage has received less attention than the other stages of the tetrapod feeding cycle because it is slow and much of what takes place is not normally visible to external observation and recording. Furthermore, it does not involve actions of the major adductor muscles, which capture the attention of most investigators. Indeed, many recent electromyographic studies on reptilian feeding (such as Throckmorton, 1980; Gorniak, Rosenberg, and Gans, 1982; Smith, 1982) do not include data on the key muscular unit of SO, the suprahyoid group. Some information on activity patterns among the suprahyoid muscles of amphibians has recently been presented in connection with studies on the feeding of *Dermophis* (Bemis et al., 1983) and *Bufo* (Gans and Gorniak, 1982a,b).

In two respects the SO-II interval may be functionally the most vital of the generalized cycle. The first has already been addressed, namely, that of properly engaging the tongue and food prior to active transport. The second concerns modulation of the chewing and transport cycles. It is well established in mammals that sensory feedback from mechanoreceptors located within the oral epithelium and periodontal membranes helps to adjust the rhythmic masticatory program to changes in food particle size, texture, mechanical resistance, and position within the oral cavity (Hendricks, 1965; Hiiemae and Crompton, 1971; Hiiemae and Thexton, 1975; Thexton, Hiiemae, and Crompton, 1980). There is mounting evidence that modulation of the feeding cycle also occurs in lizards (Throckmorton, 1978, 1980), turtles (Bramble, 1980, unpublished data), and rhynchocephalians (Gorniak, Rosenberg, and Gans, 1982) in response to differences in the physical properties of the food. The nature of tooth attachment and replacement in reptiles precludes periodontal mechanoreceptors (Throckmorton, 1980), as does the edentulous condition of turtles and birds. Hence, in these animals the perception of mechanical stimuli during feeding is probably confined to receptors in the oral epithelium or beneath the rhamphotheca (for example, Herbst's and Merkel's corpuscles in birds; Zeigler and Karten, 1973; Zeigler, Miller, and Levine, 1975).

Slow open II could be an especially sensitive period for modulatory feedback from the oral cavity. There is then maximal surface contact between the food and the oral lining, and, especially in the latter part of the period, the tongue is flattened against the undersurface of the food and is exerting an upward force on the palate, thereby increasing the potential for stimulation of oral mechanoreceptors. There is also little or no relative movement between the lingual surface and the food, thus improving the fidelity of sensory information used to establish the spatial relations of the food to surrounding oral structures. Finally, sensory input during SO-II is appropriate inasmuch as this stage immediately precedes the onset of very rapid food transport in FO and FC. Motor events are sufficiently rapid during these latter two stages to suggest that they are largely "preprogrammed" and, as such, relatively insensitive to sensory modulation.

Mechanical considerations predict that the SO-II stage should be more pronounced in an animal that is feeding on relatively large or heavy food objects than when it is ingesting light or small items. This follows from the fact that more precise fitting of the tongue will be required for effective lingual transport when the food object offers substantial inertial resistance to acceleration. Light items should require less handling time by the tongue since simple adhesive bonding suffices to hold the food securely to the tongue's surface. Experiments with tortoises (Bramble, unpublished data) and cats (Thexton, Hiiemae, and Crompton, 1980) suggest that responses to differing size, hardness, and texture of food particles are indeed expressed mainly as changes in the duration of SO-II.

A second implication of the model concerns evolutionary reversals. Because the generalized primitive feeding program centers on the tongue in food ingestion and, above all, in intraoral transport, modern tetrapod feeding mechanisms that exclude the tongue must be regarded as secondary modifications. Two general conditions seem to favor such evolutionary reversals. The first is a return to aquatic feeding and, with it, to fluid transport. To varying extents this has occurred in aquatic salamanders, a few anurans (such as *Pipa*), and many turtles. In each case there has been a reduction of the tongue and a concurrent enlargement of the hyobranchial skeleton (Shafland, 1968; Özeti and Wake, 1969; Bramble, 1973, unpublished data). The mode of inducing water flow is much like that in fishes, even though there is no opercular apparatus and the manner of cheek and jaw suspension is markedly different (see Chapter 12).

The feeding of crocodilians has also been modified by a secondary return to aquatic habits (Busby, 1982). Although they do not use suction feeding, their tongue is much simplified, largely devoid of musculature (Schumacher, 1973), and incapable of anteroposterior displacement. Therefore, the crocodilian tongue cannot contribute to intraoral transport, but it may be used to hold food against the palate (Busby, 1982).

The development of specialized inertial feeding is the second general setting in which there is secondary loss of lingual feeding. Such behavior occurs in some lizards (Frazzetta, 1962; McDowell, 1972; Smith, 1982, 1984), crocodilians, most birds, and, with great modification, in snakes. Inertial feeding is examined in more detail in a later section; here we are concerned only with those factors that favor its evolutionary origin from the generalized feeding system.

Empirical data as well as theoretical considerations suggest that specialization for inertial feeding is most likely to arise in predatory species whose diet regularly includes large prey items. Lingual transport involves surface-dependent bonding mechanisms to overcome the mass-dependent inertial resistance offered by the food. Larger, heavier food objects require disproportionately large increases in surface bonding in order to ensure effective tongue transport. Without altering the basic bonding mechanisms (adhesion, interlocking), increasing the bonding force can only result from modifications that provide greater surface contact between tongue and food. These include greater surface relief (papillation) and a relative increase in overall tongue size.

Enlargement of the tongue may, however, prove more a hindrance than an aid when attempting to ingest large, active prey. As the tongue is protracted beneath the prey item, the mandible must be depressed farther to accommodate the enlarged tongue as it shifts forward in the mouth (Fig. 13-5C). This in turn decreases the grip of the jaws on the prey and increases the chance of its escape. The risk is further increased by the fact that the SO stage is inherently one of reduced purchase on the food. Most electromyographic recordings of tetrapods having a distinct SO stage reveal little or no activity in the major adductor muscles at this time. Apparent exceptions are the caecilian *Dermophis* (Bemis, Schwenk, and Wake, 1983) and the monitor lizard *Varanus* (Smith, 1982).

Relatively large food items are also taken by tortoises, herbivorous lizards, and fruit-eating birds. In contrast to the reduced lingual structures of carnivorous reptiles and birds specializing in large prey, such herbivores tend to have massive, heavily papillated tongues, which they use extensively in food transport. Cinematographic records of tortoises and lizards eating large pieces of vegetable matter confirm that purchase on the food is sacrificed during SO. Often the lower jaw has completely disengaged the food object by the end of SO-II, leaving it held only between the palate and the tongue (Bramble, unpublished data). Thus, an evolutionary shift to inertial feeding appears linked not just to relative food size, but also to the nature of the food itself (active versus inert; surface texture).

Yet another implication for research is that more attention should be paid to the role of neuromuscular programming in trophic evolution. This difficult and neglected issue seems increasingly pertinent to any comprehensive understanding of the evolution of vertebrate feeding systems. We introduce this topic briefly to indicate some of the ways in which it may be relevant to functional and evolutionary studies of feeding behavior.

It has been proposed that the rhythmic, stereotyped motor and kinematic patterns associated with mammalian mastication are attributable to centrally located (brainstem) neural oscillators (Dellow and Lund, 1971; Thexton, 1973, 1974, 1976; Dellow, 1976). Although the existence of such pattern generators is not universally accepted, it is still reasonable to ask whether the stereotyped and rhythmic transport cycles of lower tetrapods could result from similar or even homologous neuromotor mechanisms (Bramble, 1980, unpublished data; Throckmorton, 1980; Smith, 1984). Furthermore, the many similarities shared by the transport cycles of mammals and reptiles prompt the specific question, Could the mammalian masticatory cycle have evolved from the primitive chewing cycle of reptiles with relatively little overall change in neuromotor programming? We suspect that the answer may be yes. In any event, we stress that the mammalian feeding complex should not be treated as an isolated, special case, but rather as a natural, evolved extension of reptilian feeding. The fact is that although the structural and functional problems surrounding the origin of the mammalian feeding apparatus have been extensively investigated, the question how the neuromotor control of this specialized feeding complex might (or might not) have been altered during the same evolutionary transition has never been addressed.

A demonstration that the mammalian and reptilian transport cycles are based on similar motor programming would be of considerable theoretical importance. It would suggest, for example, that the evolution of the complex mammalian masticatory system was accomplished through minimal change in the associated neuromotor mechanisms but relatively enormous alterations in the peripheral feeding structures (bones, muscles, dentition). Whether this could be a more general pattern in the functional transformation of morphological complexes is of obvious interest. Recent analyses of feeding in fishes provide mixed evidence for this possibility (Lauder, 1983a,b; Liem and Kaufman, 1984). On the other hand, basic similarities in the patterned activity of some homologous appendicular muscles between lizards and primitive mammals implies that certain aspects of mammalian locomotion may be based on neuromotor programs inherited largely intact from reptilian ancestors (Jenkins and Goslow, 1983).

The most basic implication of the model feeding cycle is that the trophic mechanisms of all modern tetrapods, no matter how unusual or specialized, represent departures from the generalized ancestral mechanism. This concept challenges morphologists to explain how such highly derived feeding behaviors as ballistic tongue projection (see below) might have originated, and how transitional stages can be characterized as regards morphological organization, mechanics, and motor programming. Is it possible, for instance, that ballistic tongue projection represents an extreme modification of the ancestral SO stage, as the current model seems to require? This would demand considerable increases in not only the extent to which the hyolingual unit is protracted during SO, but also in the relative speed with which this occurs. It is interesting, and possibly significant, that rapid tongue projection in plethodontid salamanders and chameleons is immediately preceded by a kinematic episode that bears strong resemblance to the SO-I interval of the model transport cycle (Thexton, Wake, and Wake, 1977). Additionally, the published electromyographic and kinematic data on the tongue-flipping mechanism of *Bufo* (Gans and Gorniak, 1982a,b) do not appear to contradict the hypothesis that tongue projection might be equated with a modified SO stage. The validity of this suggestion awaits further investigation, but these examples illustrate the manner in which the present model can promote the examination of new possibilities of both a practical and a theoretical nature.

Inertial Feeding

Inertial feeding is identified by the presence of a forward thrust of the head and neck at some point in each ingestive or transport cycle (Table 13-2). There are two major categories of inertial feeding, static and kinetic. Static inertial feeding (SIF) simply involves the acceleration of the head over a stationary food object. In kinetic inertial feeding (KIF) the food is accelerated rearward in a manner coordinated with the inertial thrust of the head. This form of feeding takes advantage of the dynamic (kinetic) inertia of the food object.

There are two types of static inertial feeding. Type A (SIF-A), described above, is the more common and is seen to some extent in most amniote groups. Type B (SIF-B) is unique to snakes and differs in that the food object is relatively very large and heavy. In SIF-B no effort is normally made to hold the food free of the substrate. Rather, the feeder gradually shifts its head and jaws over the stationary object, taking advantage of its static inertia and friction with the ground. Whereas SIF-A is generally characterized by rapid motor actions, those of SIF-B are typically slow.

The slow progression of the mouth over the food in SIF-B, and the fact that some portion (often most) of the oral epithelium is in contact with the food during the inertial thrust, make frictional forces of prime concern. Accordingly, either the inertia of the food object must be high, or the coefficient of friction between the mouth lining and food must be low, or both. Snakes frequently augment the inertia of the prey by pinning it to the ground with a body press or holding it with a coil as the head is forced over it. Friction is minimized by eating most prey, especially those with well-developed limbs and rough integuments, head first (Greene, 1976). The great distensibility of the snake's mouth, its thick mucous coating, a reduced tongue, and smooth, recurved teeth on the jaws and palate all act to further reduce frictional forces. Distensibility of the head is limited in primitive living snakes, but recent observations (Greene, 1983) demonstrate that some of these ophidians prey on elongate fishes, amphibians, and reptiles whose streamlined, smooth-surfaced bodies assure minimum friction during ingestion.

Several forms of kinetic inertial feeding are distinguished by the mechanism used to impart a rearward acceleration to the food during the active phase (FO and FC) of the transport cycle (Table 13-2). The most common mode is lingual-inertial feeding (LIF), in which the tongue is the transport agent. This is the feeding mechanism we attribute to the generalized tetrapod (see above). In cranioinertial feeding (CIF), the food is initially accelerated rearward by a backward jerk of the head and neck, then released to its own momentum as the head abruptly reverses direction (Gans, 1969).

Species specializing in CIF might be expected to show modification for the reduction of craniocervical mass. Kinematic profiles of lizards (Smith, 1982; Frazzetta, 1983) and birds (Zweers, 1982a) indicate that inertial feeding may require complete acceleration-deceleration and reacceleration cycles of the craniocervical complex within a few tens of milliseconds. Carnivorous lizards that regularly use this form of inertial feeding (such as *Varanus* and *Tupinambis*) have very lightly built crania. Most birds employ this feeding pattern, and their craniocervical skeleton is characteristically light. Since ancestral birds appear to have been active carnivores (Ostrom, 1974; Caple, Balda, and Willis, 1983), the feeding habits of herbivorous birds are secondary. This suggests that the inertial system, as first evolved in carnivorous birds, was sufficiently efficient (energetically) to permit its extension to the harvesting of smaller, less nutritious food packages. Although CIF is well developed in some granivores (chickens, pigeons) that eat individual food particles having high nutritive value, it appears that the best examples among birds and lizards are in species that ingest relatively large, live prey.

We note that inertial feeding is a facultative behavior for most tetrapods that use it. Pigeons, for example, utilize a type of lingual transport ("slide-and-glue," Zweers, 1982a) when eating small seeds of low mass, yet when confronted with larger, heavier seeds, immediately switch to a

Table 13-2 Comparison of selected parameters in different modes of feeding behavior in lower tetrapods.

Mode	Preferred medium	Hyoid size	Tongue size	Inertial thrust of head	Rearward acceleration of food	Speed of action	Food size relative to head size
Suction	Water	Medium-large	Small	No	Yes (water)	Fast	Small
Lingual	Air	Medium-small	Large	No	Yes (tongue)	Moderate-slow	Medium-small
Inertial-suction	Water	Large	Small	Yes	Yes (water)	Fast	Medium-large
Static inertial (type A)	Water, air	Medium-small	Small-large	Yes	No	Slow-fast	Medium-large
Static inertial (type B)	Water, air	Small	Small	Yes	No	Slow	Large
Lingual-inertial	Air	Medium-small	Medium-large	Yes	Yes (tongue)	Moderate-fast	Medium
Cranioinertial	Air	Medium-small	Medium-small	Yes	Yes (jaws)	Fast	Medium-large

form of CIF ("catch-and-throw"). Seeds of intermediate mass are transported through the mouth by combining both tactics—that is, by what is here regarded as LIF. Snakes and varanid lizards are perhaps the only tetrapods in which inertial feeding has become obligatory.

Cranial Kinesis

Cranial kinesis (intracranial mobility) has been reported in all tetrapod classes except the Mammalia. Kinetic skulls are well developed in crossopterygian fishes (Romer, 1937; Thomson, 1967), but only weakly present in a few fossil and recent amphibians (Panchen, 1964; Bolt and Wassersug, 1975; Wake and Hanken, 1982). Among amniotes significant kinesis is found in most lizards and all snakes and birds. Turtles and crocodilians have akinetic crania, as does adult *Sphenodon* (Jollie, 1960; Gorniak, Rosenberg, and Gans, 1982).

Two basic categories of kinesis are recognized (Versluys, 1912; Frazzetta, 1962; Rieppel, 1978a): metakinesis permits displacement between the dermatocranium and the underlying neurocranium, whereas mesokinesis refers to relative motion between adjacent components of the dermatocranium. Metakinesis involves components of the splanchnocranium, which develop synovial cavities at most metakinetic joints, although these appear to have been secondarily lost in snakes (Rieppel, 1978a). The joint surfaces of mesokinetic articulations are modified syndesmotic unions, and therefore lack cartilage.

Most modern lizards are said to possess amphikinetic skulls since both meta- and mesokinetic joints are present (see, for example, Frazzetta, 1962). Figure 13-6A indicates the transverse joints in the skull of a typical amphikinetic lizard. The neurocranium articulates with the dermatocranium via two pairs of lateral joints (basipterygoid and paraoccipital connections) and a single, median sliding articulation between the supraoccipital and fused parietal bones. The mesokinetic joint is a transverse hinge at the frontoparietal suture, situated above the posterodorsal edge of the orbit. Both snakes and birds also have mesokinesis, but the transverse joint lies well forward

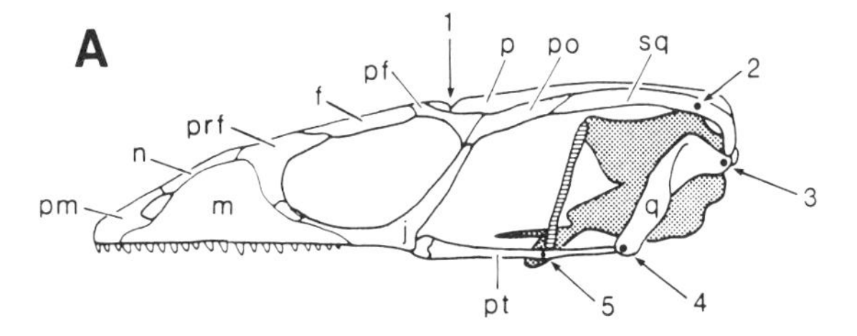

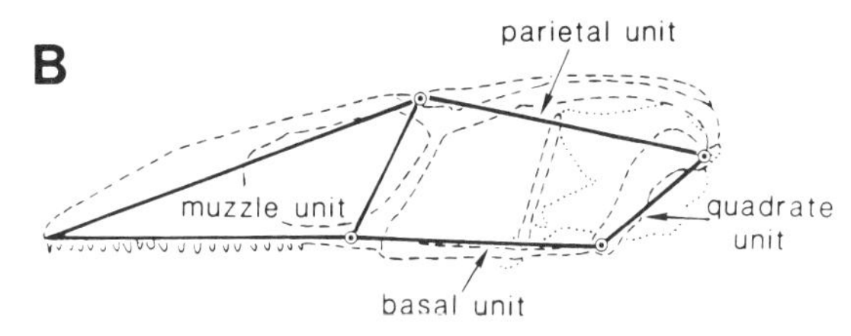

Figure 13-6 A: Lateral view of skull of the lizard *Gerrhonotus* showing cranial bones and locations of principal kinetic joints *(numbered dots)* associated with cranial kinesis. Joints: 1 = mesokinetic frontoparietal hinge; 2 = unpaired joint between neurocranium *(stippled)* and fused parietal bones; 3 = quadrate-squamosal articulation; 4 = pterygoquadrate articulation; 5 = basipterygoid articulation. Epipterygoid bone indicated by hatching. *f* = frontal; *j* = jugal; *m* = maxillary; *n* = nasal; *p* = parietal; *pf* = postfrontal; *pm* = premaxillary; *po* = postorbital; *prf* = prefrontal; *q* = quadrate; *sq* = squamosal. B: Hypothesized four-bar mechanical linkage associated with lizard skull kinesis. The four structural units are movably joined together at "pins." (Modified from Frazzetta, 1983.)

of that seen in lizards. Snakes and birds are therefore said to have prokinetic crania (Frazzetta, 1962; Bock, 1964). The prokinetic joint of birds is near the base of the upper jaw, and in snakes it is at the nasofrontal articulation.

All forms of tetrapod skull kinesis are associated with streptostyly, a condition in which the quadrate is free to move anteroposteriorly and, to a lesser extent, mediolaterally with respect to the braincase. In squamate reptiles streptostyly is correlated with the reduction and loss of the lower temporal arcade (lizards) or both the lower and upper temporal arches (snakes). Quadrate mobility in birds has been achieved at the expense of the upper arch, and the lower arch has been reduced to a slender, flexible rod linking the quadrate with the palatal complex.

A second universal feature of skull kinesis is a functional articulation between the palate and the basicranium. This is required because all forms of kinesis (including prokinesis) permit some raising and lowering of the upper jaw relative to the braincase, which in turn demands protraction and retraction of the palatal region. Thus, a basipterygoid (or palatopterygoid) joint is normally present between the medial face of the pterygoid bone and a laterally directed basipterygoid process formed chiefly or exclusively by the basisphenoid bone. A well-developed basipterygoid joint is present in crossopterygian fishes and the earliest labyrinthodont amphibians and is therefore primitive for tetrapods (Rieppel, 1978a).

That metakinesis arose early in reptilian history is indicated by its presence in such stem reptiles as *Captorhinus* (Price, 1935). The absence of metakinesis and a movable basipterygoid articulation is apparently secondary in modern chelonians, crocodilians, and *Sphenodon.* Furthermore, the ancestral Triassic turtle *Progranochelys* appears to have had a functional basipterygoid joint (Gaffney, 1983), and a lizardlike articulation between the basisphenoid and pterygoid is seen in embryonic chelonians (Rieppel, 1977). Mesokinesis, involving a straight frontoparietal hinge joint, seems to be a derived feature within the Lacertilia; such a suture is absent in the earliest known lizards (Carroll, 1977).

Lizards The cranial kinesis of lizards has been the focus of more controversy than that of snakes or birds. Modern lizards are typically amphikinetic, the degree of cranial mobility tending to vary with diet. Thus, kinesis is highly developed in carnivorous *Varanus,* but is limited to streptostyly in the herbivore *Uromastix* and is absent in adult *Iguana* (Throckmorton, 1976). Although carnivorous, chameleons are secondarily akinetic.

In 1962 Frazzetta offered the first detailed biomechanical analysis of cranial kinesis in lizards and presented what has become the most cited and debated model for the mechanism of skull movement (Figs. 13-6, 13-7). The skull is viewed as consisting of an occipital segment, which is essentially the neurocranium, and a maxillary segment encompassing most of the remainder of the

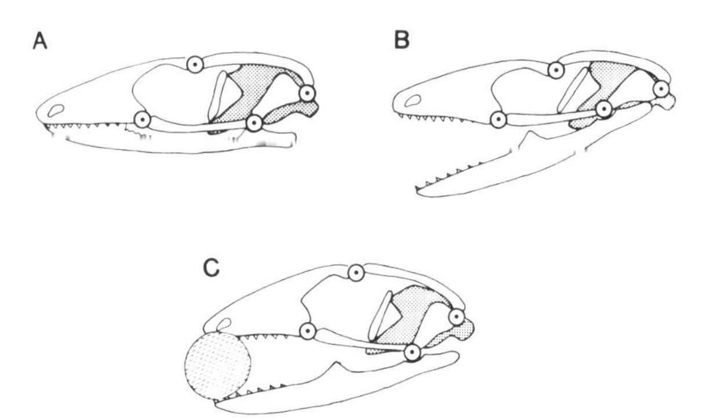

Figure 13-7 Presumed action of the mechanical model for skull kinesis in the lizard *Gerrhonotus.* A: Skull and mandible at rest (side of mandible removed to display basal unit). B: Jaws opened; quadrate and basal unit shift forward while muzzle unit is lifted. Note that frontoparietal hinge region is depressed by this action. C: Jaws biting on a food object; quadrate and basal units have shifted rearward and muzzle is flexed downward. Frontoparietal hinge is now elevated. (Modified from Frazzetta, 1983.)

cranium. Five functional units (parietal, muzzle, epipterygoid, basal, quadrate) are recognized within the maxillary segment. All but the parietal unit are paired. Of particular functional importance is the basal unit, which consists of a series of dermal bones (pterygoid, ectopterygoid, jugal, and posterior region of the palatine) forming the palatal complex.

The action of the Frazzetta model can be likened to a pinned, four-bar kinematic chain (Fig. 13-6B). The muzzle unit is protracted and elevated as the mouth opens. This involves forward rotation of the quadrate, protraction of the basal unit to which the quadrate is pinned (pterygoquadrate joint), and upward rotation of the muzzle about the mesokinetic (frontoparietal) hinge (Fig. 13-7B). Protraction of the basal unit is dependent on the contraction of two muscles, the protractor pterygoidei and the levator pterygoidei. As the jaws close, all intracranial motions are reversed (Fig. 13-7C). Frazzetta (1962) believed that the pterygoideus muscle had a key role in the retraction and depression of the muzzle unit, but Iordansky (1966, 1970) has questioned this.

Several studies provide general support for the Frazzetta model. Raising and lowering of the muzzle relative to the parietal unit has been reported during feeding in carnivorous lizards (Frazzetta, 1962, 1983; Boltt and Ewer, 1964; Rieppel, 1978b). Some of these observations are based on light cinematography, making it difficult to distinguish mesokinesis from extension and flexion at the cervicocranial joint (Iordansky, 1966; Impey, 1967). Movements generally consistent with the model have also been recorded with cineradiography (Rieppel, 1978b), but two recent reports (Throckmorton and Clarke, 1981; Smith, 1982), both utilizing cineradiography of test animals having metal implants, cast doubt on certain aspects of the Frazzetta model. They show that in several species of lizards, contrary to the requirement of the model, quadrate mobility is independent of mesokinesis. In general, however, the quadrate does swing forward as the mouth opens and backward as it closes (Fig. 13-7B,C). In some lizards (such as *Amphibolurus*) the parietal unit appears to be elevated rather than depressed, as the model predicts (Fig. 13-7B), as the mouth opens (Throckmorton and Clarke, 1981). Perhaps most surprising is the fact that in vivo strain gauge recordings have revealed little movement at the frontoparietal hinge during feeding in *Varanus.* Moreover, the direction and timing of the movements observed directly contradict the model of Frazzetta (Smith and Hylander, 1982, unpublished data).

Snakes Cranial kinesis is most extremely expressed in snakes (Gans, 1961a; Rieppel, 1980). The tooth-bearing bones of the upper and lower jaws are highly mobile and are connected to the braincase only by ligamentous and muscular unions. The braincase itself has become exceptionally solid owing to extensive secondary downgrowths of the frontal and parietal bones. Streptostyly is pronounced. The quadrate no longer contacts the otic region of the braincase, but instead articulates with the distal end of an elongated supratemporal bone. This gives the quadrate extraordinary mobility, which is further increased in many advanced snakes by the formation of a sliding joint between the supratemporal and the braincase. The lower jaws of most snakes are slender and joined at their tips by distensible connective tissues.

The most significant feature of ophidian kinesis is its association with unilateral mobility of the jaws (Gans, 1961a). This unique action is now well documented through cinematographic (Albright and Nelson, 1959a,b; Boltt and Ewer, 1964; Frazzetta, 1966; Kardong, 1977) and combined cinematographic and electromyographic recordings (Cundall and Gans, 1979; Cundall, 1983). The mechanism involves what is often described as the "pterygoid walk," in which the snake slowly works the prey into the mouth by alternately advancing and retracting the jaws. A key feature is that as the set of jaws on one side of the head is being advanced over the surface of the prey, the teeth of the other set remain fixed to the food object. The marginal tooth-bearing elements of the upper jaws normally have little to do with food transport. Nearly all the relative movement between the snake and the prey depends on carefully coordinated protractive and retractive motions of the toothed bones of the more medial palatopterygoid bars (Cundall, 1983).

The following description of feeding mechanics in snakes is based chiefly on the experimental work of Cundall and Gans (1979) and Cundall (1983). As in lizards, protraction of the palatomaxillary unit results largely from contraction of the protractors of the dorsal constrictor musculature (protractor pterygoidei, levator pterygoidei, protractor quadrati). However, in snakes opening of the mouth is assisted by long-axis rotation of the braincase presumably produced by contraction of portions of the cervicocranial musculature. As the mouth is opened the maxillary bones not only are raised and drawn forward, but also rotated laterally so that their teeth are freed of the prey. In the latter part of the opening phase, and early in the closing phase, there is considerable overlap in the activity of the depressor mandibulae and several mandibular adductors. Such antagonistic interaction is presumably important to the precise control of movements of the lower jaw and its suspensorium. The lower jaws themselves have little role in food transport, but serve mainly to control the position of the prey in the mouth and to press it upward against the teeth of the palatopterygoid bars. Closing of the mouth results in depression and some slight retraction of the toothed palatal bones. The maxillary also moves downward and rotates inward. The closing phase of the jaw cycle is accompanied by strong contraction of most adductor muscles. Among these, the pterygoideus seems to function chiefly as a protractor of the mandible rather than as a retractor of the palatopterygoid bar, as had been previously hypothesized (Boltt and Ewer, 1964; Liem et al., 1971).

Most earlier studies had assumed that the unilateral feeding cycles of snakes served to pull the food into and through the mouth (for example, Dullemeijer, 1956; Gans, 1961a, 1974). This now appears not to be the normal situation. After the palatal bones have been advanced and lowered into the prey, the braincase is pulled forward by contraction of the mandibular adductors and the retractor pterygoidei. Most of the advance of the braincase over the palatal elements of one side takes place during the opening or advance of the jaws of the opposite side (Cundall and Gans, 1979). This mode of transport may be especially advantageous when attempting to swallow prey whose mass equals or exceeds the cranial mass of the snake (Cundall, 1983). In certain instances the animal may actively push the head over the stationary prey rather than pull it forward with the movable palatal bones.

In very primitive snakes with limited unilateral excursions of the jaws, food transport seems the role of the maxillary bone as well as of the more medial palatopterygoid bar. The freeing of the maxillary from the active participation in food transport is a radical innovation of higher (colubroid) snakes and one of major evolutionary importance. It permitted the maxillary to be specialized in diverse ways for prey capture (Cundall and Greene, 1982; Cundall, 1983). The most extreme example of the division of labor between the palatal and maxillary complexes is the development of erectile maxillary fangs in several lineages of snakes. The mechanical basis of fang erection has been well studied in vipers (Kathariner, 1900; Boltt and Ewer, 1964; Kardong, 1974). The maxillary bone of such front-fanged snakes is highly modified to rotate extensively on neighboring cranial elements. The mechanism used to induce forward rotation of the maxillary, with consequent erection of the fang, is an elaboration of that used to protract the palatomaxillary arch during normal unilateral feeding. Thus, erection of the fang depends on its strategic linkage to the palatopterygoid bar and not on streptostylic movements of the quadrate, as was assumed by earlier workers (Klauber, 1939; Romer, 1956).

An overriding feature of snake feeding is its extraordinary linkage to limbless locomotion and fossoriality (Gans, 1961a, 1974, 1983; Savitzky, 1980). Small head size and the tendency to take large, active prey appears to have favored the early evolution of constricting behavior, which is widespread among primitive snakes and has probably been lost by many advanced groups (Greene and Burghardt, 1978). Constriction provides a mechanism for subduing large and potentially dangerous prey prior to swallowing. However, constrictors are often typified by relatively thick bodies and slow locomotor progression. The musculoskeletal architecture favoring powerful constriction may be incompatible with that required for rapid locomotion (Ruben, 1977), so it is possible that locomotor constraint provided much of the impetus for the development of venom delivery systems among several groups of advanced snakes as an alternative means of subduing large prey (Savitzky, 1980).

Birds The skull of all modern birds is kinetic, and in some groups (parrots, boobies) mobility is pronounced. The structural and mechanical bases of avian cranial kinesis have been extensively investigated (see reviews by Bock, 1964, and Zusi, 1967). Slow-motion cinematography has documented some gross kinematic patterns associated with feeding and drinking in a few birds (see, for example, Homberger, 1980a,b). The only extensive experimental analysis of the avian feeding mechanism comes from the exceptional work of G. A. Zweers and his associates on the mallard (Zweers, 1974; Zweers, Gerritsen, and Kranenburg-Voogd, 1977). Their studies include the anatomy, kinematics, electromyography, biomechanics, sensory and motor physiology, and behavioral attributes of feeding. This work offers the most comprehensive form-function profile of the feeding mechanism of any vertebrate, excluding man. Further information has more recently been obtained for the pigeon (Zweers, 1981, 1982a–d).

The avian skull consists of four major functional and kinematic units (Bock, 1964): the braincase, upper jaw, the bony palate plus jugal bars and attached quadrates, and the mandible (Fig. 13-8A). Streptostyly is strongly developed and its attendant structural modifications are more clearly defined than in lizards. Thus, the robust quadrate joins the squamosal region of the braincase in one or more well-formed ball-and-socket joints. The design of the quadrate-squamosal joint permits considerable anteroposterior and mediolateral displacement of the quadrate (hence jaw articulation) and also provides exceptional mechanical stability under conditions of compressive loading of the jaw joint (Bramble, 1978). Highly modified pterygoid bones function as movable struts to stabilize and guide displacements of the quadrates. The pterygoquadrate articulation of modern birds often is unusual in being a distinct ball-and-socket union. The palatal complex slides back and forth on the underside of the solid braincase; its movements are guided by a median keel, which it straddles. A basipterygoid articulation is generally absent. A few birds (anseriforms, galliforms) do possess a basipterygoid articulation (Bock, 1964), but this may not be homologous with that of other tetrapods (McDowell, 1978).

The prokinetic joint is normally developed at or near the nasofrontal junction. Most often this hinge amounts to little more than a transverse band of thin or flexible bone. However, a distinct, encapsulated transverse joint is present in some of those birds in which cranial kinesis is most pronounced (for instance, parrots). In a few birds (charadriiforms, ratites) described as having rhyncokinetic crania (Hofer, 1949), movements of the upper jaw depend on the bending of a narrow bar of bone that separates the two external nostrils. Rhyncokinesis differs from prokinesis in that it allows only the anterior region of the upper beak to be moved (Bock, 1964).

The kinetic mechanism of a typical bird skull is illustrated in Figure 13-8B. Anterior rotation of the quadrates imparts a forward push (via the jugal bars and the pterygopalatine linkages) to the base of the upper jaw. This causes the bill to rotate upward about the prokinetic joint. Displacement of the upper jaw is accompanied by protraction of the palatal complex along the underside of the braincase. Closing of the jaws is the reverse of these movements. Elevation of the upper jaw involves primarily the contraction of the protractor pterygoidei and the depressor mandibulae. Clos-

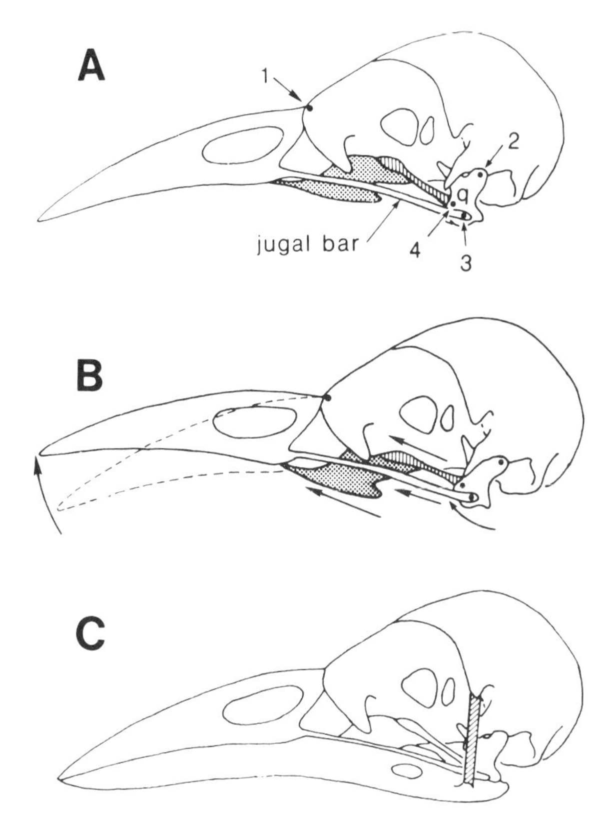

Figure 13-8 Lateral view of a crow skull illustrating the basic elements of avian cranial kinesis. A: Location of the principal kinetic joints *(numbered dots)* and associated structural elements. Joints: 1 = mesokinetic (prokinetic) nasofrontal hinge; 2 = quadrate-squamosal articulation; 3 = quadrate–jugal bar articulation; 4 = pterygoquadrate articulation. Pterygoid *(hatched)* and palatine *(stippled)* have a sliding articulation with the underside of the solid braincase. B: Mechanism of elevation of the beak. Quadrate rotates forward, resulting in forward translation of the pterygoid and palatal units as well as of the jugal bar. This thrust is resolved into upward rotation of the beak about the nasofrontal joint. For simplicity, certain small bones attached to the front of the palatines have been omitted. C: Skull and mandible articulated and at rest to show the location of the postorbital ligament *(hatched)* linking the cranium and lower jaw.

ing of the mouth is effected by the external adductor masses (including the pseudotemporalis complex) and the pterygoideus muscles. The attachments of the pterygoideus cause it to simultaneously elevate the mandible and to depress the upper jaw (Bock, 1964).

In most modern birds cranial kinesis is coupled in that movements of the two jaws are mechanically and kinematically linked (Bock, 1964; Zusi, 1967). Coupling of the jaws results either from strong ligamentous connections (postorbital and/or lacrimomandibular ligaments) between the braincase and mandible (Fig. 13-8C) and/or a special interlocking of the quadrate and articular bones at the jaw joint. Owing to these linkages, depression of the mandible must be accompanied by forward displacement of the quadrate and, hence, concurrent elevation of the beak. Among other things, coupled kinesis permits contraction of the depressor mandibulae alone to simultaneously lower the mandible and raise the upper jaw. Although the postorbital ligaments are an important mechanical component of the coupling mechanism, coordinated jaw movements persist after their surgical removal. Moreover, it is probable that the intact bird can uncouple its upper and lower jaws through voluntary adjustments in the position of the mandible (Zusi, 1967).

Streptostyly Streptostyly allows the mandible to be protracted and retracted relative to the upper jaw. Robinson (1966) believed it originated to aid the transport of food through the mouth by propalinal motions of the lower jaw, but there is no evidence that any tetrapod utilizes streptostyly in this manner. Moreover, lingual transport would be a much more direct and effective means of food manipulation, and extensive use of the tongue for food transport occurs in several species of lizards having well-developed streptostylic jaw suspension (for example, *Gerrhonotus,* Frazzetta, 1962; *Uromastix,* Throckmorton, 1976).

Theoretically, streptostyly might also function to increase gape, but in mesokinetic species it would do so only if the quadrate were retracted as the mouth opens. However, in those lizards for which reliable data exist, the quadrate usually swings anteriorly, thus effecting an actual reduction in gape (Throckmorton, 1976; Rieppel, 1978b; Smith, 1978, 1980, 1982; Throckmorton and Clarke, 1981). Conversely, streptostyly could contribute to the ingestion of relatively large food items by permitting the jaw joints to spread laterally, thereby enlarging the transverse diameter of the oropharyngeal passage (Robinson, 1966). This potential function has not been adequately explored, but there is some evidence that the quadrates in *Varanus* spread during the swallowing of bulky prey (Krebs, 1979). The value of streptostyly as a specialization for swallowing large food objects is more definitely established in snakes. Species that eat exceptionally large prey have especially elongated quadrates and supratemporal bones that further increase the potential for abduction of the jaw articulations and increased gape (Gans, 1961a, 1974). The most extreme modifications for enhanced gape are found in egg-eating snakes (Gans, 1952; Gans and Oshima, 1952; Gans, 1974; Rosenberg and Gans, 1976).

Quadrate mobility has also been implicated in jaw and muscle mechanics. Anteroposterior displacements of the quadrate can alter the length, angle of insertion, and moment arms of the mandibular adductor muscles. Rieppel (1978b) has suggested that in lizards a posterior rotation of the quadrate as the jaws close could improve the mechanical advantage of the external adductors at relatively low gape angles. Smith (1980) has pointed out that streptostyly might allow the pterygoideus musculature to contribute importantly to bite force if its operational fulcrum were located at the movable quadrate-squamosal joint rather than the jaw articulation (but see also Smith, 1982). Alternatively, a movable quadrate could provide the means of adjusting the relative contribution of the internal and external adductor muscles to bite force (Smith, 1980). Another possible role for streptostyly in lizards is that it permits some of the adductive force acting on the mandible to be relayed to the upper jaw (Iordansky, 1970). This is plausible if, by pressing the quadrate rearward, posteriorly directed translational forces from the external adductors could be transferred to the muzzle unit through the pterygoideus musculature. This model is attractive in that it offers a strategy for utilizing otherwise wasted muscular force. Such a system might be of special value to small tetrapods that feed on hard or armored invertebrate prey. A similar mechanical arrangement could operate in birds (Bas, 1957).

Among modern amphibians functional streptostyly may be restricted to caecilians. In embryonic gymnophiones a movable quadrate may aid protraction of the lower jaw, which forces the armature of the foetal dentition against the oviducal lining to stimulate secretion of nutritive fluid (Wake and Hanken, 1982).

Mesokinesis in Lizards Older hypotheses (for instance, Bradley, 1903) that mesokinesis permitted food to be clasped between the pterygoids appear to be without merit. Similarly, in lizards there is no support for the notion that mesokinesis acts in shock absorption, nor can it contribute to a stationary visual axis during feeding (Frazzetta, 1962, 1983). To the extent that mesokinesis is associated with streptostyly, it may influence both gape and jaw mechanics. The functional significance of mesokinesis in lizards, however, has been discussed mainly in terms of prehension. Frazzetta (1962) hypothesized that mesokinesis might help to orient the jaws toward the prey and subsequently to seize it. The upper and lower jaws normally contact the prey at about the same instant, presumably because the independently mobile upper jaw closes on the food at the same time the mandible does. Mesokinesis may compensate for inaccuracy in the positioning of the head during prey capture, thus eliminating the need for abrupt changes in the trajectory of the head and forebody as they deliver the jaws toward a prey item. This could be particularly important for predators that take much of their food from the ground (Frazzetta, 1983).

Mesokinesis/Prokinesis in Birds Among the many functional explanations advanced for avian skull kinesis, two seem very probable and several others much less so, but none has been critically tested. The first of those biological roles that may prove real is that of enhancing the manipulatory abilities of the jaws during feeding and other important activities (preening; nest construction). A versatile and precise grip would seem especially valuable to edentulous animals that feed upon items that are often small, hard, and smooth-surfaced (insects, seeds). The special handling properties of the bird beak are attributed to coupled jaw movements and to the fact that prokinesis permits the upper jaw to pivot about a point widely separated from that of the mandible (Zusi, 1967).

The second likely biological role for avian skull kinesis is shock absorption. Shock absorption is highly probable in birds such as woodpeckers (Beecher, 1953, 1962; Bock, 1964, 1966; Spring, 1965) and many granivorous birds (Bock, 1964), but whether this is a significant function in other types of avian feeding is uncertain.

Another potential function of avian prokinesis is that of allowing the bird to maintain a constant visual axis while the mouth is opened (Bock, 1964; Zusi, 1967). This possibility exists because the hinge lies forward of the braincase. In this regard avian mesokinesis is potentially superior to lizard mesokinesis. The ability to maintain a steady line of sight while feeding might well contribute to greater speed and accuracy when seizing small objects with the tips of the jaws. However, the fact that closure of the eyes appears to be coupled to the thrust of the head and the raising of the upper beak in some granivorous species (for example, pigeons, Zeigler et al., 1980; Zweers, 1982a) tends to undermine the visual axis hypothesis. Nonetheless, the ability of birds to raise the upper jaw independently of the cranium and the mandible appears to be exploited in a number of specialized feeding behaviors. Included are the probing and spreading of flowers and soft substrates to secure nectar and invertebrate food (Zusi, 1959) and the filter feeding of flamingoes (Jenkin, 1957). Most of these specialized trophic mechanisms are accompanied by increased size and leverage of the depressor mandibulae muscle (Zusi, 1967).

Other proposed biological roles for the kinetic skull of birds, such as increasing the gape or the speed of jaw movements or providing a passive means of maintaining the mandible in a closed position, have less support in fact or are actually controverted by experimental data (Zusi, 1967).

Projectile Mechanisms in Feeding

To secure food a predator can either search actively or wait for the prey to approach it. The latter, more passive strategy is termed "sit and wait" or "ambush feeding" (Pianka, 1966) and is often found in animals with limited locomotor capacity. Whichever method is used, in the final approach and contact the predator may propel either its entire body or just a portion of it at the food object. In the first instance, little more than a modest lunge may take place, as in salamanders, or a high-speed sprint may occur as is seen in certain lizards. Under the second option only a fraction of the body is set in significant motion. In practice this is usually the head and neck—less frequently just the hyolingual complex.

Most tetrapod projectile feeding mechanisms can be sorted into one of three groups: whole body projection (WBP); craniocervical projection (CCP); or hyolingual projection (HLP). Generalized predators frequently combine all three tactics in a single predatory act. The more specialized behaviors, CCP and HLP, are usually not accompanied by significant involvement of the rest of the body. Several factors may contribute to the development of these more restricted forms. First, the movement of only a portion of the body allows the smaller mass to be accelerated more rapidly and with less force (energy) than were the entire body projected. Second, because most of the body can remain stationary, a predator may more easily maintain crypsis. Tongue projection may also compensate for locomotor restrictions, especially in arboreal settings where rapid whole body approaches to the prey are made difficult by an unstable or discontinuous substrate. The following discussion is restricted to a consideration of the more derived CCP and HLP systems.

Craniocervical Projection Tetrapods most exemplary of the group using CCP are turtles, birds, and snakes. Snakes are different, however, in that portions of the anterior trunk region are projected as well as the head and "neck." All three groups are alike in that locomotor potential is diminished as compared with generalized tetrapods. In turtles the formation of the shell has rendered the entire trunk immobile. Birds have rigidified the trunk in connection with aerodynamic efficiency and stabilization. As in turtles, elongation and increased mobility of the cervical region in birds must be viewed largely as a compensatory response to trunk immobility. Limblessness in snakes, and associated changes in body shape, have led to increased flexibility of the trunk but also to restrictions on speed and maneuverability. It seems clear, then, that the more extreme practitioners of the CCP strategy have been forced to this pattern by locomotor constraint.

Carefully regulated, high-speed movements of the craniocervical unit typify the feeding actions of many birds. Such movements are a vital element in the well-developed inertial feeding system utilized by nearly all birds (see above). Possibly the most extreme expression of CCP is found in certain wading species (herons, egrets, storks) that use high-speed and extremely accurate strikes to spear fish and anurans. However, the ability of many herbivorous birds (including domestic fowl and pigeons) to collect seeds through a series of rapid thrusts of the head and neck is almost equally impressive (see, for example, Zweers, 1982a). Although excellent descriptive studies exist (for instance, Boas, 1929), there have so far been no detailed biomechanical studies of these projectile feeding mechanisms.

Likewise, no detailed kinematic or biomechanical analyses of the strike of snakes are available. Frazzetta (1966) described some of the strike actions of pythons from slow-motion films. These suggest not only great speed but also a remarkable ability to execute in-flight adjustments in response to escape maneuvers by the prey (rodents). Such agility is perhaps to be expected in those snakes that depend on snaring quick and agile prey with their teeth, especially when the prey have an external covering (hair, feathers) that provides little purchase. One might reasonably expect less ability to adjust the strike in species that utilize stabbing and venom injection, rather than grasping, to capture their prey.

All CCP mechanisms involve the rapid, controlled projection of a chain of vertebrae. In the cocked position the cervical column is nearly always bent into an S-shaped curve. From this

Figure 13-9 Stop action photo illustrating tongue projection in the salamander *Hydromantes italicus.* Animal is in early phase of tongue retraction. Note the sharply elevated cranium and the orientation of eyes relative to the target. Folded hypobranchial skeleton forms the slender stalk of the extended hyolingual apparatus. Photo courtesy of Gerhard Roth.

position the system is accelerated to terminal velocity as it reaches its target. The neck musculature is presumably programmed so as to effectively achieve a smooth, linear acceleration by summing nonlinear (angular) accelerations about a whole series of movable joints. The morphology, mechanics, and neuromuscular control of such projectile mechanisms are much in need of study.

Hyolingual Projection Specialized high-speed HLP mechanisms occur in several families of salamanders, most anurans, and chameleons (Fig. 13-9). Except perhaps for the primitive frog *Rhinophrynus* (Trueb and Gans, 1981, 1983), the anuran projectile system does not appear to depend on significant displacement of the hyoid skeleton. This is contrary to earlier opinions (Emerson, 1976, 1977). In chameleons hyoid movements are important to the aiming of the projectile tongue, whereas in salamanders the entire hyolingual complex is incorporated into the projection mechanism. Relatively slow but nonetheless specialized HLP mechanisms have also evolved in birds, especially in cases where the bird probes for small invertebrate food items. In woodpeckers the packing of the elongated hyoid skeleton is associated with remarkable structural adjustments, including the coiling of the epibranchial horns into the nareal passage (Burt, 1930).

Considerable structural variation in the organization of the hyoid and lingual musculature among modern frogs implies that there has been significant parallel evolution of tongue-protrusion mechanisms, which typify most species (Regal and Gans, 1976; Horton, 1982). The mechanism of tongue protrusion in frogs has long been the subject of considerable conjecture (for review, see Gans and Regal, 1976). Only recently have experimental techniques (high-speed cinerecording; electromyography) been applied to this problem (Gans and Gorniak, 1982a,b).

In the projection system of *Bufo* a fairly massive tongue becomes a ballistic projectile. As little as 150 msec may elapse between the start of the tongue projection (as determined by electromyography) and impact on the prey. Combined cine and electromyographic recordings seem to confirm that coordinated muscular contractions act, first, to stiffen the normally flaccid tongue; second, to rotate and catapult this rodlike structure over the mandibular symphysis to impact upon the target; and third, to retrieve the tongue and attached food item to the mouth. Stiffening of the tongue results mainly from contraction of extrinsic tongue musculature (genioglossus medialis). Protraction (acceleration) of the tongue mass is provided by the muscles (geniohyoideus lateralis and medialis) linking the hyolingual unit to the symphyseal region of the lower jaw. Other muscles located just behind the symphysis pivot and catapult the tongue rod as it rotates up and over the front of the mandible. Specifically, contraction of the bulbous submentalis muscle (together with the genioglossus basalis) yields a "rising fulcrum" over which the projectile unit rotates (Gans, 1961b, 1962). The upward force on the base of the tongue rod generated by these muscles, together with a downward pull on the front of the rod produced by depression of the tips of the mandibles, creates a force couple that further accelerates the tongue. Beyond the symphysis the lingual mass is carried passively by its momentum on toward the target. Because its anterior end is fixed to the jaw and the posterior end is free, the tongue automatically flips over as it leaves the mouth cavity. Accordingly, the glandular dorsal surface of the tongue is rotated ventrally where it strikes and partly encircles the prey. Bonding is primarily or exclusively adhesive. Following impact, the tongue is rapidly retracted into the mouth by the posteriorly directed pull of the hyoglossus and sternohyoideus muscles. Once the tongue and prey are well within the oral cavity, the mouth is closed, thus ending the prey capture cycle.

Unfortunately, the feeding behavior of primitive frogs has not been studied directly. Morphological data, together with a few behavioral observations, suggest that the tongue is certainly used in feeding but that it can be protruded only slightly beyond the anterior margin of the jaws (Regal and Gans, 1976). There are insufficient data to detail the transformation of such primitive feeding systems into the derived projectile mechanisms of *Bufo, Rana,* and many other living anurans.

Salamanders in five families (Hynobiidae, Dicamptodontidae, Ambystomatidae, Salamandridae, Plethodontidae) metamorphose fully and achieve a terrestrial stage in which tongue protrusion is used to capture prey. With only a few exceptions, salamanders have weak jaws and small marginal teeth. The jaws function primarily in holding prey brought to them by the tongue. All five families have palatal teeth, which become elaborated and extend far posteriorly to near the end of the parasphenoid bone in salamandrids and plethodontids. The ability to project the tongue is limited in the more primitive families but becomes great in the salamandrids and plethodontids. The tongue returns prey directly to the mouth, and small items tend to be carried far into the buccopharyngeal cavity. As the tongue moves forward to its rest position following jaw closure, the prey is taken off the tongue by the palatal dentition (Regal, 1966). In plethodontids these teeth form large, paired patches, each of which may bear as many as 100 tiny teeth.

The universal use of at least modest tongue protrusion in terrestrial salamanders argues in favor of the view that HLP is primitive for urodeles. The suction feeding modes that are used by permanently larval forms (*Necturus*) and by highly specialized, tongueless, paedomorphic species (*Cryptobranchus, Siren*), as well as by the bizarrely modified *Amphiuma,* are probably secondary departures that tell us little about the early amphibian feeding mechanism.

Tongue-protruding salamanders that have an aquatic larval stage progress smoothly from suction feeding to a tongue-protrusion phase (Özeti and Wake, 1969). Metamorphosis is gradual, and until late in the metamorphic period the larva continues to feed as before. Newly metamorphosed individuals frequently remain in the water for some time, where they continue suction feeding. However, a major change has occurred in the hypobranchial apparatus. The gills have been lost, the gill slits closed, and the distal parts of the cartilaginous or bony branchial skeleton have been resorbed. Moreover, new features are added to the remaining hypobranchial apparatus, thereby producing a bipartite adult skeleton. This skeleton consists of a pair of bladelike structures lying in the floor of the mouth. They are attached to the quadrate-squamosal region of the skull by a long rodlike extension. The rest of the hypobranchial skeleton (the articulated skeleton) includes a central basibranchial, which bears one or two pairs of anteriorly located radii. The basibranchial supports the tongue pad (partly by means of the laterally extending radii) on its anterior end. Two pair of certatobranchials (or hypobranchials) and a pair of epibranchials (or certatobranchials) are located posteriorly (Fig. 13-10). A small, disconnected urohyal may be present as an osseous inscription into which the geniohyoideous medialis and slips of the rectus cervicis superficialis insert.

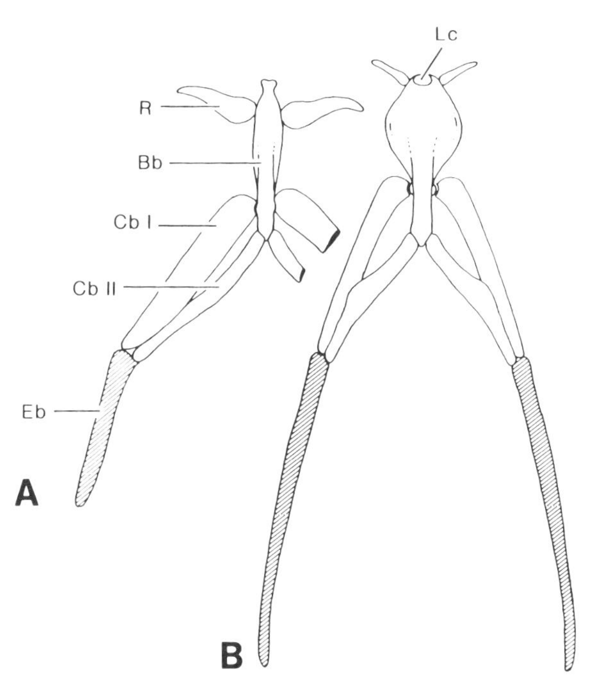

Figure 13-10 Dorsal view of the hyoid skeleton of (A) a plethodontid salamander *(Plethodon jordani)* with generalized feeding habits and (B) a plethodontid *(Eurycea longicauda)* moderately specialized for tongue projection. Notice the relatively reduced radial *(R)* and first ceratobranchial (*Cb* I) elements and the elongated epibranchials *(Eb; hatched)* of the tongue projectionist. *Bb* = basibranchial; ceratohyals and urohyals not shown. Drawings are to same scale. (Modified from Lombard and Wake, 1977.)

A large muscle present in the larva becomes elaborated in the adult as the main protractor of the tongue (Fig. 13-11). The subarcualis rectus I (SAR-I) extends from the ceratohyal posteriorly to the epibranchial, where it wraps around that element to varying degrees. In the highly specialized plethodontid salamanders this muscle becomes pinnate and forms a large bulb enclosing the tapered epibranchial. When this muscle contracts, the articulated skeleton, with the attached tongue pad, is protracted (Lombard and Wake, 1976, 1977). However, if the retractor muscles (primarily the rectus cervicis group) are simultaneously contracted in such a way that both protractors and retractors have mechanical advantage, the tongue is not protracted. Rather, the medial portions of the tongue skeleton are drawn posteriorly and the lateral, anteriorly. As a result, the hyoid skeleton buckles and the buccopharyngeal chamber expands dramatically. This is the mechanism used by metamorphosed salamanders to continue suction feeding when in water.

In the more derived members of the families Salamandridae and Plethodontidae the tongue has become so specialized that suction feeding is no longer possible. Most species of the family Plethodontidae (which contains about 65% of all living urodels) are completely terrestrial and have abandoned the aquatic larval stage. It is these species that have the most specialized feeding systems (Wake, 1982).

In order to use tongue protrusion on land, movements of the tongue must be correlated with contractions of the ventral constrictor muscles of the throat (intermandibularis and interhyoideous series). The constrictors contract to establish a relatively firm platform, which supports the ceratohyals and ensures that contraction of SAR-I leads to tongue protraction rather than buccopharyngeal expansion. In addition, longitudinal hypobranchial (geniohyoideus lateralis) or branchiomeric (subhyoideus) muscles orient the projection platform in the floor of the mouth.

In generalized salamanders tongue protraction is modest and involves little more than rolling the large, sticky tongue pad over the front of the mandible to contact prey (Larson and Guthrie, 1976). Many species have a moderate to well developed posterior tongue flap that wraps over the prey item on contact. The anterior margin of the tongue is bound to the mandibular symphysis by stout genioglossus muscles, which probably function to return the tongue to its resting position. The primary retractors are the rectus cervicis series, which attaches both to the first ceratobranchials and into the tongue pad. The pad itself contains a complex intrinsic musculature (hyoglossus, basiradialis, radioglossus, interradialis) that acts to shape the lingual surface (Özeti and Wake, 1969; Lombard and Wake, 1976, 1977).

Both salamandrids and plethodontids have diversified in their tongue-projection mechanisms. In salamandrids unique specializations have evolved in *Salamandrina* and *Chioglossa* for rotating the tongue pad itself and projecting it a considerable distance from the mouth. The first pair

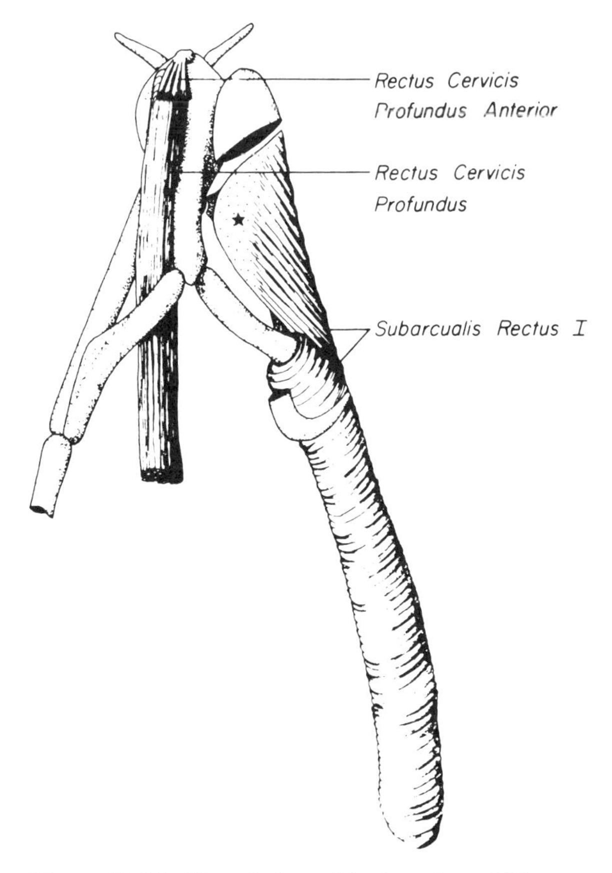

Figure 13-11 Dorsal view of the hypobranchial skeleton of the salamander *Eurycea bislineata* to show the attachments of the principal muscles of projection (subarcualis rectus I) and retraction (rectus cervicis profundus; rectus cervicis profundus anterior) of the projectile tongue. Part of the right ceratobranchial has been removed to show origin *(star)* of the subarcualis rectus I muscle on its ventral surface. This muscle passes posteriorly from its origin to envelop the posterior portions of the ceratobranchials and all of the epibranchials. The rectus cervicis profundus passes between the ceratobranchials to insert on a common myocomma with the rectus cervicis profundus anterior. The latter muscle inserts on the small lingual cartilage. (From Lombard and Wake, 1976.)

of radii are lost and the second are greatly elongated and posterolaterally oriented. Enlarged basiradialis muscles rotate the radii about the end of the ossified basibranchial. The elongated rectus cervicis muscles are folded into the tongue pad, which provides much additional length to accommodate the extensive protraction. In *Chioglossa* there is some protraction of the articulated skeleton as well (Özeti and Wake, 1969).

The most specialized of the urodele HLP systems occurs in lungless salamanders, the Plethodontidae. In these species the hypobranchial apparatus is folded and protracted from the mouth. This is made possible by the fact that the hypobranchial complex no longer functions in its generalized role as an inflator of the lungs, and thus considerable specialization has been allowed (Wake, 1982). The most extensively projectile tongues (*Hydromantes;* Fig. 13-9) and the fastest tongues (*Bolitoglossa*) occur in plethodontids without aquatic larvae. There is no new organization in these species, but their freedom from the constraints imposed by lung and gill ventilation, as well as by suction feeding, have permitted the continuation of established lines of specialization to an extreme. For example, the epibranchial and associated SAR-I muscle become exceptionally long in *Hydromantes* and some other tropical genera, in which these elements extend (under the dorsolateral skin) over the shoulder and back to the middle of the trunk. Furthermore, the main retractor of the tongue in *Hydromantes,* the rectus cervicis profundus, is thrown into loops when at rest so that it has no mechanical advantage when it first starts to contract. All the more derived tongue projectionists have lost the genioglossus muscle, thus eliminating still another barrier to extreme lingual protraction, partly through weight reduction. The larger projectile tongues of some plethodontids have sufficient mass to make them relatively slow. In *Bolitoglossa occidentalis,* however, the tongue pad is especially small and the articulated apparatus is lightly built. Tongue projection in this salamander is extremely fast, requiring only 5 to 10 msec (Thexton, Wake, and Wake, 1977).

Some electromyographic studies of *Bolitoglossa* have been conducted (Thexton, Wake, and Wake, 1977). Apparently the mandibular depressors, ventral constrictors, and the lingual protractors and retractors all contract simultaneously. The resulting action is directed and predictable because of the geometry of the peripheral structures. Thus, the length-tension curves of the protractor and retractor muscles favor instant mechanical advantage for the former. The tongue continues outward until the length-tension curves of the two sets of muscles cross, at which time the mechanical advantage shifts in favor of the retractors. This ballistic action is especially evident at great projection distance, wherein integrated electromyograms of the antagonistic muscles are the same as for short distances, but impact force on the target falls off as a function of projection distance. These results imply that tongue action is a prepatterned phenomenon unmodulated by peripheral feedback mechanisms. The mouth is opened very widely during tongue protraction such that the animal loses eye contact with the prey. But this has no effect, for *Bolitoglossa* is extraordinarily accurate and rarely misses its target.

The Chameleonidae are the only reptiles with projectile tongues. In some species the tongue can be extended more than a body-length to capture invertebrate prey (Gans, 1967). Although extremely accurate, the HLP mechanism of these lizards is considerably slower than that of many salamanders. Cinematographic data indicate that the elapsed time between the "firing" of the tongue and its impact on the target is approximately 40 msec (Zoond, 1933; Gans, 1967). Retraction of the tongue back into the mouth is much slower still (see Harkness, 1977).

The projection sequence begins with what appears to be a distinct SO stage. The hyolingual unit is protracted and elevated until the tongue protrudes noticeably beyond the margins of the partly opened jaws. Hyolingual movement is then arrested as the chameleon takes final aim by slowly orienting its head toward the prey.

Several structural specializations are associated with the HLP mechanism of chameleons (Gnanamuthu, 1930; Gans, 1967). The hyoid skeleton is unusual in having an exceptionally long, tapered lingual process (Fig. 13-12). When the system is at rest, this rod supports (front to rear) the tongue, the accelerator muscle, and the retractor muscle (Fig. 13-13A). The tongue is of modest size, slightly papillate, and richly endowed with mucous glands. The accelerator muscle encircles the anterior half of the lingual rod with a complexly interwoven array of fiber bundles. The fibers attach centrally to a hollow tendon. The tendon itself fits like a cap over the hyoid projection, thereby isolating the lingual rod from direct contact with the accelerator muscle. The posterior half of the rod is surrounded by the elongated retractor muscle, the glossohyal. To conserve on space, the retractor is folded on itself or pleated, much like the retractor musculature of the plethodontid tongue projection system (Lombard and Wake, 1977), although to a more extreme degree.

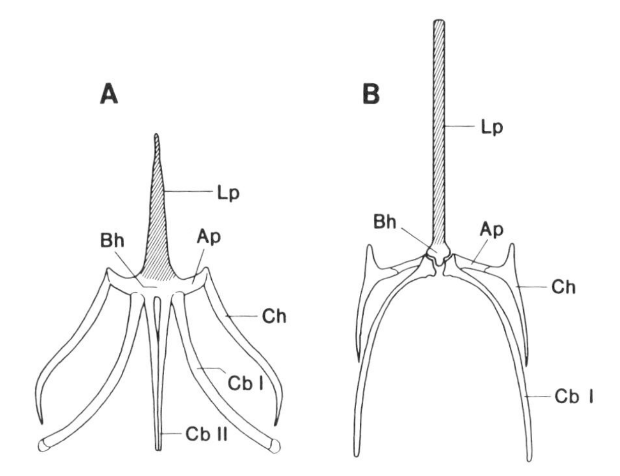

Figure 13-12 Ventral views of the hyoid skeletons of (A) a generalized lizard *(Agama agama)* and (B) a chameleon *(Chameleo namagyensis)*. The chameleon is specialized in the possession of an elongated, rodlike lingual process *(Lp)* and in lacking a second pair of ceratobranchials (*Cb* II). To facilitate comparison, the hyoid of *Chameleo* is shown in an extended position. At rest the anterior processes *(Ap)* and associated epihyals *(Eh)*, as well as the first ceratobranchials (*Cb* I), slant sharply forward from their articulations with the basihyal *(Bh)*. (Modified from Tanner and Avery, 1982.)

Cinematographic analysis (Gans, 1967) as well as electrophysiological data (Zoond, 1933) suggest that the tongue of chameleons is projected in the following manner. Contraction of the accelerator muscle compresses fluid between the lingual rod and the adjacent accelerator tendon. The accelerator muscle apparently contracts smoothly in a wave passing from front to back, thus tending to concentrate the propulsive force (fluid pressure) near the tip of the lingual rod (Fig. 13-13B). The result is that the muscle squeezes itself and

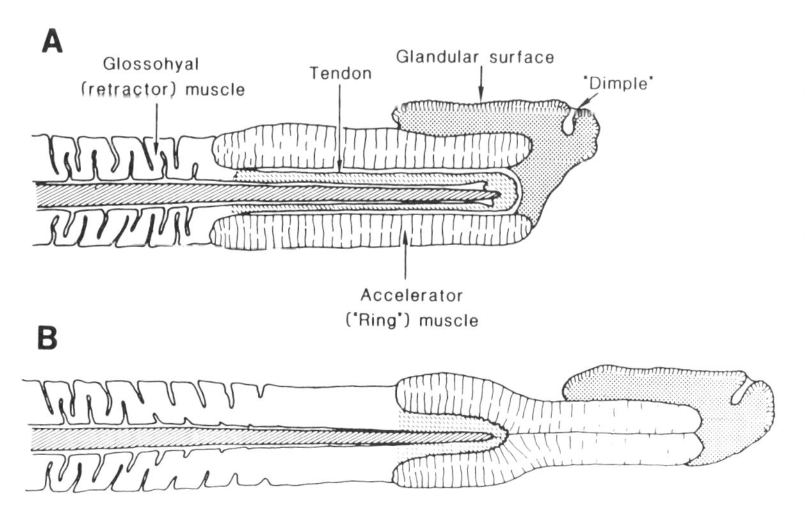

Figure 13-13 The structural and mechanical basis of tongue projection, shown by stylized sagittal sections through the anterior end of the lingual process and neighboring soft tissues of a chameleon. A: Tongue in prestrike position showing relationships of the accelerator and pleated retractor muscles as well as the hollow tendon *(broken hatching)* surrounding the tip of the lingual process *(fully hatched)*. Open spaces between tendon and adjacent rod and accelerator muscle are fluid-filled. Tendon is continuous posteriorly with glossohyal muscle. The surface of the tongue *(stippled)* is glandular and sticky and bears a distinct invagination or "dimple" near its tip. B: Tongue in the process of accelerating off the lingual process. Contraction of the accelerator muscle concentrates propulsive force near the tip of the rod. The tendon is bunching up in the same area. (Modified from Gans, 1967.)

the associated lingual mass off the anterior end of the rod at high velocity. Once clear of the rod, only momentum carries the projectile to its target. As this occurs, the accelerator tendon first slides off the lingual rod, then stretches, and finally turns inside out. The retractor muscle is simultaneously straightened as it is pulled along behind the projectile. At impact the mucous-coated tongue flattens against the target, thus establishing a strong adhesive bond. However, the presence of a conspicuous depression or "dimple" near the tip of the tongue raises the additional possibility of a suction-cup-like attachment to the target (Schwenk, 1983).

Retraction of the tongue involves contraction of the glossohyal musculature. These muscles anchor the accelerator muscle to the hyoid frame. During its relatively slow return to the mouth, the tongue and attached prey usually drop well below the chin of the lizard, there swinging pendulum-like until hoisted up to the mouth. The gape, which during projection and retraction of the tongue remains low, is generally increased as the tongue and prey pass into the oral cavity. After the food has entered the mouth, rhythmic chewing cycles provide mechanical reduction.

The elongated hyoid rod appears to be of central importance to the HLP system of chameleons in two respects. First, the length of the tapered rod establishes the length of the accelerator muscle. This in turn determines the length of time that the muscle is able to apply pressure to the tip of the lingual rod as it passes over it. The longer the thrust lasts, the higher will be the terminal velocity and the greater the momentum (and range) of the projectile. Second, the lingual rod serves as a launch rail to stabilize and to force the projectile into its proper trajectory during its brief acceleration. In this sense the lingual rod is analogous to the soft tissue guides associated with the HLP system of plethodontid salamanders (Lombard and Wake, 1976). Just as lengthening the barrel of an artillery piece will improve the accuracy of the shell fired through it, so too should lengthening of the hyoid process enhance the accuracy of the tongue fired from it.

Swallowing

Unfortunately, except in mammals the mechanisms and mechanics of tetrapod deglutition are poorly understood. That swallowing in frogs is commonly associated with forceful retraction of the eyes by the powerful retractor bulbi muscles strongly implies that such movements could be an integral part of deglutition. In theory, displacement of the large eyeballs downward through the soft tissues of the palate could push or squeeze food into the short esophagus. Although plausible, this hypothesis awaits experimental verification. The presence of large palatal vacuities in labyrinthodont amphibians has sometimes been interpreted to mean the presence of a similar eye-retracting mechanism. These observations raise the possibility that although the tongue was used for intraoral food transport in primitive tetrapods, the tongue-dependent deglutition of generalized reptiles and mammals may be a derived feature of amniotes. The only fish known to swallow out of water transport food through the pharynx with their pharyngeal jaws, not by actions of the more anterior visceral arches associated with the "tongue" (Sponder and Lauder, 1981).

Cineradiography has revealed that the swallowing mechanism of tortoises depends heavily on the tongue and that several aspects of deglutition in these primitive reptiles bear strong resemblance to those of mammals (Bramble, unpublished data) (Fig. 13-14). Swallowing begins with an SO-like kinematic episode in which the tongue is protracted and elevated against the palate to establish a seal anterior to the food mass. The tongue next expands and slides rearward along the roof of the pharynx, thereby pushing and wedging the bolus into the esophagus. During this process the hyoid apparatus spreads and rotates, thus tipping the glottis out of the path of the food. As the bolus passes over the base of the tongue, the hyoid and the root of the tongue are rapidly elevated against the trailing edge of the food. As in mammals, this phase of the swallow cycle is associated with strong contraction of the anterior suprahyoid muscles. Once past the pharynx, peristaltic contractions of the superficial constrictor musculature of the throat convey the bolus to the stomach.

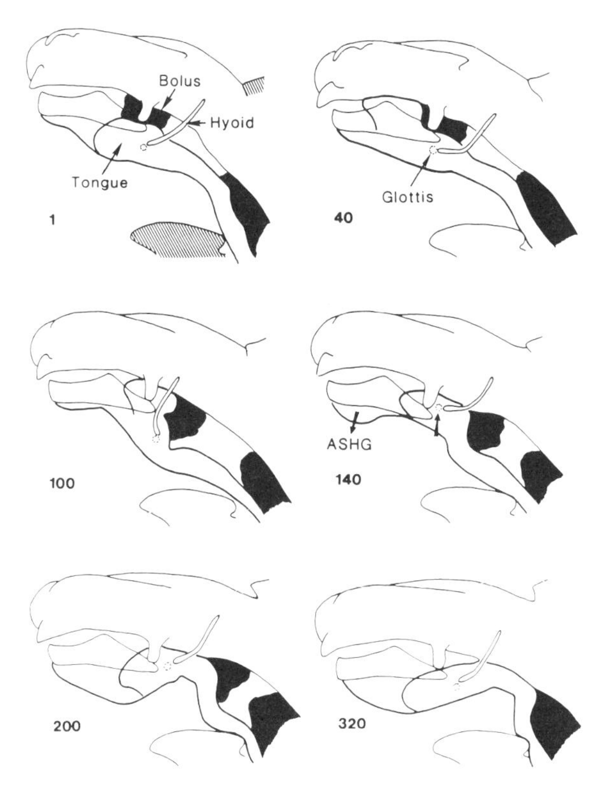

Figure 13-14 Selected tracings from a cineradiographic swallowing sequence of the tortoise *Scaptochelys agassizi*. Tongue has formed a "seal" anterior to bolus in frame 40. Tongue is driving bolus over depressed base of tongue and into esophagus in frame 100. Note reversed curvature on the spread hyoid horn and the tilted floor of pharynx below bolus. In frame 140 the hyoid raises rapidly up trailing edge of food mass as anterior suprahyoid muscles *(ASHG)* bulge behind mandibular symphysis. Floor of the oral chamber sags in frame 200 as superficial constrictor muscle tightens entrance to esophagus behind food. In frame 320 the bolus has joined another food mass in esophagus, and the pharyngeal passage reopens as tongue and hyoid descend. Framing rate is approximately 70 per second.

Similar swallowing mechanics could be expected in those reptiles having a well-developed tongue (*Sphenodon*, many lizards), but direct evidence is unavailable. It is clear, however, that different swallowing techniques are used by those amniotes in which the tongue has been reduced. In varanid lizards, for example, the hyoid apparatus is specialized to permit exceptional enlargement of the pharynx and anterior esophagus (McDowell, 1972). Large food items are first passed to the pharynx through a series of inertial feeding maneuvers and then swallowed by compressive actions of the hyoid apparatus and superficial constrictor musculature (Smith, 1982, 1984). In snakes the hyoid apparatus is vestigial (Tanner and Avery, 1982) and a special form of static inertial feeding (see above) is used to pass the prey to the rear of the pharynx. At this point peristaltic contractions take over and carry the food to the stomach (Cundall and Gans, 1979; Greene, 1983). Similar responses have been observed in the swallowing process of the limbless amphibian *Dermophis* (Bemis, Schwenk, and Wake, 1983). This form of deglutitional transport (at least in snakes) appears to recruit the anterior trunk musculature and often involves bending of the vertebral column.

The tongue of most birds is probably involved in the transfer of food from the pharynx to esophagus. However, this is not always the case. Zweers (1982a) has shown that the pigeon swallows small seeds through ratchetlike actions of the relatively large larynx. These movements (coupled to those of the hyolingual unit) serve to scrape seeds from the posterior palate and to carry them back into the rostral esophagus. Flexible, one-way valves on the surface of the larynx are vital to this scrape-and-carry mechanism (Zweers, 1981, 1982a). Larger seeds (peas) are forced into the esophagus by squeezing and wedging motions of the elevated larynx.

Conclusion

Even at this late date in the history of comparative morphological investigations there is an unfortunate tendency for workers to adopt a *Scala naturae* approach, in which amphibians, reptiles, and birds are envisioned as imperfect forms standing between fishes and mammals. Nonmammalian tetrapods, living and extinct, are enormously diverse, and many lineages have had independent histories that are older than all the living lineages of mammals. In the face of this great diversity, we have sought some general themes to give insight into the nature of the transition from water to land. In particular, we have espoused an approach that involves development of a general model for the function of tetrapod feeding systems. Such a model may be of value both for the predictions it makes and for the extent to which it encourages a broader view of the structural, functional, and historical patterns embodied in the feeding complex of all tetrapods. Too often functional morphological analyses of tetrapod feeding have cast their examples as special and isolated cases. Our hope is that the formulation of a first-order model may encourage the search for general and unifying principles.

There are many areas in which our knowledge of feeding in nonmammalian tetrapods is extremely scant. More studies are needed that combine comparative and experimental analyses. Some groups are of greater value than others for making generalizations. For example, the more primitive salamanders (families Hynobiidae, Cryptobranchidae, and Dicamptodontidae) are of key significance, for they might shed light on issues concerning the origin of tetrapod feeding and water-air transitions. Mechanical dependence on the medium can be studied more readily in turtles than in many other groups. Electromyographic analyses of the hyolingual muscles in *Sphenodon* and of the adductors in frogs would help to test the validity of our generalized model.

We are just beginning to gain an understanding of the neuromotor control of feeding mechanisms in nonmammalian tetrapods. The topic has received more formal consideration in birds (for example, Zeigler and Karten, 1973; Zeigler, Miller, and Levine, 1975, 1980; Zweers, 1980, 1982a,d) than in other groups. We have no idea how conservative central motor programming might be or to what extent it might constrain evolutionary events. To begin to test the notion that neuromotor programming is indeed conservative will require careful documentation of the normal sequencing of activity in homologous muscles during intraoral transport in a phylogenetically diverse sampling of tetrapods. Also needed are studies that coordinate kinematics, muscle activity patterns, and comparative morphology with such aspects of neurological organization as peripheral innervation, anatomy of the brain stem, and connections to higher centers.

But the study of taxa selected because they are generalized, or amenable to study, or because they display flexibility in physical habitat is not sufficient. We also need analyses of entire assemblages of living species. Ideally, such groups should include a wide array of presumably adaptive morphologies and should have been subjected to both a detailed cladistic analysis based on comparative morphological data and an analysis using modern techniques of biochemical systematics. Preferably, the group should occupy a spectrum of habitats and display a wide array of feeding modes. For example, salamanders of the family Plethodontidae have generalized species with aquatic larvae and semiterrestrial adults, but also a full array of adult morphologies. Some are permanently aquatic forms, others are strictly terrestrial with direct development and no larval stage, and the terrestrial forms range from fossorial to arboreal in habitat (Wake, 1966).

By studying such adaptive radiations the relationship between feeding morphology, trophic behavior, and ecology could be analyzed. What does it mean, for example, to be specialized as opposed to generalized in feeding morphology? Both ecomorphologies exist, often within the same lineage. Why the diversity, and why can generalists survive? Increasingly, workers are concerned with these matters and are attempting to relate feeding morphology to differences in individual fitness (see Greene, 1982; Boag and Grant, 1982; Arnold, 1983). We look forward to an integrated approach to morphological analysis of tetrapod feeding in which ecology, physiology, and phylogeny all play important roles.

Chapter 14

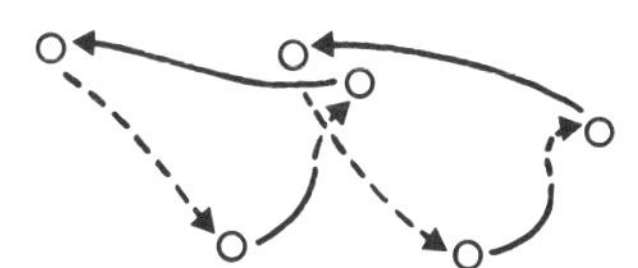

Mastication, Food Transport, and Swallowing

Karen M. Hiiemae
Alfred W. Crompton

Food has metabolic value only when the products of its digestion enter the blood stream. For most mammals a high metabolic rate depends on regular ingestion of food items, which in most cases must be mechanically broken down in the oral cavity before they can be chemically simplified in the gut. Thus, food is taken into the mouth (ingestion), processed (mastication), and then swallowed (deglutition). Enzymes in the saliva of most mammals begin to act as the material is readied for swallowing and, depending on the rate of both gastric secretion and gastric movements, continue to act for a limited time. Although some living reptiles shred or puncture food before swallowing (Throckmorton, 1976, 1980; Smith, 1982), that process can be seen as facilitating food intake and swallowing, rather than as producing the extensive mechanical reduction of food within the oral cavity that is characteristic of mammals.

Chewing, or mastication, serves two functions: first, material is reduced to a condition suitable for swallowing; second, the resulting increase in surface area facilitates the penetration of the digestive enzymes and so expedites the rate of chemical breakdown. Foods with resistant cell walls, such as grasses, require extensive mechanical breakdown before digestive enzymes are maximally effective. It is also necessary to expose the cellulose cell walls to the digestive enzymes of bacteria.

Studies of feeding mechanisms in mammals have concentrated on the morphology, or both the morphology and function, of the jaw apparatus (including teeth, mandibular joint, and elevator muscles) in relation to broad dietary habits (for example, Hiiemae and Ardran, 1968; Turnbull, 1970; Kallen and Gans, 1972; Herring and Scapino, 1974; Weijs and Dantuma, 1975; Weijs and de Jongh, 1977; Gorniak, 1977; Janis, 1979; Hylander, 1977, 1979; Fish and Mendel, 1983; Oron and Crompton, unpublished data). Further, the teeth of mammals (and other dentate vertebrates) vary widely in shape, and there are unequivocal associations between tooth form and general dietary habits. It is not, therefore, surprising that teeth and jaws have major taxonomic significance, especially in the analysis of the fossil record, and that much experimental effort has been devoted to analyzing the movements of jaws in feeding.

Mammalian mastication can be characterized (except in some highly specialized forms) as having the following features: (1) active breakdown of food is unilateral, that is, it occurs on one side of the jaw at any one time; (2) there is some element of transverse movement during food breakdown, which is minimal in carnivores and maximal in some herbivores; and (3) upper and lower molars accurately "fit" one another, although their tight occlusion is, to some extent, developed with wear. These features are found in the jaw apparatus of the earliest mammals (known from 180 million years before Pleistocene) and sharply separate them from their immediate ancestors, the mammallike reptiles.

Limited attention (for instance, Hiiemae and Ardran, 1968; Weijs, 1975, for rats; Gordon, 1984, for walruses) has been paid to the mechanisms involved in the ingestion of food or its transport and manipulation within the oral cavity. Swallowing has been extensively studied in humans but almost ignored in nonhumans. It is only recently that experimental studies have shown that the tongue and the hyoid apparatus, as well as the soft palate, have a pivotal role in the feeding process and, indeed, may have the primary role (at least in the ontogenetic sense). This chapter is, therefore, focused on the oral management of food rather than simply on the form of the teeth and jaws and the movements of the head and lower jaw in mastication.

Despite the differences in the morphology of the jaw apparatus in mammals, synchronous cinefluorographic and electromyographic studies of feeding in a variety of mammals—the American opossum *Didelphis marsupialis,* a carnivore/omnivore; the tenrec *Tenrec ecaudatus,* predominantly an insectivore; the hyrax *Procavia capensis,* a herbivore; the cat *Felis catus* and dog *Canis familiaris,* both carnivores; the pig *Sus scrofa,* an omnivore; and a higher primate, *Macaca fascicularis,* have shown that the mechanisms by which solid food is transported, processed, manipulated, and swallowed are essentially the same in all. Such differences as are found, for example, in the position of bolus formation or in the transport of food to the back of the tongue in higher primates, reflect minor, albeit significant, modifications of the basic pattern.

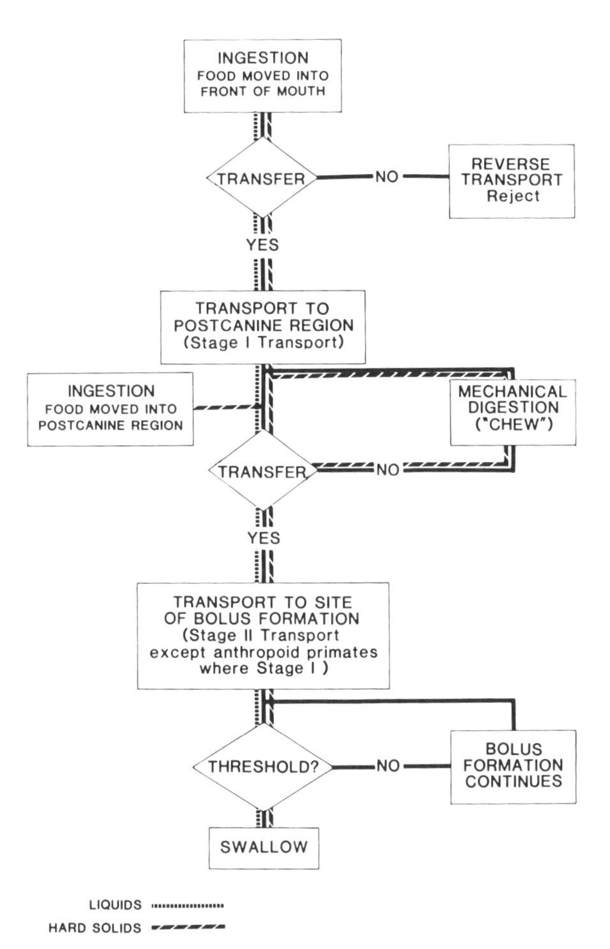

Figure 14-1 Flow chart of food processing in the mouth. Liquids and very soft foods of low viscosity are moved directly from the front of the mouth to the oropharynx for bolus formation. Solid food is ingested either through the anterior part of the mouth or, if very hard, through the postcanine region (ingestion by mastication). The number of times such food passes through the chew loop depends on its initial particle size and consistency. Stage I transport moves food posteriorly within the oral cavity; stage II transport moves food into the oropharynx (see Fig. 14-3).

Food Processing and Transport: Overview

Recent experimental studies have shown that liquid food is moved into and through the mouth in a three-stage process; solid food requires five (Fig. 14-1). Chewing, or the breakdown of food by the teeth to a condition suitable for further transport, is simply a stage, though an important one, in the more general process. The stages are sequential and usually involve different parts of the mouth in the processing of each unit, or aliquot, of food, although more than one stage can occur concurrently in relation to additional aliquots. Feeding depends on the synchronized and cyclical movements of the jaws, that is, the cranium on the cervical vertebral column (upper jaw) and the mandible on the cranium (lower jaw); the tongue base (hyoid apparatus) and the body of the tongue; as well as the soft palate (Fig. 14-2). Movements of the soft palate occur throughout the process but are highly patterned, as are those of the walls of the oropharynx in swallowing.

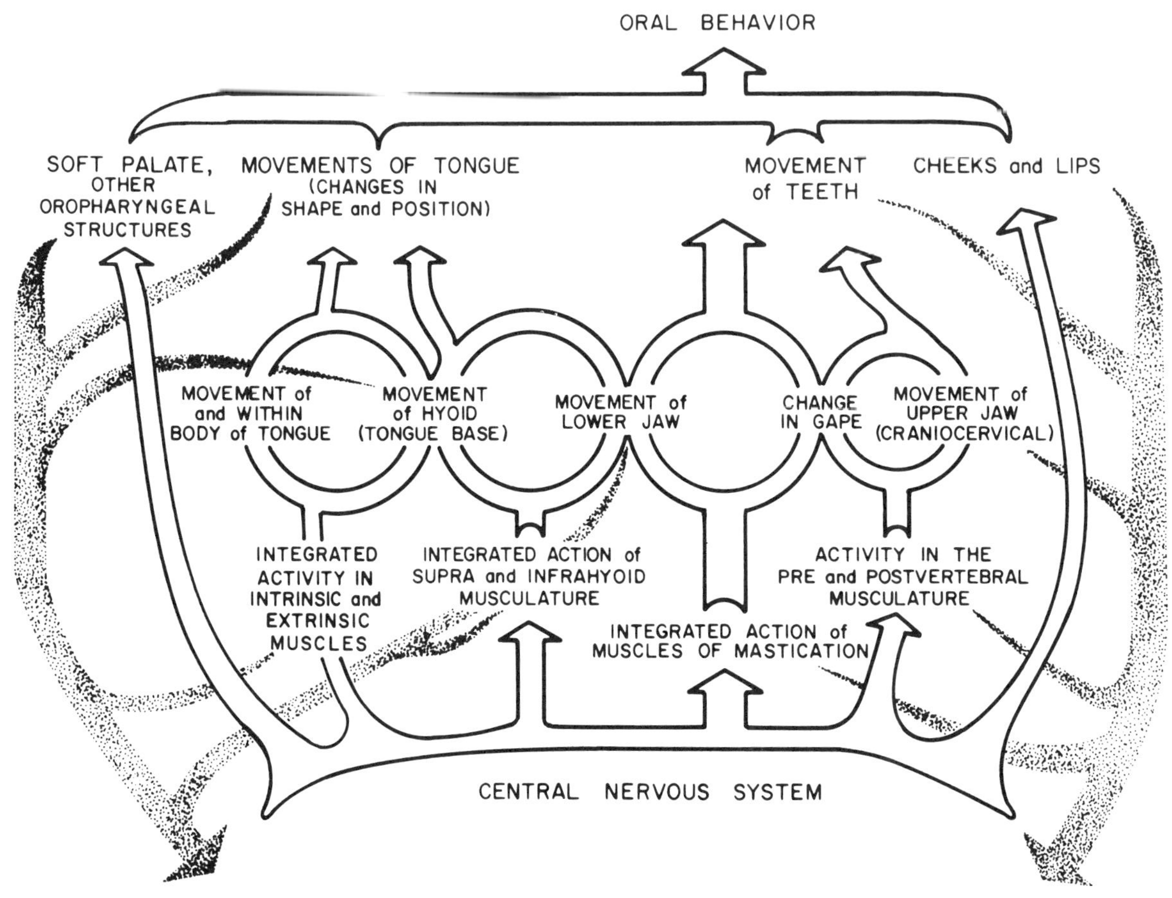

Figure 14-2 Functional relationships of the major anatomical elements in the feeding apparatus. The effector system is shown as solid arrows linking components in sequence. Dotted arrows indicate the sources of sensory input that can modulate effector activity. "Other oropharyngeal structures" include the hard palate (especially its mucosa), the epiglottis, the pharyngeal musculature, and the oropharyngeal mucosa.

Ingestion and Stage I Transport The process by which food is transferred from the external environment to the oral cavity is called ingestion. Liquids are usually lapped (a rhythmic behavior in which aliquots are collected by the tongue during jaw movements of low amplitude) or are sucked (a rhythmic reduction in intraoral pressure).

The ingestion of solids is the most variable and least well understood of the feeding processes; its pattern depends largely on the condition of the food item. Vegetation such as leaves or grasses is plucked or cut by the action of lips, tongue, and anterior teeth. Depending on size and hardness, solids may be ingested in a series of bites or gnaws, while usually, but not always, held in one or both hands. If the material is very hard, the force needed to separate a bite may require the use of the cheek teeth. Such behavior is routinely seen in primitive mammals and in carnivores feeding on bones or attempting to pull flesh from a carcass. Small food items may simply be picked up by the anterior teeth or tongue and passed into the front of the mouth or, especially in primates, manually placed in the mouth.

Stage I transport is the movement of liquid and solid food from the anterior part of the mouth back toward the cheek teeth (premolars and molars). Some rodents and primates have well-developed cheek pouches. After rapidly ingesting food, they transport it posteriorly, store it in the pouches, and remove it later for processing.

Mastication and Manipulation The particle size of the ingested food is reduced by the repeated approximation of the cutting edges and crushing surfaces of the postcanine teeth. Although usually described as "chewing," many other evocative terms have been coined to describe this process. In general, they all attempt to explain the method by which the food is reduced (cutting, tearing, crushing, chomping, grinding) and are based on visual and auditory observations of animals feeding or, alternatively, on the shape and occlusion of the molars. The actual method of reduction is a function of the physical properties of the food and the characteristics of the tooth surfaces acting on it (Lucas, 1982). As particles of food reach a size appropriate for bolus formation and swallowing, they are moved posteriorly. The residuum is then reprocessed. It is clear that there is substantial variation among mammals in the range of particle size and consistency suitable for swallowing. A ruminant artiodactyl can swallow a less thoroughly processed wad of vegetation than can a perissodactyl: the former will regurgitate the food and reprocess it; the latter must rely on an adequate level of processing before deglutition. In contrast, a carnivore swallows large lumps of barely processed material. During a chewing sequence food is moved either from one side of the oral cavity to the other or to a more distal position. This is referred to as manipulation.

Stage II Transport and Deglutition As mastication proceeds, reduced foods are moved posteriorly to form the bolus, first to a position below the soft palate and then farther to the vallecular regions and the pyriform recesses (defined below and in Fig. 14-3). This is termed stage II transport.

When a sufficient volume of liquid or solid has accumulated, the bolus is swallowed. Swallowing (deglutition) is a stereotypical behavior that occurs during feeding—every few cycles in lapping or intermittently during and, finally, at the end of a masticatory sequence. In either case the regular pattern of jaw movement is transitorily suspended as the bolus is moved from the posterior part of the oral cavity across the oropharynx to enter the esophagus or, in higher primates, from the oral cavity to the esophagus.

The process of food transport and reduction as outlined above presupposes that receptors in the oral cavity can discriminate between particles that are reduced to a size suitable for swallowing and those that are not. The concept of selection is important; if such mechanisms operate, not only must a high level of discrimination be present, but there must in addition be a translation of that discrimination into a sorting function. There is good, though somewhat circumstantial, evidence that there is an optimal particle size for at least some types of food in man (Kay and Sheine, 1979; Lucas, 1982). At present, however, no acceptable evidence exists as to how such sorting is achieved, although the tongue must be the primary organ involved.

Anatomy of the Mouth and Jaws

Major features of the jaw apparatus in mammals are correlated with the sequential pattern of internal food handling. The cheeks form a flexible external seal to the oral cavity and, together with the lips, a mobile intermittent seal (the anterior oral seal) in front of the incisors and canines. The space from the lips to the opening of the pharynx is divided into the oral cavity and the oropharynx. The hard palate is covered, to a variable extent, with transversely oriented rugae; the soft palate is highly mobile (Fig. 14-3). The oral cavity extends from the lips posteriorly to the pillars of the fauces, which flank the aperture between mouth and oropharynx and contain the palatoglossus muscles running from the external surface of the tongue into the soft palate. Anteriorly the oral cavity can be sealed off by the muscles of the lips (the orbicularis oris and associated muscles); posteriorly it can be separated from the oropharynx by a sphincter or intermediate seal formed by the contraction of the palatoglossi and a change in tongue shape.

The oropharynx has two distinct compartments in all mammals except adult *Homo sapiens;* the anterior compartment (or vallecular region)

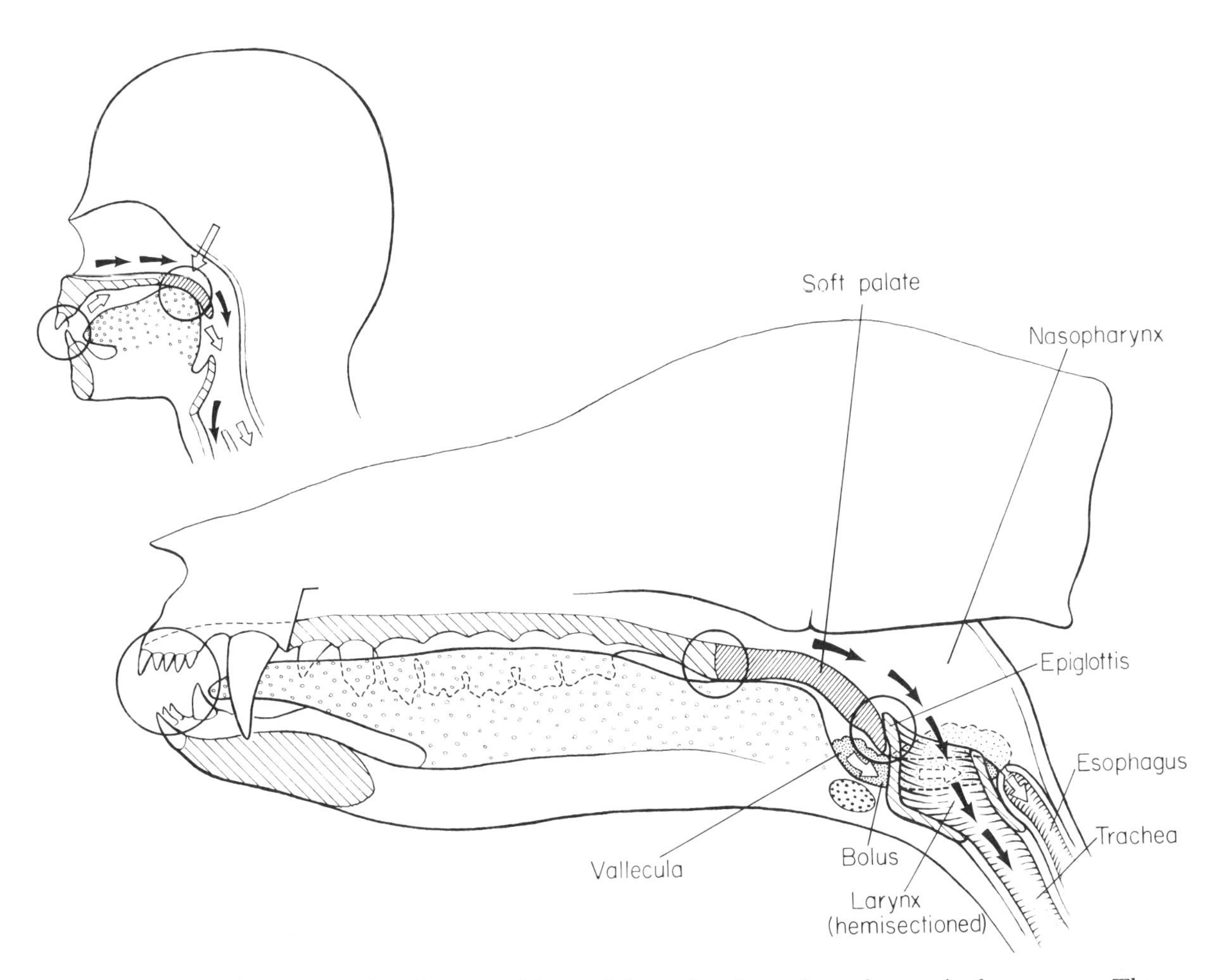

Figure 14-3 Hemisection showing the general form of the oral cavity and oropharynx in the opossum. The oral cavity extends posteriorly to the contact between the soft palate and the tongue (middle oral seal). The area between the middle and posterior oral seals is part of the oropharynx. The space between the posterior part of the tongue and the anterior surface of the epiglottis (the vallecula) communicates with a potential space, the pyriform recess, which extends laterally and posteriorly around the larynx, between it and the lateral wall of the pharynx (dotted line). In opossums a bolus accumulates in both the valleculae and the pyriform recess (shaded). The human head is shown for contrast.

extends from the pillars of the fauces in front and below the soft palate to the epiglottis posteriorly. The posterior surface of the tongue forms the floor of this compartment, which is normally closed posteriorly by the posterior oral seal formed by the overlap between the soft palate and the epiglottis (Fig. 14-3). In most mammals the anterior compartment extends distally on either side, external to the soft palate–epiglottis–larynx contact, into flaccid recesses referred to as the pyriform recesses. Food can be stored in the anterior compartment and in its extension, the pyriform recess. In anthropoid primates the pyriform recesses are reduced to furrows lying alongside the larynx and do not serve for the transient storage of a bolus. The posterior compartment of the oropharynx extends from the plane of the soft palate–epiglottis to the posterior wall of the pharynx. Except during swallowing, this posterior part is confluent with the nasopharynx above and the laryngopharynx below. Adult *Homo sapiens* have no soft palate–epiglottal seal; the posterior oral seal is between the tongue and the soft palate at the pillars of the fauces, making the oropharynx a single compartment. However, in infants a soft palate–epiglottal seal is present; milk reaches the esophagus via the pyriform recesses without disturbing the airway. During normal development humans lose the posterior oral seal, and adults rely on the intermediate seal to separate the oral cavity from the oropharynx and nasopharynx (Fig. 14-3).

Posteroventrally the base of the tongue is tied to the hyoid complex. This is a bony or osteocartilaginous structure that is linked by ligaments and muscles to the larynx, the base of the skull, the mandibular symphysis, and to the sternum and scapula. In many mammals the hyoid complex is also joined to the basicranium (presumably a primitive condition) through a system of jointed bones or cartilages (hyoid arch). The hyoid is always highly mobile.

The following general description is based on the American opossum *Didelphis virginiana,* chosen because much of our current knowledge of food transport is based on this animal. The opossum provides what is presumed to be a fairly good example of the primitive condition of the oral region. Highly evolved mammals such as higher primates, carnivores, and herbivores show considerable differences in detail (including the relative proportions of the adductor muscles and form of the digastric), but not in the general organization of the system.

Adductors In some mammals the adductors form a series of discrete muscles; in others, such as the opossum, the divisions are less clear. However, it is always possible to recognize the temporalis, masseteric, internal pterygoid, and external pterygoid muscle masses. The temporalis mass inserts on the medial and dorsal surface of the ascending ramus of the dentary (hemimandible) and arises from the dorsal region of the temporal fossa (Fig. 14-4). The masseteric mass arises from the zygomatic arch and inserts on the external and ventral surface of the mandible. The superficial fibers of this mass extend obliquely backward from the anterior region of the zygoma to the angular region of the mandible, whereas the deep fibers have a more vertical orientation and arise from the wide region on the zygomatic arch behind the origin of the superficial masseter (see Hiiemae and Jenkins, 1969, for a detailed description of these muscles in *Didelphis,* with a somewhat idiosyncratic terminology for the adductors). Although the adductors of other animals vary widely in their proportions and internal architecture (Turnbull, 1970; Weijs, 1975; Gans and Gorniak, 1978; Janis, 1979), they have very consistent sites of origin and insertion relative to one another.

Muscles Controlling the Hyoid Apparatus The hyoid apparatus consists of elements of the hyoid and first branchial arches. The ventral portion of the hyoid and first branchial arches (basihyoid and thyrohyoid, respectively) form a C-shaped structure that both cradles and is joined by muscles and ligaments to the larynx. In some mammals (ungulates and carnivores) the remaining elements of the hyoid arch form a series of cartilaginous or bony rods connecting the ventral element to the auditory capsules (the integrocornate condition). In primates, as well as opossums, lagomorphs, and tenrecs, these elements are reduced to varying degrees (the discretocornate condition) to form the lesser horn of the primate hyoid apparatus. The base of the tongue is firmly connected to the anteroventral surface of the hyoid apparatus. The epiglottis forms a mobile anterodorsal cover to the larynx. It extends upward inside the body of the C formed by the ventral portion of the hyoid apparatus, with its anterior surface facing the tongue and the posterior surface the laryngeal inlet and oropharynx.

The principal muscles that control hyoid movement are two suprahyoid muscles, the geniohyoid (mandibular symphysis to hyoid) and the stylohyoid (basicranium to ventral surface of hyoid); and two infrahyoid muscles, the sternohyoid and omohyoid (scapula to hyoid). Minor elements of other muscles contribute to the anterior suprahyoid muscles, for example, fibers of the mylohyoid and genioglossus. Some mammals, including opossums, have lost or greatly reduced the stylohyoid muscle, and its function is taken over by an equivalent muscle that parallels a larger stylopharyngeus (Fig. 14-4). A complex set of muscle sheets originating on the basicranium

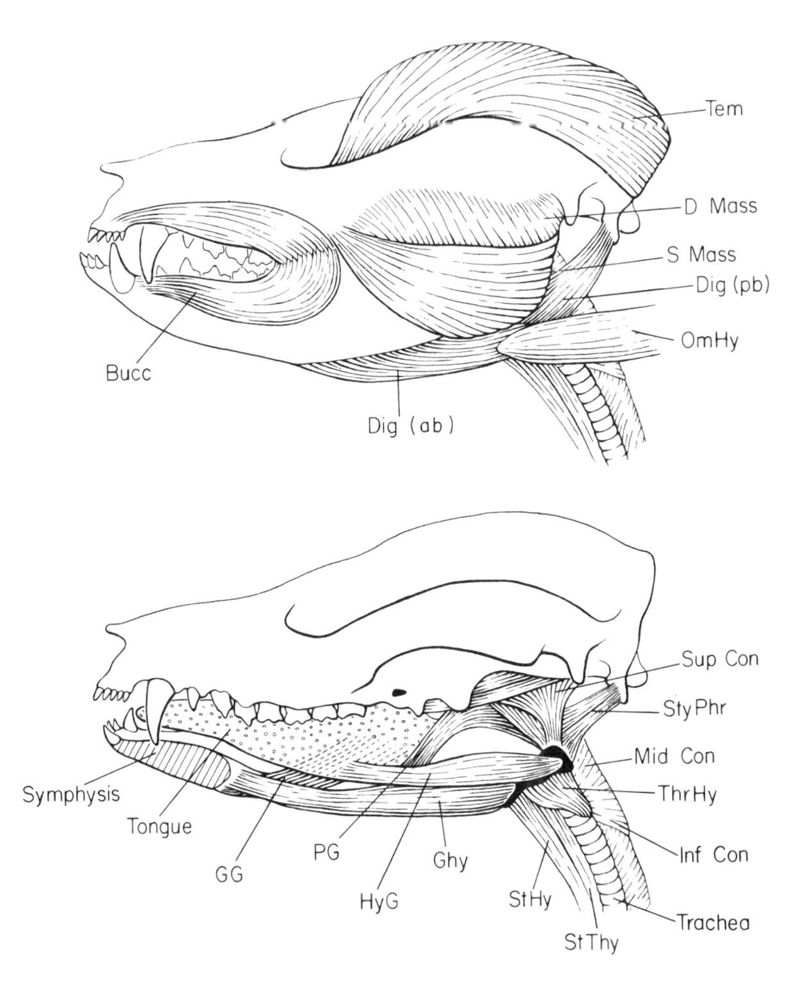

Figure 14-4 Lateral views of the head of the opossum showing structures associated with feeding. *Above:* Muscles of the jaw and hyoid. *Tem* = temporalis; *D Mass* and *S Mass* = deep and superficial masseter; *Bucc* = buccinator; *OmHy* = omohyoid; *Dig* = digastric, which has anterior *(ab)* and posterior *(pb)* bellies. Not shown are the large internal pterygoid (deep to the *Mass*), the small internal pterygoid (attaching to neck of mandibular condyle), and the mylohyoid (between the rami of the jaws). *Below:* Muscles of the tongue and pharynx. *GG* = genioglossus; *Ghy* = geniohyoid; *HyG* = hyoglossus; *PG* = palatoglossus; *StHy* = sternohyoid; *StyPhr* = stylopharyngeus; *Sup Con, Mid Con,* and *Inf Con* = anterior (or superior), middle, and posterior (or inferior) constrictors; *ThrHy* = thyrohyoid.

make up the dorsal pharyngeal wall—some of these insert directly on the dorsal surface of the thyrohyoid. The sternothyroids parallel the sternohyoids. In primates and opossums some fibers of both the anterior and posterior bellies of the digastric insert on the ventral surface of the hyoid apparatus. This appears to be the retention of a primitive condition that is lost in more specialized mammals. The hyoid-jaw link means that the hyoid muscles and the digastrics can affect jaw opening.

Muscles Controlling the Tongue The tongue can be regarded as a bag of highly specialized epithelium filled by muscle fibers. These fibers have conventionally been regarded as clearly separated into two groups of anatomically distinct muscles: the intrinsic muscles (vertical, transverse, and longitudinal) running within the substance of the tongue; and the extrinsic muscles, which have a single bony attachment external to the body of the tongue (hyoglossus, genioglossus, styloglossus, and palatoglossus). It has been argued that the extrinsic muscles regulate tongue position in space, whereas the intrinsic muscles control tongue shape (Livingstone, 1956). Such a clear and arbitrary division of function is not valid, since all the extrinsic muscles can also change the shape of the tongue; for example, the genioglossus can depress the posterior surface and the hyoglossus can shorten the posterior region of the tongue. The division of muscles into protractors and retractors (see Lowe, 1981) is also an oversimplification. The experimental evidence suggests that both groups of muscles produce changes in tongue shape and may affect minor changes in position, although major changes in position certainly depend on movement of the hyoid (that is, tongue base) resulting from differential contraction of the hyoid muscles, rather than on those of the tongue. However, we know virtually nothing about how the series of muscles making up the body of the tongue or the extrinsic muscles entering the tongue actually control tongue shape.

Muscles of the Cheeks and Lips The mouth is surrounded by a ring of muscle (orbicularis oris) that functions as a sphincter to close the lips and form the anterior oral seal. (The mobility of the lips, characteristic of man and higher primates, reflects the absence of a rhinarium. In many mammals the upper lip is tied to a rhinarium and is relatively immobile. Nevertheless, a complete anterior oral seal can be formed since the seal depends not on the overall mobility of the lips but on the ability of the animal to approximate the upper and lower lip margins even if the mouth extends far posteriorly. The buccinator, whose fibers pass anteroposteriorly and intermesh anteriorly with those of orbicularis oris (Fig. 14-4), forms the muscular lateral wall of the oral cavity and thus the bulk of the cheeks. It is attached posteriorly to the pterygomandibular raphe, which passes vertically from the pterygoid hamulus to attach to the inner surface of the mandibular body behind the molars. Its fibers then pass forward and lateral to the postcanines to form the cheek.

Muscles of the Soft Palate and Pharynx The pterygomandibular raphe also forms the anterior attachment of the anterior (superior) constrictor of the pharynx. This is one of three constrictors whose overlapping fibers form the muscular posterior and lateral walls of the pharynx. Internal to the raphe, the palatoglossus connects the soft palate and tongue, running in the folds of mucosa called the pillars of the fauces. The middle oral seal is formed between the soft palate, the pillars, and the tongue. The soft palate is highly mobile and changes shape during feeding. The movements of the palate are produced by four extrinsic muscles, the levator and tensor veli palatini, the palatoglossus, and the palatopharyngeus. The last of these helps form the palatopharyngeal sphincter during swallowing, which assists in closing off the nasopharynx from the oropharynx.

Mammalian jaws can be moved to varying degrees in all three planes, anteroposteriorly and mediolaterally as well as vertically. With the possible exception of the lateral pterygoid and the digastric, all the muscles whose sole function is to control the jaws act as adductors. They have been traditionally described (see Hiiemae, 1978) as forming a series of couples about the lower jaw. Transverse (lateromedial or ectental) movements are thought to be produced by the combined actions of the medial ptergygoids of the side on which chewing is occurring (the active side) and the posterior fibers of the temporalis on the opposite (the balancing side). This couple has been considered most important during the power stroke of mastication and the first part of opening when the jaw is moved from lateral to medial on the active side. Anteroposterior (or propalinal) movements are thought to be produced by the pterygoids (internal and external acting together) and the superficial masseters (forward movement) or the posterior fibers of temporalis (backward movement). Transverse and anteroposterior movements do not normally occur in isolation; they are combined with vertical movements in the normal cyclical activities of mastication. The only exception so far documented is the anteroposterior shift of the lower jaw to bring the incisors or molars into occlusion in rodents (Hiiemae and Ardran, 1968; Weijs and Dantuma, 1975, 1981), where the simultaneous occlusion of anterior and postcanine teeth is precluded by the geometry of the jaws.

More recent experimental work has demonstrated, however, that such biomechanical explanations of adductor muscle function are simplistic. Although "couples" may operate to produce jaw movement, electromyographic and movement analysis show that muscles, or parts thereof (and not always those predicted by their theoretical vector mechanics), are active on both sides of the jaw during the feeding cycle. Further, there is a marked and changing pattern of differential activity between corresponding muscles on the active and balancing sides. The explanation for these findings will require further studies with the level of analysis used by Weijs and de Jongh (1977) in their study of the rabbit.

The mechanics of jaw opening have been classically described as based on a downward and forward pull on the condyle, coupled with a downward and backward pull on the anterior part of the mandible from the lateral pterygoid and the digastric, respectively. This mechanically simple explanation is complicated by the variability of digastric form in mammals (du Chaine, 1914; Edgeworth, 1935) and confounded by the experimental observation that not only does jaw opening occur in two separate stages in many mammals (see below and Figs. 14-5, 14-6, 14-8), but that even in those cases where it appears to be a simple movement, it results from the complex interplay of all the muscles of the hyoid apparatus (Crompton et al., 1977). Further, when the central tendon and hyoid connection of the digastric are severed, leaving both anterior and posterior bellies free, there is no observable change in the pattern or amplitude of jaw opening. It might reasonably be argued that the opossum is a less than ideal subject from which to make generalizations on digastric function; however, the digastric in that animal does have a hyoid linkage, as it does in all primates and some other mammals. This kind of experiment should be done in animals such as cats where the digastric has no hyoid connection.

Although all these muscles can be described as belonging to specific muscle groups, it is important to appreciate that no muscle contracts in isolation. The complex cyclical movements of the jaw and tongue observed in feeding depend on a highly integrated and patterned cycle of activity in all these muscles (Fig. 14-2).

Mechanisms of Intraoral Food Handling

Until recently almost nothing was known of the mechanisms by which food was transported through or manipulated in the mouth. Our knowledge of masticatory movement was limited to naked eye observation or the analysis of cinefilms of wild or laboratory animals. Although these methods provided useful information on feeding behavior and the general pattern of jaw movements, it was difficult (even under laboratory conditions) to correlate the observations with the condition or position of the food. The only exceptions to this of which we are aware are Abd-el-Malik's observations of human tongue movement (1955) and an extraordinary film taken of intraoral food manipulation in a patient in whom the cheek and part of the left upper jaw had been surgically removed (Veterans Administration, E. L. DuBrul, personal communication).

However, in the last decade cinefluorographic and electromyographic techniques have been used to examine mandibular and oral function. Technical problems have so far limited our knowledge of tongue movements in the coronal plane to what can be seen when the mouth is open or partially open. Cinefluorographs show the movements of such radiopaque structures as bones and teeth. With careful adjustment of the radiation levels, soft tissue outlines can also be recorded. Food can be mixed with barium salt and its position then tracked. However, the most important development has been the regular use of metallic markers (Crompton and Hiiemae, 1970). The insertion of small fillings in selected teeth in both upper and lower jaws permits very accurate measurement of changes in jaw position (Hylander and Crompton, 1980); a marker sutured to the dense fascia on the anterior (ventral) surface of the body of the hyoid reflects changes in the position of that bone (Crompton et al., 1977). Similarly, the insertion of small segments of surgical wire into the muscles of the tongue below its surface led to the discovery that the mechanism of food transport involves differential longitudinal contraction and expansion of various parts of the tongue (Crompton and Sponder, 1981, 1982). Analysis of high-speed cinefluorographic film is now based on digitization of significant reference points in each frame of film. The *x-y* coordinates of each point are entered into the computer and a variety of manipulations performed (McGarrick and Thexton, 1981).

Figure 14-5 shows typical movements over time of gape, hyoid, and tongue markers for the opossum and macaque. The profiles have been smoothed by connecting the individual computer-generated points with a solid line. Lapping by the opossum is associated with high amplitudes of anterior and middle tongue markers (*ATM* and *MTM*) owing to the extensive protrusion of the tongue (see also Fig. 14-12). The spaces between the vertical lines indicate (*left to right*) the durations of first opening (*O*1), second opening (*O*2), and close (*C*) phases, which are not homologous with phases of the masticatory cycle. In eating soft food the opossum has no clear division between fast close (*FC*) and slow close (*SC*) because of the negligible resistance of the food. Long slow open (*SO*) and hyoid protraction are associated with stage I transport. In contrast to lapping, all tongue markers have nearly equal anteroposterior movement when hard food is eaten. Phases *FC, SC, SO,* and *FO* (fast open) are distinct. The *FC-SC* transition occurs as upper and lower teeth come into contact with the food.

Macaques do not lap; Figure 14-5 represents drinking from a water bottle with spout just anterior to the incisors. The amplitude of jaw movement is about the same for soft and hard food, but the profiles differ. The gape plot for soft food has no distinct phases. Movements of the markers indicate active transport. For hard food this power stroke is long and the opening phase short. Movement of food is minimal.

Correlation of data from cinefilms with electromyograms (Figs. 14-6, 14-7) show the pattern of muscle activity associated with each behavioral event. Hylander (1977) and Weijs and de Jongh (1977) have pioneered the use of strain measurement to examine the biomechanics of jaws and skulls. Single elements or rosette strain gauges are attached to the bony surface and strain levels accurately correlated with electromyograms and movement patterns (Fig. 14-6).

Behavior recorded experimentally shows that some of the processes to which the food is subjected in the oral cavity occur concurrently. Each depends on the synchronization of the activity of numerous muscle groups controlling the jaw, tongue, hyoid, soft palate, and pharynx. These muscles show modifications of their basic activity in response to changes in the size, shape, or consistency of the food and therefore in the relative movement of the structures they control during a sequence of cycles involving the opening and closing of the jaws and the coordinated movements of the hyoid apparatus, the body of the tongue, and the soft palate (Fig. 14-8).

Food is transported through the mouth, broken down by the teeth, and manipulated by the tongue and cheeks. The shape of the tongue at any moment is a function of activity in its intrinsic and extrinsic (especially genioglossus and hyoglossus) muscles. The position of the tongue is largely determined by the position (relative to the palate) of its structural base—the hyoid. The tongue is linked both directly to the symphyseal region of the lower jaw by the genioglossus and indirectly to the body of the lower jaw through the anterior suprahyoid muscles. Franks, German, and Crompton (unpublished data) have shown that the activity of the muscles that regulate tongue surface, hyoid, and lower jaw movement is coordinated. The repetitive movement of the jaws, hyoid apparatus, and tongue body are, however, linked (at least in the opossum and hyrax) only at one point in each of their cycles: the time just before the jaw leaves minimum gape. This represents the end of the power stroke, occurring after the cessation of adductor activity and at the beginning of protraction of the hyoid and of markers in different parts of the tongue. It also marks the beginning of lengthening (anteroposterior expansion as seen in lateral projection) of the posterior third of the tongue. The times at which the jaw reaches maximum gape and when the markers begin to return or the posterior tongue starts to shorten (contract) are not synchronous. It is possible for the sake of convenience to refer to jaw, hyoid, and tongue cycles; if these are defined as beginning at minimum gape, it follows that the synchronous jaw, hyoid, and tongue cycles always have the same time base. The pattern of jaw, hyoid, and tongue activity changes as the food is progressively broken down and portions swallowed. Despite the variations in behavior, it is clear that there is a basic pattern in all three systems and that this pattern is found in all mammals so far studied.

Three stages (ingestion, stage I transport, and stage II transport) have been identified in the intraoral transport of liquids, and five stages (in-

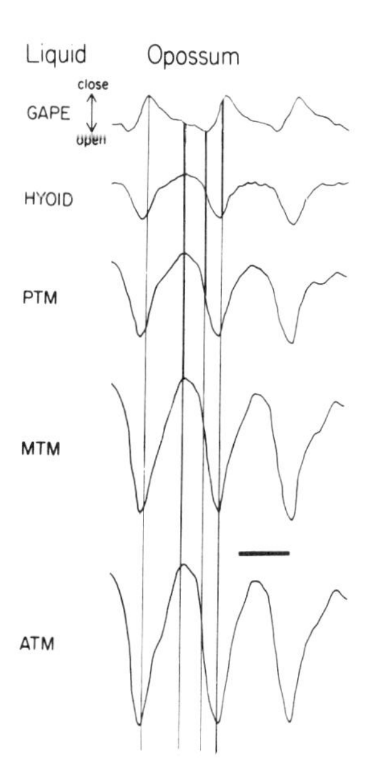
Liquid
Opossum
close
GAPE
open
HYOID
PTM
MTM
ATM

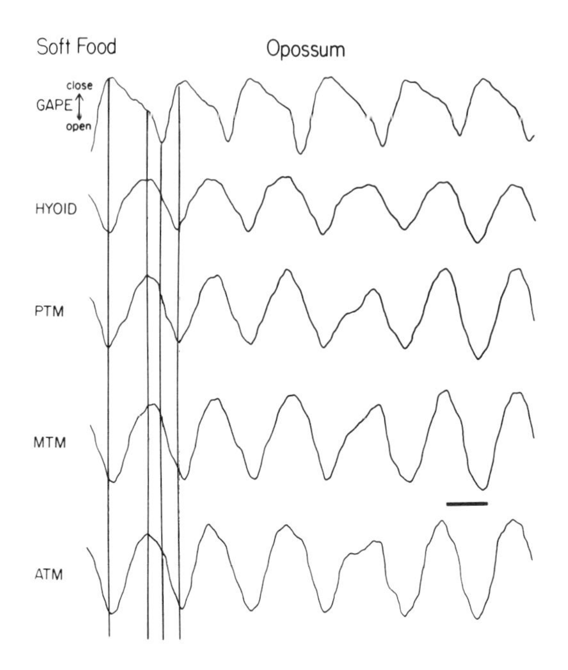
Soft Food
Opossum
close
GAPE
open
HYOID
PTM
MTM
ATM

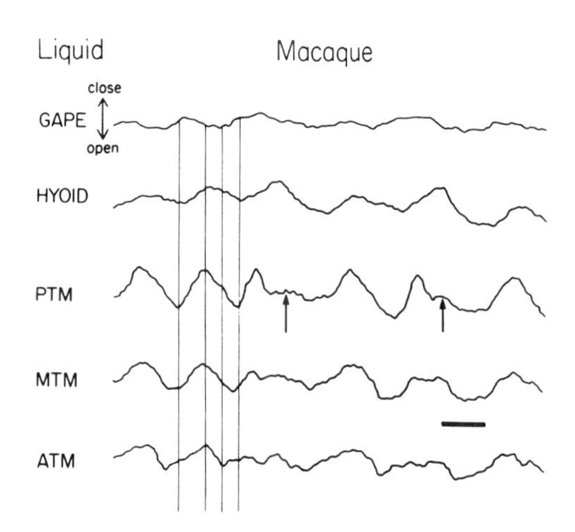
Liquid
Macaque
close
GAPE
open
HYOID
PTM
MTM
ATM

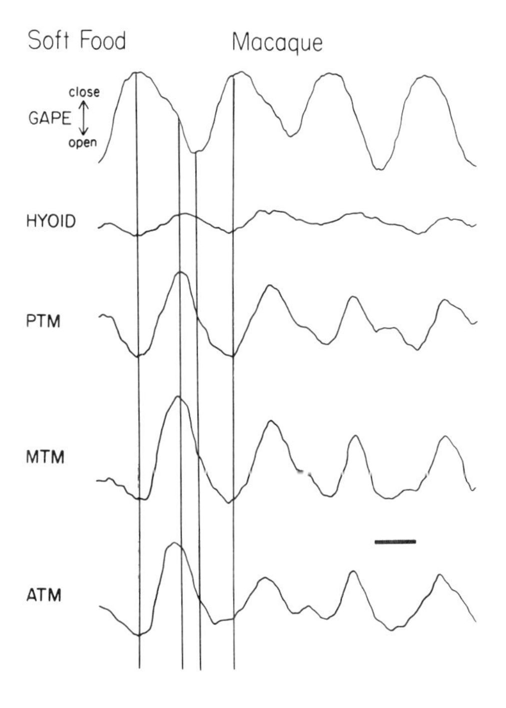
Soft Food
Macaque
close
GAPE
open
HYOID
PTM
MTM
ATM

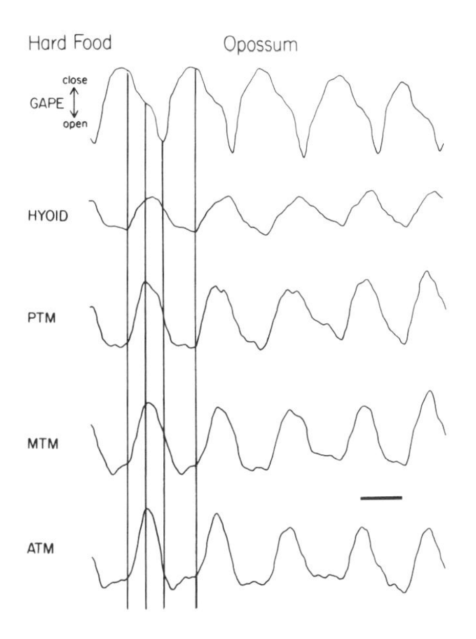

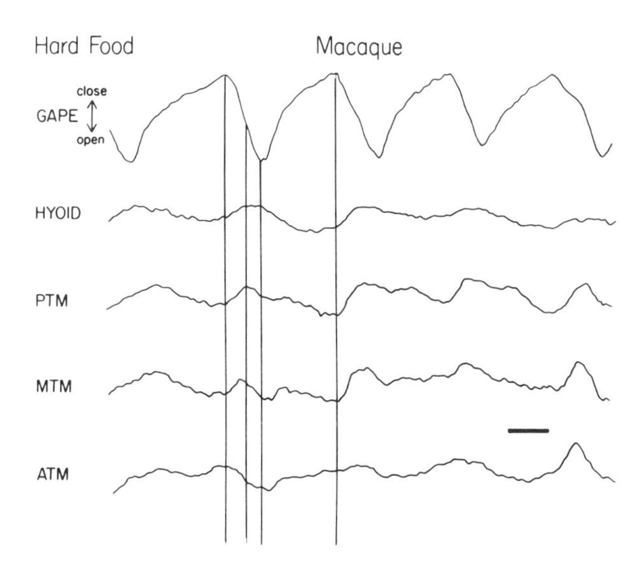

Figure 14-5 Movements of the jaw, hyoid, and tongue in the opossum and macaque when feeding on liquids and soft and hard solid foods. The heavy line shows 250 msec elapsed time. Vertical reference lines show minimum and maximum gape and *PTM* reversal (compare with Fig. 14-8). *ATM, MTM,* and *PTM* = anterior, middle, and posterior tongue markers. Arrows (macaque for liquid) show times of swallowing.

gestion, stage I transport, mastication and manipulation, stage II transport) in the handling of solids (Fig. 14-1).

Ingestion of Solid Food Solid food is either picked up and collected by the tongue and anterior teeth, often with the assistance of the lips (as in grazing herbivores), or is forcibly separated from the matrix using the anterior (but in some cases the postcanine) teeth (as in opossums, rodents, primates). Pickup or collection cycles have frequently been recorded in laboratory experiments. The jaw is opened, the food often licked and then rapidly grasped either by the anterior teeth or by the tip of the tongue and the upper anterior teeth as the jaw closes. The head is then withdrawn from the food source, elevated, and the masticatory sequence begins. Many mammals, such as rodents and primates, hold a food object (a nut, fruit, or a leafy twig) in their hands. They then cut or tear portions from it using the incisors as much for grasping as cutting; the portion of the food in the mouth is torn away from its matrix by head or hand movement. When rats gnaw on hard, brittle bariumized biscuit (Hiiemae and Ardran, 1968), pieces are chipped away by a series of repetitive cyclical chiseling movements in which the lower incisors move upward into the food (closing and power strokes) and then downward (opening stroke). With each repetition small fragments of material accumulate in the diastema immediately behind the incisors. Whatever method is used to take material into the front of the mouth, it is then transported to the postcanines for processing.

In many tetrapods, including mammals (the

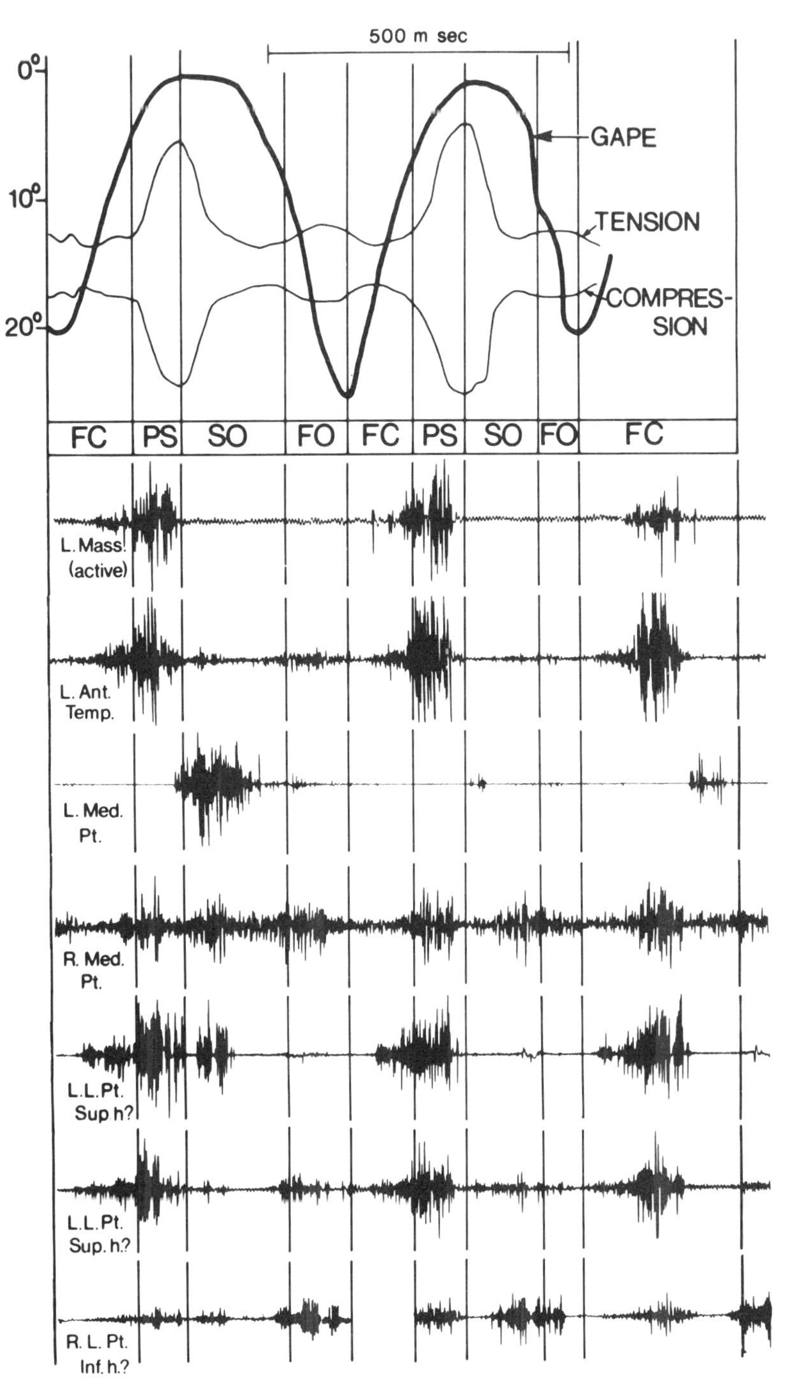

Figure 14-6 The relationships among jaw movement *(heavy line)*, analyzed from 16-mm X-ray cinefilm; strain *(fine lines)*, measured from gauges attached to the bone of the lower jaw; and electromyograms for two successive chewing cycles in a macaque. Electromyography from the left masseter and left anterior temporalis show peaks early in the power stroke. Strain in the jaw peaks at near minimum gape and tails off rapidly. Activity in the medial pterygoid during slow open is associated with lateral translation of the lower jaw. The sites of the electrodes in the lateral pterygoid were not determined by post mortem dissection. *FC* = fast close, *PS* = power stroke, *SO* = slow open, *FO* = fast open.

opossum, the tenrec, and small carnivores), food is ingested by inertial feeding (Chapter 13). Food is gripped by the incisors and the head is rapidly withdrawn and brought to a sudden stop; at the same time the jaws are rapidly opened to release the food, which then moves under its own inertia into the oral cavity. Detailed records and analyses of the path of jaw, tongue, and hyoid movements during the ingestion of solid food have not been published for any mammal, although they are

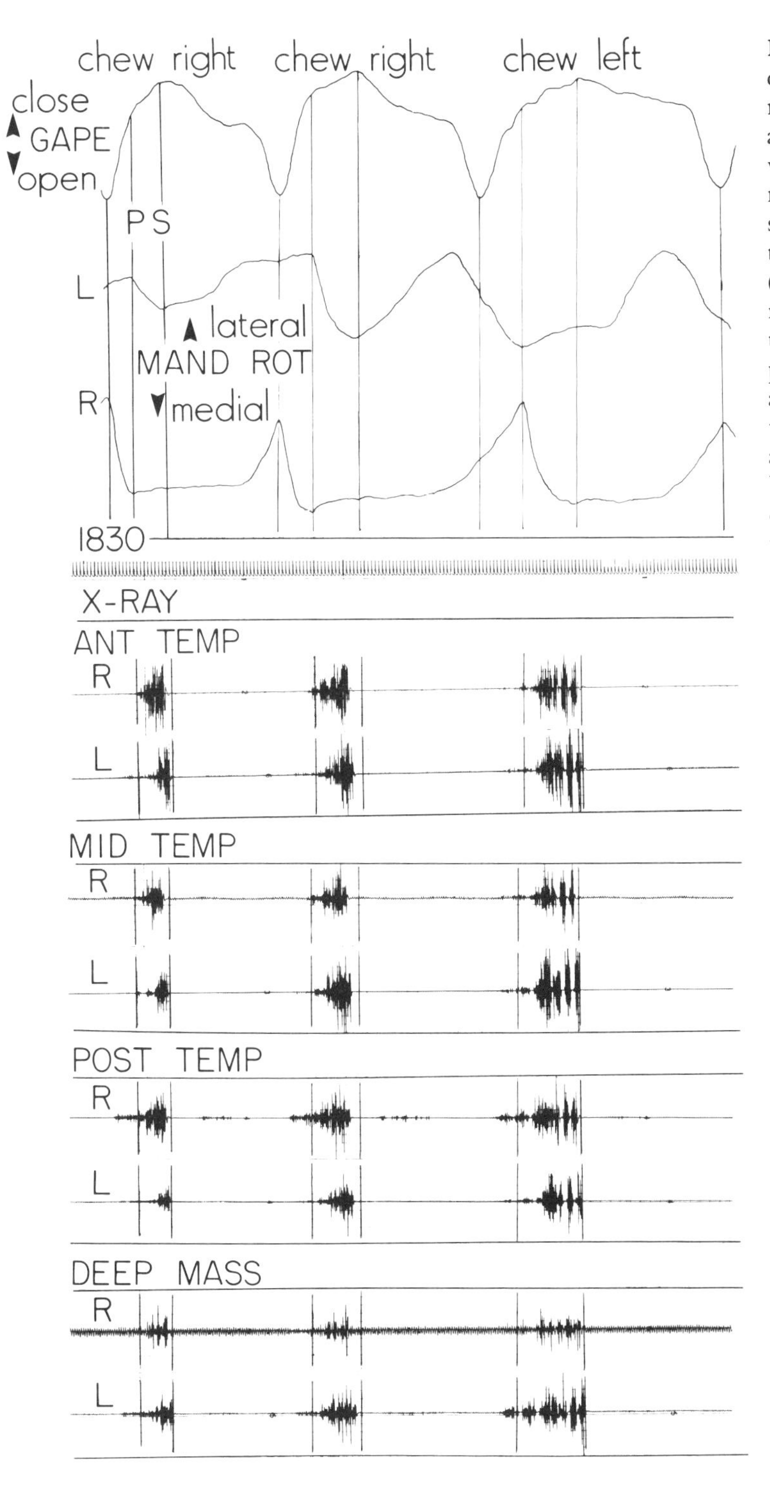

Figure 14-7 Three chewing cycles in the opossum showing the relative rotation between the active and passive sides of the mandible, which is made possible by the mobile nature of the mandibular symphysis. In the first two cycles the hemimandible on the right (active) side rotates toward the midline during *FC* and returns toward the "straight" lateral position during *PS*. These rotations are reversed in the third cycle where active and balancing sides are reversed. The left hemimandible makes reciprocal motions. The change in the electromyograms of the adductors during *PS* demonstrates the differential activity of the muscles of active and balancing sides. (Based on unpublished data of Crompton and Crabtree.)

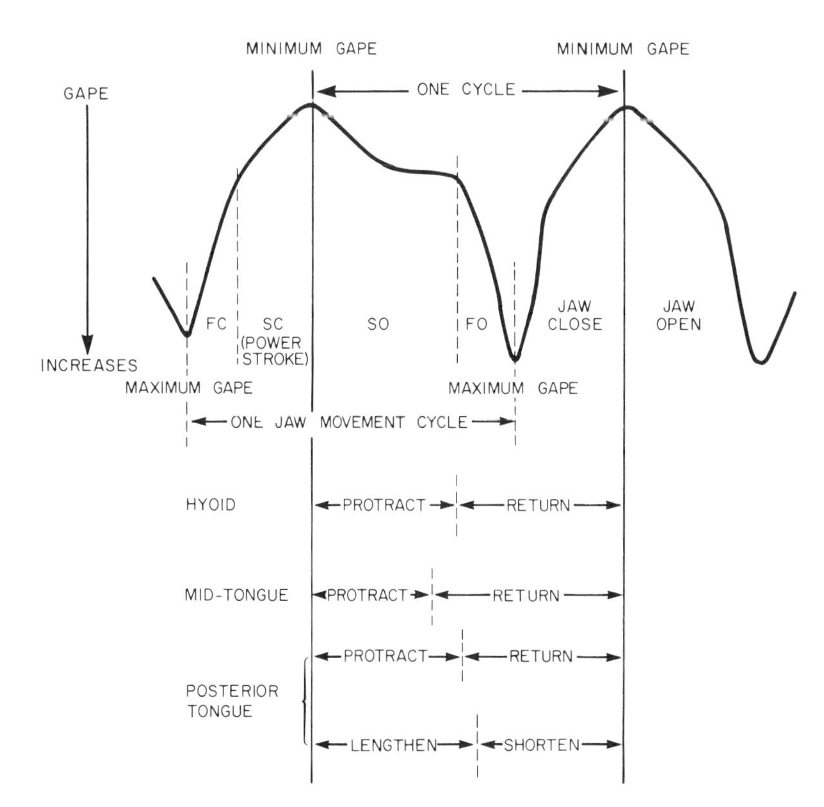

Figure 14-8 The relationship of tongue and hyoid movement to the cycle of jaw movement. Data from hyrax and opossum show that hyoid and tongue movements correlate predictably with jaw movement only at minimum gape when the hyoid and posterior two thirds of the tongue begin to protract and the posterior third to expand. *SC* = slow close; other symbols identified in Fig. 14-6.

now available for cat, opossum, tenrec, and macaque.

Mastication Although it has long been recognized that there is a general correlation between tooth form and type of diet in mammals (for instance, carnassials in carnivores, lophodont molars in herbivores), it is only recently that the mechanical properties of foods and the pattern of their comminution have been considered in the evaluation of tooth form. Following Rensburger's (1973) study of occlusion and wear patterns in fossil and recent rodents and comparable work by Janis (1979) on hyraxes, more attention has focused on the mechanics of food breakdown (Lucas, 1979, 1980) and the relationship between wear and enamel structure (for example, von Koenigswald, 1982). A new and important area is the form of microwear on the tooth surface and its value as an indicator of both direction of movement and of diet (for example, Walker, Hoeck, and Renez, 1978; Ryan, 1979; Covert and Kay, 1981; Grine, 1981; Kay and Covert, 1983; Gordon, 1984).

The molars of metatheria and eutheria are derived from the basic tribosphenic form found in the early mammals of the Cretaceous, family Aegialidontidae (Crompton and Kielan-Jaworowska, 1978). These teeth are characterized by high pointed cusps, long shearing edges (blades) on the slopes of the cusps, and a limited "mortar and pestle" system in the protocone and talonid basin (Fig. 14-9). The efficiency of this design for the breakdown of a wide variety of foods has been documented in the American opossum, which can rapidly reduce materials as hard as chicken bone and as fibrous as skin to a condition suitable for swallowing. With the expansion of mammals into new habitats in the Paleocene and Eocene, modifications of this basic design developed, producing the long, high-profile blades of the carnassials of carnivores, the parallel arrays of low-profile blades found on the ridges of the molars of rodents and herbivores, and the enhanced mortar and pestle system of the Anthropoidea (Fig. 14-10).

For food to be broken down, it must first be positioned between the occlusal surfaces of the

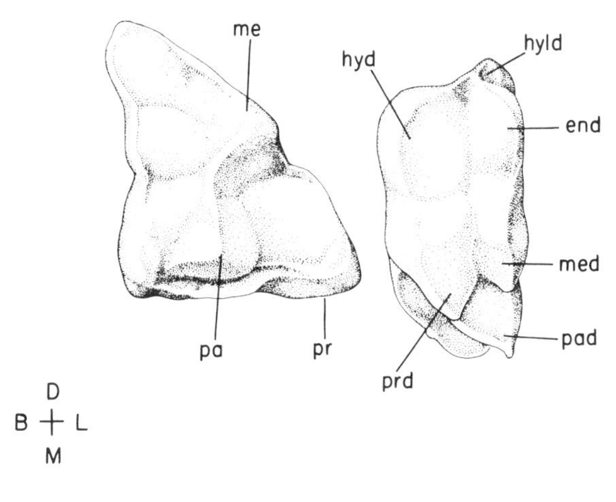

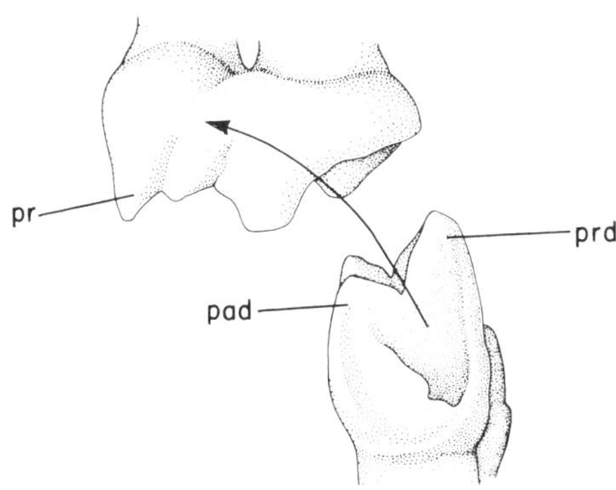

Figure 14-9 *Above:* Occlusal views of upper and lower molars of the American opossum showing the basic tribosphenic pattern with its combination of ridges (blades), cusps (pestles), and basins (mortars). *Below:* Mesial *(lower molar)* and distal *(upper molar)* profiles of the molars of the opossum to show the main matching shearing surfaces. Since the lower molar occludes between the upper molar of its own number and the molar distal to it, the upper molar in the figure has been rotated to show the distal surface, that is, the surface across which the large blade on the lower molar cuts during the power stroke. *end* = entoconid; *hyd* = hypoconid; *hyld* = hypoconulid; *me* = metacone; *med* = metaconid; *pa* = paracone; *pad* = paraconid; *pr* = protocone; *prd* = protoconid.

teeth on the working side (in most mammals mastication is unilateral). In a few animals, including man and rodents, the transverse distance between the upper and lower molar rows is similar (isognathous); bilateral mastication can occur in rodents (Hiiemae and Ardran, 1968) and some coincidental trituration may occur on the balancing side in man. Regardless of whether mastication is unilateral or bilateral, the food to be reduced must be retained between the teeth during the power stroke. The food particles are positioned by the tongue acting in conjunction with the buccinator muscles of the cheeks. Retention (see Osborn and Lumsden, 1978) is achieved either by features on the teeth (the notch in a carnassial, for example) or, for large lumps of material, by the continued action of the tongue and cheek or even a hand.

Food is broken down by the generation of internal fracture patterns. These patterns may form so rapidly and the effect be so dramatic that the material is cleanly separated in one stroke, as happens when a blade cuts through fruit or a nut shatters under pressure. Alternatively, the process may take longer or be slower, as happens in fibrous materials, which require repeated strokes to disrupt their structure. Lucas (1982, 1983) notes that fractures are generated either by blades, which are very narrow in one dimension and are usually, but not exclusively, arranged in opposing pairs (Fig. 14-10B); or by mortar and pestle combinations, where the "pestle" is a blunt convex surface and the "mortar" a reciprocal and, usually, larger concave surface (Fig. 14-10A). Both elements can be found in serial arrays or in combinations. A third pattern is a serial array of low-profile blades that act as a milling machine (Fig. 14-10C).

Foods that are hard and brittle (nuts) or turgid (fruit pulp) are most effectively crushed by a mortar and pestle system (Fig. 14-10A). Secondary fractures are propagated through brittle material. Turgid food requires that the cell walls fail, but no further comminution is needed; a mortar and pestle can crush many cells with a single stroke, blades only a few. A tight-fitting pestle increases retention but creates problems in clearing the fractured particles from the working surfaces. A mortar and pestle that do not fit tightly permit both vertical and transverse movement, first compressing the material and then rolling it (compression crush and rolling crush, respectively, Osborn and Lumsden, 1978). A spider monkey *Ateles* has molars functioning as mortars and pestles to break up fruit pulp. The ridges

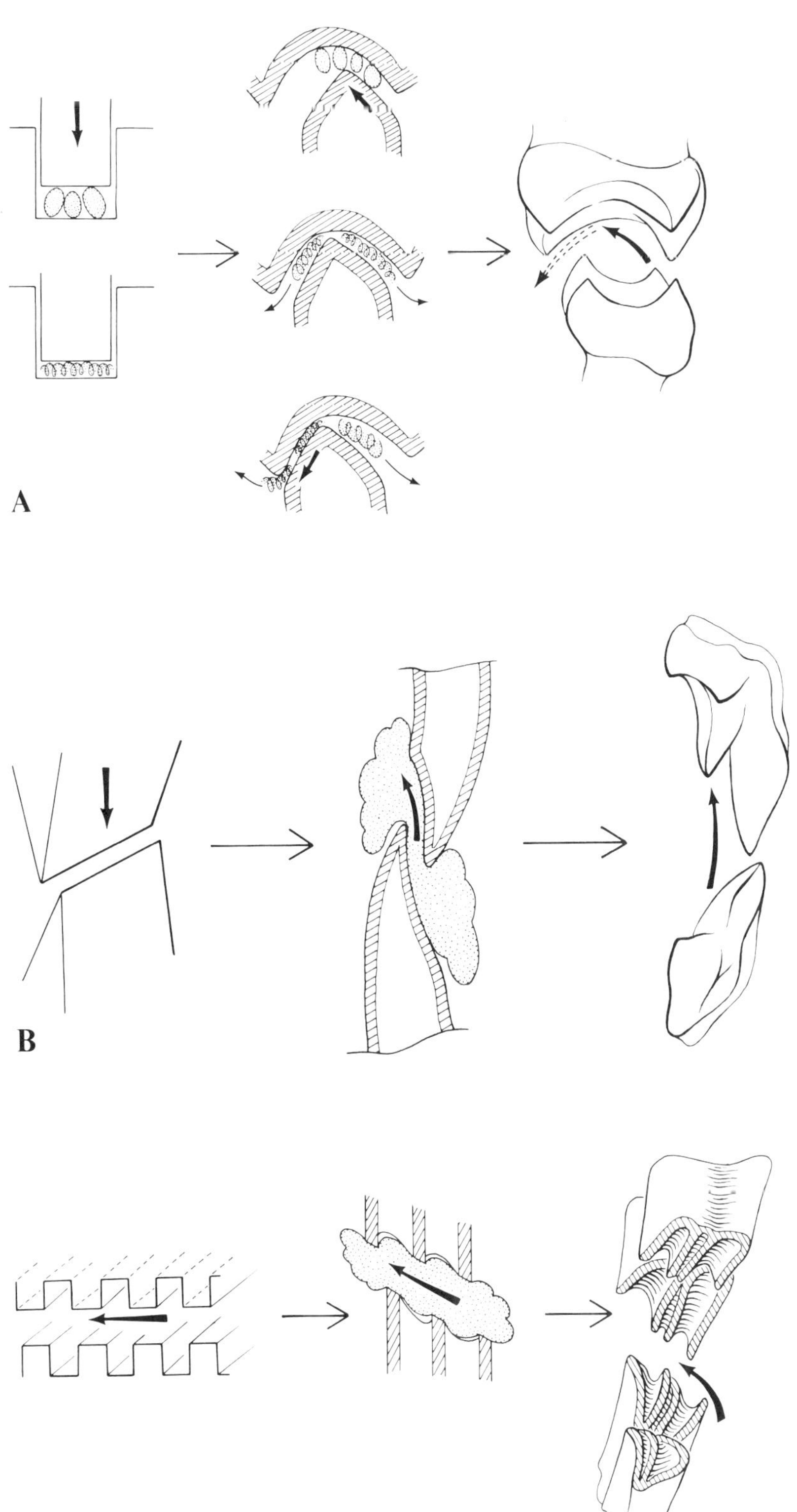

Figure 14-10 Mechanical principles of tooth design in relation to the nature of the food: hard, brittle, or turgid *(A)*; soft, tough *(B)*; or tough, fibrous *(C)*.

forming the margin of the occlusal surface also act to cut through and compartmentalize the fruit, thereby serving as the walls of a compression chamber (Kay and Hiiemae, 1976).

Foods such as muscle and skin that are soft but tough are most efficiently cut by blades (Fig. 14-10B). The simplest blades are found in the carnassials of various carnivores but are also found on the edges of cusps.

Tough, fibrous foods such as grass are best processed by the combination of compression and tearing provided by a serial array of low-profile blades acting as a milling machine (Fig. 14-10C). A system of enamel ridges and dentine troughs is formed and maintained by the different rates of wear of these materials. Such teeth are typical of ungulates, subungulates, and rodents. Although a newly erupted tooth may have simple blade structure along the slopes of its cusps, in many mammals the enamel cover is rapidly worn away, leaving two enamel edges with a softer dentine surface between. This is the most efficient condition for a parallel array of low-relief blades that are used to reduce grasses by cutting on a transverse rather than a vertical stroke. The molars of some rodents erupt with small enamel-free areas just under the cusp tips. The tips break off almost as soon as the teeth are used, producing the two enamel edges associated with horizontally oriented blades.

Many food items contain combinations of materials, each with different fracture properties. For example, many fruits have tough, fibrous skins and soft, fleshy pulps. According to Lucas (1983), insects can be considered as "fluid-filled sealed tubes," the limiting factor in their reduction being the nature of the tube. A soft-skinned larva readily bursts in a mortar and pestle; an adult with its chitinous exoskeleton is more efficiently opened by a blade. Materials such as these can be most effectively reduced by a dentition with a combination of both elements. Such an arrangement can be found both in individual teeth, for example, in the molars of prosimian primates and primitive mammals such as the opossum or, alternatively, along the tooth row, such as the combination of blades (the carnassials) and mortars and pestles (the molars) in dogs.

A third variant is the separation of function between anterior and cheek teeth, as in rodents that gnaw the fibrous shell of a nut using the incisors and then chew the kernel. Many primates peel a fruit using their incisors to grip the skin and the action of the arm and hand to pull the fruit away. The skin is then dropped and the process repeated. Once peeled, the flesh is transported to the molar region for pulping. The more homogeneous a dentition in terms of its mechanical design, the more likely it is that the animal has a diet dominated by a single type of food: examples are the grazing artiodactyls and the felid carnivores.

Jaw Closing Food breakage occurs during the power stroke as the force generated by the elevator muscles is applied through the teeth to the material between them. The power stroke is defined as beginning with tooth-food-tooth contact and ending at minimum gape, the end of the vertical movement approximating upper and lower molars. The jaws must be opened for the teeth to be repositioned for another stroke and for comminuted fragments to be collected and transported by the tongue. The upper and lower teeth are moved toward and away from each other by a combination of cranial elevation and depression and by mandibular elevation and depression relative to the moving cranium. Each gape cycle can be divided into opening and closing movements, both of which may be further divided into phases so that closing has a fast close (*FC*) and a slow close (*SC*) or power stroke phase, and opening has a slow open (*SO*) and a fast open (*FO*) phase (Fig. 14-8). The transitions between *FC* and *SC* phases and *SO* and *FO* phases may be indistinct in gape-time plots since the characteristics of jaw movement depend on the size and consistency of the food, the stage in the masticatory sequence, and the habitual amplitude of movement in each mammal.

Some mammals, notably the American opossum and carnivores, always open their mouths much more widely when eating solid food than its dimensions would appear to require (Thexton, Hiiemae, and Crompton, 1980). Other mammals, and especially herbivores, have indistinct transitions between the *FC* and *SC* phases and no clearly discernible change of rate between *SO* and *FO*. All four phases can nonetheless be distinguished on other criteria (see below). The transitions between jaw closing and opening are, however, clearly defined in behavioral records. The point at which closure ceases and opening begins (minimum gape) represents the end of food processing in that cycle and the initiation of food transport. As such it is the point in the jaw movement cycle that is highly correlated with transition points in both the hyoid cycle and the tongue surface movement cycle (Fig. 14-8). Although the power stroke ends at minimum gape, this may not coincide with full occlusion or maximum intercuspation of the teeth. The opossum and the tenrec, which have high-profiled postcanines and mobile symphyses, rotate the two halves of the lower jaw along their long axes as the mandible moves through the chewing cycle (Hiiemae and Crompton, 1971; Crompton, Oron, and Crabtree, unpublished data). In the opossum the hemimandible on the balancing side both rotates (during the power stroke) and moves laterally, whereas during the remainder of the cycle it moves medially (Fig. 14-7). In contrast, the most dorsal edge of the hemimandible on the active side rotates sharply during *FC*, stays rotated through the power stroke, and rotates sharply laterally during *FO* (T. Crabtree, personal communication).

With the exception of those few mammals (some rodents) that have lower cheek teeth and jaws wider than those of the upper jaw, chewing can take place only on one side (the active or working side) at any one time. In all mammals so far examined (some rodents excepted), the direction of lower jaw movement in the power stroke is upward (toward the upper molars) and medial (toward the midline), with a variable anterior component. During puncture crushing of large, hard objects such as bone, the power stroke ceases when the teeth are still separated; medial movement of the lower jaw on the power stroke is then only minimal (Oron and Crompton, unpublished data). So consistent a pattern among mammals with very different tooth morphologies is readily explained, given the common origin of their teeth and the basic design of the mammalian jaw apparatus. As later molar forms evolved, concomitant changes occurred in the amplitude and angulation of the power stroke. For example, the anterior component is eliminated and the medial component is very small in felids, as a scissorlike action does not require major medial movement of the carnassials; in contrast, the medial component is exaggerated in most herbivores, allowing a long lateral-to-medial traverse for the serially arranged low blades. The anterior component is elongated in many rodents, given the geometry of the dentition (Hiiemae and Ardran, 1968). The morphology of the jaw joint reflects either the predominant direction of the movement required for food breakdown (for example, the anteroposteriorly oriented condyle and glenoid fossa in rodents), or the mechanics of the system (for example, the excessive rotation of the mandible in tenrecs). The pre- and postglenoid processes of the fossa in carnivores both restrict movement and prevent dislocation.

To position the teeth for the power stroke, the lower jaw has to be moved upward and laterally from its point of widest opening. The first stages of this movement (to tooth-food-tooth or tft contact) occur without resistance, often very rapidly (*FC* phase), and are associated with low levels of activity in the adductor muscles. As soon as tft contact occurs, the rate of jaw closure slows. The rate change between *FC* and *SC* (the power stroke) and the rate at which the teeth are approximated during that stroke depend in large part on both the size and consistency of the food (Thexton, Hiiemae, and Crompton, 1980; Fig. 14-5). Activity in the elevator muscles rapidly increases on tft contact, reaching peak levels during *SC*, but ceases before or close to the point of minimum gape (Figs. 14-6, 14-7). The rapid rise in the force exerted on the food is reflected by the pattern of strain on the mandible. This peaks at or before minimum gape (Hylander and Crompton, 1980; Hylander, 1984). In animals such as the rabbit where the measurable minimum gape may not correspond exactly to maximal occlusion given the orientation and form of the tooth surfaces, the strain also peaks before completion of the power stroke (Weijs and de Jongh, 1977).

Since the action of high blades, low blades, or mortars and pestles depends on the powered approximation of their components, the power stroke is completed, strictly speaking, when the teeth reach their maximum contact (minimal separation or maximum intercuspation in occlusion) for that cycle and when the gape is therefore least. It follows that minimum gape does not always occur with the teeth in full occlusion. It also follows that minimum gape may appear to last for some time if the teeth are moving but in occlusion. Records for complete masticatory sequences (ingestion-deglutition) in cats and opossums eating hard foods show that the teeth may fail to break the food in the earliest cycles so the minimum gape is large. With successive power strokes, some firm foods are progressively reduced and the teeth move closer together with each cycle until they finally reach full occlusion at the end of the power stroke. This is true, for example, of the hyrax eating most vegetable foods, of the opossum eating apples, and of the tenrec eating cooked flesh. When irregularly shaped hard substances such as bone are chewed, a sequence of cycles with progressively reduced gapes is not observed; instead, gapes are highly variable as the object is manipulated in the mouth and engaged by the teeth in different positions. It is only when a certain consistency is reached that progressive cycles with reduced gapes are observed.

Early cycles in a masticatory sequence have been distinguished as puncture crushing and the later ones as chewing (Hiiemae, 1976). This broad categorization, although it masks the progressive change that occurs as food is reduced, does reflect both a difference in the mechanics of food breakdown and the control of the lower jaw movement. In puncture crushing the high points of the tooth crowns engage the food to break and crush it without the shearing blades being brought into play, resulting in wear predominantly on the cusp tips. In some animals food is crushed between a pestle (protocone) on the upper molars and a mortar (talonid basin) on the lower molars. Puncture crushing is less important in herbivores where small quantities of vegetation are broken down immediately at the beginning of a masticatory sequence by the low shearing blades on the molars. The opossum can generate on the order of 17 kg/mm^2 on its postcanines in a static bite; chewing forces of 5 – 15 kg are reported for human molars, with the potential to reach more than 100 kg in Inuit males (Jenkins, 1978). The predominant direction of movement in the power stroke is vertical. As soon as the particle size of the food is sufficiently small for the teeth to come into intercuspal range, the precise positioning associated with a complex occlusion can be achieved. Once this occurs, subsequent movement is largely controlled by the form of the teeth (cuspal guidance). The physical dimensions of many foods, such as blades of grass, are such that the teeth can come into very close apposition at the beginning of the first power stroke. Similarly, low-resistance material, such as fruit pulp, can be fractured in a single stroke, the teeth moving through a puncture-crushing stage and into occlusion. The precision and regularity of movement during the power stroke of chewing produces a distinctive pattern of wear facets on the surfaces of the teeth, which can be used to track relative tooth movement in a given animal. This pattern of wear has also proved valuable in tracing the phylogeny of tooth form, since the fundamental occlusal relations of the primary features of the crown in primitive mammals are maintained throughout evolution (Crompton, 1971; Crompton and Jenkins, 1979).

The movement of upper and lower teeth into the food can produce two types of wear facets. In those animals that crush food the tips of cusps are often rapidly worn away, producing cavitations or basins floored by dentine and ringed by enamel (Kay and Hiiemae, 1974). Similar cavitation of the cusps occurs when they are used as pestles. However, wear facets are also produced on the slopes of cusps where the ridges on those slopes are used for shearing. Almost all facets result from the wear on the teeth occurring during the power stroke. Originally called buccal phase facets by Mills (1967), Kay and Hiiemae (1973) used the term *phase I facets* to denote facets formed in primates as the teeth moved into centric occlusion. The few facets formed after the teeth reach centric occlusion, that is, as the jaws begin to open but while the upper and lower teeth are still in transitory contact, were called *phase II facets.* The relationship between phase I and phase II as descriptions of facet origin and the actual movements of the jaws producing them has caused some confusion. The experimental evidence shows that electromyographic activity of adductors ceases before centric occlusion is reached, although given the decay in tension (50–75 msec in macaque) characteristic of striated muscle, there is still some residual, albeit rapidly declining, force acting on the lower jaw (Fig. 14-6) during the period extending into the first part of opening. The beginning of the jaw opening in mammals is always associated with varying degrees of forward movement of the mandible (greatest in those forms in which the condyle lies high above the lower molar row). It is often also associated with a continuation of transverse movement (Janis, 1979). The residual elevator force coupled with the force from the muscles generating the medial and anterior movement could hold the teeth in contact until the cusps clear each other. According to Kay and Hiiemae (1974), phase II facets are formed during this process.

Figures 14-6 and 14-7 show a remarkable, if not unexpected, consistency in the general pattern of electromyographic activity during masticatory cycles. The adductors (temporalis, masseter, medial pterygoid) are primarily active during the *FC* and *SC* phases. There are, however, important differences in the timing and amplitude of activity between corresponding muscles on the active and balancing sides. It is these subtle differences that control jaw movements in the transverse and anteroposterior directions (Weijs and Dantuma, 1975, 1981). Some muscles, such as the medial pterygoids (contracting differentially) produce transverse (medial) movement and are therefore active when jaw movement is occurring in conjunction with mandibular depression or elevation. The force applied to soft food is largely derived from the adductors on the working side. As food hardness increases, the balancing-side muscles are progressively recruited until the two sides show almost identical levels of activity (Hylander, 1979). Hylander (1975) has shown that not only are the jaw joints of mammals loaded during the power stroke of mastication but that the greater load is on the joint of the balancing side.

Jaw Opening The jaws open slowly at first and then accelerate toward maximum gape. Where there is a distinct change of rate, two opening phases (*SO* and *FO*) can be identified (Figs. 14-5, 14-7, 14-8). A two-phase opening movement is routinely seen in the opossum, tenrec, and cat, which have high-amplitude jaw movement. Such a clear phase shift is rare in macaque, hyrax, and other herbivores, which have low-amplitude cycles. Experimental work has shown that during the opening and the following closing movements of the jaws, a complex cycle of activity is occurring in both the tongue body and the hyoid, associated with food manipulation and transport. These findings demonstrate that jaw opening is not a simple behavior dependent on the activity of the digastric, which has been classically described as the depressor of the mandible, nor on the digastric and the lateral pterygoid muscles.

The specific anatomical relations of the jaw muscles differ widely among mammals. In opossums and primates the digastric is connected to the hyoid through its intermediate tendon. In the cat, tenrec, insectivores, and most herbivores the digastric has no hyoid connection. Although it is not always possible to recognize an *SO* and *FO* phase transition in gape-time plots, two distinct phases can nevertheless be recognized from tongue and hyoid movement. At minimum gape the hyoid is at its most depressed and retracted position (Figs. 14-5, 14-8). As the jaw begins to open, the hyoid moves upward and forward. This movement is associated with a powerful contraction of the anterior suprahyoid muscles, especially the geniohyoid, and is often associated with low-level activity in the elevator muscles, which can be interpreted as regulating the rate of jaw opening to allow the hyoid to achieve its full upward and forward movement (see below). In those cases where a distinct change of rate between an *SO* and an *FO* phase is seen (opossum, cat), the hyoid reaches its maximum forward and, usually, upward position either just before or coincident with the transition from *SO* to *FO* when eating soft foods. When eating hard foods there is a pause at about this point, but the reversal to hyoid retraction may occur nearer maximum gape. Even when no such clear transition can be seen in a gape-time plot, hyoid movement reverses abruptly during opening so that at some point the hyoid starts traveling downward and backward (Figs. 14-5, 14-8). Although the digastric may show some electromyographic activity during the first part of jaw opening, its main burst occurs during the second phase (Fig. 14-9). In animals having the digastric connected to the hyoid, its activity contributes both to jaw opening and to hyoid retraction; in the absence of such a connection it acts as a primary jaw opener during later opening (*FO*). At this time its activity is aided by the combined action of the infrahyoid and geniohyoid muscles, which both depress the jaw and retract the hyoid.

In most of the mammals so far studied (see Thexton, Hiiemae, and Crompton, 1980) the amplitude of hyoid movement during jaw opening can vary widely, depending on the condition of the food (Fig. 14-5). Opening is not a simple vertical mandibular movement: the lower jaw is also moving transversely and anteroposteriorly.

Food Transport and Manipulation

To reach the stomach food has to be actively moved through the mouth, propelled across the oropharynx (deglutition), and pushed down the esophagus (peristalsis). The work involved in moving food depends primarily on its consistency. Liquids flow and some materials such as ice cream can flow and offer little resistance, but solid materials have to be manipulated. In all cases transport movements involve the tongue and hard palate (stage I transport) and the tongue and soft palate (stage II transport).

Manipulation Dogs, cats, and opossums lap liquids; pigs and *Homo sapiens* (with a straw) suck fluids. When liquids are ingested and transported through the mouth, that is, are lapped rather than sucked, the tongue acts as a conveyor belt, moving successive small quantities of liquid (aliquots) from the external source to the oropharynx for bolus formation. In sucking an aliquot is both sucked and swallowed in one cycle. In lapping such a ratio is rare; each animal normally establishes a rhythm (2 : 1, 3 : 1, or even 6 : 1) between lap (collecting stroke/ingestion) cycles and "swallow" cycles. As explained in the section on deglutition, swallowing occurs as an interpolated behavior in the *SO* or *O*1 phase of otherwise typical lap cycles. There is, therefore, no such thing as a specific swallowing cycle. The rhythm used in lapping varies not only among animals but also with the viscosity of the material lapped (Hiiemae and Abbas, 1981); once established it is very stable for that animal and food type (Hiiemae, Thexton, and Crompton, 1978; Crompton and Weijs-Boot, unpublished data).

To describe this "conveyor belt" activity (Fig. 14-11), more than one cycle has to be considered: two lap cycles and a third cycle in which both lapping and a swallow occur. In each case the jaw movement cycle of lapping has three clearly identifiable components: an initial opening movement (*O*1, Figs. 14-11, 14-12) in which the jaws are opened from minimum gape (taken as the start of the cycle) to a low-amplitude gape; a second open phase (*O*2) in which the gape is either maintained or increased, although often with some oscillation, while the now protruded tongue moves through the liquid; and last, a closing (*C*) phase in which the jaw is rapidly elevated as the tongue is retracted. Although the following description is based on lapping in the opossum, the movements of the jaw, hyoid, and tongue in the cat and dog are essentially the same. The macaque movement profiles for liquid shown in Figure 14-5 are a little different, at least for jaw movement, since the animal was taking fluid from a drop bottle. Because the tongue is barely protruded, this is an example of pure stage I transport of liquid.

At the beginning of a lap cycle (at or just before minimum gape), the hyoid begins to move upward and forward (Fig. 14-8), carrying the tongue base forward relative to the palate (Fig. 14-12, *O*1). At the same time, the posterior three quarters of the body of the tongue lengthens (expands), further assisting protraction, although the tip of the tongue contracts. Nevertheless, the resultant movement is sufficient to push the tip downward and backward into the liquid by the end of *O*1. The reversal of hyoid movement marks the junction between the *O*1 and *O*2 phases. As the hyoid begins to move backward, the posterior and middle thirds of the tongue begin to retract (Fig. 14-11, *O*2). At the same time, the tip of the tongue expands and moves forward through the liquid collecting an aliquot

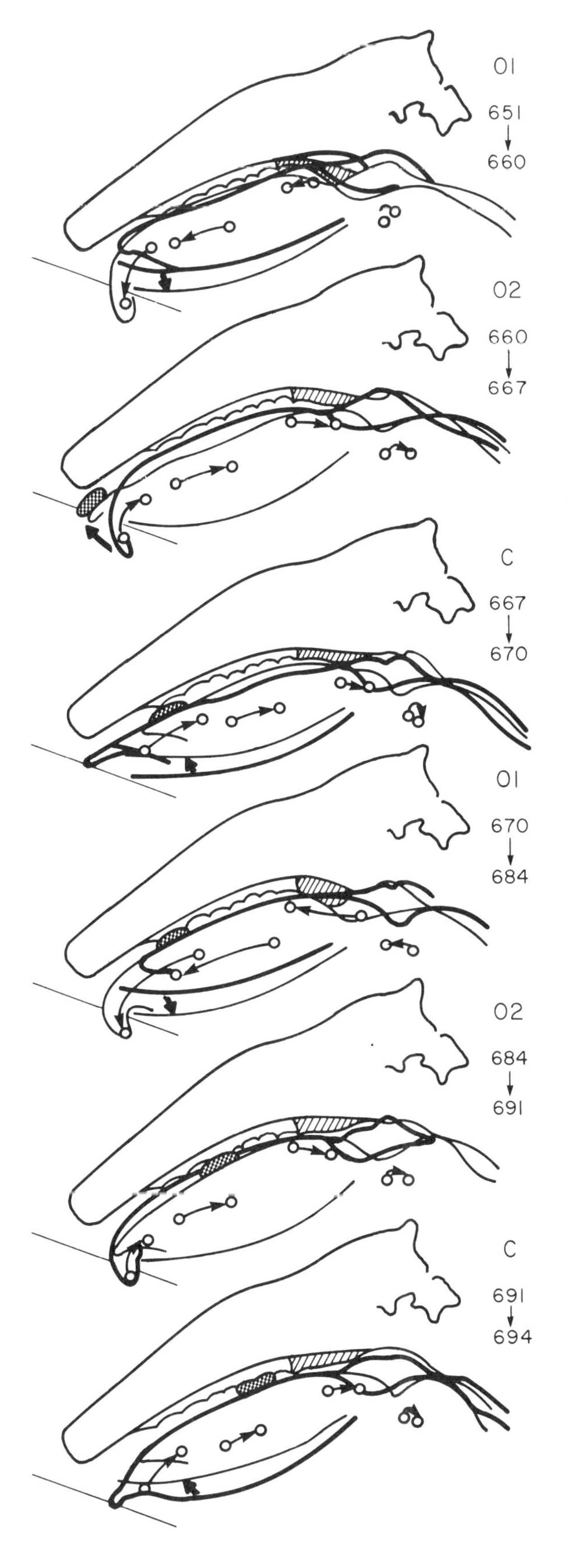

of fluid on its surface. As the tongue tip comes into alignment with the body of the tongue, the tip is sharply withdrawn into the mouth with the aliquot on its surface. As the jaws close, the posterior three quarters of the tongue continues to contract (intrinsic movement) and retract. The anterior one quarter of the tongue remains a constant length during closure, cradling the aliquot.

During the second cycle, two separate but closely related movements of liquid occur: a new aliquot is ingested and the first one is moved distally through the oral cavity (Fig. 14-11). As the jaws begin to open (*O*1), the surface of the tongue elevates, trapping the first aliquot between it and the palatal rugae. The tongue surface moves forward below the aliquot. Proximal movement of the aliquot is hindered by the shape of the rugae, which have steep anterior and gentle posterior slopes. As the tongue retracts in *O*2, its surface loses its tight apposition to the palate; the first aliquot is released and moved posteriorly by the backwardly traveling tongue surface. This distal movement of food across the palate is facilitated by the shape of the rugae. As the second aliquot is ingested in the *C* phase of the second cycle, the first aliquot is carried farther posteriorly to a position immediately in front of the hard-soft palate junction.

As a result of this conveyor belt system, repeated aliquots are added to the oropharynx to form a bolus; when it reaches an optimal size it is swallowed (Hiiemae, Thexton, and Crompton, 1978). This occurs in the middle of *O*1 of a cycle in which lapping is still taking place. Figure 14-12 shows the orbits of movement during the same cycles illustrated in Figures 14-11, 14-13, 14-14, and 14-15. Note that during *O*1 the markers move on a higher plane than during *O*2 and *C*.

Figure 14-11 Two lapping (liquid transport) cycles in the opossum. In the first cycle, the first aliquot is transported as a second is ingested *(shading)*. The fine arrows show the path of movement of the tongue and hyoid markers during each phase. The low amplitude of jaw movement in lapping is shown by the thicker arrows. The fluid level *(fine line)* is shown at an angle to simplify the figure. *O*1 = first open, *O*2 = second open, *C* = close.

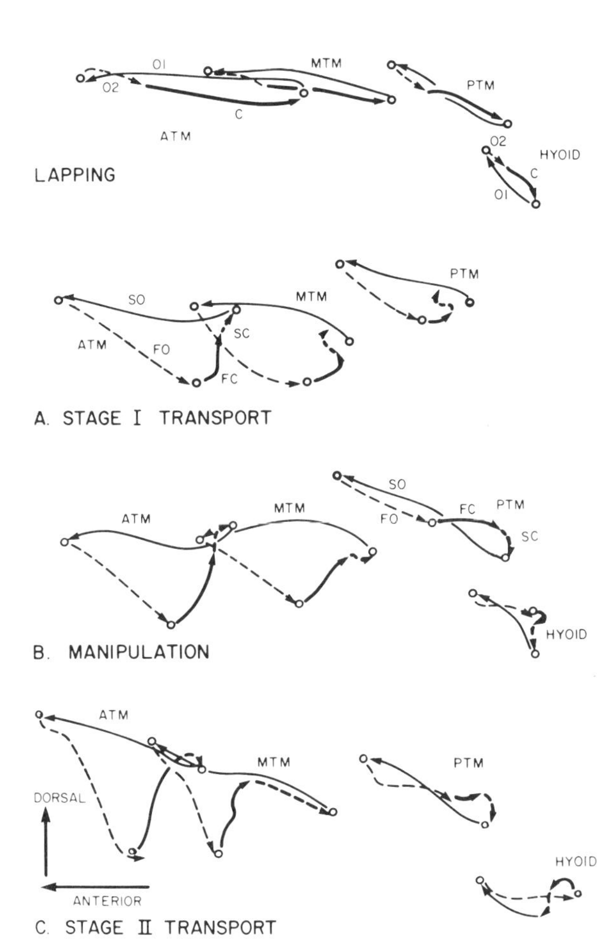

Figure 14-12 The paths of movement (orbits) of the anterior *(ATM)*, middle *(MTM)*, and posterior *(PTM)* tongue markers and the hyoid marker, all plotted relative to the palate during the cycles used to illustrate lapping and transport behaviors (Figs. 14-11, 14-13, 14-14). The effect of jaw movement on orbital profile is shown by the difference in orbital shape for the *ATM* and *MTM* between lapping, with very low-amplitude jaw movement, and manipulation or stage II transport, where the jaw is opened to a large maximum gape.

In *O1* of the third cycle the first two aliquots are trapped in their relative positions, and the tip of the tongue is protruded to pick up a third aliquot. During *O2* and *C* the tongue is withdrawn from its contact with the soft palate and drawn backward, carrying the first aliquot into the oropharynx. At the same time the second and third aliquots are also transported within the oral cavity. During *O1* of the fourth cycle the posterior tongue contacts the soft palate and maintains this contact as it slides forward, which traps the first aliquot in the oropharynx. The oropharynx is bordered by the tongue–soft palate contact, the surface of the posterior part of the tongue, the anterior wall of the epiglottis (the top of which is covered by the soft palate), and dorsally by the soft palate. In opossums, but not necessarily in cats, the oropharynx extends into a large pyriform recess lying on either side of the larynx, and it is in this recess that the major part of the bolus is held (Fig. 14-3). There is some evidence (see Thexton, 1981) that the rapid distal transport of liquids to the oropharynx in cats might be attributable to the formation of a central furrow or trough in the posterior half of the tongue, which rapidly empties from the front part backward and which has been observed in decerebrate cats, where its appearance can be initiated by palatal pressure.

Stage I Transport and Manipulation (Solids) Solid food deposited in the anterior part of the mouth is moved posteriorly by either or both of two processes: a continuation of inertial transport or stage I transport. Carnivores such as the cat and primitive mammals such as the tenrec and the tree shrew *(Tupaia)* regularly use an inertial mechanism to move food into or close to the post-canine area of the mouth.

More often, solid food is moved from the incisal to the premolar/molar area of the mouth by a mechanism essentially identical to that used for stage I transport of liquids. The only significant difference is the effect on tongue shape of the much wider jaw opening associated with the presence of an *FO* phase. Briefly, the food is held against the palate during the *SO* phase (Figs. 14-12, 14-13) as the tongue surface moves forward below it and, cradled in a depression on the tongue surface, it is carried distally as the tongue base retracts and the posterior two thirds of the tongue body contracts during *FO*. During the *FC* phase the food is transported even farther backward as the depression cradling it itself travels distally (Fig. 14-13, *FC* phase).

In summary it appears that all mammals use the same mechanism for the transport of liquids and solids within the oral cavity (that is, toward but not through the pillars of the fauces). The essen-

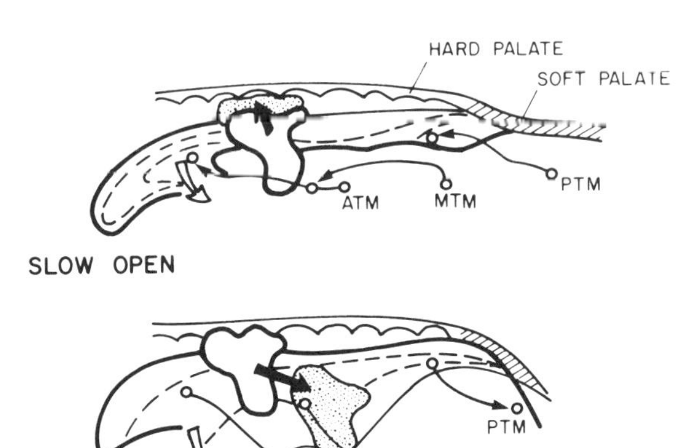

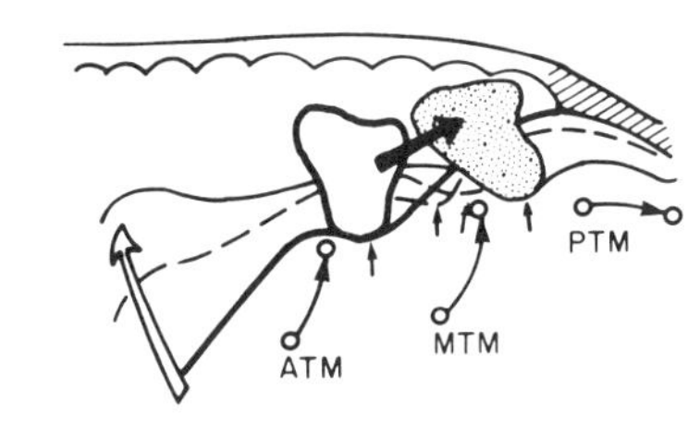

Figure 14-13 Stage I transport of solids. Large hollow arrows show direction and amplitude of jaw movement; solid black arrows show movement of food through each phase. Heavy solid line shows position and shape of the tongue and food at the beginning, and the thin solid line, their position at the end of each phase. Intermediate positions are shown by dashed lines. The position of the anterior *(ATM)*, middle *(MTM)*, posterior *(PTM)* tongue and the hyoid *(HM)* markers is shown at the beginning and end of each phase by hollow circles; their overall movement during each phase by the fine black arrows. The small black arrows in the Fast Close figure show the progressive distal movement of the base of the trough on the tongue surface where the food is held. The net distal travel of the trough being greater than that of the tongue body, the intrinsic musculature of the tongue not only forms a "wave" but propagates it distally.

tial elements of this process are: (1) the forward expansion of the body of the tongue beneath the food as the hyoid travels forward in the *SO* (or *O*1) phase; (2) the retraction of the hyoid and the body of the tongue during the following *FO* and *FC* phases, which carries food on the tongue surface distally; and (3) the distal movement of a trough (or wave) on the tongue surface, cradling an aliquot of food, at a greater rate than that of the return movement of the tongue base and body during the closing phase (*FC*/*SC*), which, in effect, carries food distally.

As solid food is broken down by the postcanines it has to be moved within the posterior part of the oral cavity: adequately triturated material has to be collected, segregated, and moved posteriorly to form a bolus while the inadequately triturated residual lumps have to be repositioned between the teeth for further reduction. Little is known of the methods used by the tongue to segregate and reposition food particles. However, the mechanisms used to move residual material anteroposteriorly within the oral cavity are known. Figure 14-12 shows the movements of the jaws, tongue, and hyoid during a masticatory cycle in which food is being manipulated. (Manipulation is defined as the movement of food within the oral cavity, as opposed to transport which has a definite directional component). As the jaws begin to open in the *SO* phase, the tongue does not, as in stage I transport, force the food against the palate where it can be held as the tongue moves forward; instead, the tongue drops slightly away from the palate and a wall forms behind the food, which is then carried forward by the anterior movement of the tongue base and internal expansion of the posterior two thirds of the tongue, both features typical of the *O*1 phase (Fig. 14-14). With the wide opening of the jaws and the retraction of the tongue seen in the *FO* and *FC* phases, the lump of food is carried backward and returned to a position in the posterior postcanine region, where it is usually positioned between the postcanine teeth. As the teeth engage the food during *SC*, the tongue continues to move and removes triturated material away from the point of bite.

Stage II Transport Stage II transport is defined as the distal movement of food through the pillars of the fauces (or the junction of hard and soft palates) to the oropharynx for bolus formation. In most mammals, but not anthropoid primates and man, a bolus is accumulated in the oropharynx (Fig. 14-3). Some mammals, including the

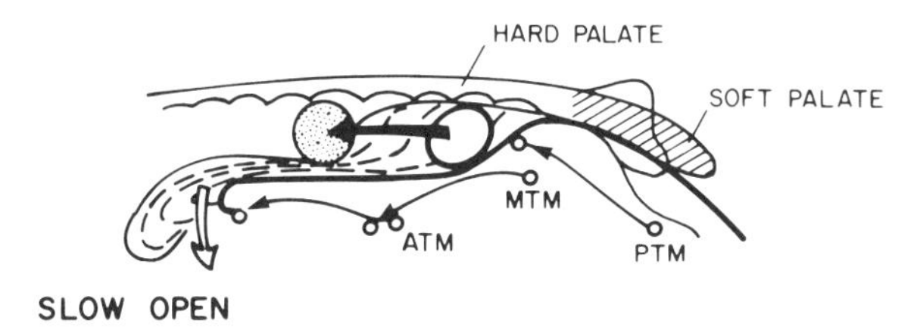

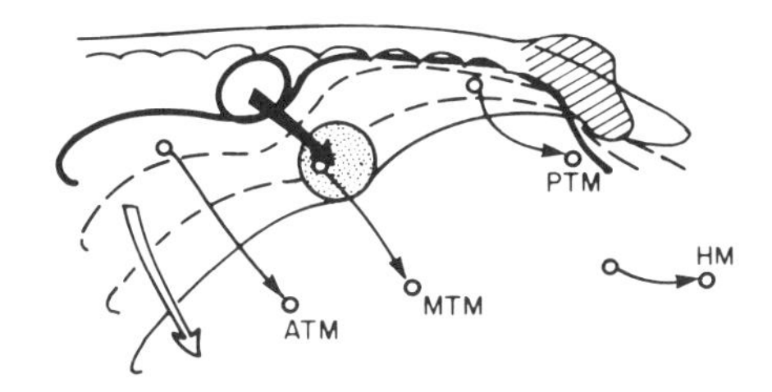

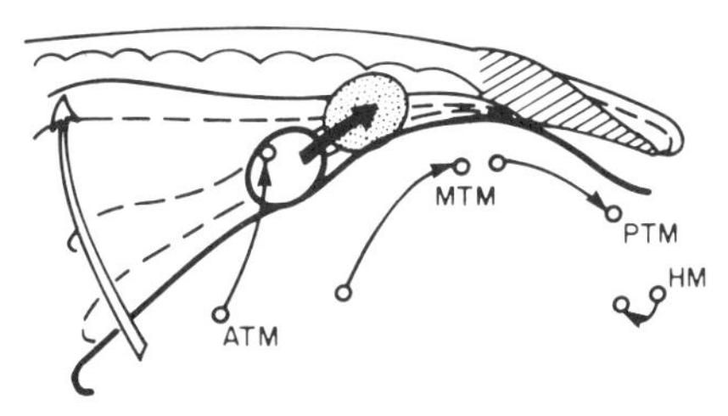

Figure 14-14 The manipulation of food during the *SO, FO,* and *FC* phases of a masticatory cycle (symbols are as in Fig. 14-13). The soft palate changes shape, in large part to maintain its contact with the tongue and thus the tongue-palate seal, as the tongue profile changes during the cycle.

opossum, have a large pyriform recess which, as an extension of the oropharynx, is the principal site for bolus formation. In swallowing, the bolus is moved from the oropharynx over the larynx to the esophagus. In all nonprimates studied (opossum, cat, dog, hyrax, tenrec, pig, and goat), stage II transport and swallowing are separate events.

The mechanism of stage II transport has been adequately documented only in opossums, so the actual mechanism may be slightly different in hyraxes, cats, and dogs, where it has also been recorded cinefluorographically. As mastication proceeds, food reduced to a consistency suitable for swallowing is separated by the tongue from the remainder of the food in the mouth and is collected in a hollow on the dorsum of the tongue below the posterior part of the hard palate (Fig. 14-13, *SO*). Food already in the oropharynx is retained in that position and prevented from moving forward by the tongue–soft palate contact. As the tongue moves forward and upward in the *SO* phase, it expands around the food collected on its dorsum, compressing and trapping it against the hard palate. With the beginning of wide jaw opening in the following *FO* phase, the middle part of the tongue rises in front of the triturated food, which is then cradled in a depression with a high anterior wall (Fig. 14-15, *FO*). During *FO* the anterior and middle thirds of the tongue are carried away from the hard palate. At the same time the soft palate flattens out so that by the *FO-FC* transition an open channel between the mouth and oropharynx has been formed. As the jaws close in the *FC* phase, the tongue not only moves upward but the depression on its surface moves upward and backward (note the position of the middle tongue marker in Fig. 14-15, *FC*) pushing the triturated food under the soft palate and toward the oropharynx to form the bolus. Fig. 14-12 illustrates a marked difference in the orbits of movement of the anterior and middle tongue markers between manipulation and stage II transport in the opossum. In the case of flesh or broken bone associated with flesh, numerous (up to six) stage II cycles are sometimes required to move food through the pillars of the fauces.

Stage II transport occurs in nonanthropoid mammals because the bolus is formed within the oropharynx but anterior to the soft palate–epiglottal seal (Fig. 14-3). Adult *Homo sapiens* has no such seal, so the bolus is formed on the dorsum of the tongue in relation to the hard palate–soft palate junction and the pillars of the fauces. In swallowing the bolus traverses the oropharynx. This part of the swallow is strictly comparable to stage II transport in nonanthropoids, but it does not occur in a separate cycle from that in which swallowing occurs.

There is, however, an interesting variation of second stage transport in macaques (Franks, Hylander, and Crompton, 1981). In these forms a soft palate–epiglottal seal is maintained except during swallowing. Food to be swallowed collects

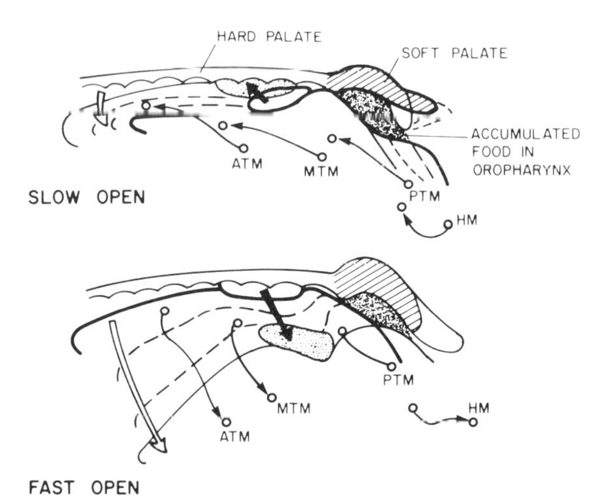

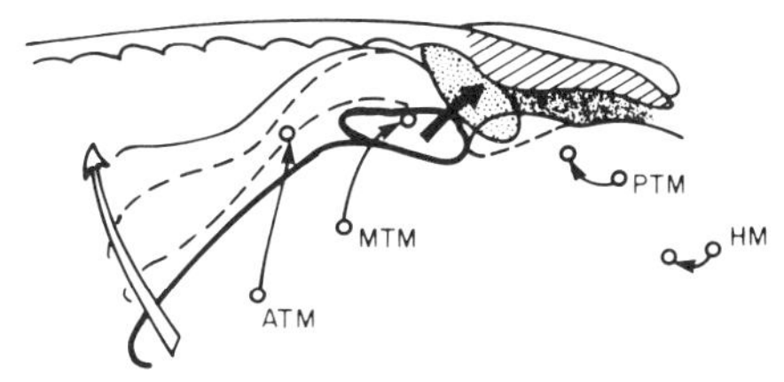

Figure 14-15 Stage II transport in which solid food is moved backward below the soft palate by a mechanism best described as "squeeze-wedge." During the *FO* phase, the contact between the palate and tongue surface (middle oral seal, Figure 14-3) travels distally, pushing food toward the valleculae. At the end of the *FO* and at the beginning of the *FC* phase, the backward and upward movement of the tongue (see *FC* phase) pushes food from the oral cavity into the oropharynx. Symbols identified in Fig. 14-13.

on the dorsum of the tongue immediately in front of the hard-soft palate junction. In the cycle preceding the swallow, a small quantity of the bolus is moved backward to lie between the back of the tongue and the soft palate (that is, in the oropharynx). However, the transport of food to this position does not take place during the *FO* and *FC* phases, as it does in nonprimates, but during the *SO* phase. As the tongue moves forward and upward during the *SO* phase, it contacts the hard palate immediately in front of that part of the bolus to be transported. As it continues to move forward, this contact point is moved progressively backward by a sequential elevation of the back of the tongue. This posterior movement of the contact point takes place despite the fact that the tongue as a whole continues to move forward. During the swallow cycle, the main bolus on the dorsum of the tongue and the small one between the tongue and the soft palate both traverse the oropharynx in the middle of the *SO* phase of jaw movement.

Deglutition Swallowing has been extensively studied in man (Miller, 1982) because isolated swallows can be elicited on command. In mammals feeding normally, swallows occur as an integral and integrated part of the masticatory sequence. A swallow can be defined as the process by which a bolus is moved across the oropharynx to enter the esophagus. This is achieved by means of a distinct series of neuromuscular events that, once initiated, appear to be little influenced by sensory feedback (Miller, 1982). However, in all the animals so far studied, including macaques, this mechanism appears as an "insert" or interpolation in the *SO* phase of an otherwise standard masticatory cycle. As the *SO* phase begins, the characteristic movements of the hyoid and tongue also begin. When a swallow is to occur, the *SO* phase is suspended: hyoid protraction as well as tongue expansion and protraction stop. The movements of the swallow are initiated and completed, after which the *SO* phase resumes. Although it is within human experience to swallow during mastication (that is, with food in the oral cavity), a situation entirely comparable to that found in other mammals, there is no evidence presently available to show the relationship between the swallow and masticatory movements nor between the swallow and food transport.

In a typical *SO* phase in the opossum, the forward and upward movement of the tongue is produced by the contraction of both bellies of the digastric, the genioglossus, geniohyoid, and mylohyoid of both sides, with relaxation of the sternohyoid and genioglossus (Crompton and Sponder, 1982; Crompton and Weijs-Boot, unpublished data). All swallows that fall within a masticatory sequence are preceded by the hyoid and tongue protraction phases of *SO*; that is, the posterior portion of the tongue and hyoid start to move forward and the geniohyoid, genioglossus, and digastrics contract. At rest the soft palate lies snugly against the epiglottis and forms a lateral wall that separates the pyriform recess from the nasopharynx (Fig. 14-3). When the soft palate is elevated in swallowing, the valleculae, pyriform recess, and the space above the larynx become confluent. In a swallow the digastric, genioglossus, geniohyoid, and the mylohyoid continue their activity, and in so doing form a tensed floor for the oral cavity and tongue base. Against this tensed floor the posterior portion of the tongue contracts and is forced backward against the food. At the same time a wave of contraction starts with the superior constrictors and passes ventrally. The combined contractions force the food through the relaxed esophageal sphincter into the esophagus. The posterior movement of the tongue appears to be caused in part by powerful activity of the hyoglossal musculature. During the swallow solid food passes over the epiglottis, which shifts backward to cover the opening of the larynx. After the swallow the tongue again moves forward to complete the remainder of the *SO* protraction phases; the remaining phases of the masticatory cycle and its successor follow in regular order. The sequential contraction of the middle and posterior constrictors after the initial contraction of the anterior constrictor forces the food down the esophagus during the completion of the *SO* and *FO* phases of jaw movement.

In macaques the swallowing mechanism seems to be identical to that described for nonprimates in that the soft palate is raised to meet the posterior pharyngeal wall, thus separating the nasal cavity from the pharynx; the posterior surface of the tongue is forced upward and backward; and, in conjunction with the superior (anterior) constrictor, forces the food through the pharynx (H. A. Franks, personal communication). Since adult man has lost the soft palate–epiglottal contact, it is not possible to store food in the pharynx prior to a swallow as this would interrupt or prevent regular breathing. During mastication, the fauces, the dorsum of the tongue, and the soft palate in the human adult form the seal that separates the oral cavity from the oropharynx. The nasopharynx and oropharynx form a continuous space so that breathing and food manipulation can take place simultaneously (Fig. 14-3). This arrangement is distinct from that of nonhumans and human neonates, where the nasopharynx and oropharynx are separated during the masticatory sequence, but where the soft palate–epiglottal seal is also present.

It is customary to divide the human swallow into two distinct phases: an oral and a pharyngeal phase. The oral phase is the period in which the food is positioned on the dorsum of the tongue in front of the fauces and corresponds to the first part of the *SO* phase immediately preceding the swallow in nonprimate mammals. During the human pharyngeal phase the bolus is transported from the dorsum of the tongue to the esophagus through the oropharynx. The retention of a preswallowed bolus between the tongue and the soft palate seen in macaques is lost.

Control of Mastication

There is a large and continuously expanding literature on aspects of oral neurophysiology (Anderson and Matthews, 1976; Sessle and Hannam, 1976; Doty, 1976; Lowe, 1981; Goldberg and Chandler, 1981; Luschei and Goldberg, 1982). Much attention centers on how the rhythmic activity is initiated and maintained. Recently, a view originally propounded in 1923 by Bremer has gained new acceptance. He argued that a *centre de correlation* (often translated as an "oscillator" or "rhythm generator") is present in the hind brain. Such a center must function by integrating sensory input from the trigeminal, facial (taste), glossopharyngeal, and vagal nerves and by activating the motor components of the same nerves as well as parts of the cervical plexus. The extent to which such a center is preprogrammed to produce a standard output, which is then modulated by peripheral sensory input, is not yet clear. The very uniform firing patterns for the jaw muscles during masticatory cycles is suggestive of a basic program, but the changes in jaw, hyoid, and tongue movement profiles within a single masticatory sequence reflect the impact of sensory input on the output of such a rhythm generator. Equally, the very stereotyped activity of swallowing and its interpolation into an otherwise conventional tongue-jaw cycle is highly suggestive of programming. It is expected that experiments using behavioral monitors (such as synchronous electromyography and movement recording) can be used to investigate this as yet poorly understood area.

The similarity of the basic feeding mechanisms in the extant mammals so far studied has led to the suggestion (Hiiemae, Thexton, and Crompton, 1978; Bramble, 1980) that the neural control system may be very conservative and so phylogenetically old. Further, the characteristics of the hyoid and tongue surface cycles and the pattern of their linkage to the jaw movement cycle suggests that the cyclical movements of the tongue may be the primary element in this complex system. It is clear that further studies should address several basic questions. First, is there a truly mammalian pattern? Although the answer seems to be yes, a careful analysis of jaw and tongue movement data from mammals with the more extreme forms of dietary specialization, such as anteaters and walruses, would offer a direct test. Second, if the tongue in its role as the active component of the food transport system is primary, then it should be possible to show the dominance of a tongue cycle not only in mammalian ontogeny but also in mammalian phylogeny. Although the vertebrates available for experimentation, such as lizards, are not those currently accepted as linearly related to mammals (Throckmorton, 1978; Smith, 1982) or fish (Lauder, 1979, 1982), experimental studies designed to address the question whether the tongue and pharyngeal structures show cyclical behavior like that of mammals could contribute valuable information even if unable to confirm the pattern of evolution of this system. Third, if, as has been suggested here, the jaw apparatus functions by virtue of three separate but correlated cycles of activity — in the jaws, the hyoid apparatus, and the body of the tongue — whose characteristics can be closely defined, then it becomes possible to design experiments to test the nature of a central pattern generator. The possibility of linkage with the system controlling respiration should also be explored. Specifically, is there a common pattern to pattern generators (such generators are reported to regulate locomotion and respiration), or is there in fact an actual linkage between the cyclical behaviors of respiration and feeding, given that phylogenetically, the latter has evolved from the former?

Chapter 15

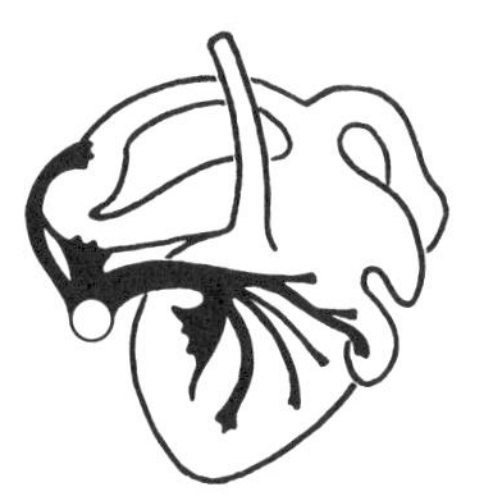

The Octavolateralis System

Richard R. Fay
Arthur N. Popper

Sound detection by animals involves a transduction of acoustic waves into a physiological response by the nervous system. In vertebrates the primary site for this transduction is the inner ear, although other peripheral structures serve to transmit and possibly modify acoustic energy to the ear. In this chapter we consider the mechanisms of acoustic transduction and detection and the different auditory mechanisms that have evolved among vertebrates to meet special acoustic needs of the various species. We also discuss the lateral line, a mechanoreceptive system of fishes and fully aquatic amphibians.

Although there is considerable interspecific variation among vertebrates for transduction of sound, several basic themes may be found among all groups. One of these is the mechanism(s) by which sound is carried to the ear. In fishes sound may stimulate the inner ear directly, whereas in tetrapods an impedance-transforming device has evolved to enable airborne sound to efficiently stimulate the fluid-filled ear. A second theme is the process by which dimensions of sound are coded or represented by the inner ear. Finally, we consider one of the major roles of the auditory system in any animal: sound localization.

Sensory Hair Cells

The central component of the octavolateralis system (ear and lateral line) of all vertebrates is the sensory hair cell (Fig. 15-1). Mechanical deformation of the stereocilia of the sensory hair cells produces changes in the conductance of the cell membrane. This triggers release of a transmitter at the synaptic junctions with the afferent neurons of the lateral line and with the eighth cranial nerves (Hudspeth and Corey, 1977; Hudspeth, 1983). Although the pathways of sound to the sensory hair cells vary among vertebrates, the basic similarity of hair cell morphology suggests that mechanisms of stimulation and physiological response of these cells are not likely to differ widely among the various inner ear and lateral line systems.

Each end organ (ear or lateral line) contains a sensory epithelium consisting of sensory and supporting cells and of accessory structures, such as an otolith, tectorial membrane, or cupula, lying close to the ciliated apical surface of the sensory cell. Each sensory cell has a ciliary bundle projecting above its apical surface toward, or in contact with, the overlying structure (Fig. 15-1). The ciliary bundle is made up of a single kinocilium located at one side of the bundle and many smaller stereocilia that are often graded in size. The kinocilium contains the typical 9 + 2 filament arrangement found in many flagellate cells, whereas the stereocilia contain actin filaments in a stiff, crystalline arrangement.

The appropriate stimulus for the sensory hair cell is a bending or shearing of its stereocilia that results in a change in membrane potential (see, for example, Hudspeth and Corey, 1977). One of the most important aspects of the hair cell's function derives from the directional polarization of the ciliary bundle. The kinocilium is placed eccentrically relative to the stereocilia, and physiological experiments have shown that a shearing of

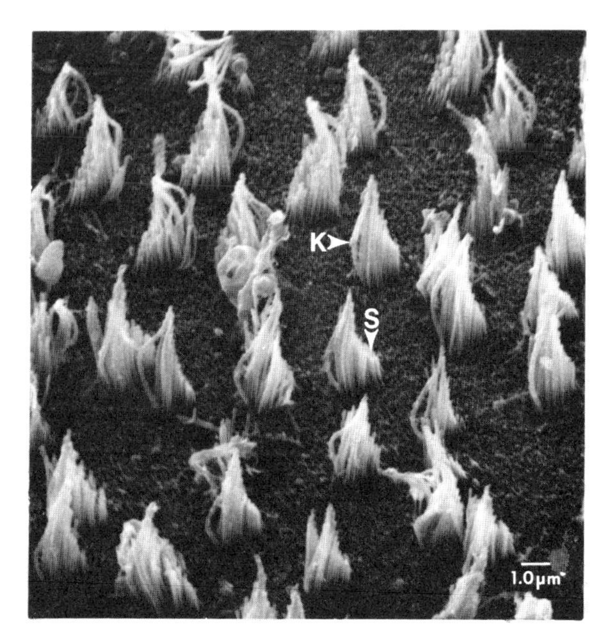

Figure 15-1 Scanning electron micrograph of ciliary bundles on the sensory hair cells of the saccule of an anabantid fish, the blue gourami. A single kinocilium *(K)* is eccentrically placed in each ciliary bundle. The more numerous stereocilia *(S)* are graded in size, the largest being closest to the kinocilium. Notice that the bundles are oriented in two directions: those located toward the top of the figure have their kinocilia on the right, whereas those on the bottom of the figure have their kinocilia on the left.

the cilia toward the kinocilium results in a depolarization of the cell membrane, whereas shearing in the opposite direction produces hyperpolarization. Bending in other directions results in a change of potential whose value is a cosine function of the bending direction (Hudspeth and Corey, 1977; Hudspeth, 1983). This physiological polarization has significance for the functioning of the octavolateralis end organs.

The Lateral Line System

The lateral line system is composed of numerous sensory units termed neuromasts, each containing sensory hair cells and supporting cells. Neuromasts are spaced along the length of the fluid-filled canals on the trunk or head of the body, although they may also be found as free end organs on the body of fishes and some aquatic amphibians. In many species the canal opens to the external environment through pores, which are generally found between pairs of neuromasts. A gelatinous mass, the cupula, lies above each neuromast, and the ciliary bundles are embedded in it.

The ciliary bundles of lateral line hair cells are oriented with their most sensitive axis along the length of the canal. About half the bundles are oriented in one direction and the rest in the opposite direction. Stimulation seems to result from liquid moving in the canal relative to the canal walls. This motion causes displacement of the cupula, thereby resulting in a shearing of cilia on the sensory hair cells. Responses of the lateral line vary depending upon the direction of the stimulus relative to the canal, thereby making the lateral line directionally sensitive.

The role(s) of the lateral line in the life of fishes and amphibians is still uncertain. This lack of clarity results from the difficulties in defining the appropriate stimulus to use in behavioral testing and in ensuring that other receptors, such as the inner ear, are not involved in any behavior associated with stimulation. Despite difficulties in studying the lateral line, it is clear that this system responds to surface waves and local water motion. However, there has been some argument whether the lateral line can be considered a low-frequency (below 100 Hz) sound detector. The trunk lateral line in the roach *Rustilus rustilus* has a vibratory threshold of 3.3×10^{-6} cm displacement at 50 Hz (its most sensitive frequency), which is considerably less sensitive than the inner ear (Sand, 1981). It appears that the trunk lateral line in the few species studied does not respond efficiently to low-frequency acoustic signals. Instead, it is more likely that it is used to detect water motions such as are produced by animals moving nearby. Removal of the lateral line effectively eliminates schooling behavior in blinded fish (reviewed in Sand, 1981).

Dijkgraaf (1963) and others have demonstrated that the head canal system is used for "distance touch," the ability to detect compression in the water mass in front of the fish as it approaches another fish or a stationary object in the environ-

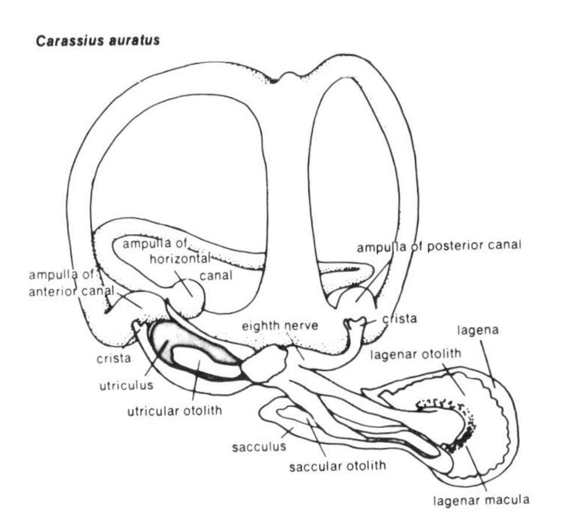

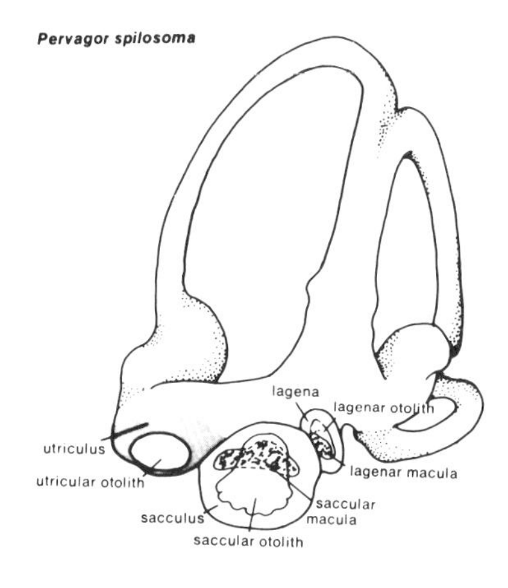

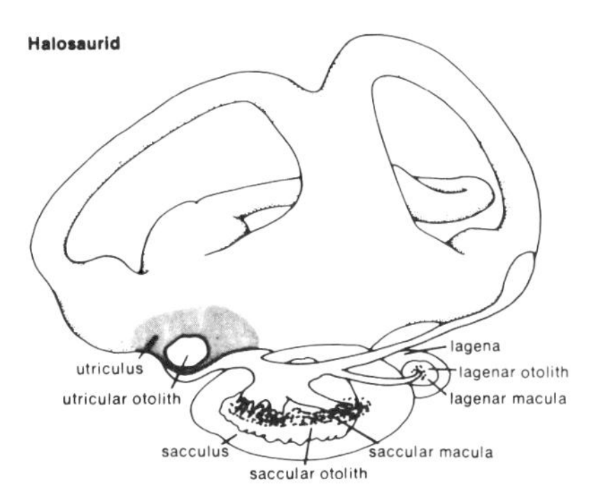

Figure 15-2 The ears of several teleosts. The *pars inferior,* which includes the semicircular canals and the utricle, do not differ substantially among species, but there is marked variation in the relative sizes and shapes of the saccular and lagenar otolith organs. Ostariophysi, represented here by the goldfish *Carassius auratus,* have lagenae that are as large as the saccule. In other than the Ostariophysi, however, the saccule is often substantially larger than the lagena. (From Popper and Coombs, 1980.)

ment. Further, the head canal system in some surface-feeding fishes detects the surface waves produced by insects moving on the water surface. These surface waves are unlike acoustic stimuli in that they are a special class of transverse waves whose propagation velocities depend on frequency.

The Auditory System of Fishes

The ears in modern fishes bear considerable resemblance to those in tetrapods, although the phylogenetic relationship among the parts of the ear in fishes and in tetrapods (and among different tetrapods) is still controversial. The fish ear has three semicircular canals and their associated ampullary regions (Fig. 15-2), as well as three otolith organs: the saccule, lagena, and utricle (Retzius, 1881; Platt and Popper, 1981; Popper and Coombs, 1982). A number of fishes (and tetrapods) have a seventh sensory region, the macula neglecta, located in the region between the utricle and the ampulla of the posterior canal (Retzius, 1881; Corwin, 1981).

Each otic end organ in fishes contains a sensory epithelium possessing numerous sensory hair cells. The sensory epithelium in the cristae of the semicircular canals and macula neglecta are overlain by a gelatinous cupula, whereas the epithelium in the otolith organs in most fishes is overlain by an otolith membrane, on top of which is a single dense calcareous otolith.

The Sensory Epithelia The sensory epithelium contains large numbers of sensory hair cells (over 200,000 in the saccule of a 16-cm long oscar *Astronotus ocellatus*), which are organized into groups based on the morphological polarization of the ciliary bundle in different epithelial regions (Fig. 15-3). The utricle and lagena each have two opposing orientation groups. The patterns in the utricle and lagena of bony fishes resemble those found in the same organs in elasmobranchs, primitive bony fishes, and most tetrapods (see, for example, Biard, 1974).

Two basic patterns of hair cell orientation are found on teleost saccules (Popper and Coombs, 1982). In Ostariophysi (catfish, goldfish, and relatives) the saccule has two groups of hair cells, one oriented dorsally (relative to the fish's body axis) and the other ventrally (Fig. 15-3). In teleosts other than the Ostariophysi, there are two additional orientation groups, generally at the rostral end, directed along the animal's horizontal axis.

The ciliary bundles on the sensory hair cells vary in length at different regions on each of the sensory epithelia. In general, the very margins of the epithelial regions of each otic organ have ciliary bundles with short stereocilia and long kinocilia. Just inside this marginal row (particularly in the saccule) are several rows of bundles with long kinocilia and long stereocilia, whereas the whole central region of most maculae contains cells with ciliary bundles of medium length. The specific positions of the different bundles varies in different organs and different species, and the functional significance of the different length bundles, and of the interspecific variation in positions, is not yet known in fishes. Physiological studies on ciliary bundles in birds and reptiles (see below) suggest that the different length ciliary bundles may be associated with responses of hair cells to different frequencies. However, a marked variation in cilia length in the bullfrog amphibian papilla does not appear to be responsible for frequency analysis known to occur along this receptor epithelium (Lewis, Leverenz, and Koyama, 1982).

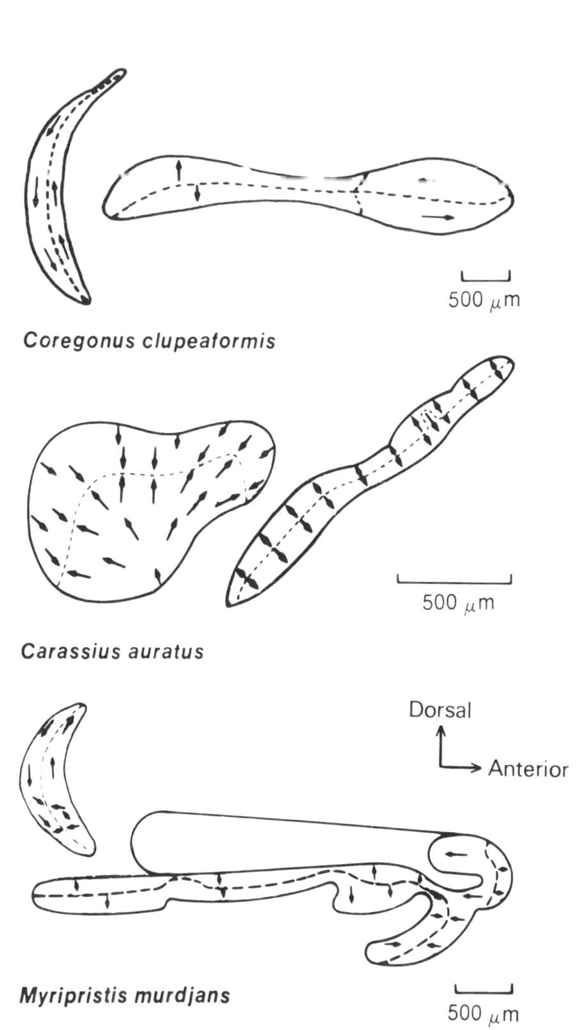

Figure 15-3 Patterns of orientation of the hair cells from the saccule *(right)* and lagena *(left)* of three teleosts. The tips of the arrows indicate the side of the ciliary bundle on which the kinocilium lies (see Fig. 15-1); dashed line indicates the approximate division between bundles oriented in different directions. In nonostariophysans there are generally four orientation groups, whereas in Ostariophysi (see *Carassius*) there are only two, oriented dorsally and ventrally. The two additional groups in the nonostariophysans are on the animal's horizontal plane and are usually found at the rostral end of the saccular macula. The precise relationship between the two horizontal groups varies somewhat among species (compare *Coregonus* and *Myripristis*). (From Popper and Coombs, 1980.)

Interspecific Variability As with the lateral line, there is substantial interspecific variability in many aspects of the otolith organs of fishes (Popper and Coombs, 1982), particularly in those aspects of the ear associated with the mechanical transduction of sound (the saccule in fishes). Interspecific variability is found in the size and shape of the otolith chamber, in the spatial patterns of different length ciliary bundles, and in the patterns of hair cell orientation, particularly of those associated with the horizontally oriented cells (Figs. 15-2, 15-3; Platt and Popper, 1981; Popper and Coombs, 1982). The functional significance of interspecific variability is not yet known, although these differences may indicate that the ear has different processing roles in various species.

Peripheral Auditory Structures Audition in fish is thought to involve the otolith organs, and in particular the saccule and lagena. However, cases are known where the utricle is the primary auditory organ (such as in the herringlike fishes, clupeids). We are becoming more aware of the possibility that each otolith organ may have multiple roles rather than being restricted to a single function such as audition or gravity reception (see, for example, Platt and Popper, 1981).

Although the inner ear is responsible for mechanical transduction, many species have other organs that are involved with the enhancement of the transmission of sound to the inner ear. The most widely known of these is the swimbladder, a gass-filled sac in the abdominal cavity of many fishes. In most species the anterior end of the swimbladder lies some distance from the cranium, whereas other species have anterior specializations of the swimbladder that project forward to connect directly to, or lie near, portions of the inner ear (some holocentrids, clupeids). These anterior diverticulae presumably provide direct acoustic coupling between the swimbladder and the inner ear. Other adaptations in the auditory system peripheral to the ear include small air bubbles that become part of the saccular wall in the Mormyriformes (elephant-nosed fishes) and pharyngeal air sacs in the Anabantidae (bubble-nest builders). Finally, an extreme specialization is encountered in the Ostariophysi, each of which has a series of bones, the Weberian ossicles, directly coupling the swimbladder to the inner ear (Fig. 15-4).

Although the specific response characteristics of each of these adaptations are not known, they appear to be associated with either improvement of sensitivity to sound pressure or to an extension of the bandwidth of frequencies detectable by fishes. Such interpretations are supported by behavioral studies indicating that fishes with anterior diverticulae and fishes with the Weberian ossicles generally (but not always) hear a wider range of frequencies, and lower amplitude sounds, than do species without such adaptations (Fig. 15-5; Fay and Popper, 1980; Platt and Popper, 1981).

Pathways for Sound to the Ear The modes of sound transmission to the ears of fishes is still an area of some discussion and experimentation. Von Frisch (1936) pointed out that since the density of fishes is about the same as water, an impinging sound would move the water mass and the fish with the same phase and amplitude. However, the air-filled swimbladder, being of very different density and compressibility than water, tends to expand and contract according to the local fluctuations in sound pressure. The moving walls of the swimbladder would then act as a secondary sound source providing sufficient particle motion to effectively stimulate the inner ear by causing motion of the otoliths relative to the sensory epithelia. Von Frisch (1936) and van Bergeijk (1967) argued that the presence of any projection of the swimbladder near the ear should enhance hearing sensitivity (see Fig. 15-5).

In the Ostariophysi the Weberian ossicles are set into motion by movements of the walls of the swimbladder, and this is communicated to the sinus impar (Fig. 15-4), a fluid-filled canal that terminates in a transverse canal projecting into both saccules. Motion of the ossicles results in motion of the fluid in the canals, which in turn causes fluid motion in the ears and motion of the otoliths against the hair cell cilia.

In addition to the swimbladder, fishes may also detect sound through direct stimulation of the ears by impinging particle motion. As the fish's body tissue moves at approximately the same am-

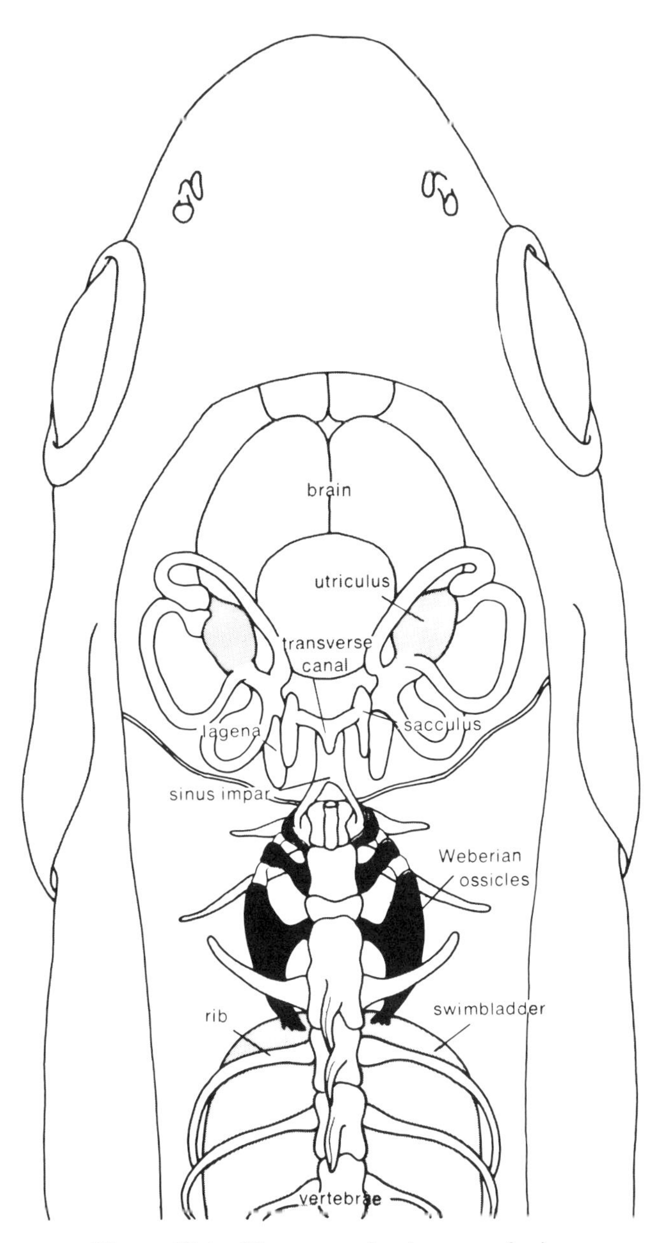

Figure 15-4 The connection between the inner ear and the swimbladder in a fish from the superorder Ostariophysi. Expansion and contraction of the swimbladder wall during acoustic stimulation result in a rocking motion of the Weberian ossicles, which leads to fluid motions in the sinus impar and transverse canal. This motion is carried to the fluids of the saccules of both ears and results in movement of the saccular otoliths relative to the sensory epithelia. (From Popper and Coombs, 1980.)

plitude as the surrounding water particles, the very dense otoliths tend to move at different amplitudes and phases owing to their greater inertia. Since the cilia of the sensory cells are attached to the sensory epithelium and contact the otolith (via the otolith membrane), a bending occurs that results in stimulation of the sensory cells.

The two major pathways for sound to the ear, the indirect (or swimbladder) path and the direct inertial path, are not mutually exclusive. The two modes of stimulation may function at different frequencies, with sound energy reaching the ear via the direct path for low frequencies and the indirect path for frequencies above several hundred Hz. It is also possible that both direct and indirect stimulation act together to affect the pattern of motion of the otoliths in some organs. Finally, some otolith organs appear to be stimulated primarily by direct signals, whereas others are primarily responsive to indirect stimulation via the swimbladder.

Acoustic Analysis in the Ear It is now clear that information about sound quality (its precise temporal waveform) is relayed directly to the central nervous system (CNS) via a "phase-locked" response of the auditory nerve fibers in which details of the waveform are preserved in the time pattern of responses, essentially as predicted by Wever (1949) in what he termed the volley principle. At the same time individual auditory fibers are somewhat frequency-selective or "tuned" (though to a broad range of frequencies) to respond best in particular frequency ranges (Fay, 1978; Fay and Popper, 1980). This means that the frequency content of a sound is represented by the identity of the fibers most active (a spatial or "labeled lines" code) as well as by the volley principle. The extent to which these representatives are used by the organism in processing acoustic information is a separate (and complex) question whose answer requires behavioral experiments. The complex variation in otolith structure and shape of the sensory epithelium suggests functional differences in different parts of the saccular epithelium, and Enger (1981) has shown that intense acoustic stimulation of the ear in the sculpin causes reproduceable spatial patterns of damage to hair cells that are frequency-

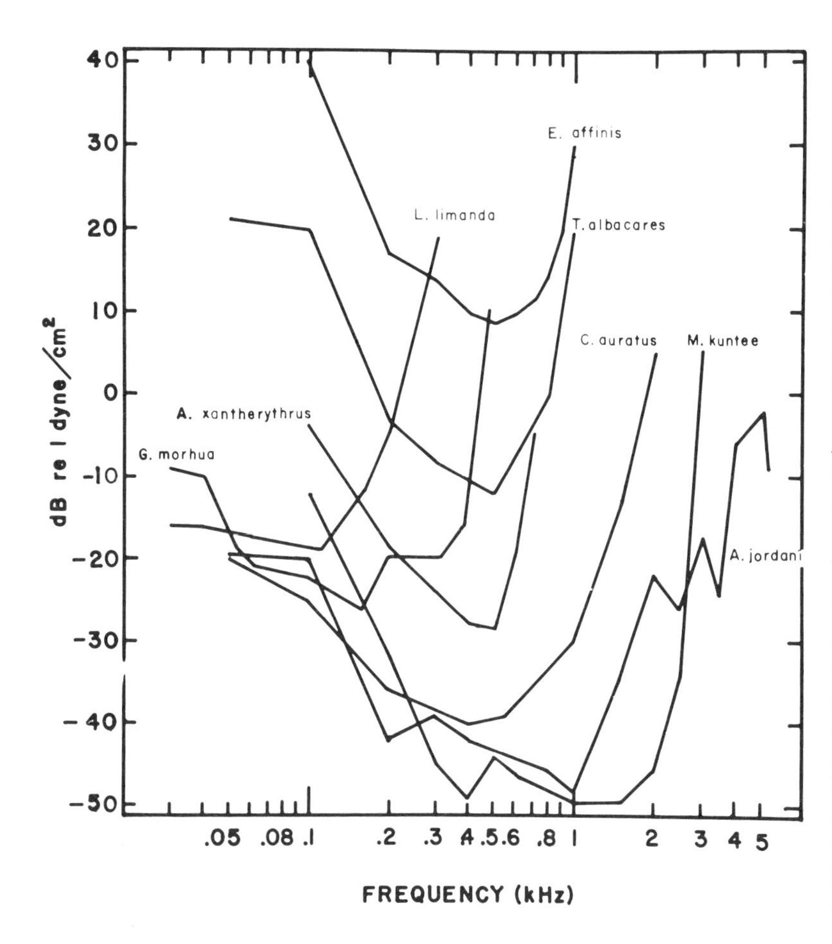

Figure 15-5 Hearing thresholds, or audiograms, for eight teleost species. The sound pressure (in decibels re: 1 microbar) necessary for minimal behaviorally conditioned responses is plotted as a function of stimulus tone frequency (1 kHz = 1000 cycles/sec). Of the three species shown with the greatest sensitivity at high frequencies, two are Ostariophysi (the goldfish *Carassius auratus* and the Mexican blind cave fish *Astyanax jordani*), and the third has a specialized swimbladder that comes into close contact with the saccule of the inner ear (a squirrelfish, *Myripristis kuntee*). The fishes having poorest high-frequency hearing (a flatfish, *Limanda limanda*) and poorest overall sensitivity (a tuna, *Euthymus affinis*) lack swimbladders. The remaining species possess a swimbladder, but do not have special adaptations for hearing (*Adioryx xantherythrus,* a squirrelfish; *Thunnus albacares,* a tuna; *Gadus morhua,* a codfish). (Data for *Carassius* from Fay, 1978; *Limanda* from Chapman and Sand, 1974; *Euthymus* from Iversen, 1969; *Thunnus* from Iversen, 1967; *Astyanax* from Popper, 1970; *Myripristis* and *Adioryx* from Coombs and Popper, 1979. Figure from Fay and Popper, 1980.)

dependent. It thus seems likely that the precise pattern of otolith motion (or orbital) would vary with different (but not yet known) signal parameters. Thus, different epithelial regions could potentially be most directly stimulated by different sounds.

Sound Localization One of the major controversies concerning fish audition has been whether fishes possess the ability to determine the position of a sound source (sound localization). Van Bergeijk (1967) argued that two independent channels of information are necessary for localization by the CNS, but that such input is not available to otolith organs because the two ears are very close together (less than 1 cm in most species) and because they are stimulated by a single structure—the swimbladder. We now know, however, that some species can determine the direction of a sound source using the inner ear and not the lateral line (reviewed in Schuijf and Buwalda, 1980). The most recent hypothesis for sound localization in fishes involves detection of the direction of the particle motion component of the sound field by the direct inertial pathway to the ear. It has been suggested that signals from different directions produce different relative motions of the otolith and that the ear has evolved to analyze the directional components of these motions. For example, the sensory cells in the saccule of nonostariophysines are oriented in two mutually perpendicular axes (Fig. 15-3). Motion of the otolith along one of the axes produces maximal stimulation in one group of optimally oriented cells and minimal stimulation of the perpendicular cells. Stimulation along any other axis produces a unique response ratio from the mutually perpendicular hair cell groups, potentially allowing the fish to derive some information about the direction of sound source. It is likely that directional information can be obtained within a single ear since it appears that the axons that innervate the horizontally oriented cells remain distinct from those innervating the vertically oriented cells until reaching the CNS, where interactions in binaural centers could potentially occur. Additional information would be obtained by the fish through comparison of inputs from ears that were oriented along somewhat different axes. Experiments on perch show that the responses from the two ears to a sound from a particular direction provide a unique ratio for each direction, again potentially providing the fish with information about the direction of a sound source (Schuijf and Buwalda, 1980).

One problem arises from this "ratio" mechanism of sound source location. Any two signals coming from 180° apart (from directly ahead or behind) produce an identical ratio at the two ears or at the mutually perpendicular hair cell groups in one ear. Behavioral experiments demonstrate that fishes can resolve this 180° ambiguity by comparing the phases of the pressure waveform received by the pathway of indirect stimulation with the particle motion waveform received through the pathway of direct stimulation (Schuijf and Buwalda, 1980).

Hearing by Elasmobranchs The ear in elasmobranchs is similar to that of bony fishes except that the macula neglecta is often very large (see Corwin, 1981). The ears are located within the chondrocranium, and a posterior canal duct (Fig. 15-6) projects upward from the saccule to the parietal fossa on the head. This fossa is filled with connective tissue and covered by a flexible membrane. Corwin has shown that the size of the macula neglecta is correlated with the feeding strategies of different species, leading him to suggest that it is associated with some aspect of prey detection.

Although sharks can detect and localize sounds from considerable distances, the pathway for sound to the ear is still not clear. Sharks and rays do not have a swimbladder, so an indirect, pressure-sensitive pathway to the ear is unlikely. It has been suggested that sound is channeled through a depression in the chondrocranium covered by taut skin, the parietal fossa, to the macula neglecta. Physiological data show that the macula neglecta is responsive to stimuli entering through this pathway (Corwin, 1981). However, the method by which a sound source could be localized by use of this path is not clear. Elasmobranchs may possibly use their lateral line system for long-distance localization, or use a behavioral strategy, such as sampling sound pressures at different positions as the animal swims, rather than an inner ear mechanism.

Entry onto Land

A major transition for the auditory system occurred when vertebrates, adapted for hearing underwater, emerged onto land and had to deal with acoustic signals in air. In order for the fluid-filled inner ears of most terrestrial vertebrates to respond to sound, a volume displacement of the fluid must occur. The problem faced by terrestrial animals is to transmit the sound energy efficiently into small fluid chambers. The terrestrial middle ear system evolved to transform airborne sound to a signal of appropriate impedance (ratio of pressure to particle velocity) to cause effective volume displacement of inner ear fluids. The system accomplishes this by using lever mechanics to trade displacement for pressure. This action depends primarily on the ratio of the area of the tympanic membrane (at the entrance to the middle ear) to that of the footplate of the stapes (at the entrance to the inner ear). A force operating on a larger area is distributed to a smaller area, thereby increasing the pressure acting over the smaller area. Force may also be increased somewhat through a lever and fulcrum system, which

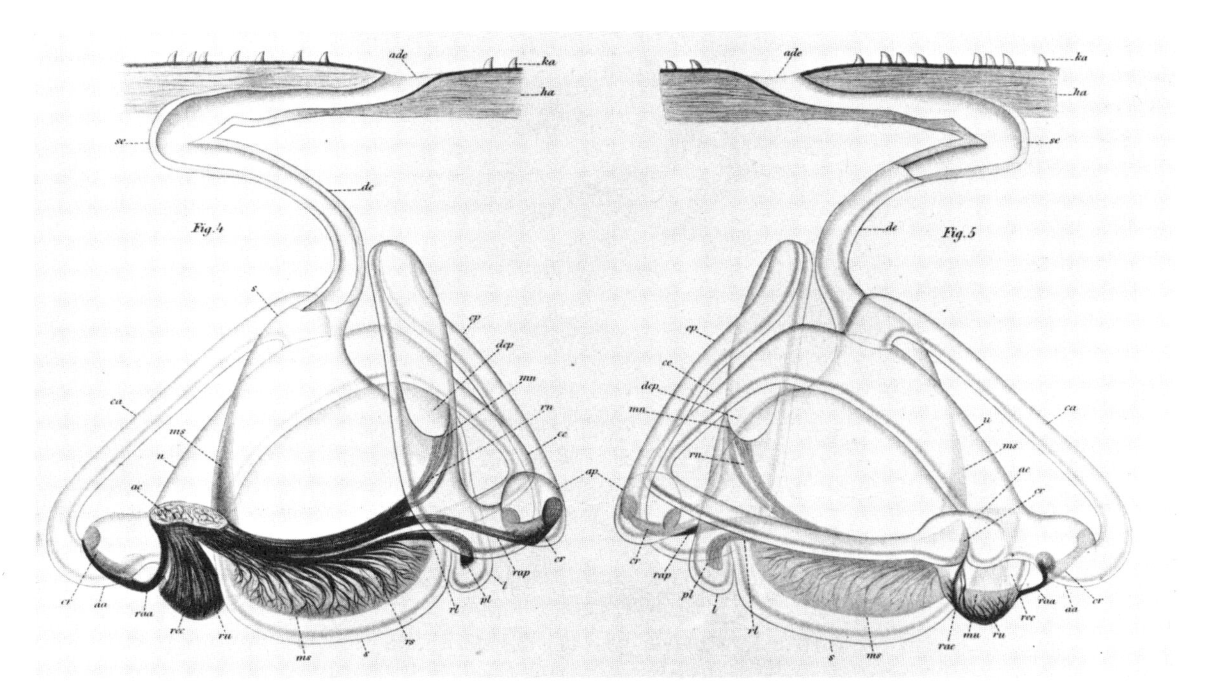

Figure 15-6 The right ear of the elasmobranch *Squatina angelus* (angel shark); medial view *(left)*; lateral view *(right)*. Note the elongate endolymphatic duct that extends to the parietal fossa on the surface of the head (see Corwin, 1981, for a complete discussion). The figure has all the labels from Retzius. The ones relevant here are: *de* = endolymphatic duct; *l* = lagenar chamber; *mn* = macula neglecta; *ms* = saccular macula; *rs* = saccular branch of the eighth nerve; *s* = saccular chamber; *u* = utricular chamber. (From Retzius, 1881).

biologically may appear quite complex (Wever and Lawrence, 1954).

The Auditory System of Amphibians

The auditory system varies somewhat among the three extant amphibian groups, the Urodela, the Anura, and the Apoda. This discussion centers on the anurans because the physiological acoustics of this group is best known.

The Middle Ear The ear in adult anurans consists of a tympanum located on the lateral side of the head, a middle ear cavity containing a single bone, the columella, and an inner ear (Fig. 15-7; see, for example, Baird, 1974; Henson, 1974; Capranica, 1976). The tympanum covers a large middle ear cavity, which connects widely to the mouth chamber by the eustacean tube. Connecting the tympanum to the oval window of the inner ear is the columella (sometimes referred to as the stapes). The columella is made up of two parts: the bony, rodlike columella, which attaches to the oval window of the inner ear; and the (often) cartilagenous extracolumella, which makes up the distal portion of the complex and attaches to the tympanum (Fig. 15-7).

A fascinating question concerns the middle ear in larval amphibians. Their auditory system resembles that of fishes in the pressure-to-displacement process (reviewed in van Bergeijk, 1967; Henson, 1974). In the common pond frog *Rana clamitans* the anterior ends of the paired lungs lie close to the ears and have a stiff "bronchial columella" terminating on the round window of the inner ear. Thus, motions of the lungs following the sound pressure waveform would be coupled to the inner ear in a way analogous to the coupling of the swimbladder to the ear via Weberian ossicles in the Ostariophysi. The oval window then appears to serve as a pressure release system for

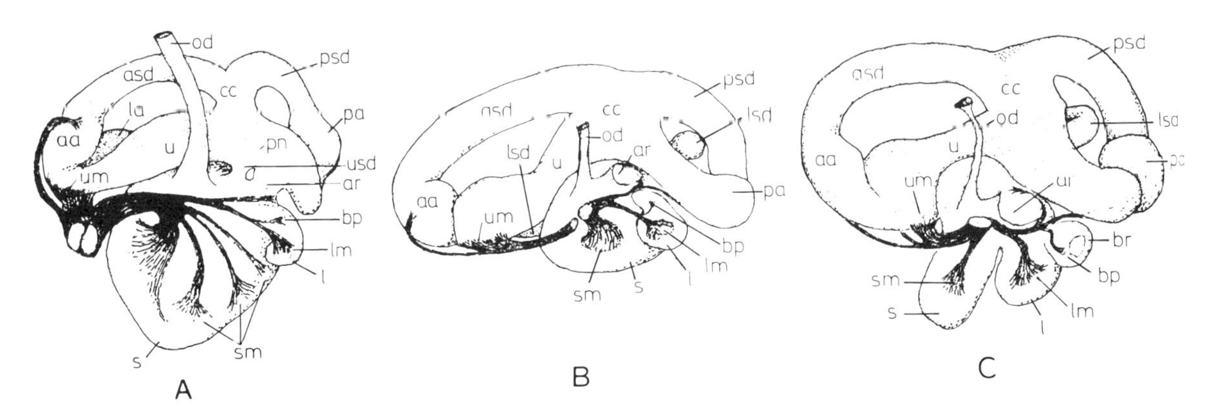

Figure 15-7 Medial sides of the inner ear of several amphibians: A, an Apodan *(Siphonops)*; B, a Urodele *(Salamandra)*; and C, an Anuran *(Bufo)*. *bp* = basilar papilla; *l* = lagena; *lm* = lagenar macula; *s* = saccular chamber; *sm* = saccular macula; *u* = utricular chamber; *um* = utricular macula. (From Baird, 1974.)

the inner ear. In larvae of other aquatic frogs, such as *Xenopus,* the bronchial sac directly contacts the round window of the inner ear in a fashion analogous to teleost species having anterior diverticula from the swimbladder.

During metamorphosis changes take place in the anuran auditory system to adapt it to sound detection in air, although there are no changes in the inner ear (van Bergeijk, 1967). The connections from the lung to the round window are lost, as is the bronchial columella in *Rana.* The tympanic membrane and columella develop during this time. It is important that the connection of the columella is to the oval window, and in the adult, the release system changes to the round window.

The impedance-matching role of the middle ear of frogs has only recently been analyzed quantitatively (Moffat and Capranica, 1978). The tympanic membrane in *Bufo americanus* and *Hyla cinerea* is a curved membranous lever that provides a gain (at frequencies within the auditory range of these species) of approximately two. There is also evidence for a lever action within the various parts of the columellar system (see Moffat and Capranica, 1978; also Manley, 1982), which may provide additional gain (of an unclear degree) to the system. Overall, the hydraulic lever system in frogs results from an areal ratio between the tympanum and the oval window of 10 : 3 for *Bufo* and 7 : 3 for *Hyla* (the values may differ depending upon how the functional area of the tympanic membrane is calculated).

At low frequencies the overall transformer ratio of the relatively simple anuran middle ear is about that of the ear of mammals with three middle ear bones, and the lever ratio of the three bones in mammals is apparently similar to that for the single bone in the frog. This leads to questions regarding the significance of the three-bone system in mammals. It is possible that the three-bone system extends the range of frequencies over which the middle ear can respond. However, the overall hearing range of a given species is probably determined by several factors, and there is no evidence that it reflects an essential "limitation" of the middle ear system.

There are other questions regarding the role of the mouth and eustacean tube in the function of the middle ear. Frogs have large eustacean tubes that are always open, with the result that the middle ear cavity is constantly open to the potential resonating chamber of the mouth (and the contralateral ear; see below). There is some controversy whether the response characteristics of the auditory system change with the opening and closing of the mouth (Moffat and Capranica, 1978; Chung, Pettigrew, and Anson, 1981).

The Inner Ear The amphibian inner ear bears some resemblance to that already encountered in fishes (Fig. 15-8; see Wever, 1973; Baird, 1974;

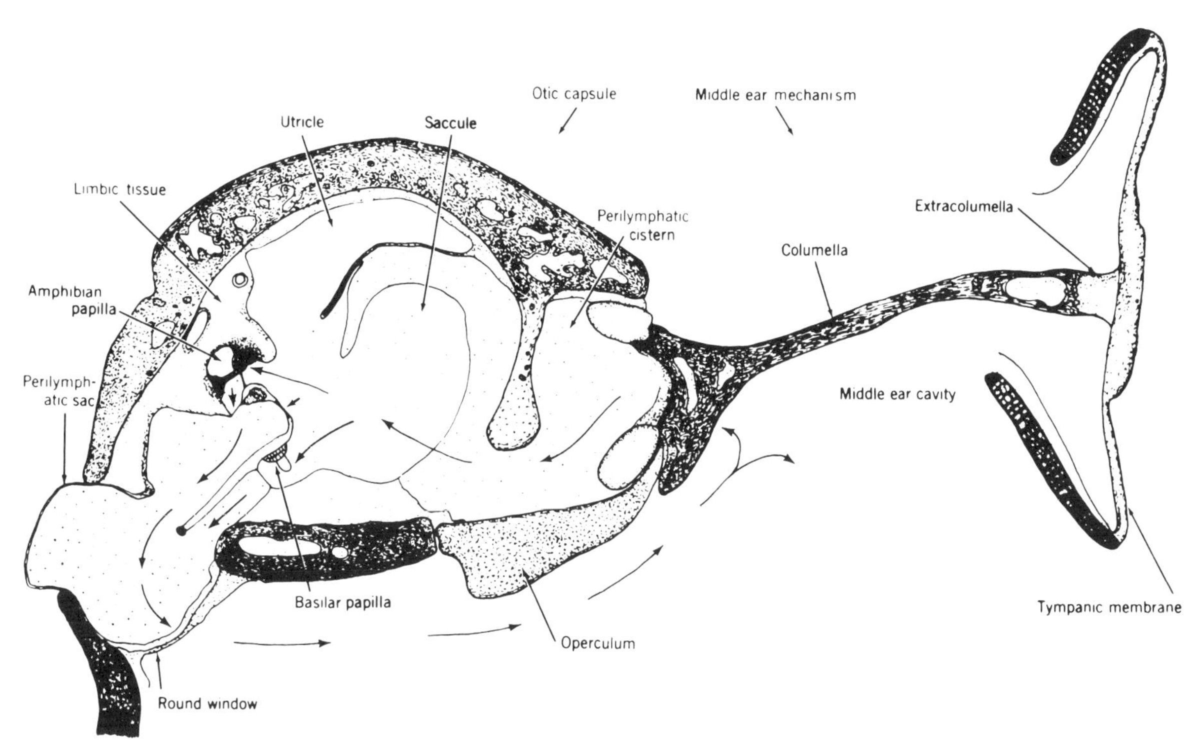

Figure 15-8 A cross section of the ear of the frog, *Rana pipiens.* The arrows indicate the pathways of fluid motion during sound stimulation. (From Wever, 1973.)

Capranica, 1976). Frogs have a lagena and utricle, which are thought to be vestibular receptors, and a saccule, which is extraordinarily vibration-sensitive and possibly sound-sensitive. Two newly evolved epithelia, the amphibian and basilar papillae, are the primary auditory organs. The amphibian papilla is found in all amphibians, whereas the basilar papilla is not always encountered. The pathway for sound to the auditory papilla in frogs is shown in Figure 15-8, which also shows the relationship of the various structures to one another. Sound enters the ear via the oval window (arrows in Fig. 15-7), passes through the saccular chamber within the periotic canal, passes the amphibian and basilar papillae, and is released from the ear by the round window.

Both amphibian and basilar papillae are firmly attached to the inner ear chambers, and each is overlaid by a tectorial mass connected to the tips of the ciliary bundles. During acoustic stimulation the passing wave imparts motion to the tectorial masses, whereas the epithelial surfaces tend to remain at rest. A shearing motion is set up over the hair cell cilia (Wever, 1978, 1981). This is essentially the stimulation mechanism found in fishes for auditory input from the swimbladder, which brings the otolith into motion over relatively stationary hair cells. The tectorial mass remains stationary in reptiles, birds, and mammals, and the epithelial surface moves relative to the mass. Such major differences in the way that the sensory cells are stimulated (along with other aspects of inner ear structure and function) have led Wever (1978, 1981) and others to propose that the auditory regions of the ear have evolved several times within the vertebrates and that there may not be homology among the auditory regions.

Behavioral and physiological studies have revealed that many frog species can detect sounds up to several thousand Hz and that many species have two regions of best sensitivity, one usually

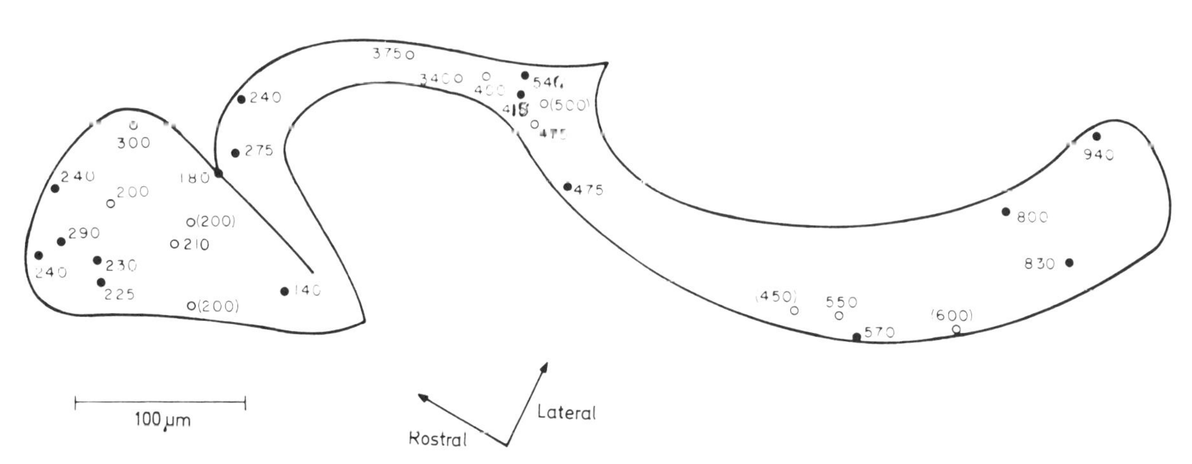

Figure 15-9 Outline of the sensory epithelium of the amphibian papilla of the bullfrog showing the sound frequency that is most excitatory at different points. These data were determined by measuring best excitatory frequencies of individual auditory neurons from the papilla and filling the neurons with a dye to trace their origin to hair cells on the surface of the epithelium. They demonstrate that there is some tonotopic organization to the amphibian papilla in the bullfrog. (From Lewis, Leverenz, and Koyama, 1982.)

below 1000 Hz and the other above 1000 Hz (Capranica, 1976). In order for a female frog to respond to the call of a male, she must hear a sound that has both high- and low-frequency energy. Physiological studies have demonstrated that the amphibian papilla responds to low frequencies, whereas the higher frequencies (above 1000 Hz) are detected by the basilar papilla (Capranica, 1976). The amphibian papilla is tonotopically organized so that the elongate caudal region responds to frequencies of 800 to 900 Hz, the middle region to 200 to 500 Hz, and the rostral responsive to sounds from 100 to 300 Hz (Fig. 15-9; Lewis, Leverenz, and Koyama, 1982). This spatial representation of frequencies does not occur in the basilar papilla.

The mechanical basis for the tonotopic organization of the amphibian papilla is not clear, although there is some variation in the thickness of the tectorial mass overlying different parts of the papilla (Lewis, Leverenz, and Koyama, 1982). The thicker region of the mass lies over the rostral, low-frequency end of the papilla, and the thinner region lies over the caudal, higher-frequency end. These authors point out that the high-frequency end of the amphibian papilla is that closest to the middle ear, a pattern that is similar to the high-frequency end of the mammalian cochlea and the avian basilar papilla. They also propose a traveling wave for the amphibian papilla, as in the cochlea, but argue that the mechanical support for such a wave in the amphibian papilla would be associated with spatially distributed mass elements in the tectorium. In contrast, the traveling wave in mammals is based upon graded mechanical properties of the basilar membrane and organ of Corti (see below).

Sound Localization With anuran amphibians, as with other vertebrates, sound localization is a major aspect of acoustic behavior. Although there are data demonstrating that frogs are able to locate the source of a sound (such as a calling male) and other data demonstrating binaural interactions and processing of directional information in the CNS, little is known about the contributions of the peripheral auditory structures to localization. Data from a number of laboratories (see, for example, Chung, Pettigrew, and Anson, 1981; Feng, 1981) indicate that the ear and associated structures, such as the mouth cavity, provide directional cues to the animal, and that these structures may enhance phase, time, and intensity differences between signals at the two ears (as suggested for birds; see below). A single ear may have frequency-dependent directional characteristics. However, there is still disagreement over the role of the eustacean tube, the mouth cavity, and the interconnections between the two ears in providing the animal with information about the location of a sound source.

The Auditory System of Reptiles

Much of what we know about the structure of reptilian auditory systems comes from the work of Wever and his colleagues and has been brought together in a monumental volume (Wever, 1978) from which much of the following discussion is derived. Significant variation exists both within and among reptilian orders in the degree of elaboration of middle ear structures (and thus in the efficiency and bandwidth of sound conduction to the inner ear) and in the complexity of inner ear structures (basilar papilla) serving hearing (also see Miller, 1980).

An outer ear is absent in turtles, snakes, amphisbaenids, and *Sphenodon.* In many lizards the tympanic membrane is very near the head surface; in geckonid lizards and crocodilians it lies at the end of a very short canal (external auditory meatus) that may be closed off by "ear lids" in some species.

The Middle Ear Figure 15-10 shows the major structures of the middle and inner ear of a lizard. The thin tympanic membrane (Fig. 15-10B) is supported at the edge primarily by the quadrate bone, and (as in anurans) it communicates mechanically with the inner ear via the extracolumella and the thin, rodlike columella. As in other terrestrial vertebrates, the impedance-matching function of the middle ear is accomplished through two principles: the ratio between the areas of the tympanic membrane and the columellar footplate at the oval window; and a lever system that trades force for displacement as motion proceeds from the tympanic membrane to the inner ear. The areal ratio varies among surveyed species from about 1:13 in the lizard *Crotaphytus* to 1:60 in a gecko. The lever process may reduce displacement and amplify force by approximately two. At frequencies below 2 kHz the combination of these two processes produces sound transmission to the inner ear that is at least as efficient as that observed in mammals. However, efficiency declines relatively rapidly for the reptilian ear (and for most other columellar ears found among amphibians and birds) above several kHz (Fig. 15-11). Manley (1982) speculates that this loss at high frequency could be due to the greater flexibility of the extracolummella and columella, which may be a protective adaptation for an otherwise unprotected middle ear. As in amphibians and birds the air-filled middle ear cavities of most reptiles communicate across the head via the pharynx.

The Inner Ear The inner ears of reptiles vary among species in a number of features. Primary among these are the "mobilization" principle employed (Wever, 1978), the length of the basilar papilla and the ratio of its length to width, the mechanisms by which hair cell cilia are restrained and stimulated, and the patterns of hair cell orientation.

The ear pictured in Fig. 15-10B illustrates the "conventional" mobilization principle, which is essentially the same as that found among the birds and mammals. Fluid motion set up in the cochlea by pistonlike movements of the columella pass through the basilar papilla on the basilar membrane, and is then "released" by corresponding motions at the air-bounded round window. A second pathway for fluid mobilization, which operates in turtles, snakes, *Sphenodon,* amphisbaenids, and a few lizards, has been termed *reentrant* (Wever, 1978). In this case, the round window is either absent or is bounded by fluid on both sides, rendering it essentially immobile. Mobilization is accomplished through a "closed circuit" fluid pathway leading from the front of the columellar footplate, through the perilymphatic and then the endolymphatic fluid chambers, and then back to the perilymphatic fluid at the back of the columellar footplate. In general, ears of the conventional type may be highly sensitive and have a relatively wide bandwidth compared with the reentrant type, which tend to have very poor high-frequency sensitivity.

The reptilian cochlea is composed of the perilymphatic chambers, scalae vestibuli and tympani, and the intervening endolymphatic chamber, the scala media. Although these chambers may be somewhat elongated (with

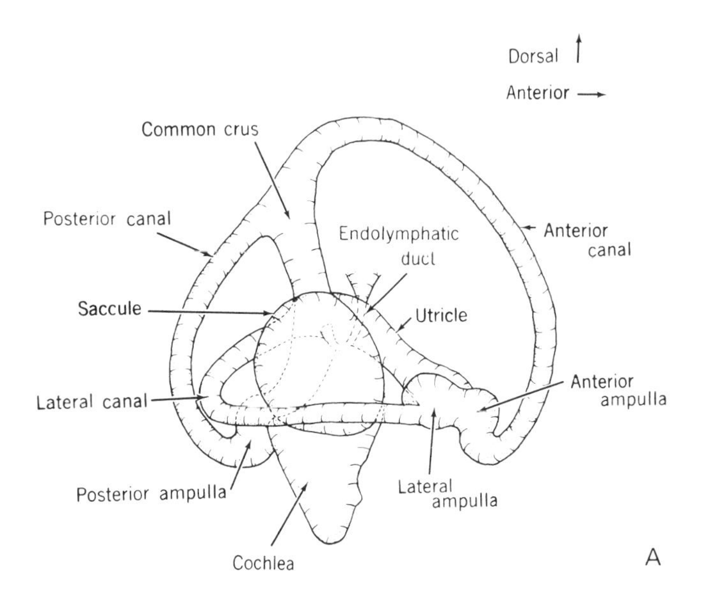

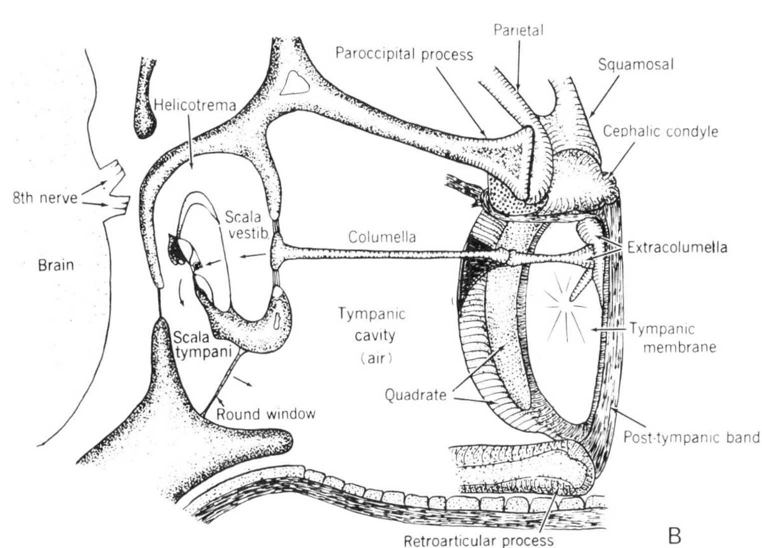

Figure 15-10 Two views of the inner ear of a lizard. A: The general form of the membranous labyrinth of *Iguana tuberculata* (the right ear in lateral view). B: A somewhat schematic cross section showing the relations of the middle and inner ear structures in *Sceloporus magister* (the right ear is viewed from behind). (Both from Wever, 1978.)

shapes depending on species), they are not coiled as in mammals. The basilar papilla sits on top of the basilar membrane, which separates the scala media and scala tympani, and contains supporting cells and hair cell receptors. The length of the basilar papilla ranges from 0.15 mm (in amphisbaenids, which have a total of 38 hair cells; Miller, 1980), to over 4 mm in crocodilians and large lizards. The longer basilar papillae with a greater number of hair cells are associated with good high-frequency responses (although not necessarily with greater sensitivity) and with relatively more highly tuned (frequency-selective) responses from single fibers of the auditory nerve (Turner, 1980; Turner, Muraski, and Nielsen, 1981).

There are three basic patterns of hair cell orientation in the reptilian ear (Fig. 15-12). In some regions of the basilar papilla all hair cells are oriented in the same direction (with the kinocilium oriented away from the auditory nerve). In other regions groups oppose each other across a

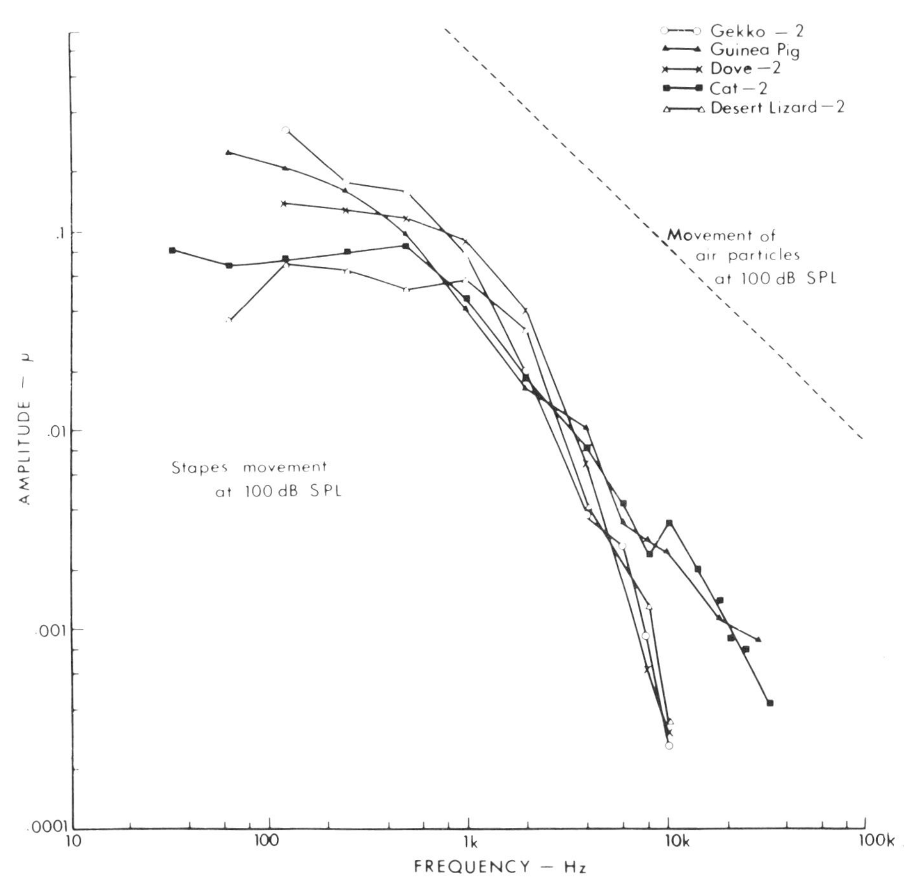

Figure 15-11 Frequency response functions for the middle ears of two lizards, a bird, and two mammals. The displacement amplitude of the columellar footplate is plotted as a function of frequency for a sound stimulus of 100 dB re 0.0002 dynes/cm². Above about 800 Hz, the functions tend to parallel the displacement amplitude of air particles, with an approximately 25 dB frictional loss. However, the loss for the nonmammalian species tends to grow above 2500 Hz where the mammals have a relative superiority. (From Johnstone and Taylor, 1971.)

midpapillary axis in a "bidirectional" pattern. In still other regions of bidirectional pattern hair cells of opposing orientations are intermixed.

Several mechanisms of hair cell restraint and stimulation have been identified by Wever (1978) (Fig. 15-13). Variations of the familiar tectorial membrane are found in a wide variety of reptilian ears. These seem to be arranged (as in mammals) so that stimulation of hair cells is probably proportional to displacement of the basilar membrane. In addition, in regions of the basilar papilla of some species massive structures termed *sallets* and *culmen* are found, which seem to be analogous to otoliths of the fish ear. These appear to stimulate underlying hair cell cilia through a combination of inertial and frictional forces, thereby stimulating hair cells in proportion to the acceleration or velocity of the basilar membrane. Finally, the cilia of some hair cells are free-standing, as are the inner hair cells of the mammalian cochlea, and therefore stimulation would probably be in proportion to the velocity of the basilar membrane. Two or more of these ciliary restraint systems may occur in a single ear (as in the gekonids and some lacertids).

The functional significance of species variation in patterns of hair cell orientation and in the mechanisms of ciliary restraint is not known.

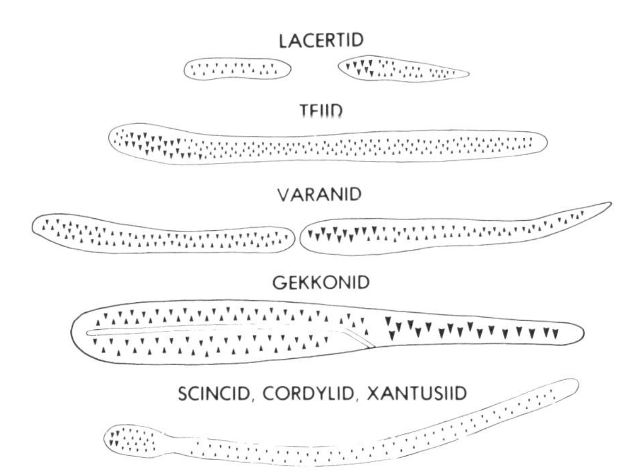

Figure 15-12 Orientation patterns of the hair cells and shapes of auditory papillae in several families of lizards. The arrows indicate the direction of the kinocilium (as in Fig. 15-3). In many cases lizards have hair cells oriented in two directions. Large arrowheads indicate areas of unidirectionally oriented hair cells. (From Miller, 1980.)

Wever (1978) reasoned that the sallet and culmen systems would probably be most effective at selected high frequencies and could contribute to high sensitivity through summation (a large number of hair cells underlying a single sallet or culmen would essentially respond in phase and thus efficiently sum their response). The tectorial and free-standing cilia, on the other hand, are more in a position to respond independently and may thus lead to greater frequency selectivity, though at the expense of sensitivity.

Although multiple orientation patterns of hair cells are possible adaptations for directional hearing in the fishes, it is not clear how they could be similarly utilized in the reptilian ear. One of the general consequences of bidirectional patterns of hair cells, however, is that the acoustic waveform would be coded in phase-locked response in the auditory nerve in greater detail and density than would be the case with hair cells oriented in only one direction (as in mammals and birds). Thus, movement of the basement membrane in both directions would be coded with great fidelity. Another consequence of this arrangement is that time intervals between responses evoked in fibers innervating hair cells of opposite orientation would reflect waveform periodicities (frequency content), which could be "analyzed" simply as time-of-arrival differences between spikes arriving at central sites from pairs of independent neural inputs.

The mechanical response of the basilar membrane of the alligator lizard, *Gerrhonotus* (Weiss et al., 1976), showed that all parts of the membrane were essentially identically tuned in a frequency range related to the best frequency of response of the middle ear. It is thus unlikely that a frequency-selective traveling wave exists in the ear of this species as it does in mammals. This finding is in contrast to the observation that single auditory nerve fibers of the same species respond to a narrow band of frequencies, and that there is a continuous distribution of these best frequencies throughout the range of basilar membrane response. The difference in results of these two experiments suggests that factors other than tuning of the basilar membrane lead to peripheral frequency selectivity in reptiles and possibly in other classes as well. A possible candidate is the "micromechanics" of the hair cell stereocilia. Turner, Muraski, and Nielsen (1981) have shown a robust correlation between length of cilia and best frequency in auditory nerve fibers of the granite spiny lizard. Thus, frequency selectivity appears to be determined at the hair cell level (also Khanna, 1983; Saunders and Dear, 1983). A second example comes from recordings from the hair cells of the ear of the terrapin *(Pseudemys scripta)* by Crawford and Fettiplace (1980), showing frequency selectivity or tuning that appears to arise from the filtering properties of the receptor cell membrane.

These studies, together with numerous single-unit studies of information coding in the reptilian auditory nerve (see reviews by Turner, 1980, and Manley, 1982) show that, whatever the underlying mechanisms, sound frequency is analyzed at the periphery according to a place principle. Unfortunately, we have no clear idea how these physiological data are related to hearing in reptiles: whether, for example, this place-principle is a mechanism for qualitative analysis of sound, or whether it simply serves to gain a favorable signal-to-noise ratio for sound detection. These kinds of questions require a behavioral analysis, and in contrast to what is known about some

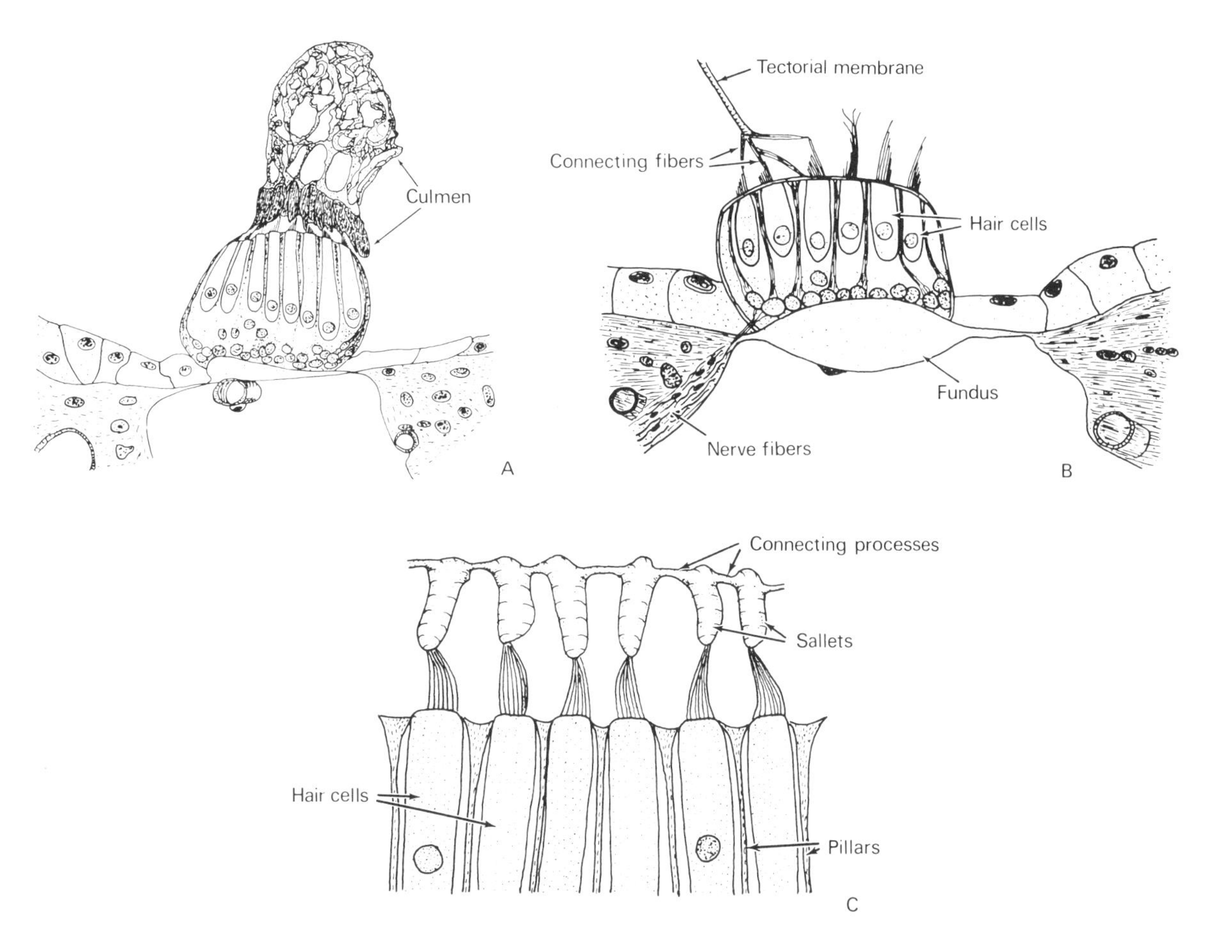

Figure 15-13 Restraint systems of hair cells found in the ears of various lizards: A, the culmen structure, which may provide inertial and frictional ciliary restraint in a skink (from Wever 1970); B, tectorial restraint and free-standing cilia in an iguana (from Wever 1967); C, longitudinal view of the basilar papilla with overlying sallets in a gecko (from Wever, 1978).

fishes and amphibians, little data of this kind exist for the reptiles.

The Auditory Systems of Birds

In contrast to reptiles, the hearing capacities of birds are well known. Data on morphology (for example, Smith, 1982), physiology (for example, Sachs, Woolf, and Sinnott, 1980), and on hearing capabilities (Dooling, 1973, 1982), including localization ability (Knudsen, 1980), tend to show that birds are well adapted for hearing and do not have the variation in peripheral morphology seen in fishes, amphibians, and reptiles. In this sense, and in terms of both physiology and psychophysics, they are more mammallike.

The Outer and Middle Ears The outer ears of birds are generally quite inconspicuous, consisting of a simple tubelike canal or external auditory meatus ending at the tympanium. The canal is usually covered by feathers, which are often specialized to reduce turbulence, and hence noise, in flight (Schwartzkopf, 1973). In some owls specialized ridges of skin run behind and in front of the ear canal openings and support the specialized feathers that make up the "facial ruff," a dense parabaloid reflector that collects sound energy and "funnels" it to the ear (Konishi, 1973). The right and left ear (and related skull features) of some owl species are asymmetrical in terms of the size of the ear openings, the vertical position of the openings, and in the vertical positions and shapes of preaural flaps. This asymmetry is a special adaptation for localization in the vertical plane (discussed below), which has not been observed in other vertebrates.

The avian middle ear is a large air-filled cavity that communicates widely with the pharyngeal space and with the opposite middle ear cavity. Birds have this feature in common with amphibians and reptiles. It functions in acoustic localization by producing interaural time or phase differences that are larger than would occur without such coupling. The tympanic membrane is more delicate than that found in the reptiles, but the general plan of the middle ear is similar. The thin columella attaches to the tympanic membrane via an extracolumella (the precise form of which is species-specific). Areal ratios between the tympanic membrane and the columellar footplate may be as large as 40:1 in an owl. The overall efficiency of impedance matching of the avian middle ear is similar to that of other terrestrial vertebrates (including mammals) in the frequency region below 2 to 4 kHz, but like both amphibians and reptiles, efficiency declines above this range relative to the mammals (see Fig. 15-11).

The Inner Ear The auditory portion of the avian inner ear is termed the cochlea (or basilar papilla) because of its similarities with the mammalian ear. The cochlea is a somewhat bent tubular structure divided longitudinally by the basilar membrane and the overlying basilar papilla and by the tegmentum vasculosum, which together enclose the endolymphatic cochlear duct (Fig. 15-14). The cochlear duct ends at the lagena, the function of which is not understood in birds.

The basilar membrane is short (3–5 mm in most species; considerably longer in some nocturnal owls) but wide, with 20 or more hair cells across the basilar membrane with their cilia (numbering up to 100 per cell) apparently firmly anchored in the overlying tectorial membrane. The properties of the hair cells vary across the basilar membrane. The "inner" (anteriorly placed over the inner, or nervous, cartilage) hair cell types are more elongated, and the "outer" types, quite short and squat with significantly more stereocilia projecting from a substantial cuticular plate (see Fig. 15-14). The existence of kinocilia on the hair cells is variable in different regions of the basilar papilla, and these patterns may vary among species (Smith, 1982). However, as in mammals but unlike fishes, amphibians, and reptiles, all kinocilia are oriented in the same direction (away from the origin of the tectorial membrane).

The mechanical response of the avian inner ear is only beginning to be understood. Von Bekesy (1960) measured the positions on the chicken basilar membrane at which displacement maxima occurred for tones of different frequency, and found that higher frequencies give best stimulation at the narrower end of the membrane (the basal end nearest the columella). The point of maximum displacement moves toward the apex for lower frequencies. These observations suggest the existence of a traveling wave on the avian basilar membrane. Single-unit neurophysiological measurements from the avian auditory nerve show a degree of frequency resolution that is at least as acute as that found in some mammals (Sachs, Woolf, and Sinnott, 1980). The question of the origin of sharp frequency resolution at the level of the auditory nerve, compared with the presumed displacement of the basilar membrane, is at the center of some recent, exciting research on the avian, as well as on the reptilian and mammalian, ear. Saunders and Dear (1983) point out that the height of stereocilia varies exponentially along the length of the basilar membrane in the chick (as well as in the reptile and mammal species studied) such that the shortest are found in the high-frequency region (base) and the tallest in the low-frequency region. They conclude that the frequency selectivity of hair cells (and of the nerve fibers innervating them) may be due in part to resonance of the stereocilia.

In general, the physiology of the avian auditory nerve shows many similarities to that of mammals, and these similarities are reflected in the hearing capabilities of birds (Stebbins, 1970; Dooling, Mulligan and Miller, 1971; Dooling, 1973, 1982). The absolute auditory sensitivity of birds is generally within 10 dB of that of man at

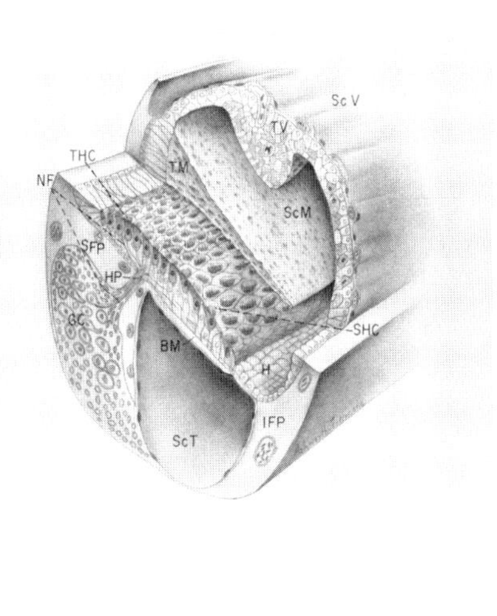

Figure 15-14 *Left:* Labyrinth and columellar apparatus of the chicken. The bone surrounding the cochlea is drawn as though transparent in order to illustrate the basilar papilla (sometimes referred to as a cochlea) and the lagenar macula. (From Tanaka and Smith, 1978). *Right:* Cross section of the pigeon's cochlear duct, including the scala media *(ScM)*, scala tympani *(ScT)*, scala vestibuli *(ScV)*, ganglion cells of the eighth nerve *(GC)*, short hair cells *(SHC)*, tall hair cells *(THC)*, and tectorial membrane *(TM)*. (From Takasaka and Smith, 1971.)

about 2 kHz, where birds are most sensitive. At frequencies below 100 Hz (down to 0.05 Hz) (Fig. 15-15), the pigeon may be as much as 50 dB more sensitive than man. Except for some of the owls, which hear with a greater sensitivity and in a bandwidth approaching that of man (Konishi, 1973), birds generally have poorer high-frequency hearing than man and differ little from one another in sensitivity and bandwidth. Passerines tend to hear somewhat better at high frequencies (3 to 10 kHz) and somewhat poorer at low frequencies (below 2 kHz) than nonpasserines (Dooling, 1982).

The ability of birds to discriminate changes in sound intensity and frequency has been measured in several species and found to be comparable to those of other vertebrates, including the mammals. Measures of temporal resolution in birds (primarily on the parakeet) show sensitivities that are typically "vertebrate" and similar to that measured for man (Dooling, 1982). Thus, the avian auditory system, in general, falls solidly into the vertebrate (including mammalian) pattern. There is little evidence for the hypotheses that birds should show unusually acute time-, intensity-, and frequency-resolving power.

Sound Localization Although not many species have been quantitatively tested, birds in general are able to locate the source of many sounds with considerable accuracy. The basis for this is undoubtedly (although not exclusively) the differences in the acoustic waveform that exist at the two ears of birds, as well as of all other terrestrial animals (mechanisms of sound localization have

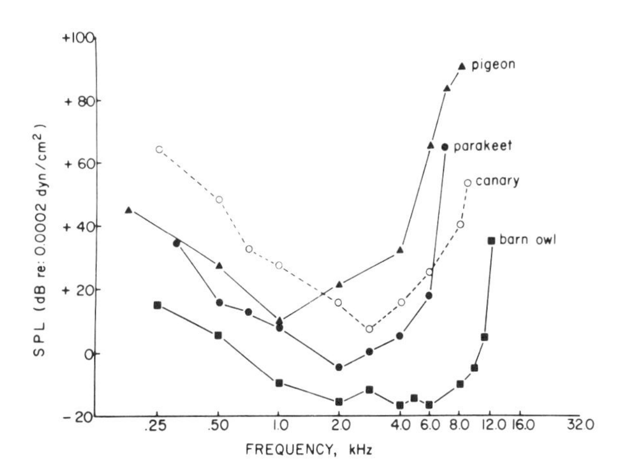

Figure 15-15 Auditory thresholds for four representative birds measured using behavioral methods. The barn owl shows unusually good high-frequency hearing and absolute sensitivity. (Pigeon from Stebbins, 1970; parakeet from Dooling, 1973; canary from Dooling, Mulligan, and Miller, 1971; barn owl from Konishi, 1973. Figure from Smith, 1982.)

been reviewed by Gourevitch, 1980, and Knudsen, 1980). These differences are thought to be in (1) time of arrival (for example, of a transient or "click"); (2) overall time delay or phase difference for ongoing sounds (reflecting the difference in distance between the sound source and the two ears enhanced by connections between the two middle ear cavities); (3) overall differences of intensity caused by the head's "shadowing" the far ear from the impinging sound; and (4) interaural differences in the shapes of spectra arriving at the two ears (the head may be more or less effective in shadowing in different frequency bands). In order for the time differences to be used by the organism, the neurally coded output of the ear must, with great accuracy, represent temporal details of the sound waveform. In order for frequency-specific interaural differences in intensity to be used as a cue, a frequency analysis of sound (an independent coding of sound in different frequency bands) must take place in the ear. Physiological studies show that both time- and frequency-domain coding are at least as acute in the avian eighth nerve as in that of mammals and other vertebrates (Sachs, Woolf, and Sinnott, 1980).

The use of these cues in behavior has only been documented in detail for one avian species, the barn owl *Tyto alba* (work done primarily by Konishi and colleagues, as recently reviewed by Knudsen, 1980). This animal is a nocturnal raptor that depends upon hearing to locate prey both in azimuth (the horizontal plane) and in elevation. The basis for vertical localization in this species is the unequal elevation of the two ear openings in the vertical plane, so that sounds at different elevations produce interaural differences in time and spectrum. Birds (and other terrestrial animals) without such an asymmetry must rely on head movements to provide good information about sound source elevation. Behavioral and physical analyses have shown that the barn owl determines the position of a sound source in both azimuth and elevation instantly with the arrival of a sound, and that it uses all the potential binaural cues listed above such that each direction in space has associated with it a unique pattern of interaural differences in time and frequency. Physiological analyses of neurons in the auditory midbrain of the owl reveals representations of auditory space that are "place-mapped" in neural tissue. This map is the result not of the isomorphic projection of a receptor surface, but of a neural computation of auditory space based on differences in time and frequency-spectrum input to the two ears.

The Auditory System of Mammals

Mammals are quite diverse in some aspects of the organization of the auditory system (such as the forms of the outer ear), but apparently quite homogeneous in others (see, for example, Stebbins, 1980). The primary characteristics of mammalian auditory systems that set them apart from the other vertebrates are (1) the existence of the pinna or external ear (auricle), which varies widely in form among species; (2) adaptations for hearing at high frequencies (in the range from 4 kHz to over 100 kHz in some specialized animals); (3) mechanisms for sound localization that make use of rather large distances between the ears (head size) in some species; (4) the existence of the three-boned ossicular system; (5) the presence of the organ of Corti with its clearly defined inner and outer hair cells; and (6) the structural and functional uniformity of the cellular elements of the cochlea.

These and other characteristics of the mammals have led many to the assumption that capacities for processing auditory information are most highly developed among mammals, and especially so among primates (Fig. 15-16). The evidence for this view is not overwhelming. Although mammals as a group are specially adapted for good high-frequency sensitivity, the capacities of mammals for intensity and frequency resolution, and other aspects of hearing such as directional hearing, are no more acute than they are in many other vertebrates. It seems, however, that man's auditory system outperforms those of other mammals and other vertebrates in basic functions such as intensity and frequency resolution and directional hearing.

The pinna functions in directional hearing and in the funneling of sound to the middle ear. In some species it appears to function also as a frequency-dependent acoustic delay and attenua-

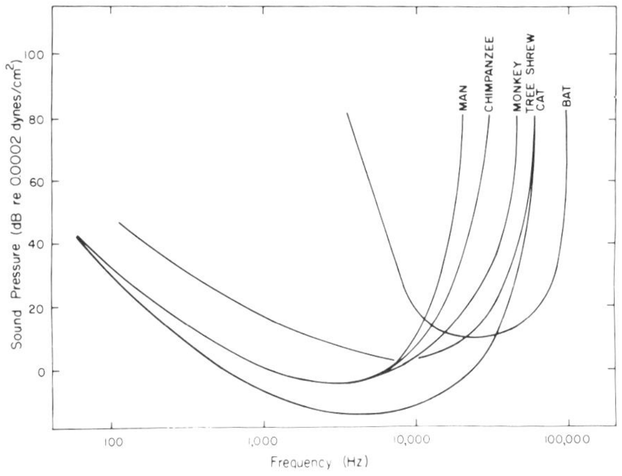

Figure 15-16 Auditory thresholds for six mammalian species, including man, measured using behavioral methods. (From Stebbins, 1980.)

tion system so that the location, and possibly, the distance of a sound source relative to the ear is coded. However, there are few detailed analyses of how a pinna of particular shape can affect sound transmission to the ear. The external auditory meatus of mammals is essentially a tube, partially bony and partially cartilagenous, whose length (generally proportional to head size) may determine a resonant frequency.

The Middle Ear The mammalian middle ear is comprised of the tympanic membrane and the three ossicles (the malleus, incus, and stapes) enclosed in an air-filled cavity whose size, and therefore resonant frequency, is species-specific and not necessarily allometric. The stapes is analogous, and possibly homologous, to the columella of amphibians, reptiles, and birds. A detailed description of the complex manner of sound transmission through the mammalian middle ear is beyond the scope of this chapter (see Wever and Lawrence, 1954, for a classical and detailed treatment, and Guinan and Peake, 1967, for more recent measurements). It is sufficient here to note that a rise in sound pressure at the outer face of the conical tympanic membrane causes it to move inward (in a complex, frequency-dependent pattern). The displaced tympanic membrane in turn moves the manubrial arm of the malleus inward, the articulating process of the incus inward, and finally the stapes in a pistonlike way in and out of the oval window of the bony labyrinth. This causes a rise in pressure in the perilymphatic vestibule and the scala vestibuli, which is communicated across the scala media and the helicotrema to the scala tympani and finally back to the membrane of the air-bounded round window. Here the pressure is released and the cochlear fluids are set in motion (Wever, 1978). This process is illustrated in a set of schematic representations presented by Kiang (1975) (Fig. 15-17). As in other vertebrates the mammalian middle ear is an efficient impedance transformer by virtue of the areal ratio between the effective area of the tympanic membrane and the area of the footplate of the stapes and by the displacement-reducing lever action occurring between the manubrial arm of the malleus and the articulating process of the incus.

Significant species differences exist in the overall efficiency of sound transmission by the middle ear and in the frequency ranges that are transmitted best. In general, the middle ear system has a frequency point above which efficiency declines. This point is determined primarily by the mass of the middle ear system, which in turn is related to the size and weight of the animal. The smallest mammals (rodents and bats) with the lightest middle ear systems efficiently transmit sound to the ear up to several thousand Hz, whereas in larger mammals (cats and man), sensitivity begins to decline at about 1300 Hz and 1000 Hz respectively. Small mammals, which hear best at high frequencies, generally have poor low-frequency hearing, but certain desert rodents from arid regions (the kangaroo rat and the chinchilla) are exceptions, having large middle ear air spaces (bullae) and excellent low-frequency hearing. In several burrowing mammals (moles and molelike mammals) what appear to be specializations for hearing low-frequency substrate vibration have been noted in the organization of the middle ear (Henson, 1974). The right and left middle ear cavities are separated by thin bone in some species and are actually joined in others. In addition, these animals have rather massive ossicles, which presumably increase sensitivity to substrate-borne vibration through their great inertia. In this type of "bone conduction" hearing the ossi-

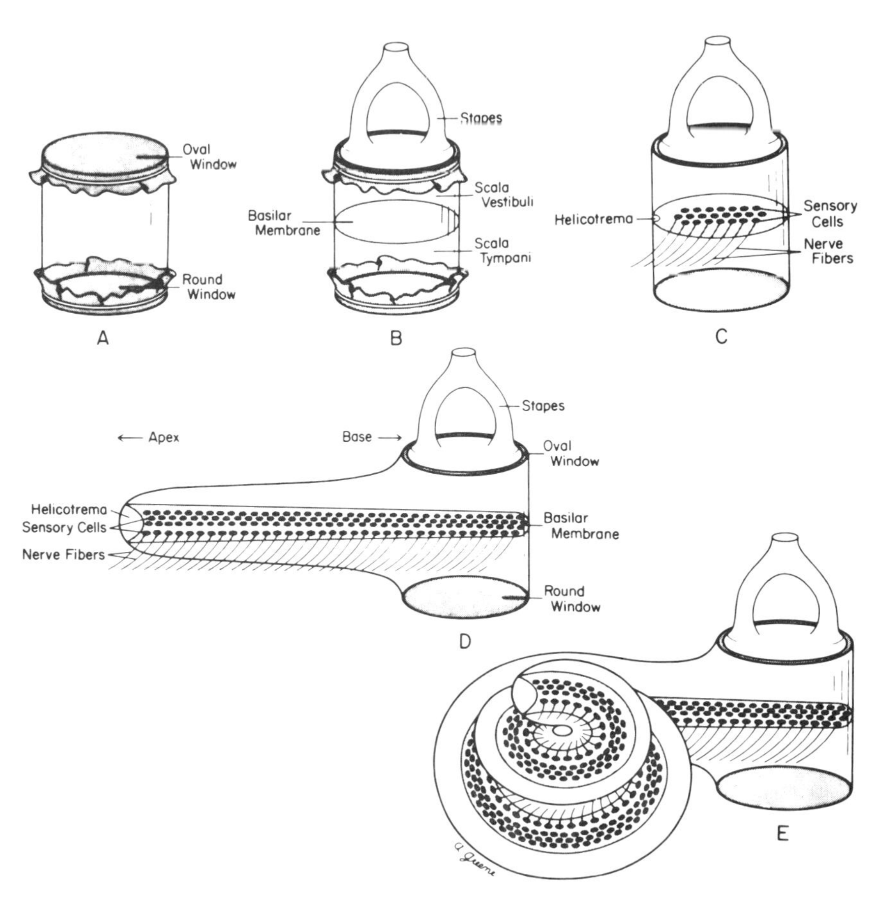

Figure 15-17 Overall functional organization of the cochlea of mammals (and, in regard to some aspects, of the reptilian and avian ears as well). Drawing A is a fluid-filled tube with air-bounded flexible membranes at each end. Drawing B shows that sound is put into this system via the stapes in the oval window, and the basilar membrane with its sensory cells can be viewed as spanning the vibration pathway between input and output. Drawing C can be viewed as representing the simplest of reptilian ears, whereas D is more like the more highly differentiated lizards, crocodilians, alligators, and birds. The coiling shown in E is only found among the mammals. (From Kiang, 1975, by permission of Raven Press, New York.)

cles may function analogously to the otoliths of fishes by providing an inertial reference for the detection of relative skull movement. Another example of a specialized inertial mechanism for detecting ground vibration may be seen in the enormous "vestibular" otoliths of the burrowing tortoise. Unfortunately, there are little behavioral, physiological, or biomechanical data on the function of these specialized ears. Particularly deserving experimental attention are questions of the mechanisms by which these burrowing animals may locate the source of sound or vibration in their usual environments.

The Inner Ear The auditory portion of the mammalian inner ear (cochlea) is a membranous tube coiled around a central core (modiolus), which contains the auditory nerve and is enclosed in the petrous portion of the temporal bone (Fig. 15-18). The tube is divided longitudinally into the scalae tympani, media, and vestibuli. Scala media, containing endolymph and the organ of Corti, is separated from the perilymphatic scala vestibuli by Reissner's membrane and from the perilymphatic scala tympani by the basilar membrane.

The basilar membrane supports the organ of

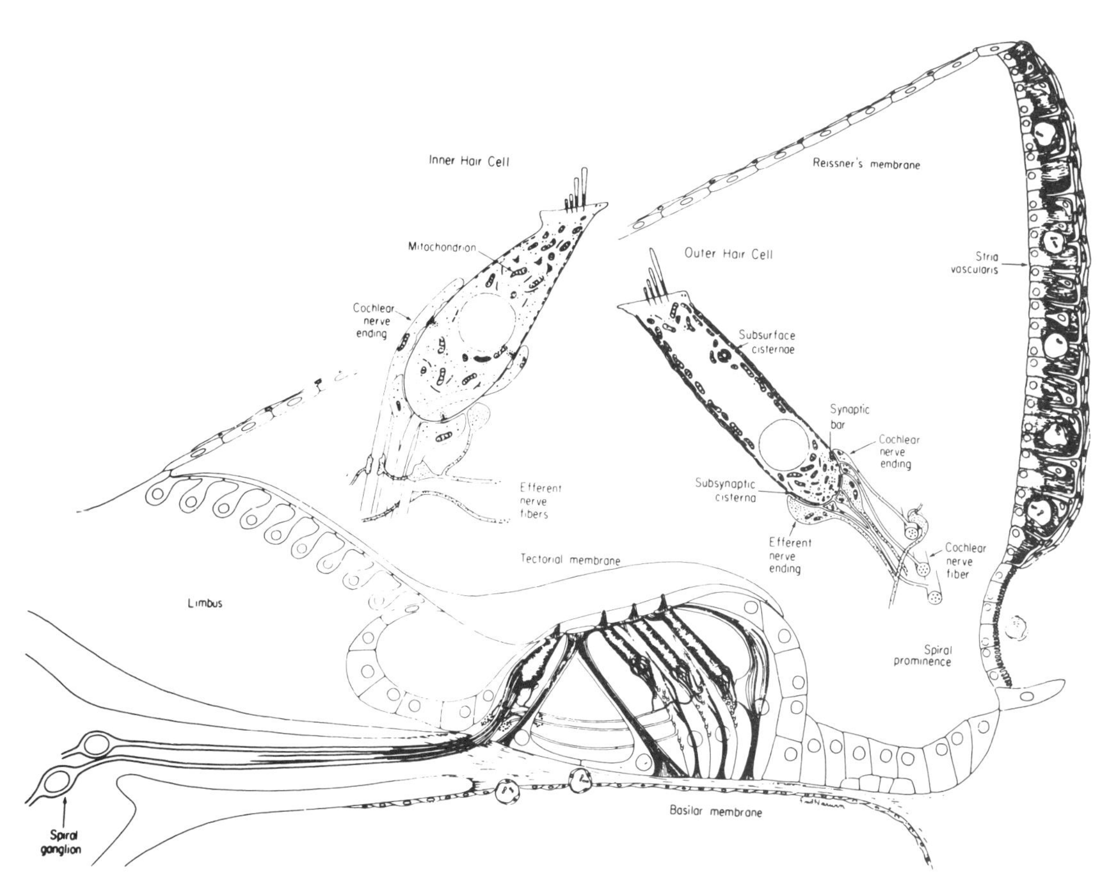

Figure 15-18 Cross section of the cochlear duct (scala media) of the guinea pig, illustrating the major morphological differences between inner and outer hair cells. (From Smith, 1975, by permission of Raven Press, New York.)

Corti and runs the length of the cochlea. The organ of Corti is composed of supporting and sensory hair cells and is overlaid by the tectorial membrane, which originates from the modiolar side at the limbus (Fig. 15-18). There is a single row of inner hair cells on the modiolar side of the tunnel and three or four rows of outer hair cells on the opposite side. Although the inner hair cells are surrounded and enclosed by other cellular elements, the outer ones are supported only at their bases, by Deiter's cells, and at their apical ends by the reticular lamina. About 95% of the afferent fibers synapse with a single (nearest) inner hair cell (each hair cell is multiply innervated). The remaining 5% may innervate as many as 10 outer hair cells along the way.

Cochlear hair cells are presumably stimulated by a radial shearing motion between the tectorial membrane and the reticular lamina that deforms the hair cell cilia. This shearing motion is set up when the basilar membrane and its organ of Corti move upward (toward the scala media) and downward in response to outward and inward motions of the stapes in its oval window. The direction and magnitude of the shearing forces are determined by the complex geometry of the cochlear cross section, and this may vary from base to apex and possibly among species.

Perhaps the most remarkable aspect of the mechanical response of the mammalian cochlea is the frequency-dependent spatial and temporal patterns of motion occurring along its length during sound stimulation. These patterns of motion were first observed and quantitatively analyzed in cadavers by von Bekesy 1960 and have only recently been observed in living mammals. The type of motion observed has been termed the "traveling wave" since the motion of each successive point on the membrane (from base to apex) is delayed somewhat relative to the point just basal. A wave thus progresses along the cochlea from base to apex. The peak amplitude of the wave grows gradually to a maximum that occurs at a frequency-dependent place, and then diminishes over a short, apical distance (Fig. 15-19). This is due primarily to a progressive reduction in the stiffness, and an increase in the width, of the basilar membrane and organ of Corti from apex to base. Consequently, both wavelength and velocity of propagation decrease toward the apex. This traveling wave is not simply propagated within the membrane and organ of Corti, but is also carried by both the fluid systems and cellular elements of the cochlea.

Considerable recent interest has focused on the question of the sharpness of tuning (or the frequency selectivity of the traveling wave) on the basilar membrane, on the relations between measures of mechanical tuning and that observed in single auditory nerve fibers, and on the mechanisms that may underlie tuning in the auditory periphery. It has long been known that there is a discrepancy between the very sharp tuning of auditory nerve fibers innervating the cochlea (all recordings to date seem to be from larger diameter neurons innervating inner hair cells) and the classical measurements of basilar membrane motion (Bekesy, 1960). This, and other considerations, led to the concept of a "second filter," or some frequency-selective mechanism in addition to motion of the basilar membrane interposed between membrane displacement and nerve fiber response. Intracellular recordings of potentials in inner hair cells in a mammal (Russell and Sellick, 1978) showed that tuning is as sharp as that observed in nerve fibers, which led to the notion that the second filter must operate at the level of the hair cell, or earlier.

The most recent measurements of motion of the basilar membrane (Khanna, 1983) reveal that some aspects of earlier measurements may potentially have been in error. If great care is taken not to damage the physiological integrity of the coch-

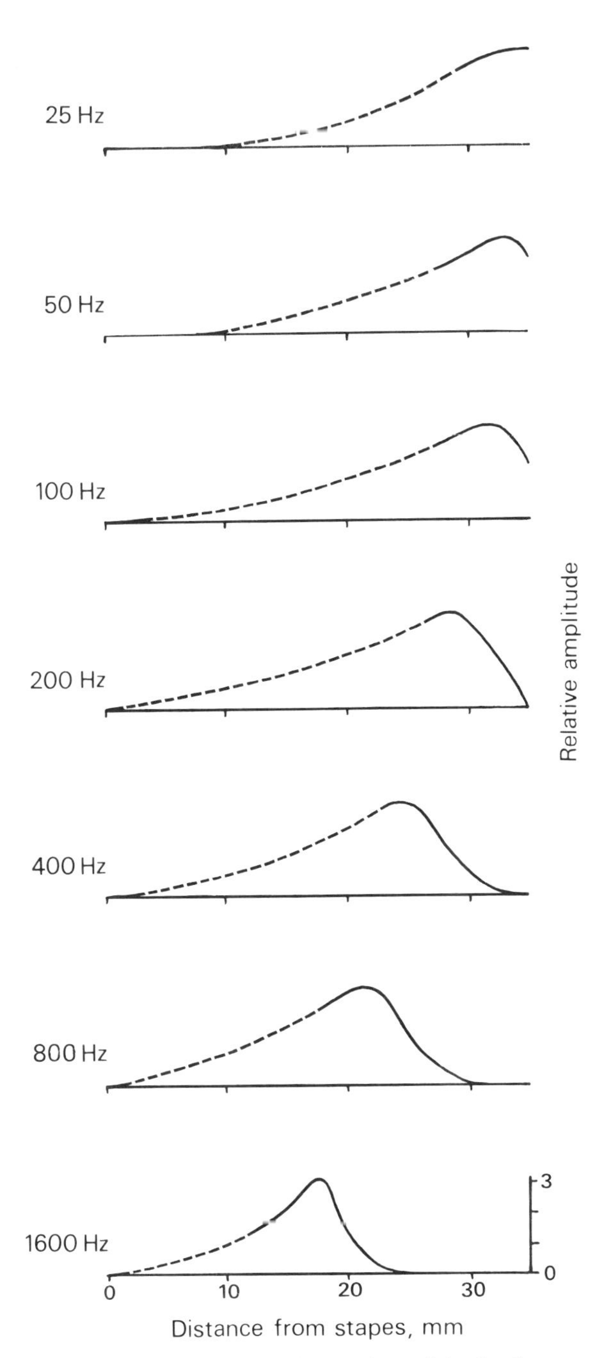

Figure 15-19 Patterns of vibration of the basilar membrane of the human cadaver deduced from the amplitude of vibration at a given place as a function of frequency at that place. These are the envelopes of the traveling wave, which shows a frequency-dependent place of maximum excitation. (Redrawn from von Bekesy, 1960.)

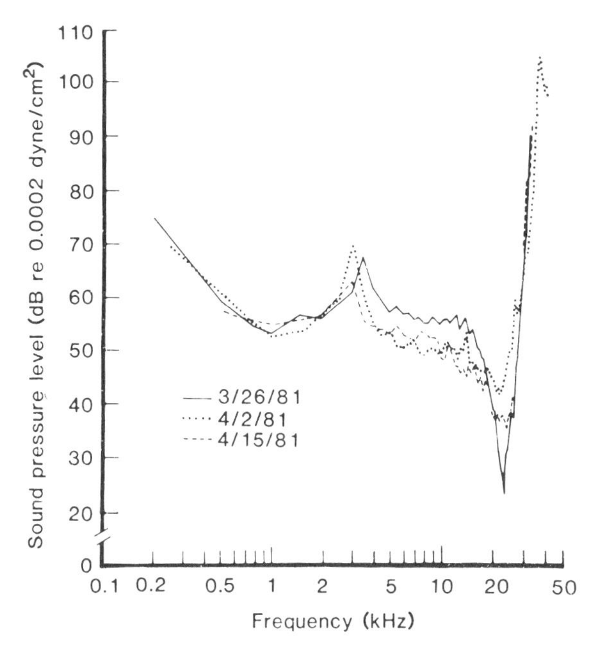

Figure 15-20 The level of sound pressure at the tympanic membrane of the middle ear necessary to obtain a certain displacement of the basilar membrane at different frequencies. These data show that less sound pressure is needed to stimulate the basilar membrane at 20 kHz than at any other frequency. The data help to show that different regions of the basilar membrane are "tuned" to different frequencies. (From Khanna and Leonard, 1982, *Science* vol. 215, pp. 305–306; copyright 1982 by the American Association for the Advancement of Science.)

lea during mechanical measurement, the degree of mechanical tuning approaches that seen in the nerve response (Fig. 15-20). If the cochlea is damaged in almost any way, the mechanical tuning deteriorates to levels characteristic of the classical measurements. The physiological vulnerability of sharp tuning has led to hypotheses that the motion of the basilar membrane is not simply a passive mechanical response, but rather a result of amplification that depends on metabolic energy. Thus, although the concept of the second filter survives, the question now centers on how this cellular energy couples back into the mechanical response. These considerations, in combination with other hypotheses regarding electrical tuning in hair cells (Crawford and Fettiplace, 1980) and resonance of stereocilia (Saunders and Dear, 1983), have begun to change in profound ways our general view of tuning mechanics in the vertebrate cochlea.

Sound Localization The ability to locate the source of a sound is perhaps the most useful aspect of hearing for an organism's survival. Thus, the evolution of various aspects of ear morphology and physiology may best be understood from the standpoint of adaptations to optimize sound localization (Masterton, Heffner, and Ravizza, 1969). Among the mammals the cues for sound localization include those listed earlier for the birds (interaural differences in time and frequency spectrum), as well as the cues to be gained from independent mobility of the pinnas. Owing to the complex and species-specific shapes of mammalian external ears, however, monaural cues for sound direction may be relatively more important for the mammals (although not well understood) than for the birds and other vertebrates. These cues essentially derive from subtle differences in the frequency spectrum of sound reaching the tympanum as it is complexly "filtered" by the structures of the pinna. Since we know that for humans these effects help to prevent fore-aft confusions and provide cues regarding the elevation of a sound source, they may be at least as useful for larger-eared mammals. There are little data on this point.

Classical theory holds that binaural directional cues are essentially interaural differences in time reception at low frequencies (below 1500 Hz), where robust neural synchronization to the stimulus waveform is possible, and are interaural differences in spectrum shape at higher frequencies, where synchronization deteriorates and the head becomes a more effective sound reflector (thus creating a "shadow"). The remarkable ability of mammals to hear at high frequencies is an evolutionary adaptation for localization, since the earliest mammals were quite small, and a small head casts a good sound shadow only for sounds of high frequency (and short wavelength). In fact, there is a remarkably good correlation between head size and the frequency range of hearing among the mammals.

Recent research has somewhat refined the classical view of cues for localizing sound to include the idea that time differences may be useful even for high-frequency sounds to the extent that these sounds have a characteristic temporal fluctuation in intensity (envelope structure) that arrives at the ears at different times. Finally, it appears that there may be species differences in the upper frequency limit for the cue of interaural phase difference, with cats and monkeys able to utilize this information at higher frequencies than man (Gourevitch, 1980). In general, however, man is the most sensitive mammal tested so far, with an ability to discriminate sound sources only 1° apart in the horizontal plane, compared with a monkey's best performance of about 8°.

Chapter 16

The Vertebrate Eye

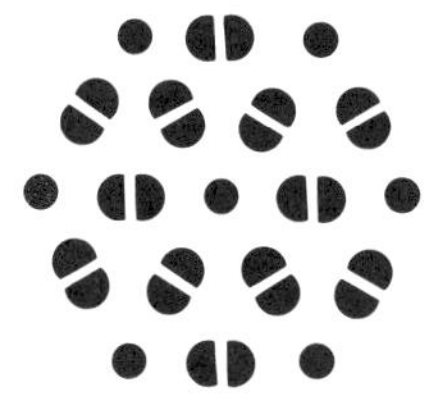

Joseph S. Levine

Green leaves, red berries, brown algae, and the iridescent plumage of tropical birds all differ from one another in their absorption and reflection of the sun's electromagnetic energy. In order to discriminate among these objects, animals have evolved visual systems that detect and report subtle differences in patterns of reflected energy reaching the eye. Eyes exhibit such elegance of design that Darwin discussed them separately in a section of *The Origin of Species* entitled "Difficulties of the theory: organs of extreme perfection and complication." As Darwin noted, each major group of animals has evolved a visual system exhibiting unique adaptations to specific visual tasks. Despite the obvious differences in both gross anatomy and fine structure among these diverse visual systems, they are all fashioned around an unvarying set of physical and perceptual requirements, and they exhibit remarkable similarities in information processing.

A comprehensive examination of *The Vertebrate Eye and its Adaptive Radiation* occupied Gordon Walls (1942) for many years and numerous pages. A more recent text on the *Physiology of the Eye* (Davson, 1980) is equally impressive in length and scope. It is neither necessary nor possible here to cover the entire story of structure and function in the vertebrate eye; the reader will be referred to other sources for more detailed information. Instead, this chapter serves to introduce the reader to a few of the functional requirements faced by vertebrate visual systems, and to examine the manner in which those requirements are expressed in the gross anatomy of the eye and the cellular organization of the retina.

Light and the Visual Environment

Light as a Physical Stimulus Light is the term given to that portion of the electromagnetic spectrum that functions as an adequate stimulus for vision in humans. Our visible spectrum ranges in wavelength from about 400 nanometers (nm), which we perceive as violet, to roughly 700 nm, which we perceive as deep red. Newton demonstrated that "white" sunlight contains all the colors of the spectrum and set the stage for the development of modern visual science by uncovering the mathematical relationships among light rays of different wavelengths (Newton, 1730). The wave properties Newton examined with his elegant experiments are demonstrated in the optical phenomena of propagation, reflection, refraction, scattering, and interference.

These observations—which described the wavelike properties of light—conflicted with observations on the interaction of light with matter that hinted at light's "corpuscular" (particulate) character. Modern quantum mechanical theory recognizes both sets of properties, describing light simultaneously in terms of photons (discrete packets of energy) and in terms of either wavelength or frequency.

The energy of a photon is interrelated with its associated wavelength and frequency by the equation

$$E = hv = \frac{hc}{L},$$

where h = Plank's constant, c = the speed of light, v = frequency, and L = wavelength.

Because the eye needs to focus light, it is worthwhile to recall that the angle of refraction of light incident upon an interface between two media is described by Snell's law:

$$\frac{\sin \theta 1}{\sin \theta 2} = n2 - n1,$$

where $\theta 1$ is the angle between the incident ray and the normal to the interface, $\theta 2$ is the angle between the refracted ray and the normal, and $n2 - n1$ is a constant obtained by taking the difference between the index of refraction of medium 2 and that of medium 1.

It is commonly assumed that light rays from distant objects arrive at the eye's first optically important interface parallel to one another. The extent to which these rays are refracted at that first interface is therefore a function of the difference in refractive index between the two media and of the radius of curvature of the optically important surfaces. In any given medium the denser the components of the refractive apparatus and the smaller their radius of curvature, the more strongly incoming light is bent during its passage to the retina.

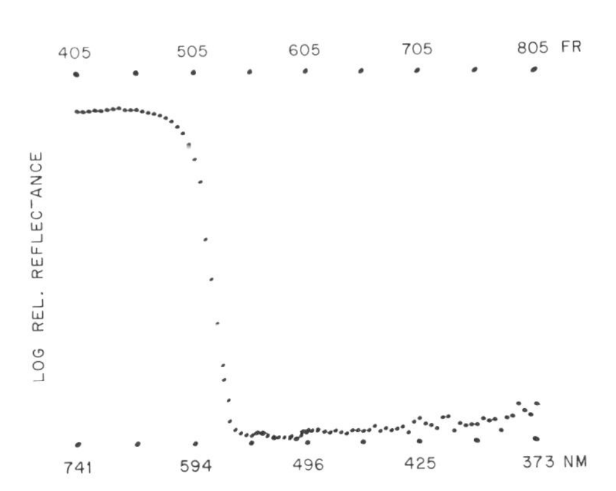

Figure 16-1 Reflectance spectrum of a typical red carotenoid pigment: the percentage of incident light reflected at each point in the visible spectrum plotted on a linear frequency scale. Labels on upper row of dots are in Fresnels (also called Terahertz); labels on lower row are in nanometers. (Modified from Levine, Lobel, and MacNichol, 1980.)

Illumination and Reflectance With the exception of those few organisms and naturally occurring compounds that are luminescent, objects are visible by virtue of light they reflect from an external source to the eye of an observer. White objects reflect a high percentage of all visible wavelengths, black objects reflect very little visible light, and grey objects reflect intermediate amounts. Most natural objects, however, do not exhibit flat reflectance curves, but reflect a greater percentage of certain wavelengths than others. These objects we call colored (Fig. 16-1). It should be noted that these statements regarding brightness and color refer only to natural objects in the environment. Under laboratory conditions sensations of white, black, and various colors can be created by numerous and varied kinds of stimuli. Under all conditions, the presence of other objects in the vicinity can have significant effects on the perception of lightness and color.

Reflectance curves provide information only about the percentage of light of any wavelength that the object under consideration reflects. The energy spectrum actually emanating from an object at any given time is the product of the object's reflectance curve and the energy spectrum of the light incident upon it. The light energy actually reaching an observing eye from an object will resemble its reflectance spectrum only if the object is illuminated by a standard white light source. At any other time the light reflected is directly dependent upon the spectral nature of the ambient illumination. (For a more detailed discussion of these phenomena, see Land, 1977, and Cornsweet, 1970.)

Light in Natural Environments Sunlight is the primary source of light for vision on earth. The emission spectrum of the sun—which exhibits a color temperature of roughly 6000° C—is modified before it reaches the earth's surface by the absorption of infrared and ultraviolet light by water vapor and ozone in the atmosphere (Dartnal, 1975).

The spectral content of ambient light changes dramatically between midday and twilight, as shown by McFarland and Munz (1975a,b). Furthermore, light intensity drops by roughly 10 log units (orders of magnitude, or decades) between bright midday sun and starlight on a moonless night. Thus, as pointed out by Land (1977), we live in a world of changing illumination that constantly alters both the intensity and spectral distribution of the stimuli presented to our eyes by objects around us. Our visual systems are confronted with the formidable task of providing constant, reliable information about our surroundings under these dramatically changing circumstances.

The visual systems that have evolved to deal with these problems are extremely complex. The comparative approach — examining the evolutionary adaptations of visual systems faced with different demands — is useful in helping to understand the design and operation of those systems. Vertebrates conveniently provide us with visual systems that have adapted to drastically different operating requirements. Diurnal and nocturnal animals live in different visual worlds, but most terrestrial animals are exposed to illumination that varies relatively little in spectral content. Because passage through short distances of clean air affects sunlight very little, a bird flying at a height of 20 m experiences the same light environment as a mouse scurrying along the ground below. Differences in the spectral quality of illumination do exist in some terrestrial habitats, such as forests, but these variations are relatively minor.

Aquatic animals, on the other hand, must cope with a medium that interacts strongly with light. Water also carries dissolved and suspended materials that further affect both the intensity and spectral characteristics of light passing through it. Bodies of organically pure, fresh and salt water absorb ultraviolet, infrared, and red light much more strongly than light of blue and green wavelengths (Lythgoe, 1979). Blue light of wavelengths near 480 nm penetrates such water most readily, and it is the transmission of this group of wavelengths — along with substantial scattering of shortwave light — that imparts to the open ocean and pristine lakes their characteristic blue color.

Most freshwater lakes and coastal marine waters contain various dissolved organic substances called *Gelbstoffe* ("yellow substances"), associated both with animal wastes and with the breakdown of chloroplasts (Yentsch and Reichert, 1962; Lythgoe, 1979). Yellow substances absorb ultraviolet and blue light very strongly, and, along with the chlorophylls of phytoplankton common to these waters, shift the transmission maximum of the medium to about 550 nm. Such waters appear green to human observers.

Still another visual environment is created in marshes, swamps, and certain tropical river systems surrounded by decaying vegetation. In addition to *Gelbstoffe,* these waters contain varying amounts of tannins, lignins, and other products of plant decomposition, which absorb light strongly across much of the visible spectrum. Depending on the precise nature and concentration of these compounds, they can impart to the water a range of hues from the color of light tea to a dark reddish brown, indicating spectral distributions with energy maxima ranging from 600 nm to more than 700 nm (Kinney, Luria, and Weitzman, 1967). Our visual systems are relatively insensitive to light in this range, and we have anthropocentrically labeled them *blackwaters.*

The deeper one descends in any of these three water types, the dimmer the illumination becomes and the more filtered and tinted the light is by the water above. Major differences in both spectral quality and intensity therefore confront aquatic animals living only a short distance apart. Species like the common guppy *Poecilia reticulata,* which restrict their activities to the very surface of the streams in which they live, encounter light that covers a broad range of wavelengths and may be several orders of magnitude more intense than the light surrounding catfishes like *Corydoras,* which meander along in the mud at the bottom of the same streams (Levine and MacNichol, 1979). In the marine environment, shallow-water, coral reef species inhabit environments where bright, broad-band illumination is available, whereas mesopelagic animals live in a world of perpetual blue twilight (McFarland and Munz, 1975a).

In discussing the functional morphology of the vertebrate eye, therefore, it is instructive to observe the similarities and differences among the eyes of terrestrial vertebrates, aquatic species living near the surface in various types of water, and those living at greater depths.

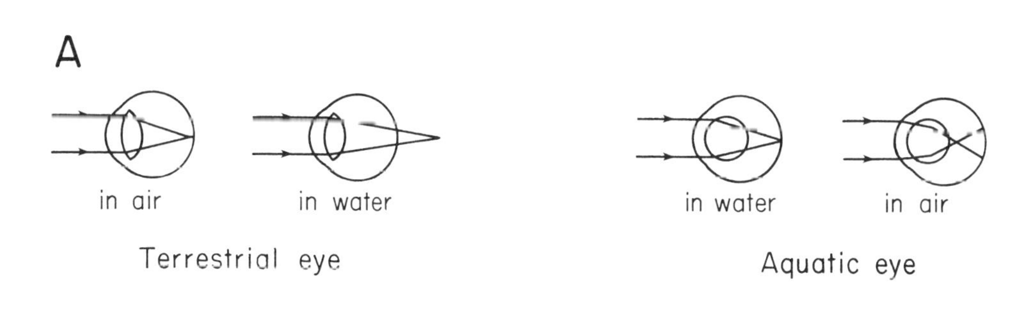
A
in air
in water
Terrestrial eye
in water
in air
Aquatic eye

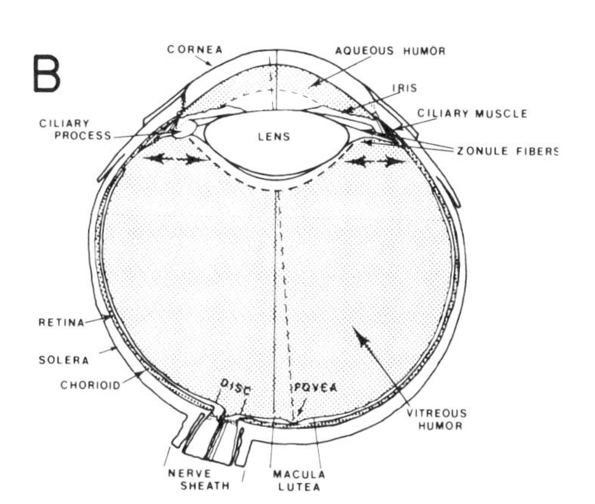
B
CORNEA
AQUEOUS HUMOR
IRIS
CILIARY MUSCLE
CILIARY PROCESS
LENS
ZONULE FIBERS
RETINA
SOLERA
CHORIOID
DISC
FOVEA
VITREOUS HUMOR
NERVE SHEATH
MACULA LUTEA

Terrestrial eye

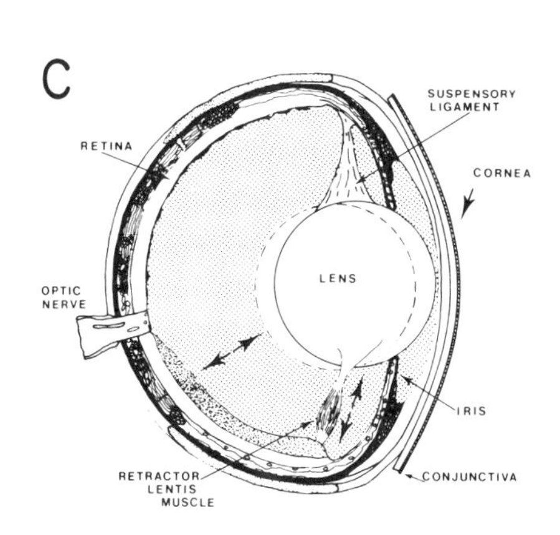
C
SUSPENSORY LIGAMENT
RETINA
CORNEA
LENS
OPTIC NERVE
IRIS
RETRACTOR LENTIS MUSCLE
CONJUNCTIVA
Aquatic eye

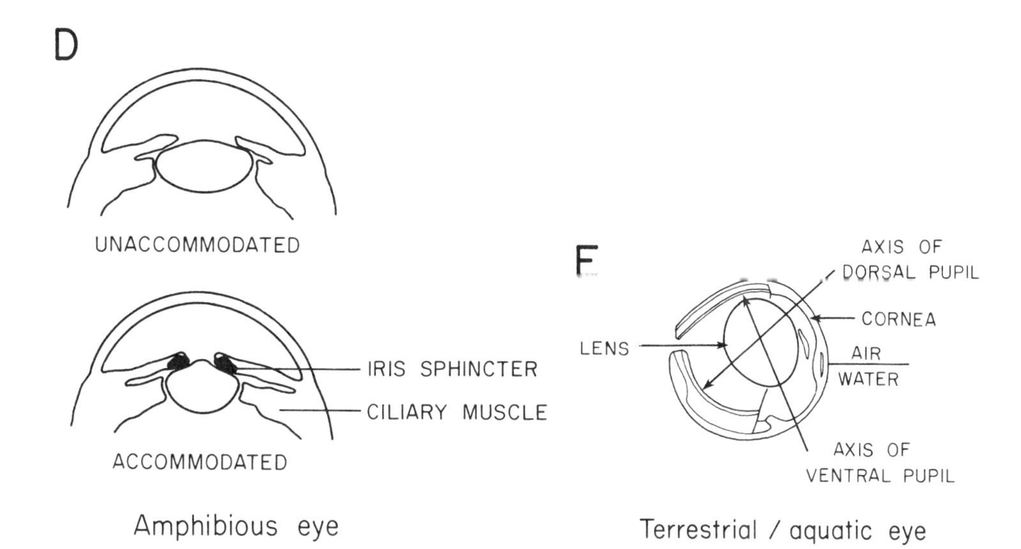
D
UNACCOMMODATED
IRIS SPHINCTER
CILIARY MUSCLE
ACCOMMODATED
Amphibious eye
E
AXIS OF DORSAL PUPIL
CORNEA
LENS
AIR
WATER
AXIS OF VENTRAL PUPIL
Terrestrial / aquatic eye

Figure 16-2 A: Comparison of the refractive abilities of aquatic and terrestrial eyes in air and water. Since the difference in refractive index between the cornea and the external environment is far greater in air than in water, much more refraction occurs at an air-cornea interface than at a water-cornea interface. (After Sivak and Millodot, 1977; with permission of Springer-Verlag.) B: Cross section of human eye. Note the lens-shaped cornea, the oval lens, and the ciliary muscles that change the shape of the lens for focusing. (Modified from Walls, 1942.) C: Cross section of typical teleost eye. Note the round, solid lens that protrudes almost to the point of touching the cornea. This high-density lens cannot be deformed easily, and is focused instead by the retractor lensis (or lentis) muscle, which moves the lens back and forth with respect to the retina. (Modified from Walls, 1942.) D: Cross section of a truly amphibious eye with a dual focusing mechanism of the type found among reptiles, birds, and some amphibious mammals. Note that in addition to the action of the ciliary muscle, the functional part of the lens is deformed by strong pressure from the iris sphincter. (Redrawn from Sivak, 1978, after Walls, 1942.) E: Cross section of the eye of the fish *Anableps anableps,* showing split pupil and differing lens curvatures for vision in air and water. (After Walls, 1942.)

Gross Anatomy and Visual Environment

Refractive Apparatus in Air and Water People who open their eyes underwater experience instant hyperopia (farsightedness). Why should an eye designed to operate in air become useless when submerged? The density of the operating environment is critical for any device that needs to bend light in order to focus it because, as we saw earlier, refraction at an interface between two media depends both on the curvature of the interface and on the difference between the refractive indexes of the two media. The optically important interfaces in the vertebrate eye occur where the external medium—air for terrestrial eyes, water for aquatic ones—abuts the cornea, and at the front and back edges of the lens where it meets the aqueous and vitreous humor, respectively (Fig. 16-2A).

The optical density of air is 1.0003, whereas that of water is 1.33. The difference in density—and hence in refractive index—between air and the typical vertebrate cornea and aqueous humor is substantial. In terrestrial situations, therefore, a good deal of refraction—and hence light gathering and focusing—can be effected by a cornea that is properly curved to act as a lens. The precise shape of the cornea is critical to terrestrial animals, and even minor aberrations in corneal curvature can lead to distortions in the visual image known as astigmatism. Very little additional refractive power is actually needed to form a clear image on the retina; correction for close-up and long-distance vision can be accomplished by a lenticular lens not much denser than the aqueous and vitreous humor in which it operates. Lenses of this density are malleable and are focused by contraction and expansion of ciliary muscles that ring the lens and change its shape (Figure 16-2B). (For a detailed and superbly illustrated treatment of human lens structure and development, see Warwick and Williams, 1973.)

In an aquatic situation, on the other hand, the refracting power of the cornea is severely diminished by the high density of the surrounding water. Even if an aquatic cornea is properly curved, therefore, it cannot focus light the way it could in air (Sivak, 1978). The shape of the cornea is thus of little consequence in totally aquatic animals, and great variation exists among them. In order to compensate for the loss of corneal refractive power, the lenses of underwater eyes must accomplish substantially more refraction, both to gather light and to project it accurately onto the visual cells of the retina. Typical terrestrial lenses, with their lower refractive index and finite flexibility, cannot undergo the radical changes in curvature required to make this necessary compensation. The optical apparatus in the eyes of most terrestrial vertebrates is thus unable to bring an image into focus on the plane of the retina when the eye is forced to operate in an aquatic situation, and hyperopia results (Sivak, 1978).

The majority of aquatic vertebrates (and several cephalopods) have solved this problem by developing nearly round, crystalline lenses of high refractive index. The shape of these basically solid lenses cannot be changed to any great degree, so accommodation is accomplished by changing the position of the lens through contraction of the retractor lensis muscle and the opposing pull of the suspensory ligament (Figure 16-2C).

Animals that have need for acute vision in both air and water have overcome the refractive difficulties brought on by the air-water transition in a variety of ways. Flying fishes and penguins, for example, have corneas that are almost perfectly flat, so the difference between their refractive performance in air and water is minimal. Certain amphibious reptiles and birds possess lenses of extreme flexibility that are shaped during accommodation both by large ciliary muscles and by an unusually strong iris sphincter that deforms the anterior surface of the lens (Figure 16-2D). A few odd creatures, like the surface-dwelling fish *Anableps anableps,* manage to see simultaneously in air and in water by employing eccentrically shaped lenses and an hourglass-shaped pupillary slit to bring separate images of above- and below-water surroundings into focus on different regions of the retina (Sivak, 1978; Figure 16-2E).

Eyes for Low-Light Situations The gross morphology of the vertebrate eye responds dramatically to the low levels of light encountered by nocturnal species and by aquatic species living at great depths in the sea. The first noticeable changes are increases in pupillary opening and total eye size, which improve the light-gathering ability of the optical apparatus. Note, for example, the large eyes of nocturnal terrestrial vertebrates and the irises of cats that can increase their pupillary openings from narrow slits in bright light to full circles in semidarkness. The trend toward large eyes and pupillary openings is carried to the extreme in many mesopelagic fishes, in which the eyes have enlarged until there is barely enough room for them among the other structures of the neurocranium. In many deep-sea fishes a normally proportioned eye with a sufficiently large pupil could not be accommodated in the head, and the pupillary opening and lens have continued to expand out of all proportion to the rest of the eye. The tubular eyes thus produced (Fig. 16-3A) exhibit outstanding light-gathering ability, but are restricted to very limited fields of view. In the abyssal depths, where practically the only light available is the dim, blue glow of bioluminescence, these tubular eyes point either forward or upward, depending, apparently, on the behavioral requirements of the particular species (Figure 16-3B).

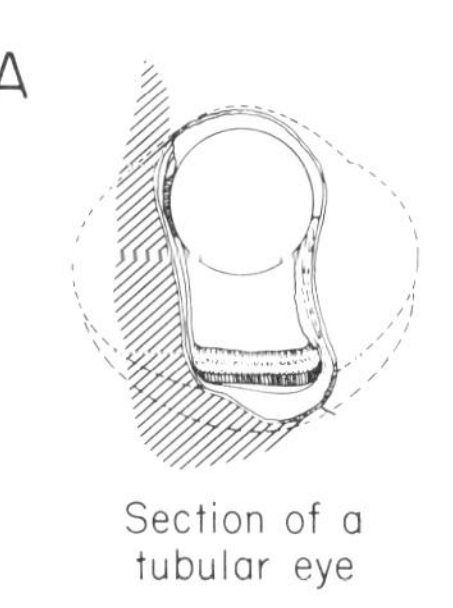

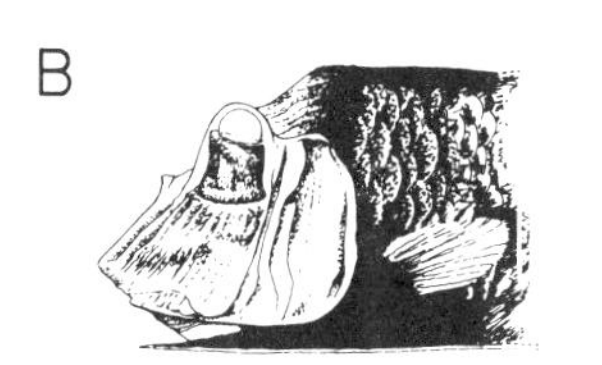
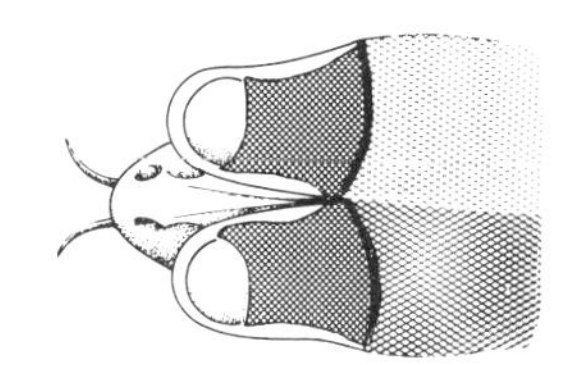

Figure 16-3 A: Tubular eye of *Argyropelecus,* a deep-sea fish. The effective aperture of this eye is equal to that of a much larger, normally shaped eye. The large, round lens, unobstructed by an iris, gathers light with great efficiency. (Modified from Walls, 1942.) B: Tubular eyes of various deep-sea fishes may point in various directions, depending upon the visual requirements of the species. (Modified from Walls, 1942.)

Retinal Structure and Visual Environment

The vertebrate retina is formed embryologically from two lateral diverticula of the forebrain known as optic vesicles, which, like the ventricles of the brain, are lined with ciliated neuroepithelial cells (Davson, 1980). When the optic vesicles invaginate during development to form the eyecup, these two ciliated layers interdigitate with each other and differentiate into the collumnar, pigment-bearing retinal epithelium and the neural or sensory retina (Fig. 16-4). Both these

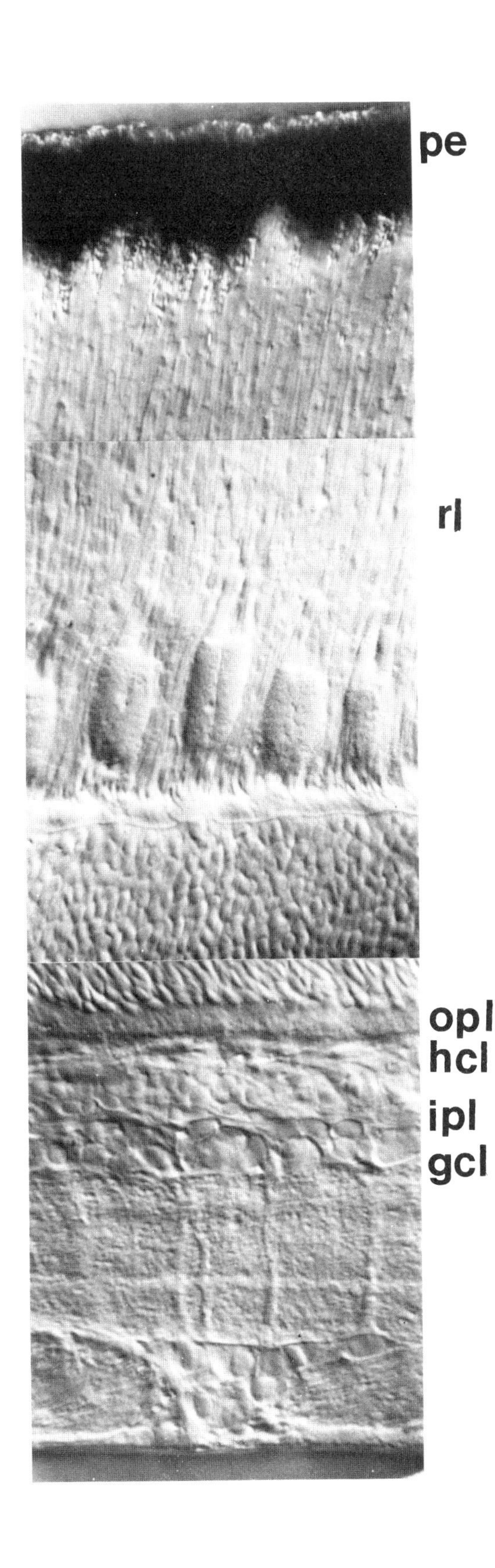

layers have evolved into varied forms, and both have undergone specialization in species with different visual requirements.

The Retinal Epithelium The pigmented epithelium serves the eye both by providing physical and physiological support to the photoreceptors (Davson, 1980) and by helping the retina to function under a wide range of changes in light intensity. In diurnal animals active in bright light, processes of the epithelial cells wrap tightly around the photoreceptors, shielding them from stray light with densely packed, light-absorbing melanin granules. A substantial amount of light is not absorbed during a single passage through the neural retina, and random scattering of this light would seriously degrade visual acuity. The pigmented epithelium of diurnal animals absorbs most of this "surplus" light, preventing backscattering and preserving acuity.

For nocturnal animals like cats, on the other hand, photons are few and absolute sensitivity is more important than acuity. Under these conditions a reflective backing to the retina can double the eye's sensitivity to light by bouncing back through the retina light that was not absorbed on the first pass. This second trip gives the photosensitive pigments in the receptor cells a second chance to absorb sufficient photons. The retinal epithelium and its supporting layers can provide any of three types of reflecting layers or tapeta lucida (Rodieck, 1973), described as a retinal tapetum, tapetum cellulosum, or tapetum fibrosum, depending on location and structure. Tapeta are responsible for the "eye shine" observed when nocturnal animals are illuminated by automobile headlights or spotlights in the dark.

In many lower vertebrates diel changes from light to dark are accompanied by migration of pigment granules within the epithelial cells. In bright light the granules descend in the epithelial

Figure 16-4 Section of the retina of a typical vertebrate eye, showing both the neural retina and the pigment epithelium. *pe* = pigment epithelium; *rl* = receptor layer; *opl* = outer plexiform layer; *hcl* = horizontal cell layer; *ipl* = inner plexiform layer; *gcl* = ganglion cell layer.

processes, shielding the photoreceptors as described above. At the onset of darkness the granules are withdrawn back into the epithelial cell bodies, decreasing the shielding effect. The precise mechanism by which these pigment migrations are mediated is not known, but it is clear that absorption of light in the photoreceptor layer is an important local controlling factor (Ali, 1959, 1975).

The Photoreceptor Layer The sensory retina consists of the photoreceptors, several classes of interneurons, and glial cells, all of which are arranged into clearly recognizable layers, and each of which can demonstrate evolutionary adaptations to ambient light conditions. The photoreceptor layer in many vertebrates contains two main types of photoreceptor cells: rods, which are sensitive enough to signal the absorption of a single photon, and cones, which provide neural signals only when they absorb significantly more quanta. Cones are generally associated with providing high acuity (resolution) and color vision, whereas rods are generally associated with high sensitivity, low acuity, and achromatic vision.

Photoreceptor cells owe their light sensitivity to a class of photosensitive chemicals known as visual pigments (see below) that are contained in specialized regions known as outer segments. When visual pigment molecules absorb light, they undergo a series of configurational alterations that trigger electrochemical changes in the receptor cells that contain them. The mechanism by which this electrical change is effected is still not understood (Davson, 1980; Fein and Szuts, 1982).

Photoreceptors do not generate the classic all-or-nothing action potentials that characterize motor neurons. Instead, they exhibit rapid, graded potentials—in vertebrate photoreceptors always in the negative or hyperpolarizing direction—which are the first in the chain of events that ultimately results in vision.

A retina usually has one class of rods (although some amphibians have two; Walls, 1942), which are neurally wired so that signals from numerous receptors are pooled as information travels centrally in the nervous system. Such pooling, or summation, allows numerous small stimuli to add up among the interneurons, greatly increasing the absolute sensitivity of the rod system as a whole. The increased sensitivity afforded by this arrangement has its price; the visual acuity (resolving power) of the rod system is low.

Diurnal vertebrates may possess as many as four classes of cone cells (humans have three), each of which contains a spectrally distinct visual pigment. As described below, interaction of the signals originating from different spectral classes of cone cells enables organisms to discriminate among objects on the basis of color as well as brightness. Although extensive interaction among cone signals is necessary for color vision, the cone system typically exhibits substantially less summation than the rod system, resulting in higher acuity with lower absolute sensitivity.

Visual Pigment Content and Receptor Morphology Microspectrophotometric studies (see below) have confirmed that in many (although not all) vertebrates, cone classes containing different visual pigments are morphologically distinct from one another. These physically different cone types are often organized in one of a number of regular, repeating mosaic patterns (Figure 16-5; for a variety of examples, see Wagner, 1972). The most detailed studies published to date on cone morphology and arrangement vis-à-vis visual pigment content have been performed on goldfish (Figure 16-8; Stell and Harosi, 1976; Marc and Sperling, 1976; Marc, 1977). Other investigations on the relationships among cone morphology, retinal mosaic patterns, and visual pigment content have been made in several cichlid fishes (Ali, Harosi, and Wagner, 1978; Levine et al., 1979), a few primates (Marc and Sperling, 1977), and turtles (Baylor and Hodgkin, 1973; Richter and Simon, 1974).

Little is known regarding the functional significance of morphological differences among cone types. Members of the double cones common in many fish species, as well as in several frogs and birds, contain different visual pigments. The electrical responses of these double cones to light of different wavelengths indicates that the signals from each member of the pair interact strongly with those of the other member (Richter and

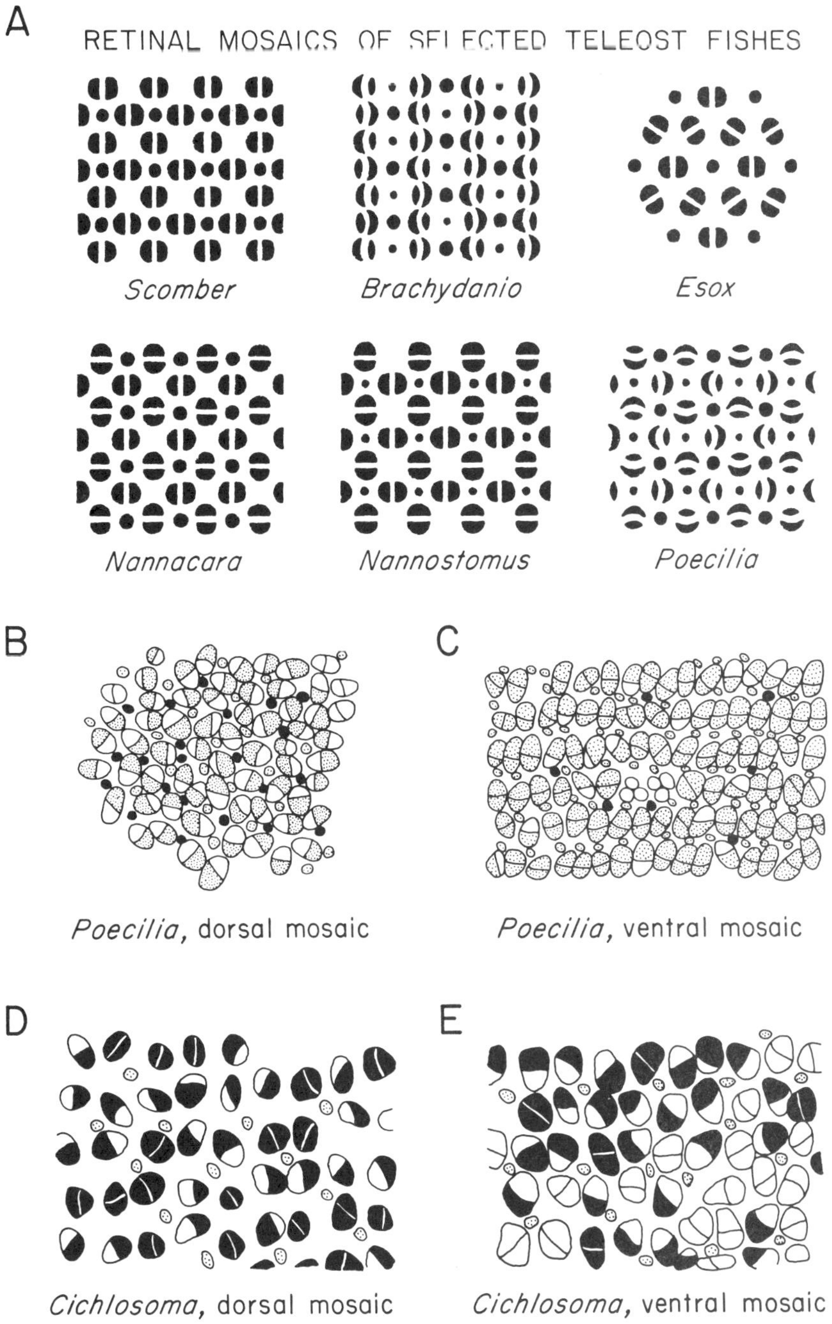

Figure 16-5 A: Diagrammatic representation of the typical cone mosaic patterns found in a variety of teleost fishes. (Modified from Wagner, 1972.) B: In many species the "typical" mosaic pattern varies significantly from one part of the retina to another. In this end-on view of the actual photoreceptor mosaic from the dorsal retina of the guppy *Poecilia reticulata,* several of the mosaic units shown in A are visible. The visual pigments that can be assigned to these cells on morphological grounds are indicated as follows: dotted cells = max 545 nm; clear cells = max 468 nm; black cells = max 410 nm. C: Photoreceptor mosaic from the ventral retina of *P. reticulata.* Visual pigment content of cells indicated as in B. D: Even in retinas where the members of the paired cones are morphologically indistinguishable, the chromatic arrangement of cells can change from one part to another. This end-on view of the photoreceptor mosaic from the dorsal retina of *Cichlasoma longimanus* shows the chromatic organization of the cells as determined by a vital staining technique. Visual pigment contents of these cells as follows: dotted cells = max 455 nm; black cells = max 579 nm; clear cells = max 531 nm. E: Chromatic organization of the ventral retina of *C. longimanus.* Key to pigment contents as in D. (Modified from Levine et al., 1980.)

Simon, 1974). As we will discuss in more detail, interaction between signals from cells containing different pigment types is essential in color vision. Sampling of the same image space by members of double cones containing different pigments certainly enables these cells to effectively subserve color discrimination, but this type of interaction is accomplished in many other animals (like primates) that have only single cones. Double cones are thus neither a unique result of this functional requirement nor essential for good color vision.

Many aquatic vertebrates possess twin cones: paired cones in which both members are morphologically identical and possess the same visual pigment. Twin cones are found in species from a variety of habitats, and the literature contains both suggestions that twin cones increase sensitivity in dim light and assertions that they are somehow useful in extremely bright light. Experiments that provide quantitative evidence for either function have not yet been done. Various hypotheses regarding the significance of differences in length among cone types in many animals are similarly intriguing but ultimately unsatisfying.

The significance of intraretinal and interspecific differences in retinal mosaic organization is also obscure. In various species different cell and pigment types are present in different proportions (see, for example, Marc and Sperling, 1977), and in several species studied these proportions vary from one part of the retina to another (Levine et al., 1979; Levine, Lobel, and MacNichol, 1980). In some cases, sensitivity may be the critical determining factor, whereas in others the situation is undoubtedly more complex (Hashimoto and Inokuchi, 1981). Studies relating photoreceptor morphology and organization to the visually important behaviors they mediate are urgently needed.

Retinal Structure and Environmental Light Regimes The photoreceptor cells and their patterns of organization vary markedly between diurnal species and species that restrict their activity either to nocturnal periods or to dim habitats. In diurnal shallow-water fishes, both rods and cones are slender, tightly packed, and very numerous (Fig. 16-6A). In teleosts paired cones and single cones are usually both present, and they are arranged in regular, geometrically uniform mosaic patterns (Fig. 16-6B). Rods are usually forced to a peripheral level in the light-adapted retinas of such species.

Fishes that must use their eyes at greater depths, or during twilight periods in shallow water, have cone cells that are enormously enlarged compared with those of the previous group. Not only do the cells themselves enlarge, but the retinal mosaic also expands and its geometric regularity deteriorates. Single cones become more randomly distributed and triple cones (clusters of three)—whose visual pigments and function are still unknown—intrude into the pattern unpredictably (Figure 16-6C,D). The wide spaces left among the cones are tightly packed with numerous rods, which play a critical role in low-light vision.

Why are the cones in these species so large? The physical correlate of light intensity is photon flux density: the number of photons that pass through a given unit area per second. As light intensity drops, the number of photons striking a cell of a given size also diminishes. Since a certain number of photons must be absorbed in a cone in order to cause the cell to produce a neural signal, increasing the cell's cross-sectional area increases its chances of catching sufficient photons at low flux densities. Although this reasoning is straightforward, the experiments necessary to test the quantitative contribution of enlarged cones to visual sensitivity have not yet been performed.

Below certain levels of ambient illumination, however, there is not enough light for even enlarged cones to function efficiently, and in many deep sea or totally nocturnal species cones are either absent or vestigial (Figure 16-6E). These retinas are composed entirely of rods. Figures 16-5E and F show the retina of a species in which rods totally dominate the photoreceptor layer and have multiplied into several bands. Since only a portion of the light that strikes a visual receptor is absorbed while passing through the cell, the existence of several layers of rods increases the total light-absorbing potential of the retina substantially.

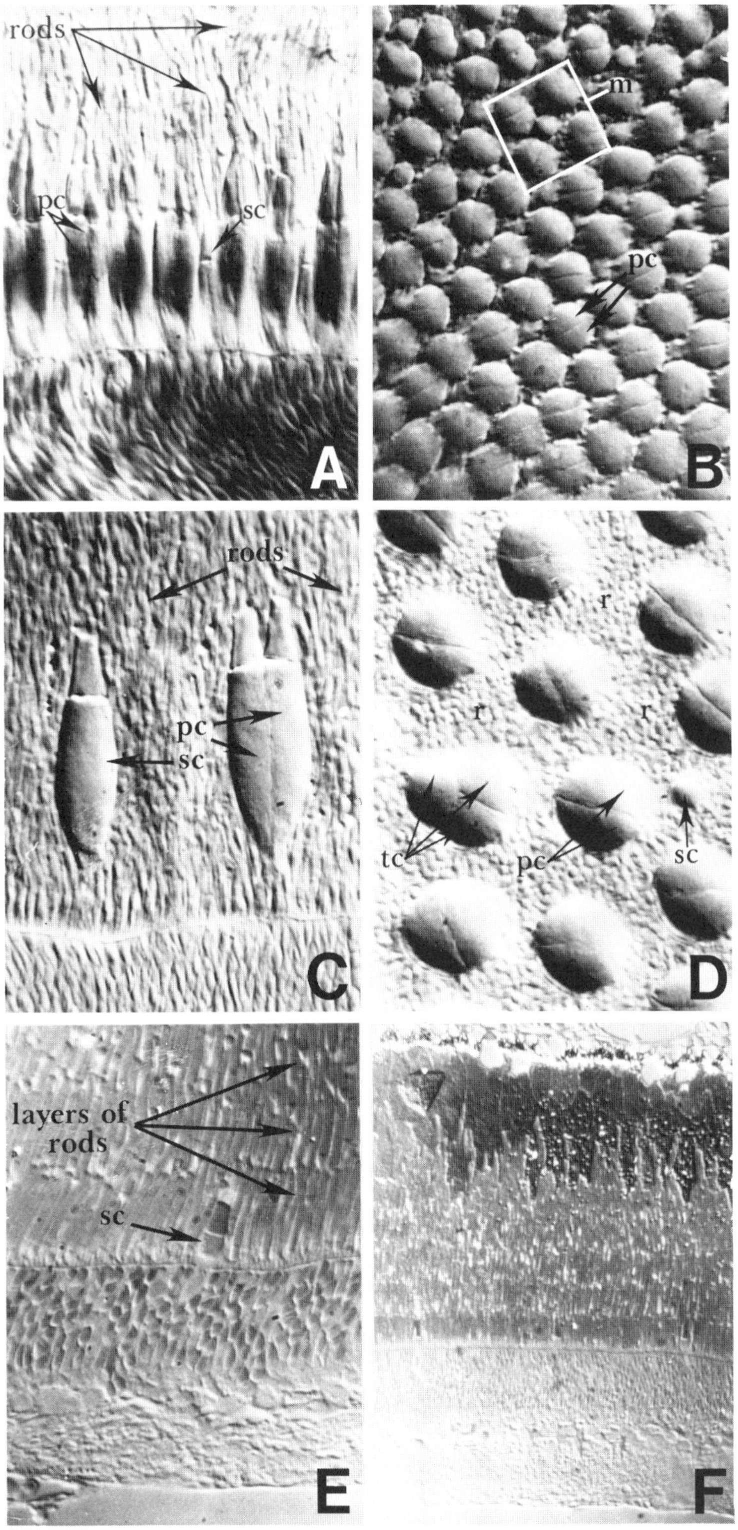

Figure 16-6 A: Longitudinal section of the retina of the mackerel *Scomber scombrus* showing slender, tightly packed paired cones *(pc)*, single cones *(sc)*, and rods. B: Cross section of the retina of the wolf-fish *Anarhichas lupus* retina, showing regular, repeating mosaic unit *(m)* of one single cone and four paired cones, with a few rods visible in the intercone spaces. C: Longitudinal section of the retina of the cusk *Brosme brosme* retina at same magnification as A, showing greatly enlarged, widely spaced cones and numerous long rods. D: Cross section of cusk retina at same magnification, showing decreased regularity in spacing and arrangements of cones. Note the presence here of a triple cone *(tc)* in addition to the expected single and paired cones. In this species the numerous rods completely fill the spaces between the cones. E: Longitudinal section from the retina of the still deeper-dwelling fawn cusk eel *Lepophidium cervinum*, still at the same magnification, showing a single, small cone and a retina overwhelmingly dominated by rods. F: The cusk eel retina at lower magnification, showing the tiered arrangement of rods. (From Levine, 1980.)

Visual Pigments and Visual Function The visual pigments contained in photoreceptors consist of a protein fraction (opsin) conjugated with a light-absorbing chromophore derived from either vitamin A-1 or A-2 (Lythgoe, 1979; Davson, 1980). All visual pigments are capable of absorbing light of most visible wavelengths to some degree, but each particular visual pigment is maximally sensitive to light of certain wavelengths (Fig. 16-7). For convenience in discussion, visual pigments are often referred to in terms of their wavelengths of maximum sensitivity or Lmax. Thus, an animal may be said to possess either blue-, green-, and red-sensitive pigments or pigments of Lmax 465, 550, and 625 nm.

Until the late 1960s analysis of visual pigments required large quantities of material extracted from numerous retinas. Dependence on this technique posed several problems. First, in duplex (rod and cone) retinas rods almost always outnumber cones, so cone pigments are literally buried beneath large quantities of rod pigments. Cone pigments are also more labile in such extractions, and the techniques required for analysis of mixtures containing three or four different light-sensitive compounds are difficult (Munz and McFarland, 1975).

The development of microspectrophotometry enabled the examination of visual pigments in situ within the outer segments that contain them. Recent refinements in this technique (MacNichol, 1978) have allowed the accumulation of much accurate data on cone pigments in a variety of animals (Levine and MacNichol, 1979; Lythgoe, 1979).

One visual pigment in a retina is sufficient to enable an organism to see, but single-pigment visual systems confine their bearers to a monochrome world in which all objects appear in black, white, or shades of grey, as they do on a black and white television screen. These monochromats can thus differentiate objects only by detecting differences in their relative brightnesses. In order to discriminate among objects on the basis of color as well as brightness, an organism requires a minimum of two visual pigments with different spectral sensitivities. Humans, and many other diurnal animals with well-developed visual systems (including many fishes, birds, and reptiles),

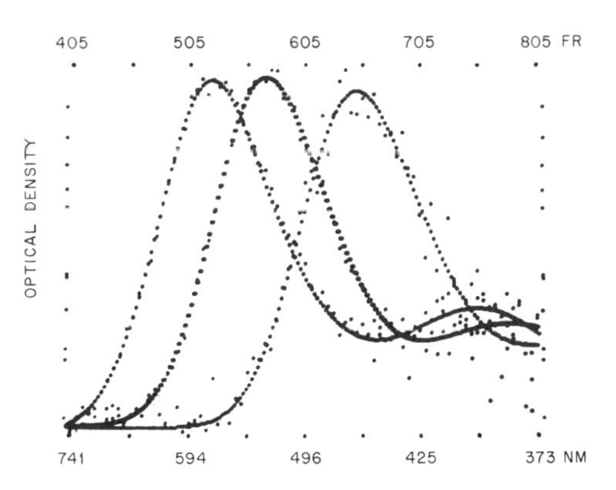

Figure 16-7 Absorption spectra of the three cone visual pigments from the cichlid fish *Cichlasoma longimanus,* plotted as normalized optical density against a linear frequency scale. Labels as in Figure 16-1. Note that although each curve exhibits a well-defined absorption maximum, each pigment absorbs light over a broad spectral range. (After Levine et al., 1979.)

have four different visual pigments; one in the rods and one each in three different classes of cones. As it is primarily the three-pigment cone system that is involved in color vision, such visual systems are called trichromatic.

Visual Information Processing and Retinal Neuroanatomy

But to determine more absolutely . . . by what modes or actions [light] produceth in our minds the phantasms of colors is not so easie.
—Isaac Newton

The detection of light by photoreceptors is merely the first step in the long series of neural operations that ultimately result in vision. The complexity of these operations is mirrored in the intricate neural networks of the retina and higher visual centers. Visual information processing is far more complex than most introductory science courses and texts lead students to believe. The simple explanation of trichromatic vision usually provided teaches that the retina, with its three visual pigments, functions more or less like a color television screen in reverse. Each retinal area is composed of a mosaic of red-, green-, and blue-sensitive cones that respond to light of different wavelengths. When the retina detects a certain amount of long-wave light emanating from a particular region of the visual field, we are told, it designates that area as "red." Where shortwave light predominates it will generate a sensation of "blue."

This kind of simplistically designed visual system would fail at a number of critical visual tasks in the real world. As mentioned earlier, natural illumination changes dramatically in intensity and spectral distribution from time to time and from one place to another. Differences in illumination mean differences in the absolute amounts of light of various wavelengths reflected to observing eyes from the same objects at different times. The simple visual system described above would regularly confuse green objects seen in predominantly red light with orange or red objects seen in predominantly green light. (For a detailed discussion of confusions possible for a monochromat, see Cornsweet, 1970). For organisms that rely heavily on vision for accurate information about objects around them, this kind of ambiguity is unacceptable. But how can a visual system cope with these problems? Our knowledge in these areas is still incomplete, but a comparative study of visual systems provides us with several insights into the function and evolution of vertebrate visual systems.

Surveys of visual pigments covering a variety of species show that wherever animals are active in well-lit situations, they have evolved multipigment visual systems theoretically capable of color vision. This is a simple—but powerful—affirmation of the importance of color vision. Single-pigment (and necessarily color blind) visual systems are found only in species that are strictly nocturnal (or whose recent ancestors have been), in species from significant depths in the sea, and in a few aquatic vertebrates whose eyes are used only for spotting silhouettes of prey from below. In these situations, dim, spectrally restricted light makes maximum sensitivity to available illumination the most important characteristic of the visual system. To maximize this sensitivity, the absorption maxima (Lmax) of these species's single visual pigments are invariably closely matched to the wavelengths of light that predominate in their environment.

To understand why matching is necessary, imagine being seated in a dimly lit theater and being obliged to wear a pair of colored sunglasses whose lenses transmit light in a fashion that mimics the spectral sensitivity of a single visual pigment. If the stage is bathed in blue light and you don a pair of sunglasses tinted blue, you will be able to see reasonably well. Because the lenses transmit blue light, the sensitivity of your eyes is spectrally matched to the available illumination. If, however, you choose a pair of green or rose-colored glasses, which block blue light, the scene would go dark, and you would miss the show.

Outside the darkened theater in bright sunlight, however, one has more options. Under bright, broad-spectrum illumination almost any color can function adequately, because there is plenty of light available at most wavelengths. The scene will be clear, but in monochrome—all in shades of blue, or red, or green, depending on the choice of lenses. You might decide to take along several pairs of glasses and use the best one for each task you want to perform. A red filter, for example, might be useful for studying color variation in roses.

It would be bothersome to switch glasses all the time (and certainly not trivial for animals to change their visual pigments), and it might be more efficient to wear a different color lens over each eye, or even composite lenses made of two or three color strips. By peering through these different filters one after another and comparing the appearance of your surroundings as seen through each strip, you would be able to make more accurate visual discriminations than with glasses of any single color. This is the selective advantage that has probably caused so many animals to develop multipigment visual systems.

Virtually all diurnal vertebrates that inhabit brightly lit environments have met the minimum criteria for color vision by evolving two or more spectrally distinct classes of cone cells, as described above. But color vision is not guaranteed by the existence of spectrally different cones; in order to process color information the nervous system of an animal must be able to dynamically compare and contrast the signals generated by those cones. In organisms where visual pigments with different Lmax values can function at the

A

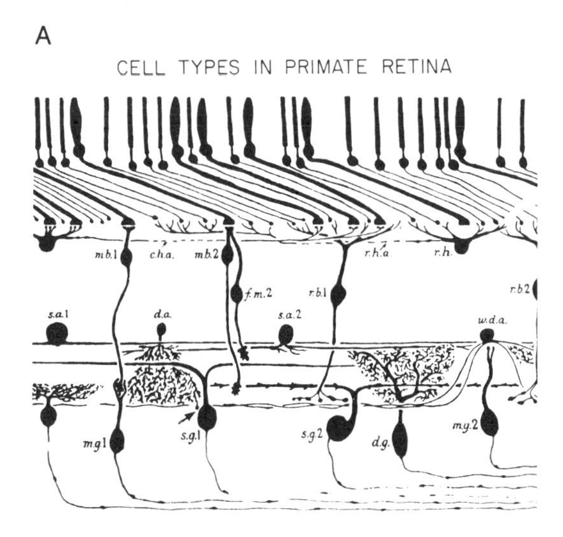

B

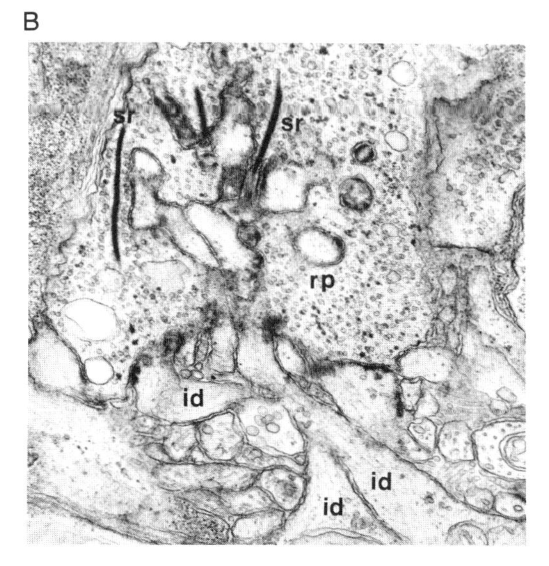

Figure 16-8 A: Photoreceptors and retinal interneurons identified in the primate retina by Boycott and Dowling (1969). Note the extensive lateral spread of several interneuron types. Cell classes: *ch, rh* = horizontal cells; *mb*1, *mb*2, *fb*, *rb*1, *rb*2, *fm*1, *fm*2 = subclasses of bipolar cells; *sda, sa*1, *sa*2, *da, wda* = subclasses of amacrine cells; *sd*1, *sd*2, *sg*1, *sg*2, *mg*1, *mg*2 = subclasses of ganglion cells. B: Electron micrograph of receptor pedicel *(rp)* showing the participation of dendrites from several interneurons *(id)* in the complex synapse. Note the synaptic ribbons *(sr)* associated with the complex.

same time, complex neural networks perform a variety of operations on the signals originating in the various receptor classes. Although our understanding of this process is far from complete, enough data exist to permit linking morphologically identifiable classes of neurons with their role in visual information processing.

Retinal Interneurons The neural signals generated by the photoreceptors are processed extensively by several layers of retinal interneurons (Fig. 16-8) whose connections are concentrated in two pronounced synaptic layers, the outer and inner plexiform layers. The meticulous investigations of Tartuferi (1887) and Cajal (1933) revealed the existence of at least four classes of interneurons tied together by a network of interconnecting dendrites.

A thorough examination of the morphology, physiology, and visual functions of all retinal interneurons would fill a good-sized volume, and readers are referred to such volumes (Rodieck, 1973; Davson, 1980) for comprehensive treatments of retinal neuroanatomy and physiology. Here we will briefly describe the major cell types, and then concentrate on the results of recent functional and anatomical studies on a few cell classes.

Anatomy of Retinal Interneurons In the outer plexiform layer the bases or pedicles of the photoreceptor cells make functional connections with dendritic processes from both bipolar cells, which carry information onward toward the rest of the central nervous system, and horizontal cells, whose processes carry information laterally across the retina. In the inner plexiform layer, the bipolar cells make synaptic connections with the ganglion cells, whose long axons leave the eye through the blind spot or optic disk to form the optic nerve, and with the amacrine cells, whose processes ramify laterally in the inner plexiform layer. For excellent reviews of retinal anatomy, see Rodieck (1973), Dowling (1974, 1979), and Davson (1980).

The relative numbers of these cell types found in different species offer some insights into retinal function. In diurnal species, for example, all

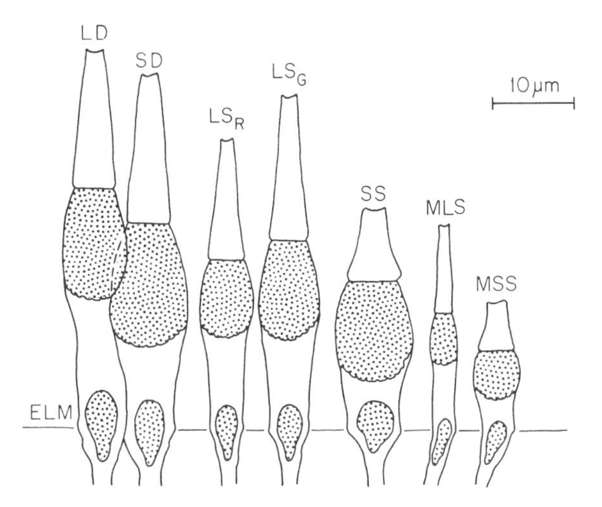

Figure 16-9 Morphological classes of cones in the goldfish *Carassius auratus,* after Stell and Harosi (1976). Morphological identifications and assignments of red-, green-, and blue-sensitive pigments to these cells are: *LD* = long member of double (red); *SD* = short member of double (green); *LS* = long single (red or green); *SS* = short single (blue); *MLS* = miniature long single (red); *MSS* = miniature short single (blue).

classes of interneurons are well represented. In humans and other primates single cone cells often "feed" into single midget bipolars, which receive their primary input from that cone, and which in turn attach to their own "private" midget ganglion cells (Davson, 1980). Large numbers of these midget bipolars and ganglion cells produce thick plexiform, nuclear, and ganglion cell layers and indicate low summation and high acuity. In nocturnal and deep-sea species, on the other hand, the bipolar and ganglion cell layers are much thinner. In these cases large numbers of receptors converge on small numbers of interneurons, emphasizing sensitivity at the expense of acuity.

Connectivity among Retinal Interneurons After Marc and Sperling (1976) and Stell and Harosi (1976) reported the correlation between cone structure and visual pigment content in goldfish (Fig. 16-9), the way was cleared for anatomical studies to elucidate connections among horizontal cell types, bipolar cell types, and photoreceptors of known spectral class. Stell and his colleagues (Stell and Lightfoot, 1975; Stell, Ishida, and Lightfoot, 1977; Ishida, Stell, and Lightfoot, 1980) performed such studies, using a variety of neuroanatomical techniques including Golgi impregnation and specific staining of junctional ultrastructure for electron microscopy. Horizontal cells that contact cones in goldfish were divided into three categories — labeled H1, H2, and H3 on Figure 16-10 — on the basis of distance from the outer synaptic layer, the extent of their dendritic spread, and the density of their contacts with cones. Stell and Lightfoot (1975) determined that H1 cells contact all three cone types, whereas H2 cells contact only green and blue cones, and H3 cells contact only blue cones.

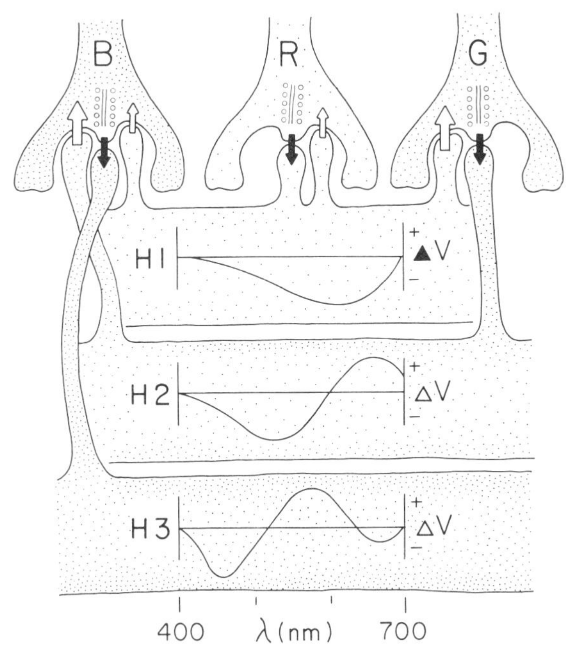

Figure 16-10 Functional connections of three horizontal cell classes (*H1*, *H2*, *H3*) in the goldfish, after Stell et al. (1975), as determined by joint electrophysiological and morphological studies. *B* = blue; *R* = red, *G* = green. Arrows indicate both direction and polarity of information transfer: black arrows indicate information transfer conserving response polarity; white arrows indicate transfer with inverted polarity. *λ (nm)* = wavelength; *ΔV* = change in resting potential.

A

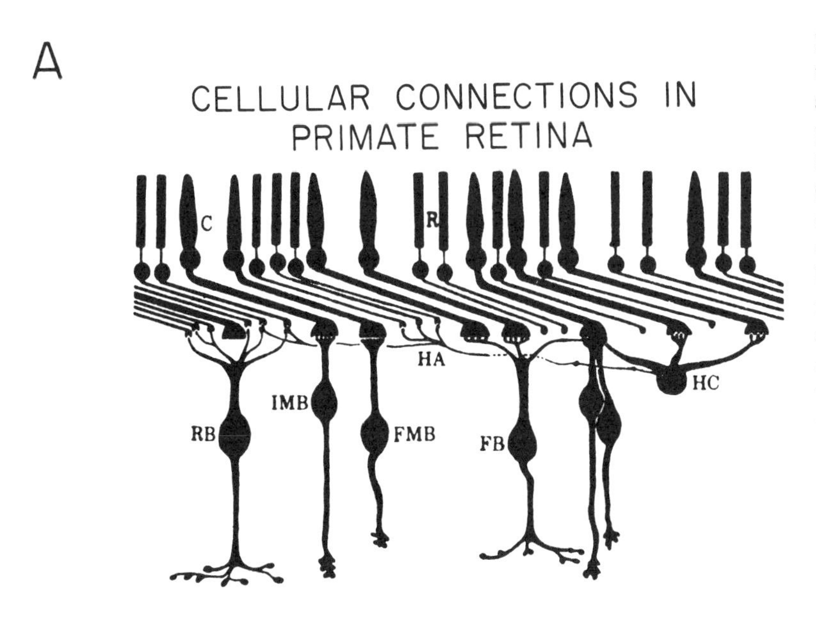

B

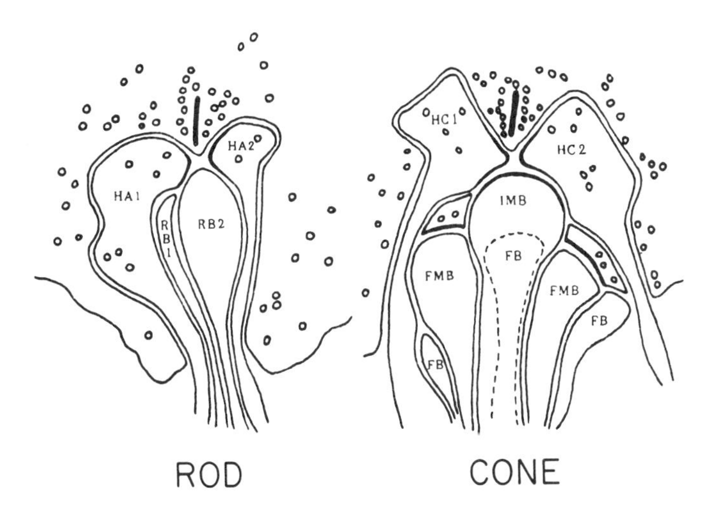

Figure 16-11 A: Diagrammatic representation of connections among horizontal, bipolar, and photoreceptor cells in the primate retina. Note that midget bipolars *(IMB, FMB)* connect to one and only one cone, but send many dendrites to that cone. Rod bipolars *(RB)* contact numerous rods, and flat bipolars *(FB)* contact several cones. Note that with the exception of the third cone from the right, this figure does not show the overlapping of various horizontal cells *(HC)* and bipolars onto each cone. (From Kolb, 1970.) B: Arrangement of bipolar (*RB*1, *RB*2, *FMB, FB, IMB*) and horizontal cell (*HA*1, *HA*2, *HC*1, *HC*2) dendrites in synaptic invaginations of rod *(left)* and cone *(right)* in primate retina. Note the close apposition between bipolar dendrites and H-cell dendrites within the synapse. (From Kolb, 1970.)

Lest the schematic nature of Figure 16-10 give a false impression, it is important to recall that bipolar cell dendrites also participate in the complex synapses of cone pedicels (Fig. 16-11). There are often two or more classes of bipolar cells in each retina, with morphology and terminology varying from species to species. Among classes that have been named on the basis of their anatomy are the giant (rod) and small (cone) bipolars of the goldfish (Stell, 1967), as well as the flat midget bipolars, invaginating midget bipolars, flat bipolars, and rod bipolars of primates (Kolb, 1970). As shown in Figure 16-11B, the dendrites of bipolar cells are in intimate contact not only with the presynaptic membrane of the photoreceptor pedicel, but with extensive regions of horizontal cell dendrites as well. This complicated, three-way physical contact among photoreceptors, horizontal cells, and bipolar cells provides a structural basis for understanding the first steps in visual processing discussed below.

The Link between Anatomy and Function Histological descriptions of vertebrate retinal interneurons raised questions about the relationship between structure and function in these cells. All retinal cell types can be subdivided into several groups on the bases of morphology and location, but the functional significance of these structural differences remained unclear. Horizontal cells, for example, were named for their extensively branching dendritic networks that traverse the retina horizontally (Fig. 16-12), but the function of the lateral information flow that these neurons ostensibly facilitate was unknown.

Even before the anatomical work discussed above was finished, electrophysiological experiments had examined the response characteristics of several classes of retinal cells. But because physiologists had no way of determining which anatomical cell classes they were recording from, the anatomical and electrophysiological information on the retina remained two separate bodies of knowledge that could not be correlated.

This problem was solved in several steps through the work of Tomita (1965), Werblin and Dowling (1969), and Kaneko (1970). These physiologists and others in their laboratories developed and applied techniques of iontophoretic dye injection that made it possible to stain and recover in histological preparations cells whose response properties had been examined with microelectrodes. These dyes, like Niagara Sky Blue, Procion Yellow, and the very strongly fluorescent Lucifer Yellow are electrically charged and can thus be injected by the application of the proper current through the microelectrode. The latter two dyes diffuse particularly well in the dendritic branches of cells into which they are injected, and their bright fluorescence is clearly visible in the fluorescence microscope. Armed with these techniques research conducted over the last several years has made major strides toward making functional sense out of retinal anatomy.

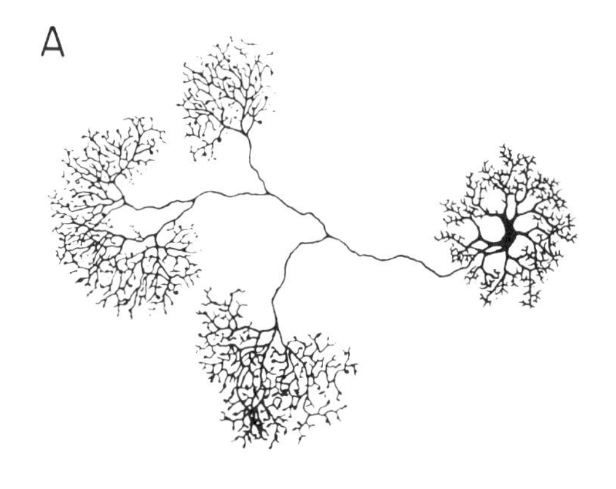

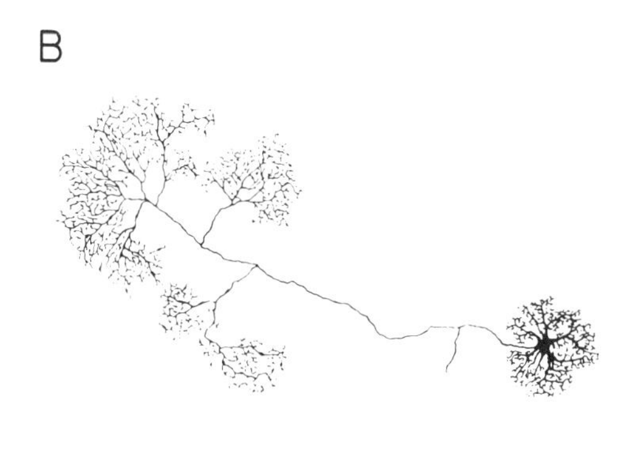

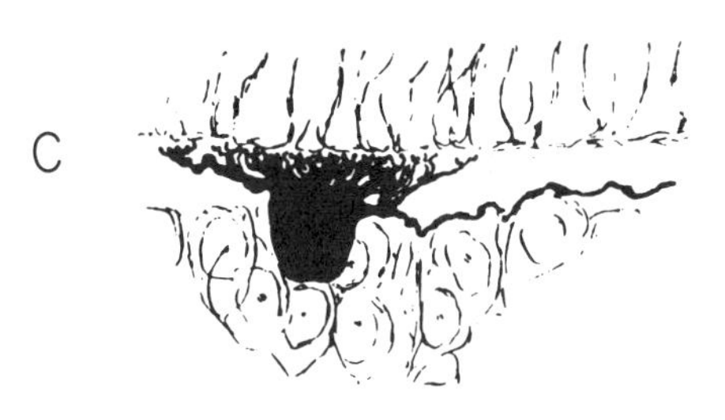

Figure 16-12 A, B: Spreading dendritic arborizations of horizontal cells from the calf retina, revealed by Golgi staining and shown in horizontal section. (From Rodieck, 1973, after Marenghi, 1901.) C: View of horizontal cell dendrites in transverse section of Golgi-stained monkey retina. (From Rodieck, 1973, modified from Polyak, 1941.)

Horizontal Cells: The First Keys to Color Processing In the mid-1950s physiologist Gunnar Svaetichin, working in Venezuela, reported a set of singular responses from a class of cells (believed at the time to be photoreceptors) in the retinas of fishes (Svaetichin, 1956). Edward F. MacNichol, Jr., then of Johns Hopkins University, joined Svaetichin in Caracas to further investigate these responses from what were later identified as horizontal cells (MacNichol, Feinberg, and Harosi, 1973). Like photoreceptor cells, horizontal cells exhibited graded changes in their membrane potentials when stimulated. But unlike any other neurons previously described, the interior of horizontal cells could become either more or less negative with respect to the outside; that is to say they could either hyperpolarize or depolarize.

The most intriguing part of the discovery, however, was that the responses of a number of horizontal cells were clearly biphasic, becoming more negative in response to certain wavelengths and less negative in response to others. An interneuron exhibiting both hyperpolarizing responses to green light and depolarizing responses to red light (Fig. 16-13) was an exciting find indeed. This class of horizontal cells was somehow receiving input of opposite sign from red- and green-sensitive photoreceptors, demonstrating what is called opponent processing. Some horizontal cells even appeared to combine input from three receptor types.

Lipetz (1978) devised a model that combined the information on horizontal cell morphology with data on the functional connections among the dendrites in photoreceptor pedicles and the response characteristics of the horizontal cells. This model, incorporated in Figure 16-10, hypothesized that electrical signals passed from photoreceptors into one class of horizontal cells could be passed with a change in sign to another class of horizontal cells through a photoreceptor synapse in which both cells participated. Although this model did not fully explain all aspects of these interactions, it represented a major conceptual step forward in our understanding of retinal structure and function. More recent experimental work by Stell, Kretz, and Lightfoot (1982) and others has confirmed and extended this model for horizontal cell function in goldfish. Many recent experiments have provided quantitative electrophysiological data on chromatic interaction and affirmed that the multifaceted photoreceptor synapses are the locations where feedback from one horizontal cell type to another takes place.

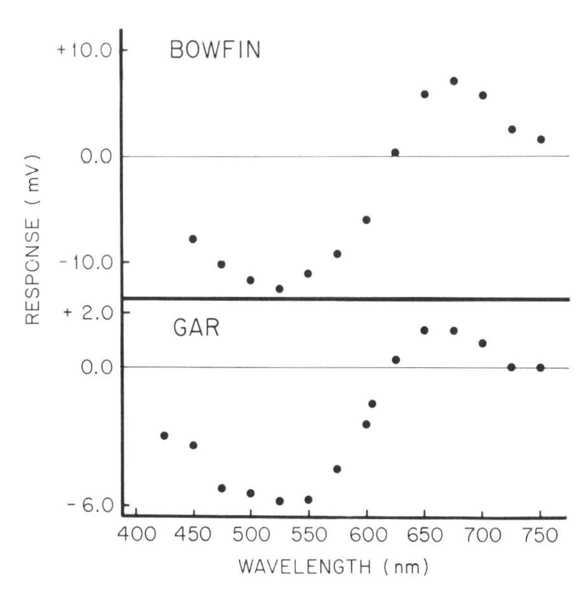

Figure 16-13 Biphasic responses from horizontal cells of the type originally obtained by Svaetichin (1956). These records were obtained from bowfin and gar retinas by Burkhardt et al. (1983).

Bipolar Cells: The Next Link in the Chain Although we cannot go into details of bipolar cell physiology here, it is worth mentioning that these cells exhibit response properties that show up repeatedly in higher visual centers. The work of Kuffler (1953) on cat retinas showed that presenting small, round light stimuli to certain parts of the retina might evoke one kind of response from a given bipolar cell, whereas the presentation of the same spot in a slightly different location might make the same bipolar cell respond in the opposite fashion. Further investigation showed that bipolar cell receptive fields are characteristically arranged with two antagonistic, overlapping areas; a circular center and an annular surround (Fig. 16-14A). If a spot in the center produced an "on" response in the cell, a spot in the surround would produce an "off" response, and vice versa.

Still further investigations with colored stimuli revealed that in some bipolar cells, this center-surround organization is combined with color opponency, to provide, for example, a cell with a

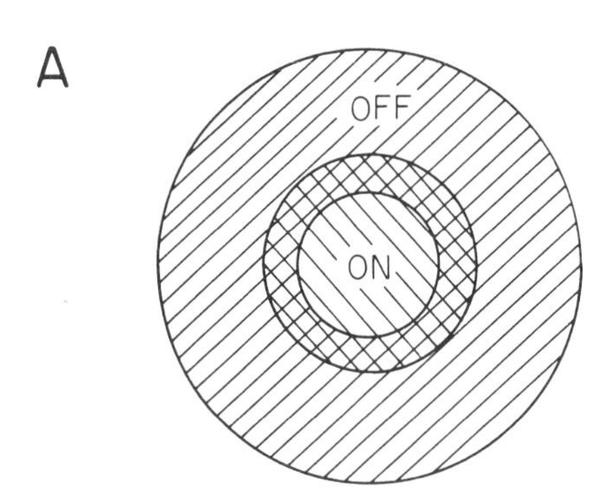

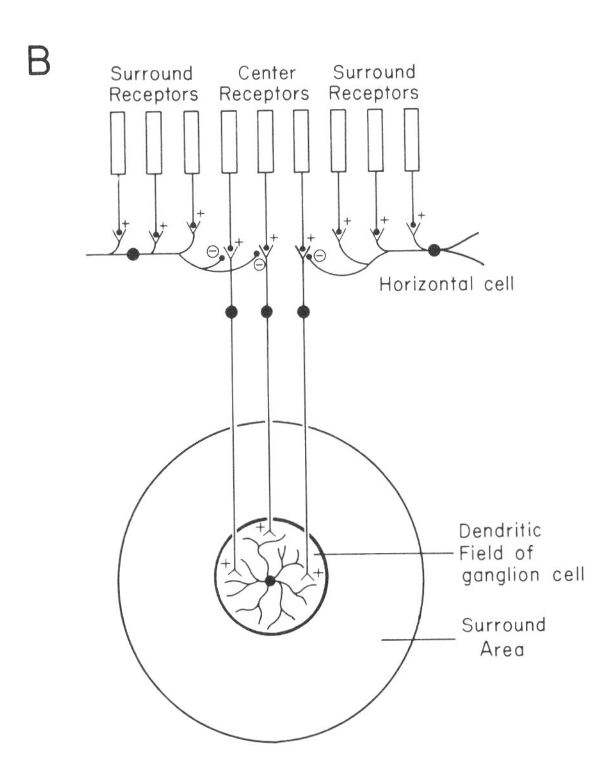

Figure 16-14 A: Schematic representation of the type of center-surround organization found in both bipolar and horizontal cells in many retinas. B: Schematic wiring diagram to explain the center-surround organization of a bipolar cell and an associated ganglion cell. Note that the central response is provided by direct, positive connections between the central receptors and the bipolar cells, whereas the antagonistic surround response is generated through negative input from horizontal cells. (After Davson, 1980.)

red "on" center and a green "off" surround (MacNichol, Feinberg, and Harosi, 1973). A functional-anatomical model incorporating what we know about photoreceptor synapses and bipolar cell responses is shown in Figure 16-14B, which also makes the point that the output of bipolar cells serves as the input to ganglion cells.

What might be the functional utility of the center-surround organization? For many of these cells, the effects of the antagonistic center and surround areas of the receptive field balance each other out; uniform illumination of the entire receptive field produce little net effect on the cell. But a spatial discontinuity of the sort produced by an edge or a bar differentially activates center and surround, giving rise to a signal that enhances the perception of form.

For the fascinating story of the continuing processing of color and form that occurs at the ganglion cell level and in the higher visual centers of the brain, the reader is referred to Cornsweet (1970) and Davson (1980).

Retinal Information Processing and Color Vision Theory Electrophysiological studies on horizontal cell and bipolar cell connections and color processing in the retina support a synthesis of two major, previously competing theories of color vision. The 200-year-old trichromatic theory—first proposed by the English physiologist Thomas Young and later extended by the German physicist Herman Ludwig von Helmholtz—suggested that human color vision was based on three photochemically decomposable substances located in the fibers of the optic nerves. This extraordinarily prescient hypothesis could not, however, explain how color mixing took place. Ewald Hering, a nineteenth-century German physiologist, argued from psychophysical experiments that the basic sensations of vision were arranged in opponent pairs: red-green, blue-yellow, and black-white. The combination of anatomical and physiological data provided by Stell, Svaetichin, and their coworkers has shown that three types of photoreceptors were indeed hooked up in antagonistic or opponent fashion.

How widespread is this kind of visual information processing? It is now apparent that opponent information processing occurs at some level in the visual system, not only in many fishes but in a large number of other vertebrates, including primates. The location of the neural hookups that accomplish this processing varies from species to species, occurring more peripherally in lower vertebrates than in primates, but the general phenomenon is widespread.

Dwight Burkhardt and his colleagues at the Vision Laboratory of the University of Minnesota, working in collaboration with the Laboratory of Sensory Physiology, found opponent processing in the retina of the bowfin *Amia calva,* one of the most primitive of all living fishes (Burkhardt, Gottesman, and Levine, 1983). The presence of opponent processing in *Amia,* a species left over from the age of dinosaurs, as well as in modern fishes, turtles, and scores of higher vertebrates (Daw, 1973; Davson, 1980) indicates that this basic mechanism of color vision may have originated in a common vertebrate ancestor eons ago, and was preserved, albeit with some variations, in all the major lines of vertebrate evolution. The alternative explanation — that opponent processing has appeared several times in independent evolutionary lines — makes an even stronger case for the unique applicability of this type of processing to common visual problems. If the system evolved only once, millennia ago, and has been kept fundamentally unchanged ever since, or if it has evolved independently in several different branches of the evolutionary tree, it must be a good way of doing things.

What are the characteristics of a system based on opponent processing, and what crucial benefits accrue from its operations? Although the responses of an opponent system are ultimately based on the stimulation of photoreceptors in the retinal mosaic, a signal originating in any cone has interacted with signals from at least one, and sometime two other receptor types by the time it has left the retina (Lipetz, 1978; Dowling, 1979). Furthermore, extensive horizontal information transfer in the retina ensures that these interacting networks compare signals from large numbers of cones, so that the network as a whole responds to stimuli spread out over substantial areas of the visual field (Daw, 1973).

Edwin Land and his coworkers at the Polaroid Vision Laboratories have hypothesized that in order to function accurately, the color processing system must assess the long-wave, middle-wave, and shortwave light falling on large areas of the retina and compare the relative amounts of energy present at these wavelengths (Land, 1977). Only after these calculations are complete does the system assign colors to the objects in the visual field. That way, if the overall color balance of the illumination changes — for example, if the ambient illumination suddenly became much more blue — the system takes note of the extra blue light and subtracts it before deciding that a ripe red apple has turned purple. Such a system could not work without extensive lateral information processing, some of which occurs in the retina, and some of which may occur in higher centers all the way up to the visual cortex. This working unit Land calls the "Retinex." Such a system is of obvious value for humans and is even more valuable to fishes, which experience major changes in illumination whenever they swim vertically.

Additionally, behavioral experiments during the last two decades have shown that certain species of freshwater fishes have an unusually high sensitivity to light of wavelength 660 – 680 nm. Their visual systems respond as if they possessed a visual pigment with its maximum wavelength (λmax) out in the far red, at about 680 nm (Naka and Rushton, 1966; Northmore and Yager, 1975). But the pigment actually measured in their photoreceptors has a λmax of only 625 nm, and no visual pigment with a λmax greater than 625 nm has ever been found in *any* species. For fishes living in water where long-wave light predominates, such a mechanism might clearly be valuable, but if its source is not in the cones and their pigments, where is it?

A similar long-wave-sensitive mechanism is also present in the human visual system, according to a new series of demonstrations in the Polaroid

Vision Laboratories. Edwin Land's studies there have confirmed that our visual system—whose red-sensitive pigment has a λmax of no longer than 567 nm—actually responds as though it had a far-red pigment with a λmax around 630 nm. This long-wave Retinex "mechanism" and the "red" in standard color-mixing functions (Cornsweet, 1970) are both spectrally located far beyond our red visual pigment.

To several researchers the 680 nm "mechanism" in the goldfish and the 630 nm "phenomenon" in humans appear to result from similar color-opponent processing, although that processing does not necessarily occur in the same portions of the visual system. As demonstrated mathematically by Sirovich and Abramov (1977), opponent interaction between mechanisms based upon two different cone action spectra can generate a response function with peak sensitivity at a much longer wavelength than either of the two contributory cone pigments. These response functions closely resemble the electrical behavior of certain classes of horizontal cells in teleosts and other cells from higher visual centers in primates.

Morphological and electrophysiological studies of higher visual centers in vertebrates still have a long way to go before they reach a level of sophistication that will allow us to piece together the known components into a working model of the visual system. The continuing use of the comparative method, with an emphasis on correlating structure and function in the visual system, is invaluable in furthering our understanding of this magnificently flexible assemblage of neurons.

Chapter 17

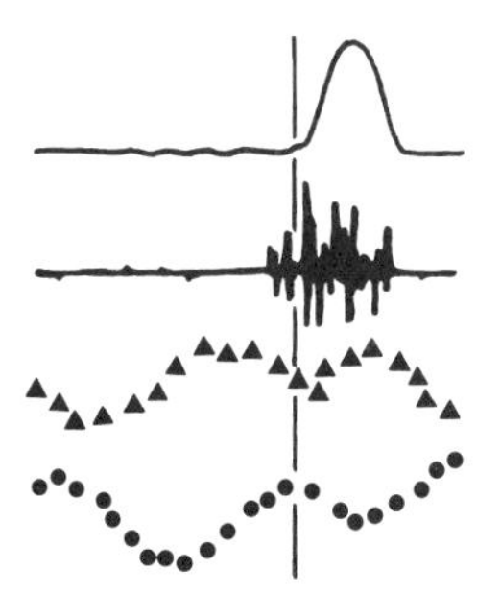

Neural Control of Locomotion

George E. Goslow, Jr.

The split-second timing necessary for survival in a vertebrate is epitomized by the midflight adjustments of a raptorial bird as it descends upon its prey. Such precise maneuvers require a complex interplay of the sensory and motor systems that ongoing research the world over is just beginning to unravel. Man's interest in animal movement seems almost innate, and published accounts stem from early history. Aristotle in his *De motu animalium* (Peck and Forester, 1937) struggled at length to understand and explain the diversity of appendage morphology and modes of progression of animals. Leonardo da Vinci's quantification of animal movements through his remarkable insight into functional anatomy 500 years ago will undoubtedly stand for hundreds of years to come.

The study of adaptation as it relates to the neural control of vertebrate locomotion is in its infancy. Although the general anatomical and physiological features of the neural control apparatus are known, few studies exist of its components as they relate to an organism's fitness. Such studies remain for future researchers.

The work of Bernstein, Marey, Philippson, Sherrington, and Graham Brown in the late 1800s and early 1900s (see Wetzel and Stuart, 1976; Grillner, 1981); established that the neural control of locomotion results from the interplay of three major elements. This tripartite view for neural control is summarized in Figure 17-1 for a cheetah in midflight. A central pattern generator (or CPG, a neural network within the spinal cord capable of generating cyclic patterns independent of supraspinal or afferent influence) for each limb is influenced by peripheral input from a variety of receptors from muscles, joints, and skin, as well as descending input from the higher supraspinal centers. Through peripheral output pathways to the trunk or limb muscles, appropriate muscle contractions for the task at hand ensue.

This account begins with a discussion of the neural control of muscle that includes the composition and recruitment of motor units and the transducing properties of their receptors. Muscle contraction results in forces that stabilize and move the organism. Thus, the second consideration is the kinematics and kinetics of these movements to illuminate the general pattern of the peripheral output systems. Third, evidence for the existence of CPGs within the spinal cord is presented, followed by a discussion of the influences on the CPG of afferent input for fishes and cats. Experiments that provide insight into the supraspinal pathways involved in the neural control program are then presented, followed by areas that appear fruitful for future studies.

The domestic cat, because of its size and availability, has historically been the mammal of choice for various anatomical, physiological, and behavioral studies. This discussion is skewed toward two hind-limb extensor muscles of the cat, the soleus (Sol) and the medial gastrocnemius (MG). Although we know bits and pieces about other muscles, the cat Sol and MG are by far the best known and most completely understood muscles of the tetrapod limb. The reasons for this are several. First, the dorsal and ventral roots of the spinal nerves in the lumbosacral area of the cat are long. This enables one to record and stim-

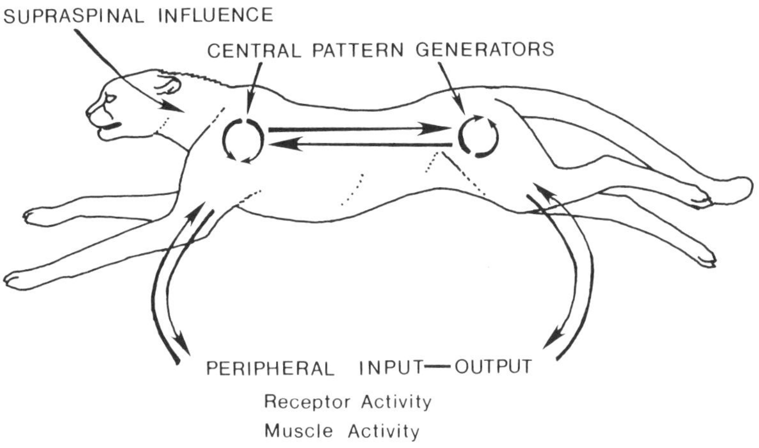

Figure 17-1 Basic components in the neural control of locomotion. The model is tripartite, recognizing supraspinal influence, central pattern generation, and peripheral input-output. CPGs (one for each limb) capable of rhythmic activity are influenced by peripheral input from a variety of somatosensory receptors as well as descending influence from supraspinal centers. Through peripheral output pathways, appropriate muscles are contracted to perform the task at hand.

ulate them relatively easily. Second, the tendons of insertion of Sol and MG are distinct and can easily be cut and attached to a recording transducer without compromising their blood supply. Third, the two muscles act synergistically for extension of the ankle but differ in the composition of their types of fibers; the Sol is homogeneous for slow-twitch, high-endurance fibers, whereas the MG consists of fibers heterogenous for these properties.

Composition of Single Muscles

Muscle fibers of vertebrates fall into two major categories, tonic and twitch. Although the two types are generally separated on the basis of a collection of anatomical and physiological properties, their distinction is not always clear (Hess, 1970). It is known that twitch fibers, at least, are organized into distinct, functionally independent components known as motor units (Fig. 17-2). A motor unit consists of a neural component, the motor neuron (motoneuron), which includes the dendrites and cell body in the central nervous system (CNS), the axon in the peripheral nerve, and the axon collaterals in the muscle. A motor unit also includes the collection of muscle fibers that it innervates (the muscle unit). Usually, but not always, the axon collaterals are in one muscle. The number of muscle fibers innervated by a single motoneuron (innervation ratio) varies from a few to several hundred, presumably depending on the task required of the muscle. The cell bodies of a population of motoneurons to a particular muscle are located in motoneuron pools. Within a muscle's pool, or across pools for groups of muscles, motoneurons are activated (recruited) appropriately to complete the locomotor task. Two reviews of mammalian motor units as related to neural control are available (Burke, 1981; Stuart and Enoka, 1983).

Tonic and Twitch Muscle Tonic fibers are found in the appendicular muscles of all classes of vertebrates with the exception of mammals. These fibers generally do not generate significant force in response to the transmitter released by a single motoneuron action potential. The transmitter causes a local, nonpropagated muscle action potential that results in relatively slow force production. Tonic fibers are multiply innervated, which is apparently necessary to get total fiber activation in a reasonable period of time. Tonic fibers are able to develop and maintain isometric tension economically, which strongly implicates them for the maintenance of sustained force.

Twitch (phasic) fibers are found in the somatic muscles of all classes of vertebrates. A twitch fiber's response to a single propagated nerve action potential is a single propagated muscle action potential and relatively rapid force development to give the classical physiological "twitch" response. All the fibers of a muscle unit respond within 40 μsec of one another such that their contraction is essentially synchronous.

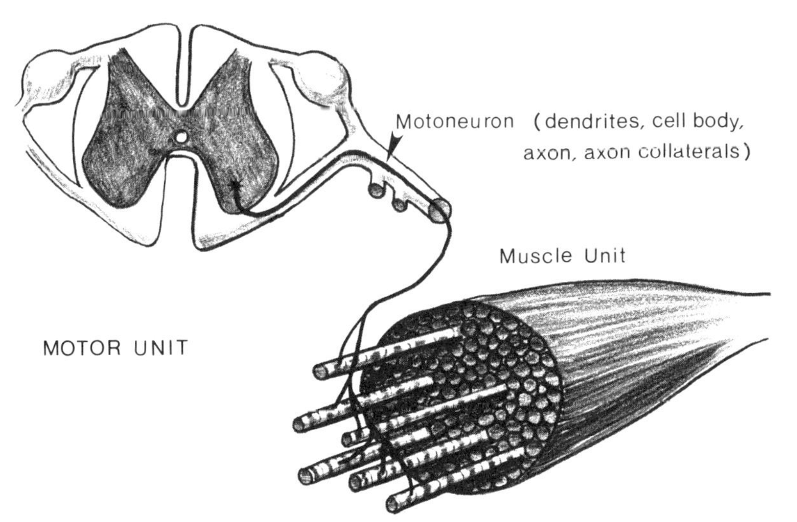

Figure 17-2 The motor unit, which is the smallest functional neuromuscular entity of twitch muscle. It consists of a neural component, the motoneuron, as well as the collection of muscle fibers innervated by that motoneuron, the muscle unit.

We are familiar with the light and dark meat of our Thanksgiving turkeys, and we recall that some muscles of the cat, rabbit, or mink appear darker than others. These differences were recognized years ago, and it was noted that the contraction time of the darker muscle was slower than that of the lighter muscle. Thus, the terms *red* and *white,* and *slow twitch* and *fast twitch* were coined to describe these muscles. We now know that red color relates to the density of the capillaries, pigmental myoglobin, and mitochondria (site of oxidative enzymes) within the muscle, and therefore relates to the muscle's capacity for aerobic contraction and endurance. These terms are still in general use, but note that there is an unfortunate implication of the terms *fast* and *slow.* Because of the early association of "slow" muscle with "red" color, it is often implied that fast twitch muscle is fatigable, whereas slow twitch muscle is resistant to fatigue. The latter appears to be true, but the *former is not;* a high percentage of vertebrate twitch fibers are "fast" but nonfatigable (red).

The limb muscles of most quadrupeds are a mixture of different types of twitch fibers, but some muscles may be predominantly one type or another. Such muscles are useful models for the study of various morphological, physiological, or biochemical properties of single types of fiber and, further, lend themselves to analyses of the energetics of contraction.

Types of Motor Units The different kinds of motor units within a muscle can be studied electrophysiologically. This is done by functionally isolating and stimulating a motoneuron or its axon and recording its biophysical properties along with the mechanical properties of the muscle unit. The experimental procedure involves anesthetization of the animal and surgical exposure of the spinal cord and motor roots of the spinal nerves. The muscle to be studied is freed from the surrounding connective tissue and its tendon attached to a transducer capable of measuring small forces. From such a preparation (Fig. 17-3) the properties of motoneurons, such as their reflex and descending connections, their size (surface area), and their sensitivity to discharge and conduction velocity, can be measured. In addition, various muscle unit properties such as twitch contraction time, unit size as reflected by absolute twitch or force of tetanic contraction, and resistance to fatigue can be measured. Such studies of the cat MG and the few other mixed muscles studied have resulted in a description of four basic types of motor units that relate to motor control (Table 17-1). In cat MG the four types of motor units are distinguished as

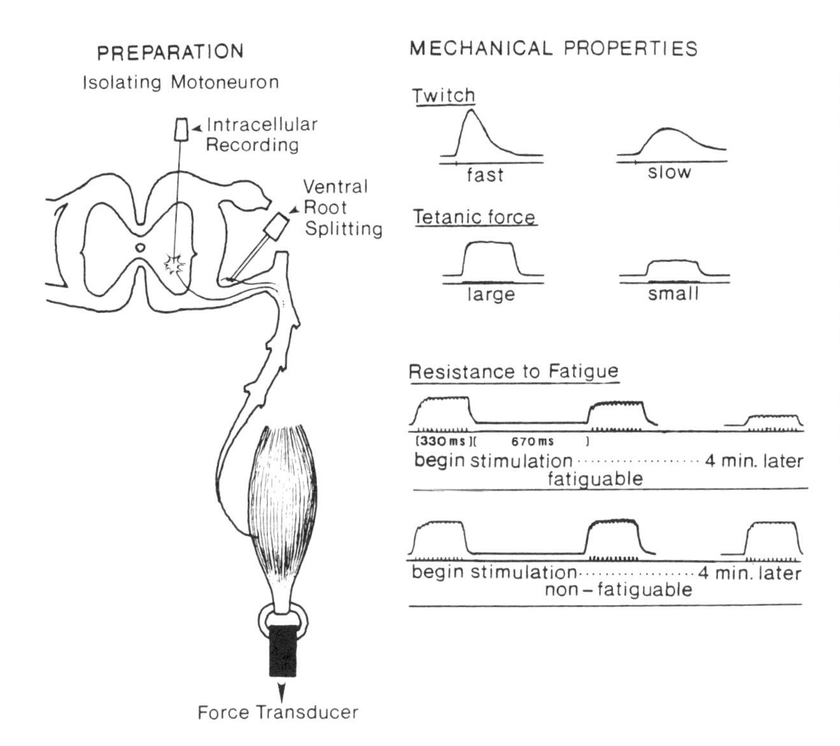

Figure 17-3 Experimental arrangement for the study of individual motor units. *Left:* Motoneurons are functionally isolated either by impaling the cell body with a microelectrode or by teasing the ventral root into filaments. All muscle nerves other than the one to the muscle under study are cut. The muscle's tendon is attached to a force transducer. *Right:* The mechanical properties of twitch contraction time, maximal force, and resistance to fatigue of each isolated unit can be measured.

follows even though there is some overlap: (1) fast and slow units can be differentiated by relative contraction times in general; and (2) three fast-twitch types can be distinguished on the basis of their response to sustained stimulation designed to determine fatigability or endurance.

This makes for four unit designations. *Slow twitch* (S) units are characterized by a relatively small motoneuron; they make up an estimated 20% of the unit population of MG. These units have a relatively slow contraction time, small force, and the highest resistance to fatigue. Experiments with whole hamster muscle reveal that S units are considerably more economical at developing and maintaining isometric tension and slow isotonic movements than are fast twitch populations (see Goldspink, 1977). Events at the cross-bridge level appear responsible for this difference in economy. *Fast twitch, resistant to fatigue* (FR) units make up an estimated 30% of the unit population in cat MG. FR units tend to be intermediate in size between the S and FF units, possess fibers with relatively fast contraction times, and are resistant to fatigue. *Fast twitch, fatigable* (FF) units possess a relatively large motoneuron and make up an estimated 42% of the unit population in cat MG. Contractile characteristics of FF muscle units include a relatively fast contraction time, large force, and very little resistance to fatigue. Experiments with whole muscles that are primarily fast twitch reveal that these types of fibers are adapted for high power output and are reasonably efficient for producing work. *Fast twitch, intermediate* (FI) units make up a small percentage of cat MG (estimated at less than 10%). FI units produce forces and possess a resistance to fatigue that are intermediate between those of FF and FR units.

Muscle Fiber Histochemistry Presumably, muscle units of different contractile parameters possess characteristic biochemical properties, which must relate to function. Various histochemical techniques, which are an indirect approximation of fiber biochemistry, have been devised for the identification of "fiber types." This requires making a thin section of muscle and incubating it with an appropriate substrate. Two

Table 17-1 Mechanical properties of motor units in cat muscles.

Motor unit type	Relative twitch contraction time	Relative tetanic force output	Relative resistance to fatigue
S	Slow	Small	Very high
FR	Fast	Medium	High
FI	Fast	Intermediate	Intermediate
FF	Fast	Large	Low

words of caution: (1) the amount of *functional* information that can be extracted from histochemistry is limited and filled with pitfalls; and (2) this area is changing very rapidly with the advent of new biochemical techniques (for example, Nemeth, Pette, and Vrbova, 1981).

Differences in twitch time are thought to reflect the existence of at least two kinds of myosin light chain molecules. Two fiber types can usually be differentiated by noting their adenosine triphosphatase (ATPase) activity after they have been soaked in an alkaline solution. The myofibrillar actomyosin of the fast type is reasonably alkali-stable. Thus, if exposed to a pH of 10.4, for example, mammalian fast fibers stain darkly, whereas slow fibers show no activity. If the tissue is first exposed to an acid pH (not so commonly done), the stain is reversed, but not all fibers will consistently reverse at the same pH. These observations are not well understood. In fact, the response of fibers to acid pH has recently been shown to be time- and temperature-dependent; an important consideration for fiber typing (see below; Gollnick, Parsons, and Oakley, 1983). One can also test for any number of oxidative enzymes, such as nicotinamide adenine dinucleotide dehydrogenase NADH-D) or succinic dehydrogenase (SDH). Fibers with a higher capacity for aerobic activity stain darkly (mitochondria and associated oxidative enzymes are abundant), whereas fibers designed for anaerobic activity do not. Thus staining for ATPase activity and an oxidative enzyme (alternately in serial sections) is a quick way of establishing the approximate fiber type composition of a muscle. Figure 17-4 illustrates serial sections from the horse gluteus medius muscle stained for NADH-D (17-4B) and alkaline-preincubated myosin ATPase (17-4C).

The physiological properties of histochemically typed fibers have been determined for only a few muscles of a few mammalian species. The procedure requires functional isolation of a motor unit, physiological testing of the unit, and stimulation over a long period of time to force the unit's fibers to use stored glycogen. Serial sections, stained appropriately to match the glycogen-depleted fibers typed physiologically with those typed histochemically, allow the correlation. This same technique is used to map the distribution of a muscle unit within a muscle. Studies within the cat MG and Sol resulted in the following motor unit–histochemical fiber type equivalents: S (slow twitch) = SO (slow twitch, oxidative); FR (fast twitch, resistant to fatigue) = FOG (fast twitch, oxidative glycolytic); FF (fast twitch, fatigable) = FG (fast twitch, glycolytic); FI (fast twitch, intermediate) = FI (fast twitch, intermediate). (See Burke and Edgerton, 1975.) Published accounts often do not recognize the histochemical FI type.

Cautions and Limitations The existence of tonic and twitch fibers among all vertebrate classes has been recognized for quite some time. The morphological and physiological differences between them across the few vertebrate species that have been examined are not consistent, however (Hess, 1970). Further, little work has been done with the tonic muscles to elucidate their functional role in amphibians, reptiles, and birds.

The physiological properties of motor units have been studied primarily in domestic species and humans. The motor unit types described here for the cat MG are not necessarily the same as those that have been found or will be found in other species (Burke, 1981). Further, intermuscular differences may exist within a species. For example, the type S motor unit of cat Sol, which is

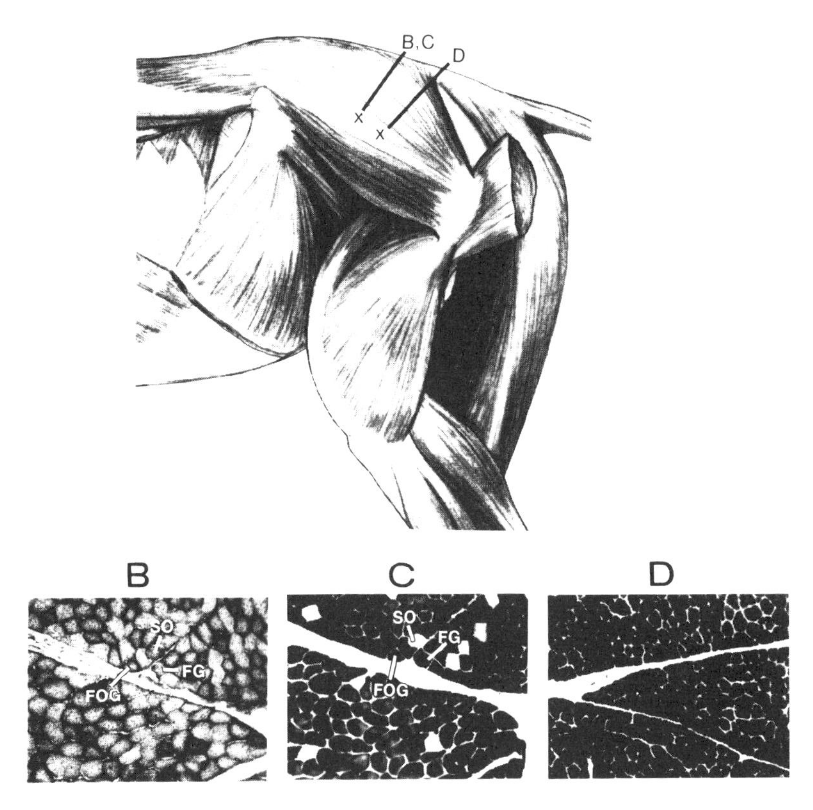

Figure 17-4 Histochemical reactions. *Above:* Approximate location of two biopsy samples (B,C and D) taken 6.5 cm apart from horse gluteus medius muscle. *Below:* Sample B stained for NADH-D and C stained for alkaline preincubated myosine ATPase. Fiber types from the two serial sections are: *SO* = slow twitch, oxidative; *FOG* = fast twitch, oxidative glycolytic; *FG* = fast twitch, glycolytic. Sample D also stained for ATPase. Note the absence of *SO* fibers in D and their presence in C. Fiber types may not be uniform throughout a muscle. (Courtesy of G. Bruce, Northern Arizona University.)

a homogeneous muscle, differs in some respects from the type S unit of cat MG. This may simply be the result of slight differences of biochemistry associated with the evolutionary isolation of these two ankle extensors, or it may reflect the functional roles of the muscles.

Distinct histochemical fiber types may exist in one species, whereas in a second one histochemical type may subtly grade into another. Further, fiber types may show plasticity and transition within limits under a variety of experimental, training, or disease states. As a result, a variety of schemes to classify or identify specific fiber types have been proposed and are in use. Although it is attractive to consider many of these identified fiber types as equivalent across species, this is a risky and ill-advised assumption. Histochemistry is useful for a "coarse-grain" view of fiber types, but it is unreliable for the detection of relatively small differences. This is particularly true for the ATPase stain, where careful control of various steps in the technique is required (Nemeth, Hoffer, and Pette, 1979).

Sampling of a muscle for histochemical fiber types must be done in such a way as to get an accurate estimate of the fiber population. We know, for example, that within some heterogeneous muscles certain kinds of fiber types may be found in distinct anatomical areas of the muscle (Gonyea, Marushia, and Dixon, 1981). Thus if only a small portion of a muscle is sampled (with a biopsy needle), a misrepresentation of the muscle's true fiber composition can result. Figure 17-4 illustrates this phenomenon for the horse gluteus medius. Figures 17-4C and 17-4D are samples taken 6.5 cm apart and stained for myosin ATPase (alkaline preincubation). Note in C the presence of some SO fibers; in D, their absence.

Orderly Recruitment and the Size Principle

As the force requirements of a muscle or limb increase during locomotion, greater activity from the motor units is necessary. Evidence indicates that force in a muscle may be increased by two primary mechanisms: recruitment of additional motor units (increasing the number of active motor units); and/or rate coding (increasing the activation rate of already recruited units). There are some limited data that suggest that the interplay of these two parameters is related to muscle function and task to be accomplished (Stuart and Enoka, 1983). Of primary concern here is the order of muscle and motor unit recruitment during locomotion.

Intuitively, it seems most logical for force recruitment to follow a progressive order such that motor units that produce small, high-endurance forces are followed by motor units that produce relatively larger, low-endurance forces. Kinesiological data reviewed below support this hypothesis.

Fishes Among those vertebrates that possess both tonic and twitch fibers, it is probable that tonic fibers are recruited for relatively slow, sustained activities, whereas twitch fibers are recruited in addition for more powerful and forceful movements. Strong support for this assumption in fishes was initially provided in a study of the trunk musculature of several species of elasmobranchs (Bone, 1966). In elasmobranchs, as in many other fish species, tonic fibers (red) form a thin lateral superficial sheet just under the skin, whereas twitch fibers (white) make up the underlying mass of the myotome (Figure 17-5). By histological and biochemical analyses, Bone showed the tonic fibers to be relatively small and richly endowed with glycogen, fat, and myoglobin. The twitch fibers are relatively larger and contain much less glycogen, fat, and myoglobin. These fibers in all probability consume energy rapidly but subsequently demand a period of rest for metabolite replenishment. Bone described a third, "transition" muscle type, the pink fibers, which are apparently twitch fibers with metabolic properties intermediate to the other two fiber types. These fibers seem analogous if not homologous to the SO fiber type of mammals.

Bone (1966) placed electrodes in pure tonic and deepest twitch fibers and recorded muscle activity during slow and fast swimming movements from a "spinal" shark (a shark in which the spinal cord had been severed). Such a preparation is capable of generating swimming movements through an interplay of the CPGs and afferent inflow. Results of the study indicated that the tonic fibers are recruited first for slow, sustained swimming movements and that the deep twitch fibers are subsequently recruited for rapid, forceful movements.

Subsequent studies of a variety of free-swimming teleosts show that during sustained swimming in some species, tonic fibers are recruited first and the pink twitch fibers next; as swimming speed increases the white twitch fibers become involved (Goldspink, 1977). For other species, such as the striped bass and bluefish, however, it has been shown that tonic fibers are used for all sustainable speeds and that the white twitch fibers are only recruited for high-speed burst swimming above maximum sustainable speeds (Freadman, 1979).

Mammals For information on the precise order of recruitment within motor unit populations we must turn to studies of twitch fibers in mammals. As long ago as 1929 Denny-Brown, a student of Sherrington's, showed that the slow twitch Sol muscle was recruited earlier in various stretch reflexes than its faster contracting synergists, the gastrocnemii. Thus, one way that force is graded is by the serial recruitment of synergistic muscles as the task becomes more demanding. Denny-Brown further observed that the deeper red portion of a muscle was often active before its more superficial white component. Hence he proposed that there is a stable and predictable order of motor unit recruitment that depends on a scaling of excitatory (or inhibitory) synaptic input. This idea is intuitively attractive because it gives order to the way the neuromuscular system is designed. In general, our view of recruitment has not changed from that of Denny-Brown. However, controversy exists relative to the mechanism by which motoneurons (motor units) are recruited, as well as to the unit parameter that is actually ordered (that is, amplitude of postsynaptic excitatory potential, size of motoneuron, force output of the unit).

The role that E. Henneman has played in the

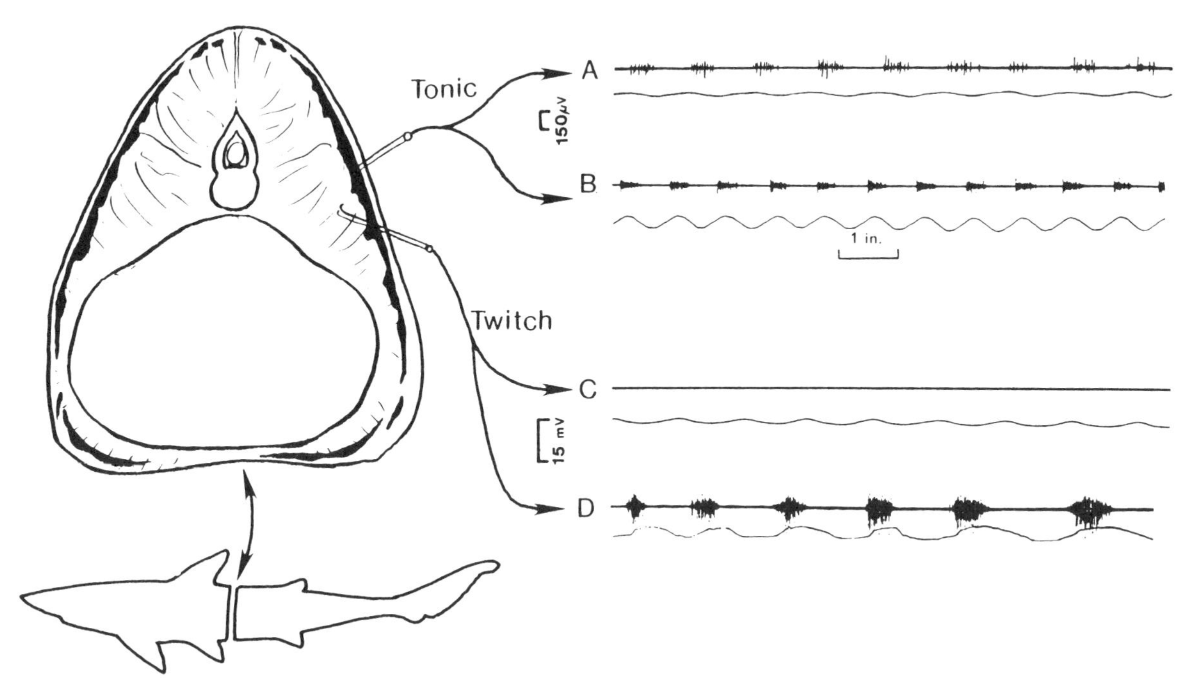

Figure 17-5 Muscle recruitment of tonic and twitch muscle in the shark. *Top:* Electromyographic activity; *bottom:* tail excursion. Tonic muscle illustrates burst activity at slow and moderate swimming speeds (A,B); twitch muscle is inactive during slow swimming but is recruited for fast swimming (C,D). (From Bone, 1966.)

understanding of motor units cannot be overstated (Stuart and Enoka, 1983). In 1957 and subsequently, he postulated a mechanism by which motor units are recruited based on size of motoneuron (Henneman, Somjen, and Carpenter, 1965). According to his size principle, the size of the motoneuron (that is, the surface area of its cell body and dendrites) is the primary property that determines recruitment: small ones first, ever larger ones following. The mechanism by which this ordering is accomplished is controversial. The relative importance of extrinsic (relating to synaptic organization) versus intrinsic (relating to cell properties) is currently being examined.

An important assumption made early on in tests of the size principle is that there is a close correlation between size of cell body and size of axon; that is, relatively small cell bodies possess axons of small diameter, whereas relatively large cell bodies possess axons of large diameter. Since it is known that the diameter of an axon determines its conduction velocity (CV), measurement of axonal CV has been used in numerous studies as an indicator of axon size, and hence, cell body size.

From the premises of the size principle, it is expected that there exists a close correlation of cell body size (hence axonal CV) and a variety of properties of the muscle unit, including twitch contraction time, peak tetanic force output, and fatigability. These correlations have been sought for select hind-limb muscles of the cat, rat, baboon, and skunk. Results are controversial in that correlations are strong for entire populations of motor units in small foot muscles of the cat (Bessou, Emonot-Denand, and Laporte, 1963), but weak for intermediate and large cat hind-limb extensor muscles (MG, lateral gastrocnemius, plantaris, tibialis posterior) and flexor muscles (tibialis anterior, extensor digitorum longus). Motoneuron size alone (as reflected by CV) does not explain the differences in the force output and contraction time of each individual S, FR, FI, and FF motor unit within a given motor pool. It is conceded, however, that this is a controversial and highly active area of research. For now, the size principle is an important and testable hypothesis.

Controversy exists as to what property of the motor unit is precisely ordered during recruitment. Some definitions relating to afferent-motoneuron interaction are in order. The largest afferent fiber from a muscle spindle, the Ia, enters the spinal cord and projects to a variety of neuron types reviewed later. Of immediate relevance is that Ias may synapse directly onto a motoneuron. Such a connection is monosynaptic. The synaptic connection between a Ia fiber and any postsynaptic neuron is always excitatory in that a presynaptic terminal of the Ia releases transmitter, which causes a depolarization, or excitation, of the postsynaptic cell. The depolarization is known as an excitatory post-synaptic potential (EPSP). A synapse from a Ia may produce in the motoneuron a range of EPSP intensities dependent upon various presynaptic membrane, transmitter, and postsynaptic membrane properties. If the EPSP is large enough, the threshold of the motoneuron for firing will be reached and a propagated action potential will result. If a motoneuron's muscle unit is in the same muscle from which the spindle's Ia originated, the connection is said to be homonymous. Heteronymous connections are those in which the synapse of the Ia is to motoneurons of muscles that are synergistic to the muscle of the spindle's origin.

Data from the laboratory of Burke relate to the excitatory effects of Ia afferents from spindles on specific types of motor units. His group has shown, for select cat hind-limb muscles, that homonymous and heteronymous Ia EPSPs are larger, on the average, in motoneurons of type S units than in type F units. Further, these Ia EPSP amplitudes scale rather precisely with motor unit type. In other words, when considering a population of motoneurons within a single motoneuron pool, Ia EPSPs are relatively larger in the motoneurons of small-force-generating, nonfatigable units, and smaller in large-force-generating, fatigable units. Accordingly, Burke (1979) proposed a recruitment model for the cat MG based on an ordering of monosynaptic Ia EPSP size (Fig. 17-6A). The figure shows the cumulative MG output force that would be developed as units are sequentially recruited according to decreasing Ia EPSP size, plotted incrementally. When ordered in this way, the first-recruited half of the MG population consists almost entirely of small, fatigue-resistant units (types S, FR, and FI), which together produce only about 25% of the maximum force available from the entire MG population. The remaining 75% of the muscle's force is produced primarily by the higher threshold FF units, which produce, on the average, relatively large tetanic forces.

A second model is that of Sypert and Munson (1981). They propose that motor units are progressively ordered according to motor unit type, that is, S < FR < FI < FF. The basis of their argument is that the type of a motor unit is determined by motoneuron threshold and amount of synaptic drive. Important to their model is the assumption that within each type of unit, ordering is random with respect to force output. Figure 17-6B shows, for a population of 43 cat MG units, a comparison of the profile of cumulative force production by increasing force outputs, type of motor unit, and recruitment according to progressively increasing axonal CV. As implied earlier, the CV recruitment profile is not a smooth one owing to the poor correlation in this large muscle between the CV of a motoneuron and the tetanic force output of its muscle unit. In contrast, both the other two profiles provide the intuitively satisfying fine gradation of muscle force, which has led Sypert and Munson to propose that motor unit type is what is ordered in recruitment. As pointed out by Stuart and Enoka (1983), however, if recruitment order within each motor unit type is random, as proposed, orderly recruitment within the cat Sol muscle (100% S units) should not occur. Binder et al. (1983) have recently shown that threshold differences of motoneuron pairs within this muscle do, however, exist.

Sensory Feedback from Limb Afferents

Historically, nerve fibers were divided into A, B, and C types and further subdivided into groups on the basis of conduction velocity. Tables 17-2 and 17-3 provide a useful summary.

Numerous somatosensory receptors play a role in the execution of smooth movement. Primary

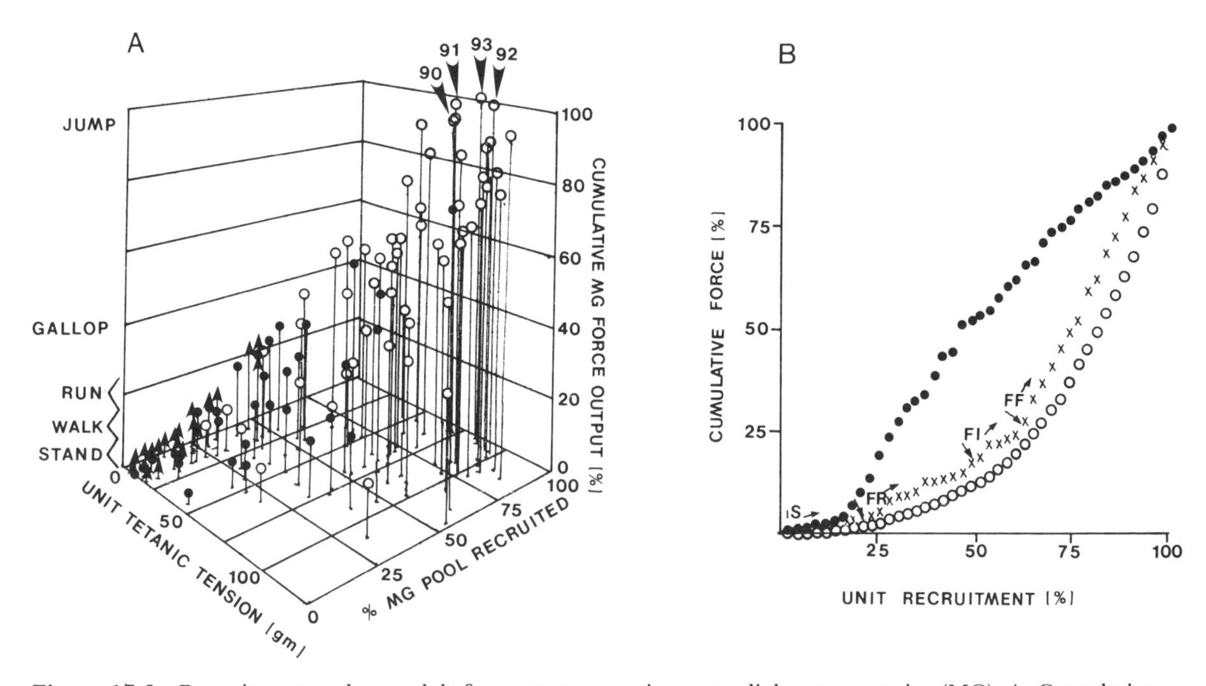

Figure 17-6 Recruitment order models for motoneurons in cat medial gastrocnemius (MG). A: Cumulative MG output force *(vertical ordinate)* that would be developed as 93 units pooled from several experiments are sequentially recruited *(from left to right)* according to decreasing size of the excitatory postsynaptic potential (EPSP) measured during maximal homonymous Ia stimulation. Each unit's individual tetanic tension is added to the cumulative; the last four units recruited are numbered 90–93. Each point represents a different unit, with unit type denoted by closed triangles *(S)*, closed circles *(FR)*, and open circles *(FF)*. (From Burke, 1979.) B: Recruitment of 43 MG units according to cumulative force output *(open circles)*, motor unit type but randomly within each type *(crosses)*, and progressively increasing axonal conduction velocity *(closed circles)*. Recruitment by cumulative force is considered ideal. Note the similarity of the type recruitment line to the force line. (From Sypert and Munson, 1981.)

Table 17-2 Afferent fiber types and functions.

Type	Group	Diameter (μm)	Conduction velocity (m/sec)	Subgroup	Source
A	I	12–20	72–120	Ia	Primary endings of muscle spindles
				Ib	Golgi tendon organs
	II	6–12	36–72	Muscle	Secondary endings of muscle spindles
				Skin	Pacinian corpuscles, touch receptors
	III	1–6	6–36	Muscle	Pressure, pain endings
				Skin	Touch, temperature, and pain receptors
C	IV	<1	0.5–2	Muscle	Pain receptors
				Skin	Touch, pain, and temperature receptors

Source: Willis and Grossman, 1973.

Table 17-3 Efferent fiber types and targets.

Type	Group	Diameter (μm)	Conduction velocity (m/sec)	Target
A	Alpha	12–20	72–120	Extrafusal skeletal muscle fibers
	Gamma	2–8	12–48	Intrafusal muscle fibers
B		3	3–15	Preganglionic autonomic fibers and some postganglionic fibers
C		<1	0.5–2	Preganglionic and postganglionic and autonomic fibers

Source: Willis and Grossman, 1973.

are the cutaneous receptors of the skin and the proprioceptors of the joints and muscles. Physiologists have worked to characterize each anatomical type and to learn what stimuli cause each to discharge and to what extent. Such studies have provided much valuable information, but the role of individual afferent species in locomotor control remains elusive. This is due, in part, to the fact that even within a local population of receptors of the same type there can be substantial variation in their threshold of response to identical stimuli.

The central connections of various afferent species are most easily studied by intracellular recording. A microelectrode (tip diameter ≅ 1 μm) placed within a cell allows for the measurement of any excitatory or inhibitory effects produced by the afferent input being tested. The afferent may be activated in any variety of natural or artificial ways. All afferents terminate on interneurons within the gray matter of the spinal cord. In addition, afferents from muscle spindles (but no other limb receptors) have collaterals that terminate directly on selected motoneurons. Interneurons may vary in size from very small cells with local projections to large tract cells whose axons project to distant centers. In general, primary afferents from the limb make excitatory connections with interneurons and motoneurons. In contrast, the terminations of interneurons onto other interneurons or motoneurons may be excitatory or inhibitory.

Anatomy and Central Connections Pressure receptors are of particular interest to us as many of them possess myelinated fibers of large enough size to conduct action potentials fast enough to be involved in postural adjustments. Recall that cutaneous receptors and the mechanoreceptors found in joint capsules and associated fascia vary from free endings simply intertwined among subsurface skin cells to organized encapsulated corpuscles. Cutaneous and joint afferents enter the spinal cord and excite interneurons of the dorsal horn. These interneurons act on the motoneurons, either directly or through relay pathways involving other interneurons. The reflex effects of joint receptors has been difficult to determine (for details, see Tracey, 1980).

The simplest proprioceptor is no more than a spray of free nerve endings, as found within the myoseptal connective tissue of agnaths. Recent studies (though limited) of relatively unspecialized free nerve endings found in the connective tissue of highly mobile structures of teleosts (hyoid barbels, modified ray fins) suggest that such free nerve endings may have given rise to both the encapsulated Golgi tendon organs of birds and mammals and the variety of proprioceptors of dipnoans, urodeles, anurans, and reptiles (Ono, 1982). The first report of an encapsulated spindle in teleosts has been published by Maeda, Miyoshi, and Toh (1983).

The most comprehensively studied muscle proprioceptors in the context of postural control are the mammalian Golgi tendon organ and the muscle spindle. These receptors are abundant and often possess low thresholds of stimulation, suggesting that they are important in moment to moment adjustments during locomotion. Surprisingly, we know little about their particular contributions.

Golgi tendon organs of the mammal are the most extensively studied tendon organ among vertebrates. The structure consists of collagenous fascicles and nerve endings encased in a spindle-shaped capsule of connective tissue. Few tendon organs are found within the tendon proper; they are instead associated with aponeuroses of the muscle belly. The collagenous fascicles connect at one end of each tendon organ with an aponeurosis and at the other with 5–25 muscle fibers. These fibers are considered to be "in series" with the tendon organ and may all belong to different motor units. Fibers adjacent to the capsule are said to be "in parallel." The sensory neuron of the tendon organ is a heavily myelinated Ib fiber. Collaterals of the Ib axon are laced throughout the collagenous fascicles, which appears to provide a physical arrangement conducive for stimulation. That is, upon contraction of the in-series muscle fibers, the collagenous fascicles straighten, which results in a pinch of the Ib collaterals. The Ib endings enter the dorsal gray matter and project to interneurons, which in turn send inhibitory connections to homonymous motoneurons as well as to heteronymous synergistic motoneurons. These afferents also project excitatory connections to antagonists.

The spindle fiber is anatomically and functionally complex and reaches its zenith in the mammal. It is an encapsulated receptor possessing three kinds of muscle fibers within the capsule (intrafusal fibers); two kinds of afferent nerve fibers, a large primary fiber (Ia) and a somewhat smaller secondary fiber (spindle group II); and a variable motor supply (the fusimotor system) to the intrafusal fibers by relatively small gamma and (at least in some cases) beta axons. (A beta motoneuron innervates simultaneously an intrafusal and an extrafusal muscle fiber.) Even within the same muscle, spindles may differ in their structural components. For example, a spindle may have one or several secondary endings, or none at all. As a rule, the complexity of the spindle increases from anuran amphibians to mammals as follows: (1) the number of intrafusal fibers within the capsule increases; (2) part of the motor supply to the intrafusal fibers becomes separate from the motor system to the extrafusal fibers (in anurans and reptiles the fusimotor innervation is derived from branches of neurons to the extrafusal muscle fibers); and (3) the primary sensory ending found in the anuran spindle becomes, in mammals, distinctly annulospiral in its innervation pattern and is joined by one or more secondary sensory endings.

Spindle afferents possess complex connections to motoneurons and, in addition, have collaterals to supraspinal centers. The spindle Ia and group II fibers have monosynaptic excitatory connections with their homonymous motoneurons and some motoneurons supplying synergists (Fig. 17-7). In addition, some evidence suggests that Ia afferents and Ib afferents have a variety of polysynaptic connections that may be excitatory and inhibitory to the *same* motoneuron. These alternative pathways to motoneurons are available to the nervous system such that the nature of the reflex output is adjusted to the task at hand (Binder et al., 1982). However, proving this point is a formidable one for present-day physiologists.

The intrafusal fibers of a spindle are supplied by either gamma or beta motoneurons, the fusimotor system (described in general texts such as Willis and Grossman, 1973). This system can be used to change the sensitivity of a spindle at a given length change.

Receptor Transducing Properties Cutaneous receptors are sensitive to mechanical stimuli and demonstrate a variety of thresholds dependent somewhat on anatomical design and location. Upon stimulation some of them produce an impulse train that only slowly decreases with time (slowly adapting). Others will adapt quickly to receptor deformation. The transducing properties of joint receptors, particularly as related to locomotion, has been difficult to determine with single-unit recording techniques. Some studies have shown that joint receptor activity is closely related to joint angle as well as to other factors such as limb loading and associated muscle forces around the joint; other studies have shown that joint receptors respond vigorously only at the extremes of joint position.

The transducing properties of muscle spindles and Golgi tendon organs have been intensely studied in isolated preparations. Reviews are available for muscle spindles (Matthews, 1981),

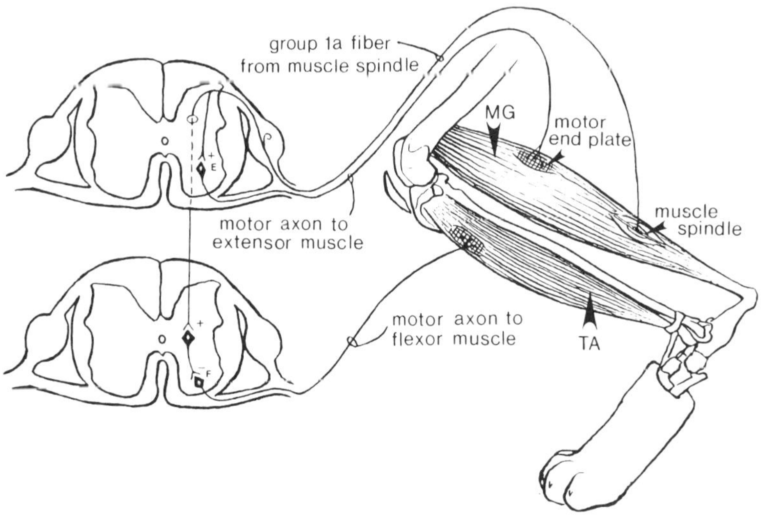

Figure 17-7 Pathway of the spindle Ia, shown as the simplest intersegmental connections between spindle Ia and the motoneurons of extensor (medial gastrocnemius, *MG*) and flexor (tibialis anterior, *TA*) muscles. Group Ia afferent fibers from the muscle spindle enter the cord through the dorsal root, traverse to the gray matter, and end synaptically on alpha motoneurons to the homonymous muscle and to the synergistic muscles. The inhibitory component is disynaptic.

Golgi tendon organs (Proske, 1981), and both (Hasan and Stuart, 1984). The number of impulses per second may be monitored from a given afferent while the muscle is held at successively different but constant lengths, thus testing its "static" sensitivity. During change in muscle length, an increase in rate of discharge may be seen that corresponds to the velocity of change. This reflects a "dynamic" sensitivity.

Golgi tendon organs may be activated by the stretch of whole muscle, but probably more significant to neural control is their remarkable sensitivity to contractions of individual motor units. Recall that several extrafusal fibers connect to each Golgi tendon organ capsule. It has been estimated that in cat MG, for example, each tendon organ is capable of sampling the forces produced by 10–15 motor units. Furthermore, in cat Sol tendon organs can respond to forces as small as 4 mg (Binder et al., 1977). Under static conditions the primary and secondary endings discharge at similar rates, although the discharge of the secondary endings is more regular. The group IIs are less sensitive than the primary endings to any kind of dynamic stimulus. Perhaps the CNS can use the signals from these two endings to determine both the amplitude (group II) and the velocity (Ia) of the muscle perturbations, but this is not clear.

Consideration of the isolated properties of receptors is fundamental to the eventual understanding of their significance, but a clear view of their central interactions and influence on the motoneuron of an intact, sprinting animal is another matter. Various approaches have been taken. Studies concerned with the deprivation of afferent input and with the discharge profile of select afferent species in reduced preparations (controlled locomotion) and conscious animals (natural locomotion) have provided insight into patterns of afferent discharge.

In 1966 a group of Russian scientists published a paper that has subsequently had great impact on our general understanding of controlled locomotion (Shik, Severin, and Orlovsky, 1966). Cats that have had the midbrain sectioned walk on a moving treadmill provided there is repetitive electrical stimulation of the caudal midbrain (Fig. 17-8; see below). Such a preparation allows for the simultaneous recording of muscle afferents, muscle efferents, electromyograms, and a variety of kinesiological parameters.

Early findings of this group were highly significant. They found, first, that spindle afferents (unidentified as to Ia or group II) from extensor muscles were active during the time when the foot was on the ground. During this time the extensors were contracting, which would be expected to unload an "in parallel" spindle fiber. On the basis of these data it was concluded that the alpha motoneurons to the extrafusal fibers and the gamma motoneurons to the intrafusal

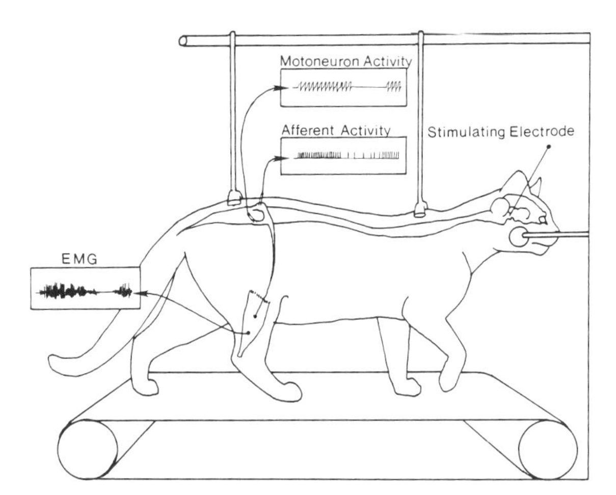

Figure 17-8 Controlled locomotion of a cat whose midbrain has been sectioned, studied on a treadmill. Repetitive stimulation of various locomotor regions in the midbrain results in descending activity, which excites the central pattern generators. As the animal walks, trots, or gallops, simultaneous records from isolated motoneurons *(top)*, afferents *(middle)*, or hind-limb muscles *(bottom)* may be made.

fibers must be active simultaneously. This is known as alpha-gamma coactivation. The point had long been contested. Second, the spindle afferents were not particularly sensitive to passive stretch of whole muscle, even at the fastest gaits. Third, Golgi tendon organs from extensor muscles fired most intensely during the stance phase.

The technically demanding task of chronically recording from various afferent species in conscious cats has been accomplished (Prochazka, Westerman, and Ziccone, 1976; Loeb, Bak, and Duysens, 1977). Though limited, these studies have provided further insight into the discharge patterns of afferents and may be summarized as follows (Loeb, 1981).

(1) Cutaneous mechanoreceptors gave highly stereotyped responses. Most low-threshold receptors were activated by any direct contact between the receptive field and any object. Of significance, however, was the finding that stretch-sensitive receptors and receptors located near skin folds were activated by the normal movements of the skin during walking. This observation may have relevance to the onset of extensor electromyographic activity prior to foot placement. The importance of any kind of peripheral input for this event had formerly been discounted.

(2) Though only eight units of joint receptors were studied, the data are thought provoking. Some of these units discharged only rarely during locomotion and were primarily activated only by axial rotations of the joint rather than by flexion-extension. Two knee joint receptors showed intense activity during the time the foot was on the ground, but were relatively quiet during the larger knee excursions as the limb was swung forward prior to placement. Apparently several factors, including joint angle, determine the excitability of these afferents.

(3) The patterns of activity of Golgi tendon organs appeared to closely follow the general electromyographic activity of the parent muscle.

(4) The discharge of muscle spindles was highly variable, depending on several factors. First, activation of the fusimotor system while the extrafusal fibers were either shortening or lengthening did seem to occur, but fusimotor activity could be modulated according to the momentary action of the muscle. Second, it was demonstrated that firing differences between spindles may relate to the kinesiological role of the muscle from which the afferent originates. These investigators emphasize that the kinesiological details of muscle function must be known before generalizations about spindle function emerge.

Much can be learned by studying the individual firing patterns of isolated receptors, but it is probably their ensemble effect on the CPGs and motoneurons that is important for quality locomotion (Wetzel and Stuart, 1976).

Organization of Muscle Receptors Data reviewed here suggest that Golgi tendon organs and spindles are sensitive to small intramuscular forces. If this is true, and if the S and FR motor unit types are the first to be recruited within a muscle, then one might expect a high population of spindles and Golgi tendon organs to be associated with SO and FOG fibers (or their equivalents). This hypothesis has been supported by Botterman, Binder, and Stuart (1978) and is still being tested.

The density within muscles of spindles (but to a

far lesser extent tendon organs) has been determined for only a few mammalian and avian muscles. Specific data for the cat, summarized by Botterman and associates (see their table 1 and figure 3), suggest that both types of receptor are plentiful in muscles that develop relatively substantial force and are subject to unpredictable loads, and that spindle densities correspond to some degree with the presumed role of the muscle in the control of postural stability or fine movement. These receptors tend to be closely associated with oxidative (low-force, low functional threshold) muscle fibers, particularly in "compartmentalized" muscles.

Maier has studied both mammalian (1979) and avian (1981) systems with an eye toward the receptor-muscle relationship. His findings for spindle densities in three heads of the pigeon gastrocnemius do not support some of the general relations described above. He found that densities in pars interna and pars externa were not significantly different, even though the contraction times and fatigue indexes of the two muscles differ significantly. Furthermore, although the pars interna and pars media were similar in these two physiological properties, their spindle densities were significantly different. It is evident that more comparative work is needed before the distribution of muscle receptors can be related with confidence to muscle function.

The presence of "compartments" of fiber types, coupled with uneven distributions within a muscle, suggests that functional subcompartments of whole muscle exist. Cameron et al. (1980) studied in cat MG receptor sensitivity to contractions of intramuscular compartments as defined by those muscle fibers innervated by a branch of the muscle nerve. They found that in general, Golgi tendon organs and spindles are more sensitive to contractions of the intramuscular compartment in which they are located than to adjacent intramuscular compartments. Thus muscle receptors may generate a "sensor partitioning" of their parent muscle and provide for an intramuscular localization of stretch reflexes. This is consistent with our knowledge that Ia synaptic connections are stronger for homonymous than for heteronymous motoneurons.

Evidence that motoneurons that serve a particular intramuscular compartment are functionally localized within the spinal cord has been reported for cat biceps femoris (Botterman et al., 1983). Three distinct nerve branches innervate the anterior, middle, and posterior regions of the biceps femoris. Further, the anterior fibers may act primarily in hip extension, whereas the posterior fibers may act in flexion of the knee. Intracellular recordings revealed that the anterior, middle, and posterior biceps nerve branches contributed larger EPSPs to their "own" motoneurons than to those supplying the other intramuscular compartments.

In summary, these preliminary findings suggest the existence of an orderly organization of peripheral neuromuscular components and their central counterparts for the neural control of locomotion, but it is too soon for general patterns to be evident. Among other things, it remains to be determined if the histochemical "compartmentalization" noted earlier has a physiological counterpart.

Kinematics and Kinetics

Requisite to our understanding of the neural control of movement is accurate description of the movements themselves. Morphologists armed with cinematographic and cineradiographic cameras are able to make these kinematic analyses of moving animals. The forces and energy requirements associated with movement are also fundamental, and cinematography is often combined with electromyography to delineate the timing of muscle contraction; with force plates, bone strain gauges, and force transducers placed directly onto muscle tendons to monitor relevant limb, segment, and muscle forces; and with respiratory gas analysis to measure cost of energy (see Chapter 10). Several kinesiological and kinetic parameters of the cat hind limb during locomotion have been determined, and they can be used to illustrate these kinds of measurements.

The architectural arrangement of muscle fibers relates to neural control in that it influences force production and velocity of shortening. Recall that some muscles possess long fibers which, in some cases, extend from one end of the muscle to the other. They also tend to parallel the line of pull and are thus known as parallel-fibered muscles. The cat Sol possesses relatively long parallel fibers, compared with the MG. Muscles such as the MG, which contain short fibers variously oriented within the muscle belly, are pinnately fibered. Given two muscles of the same anatomical volume, mofibril density, and intrinsic rate of shortening, differences in fiber architecture will result in quite different performance.

Consider maximum force production. Recall that at a given length the isometric force a muscle can develop relates to the cross-sectional area of the contractile component. Pinnation increases the number of fibers acting on a tendon and, all other factors being equal, a pinnately fibered muscle is capable of generating more maximum force than a parallel-fibered muscle of equal weight or volume. Although the pinnately fibered MG weighs but three times the parallel-fibered Sol, it produces five times more maximal force. It should also be recalled, however, that the isometric force a muscle can produce at a given length relates to the relationship of its contractile proteins. At the sarcomere level there exists an optimal degree of overlap of the thick and thin filaments for maximum active tension or force development. At overlaps that are less or greater than optimal, active force production drops off. Thus, for a whole muscle or single fiber there exists a relationship between length and maximal active tension. Experimentally, one is able to determine this relationship and plot it as a length–active tension curve. The last general point to keep in mind is that the rate at which a muscle or muscle fiber can shorten is dependent upon two factors: the intrinsic rate of shortening and the number of sarcomeres in series. The intrinsic rate of shortening is simply the rate at which the myosin cross-bridges work.

These general considerations of muscle fiber architecture, length-force relationships, and rate of shortening suggest that a parallel-fibered muscle can contract through a greater distance at a faster rate than a pinnately fibered one of similar length (and similar fiber type). They also suggest that the absolute length over which a parallel-fibered muscle can maintain near maximal tension is greater than that of a pinnately fibered muscle of equal length. The fact that the parallel-fibered muscle has more sarcomeres in series in a myofibril supports these predictions. For example, a relatively small length change in MG markedly influences the degree of thick and thin filament overlap for each sarcomere of its relatively short fibers. As a result, the ascending and descending arms of the length–active tension curve of this muscle are steep compared with those of its parallel-fibered synergist, Sol. In addition to the examples given below, see Sacks and Roy (1982) for an extensive discussion of muscle fiber architecture and the neural control of the cat hind limb.

The physical condition of a muscle at various times throughout the movement phases is also an important consideration. For example, it is important to know parameters of a muscle relative to its contractile state (active versus inactive) and length (static or constant length; dynamic or changing length). Also important if length is changing are the direction (shortening versus lengthening) and rate of change simultaneous with the muscle's contractile state of activation. If a muscle is shortening while it is electrically active, this is said to be active shortening (concentric contraction). This may be contrasted to passive shortening (no electrical activity). On the other hand, a muscle may be stretched or lengthened while it is either electrically active (active lengthening; eccentric contraction) or electrically inactive (passive lengthening).

Cat Hind Limb Various schemes have been used to describe the step cycle, or movements and epochs of the limb during stepping. The description that has proven most useful is that of Philippson (1905). It recognizes the two basic time periods when the foot is either off the ground (swing, recovery, or protraction) or on the ground (stance, support or propulsion), and four subphases (F, E^1, E^2, and E^3), as illustrated in Figure 17-9. Sometimes the ankle or knee joint begins its flexion or extension 50–100 msec earlier or later than the other, but these two joints move more or less synchronously throughout the step cycle. The hip, on the other hand, flexes in phase with the ankle and knee in F, remains stable in E^1 and E^2, and returns to an in-phase pattern as

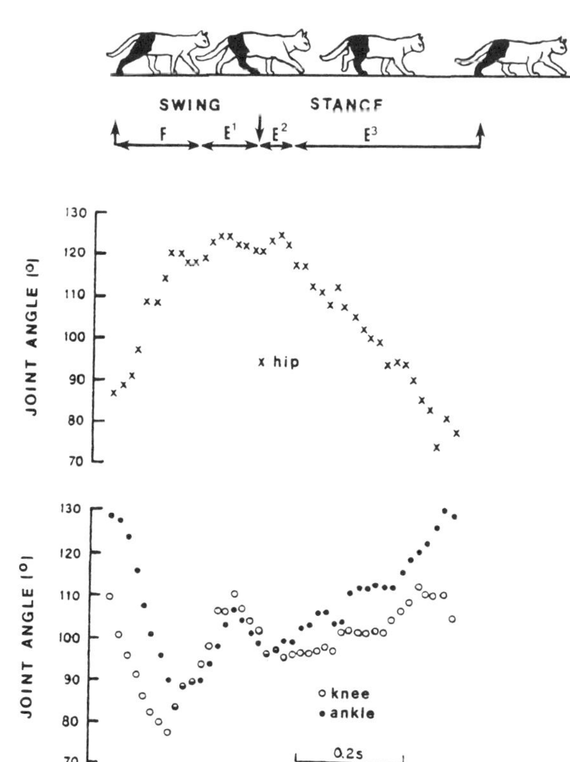

Figure 17-9 The step cycle of a single hind-limb of the cat during walking. Upward arrows indicate foot lift-off; the downward arrow, touch-down. The flexion *(F)* phase begins as the foot is thrust off the ground, at which time flexion begins at the hip (ischio-femoral angle), knee, and ankle joints. This phase ends as extension (E^1) begins in the knee and ankle. During the second extension phase (E^2) the knee, ankle, and metatarsophalangeal joints *(not shown)* yield (flex) under the animal, but the hip remains stable. The third extension phase (E^3) is characterized by progressive extension of the hip, knee, and ankle joints. (From Goslow, Reinking, and Stuart, 1973a.)

the three joints extend in E^3. These phase relationships have relevance to the neural control of the limb.

As the animal moves faster, the swing phase remains nearly constant, whereas the stance phase shortens (Fig. 17-10). Limb movements and joint angles can be coupled with electromyographic recordings and appropriate lever-arm measurements to provide a dynamic portrayal for selected muscles.

Muscle length change may be determined from joint angle change and lever-arm measurements. Such a joint angle–muscle profile is shown for the cat Sol in Figure 17-11. Anatomically, the Sol is relatively simple in that it crosses only one joint, the ankle. From Figure 17-11A one sees that two stretches of quite different natures are imposed on the Sol during the step cycle. The stretch in the *F* phase is more or less passive, whereas the E^2 stretch is an active lengthening. Furthermore, Figure 17-11B presents changes in muscle length in terms of the extent and rate of passive and active stretch, and 17-11C shows a "working range" of muscle length. This is done by relating maximum to minimum in situ lengths to the length associated with quiet standing.

Further information about the interrelationships of limb movement and muscle design is derived by correlating the length–active tension profile of a muscle to its activity pattern and excursion. Figure 17-12 is from a study of the cat MG done by Stephens, Reinking, and Stuart (1975). It illustrates the length–active tension curve as it relates to change in the angle of the ankle joint during walking and galloping. The

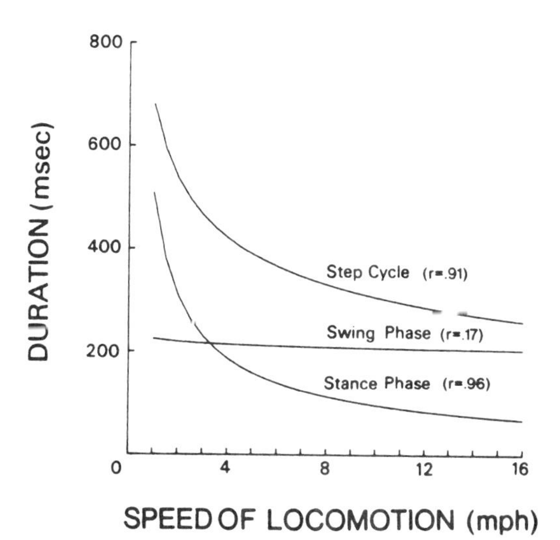

Figure 17-10 The duration of the entire step cycle and its components of swing and stance for cats at varying forward speeds. As forward speed increases, the swing phase remains relatively constant but the stance phase shortens considerably. (From Goslow, Reinking, and Stuart, 1973a.)

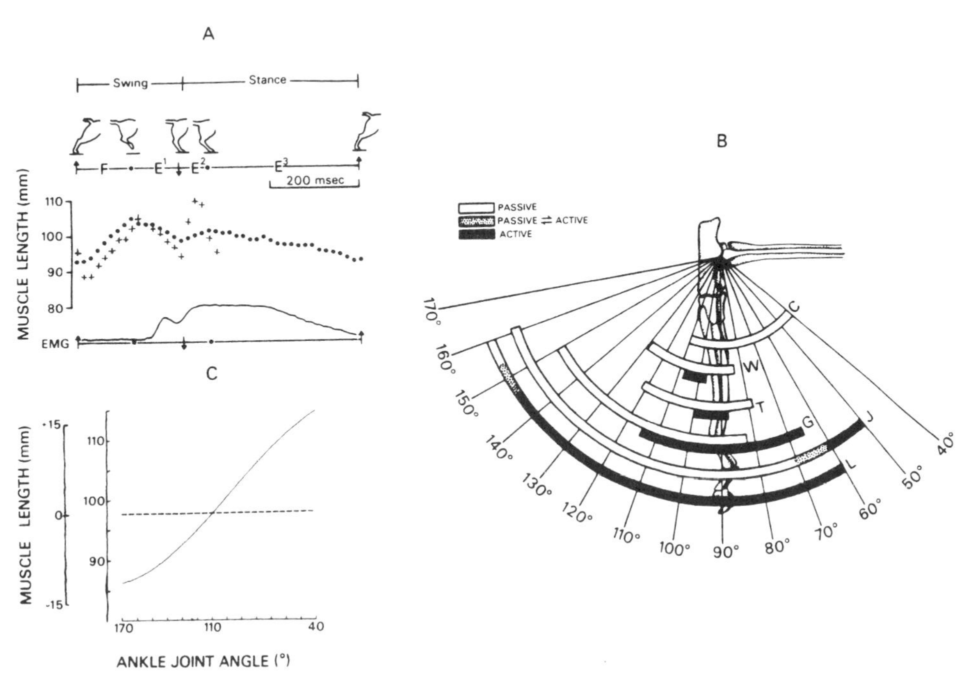

Figure 17-11 The extents of stretch for the soleus (Sol) of a cat. A: Upper sketches show cat step cycle; middle plot shows muscle length during a 1.5 mph walk *(solid circles)* and a 16.3 mph gallop *(crosses)*; lower sketch shows an integrated electromyographic profile. B: Range of ankle dorsiflexions as the cat crouches *(C)*; walks *(W)*; trots *(T)*; gallops *(G)*; completes a vertical jump *(J)*; and lands *(L)* from a 3-foot fall. The locomotor *(W,T,G)* stretches are for both *F (open bars)* and E^2 *(solid bars)* phases of the step cycle. C: A conversion graph for relating ankle angle to Sol length. The dotted horizontal line is standing length (when ankle angle ≅ 110°). (From Goslow, Reinking, and Stuart, 1973b.)

figure shows a variety of adaptations of the musculoskeletal system. First, note that the peak of active tetanic tension for MG is near the longest lengths associated with natural movements. Second, the authors made individual length–active tension curves for S, FR, and FF motor units and found that the mean optimal length for force production of the S and small FR units corresponds to those lengths that are at or near standing length. Figure 17-12 emphasizes that increased demands on MG (as in the change from standing to walking to galloping) results in an increase in muscle "usage" (product of displacement and time) at lengths greater than the standing length. These observations appear, then, to offer functional significance to not only the length–active tension of whole muscle but also of their constituent motor units. Third, it can be seen that in normal locomotion the muscle displays electrical activity during shortening and lengthening contractions.

The relation between force and velocity of a contractior relates importantly to the architecture of muscle fibers. The cat Sol possesses relatively long parallel fibers, whereas the complexly pinnate MG possesses shorter fibers. Spector et al. (1980) reported for the cat a nearly threefold greater maximum isotonic shortening velocity (Vmax) of the MG sarcomeres relative to Sol sarcomeres. These differences are presumed to relate to biochemistry at the fiber level. However, and more relevant to the animal's locomotor performance, the Vmax developed by the MG at the Achilles tendon is only 1.5 times that of the Sol owing to the influences of these muscles' specific fiber architectures. Force-velocity curves for the Sol and MG, determined at the Achilles tendon, are reproduced from these authors in Figure

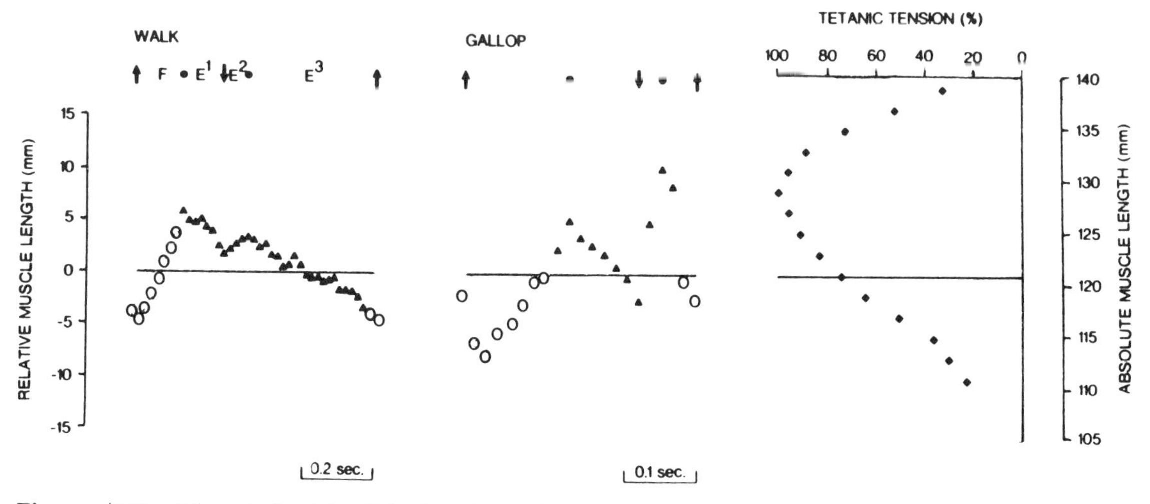

Figure 17-12 The relationship of the length–active tension curve of the cat medial gastrocnemius (MG) to its length and electromyographic activity (EMG) during the walk and trot. The right side of the figure shows the mean length–active tension curve for a sample of MG muscles. Length of the muscle is expressed in absolute units *(right)* and units relative to the standing length *(left)*. EMG activity during stepping is shown by closed triangles; circles show inactivity. The peak of active tetanic tension is near the longest lengths associated with natural movements. (From Stephens, Reinking, and Stuart, 1975.)

17-13. They concluded that at high speeds of locomotion and jumping, the extension velocities of the ankle exceed the maximum shortening velocity of the Sol. The MG, however, which is capable of producing nearly five times the force of the Sol, is capable of shortening velocities consistent with observed speeds of ankle extension.

The final piece of evidence necessary to profile a muscle is knowledge of the force of each muscle through the step cycle. In a study by Walmsley, Hodgson, and Burke (1978) force transducers designed for chronic implantation were placed on the individual tendons of the Sol and MG in the same hind limb of adult cats. Kinesiological and electromyographic data were also recorded during standing, treadmill locomotion, and jumping (Fig. 17-14). Several findings are of particular significance here. First, during standing, the Sol force was near that seen in locomotion, whereas the MG produced only small forces (approximately 5 N). Second, they found that throughout a wide range of locomotor speeds (from slow walk to moderate run) the Sol develops approximately the same peak force (about 20 N), whereas the MG produces forces over a threefold range (from about 6 to 20 N). Third, during stepping the peak forces of both the Sol and the MG occur before

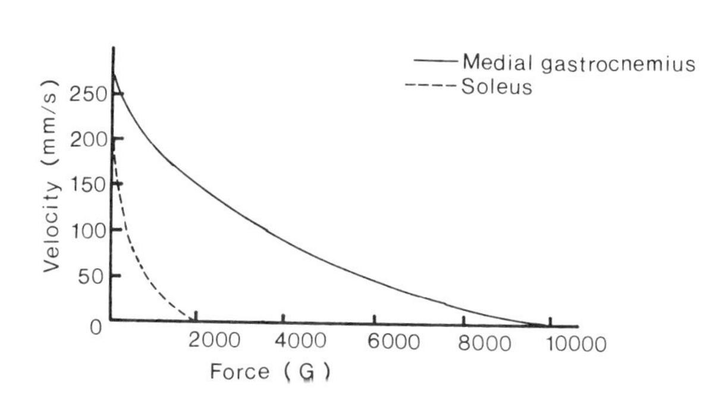

Figure 17-13 Force-velocity relationships of cat soleus (Sol) and medial gastrocnemius (MG) muscles as measured at the Achilles tendon. The maximum rate of isotonic shortening *(ordinate)* is dependent, in part, on the load or force developed *(abscissa)*. The maximum sarcomere shortening rate of the parallel-fibered Sol is one third that of the pinnately-fibered MG. When fiber length and angle of pinnation are accounted for, however, the differential in maximum velocities produced at the Achilles tendon is not as great. At high speeds of locomotion the extension velocities of the ankle exceed the maximum shortening velocity of the Sol but not the MG. (From Spector et al., 1980.)

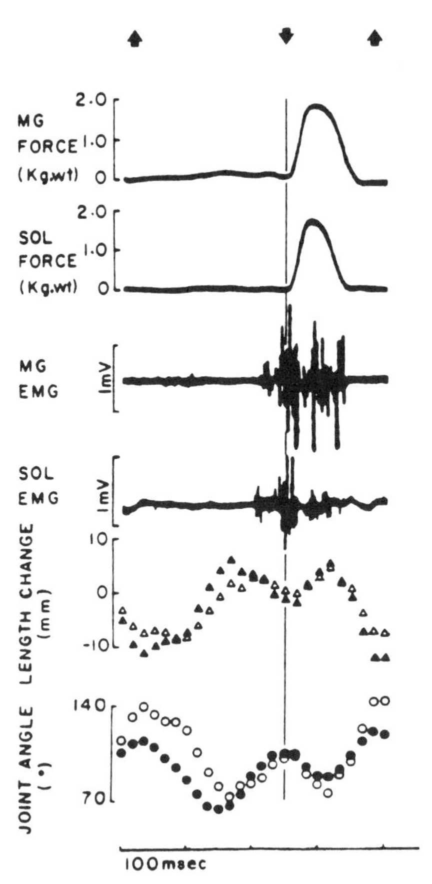

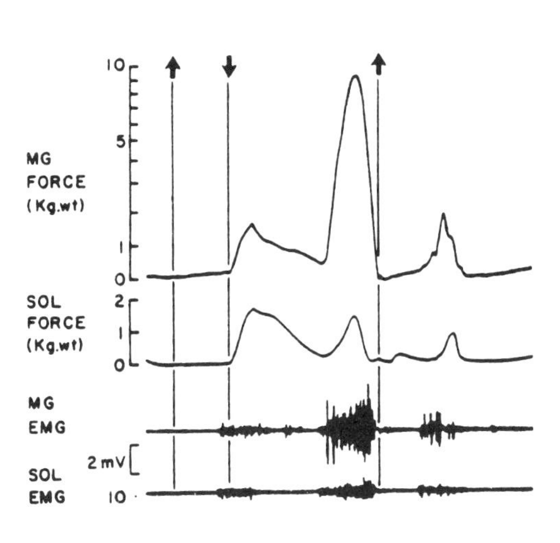

Figure 17-14 Records from an adult cat with force gauges attached to the soleus (Sol) and medial gastrocnemius (MG) tendons. *Left:* Records for running at a moderate treadmill speed; upward and downward arrows identified in Fig. 9. At the bottom are the joint angles of the knee *(closed circles)* and ankle *(open circles)*. Estimates of relative length changes for Sol *(closed triangles)* and MG *(open triangles)* are also given, using standing length as a reference. *Right:* EMG and force recordings from the Sol and MG tendons during a vertical jump of 120 cm. During running the peak forces of the Sol and MG are about the same, whereas in the jump, although the Sol continues to produce about 2 kg wt, the MG force is about five times its force during running. (From Walmsley, Hodgson, and Burke, 1979.)

the end of E^2, as both muscles undergo active lengthening. Fourth, for vertical jumping (Fig. 17-14) MG force greatly increases (to 90 N for medium-sized cats) above the forces seen during stepping. They concluded from these data that the division of labor between the absolute force levels produced by the two muscles during posture, stepping, and jumping appear consistent with their differing motor unit populations.

Nonmammalian Vertebrates There are some important and precise data on the kinematics of normal locomotion for other vertebrate classes, which suggest a valuable evolutionary framework in which to view vertebrate neural control. Some kinematics of fish locomotion will be presented later in this account. Of special note here is that most fishes use a "traveling wave" mode of vertebral undulation. In this mode a wave of contraction in the axial musculature on one side of the body begins at the anterior end and travels posteriorly down the body. Details may be found in Chapter 7, Gray (1968), and Grillner (1981). Kinematic studies of the anuran limb often relate to jumping (see Chapter 4). Edwards (1976) has done the most complete cinematographic study of urodele locomotion (48 species). Slow movements (walking) are characterized by a "standing wave" mode of vertebral undulation in which points on the vertebral column having no lateral movement (nodes) alternate with regions of large lateral movement. The nodes correspond to the points of attachment of the pectoral and pelvic girdles. Of interest is that trotting salamanders employ the traveling wave mode associated with fishes. These findings, when coupled with various embryological studies (below), provide evidence for the evolution of the neural control of tetrapod locomotion. In addition, Edwards studied the influence of body size and rate of travel on gait selection, the three methods of forward propulsion, and the probable muscular contributions employed for movement.

The kinematics of limbless locomotion among reptiles has received some attention (Chapter 9). Studies of terrestrial and aquatic locomotion in turtles (see Walker, 1979) provided an important foundation for subsequent studies in which the turtle was used as a vertebrate model (Stein, 1978; see below). In a cineradiographic and electromyographic analysis of the shoulder of a varanid lizard there is a discussion of the contribution to locomotion made by homologous reptilian and mammalian muscles (Jenkins and Goslow, 1983).

Although kinematic studies of wing movements have been done, little work relating to the neural control of adult bird locomotion has been pursued. Studies of the topographic relationship of neuron location in the spinal cord relative to muscle location in the hind limb are important exceptions. Such studies demonstrate an approach fundamental to our understanding of comparative neural control (Landmesser, 1976).

Central Pattern Generators

Evidence for Fishes Evidence for the existence of CPGs within the spinal cord of fishes comes from studies of the lamprey and the dogfish. The spinal cord is isolated from supraspinal influences by transection to produce a "spinal animal" or a "spinal preparation." Undulatory movements, as in swimming, are seen in a spinalized dogfish. Early studies illustrating this phenomenon were done by Sir James Gray. The movements seen, however, may excite many somatoreceptors, which in turn may feed back into the spinal cord, excite various neurons, and strongly influence further movements through feedback mechanisms. The most direct way to prevent afferent inflow in the spinalized fish is to section the dorsal roots, which requires extensive surgery. A second procedure commonly employed is to administer curare intravenously to block the neuromuscular junction and paralyze the animal. When either of these procedures is employed and recording electrodes are placed on one or more ventral roots of the spinal fish, rhythmic efferent outflow of constant phase coupling and frequency is seen (Fig. 17-15). No muscles are actually contracting, but the outflow impulses from the spinal cord are suggestive of those seen in movement. This preparation is said to illustrate "fictive" (imaginary, or feigned) locomotion. True fictive locomotion requires that efferent outflow patterns be bilateral, which mimics that seen in normal locomotion. Since this cyclic pattern of outflow cannot be the product of supraspinal control (prevented by spinalization) nor result from the phasic flow of afferent information (prevented by dorsal root section or neuromuscular block), it must result from an innate CPG(s) located in the spinal cord. If the cord is sectioned into several transverse pieces, each piece produces phasic efferent outflow. Thus, the dogfish spinal cord is thought to possess segmental pattern generators.

The lamprey is an ideal model for the study of locomotion, and early studies of this preparation are being extended by Grillner and his colleagues (Grillner, 1981). A preparation of the spinal lamprey similar to that described for the dogfish does not produce spontaneous fictive locomotion. The reason for this is not clear. However, if the spinal neurons of the cord are given some sort of "background" stimulation by way of chemicals, fictive locomotion ensues. The chemicals usually used are either precursors to normal neural transmitters, membrane-receptor stimulators, or transmitters. Although the drugs obviously introduce another variable, their use has been instructive.

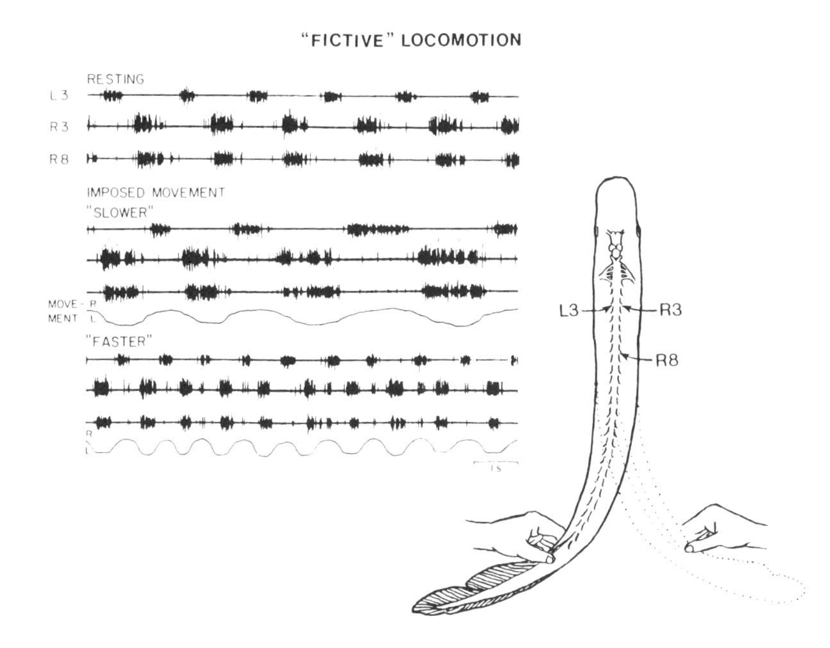

Figure 17-15 "Fictive" locomotion in an isolated notochord-spinal cord preparation of the lamprey. *Right:* Locomotor bursts were recorded in segments 3 and 8 with suction electrodes. *Top left:* Bursts from the ventral roots were recorded from the stationary preparation, suggesting "fictive" locomotion. Note the unchanging but constant frequency and the appropriate phase lag from *L*3 to *R*3. *Middle left:* After fixation of the notocord rostrally, movement of the preparation to the right and left at frequency that was either slower *(middle)* or faster *(bottom)* than resting frequency resulted in an appropriate change in efferent outflow. (From Anderson, Forssberg, and Grillner, 1981.)

An isolated lamprey spinal cord bathed in a solution of various amino acids produces an efferent outflow of constant phase coupling and frequency similar to that seen for the dogfish. These studies confirm that the neuronal circuitry for swimming of the dogfish and lamprey resides in the spinal cord. Presumably this is true for other fishes as well.

Evidence for Mammals As long ago as 1906 Sherrington showed that a dog or cat displays rhythmic, alternating hind-limb movements after spinal cord transection at the thoracic level. However, clear evidence for the general presence of a single series of CPGs in tetrapods has only recently been established. Extending and refining studies of earlier workers (such as Wetzel and Stuart, 1976), Grillner and his colleagues have clearly shown that the mammalian spinal cord is capable of generating a phasic efferent outflow to each limb (Grillner, 1981). The preparation that conclusively demonstrates this is the spinal cat, in which an intravenous injection is given of selected pharmacological agents to provide background stimulation and curare to prevent movement. In such a preparation, as for fishes, rhythmic activity from the CPG can be recorded from ventral root filaments.

Influences of Afferents on the CGP By returning to the isolated notochord–spinal cord lamprey preparation, we see the ensemble effects of somatosensory inflow on the rhythmic influence of the CPG network. Recall that in this isolated in vitro preparation, fictive locomotion results from a constant efferent outflow of constant frequency. However, if the notochord–spinal cord complex is manually moved back and forth, as in swimming at either a faster or slower rate (Fig. 17-15) than that corresponding to the normal frequency of the CPGs, then the frequency of efferent outflow changes to correlate with the imposed movements. Apparently, sensory feedback to the spinal generators influences the frequency of motor output to adjust to the needs of the fish. Under normal conditions the sensory rhythm will be identical, or almost identical, to the motor frequency. But if something occurs in the environment that does not allow the body of the fish to follow the basic motor rhythm, the sensory input can inform the motor rhythm. It is suggested that in the case of the lamprey, as yet unidentified receptors in the spinal cord and notochordal sheaths are responsible for this peripheral feedback.

Similar influence of peripheral feedback on the frequency of the CPG is seen in the spinalized, curarised dogfish. The superimposed movements themselves may be of much smaller amplitude than those of the intact, swimming fish. Positive feedback from the side toward which the tail is moving seems to be important. During a movement to the right, peripheral input reinforces efferent outflow to the right and simultaneously inhibits efferent outflow to the left. This has been termed a positive directional sensitivity response. On the other hand, when the trunk is bent to the normal extreme left or right, a reflex reversal occurs, in that instead of a reinforced efferent outflow toward the side of movement, there is inhibition; and instead of inhibition to the side away from the movement, there is excitation. This directional switch phenomenon can be related clearly to the control of flexion and extension of the tetrapod limb.

Similar experiments showing the influence of peripheral somatosensory input on the CPGs have been done in the cat by Grillner and Rossignol (1978). As in the intact cat, the hind-limb step cycle of a "low spinal" cat (cord sectioned posterior to forelimbs) walking or running on a treadmill will adapt to the speed of the treadmill. As the treadmill belt speed is increased, the swing phase remains relatively constant but the stance phase becomes shorter, as in normal locomotion. Thus, there is some sort of reflex control modulating the CPG to change the stride length. If one hind limb of a spinal cat is stopped during the first or the middle part of its extension in the stance phase, the initiation of flexion is prevented even though the other limbs continue their normal cycles. In such a fixed limb, extensor activity continues indefinitely until the limb is permitted to extend farther (Fig. 17-16). The figure shows dramatically the influence of limb position on efferent output. When the limb has reached a hip angle near its most caudal position in the normal step cycle, flexion suddenly ensues. Analogous to the fish experiments, a directional switch is initiated at the extreme of hip extension; the joint angles of the lower limb do not appear to be so important for the extension-flexion transition. The specific receptors facilitating this peripheral influence have not been identified, but they are presumably located in and around the tissue and muscles of the hip joint. Some studies suggest that a decrease in the load on the limb at the end of the stance phase acts to stimulate the beginning of the flexion (swing) phase.

Studies of fictive locomotion in spinalized, curarized cats have provided data on the general effects of somatosensory input on CPGs that are similar to those for fishes. First, repetitive movements back and forth of the hind limb (simulating flexion and extension) resulted in efferent bursts that were either faster or slower than the spontaneous burst recorded at rest. Second, flexor activity occurred during the imposed flexion, and extensor activity occurred during the imposed extension. Third, the entrainment pattern of efferent bursts can be induced by movements of much smaller amplitude than those normally seen in the intact animal.

Finally, somatosensory influence on the CPG interneuron machinery is dependent on the locomotor phase that exists at the instant of input. A second set of experiments on spinalized but not curarised dogfish and cats illustrates. As the animals were moving in response to their spinal generators, a tactile stimulus was applied to the skin's surface. In a swimming spinal dogfish, touching the tail (either side) elicited an exaggerated movement to the right if the tail was starting to move to the right; the same stimulus elicited an exaggerated movement to the left if the tail was starting to move to the left. In a spinalized walking cat, touching the dorsum of the foot caused enhanced extension as if to step over something if the limb was already flexed; the same stimulus enhanced extension if the limb was already extended. These responses, termed phase-dependent reflex reversals, illustrate that the reflex is dynamically changing as the animal moves. Apparently, somatosensory input can be directed along different pathways according to the state of the CPG.

Numerous deafferentation studies have been done on a variety of species to evaluate the contribution of afferents. Suffice it to say that although an animal may be able to move or even use a limb in locomotion, the quality and precision of that limb is distinctly impaired following deafferentation. In general, the effect on fishes and amphibians is less debilitating than on mammals. Most deafferentation studies focus on what the animal is able to do, rather than on what elements of

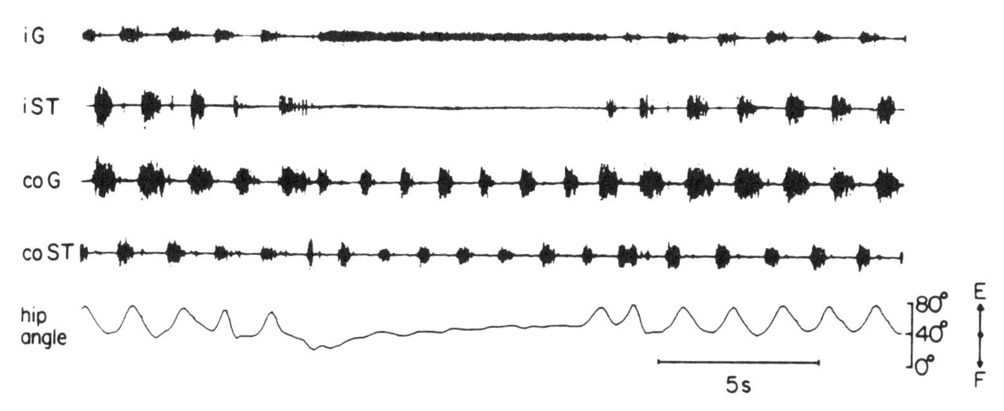

Figure 17-16 Effect of impaired limb movement of a spinal cat on a treadmill, shown by electromyograms of the ipsilateral *(i)* and contralateral *(co)* gastrocnemius *(G)* and semitendinosus *(ST)*, as well as by successive changes in the hip angle during stepping. During the stance period the ipsilateral limb was held and prevented from continuing its movement. The *iG* remained continuously active and the *iST* remained silent, while the contralateral limb continued to step. These data suggest that peripheral input at hip angles associated with the stance phase provides positive feedback to the extensor motoneurons. (From Grillner and Rossignol, 1978.)

refined locomotion are lost; the latter observations reveal a great deal about the necessity of afferent input. A problem with the deafferentation model is that it is not selective to just the periphery. As a result of the loss of afferent input, there is a change in the synaptic characteristics of the cells of the CPGs and motoneurons.

Supraspinal Influences

Much of what is known about the supraspinal control of locomotion is extensively reviewed in Grillner (1981), Shik and Orlovsky (1976), and Wetzel and Stuart (1976). Most of our information regarding supraspinal centers and pathways comes from study of reduced preparations. In fact, removal of many higher centers is necessary in order to isolate particular areas of the CNS for study.

The general aspects of supraspinal control have been outlined using the cat as a model. Methods of study have included transection at progressively lower levels of the brainstem, in conjunction with electrical stimulation (to induce controlled locomotion), and electrical stimulation of and recording from descending tracts.

Mesencephalic Locomotor Region A relatively autonomous stepping program can be initiated and sustained by way of descending impulses. The dominant descending cells appear to be reticulospinal neurons, which project to the motoneurons either monosynaptically or via interneurons. The cell bodies of these neurons are located in the caudal reticular nucleus of the pons and in the giant cell reticular nucleus of the medulla (Fig. 17-17). Apparently, these cells can induce locomotion by way of a number of suprapontine centers. If the neuraxis of a cat is severed on a plane that passes rostral to the superior colliculus and rostral to the mammillary body (Fig. 17-17A), the animal will show spontaneous stepping activity, although the activity is not purposeful. This section represents the caudalmost level that the brain can be severed and still show spontaneous locomotor activity. A section at level B, which courses caudal to the mammillary body, is that used by Shik, Severin, and Orlovsky (1966). This is often called the mesencephalic preparation. The animal walks if the mesencephalic locomotor region (MLR) under the inferior colliculus is stimulated electrically or if the cat is allowed to recover for one or two weeks. The MLR is considered to be the most important of the suprapontine locomotor regions.

When the MLR of the mesencephalic preparation is stimulated, the cat stands up and begins stepping. If the treadmill belt is in motion, the cat may accelerate the belt. Stimulation is typically in

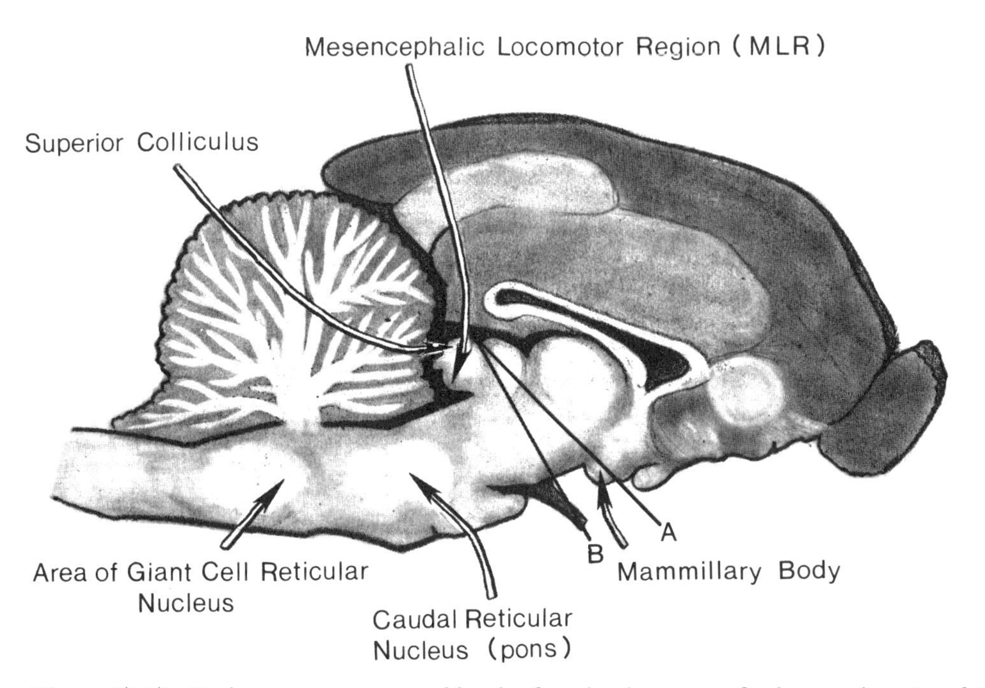

Figure 17-17 Brainstem structures and levels of section important for locomotion. *A* and *B* are levels of section designated as "premammilary" and "postmammilary" preparations, respectively. A premammilary preparation displays spontaneous locomotion within 24 hours of surgery and if placed on a moving treadmill belt will conform to the imposed velocity. The postmammilary preparation excludes some caudal hypothalamic and subthalamic tissue, and an adult acute cat so sectioned will not walk spontaneously. However, controlled stepping on a moving treadmill can be initiated by repetitive electrical stimulation of the mesencephalic locomotor region (MLR). Shown are the approximate positions of caudal reticular and giant cell reticular neurons of the pons and medulla thought to be of major importance for "switching on" and sustaining the spinal stepping circuitry.

the form of 1-msec shocks at 25 to 50 Hz. Stronger stimulation is required to initiate forelimb movements than hind-limb movements. The characteristics of stimuli that produce locomotion illustrate to some degree the nature of the descending control signals. Within and around the MLR, stimulus rates of 10 Hz evoke movements that are phase-locked to each individual shock. Stimulus rates from 25 to 500 Hz within the MLR are equally effective for evoking locomotion. Finally, within the locomotor region an increase in stimulus intensity, but not rate, leads to faster and more vigorous locomotion, including gait conversion from walking to trotting to galloping. To emphasize, stimulation of the MLR apparently creates enough "background excitability" among the reticulospinal cells to initiate and maintain stepping.

Descending Tracts Involved in Locomotion
Once the MLR had been located, it was logical to determine which descending pathways were involved during controlled locomotion. Orlovsky (1972) made some important observations to this end by not only electrically stimulating but also recording from four descending pathways during controlled locomotion of the mesencephalic cat. The pathways studied were the lateral reticulospinal tract, the lateral vestibulospinal tract, the rubrospinal tract, and the corticospinal (pyramidal) tract neurons. All four of these tracts project directly to the motoneurons. Orlovsky not only monitored the levels of sustained discharges from these pathways, but he also noted how these discharges varied with the different phases of the step cycle. He stimulated each tract at different times in the step cycle (during *F* or during E^3 and

so on) and noted whether phase-modulated effects could be observed.

It appears, then, that there are essentially two types of descending control during stepping. The more generalized one is to activate the spinal centers (CPGs) that control muscle activity of each of the four limbs. The second type is the facilitation and adjustment of the degree of motor unit activity in appropriate phases of the step. Certainly other portions of the brain are involved in locomotion; their absolute necessity for the basic stepping rhythm, however, appears minimal. Comparable data on the supraspinal contributions to locomotion are simply not available for other vertebrate groups.

Interlimb Coordination

Evidence from a variety of fronts indicates that in tetrapods each of the four limbs possesses its own CPG. The timing of the movements of the four limbs for appropriate stepping is thought to result from spinal interconnections among the four CPGs. Two problems have plagued attempts to characterize these neuronal interconnections: although ascending and descending propriospinal pathways within the spinal cord have been implicated, they are so intertwined and diffused with a variety of other spinal pathways that efforts to selectively destroy them have not been satisfactory; in addition, the interlimb timing patterns (or phase relationships) are variable, dependent on the gait employed as well as on rapid adjustments to the terrain. Thus, the capabilities within the animal to handle a variable environment have frustrated attempts to document the functional constraints of the interconnections of the CPGs.

The study of interlimb coordination in vertebrates is still in its infancy, but several laboratories are currently exploring this important topic. Evidence that supraspinal influence is not necessary for interlimb coordination has been obtained by Stein (1978) and his colleagues. When a turtle is spinalized by transection at the first cervical segment, no swimming movements can be observed. If the cut end of the spinal cord is stimulated by electrical impulses delivered to the dorsolateral funiculus (descending pathway), swimming movements of the ipsilateral pair of limbs in the proper sequence (as reflected by electromyograms of selected limb muscles and by limb movement) are produced. The cycle period of the swimming movements can be altered by changing the frequency of electrical pulse stimulation, but interlimb phasing is properly regulated. Thus, supraspinal adjustment is not necessary to maintain interlimb coordination; spinal cord pathways that interconnect the limb CPGs are sufficient.

Studies done earlier of interlimb movements of the cat during a variety of locomotor acts resulted in a hypothesis of programmed interlimb coordination (Miller and VanDerBerg, 1973). Based on measurements of joint angles as well as limb electromyogram patterns, a fairly tight coupling of the different limbs was proposed. In contrast, Stuart et al. (1973) stressed the variability in interlimb timings and suggested that the interlimb coordinating mechanisms are facultative (dependent on the situation), which, of course, precludes a tight coupling hypothesis. A middle-of-the-road position has been taken by English and his colleagues following a series of studies of interlimb coordination for the normal cat as well as for cats with specific spinal cord lesions (English and Lennard, 1982).

Figure 17-18A shows the electromyograms of the lateral triceps muscles of the two forelimbs (English, 1979). Duration of step is measured from the cessation of one burst to the next, and burst duration is from beginning to end of a selected signal. A duty factor, calculated for each muscle, is the ratio of burst duration to step duration (Fig. 17-18B). The temporal spacing (phase interval) of different limbs is expressed as the latency or lag time of the cessation of electromyographic activity in the two limbs. Phase intervals are expressed in degrees of a circle to emphasize their continuous distribution. The distance from the center along a radius expresses the duty factor (for example, a cross coinciding with the first inner circle means the muscle was active 25% of the total step duration), and the position on the circumference shows the corresponding phase interval. In a perfectly synchronized trot, for instance, the forelimbs are 180° out of phase. If one limb leads or lags the other, points appear to the right or left of the 180° meridian. If the forelimbs

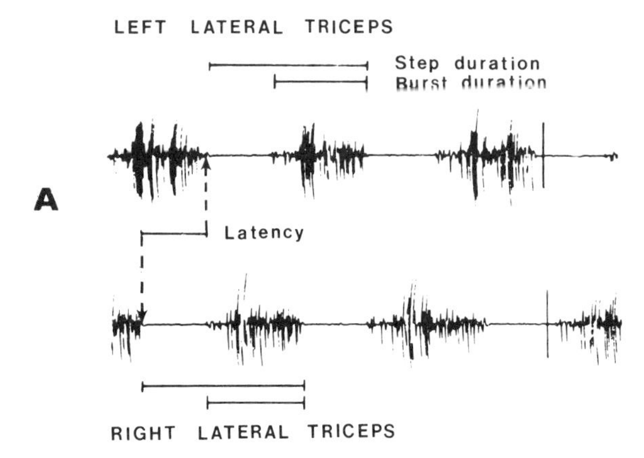

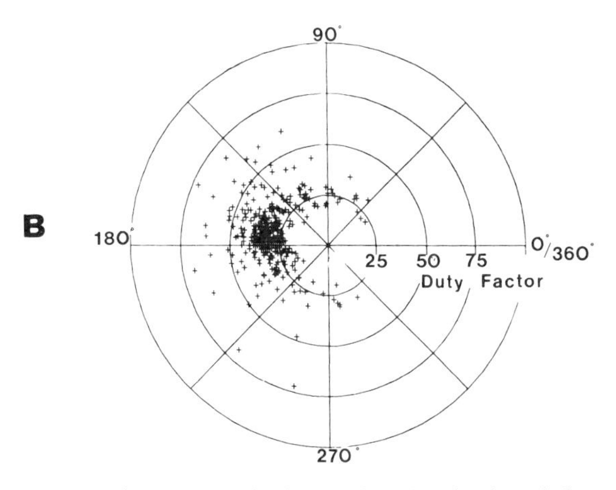

Figure 17-18 Interlimb coordination in the adult cat: a scheme for expressing limb linkage and variability. A: Electromyograms of the lateral triceps during stepping. Burst and step duration is determined as indicated. The ratio of burst duration to step duration is used to calculate a "duty factor." The "phase interval," or latency of the cessation of EMG activity in the two limbs, is expressed as a function of the step duration of the first limb used in the calculations. Phase intervals are expressed in degrees. B: Patterns of coordination for the two forelimb step cycles. The distance from the center is the "duty factor," and the position on the circumference the "phase interval." (From English, 1979.)

were precisely in phase, as in a pace, points would appear on the 0–360° meridian. Such an analysis can be done for any two limbs at a variety of gaits and speeds. The interlimb changes associated with gait transition, which are often ignored, have also been examined. This method of expressing interlimb coordination shows not only linkage patterns but variability, an important element of the system.

It is evident that the modes of stepping should be defined in terms of the relative strengths of coupling between the neural elements generating the movements. The strength of coupling between a pair of limbs is a state-dependent variable that includes the net effects of all the converging inputs (supraspinal, peripheral, intersegmental) to the pattern generator of each limb.

General Remarks and Future Study

At this point in an essay such as this, a general model for the CPG might be expected, or at least a satisfactory model for the neural control of alternating muscle movement or interlimb coordination. Unfortunately, so little is known about the interneuronal machinery of the CPG that current models must be incomplete and tentative. Two of the more useful are the Graham-Brown half-center hypothesis and the ring hypothesis (Shick and Orlovsky, 1976). Each of these models is formulated around the concept of alternating flexor-extensor activity during the step cycle, a notion that now appears too simplistic.

Vertebrate morphology has a clear mission in this field; extensive comparative data couched in evolutionary and developmental frameworks are needed both within and across the major vertebrate classes. Work with the intact, unimpeded animal as it moves through its environment is desirable to avoid the pitfalls and limitations of working with reduced preparations. Nevertheless, the benefits of observing one or more components of the neuromuscular or musculoskeletal system in relative isolation are great. These components can eventually be integrated to illustrate, in part at least, the whole. Ongoing studies of primitive tetrapods will illustrate.

Recall that Edwards (1977) noted that salamanders use standing wave undulations when walking and (like fishes) use traveling wave undulation when trotting. Data from embryological studies of salamanders (Brandle and Szekely, 1973) suggest that traveling waves are mediated by supraspinal impulses that travel to the motoneurons via the reticulospinal pathways. The limbs are more or less passively carried along by the traveling waves, resulting in a trot. In contrast, the walking gait associated with a standing wave is thought to be generated by each limb's individual "limb-moving generator" (Szekely, 1976). Based on these observations as well as on observations of the limb movements of "walking" angler fishes, Edwards (1977) has speculated that a traveling wave mode of progression was primitive for protoamphibia and that a standing wave mode was secondary. One would expect the organization of the fore- and hind-limb motoneuron pools to reflect this historical development. Edwards is currently testing this hypothesis using cinematography, electromyography, and motoneuron labeling.

Be careful, there is something seductive about the study of the neural control of locomotion for morphologists. Sir James Gray began his vertebrate studies by analyzing the kinetics of support and movement and ended his investigations pondering the role of afferents in limb movement. Gambaryan spent years analyzing the musculoskeletal adaptations for the locomotion of ungulates, carnivores, and rodents before working with Shik, Severin, and Orlovsky to analyze electromyographic patterns in the mesencephalic cat. English and Gonyea studied forelimb musculoskeletal adaptations in sea lions and saber-toothed cats, respectively, before moving into the neural control arena. And I studied adaptations of the hind limb of raptorial birds for striking and capturing prey before speculating on the properties of motor units of such a mundane animal as the domestic cat. I admit to you, however, that when I glimpse the rapid flight of a goshawk in pursuit of its prey, or the tilting of a falcon wing high in the sky, my heart quickens with the excitement of witnessing such an exquisite expression of the neural control apparatus. And as I watch in wonder, I think maybe someday we will understand.

Chapter 18

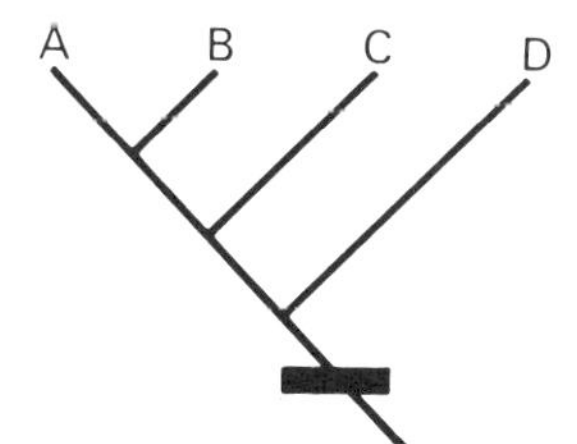

Morphology: Current Approaches and Concepts

Karel F. Liem
David B. Wake

The recent renaissance of morphology, of which this book is one example, does not result only from the use of powerful modern experimental tools and an in-depth understanding of the nature of tissues and cells. Biologists have been drawn to the field because forms and functions are what work in nature. Forms and functions are at the center of biology. It is clear from the preceding chapters that no central principles have as yet emerged and that there are few commonly accepted long-term goals. Many parallel approaches are discernible, but there is little cohesion to the field. Some functional morphologists ask, How does it work? and seek proximate answers. With ever more sophisticated tools and experimental approaches they move from one finding to the next question, inexorably adding to a valuable empirical foundation. Some functional morphologists explicitly omit the historical component of biological systems and assume that an optimal equilibrium exists between the organism and the environment. They contribute by finding general structural principles and systems that are necessary to meet particular functional demands imposed by the environment. Such reductionist approaches are of interest because of the well-defined mechanical theory underlying the interpretation of design. Other morphologists ask, How has it come to work? It is this group that tries to explain form and function in an evolutionary perspective.

The task of the morphologist is to analyze and explain the diversity of structure and function exhibited by organisms. Although morphologists have long used the comparative method, and in fact the comparative method may be morphology's outstanding contribution to science in general, modern morphology also relies heavily on the experimental method. We define functional and evolutionary morphology as that field of biology that studies the evolution of form and function by combining comparative and experimental methods of analysis. Morphologists have analyzed form and function in two ways: equilibrium and historical analysis (Lauder, 1982). Equilibrium analysis is by far the dominant approach to the study of design in functional morphology. Organisms are considered to be in optimal equilibrium with the environment, so present-day environmental correlates of structure are sought. An equilibrium approach may also have a time axis in that structural changes of organisms are described as adaptive modifications to changing environments. Historical analysis, in contrast, is not concerned with the relationship between form and the environment (extrinsic factors), but focuses instead on the evolutionary transformation of intrinsic organizational features (Lauder, 1981). The historical analysis of design is a phylogenetic approach, which reveals the historical pattern by which any combination of structural features was constructed. If a complete explanation is ever to be given for the enormous diversity in form and function of organisms, then a synthesis of both extrinsic environmental influences and intrinsic structural and functional properties must be achieved within a phylogenetic or historical framework.

Extrinsic Determinants of Design

We follow Lauder (1982) in defining *design* as the organization of biological structure in relation to a hypothesized function. Virtually all morphologists accept that evolution has occurred and believe that natural selection has molded the organism and its design (Bock, 1980). There is little doubt that the environment poses at least general constraints on design (Lewontin, 1978). The analysis of extrinsic determinants is essential for understanding the relationships among structure, function, and limits to design.

Structural Analysis A purely structural analysis results in a comprehensive description of the form of structural elements and their relationships to one another, the nature of their building materials, and their distribution in different taxa. In an equilibrium approach explanations for particular structural features are extrinsic in that factors external to the organism (resource abundance, competition, habitat) are thought to be the primary molding forces. During such a structural analysis many workers search for environmental correlates to see whether a particular structural feature, identified in many independent phyletic lines, is always associated with a particular habit or habitat. If so, we may hypothesize that the structure is an adaptation, that is, a feature of an organism that contributes to its ability to survive and reproduce. Thus, an explanation of specific (unique) features is given in relation to certain extrinsic environmental factors. Structural analysis can then proceed by comparing individuals in different genera, families, or orders. Selective forces on form may be hypothesized by studying large-scale differences in form and by proposing environmental factors that could account for them. Analysis at this general level is often assumed to provide insights into selective processes acting at the intraspecific level.

Structural analysis also provides probability statements concerning function and the mechanical groupings of elements. However, the derivation of function from structure is generally invalid (or at least incomplete, or unreliable) because it is based on the fundamentally weak assumption that a given structure is molded by a single, overriding current function. Most morphologists agree that patterns of structure are related not only to the exigencies of the present environment, but also to past sequences of phyletic and environmental change. Structural elements often perform a multiplicity of functions, and each element or composite unit must serve these simultaneously or in sequence.

Finally, structural analysis often generates nested sets of uniquely derived structural features with which organisms can be ordered into clusters, which in turn may be interpreted as a phylogeny, reflecting genealogical relationships. The phylogenetic pattern of relationships and the distribution of structural features specified by accepting any particular phylogeny may then form a basis for interpreting functional patterns. A phylogeny may also provide a very general test of the correlation between structure and environment. Congruence of the pattern of historical change in environment with the pattern predicted from phylogeny would corroborate a proposed relationship between the structural feature and the environment (Lauder, 1981).

Functional Analysis Functions can be analyzed with great precision by elegant experimental methods, as is evident from the preceding chapters. In its simplest form functional analysis can elucidate function and answer the question, How does it work? It has provided us with a basic understanding both of how extrinsic factors limit biological design and of the mechanical principles used in constructing biological systems (see, for example, Wainwright et al., 1975; Wainwright, 1980). Functional analysis is often used as a basis for constructing optimality models, which treat designs at equilibrium with external molding forces. In the construction of optimality models the structural specification is considered to be the paradigm for the proposed function: as noted by Rudwick (1964), the properties of available anatomical materials define the structure that would give optimal efficiency in the performance of a function. Such mechanical optima (paradigms) are held to exist because they provide a competitive, or selective, advantage.

Several morphologists accept the principles of optimality and explain the nature and emergence of a particular design by comparing the actual structural and functional pattern with the optimum expressed in engineering terms. Congruence of the actual design with the optimal model is then considered a proof that the effect of natural selection is an optimization of structure and that phyletic sequences are governed by principles of optimality (see, for example, Dullemeijer, 1974). Of course it is impossible to falsify the optimality hypothesis by the use of the paradigm method since it cannot show that a particular structure had no function or was nonadaptive. Several morphologists (such as Gans, 1966, 1974; Gould, 1970, 1980; Liem, 1980) have pointed out the potential pitfalls of the use of optimization models. Structures often have several functions, and thus do not fit any one paradigm. Because optimality models are made on the assumption that organisms are at an optimal equilibrium with current environmental demands, they fail to explain a design in which intrinsic historical factors have been important determinants of form. It is not always sufficient to view organisms as being in equilibrium with the environment. Thus, in some instances the analysis of present-day environmental factors alone cannot be used to understand functional design, which is often a compromise between current demand and past history.

Functional analysis is a powerful method for determining the functional and mechanical relationships among structures that are integrated into functional components. Structural elements interact with one another by means of all or any combination of their following characteristics: position, shape, structure, size, function, and mechanics. Whereas interactions in regard to position, shape, structure, and size can be determined by comparative structural and ontogenetic analyses, functional and mechanical linkages or couplings can only be analyzed by experimental methods. Thus, structural and functional analysis can lead to an understanding of the mutual influences that structural elements exert upon one another.

The results of such an analysis are diagrammed in Figure 18-1, which illustrates the concept of a network of interacting constraints with an example of the jaw protrusion apparatus in a generalized percoid fish (Liem, 1980). The structural elements are depicted in rectangles (bony elements) and parallelograms (muscular elements), which are connected by arrows indicating influence. The direction of the arrows, and thus of the influence, is found by logical derivation, constructional analysis, and functional analysis. Outgoing arrows represent the degree of freedom and the ingoing ones the degree of dependence. Networks of interacting constraints are constructed in relation to a function, which is indicated by a three-dimensional block. Upper jaw protrusion is the realized activity (*r*). An arrow is pointed from the maxilla to the premaxilla expressing the dominance of the maxilla over the premaxilla. The premaxilla is influenced in respect to three characteristics: structure and shape

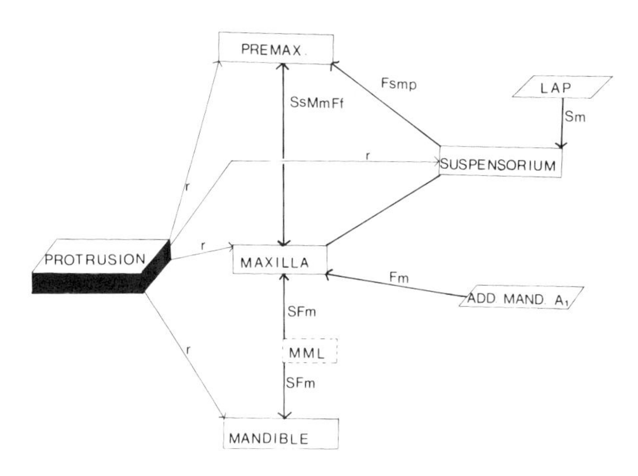

Figure 18-1 Pattern of mutual influences among structural elements involved in upper jaw protrusion in a generalized percoid fish. Solid-line rectangles stand for bony elements; broken-line rectangles stand for ligamentous elements; and parallelograms stand for muscles. Heavy arrows run from the influencing to the influenced element. The light arrows *(r)* indicate realization. The capital letters indicate the characteristic that is influenced, the small letters the influencing characteristic. *f* = function; *m* = mechanics; *p* = position; *s* = shape and structure; *r* = relation between form and function; *ADD. MAND.* A_1 = adductor mandibulae part A_1; *LAP* = levator arcus palatini; *PREMAX.* = premaxilla. (From Liem, 1980.)

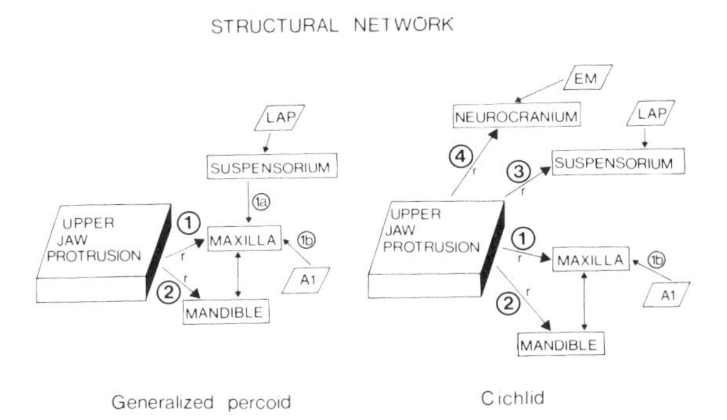

Figure 18-2 Structural networks in the head of a generalized percoid and cichlid fish. In generalized percoids two mechanical pathways mediate protrusion: mandibular depression and maxillary rotation (1, 2). Suspensorial movement influences the upper jaw by an intermediary articulation with the maxilla (pathway 1*a*). In cichlids, suspensorial movement can effect jaw protrusion independently of the maxilla, so the suspensorium has been mechanically decoupled from the maxilla (3). A consequence of the decoupling is the increase in number of mechanical pathways controlling upper jaw protrusion (1–4). A_1 = part A_1 of adductor mandibulae; *EM* = epaxial muscles; *LAP* = levator arcus palatini; *r* = realization of the function of upper jaw protrusion by the indicated pathway; 1*b* = mandibular depression pathway. (From Lauder, 1981.)

(*S*), mechanics (*M*), and function (*F*). The influencing characteristics of the maxilla are shape and structure (*s*), mechanics (*m*), function (*f*), and position (*p*). The mutual influences between the premaxilla and maxilla are determined by their anatomical relations and mechanical linkages. The anterodorsal component of the suspensorium (palatopterygoid arch) can press on the head of the maxilla, which in turn causes protrusion of the premaxilla. Thus, the functional characteristic (*F*) of the premaxilla is influenced by the structure (*s*), mechanics (*m*), and position (*p*) of the suspensorium. To summarize, in generalized percoids the network of interacting constraints underlying jaw protrusion is characterized by multiple constraints of the maxilla and suspensorium acting on the premaxilla. This general methodology has been used successfully for different groups of organisms (Dullemeijer, 1980). Lauder (1981) offers an alternative method for the construction of networks of interacting constraints (Fig. 18-2). He restricts the network to structural and mechanical constraints, thereby emphasizing biomechanical couplings and mechanical pathways by which jaw protrusion is effected.

Functional and structural analyses allow the identification of a structural and functional network. A network of interacting constraints allows us to understand the intrinsic organizational properties of various components of the organism. The occurrence of biological constraint has been ascribed an increasingly important role in evolutionary biology (Gans, 1966; Seilacher, 1973; Riedl, 1978; Gould and Lewontin, 1979; Gould, 1980; Liem, 1980; Lauder, 1981; Ostert and Alberch, 1982; Reif, 1982), yet it has remained an abstract concept that has not been described either qualitatively or quantitatively. By constructing networks of interacting constraints for various species, morphologists are in a very good position to offer a qualitative description of the concept of biological constraint, thereby providing a useful and concrete methodology for analyzing its role in evolutionary biology. Morphologists may reveal that a variety of constraints gives direction, or at least bias, to the evolutionary change of each system studied (Wake, 1982).

Experimental Ecological Analysis It is essential to include an in-depth experimental study of the ecology of an organism in order to explain the extrinsic determinants of design. Ideally, the performance of the particular function under study should be measured. Advocates of the supremacy of equilibrium analysis will search in the field for perfect or optimal correlations between "specialized" design and "specialized" performance. They search for proximate answers and have added greatly to our rapidly growing empirical foundation illustrating the perfect fit between functional design and optimal performance in the wild. However, perfect fit has been demonstrated neither experimentally nor quantitatively. Instead, the perfect fit has been speculatively invoked by ecologists with insufficient data on design and by morphologists with inadequate experimental ecological evidence. As a result, design is provided with an adaptive explanation by finding the problem to which it is a solution! Such a tautological approach presupposes constancy of both design and environment. Thus, an originally evolutionary argument is turned into a remarkably pre-Darwinian static view of the harmony of nature.

Experimental ecological analysis must take into account that the external environment is constantly changing with respect to variable organisms. Thus, the dynamic relationship between design and environment enables the organism to maintain its state of relative adaptation rather than to optimize it. The evolutionary trajectory of vertebrate species is largely determined by occasional "bottlenecks" of intense selection during a small portion of their history (Wiens, 1977; Boag and Grant, 1981). For example, Boag and Grant demonstrated the highest level of natural selection ever measured in vertebrates. During a period of intense drought, large-beaked members of a species of Darwin's finches had a strong advantage over others because the design of their trophic apparatus was well adapted to open the large seeds that came to predominate as the drought progressed. Other species could not exploit the large seeds. Whereas morphological literature was cited in passing, the only morphological data presented were simple multivariate statistics of external bill dimensions. There is a need for morphologists to delve into the functional significance of the morphological differences among individuals and to apply the results to the varying environments and adaptive radiation of these finches.

Recently the importance of long-term ecological analysis, conducted under varying conditions in different geographical areas, has been shown by McKaye and Marsh (1983) and Ribbink et al. (1983) for cichlid fishes. Functionally herbivorous species exhibit considerable intraspecific variation under controlled laboratory conditions (Liem, 1980). Direct laboratory investigation revealed a repertoire of as many as eight modes of feeding. In the field these species are not only remarkable specialists in a narrow sense (that is, under particular conditions in a restricted study area), but they are also jacks-of-all-trades when other highly nutritive food resources become abundant (McKaye and Marsh, 1983; Ribbink, 1983). Thus, intraspecific feeding repertoires (intraspecific variation) as experimentally demonstrated in the laboratory have been confirmed in long-term field studies focusing on variability in various environments. The capacity of the specialized herbivorous cichlid species to exploit available resources (such as algae) when rich ones are scarce, and to switch to the nutritionally rich ones (invertebrates and their larvae) when they become abundant, ensures the survival of the species under varying environmental conditions. The important concept to emerge from these combined laboratory and field studies is that the extremely specialized design of the feeding apparatus of many herbivorous cichlid fishes does not necessarily increase efficiency of feeding upon preferred food; rather, it enhances the use of a secondary, less preferred food that serves as a refugium during "ecological crunches" (Wiens, 1977). Liem and Kaufman (1984) have studied a highly polymorphic cichlid species by measuring the effects of morphological and functional variation on performance under controlled laboratory conditions and by determining the effects of performance on fitness in the field. They have shown that a different set of external factors operate on trophic design when food is varied and abundant than when food is scarce. Even though a combined experimental laboratory and field research program is difficult and time consuming, we must have such studies to fully understand the extrinsic determinants of design.

Arnold (1983) has attempted to show how one might demonstrate adaptive significance of differing morphologies within a species. He was concerned with variation in dimensions of bones forming the jaws of snakes relative to the size and other characteristics of their prey. He broke the task into two parts: measurement of the effect of morphological variation on performance and measurement of the effects of differences in performance on fitness. Arnold outlined a specific program of research involving both laboratory and field components.

Intrinsic Determinants of Design

As we have seen above, optimal models explicitly neglect the historical component of biological systems and treat the organism-environment relationship as though the system were in maximal equilibrium. Yet the problem of biological design is fundamentally a historical one (Lauder, 1981, 1982a,b). The past history of an organism limits and determines future directions of structural and functional change. Most studies have paid only lip service to the existence of historical constraints. Few attempts have been made to provide a sound methodology for understanding the influence of intrinsic (historical) factors on biological design. The analysis of intrinsic determinants of form complements equilibrium analyses by using a different explanatory framework and by focusing on a distinct level of morphological novelty. Two assumptions underlie historical analysis of form: first, organisms possess features that may be ordered into nested sets (a cladogram or phylogenetic tree); and second, this nested pattern is a reflection of a historical process of descent with modification.

Phylogenetic Analysis In general, similarities of specialized characters among taxa are found, and these shared derived characters are ordered into the most parsimonious nested set. Thus a genealogical hypothesis is made (see, for example, Eldredge and Cracraft, 1980). Hypotheses regarding character homology can then be derived from the cladogram and are considered to be synapomorphies (Fig. 18-3). Characters that are not congruent with the preferred hypothesis are convergences. It is only by having a corroborated hypothesis of phylogeny that testable generalizations about historical changes of form can be made (Lauder, 1981, 1982a,b).

Historical Analysis of Design Networks of interacting constraints derived from equilibrium analysis can now be studied in a historical context. The historical analysis deals with the following questions: What happens to structurally adjacent elements when one of the components changes functionally and/or structurally? Can a level of functional integration be recognized that correlates with the rigid maintenance of character-complex identity despite the change of adjacent elements? Is there a correlation between diversity in form and certain general features of the network of interacting constraints?

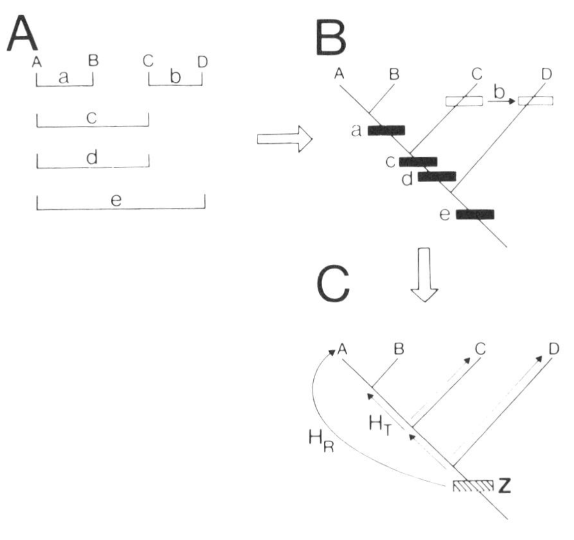

Figure 18-3 Method for testing historical hypotheses about the transformation of design. Nested sets of similarities (A) are ordered into the most parsimonious cladogram (B). Congruent characters *c* and *d* are recognized as homologies for the taxa they define, whereas the incongruent character *b* is a convergence. The corroborated cladogram is a hypothesis of the phylogenetic relationships (C) of the terminal taxa *(A-D)*. Transformational hypotheses (H_T in C) concern the historical sequence of structural and functional change. Relational hypotheses (H_R in C) involve correlations between aspects of the primitive network (*Z* in C) and morphological diversity (both within and between terminal taxa). Transformational hypotheses concern historical pathways of change in design, whereas relational hypotheses concern historical consequences. (From Lauder, 1981.)

Lauder (1981) offers an example of a historical analysis. In the teleostean fish family Cichlidae (Fig. 18-2) an increase in the number of kinematic pathways governing movements of jaws is thought to relate to an increase in diversity of jaw form (Liem, 1980). Thus, in the primitively coupled linkage system the number of mechanical pathways controlling jaw movements is limited, whereas decoupled systems have increased the number of mechanical pathways. Taxa with coupled (constrained) systems would have a lower morphological diversity than those with decoupled systems. This circumstance provides an interpretation for the correlation (H_R in Fig. 18-3) between decoupled elements of the primitive network (Z in Fig. 18-3) and morphological diversity in terminal taxa. The hypothesis is that there is an inverse correlation between morphological diversity in a lineage and the number of constraints in the network (Liem, 1980). This hypothesis may be refuted if the two related lineages have different networks of constraints but similar patterns of morphological diversity. Tests of this hypothesis should involve comparisons of the networks among related monophyletic lineages and their respective outgroups. Thus, the independently derived genealogy furnishes a powerful tool for the analysis of form and function, because the historical track (Fig. 18-3, H_T) taken by networks in organisms can be used for testing hypotheses of constraint, canalization, or versatility (Fig. 18-3, H_R).

A similar hypothesis by Vermeij (1973) also relates morphological diversity to complexity of design and can also be tested by a historical analysis. Complexity of design is defined as the number of independent parameters controlling form; the greater the number of controlling parameters (or descriptors of form), the more complex the design. The analysis should show that primitive members of morphologically diverse monophyletic lineages possess a greater complexity than closely related lineages whose primitive members have a less complex design. However, if these related lineages show similar patterns of morphological diversity to the taxa with a primitively more complex design, the hypothesis is refuted. In order to determine the generality of the phenomenon, the tests should be conducted on a wide range of taxa differing in general body plan. This historical analysis can show how features of the network of interacting constraints canalize form and restrict or increase the potential structural and functional diversity of descendant taxa. The possibilities of optimization, the sensitivity of the phenotypic traits to environmental factors, and the correlated evolutionary responses of different traits were often related to this network of interacting constraints.

Analysis of Biological Repetition Equilibrium analysis often reveals the existence of redundant or repetitive functional systems. For example, in a cichlid fish upper jaw protrusion can be accomplished in at least four ways (Fig. 18-2): lifting the neurocranium; abducting the suspensorium; lowering of the mandible; or swinging of the maxilla. It is possible that initially redundant functional systems have played an important role in the emergence of structural and functional diversity within lineages by allowing new functions to appear without disrupting the primitive function, which is retained by the remaining redundant systems. Likewise, repetition of structural elements allows independent specialization of some segments, whereas other segments retain the primitive structural and functional associations (Lauder, 1981). The consequence of redundant functions and repeated structural elements for the subsequent morphological history of a taxon can also be analyzed with a historical hypothesis as a basis. We can test the association between the presence of primitively duplicated functions or structures and the pattern of structural or functional diversity in the terminal taxa of the clade. Repeated elements and redundant functional systems may provide the organism with constructional flexibility by allowing one component of a system to be modified without disrupting other elements (Liem, 1980; Lauder, 1981). Accordingly, the historical effect of structural repetition and functional redundancy is structural and functional diversity.

Ontogenetic Analysis Attempts to learn something from development about the evolution of morphology predate Darwin (Gould, 1977). The parallelism between ontogeny and phylogeny has been a major theme in evolutionary investigations, but the entire field underwent an eclipse during the period of the great evolutionary synthesis of the 1930s to the 1960s (Provine and Mayr, 1980). Recently, there has been a rebirth of interest in ontogenetic analysis in studies of morphological evolution. There have been major international conferences devoted to the topic of development and evolution (Bonner, 1982; Goodwin, Holder, and Whyle, 1983), as well as a great increase in publications relating to the topic.

Evolution tends to follow certain pathways, to "run in grooves," or to follow avenues of least resistance. Not all possible morphologies exist; there are "gaps" in morphological space and not all conceivable forms exist now, or can exist. This truism has led to the conception of constraints on evolution (see above). The major constraints may be developmental in nature. For example, the vagaries of history have played an important role in evolutionary diversification. The origin of the neural crest developmental system in vertebrates can be viewed as a historical "accident" at one level. But at another it is an "opportunity," and at yet another it is a developmental constraint—something that sets bounds on the possible developmental states and their morphological expression in the lineage.

As far as morphological change is concerned, evolution acts by altering development. The particular aspect of development that concerns us is epigenetics—the control of gene expression by the microenvironments encountered by cells during development, or, more generally, the mechanisms by which genes express their phenotypic effects (Hall, 1983). Vertebrate development is hierarchical. A highly integrated, but largely particulate, genome both produces proteins and controls the pattern of their production. The proteins and enzymes produced undergo a self-assembly so that a definable structure arises, following physiochemical laws. The resultant cells differentiate to form epithelial or mesenchymal tissues. This minimal differentiation has profound effects: each cell type can do only certain things, and the different cells interact in particular ways to produce form. There is a limited array of autonomous cell activities, such as adhesion, protrusion, contraction, locomotion, division, assembly of cytoskeleton, transport, communication, and, of course, death. But cells also integrate and coordinate; they have a social behavior (such as miosis); they are responsive to gradients (such as in diffusion). And as differentiated cell types they display characteristic properties—mechanical properties, metabolic coupling, local autonomy, and so on (see Wessells, 1982, for elaboration of these points). Cells form tissues, and the tissues interact in complex and often very specific ways, as in the extraordinarily varied manner of the neural crest system (Gans and Northcutt, 1983; Maderson, 1983). Epigenetic controls such as tissue interactions of the neural crest must have arisen very early in the evolutionary history of vertebrates, and they have been preserved with amazing fidelity. In fact, it may be the characteristics of the epigenetic system of organisms that are responsible for one of the persistent problems in evolutionary biology—the unity of type. Simply speaking, this is the concept of a "Bauplan," or a common theme that characterizes the developmental and adult morphology of entire taxa. An illustrative example is the relatively stable general morphological theme of frogs, which despite aquatic, arboreal, burrowing, and other specialized forms, is amazingly conservative, at least as compared with mammals (Cherry et al., 1978, 1982).

The inductive signals and the competence of cells to respond to signals of different kinds seem to have remained stable over vast periods of evolutionary time, but yet, great diversity of form can come about by subtle changes in elements of epigenetic control (for example, position of epithelial and mesenchymal cells in early development, migration rates of neural crest cells, timing of cell migration, onset and termination of induction, length of time inducers retain inductive activity; Hall, 1983). The pattern of feedback between genetic and epigenetic levels is illustrated in Figure 18-4. Oster and Alberch (1982:453) describe the pattern as follows: "(1) Genes code for informational and structural macromolecules. (2) These molecules then self-assemble to form characteristic cell properties: specific adhesion sites, cytoskeletal architecture, etc. (3) Early in embryogenesis blastomeres differentiate into two broad classes of cells: (a) epithelia, characterized by cuboidal cells which are arrayed in layers and with many intercellular junctions of various types; (b) mesenchymal cells characterized by amorphous morphology and generally motile. (4) Because of its macromolecular architecture and differential gene activity, each cell type is able to carry out characteristic cellular processes: e.g., epithelia secrete a basal lamina, while mesenchymal cells secrete extracellular matrix material—which can "contact guide" other motile cells. Both types can execute characteristic cell shape changes via modulation of their cytoskeleton; each can secrete and receive signal and recogni-

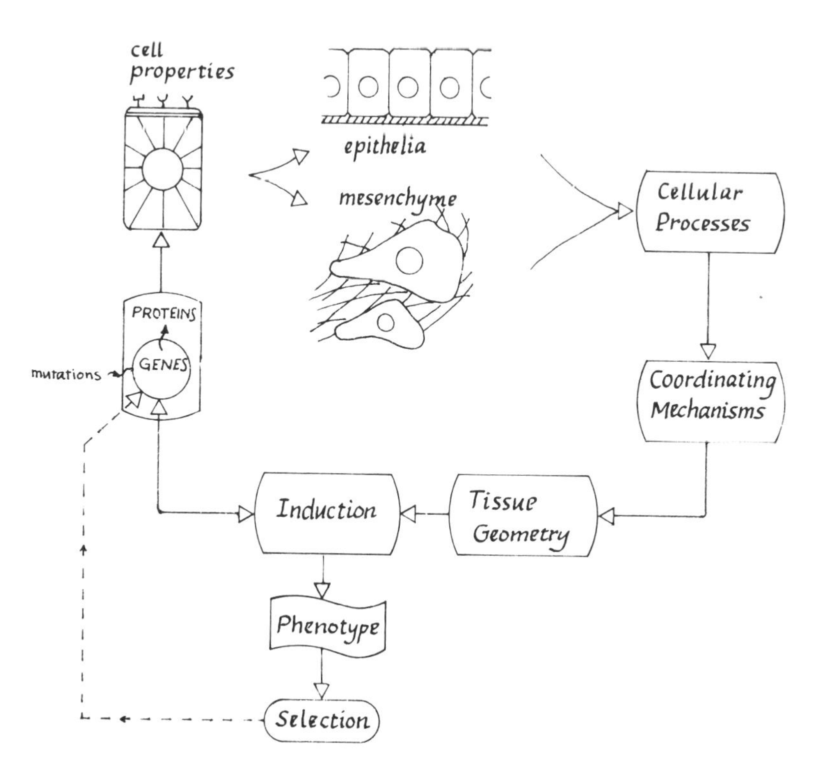

Figure 18-4 Feedback loop between genetic and epigenetic levels, which characterize all developmental programs. (From Oster and Alberch, 1982.)

tion molecules, etc. (5) The various cellular processes are coordinated by various interaction mechanisms so as to orchestrate a coherent morphogenetic unfolding. These include chemical and mechanical interactions between cells, cellular clocks, contact inhibition and guidance, etc. (6) The outcome of (4) and (5) is a particular tissue geometry characterized by specific juxtapositions of epithelia and mesenchymal cell populations. (7) When cell populations with different histories are juxtaposed, the phenomena of tissue induction come into play. That is, intercellular communication between different tissue types initiates different patterns of genome activity leading to new cell types. Thus the feedback loop between epigenetic processes and gene activity is closed, leading to further rounds of morphogenesis which culminate in the formation of a particular tissue or organ phenotype. (8) Depending on the functional adaptation of this design, selection (9) acts in the usual way to amplify those gene combinations which produced the more adaptive phenotype. Thus there is a higher level feedback on gene frequency which acts at the ecological level; this is the province of population genetics, which deals essentially with steps (1), (8) and (9), relegating steps (2)–(7) to phenomenologically fixed parameters."

A disturbing feature of epigenetic controls on morphological diversification is that there are seemingly more than adequate amounts of genetic variation and possibilities for variation. The genetic variation may be disturbingly nonspecific so that any of a whole class of genetic changes that affect cell surface phenomena might have essentially identical phenotypic outcomes. For example, at least seven so-called point mutations result in limblessness in chickens. As expected, all of them have a host of pleiotropic effects, rendering the phenotypes lethal and demonstrating that some fundamental cellular character has been modified. They are not "limbless" mutants, but modifications of the ways in which cells communicate or interact with one another to produce, among an array of other phenomena, chicks that lack limbs. Perhaps they all indirectly affect the number of cells in the embryonic apical epidermal ridge of the limb bud, but the ways in which the number is changed might be diverse. This example points out that some basic parameter, such as rate of cell division in one part of an embryo relative to another, can be amplified during development so as to have a major effect on phenotype.

The developmental control parameters that have most commonly been implicated as having significant evolutionary effects are those having to do with rates and timing. Since the publication of Gould's (1977) treatise on ontogeny and phylogeny there has been renewed interest in heterochrony, or shifts in developmental timing during phylogeny, as a source of morphological novelties. Heterochrony is appealing because it provides a relatively simple mechanism whereby major morphological change can be accomplished. The most celebrated, and the oldest, case is that of the Axolotl, which is a permanently larval but reproductively mature salamander of the genus *Ambystoma* occurring on the Mexican Plateau. But heterochrony is a phenomenon of potentially broad relevance in evolutionary morphology, and it has been implicated in many groups, from the evolution of arboreal salamanders in the lowland tropics (genus *Bolitoglossa;* Wake, 1966; Wake and Brame, 1969; Alberch and Alberch, 1981) to the evolution of humans in relation to great apes (Gould, 1977).

In recent years we have come to have a deep understanding of the fundamental cellular and tissue relationships in vertebrate development. Experiments involving recombinations of different-aged epithelia and mesenchyme in tissue culture (see Hall, 1983; Maderson, 1983) have been especially instructive. For example, the early induction of mesenchyme for chondrogenesis involves position and timing for successful completion. The site of the tissue is under epigenetic control—it is fixed by the stationary epithelium and not by the migrating mesenchyme. The timing of the interactions is also under epithelial control. Timing, position, and degree of contact all play important roles in mandibular development, and Hall (1983) has suggested that the search for regulation of gene expression in development and evolution should involve investigations of epigenetic associations and how they activate the genomes of adjacent cells. Perhaps the most interesting of the several recent experiments is the demonstration by Kollar and Fisher (1980) that the recombination of mouse dental mesenchyme with chick mandibular arch epithelium elicits production of enamel by the chick cells, even though birds have not produced enamel for nearly a hundred million years!

A particularly instructive demonstration relating to both developmental constraints and heterochrony is the array of experiments by Alberch and Gale (1983). They examined the rate of cell proliferation and its effects on production of limb morphology in amphibians. Frogs and salamanders are typically five-toed, but in both groups there are unrelated lineages that have only four toes. Four-toed frogs have lost the first toe, but four-toed salamanders have always lost the fifth. Embryos in the limb bud stage of normally five-toed species of the frog genus *Xenopus* and of the salamander genus *Ambystoma* were used in the experiments. Limb buds on one side of the embryo were bathed in colchicine, a mitotic inhibitor, while those on the other were used as controls. In a significant percentage of cases the treated limb buds developed but four toes, always with the predicted arrangement—the first missing in frogs and the fifth in salamanders. Perhaps the loss of toes phylogenetically results simply from the reduction in rate of cell division at a crucial time in development, but at different sites in the two groups of amphibians.

In 1979 Alberch et al. formalized the concept of heterochrony as a mechanism for morphological evolution. A primary focus was on four parameters of development: onset of growth of the feature under consideration, offset of growth of that feature, a growth rate, and an initial size. The formulation was sufficiently general so that growth could follow any of a number of specific growth laws (isometry, positive or negative allometry, Gompertz function, and so on; see Chapter 2).

In order to understand how the above parameters relate to morphological evolution one must return to the nineteenth century and recall the fundamental contribution of two great biologists, von Baer and Haeckel. Von Baer is responsible for the seemingly simple but profound generalization that during ontogeny organisms pass from states of general similarity to specialization. Thus, early stages in the ontogeny of a group (for instance, the vertebrates) are more similar than are later stages. Haeckel's famous aphorism—ontogeny recapitulates phylogeny—is related in a direct but somewhat convoluted way to von Baer's generalization (for an extended account, see Gould, 1977). For many years Haeckel's views dominated biology, but they simply died out in the twentieth century. It became obvious that he was not entirely right; but, he was about half right! Haeckel argued that during their ontogeny organisms pass through morphological stages equivalent in a general way to the adult stages of forms in their ancestry. In his view evolution was the result of terminal addition or extensions of existing conditions. Haeckel realized that there would be insufficient time for organisms to preserve perfectly all their phylogeny in their ontogeny. Accordingly, he postulated a condensation of the ontogeny, which would permit at least a general repetition of the phylogenetic history of the lineage. Today we realize that Haeckel badly overstated the matter, but we also see a general kind of parallel between ontogeny and phylogeny.

In order to analyze ontogenetic-phylogenetic interactions, a well-developed phylogenetic hypothesis is essential (Fink, 1982; Lauder, 1982a). One must recognize that in ontogenetic analyses all is relative. One asks if a descendant, relative to an ancestor, demonstrates recapitulation and terminal addition, or if some other condition holds. There is much controversy concerning the ideal way in which to formulate phylogenetic hypotheses, but it is becoming generally recognized that morphological features are well suited for cladistic analysis (Hennig, 1966). An account of the application of these principles is found in Eldredge and Cracraft (1980) and Wiley (1981), and examples in vertebrate morphology are provided by Lauder and Liem (1983).

We now know that phylogeny may recapitulate, but that it also may follow a reverse direction. Alberch et al. (1979) used the term *peramorphosis* ("shapes beyond") for morphological expressions of recapitulation, such as when a derived species simply adds one more step onto the normal ontogenetic progression of its ancestral or equivalent species. The contrasting morphological expression, in essence a reversal of recapitulation, is *paedomorphosis* ("juvenile shapes"), in which the descendant morphology is more similar to juvenile stages of the ancestor than to adult stages.

Now the control parameters of development mentioned earlier gain special meaning. If one considers all else as being constant and changes the parameters one at a time, one can make predictions concerning direction of morphological evolution. For example, a given peramorphic change might come about as a result of a simple increase in the rate of development (acceleration), or it might result from development at the same rate but extended in time (having a later offset, or hypermorphosis). Paedomorphic change might result from a decrease in the rate of development (neoteny) or from a general truncation of development (having an early offset, or progenesis). There are also other possibilities, but the point is that one can now generate testable hypotheses concerning the role of ontogenetic change in evolution. Further, an evolutionary change in ontogeny might affect only one part of the organism (dissociation), or it might affect the entire organism (as is often the case in progenesis, where descendants are frequently dwarfed; see Alberch and Alberch, 1981). Crucial to the analytical phase is the recognition of specific ontogenetic trajectories followed during the development of either whole organisms or organs. It is the pattern of the ontogenetic trajectory that is modified to give morphological change. To understand ontogenetic trajectories one must understand epigenetics and developmental dynamics. One of the most intriguing aspects involves the complex nonlinear interactions during development. Small changes in developmental control parameters can be amplified during ontogeny to give dramatically different morphological results. It is this final point that has led to the recent excitement in analyzing ontogeny in relation to evolutionary morphology and phylogeny.

Conclusions

The analysis of environmental factors may reveal the limits imposed by the environment on biological design, but the intrinsic phylogenetic and ontogenetic components of design may limit the directions of structural and functional evolution (see, for example, Hickman, 1980; Reif, 1982; Wake, 1982; Wake et al., 1983). Variation in morphology, as manifested by different patterns of development among individuals of a species, is strongly directional. Directional changes emerge as a result of ontogenetic patterns, networks of interacting constraints, complexity of design, and structural and functional repetitions. Directional evolutionary change in structure and function may be due to extrinsic determinants as well as to inherited constraints of body plan. Thus, when a potential morphospace remains unoccupied, the explanation may not lie with the absence of appropriate selection pressures but rather with historical and ontogenetic constraints on possibilities for morphological change.

We have discussed a testable research program to understand the factors controlling biological design. We agree with Lauder (1982a,b) that in order to understand morphology, a synthesis of both extrinsic and intrinsic determinants of design must be achieved within a phylogenetic (historical) framework. In order to achieve such a goal, it will be increasingly important for morphologists to work as systematists, evolutionary biologists, biomechanical engineers, developmental biologists, population biologists, experimental and community ecologists, mathematical biologists, and even molecular biologists. "By incorporating experimental approaches and solid theory, and by making the comparative method ever more sophisticated, morphologists will continue to occupy a center stage in modern biology" (Wake, 1982:617).

References

1. Functional Adaptation in Skeletal Structures

Alexander, R. McN. 1981. Factors of safety in the structure of animals. *Sci. Prog.* 67:109–130.

Alexander, R. McN., A. S. Jayes, G. M. O. Maloiy, and E. M. Wathuta. 1979. Allometry of the limb bones of mammals from shrews to elephant. *J. Zool., Lond.* 189:305–314.

Bacon, G. E., P. J. Bacon, and R. K. Griffiths. 1980. Orientation of apatite crystals in relation to muscle attachment in the mandible. *J. Biomech.* 13:725–729.

Biewener, A. A. 1982. Bone strength in small mammals and bipedal birds. *J. Exp. Biol.* 98:289–301.

Biewener, A. A., J. Thomason, A. Goodship, and L. E. Lanyon. 1983. Bone stress in the horse forelimb during locomotion at different gaits: a comparison of two experimental methods. *J. Biomech.* 16:565–576.

Bouvier, M., and W. L. Hylander. 1981a. Effect of bone strain on cortical bone structure in macaques. *J. Morph.* 167:1–12.

——— 1981b. The relationship between split line orientation and in vivo bone strain in galago and macaque mandibles. *Am. J. Phys. Anthrop.* 56:147–156.

Burstein, A. H., J. P. Currey, V. H. Frankel, and D. T. Reilly. 1972. The ultimate properties of bone tissue: the effects of yielding. *J. Biomech.* 5:35–44.

Carter, D. R., W. C. Hayes, and D. J. Schuman. 1976. Fatigue life of compact bone. II. Effects of microstructure and density. *J. Biomech.* 9:211–218.

Carter, D. R., W. E. Caler, D. M. Spengler, and V. H. Frankel. 1981a. Cortical bone fatigue: the effect of strain range, stress range, and elastic modulus. *Trans. Orth. Res. Soc.* 27:44.

Carter, D. R., W. H. Harris, R. Vasu, and W. E. Caler. 1981b. The mechanical and biological response of cortical bone to in vivo strain histories. In *Mechanical properties of bone,* ed. S. Cowin. *ASME Publ. AMD* 45:81–92.

Churches, A. E., and C. R. Howlett. 1981. The response of mature cortical bone to controlled time varying loading. In *Mechanical properties of bone,* ed. S. Cowin. *ASME Publ. AMD* 45:69–80.

Churches, A. E., C. R. Howlett, and G. W. Ward. 1980. Bone reaction to surgical drilling and pinning. *J. Biomech.* 13:203–209.

Churches, A. E., C. R. Howlett, K. J. Waldron, and G. W. Ward. 1979. The response of living bone to controlled time varying loading: method and preliminary results. *J. Biomech.* 12:35–45.

Currey, J. D. 1959. Differences in tensile strength of bone of different histological types. *J. Anat.* 93:87–95.

——— 1969. The mechanical consequence of variation in the mineral content of bone. *J. Biomech.* 2:1–11.

——— 1975. The effect of strain rate, reconstruction, and mineral content on some mechanical properties of bovine bone. *J. Biomech.* 8:81–86.

——— 1979. Mechanical properties of bone tissue with greatly differing functions. *J. Biomech.* 12:313–319.

——— 1982. "Osteons" in biomechanical literature. *J. Biomech.* 15:717–718.

Donaldson, C. L., S. B. Hulley, J. M. Vogel, R. S. Hattner, J. H. Bayers, and D. E. McMillan. 1970. Effect of prolonged bed rest on bone mineral. *Metabolism* 19:1071–1084.

Engesaeter, L. B., A. Ekeland, and N. Langeland. 1979. Methods for testing the mechanical properties of the rat femur. *Acta Orth. Scand.* 49:512–518.

Frost, H. M. 1973. *Bone modelling and skeletal modelling*

errors. Orthopaedic Lectures, 4. Springfield, Ill.: Charles C Thomas.

Goodship, A. E., L. E. Lanyon, and H. MacFie. 1979. Functional adaptation of bone to increased stress. *J. Bone Jt. Surg.* 61A:539–546.

Hayes, W. C., and B. Snyder. 1981. Toward a quantitative formulation of Wolff's law in trabecular bone. In *Mechanical properties of bone,* ed. S. Cowin. *ASME Publ. AMD* 45:43–68.

Hert, J. 1969. Acceleration of the growth after decrease of load on epiphyseal plates by means of spring distractors. *Folia Morph.* (Prague) 17:194–204.

Hert, J., M. Liskova, and J. L. Landa. 1971. Reaction of bone to mechanical stimuli. I. Continuous and intermittent loading of tibia in rabbit. *Folia Morph.* (Prague) 19:290–317.

Hert, J., A. Skelenska, and M. Liskova. 1971. Reaction of bone to mechanical stimuli. V. Effect of intermittent stress on the rabbit tibia after resection of the peripheral nerves. *Folia Morph.* (Prague) 19:378–387.

Hert, J., P. Kucera, M. Vavra, and V. Volenik. 1965. Comparison of the mechanical properties of both primary and Haversian bone tissue. *Acta Anat.* 61:412–423.

Hillman, J. R., R. W. Davies, and Y. Z. Abdelbaki. 1973. Cyclic bone remodelling in deer. *Calcified Tissue Res.* 12:323–330.

Holmdahl, D. E., and B. E. Ingelmark. 1948. Der Bau des Gelenkknorpels unter verschiedenen funktionellen Verhaltnissen. Acta Anat. 6, fasc. 4:310–375.

Howell, J. A. 1917. An experimental study of the effect of stress and strain on bone development. *Anat. Record* 13:233–252.

Jaworski, Z. F. G. 1981. Physiology and pathology of bone remodeling. *Orth. clinics N. A.* 12:485–512.

Jones, H. H., J. D. Priest, W. C. Hayes, C. C. Tichenor, and A. A. Nagel. 1977. Humeral hypertrophy in response to exercise. *J. Bone Jt. Surg.* 59A:204–208.

King, A. I., and F. G. Evans. 1967. Analysis of fatigue strength of human compact bone by the Weibull method. In Intl. Conf. Med. and Biol. Engr., 7th, Stockholm. Digest, ed. B. Jacobson. Stockholm: Almquist and Wiksell.

Koch, J. C. 1917. The laws of bone architecture. *Am. J. Anat.,* 21:177–298.

Lafferty, J. F., and P. U. Raju. 1979. The influence of stress frequency on the fatigue strength of cortical bone. *J. Biomed. Engr.* 101:1120–31.

Lanyon, L. E. 1973. Analysis of surface bone strain in the calcaneus of sheep during normal locomotion. *J. Biomech.* 6:41–49.

——— 1974. Experimental support for the trajectoral theory of bone structure. *J. Bone Jt. Surg.* 56B:160–166.

——— 1980. The influence of function on the development of bone curvature: an experimental study on the rat tibia. *J. Zool., Lond.* 192:457–466.

Lanyon, L. E., and D. C. Baggott. 1976. Mechanical function as an influence on the structure and form of bone. *J. Bone Jt. Surg.* 58B:436–443.

Lanyon, L. E., and S. Bourne. 1979. The influences of mechanical function on the development and remodelling of the tibia. *J. Bone Jt. Surg.* 61A:539–546.

Lanyon, L. E., P. T. Magee, and D. G. Baggott. 1979. The relationship of functional stress and strain to the processes of bone remodelling. *J. Biomech.* 12:593–600.

Lanyon, L. E., A. E. Goodship, C. Pye, and H. McFie. 1982. Mechanically adaptive bone remodelling: a quantitative study on functional adaptation in the radius following ulna osteotomy in sheep. *J. Biomech.* 15:141–154.

Lanyon, L. E., W. G. J. Hampson, A. G. Goodship, and J. S. Shah. 1975. Bone deformation recorded in vivo from strain gauges attached to the human tibial shaft. *Acta Orth. Scand.* 46:256–268.

Lincoln, G. A. 1975. An effect of the epididymis on the growth of antlers of castrated red deer. *J. Repro. Fertil.* 42:159–161.

Mazess, R. B. 1982. On aging bone loss. *Clin. Orth. and Rel. Res.* 165:239–252.

Montoye, H. J., E. L. Smith, D. F. Fardon, and E. I. Howley. 1980. Bone mineral in senior tennis players. *Scand. J. Sp. Sci.* 2:26–32.

Morey, E. R., and D. J. Baylink. 1978. Inhibition of bone formation during space flight. *Science* 201:1138–41.

Murray, P. D. F. 1936. *Bones: a study of the development and structure of the vertebrate skeleton.* London: Cambridge University Press.

O'Connor, J. A., A. G. Goodship, C. T. Rubin, and L. E. Lanyon. 1981. The effect of externally applied loads on bone remodelling in the radius of the sheep. In *Mechanical factors and the skeleton,* ed. I. A. F. Stokes. London: John Libbey.

Oxnard, C. E. 1972. Tensile forces in skeletal structures. *J. Morph.* 134:425–436.

Radin, E. L., S. Simon, R. M. Rose, and I. L. Paul. 1979. *Practical biomechanics for the orthopaedic surgeon.* New York: Wiley.

Ralis, Z. A., H. M. Ralis, M. Randall, G. Watkins, and P. D. Blake. 1976. Changes in shape, ossification and quality of bones in children with spina bifida. *Devel. Med. and Child Neurol.* suppl. 37:29–41.

Rasmussen, P. 1977. Calcium deficiency, pregnancy,

and lactation in rats: microscopic and microradiographic observations in bones. *Calcified Tissue Res.* 23:179–184.

Rubin, C. T., and L. E. Lanyon. 1981. Bone remodelling in response to applied dynamic loads. *J. Bone Jt. Surg.* 5:237–238.

——— 1982a. Peak functional strain and fatigue properties in bone. *Trans. Orth. Res. Soc.* 7:83.

——— 1982b. Limb mechanics as a function of speed and gait: a study of functional strains in the radius and tibia of horse and dog. *J. Exp. Biol.* 101:187–211.

——— 1983a. Regulation of bone mass by peak strain magnitude. *Trans. Orth. Res. Soc.* 8:70.

——— 1983b. Dynamic strain similarity in vertebrates: an alternative to allometric limb bone scaling. *J. Theor. Biol.* 107:321–327.

——— 1983c. Regulation of bone formation by applied dynamic loads. *J. Bone Jt. Surg.* 66-A:397–402.

Sedlin, E. D., and C. Hirsch. 1966. Factors affecting the determination of the physical properties of femoral cortical bone. *Acta Orth. Scand.* 37:29–48.

Seireg, A., and W. Kempke. 1969. Behaviour of in vivo bone under cyclic loading. *J. Biomech.* 2:455–461.

Stinchfield, A. J., J. A. Reidy, and J. S. Barr. 1949. Prediction of unequal growth of the lower extremities in anterior poliomyelitis. *J. Bone Jt. Surg.* 31A:478–486.

Swanson, S. A., M. A. Freeman, and W. H. Day. 1971. The fatigue properties of human cortical bone. *Med. and Biol. Engr.* 9:23–32.

Tilton, F. E., J. C. Degioanni, and V. S. Schneider. 1980. Long-term followup of Skylab bone demineralization. *Aviation, Space, and Environ. Med.* 51:1209–13.

Uhthoff, H. K., and Z. F. G. Jaworski. 1978. Bone loss in response to long-term immobilisation. *J. Bone Jt. Surg.* 60B:420–429.

Vincent, J. F. V. 1980. Insect cuticle: a paradigm for natural composites. In *Mechanical properties of biological materials,* ed. J. F. V. Vincent and J. D. Currey. *Symp. Soc. Exp. Biol.* 34:183–210.

Wainwright, S. A., W. D. Biggs, J. D. Currey, and J. M. Gosline. 1976. *Mechanical design in organisms.* Princeton: Princeton University Press.

Wolff, J. 1870. Über die innere Architektur der Knochen und ihre Bedeutung für die Frage vom Knochenwachstum. *Arch. Anat. Physiol.* 50:389–453.

Woo, S. L-Y., M. A. Gomez, D. Amiel, N. G. Cobb, W. C. Hayes, and W. H. Akeson. 1981a. The effect of short and long term exercise training on cortical bone hypertrophy. *Trans. Orth. Res. Soc.* 6:63.

Woo, S. L-Y., S. C. Kuei, D. Amiel, M. A. Gomez, W. C. Hayes, F. C. White, and W. H. Akeson. 1981b. The effect of prolonged physical training on the properties of long bone: a study of Wolff's law. *J. Bone Jt. Surg.* 63A:780–787.

Yamada, H. 1970. *The strength of biological material.* Baltimore: Williams and Wilkins.

2. Body Support, Scaling, and Allometry

Aiello, L. C. 1981. The allometry of primate body proportions. *Symp. Zool. Soc. Lond.* 48:331–358.

Alexander, R. McN. 1968. *Animal mechanics.* London: Sidgwick and Jackson.

——— 1977. Terrestrial locomotion. In *Mechanics and energetics of animal locomotion,* ed. R. McN. Alexander and G. Goldspink. London: Chapman and Hall.

——— 1982. *Locomotion of animals.* Glasgow: Blackie.

Alexander, R. McN., V. A. Langman, and A. S. Jayes. 1977. Fast locomotion of some African ungulates. *J. Zool., Lond.* 183:291–300.

Alexander, R. McN., A. S. Jayes, G. M. O. Maloiy, and E. M. Wathuta. 1979. Allometry of limb bones of mammals from shrews *(Sorex)* to elephants *(Loxodonta). J. Zool., Lond.* 189:305–314.

——— 1981. Allometry of the leg muscles of mammals. *J. Zool., Lond.* 194:539–552.

Biewener, A., R. McN. Alexander, and N. C. Heglund. 1981. Elastic energy storage in the hopping of kangaroo rats *(Dipodomys spectabilis). J. Zool., Lond.* 195:369–383.

Economos, A. C. 1983. Elastic and/or geometric similarity in mammalian design? *J. Theor. Biol.* 103:167–172.

Greenewalt, C. H. 1975. The flight of birds. *Trans. Am. Phil. Soc.* 65:1–67.

Hill, A. V. 1950. The dimensions of animals and their muscular dynamics. *Sci. Prog.* 38:209–230.

LeCren, E. D. 1951. The length-weight relationship and seasonal cycle in gonad weight and condition in the perch *(Perca fluviatilis). J. Anim. Ecol.* 20:201–219.

Maloiy, G. M. O., R. McN. Alexander, R. Njau, and A. S. Jayes. 1979. Allometry of the legs of running birds. *J. Zool., Lond.* 187:161–167.

McMahon, T. A. 1973. Size and shape in biology. *Science* (N.Y.) 179:1201–4.

——— 1975. Using body size to understand the structural design of animals: quadrapedal locomotion. *J. Appl. Physiol.* 39:619–627.

Norberg, U. M. 1981. Allometry of bat wings and legs and comparison with bird wings. *Phil. Trans. Roy. Soc. Lond. B* 292:359–398.

Pedley, T. J., ed. 1977. *Scale effects in animal locomotion.* London: Academic Press.

Pennycuick, C. J. 1975. On the running of the gnu *(Connochaetes taurinus)* and other animals. *J. Exp. Biol.* 63:775–800.

Prange, H. D., J. F. Anderson, and H. Rahn. 1979. Scaling of skeletal mass to body mass in birds and mammals. *Am. Nat.* 113:103–122.

Radinsky, L. 1978. Evolution of brain size in carnivores and ungulates. *Am. Nat.* 112:815–831.

Rashevsky, N. 1962. *Mathematical biophysics.* New York: Dover.

3. Walking and Running

Alexander, R. McN. 1974. The mechanics of jumping by a dog *(Canis familiaris). J. Zool., Lond.* 173:549–573.

——— 1980. Optimum walking techniques for quadrupeds and bipeds. *J. Zool., Lond.* 192:97–117.

Alexander, R. McN., and H. C. Bennet-Clark. 1977. Storage of elastic strain energy in muscle and other tissues. *Nature* 265:114–117.

Alexander, R. McN., G. M. O. Maloiy, R. F. Ker, A. S. Jayes, and C. N. Warui. 1982. The role of tendon elasticity in the locomotion of the camel *(Camelus dromedarius). J. Zool., Lond.* 198:293–313.

Bartholomew, G. A., Jr., and H. H. Caswell. 1951. Locomotion in kangaroo rats and its adaptive significance. *J. Mammal.* 32:155–168.

Biewener, A., R. McN. Alexander, and N. C. Heglund. 1981. Elastic energy storage in the hopping of kangaroo rats *(Dipodomys spectabilis). J. Zool., Lond.* 195:369–383.

Bock, W. J. 1968. Mechanics of one- and two-joint muscles. *Am. Mus. Novit.* 2319:1–45.

——— 1974. The avian skeletomuscular system. In *Avian biology,* ed. D. S. Farner, J. R. King, and K. C. Parkes, vol. 4. New York: Academic Press.

Bramble, D. M., and D. R. Carrier. 1983. Running and breathing in mammals. *Science* 219:251–256.

Brinkman, D. 1980. Structural correlates of tarsal and metatarsal functioning in *Iguana* (Lacertilia; Iguanidae) and other lizards. *Can. J. Zool.* 58:277–289.

——— 1981. The hind limb step cycle of *Iguana* and primitive reptiles. *J. Zool., Lond.* 181:91–103.

Butler, D. K., Jr. 1974. *The principles of horseshoeing.* Publ. by author: P.O. box 370, Maryville, Mo. 64468.

Camp, C. L., and N. Smith. 1942. Phylogeny and functions of the digital ligaments of the horse. *Mem. Univ. Calif.* 13:69–124.

Cavagna, G. A., N. C. Heglund, and C. R. Taylor. 1977. Mechanical work in terrestrial locomotion: two basic mechanisms for minimizing energy expenditure. *Am. J. Physiol.: Regulatory, Integrative and Comparative Physiol.* 2:R243–R261.

Charig, A. J. 1972. The evolution of the archosaur pelvis and hind limb: an explanation in functional terms. In *Studies in vertebrate evolution,* ed. K. A. Joysey and T. S. Kemp. New York: Winchester Press.

Coombs, W. P., Jr. 1978. Theoretical aspects of cursorial adaptations in dinosaurs. *Q. Rev. Biol.* 53:393–418.

Cracraft, J. 1971. The functional morphology of the hind limb of the domestic pigeon, *Columbia livia. Bull. Am. Mus. Nat. Hist.* 144:173–268.

Dawson, T. J., and C. R. Taylor. 1973. Energetic cost of locomotion in kangaroos. *Nature* 246:313–314.

Edwards, J. 1976. A comparative study of locomotion in terrestrial salamanders. Ann Arbor: University Microfilms International.

English, A. W. 1978. Functional analysis of the shoulder girdle of cats during locomotion. *J. Morph.* 156:279–292.

——— 1980. The functions of the lumbar spine during stepping in the cat. *J. Morph.* 165:55–66.

Fedak, M. A., N. C. Heglund, and C. R. Taylor. 1982. Energetics and mechanics of terrestrial locomotion. II. Kinetic energy changes of the limbs and body as a function of speed and body size in birds and mammals. *J. Exp. Biol.* 79:23–40.

Frohlich, C. 1980. The physics of somersaulting and twisting. *Sci. Am.* 242:155–164.

Gambaryan, P. P. 1974. *How mammals run: anatomical adaptations.* New York: Wiley. (Orig. publ. in Russian, 1972.)

Gambaryan, P. P., G. N. Orlovskii, T. Y. Protopopova, F. V. Severin, and M. L. Shik. 1971. The activity of muscles during different forms of locomotion of cats and the adaptive function of the (hindlimb) musculature in the family Felidae. (In Russian.) *Trudy Zool. Inst. Akad. Nauk, USSR* 48:220–239.

Goldspink, G. 1981. The use of muscles during flying, swimming, and running from the point of view of energy saving. In *Vertebrate locomotion,* ed. M. H. Day. *Symp. Zool. Soc. Lond.,* 48. New York: Academic Press.

Gray, J. 1944. Studies on the mechanics of the tetrapod skeleton. *J. Exp. Biol.* 20:88–116.

——— 1968. *Animal locomotion.* New York: Norton.

Hildebrand, M. 1959. Motions of the running cheetah and horse. *J. Mammal.* 40:481–495.

——— 1961. Further studies on locomotion of the cheetah. *J. Mammal.* 42:84–91.

——— 1976. Analysis of tetrapod gaits: general considerations and symmetrical gaits. In *Neural control of locomotion,* ed. R. M. Herman et al. Advances in Behavioral Biology, 18. New York: Plenum Press.

——— 1977. Analysis of asymmetrical gaits. *J. Mammal.* 58:131–156.

——— 1980. The adaptive significance of tetrapod gait selection. *Am. Zool.* 20:255–267.

Howell, A. B. 1944. *Speed in animals.* Chicago: University of Chicago Press.

Jayes, A. S., and R. McN. Alexander. 1978. Mechanics of locomotion of dog *(Canis familiaris)* and sheep *(Ovis aries). J. Zool., Lond.* 185:289–308.

——— 1980. The gaits of chelonians: walking techniques for very low speeds. *J. Zool., Lond.* 191:353–378.

Jenkins, F. A., Jr. 1970. Limb movements in a monotreme *(Tachyglossus aculeatus):* a cineradiographic analysis. *Science* 168:1473–75.

——— 1971. Limb posture and locomotion in the Virginia opossum *(Didelphis marsupialis)* and in other non-cursorial mammals. *J. Zool., Lond.* 165:303–315.

——— 1972. Chimpanzee bipedalism: cineradiographic analysis and implications for the evolution of gait. *Science* 178:877–879.

——— 1974. The movement of the shoulder in claviculate and aclaviculate mammals. *J. Morph.* 144:71–83.

Jenkins, F. A., Jr., and S. M. Camazine. 1977. Hip structure and locomotion in ambulatory and cursorial carnivores. *J. Zool., Lond.* 181:351–370.

Jenkins, F. A., Jr., and G. E. Goslow, Jr. 1983. The functional anatomy of the shoulder of the savannah monitor lizard *(Varanus exanthematicus). J. Morph.* 175:195–216.

Jenkins, F. A., Jr., and W. A. Weijs. 1979. The functional anatomy of the shoulder in the Virginia opossum *(Didelphis virginiana). J. Zool., Lond.* 188:379–410.

Kummer, B. 1959. *Bauprinzipien des Säugerskeletes.* Stuttgart: Georg Thieme.

Manter, J. T. 1938. The dynamics of quadrupedal walking. *J. Exp. Biol.* 15:522–540.

Maynard Smith, J., and R. J. G. Savage. 1956. Some locomotory adaptations in mammals. *J. Linn. Soc. Zool.* 42:603–622.

Mochon, S., and T. A. McMahon. 1980. Ballistic walking. *J. Biomech.* 13:49–57.

Rewcastle, S. C. 1981. Stance and gait in tetrapods: an evolutionary scenario. In *Vertebrate locomotion,* ed. M. H. Day. *Symp. Zool. Soc. Lond.* 48. New York: Academic Press.

Rieser, G. D. 1977. A functional analysis of bipedalism in lizards. Ph.D. diss., University of California at Davis.

Schryver, H. F., D. L. Bartel, N. Langrana, and J. E. Lowe. 1978. Locomotion in the horse: kinematics and external and internal forces in the normal equine digit in the walk and trot. *Am. J. Vet. Res.* 39:1728–33.

Snyder, R. C. 1962. Adaptations for bipedalism of lizards. *Am. Zool.* 2:191–203.

Sukhanov, V. B. 1974. *General system of symmetrical locomotion of terrestrial vertebrates and some features of movement of lower tetrapods.* Washington, D.C.: Smithsonian Institution and National Science Foundation. (Orig. publ. in Russian, 1968.)

Taylor, C. R. 1978. Why change gaits? Recruitment of muscles and muscle fibers as a function of speed and gait. *Am. Zool.* 18:153–161.

Taylor, C. R., A. Shkolnik, R. Dmi'el, D. Baharav, and A. Borut. 1974. Running in cheetahs, gazelles, and goats: energy cost and limb configuration. *Am. J. Physiol.* 227:848–850.

Walker, W. F., Jr. 1971. A structural and functional analysis of walking in the turtle, *Chrysemys picta marginata, J. Morph.* 134:195–213.

Wentink, G. H. 1979. Dynamics of the hind limb at walk in horse and dog. *Anat. Embryol.* 155:179–190.

Yalden, D. W. 1970. The functional morphology of the carpal bones in carnivores. *Acta Anat.* 77:481–500.

——— 1971. The functional morphology of the carpus in ungulate mammals. *Acta Anat.* 78:461–487.

Zug, G. R. 1971. Buoyancy, locomotion, morphology of pelvic girdle and hindlimb, and systematics of cryptodiran turtles. *Univ. Mich. Mus. Zool.* Misc. Publ. 142.

4. Jumping and Leaping

Alexander, R. McN. 1974. The mechanics of jumping by a dog *(Canis familiaris). J. Zool., Lond.* 173:549–573.

Alexander, R. McN., and H. Bennet-Clark. 1977. Storage of elastic strain energy in muscle and other tissues. *Nature* 265:114–117.

Alexander, R. McN., and A. Vernon. 1975. The mechanics of hopping by kangaroos (Macropodidae). *J. Zool., Lond.* 177:265–303.

Alexander, R. McN., A. S. Jayes, G. M. O. Maloiy, and E. M. Wathuta. 1981. Allometry of the leg muscles of mammals. *J. Zool., Lond.* 194:539–552.

Badoux, D. 1965. Some notes on the functional anatomy of *Macropus giganteus* with general remarks on the mechanics of bipedal leaping. *Acta Anat.* 62:418–433.

Bartholomew, G., and H. Caswell. 1951. Locomotion in kangaroo rats and its adaptive significance. *J. Mammal.* 32:155–169.

Bauschulte, C. 1972. Morphologische und biomechanische Grundlagen einer funktionellen Analyse der Muskeln der Hinterextremität (Untersuchung an quadrupeden Affen und Kanguruhs). *Z. Anat. Entwickl. Gesch.* 138:167–214.

Bennet-Clark, H. 1977. Scale effects in jumping animals. In *Scale effects in animal locomotion,* ed. T. Pedley. New York: Academic Press.

Biewener, A., R. McN. Alexander, and N. Heglund. 1981. Elastic energy storage in the hopping of kangaroo rats *(Dipodomys spectabilis). J. Zool., Lond.* 195:369–383.

Calow, L., and R. McN. Alexander. 1973. A mechanical analysis of a hind leg of a frog *(Rana temporaria). J. Zool., Lond.* 171:293–321.

Dobrowolska, H. 1973. Body part proportions in relation to the mode of locomotion in anurans. *Zoologica Poloniae* 23:59–108.

Dunlap, D. 1960. The comparative myology of the pelvic appendage in the Salientia. *J. Morph.* 106:1–76.

Emerson, S. 1978. Allometry and jumping in frogs: helping the twain to meet. *Evol.* 32:551–564.

Emerson, S., and H. J. de Jongh. 1980. Muscle activity at the ilio-sacral articulation of frogs. *J. Morph.* 166:129–144.

Fokin, I. 1978. *The locomotion and morphology of the locomotory organs in jerboas.* (In Russian.) Leningrad: Academy of Sciences of USSR, Zoology Institute.

Gambaryan, P. 1974. *How mammals run: anatomical adaptations,* trans. Israel Program for Scientific Translations. New York: Wiley.

Hall-Craggs, E. 1965. An analysis of the jump of the lesser galago *Galago senegalensis. J. Zool., Lond.* 147:20–29.

Hatt, R. 1932. The vertebral columns of richochetal rodents. *Bull. Amer. Mus. Nat. Hist.* 63:599–738.

Hill, A. V. 1950. The dimensions of animals and their muscular dynamics. *Sci. Prog.* 38:209–230.

Hirsch, W. 1931. Zur physiologischen Mechanik des Froschsprunges. *Z. vergl. Physiol.* 15:1–49.

Howell, A. 1932. The saltatorial rodent *Dipodomys:* the functional and comparative anatomy of its muscular and osseous systems. *Proc. Am. Acad. Arts Sci.* 67:377–536.

Jouffroy, F., and J. Lessertisseur. 1979. Relationships between limb morphology and locomotor adaptations among prosimians: an osteometric study. In *Environment, behavior and morphology: dynamic interactions in primates,* ed. M. Morbeck, H. Preuschoft, and N. Gomberg. New York: Fisher Press.

Kenagy, G. 1973. Daily and seasonal patterns of activity and energetics in a heteromyid rodent community. *Ecol.* 54:1201–19.

Klingener, D. 1964. The comparative myology of four dipodoid rodents (genus *Zapus, Napaeozapus, Sicista,* and *Jaculus*). *Univ. Mich. Mus. Zool. Misc. Publ.* 124:100.

Maynard Smith, J., and R. Savage. 1956. Some locomotory adaptations in mammals. *J. Linn. Soc. Zool.* 42:603–622.

McArdle, J. 1981. Functional morphology of the hip and thigh of the Lorisiformes. *Contr. Primat.* 17:1–132.

Oxnard, C., R. German, F. Jouffroy, and J. Lessertisseur. 1981. A morphometric study of limb proportions in leaping prosimians. *Am. J. Phys. Anthrop.* 54:421–430.

Pinkham, C. 1976. A biomechanical analysis of leaping in the kangaroo rat and the Mexican spiny pocket mouse. *Utah Acad. Proc.* 53:1–15.

Rand, A. S. 1952. Jumping ability of certain anurans with notes on endurance. *Copeia* 1952:15–20.

Sperry, D. 1981. Fiber type composition and postmetamorphic growth of anuran hindlimb muscles. *J. Morph.* 170:321–345.

Stokely, P., and J. Berberian. 1953. On the jumping ability of frogs. *Copeia* 1953:187.

Wermel, J. 1934. Über die Körperproportionen der Wirbeltiere und ihre funktionelle Bedeutung. Biometrische Übungen. Zweiter Aufsatz. Die Extremitätenproportionen und der Sprung den Salientia. *Z. Anat. Entwickl. Gesch.* 103:645–659.

Zug, G. 1972. Anuran locomotion: structure and function. I. Preliminary observations on the relation between jumping and osteometrics of the appendicular and postaxial skeleton. *Copeia* 1972:613–624.

——— 1978. Anural locomotion: structure and function. II. Jumping performance of semiaquatic, terrestrial and arboreal frogs. *Smith. Contr. Zool.* 276: 1–31.

5. Climbing

Alberch, P. 1981. Convergence and parallelism in foot morphology in the Neotropical salamander genus *Bolitoglossa.* I. Function. *Evol.* 35:84–100.

Andrews, D. H., and R. J. Kokes. 1962. *Fundamental chemistry.* New York: Wiley.

Barnett, C. H., and J. R. Napier. 1953. The form and mobility of the fibula in metatherian mammals. *J. Anat.* 87:207–213.

Benton, R. S. 1967. Morphological evidence for adaptation within the epaxial region of the primates. In

The baboon in medical research, ed. H. Vagtborg, vol. 2. Austin: University of Texas Press.

Biegert, J. 1961. Volarhaut der Hände und Füsse. In *Primatologia,* ed. H. Hofer, A. H. Schultz, and D. Starck, vol. 2, pt. 1, no. 3. Basel: Karger.

——— 1963. The evaluation of characteristics of the skull, hands, and feet for primate taxonomy. In *Classification and human evolution,* ed. S. L. Washburn. Chicago: Aldine.

Biegert, J., and R. Maurer. 1972. Rumpfskelettlänge, Allometrien und Körperproportionen bei catarrhinen Primaten. *Folia Primatol.* 17:142–156.

Biewener, A. A. 1982. Bone strength in small mammals and bipedal birds: do safety factors change with body size? *J. Exp. Biol.* 98:289–301.

Bishop, A. 1964. Use of the hand in lower primates. In *Evolutionary and genetic biology of primates,* ed. J. Buettner-Janusch, vol. 2. New York: Academic Press.

Bock, W. J., and W. D. Miller. 1959. The scansorial foot of the woodpeckers, with comments on the evolution of perching and climbing feet in birds. *Am. Mus. Novit.* 1931:1–45.

Böker, H. 1932. Beobachten und Untersuchungen an Säugetieren während einer biologisch-anatomischen Forschungreise nach Brasilien im Jahre 1928. *Gegenbaurs morph. Jahrb.* 70:1–66.

Bowden, F. P., and D. Tabor. 1973. *Friction: an introduction to tribology.* Garden City, N.Y.: Anchor Press/ Doubleday.

Cartmill, M. 1974. Pads and claws in arboreal locomotion. In *Primate locomotion,* ed. F. A. Jenkins, Jr. New York: Academic Press.

——— 1975. Primate evolution: analysis of trends. *Science* 189:229–230.

——— 1979. The volar skin of primates: its frictional characteristics and their functional significance. *Am. J. Phys. Anthrop.* 50:497–510.

Cartmill, M., and K. Milton. 1977. The lorisiform wrist joint and the evolution of "brachiating" adaptations in the Hominoidea. *Am. J. Phys. Anthrop.* 47:249–272.

Charles-Dominique, P. 1977. *Ecology and behaviour of nocturnal primates.* New York: Columbia University Press.

Coimbra-Filho, A. F., and R. A. Mittermeier. 1976. Exudate-eating and tree-gouging in marmosets. *Nature* 262:630.

Cummins, H., and C. Midlo. 1943. *Finger prints, palms and soles: an introduction to dermatoglyphics.* New York: Dover (reprinted, 1961).

Dankmeijer, J. 1938. Zur biologischen Anatomie der Hautleisten bei den Beuteltieren. *Gegenbaurs morph. Jahrb.* 82:293–312.

Dunn, E. R. 1931. The disc-winged bat *(Thyroptera)* in Panama. *J. Mammal.* 12:429–430.

Emerson, S. B., and D. Diehl. 1980. Toe pad morphology and mechanisms of sticking in frogs. *Biol. J. Linn. Soc. Lond.* 13:199–216.

Emmons, L. H. 1981. Morphological, ecological, and behavioral adaptations for arboreal browsing in *Dactylomys dactylinus* (Rodentia, Echimyidae). *J. Mammal.* 62:183–189.

Erikson, G. E. 1963. Brachiation in New World monkeys and in anthropoid apes. *Symp. Zool. Soc. Lond.* 10:135–164.

Fleagle, J. G., and R. A. Mittermeier. 1980. Locomotor behavior, body size, and comparative ecology of seven Surinam monkeys. *Am. J. Phys. Anthrop.* 52:301–314.

Flower, W. H. 1885. *An introduction to the osteology of the Mammalia.* London: Macmillan.

Ganslosser, U. 1980. Vergleichende Untersuchungen zur Kletterfähigkeit einiger Baumkanguruh-Arten (*Dendrolagus,* Marsupialia). *Zool. Anzeiger* 205:43–66.

Garber, P. A. 1980. Locomotor behavior and feeding ecology of the Panamanian tamarin (*Saguinus oedipus geoffroyi,* Callitrichidae, Primates). *Int. J. Primatol.* 1:185–201.

Green, D. M. 1981. Adhesion and the toe-pads of treefrogs. *Copeia* 1981:790–796.

Green, D. M., and P. Alberch. 1981. Interdigital webbing and skin morphology in the Neotropical salamander genus *Bolitoglossa* (Amphibia, Plethodontidae). *J. Morph.* 170:273–282.

Haines, R. W. 1955. The anatomy of the hand of certain insectivores. *Proc. Zool. Soc. Lond.* 125:761–777.

Hall-Craggs, E. C. B. 1966. Rotational movements in the foot of *Galago senegalensis. Anat. Record* 154:287–294.

Hershkovitz, P. 1970. Notes on Tertiary platyrrhine monkeys and description of a new genus from the Miocene of Columbia. *Folia Primatol.* 12:1–37.

——— 1977. *Living New World monkeys (Platyrrhini), with an introduction to primates,* vol. 1. Chicago: University of Chicago Press.

Hildebrand, M. 1978. Insertions and functions of certain flexor muscles in the hind leg of rodents. *J. Morph.* 155:111–122.

——— 1982. *Analysis of vertebrate structure,* 2nd ed. New York: Wiley.

Hill, W. C. O. 1953. *Primates: comparative anatomy and taxonomy.* I. *Strepsirhini.* Edinburgh: Edinburgh University Press.

Hiller, U. 1968. Untersuchungen zum Feinbau und zur Funktion der Haftborsten von Reptilien. *Z. Morph. Tiere* 62:307–362.

Jenkins, F. A., Jr. 1970. Anatomy and function of expanded ribs in certain edentates and primates. *J. Mammal.* 51:288–301.

——— 1971. The postcranial skeleton of African cynodonts: problems in the early evolution of the mammalian postcranial skeleton. *Peabody Mus. Bull.* 36:1–216.

——— 1974. Tree shrew locomotion and the origins of primate arborealism. In *Primate locomotion,* ed. F. A. Jenkins, Jr. New York: Academic Press.

——— 1981. Wrist rotation in primates: a critical adaptation for brachiators. *Symp. Zool. Soc. Lond.* 48:429–451.

Jenkins, F. A., Jr., and F. R. Parrington. 1976. The postcranial skeletons of the Triassic mammals *Eozostrodon, Megazostrodon* and *Erythrotherium. Phil. Trans. Roy. Soc. Lond. B* 273:387–431.

Jensen, A., and H. H. Chenoweth. 1967. *Applied strength of materials.* New York: McGraw-Hill.

Johnson, R. G. 1955. The adaptive and phylogenetic significance of vertebral form in snakes. *Evol.* 9:367–388.

Jouffroy, F. K., and J. Lessertisseur. 1959. Reflexions sur les muscles contracteurs des doigts et des orteils *(contrahentes digitorum)* chez les Primates. *Ann. Sci. Nat. Zool.,* sér. 12, 1:211–235.

Jullien, R. 1968. Evolution des supports osseux et musculaires de la prehension au pied des *Dendromurinae* arboricoles. *Mammalia* 32:276–306.

Jungers, W. L. 1978. The functional significance of skeletal allometry in *Megaladapis* in comparison to living prosimians. *Am. J. Phys. Anthrop.* 49:303–314.

——— 1985. Allometry of the limbs in primates. In *Size and scaling in primate biology,* ed. W. L. Jungers. New York: Plenum.

Kay, R. F. 1973. Mastication, molar tooth structure, and diet in primates. Ph.D. diss., Yale University.

Keith, A. 1923. Man's posture: its evolution and disorders. *Brit. Med. J.* 1:451–454, 499–502, 545–548, 587–590, 624–626, 669–672.

Kidd, W. 1907. *The sense of touch in mammals and birds, with special reference to the papillary ridges.* London: Adam and Charles Black.

Kingdon, J. 1971. *East African mammals: an atlas of evolution in Africa,* vol. 1. London: Academic Press.

Kinzey, W. G., A. L. Rosenberger, and M. Ramirez. 1975. Vertical clinging and leaping in a Neotropical anthropoid. *Nature* 255:327–328.

Le Gros Clark, W. E. 1936. The problem of the claw in primates. *Proc. Zool. Soc. Lond.* 1936:1–24.

——— 1971. *The antecedents of man: an introduction to the evolution of the primates,* 3rd ed. Edinburgh: Edinburgh University Press.

Lewis, O. J. 1972. Evolution of the hominoid wrist. In *The functional and evolutionary biology of primates,* ed. R. H. Tuttle. Chicago: Aldine-Atherton.

Leyhausen, P. 1963. Über südamerikanische Pardelkatzen. *Z. Tierpsychol.* 20:627–640.

Loveridge, A. 1956. *Forest safari.* London: Lutterworth Press.

Mendel, F. C. 1979. The wrist joint of two-toed sloths and its relevance to brachiating adaptations in the Hominoidea. *J. Morph.* 162:413–424.

Musser, G. G. 1972. The species of *Hapalomys* (Rodentia, Muridae). *Am. Mus. Novit.* 2503:1–27.

Napier, J. R. 1967. Evolutionary aspects of primate locomotion. *Am. J. Phys. Anthrop.* 27:333–342.

Noble, G. K. 1931. *The biology of the Amphibia.* New York: McGraw-Hill.

Noble, G. K., and R. Jaeckle. 1928. The digital pads of the tree frogs: a study of the phylogenesis of an adaptive structure. *J. Morph.* 45:259–292.

Peterka, H. E. 1937. A study of the myology and osteology of tree sciurids with regard to adaptation to arboreal, glissant and fossorial habits. *Trans. Kans. Acad. Sci.* 39:313–332.

Peterson, J. A., and E. E. Williams. 1981. A case history in retrograde evolution: the *onca* lineage in anoline lizards. II. Subdigital fine structure. *Bull. Mus. Comp. Zool.* 149:215–268.

Pocock, R. I. 1922. On the external characters of the beaver (Castoridae) and of some squirrels (Sciuridae). *Proc. Zool. Soc. Lond.* 1922:1171–212.

Richardson, F. Adaptive modifications for tree-trunk foraging in birds. *Univ. Calif. Publ. Zool.* 46:317–368.

Rosenberger, A. L. 1979. Phylogeny, evolution and classification of New World monkeys (Platyrrhini, Primates). Ph.D. diss., City University of New York.

Ruibal, R., and V. Ernst. 1965. The structure of the digital setae of lizards. *J. Morph.* 117:271–294.

Russell, A. P. 1975. A contribution to the functional anatomy of the foot of the tokay, *Gekko gecko* (Reptilia, Gekkonidae). *J. Zool., Lond.* 176:437–476.

Russell, D. E. 1964. Les mammifères paléocènes d'Europe. *Mém. Mus. Natl. d'Hist. Nat.* (Paris), sér. C, 13:1–321.

Schliemann, H., and Rehn, C., 1980. Zur Kenntnis der Haftorgane von *Eudiscopus denticulus* (Osgood, 1932) (Mammalia, Microchiroptera, Vespertilionidae). *Z. Säugetierkunde* 45:29–39.

Schultz, A. H. 1961. Vertebral column and thorax. In *Primatologia,* ed. H. Hofer, A. H. Schultz, and D. Starck, vol. 4, no. 4. Basel: Karger.

Sclater, W. L. 1901. *The mammals of South Africa.* London: Porter.

Szalay, F. S. 1972. Paleobiology of the earliest pri-

mates. In *The functional and evolutionary biology of primates,* ed. R. H. Tuttle. Chicago: Aldine Atherton.
——— 1981. Functional analysis and the practice of the phylogenetic method as reflected by some mammalian studies. *Am. Zool.* 21:37–45.
Szalay, F. S., and M. Dagosto. 1980. Locomotor adaptations as reflected in the humerus of Paleogene primates. *Folia Primatol.* 34:1–45.
Szalay, F. S., and G. Drawhorn. 1980. Evolution and diversification of the Archonta in an arboreal milieu. In *Comparative biology and evolutionary relationships of tree shrews,* ed. W. P. Luckett. New York: Plenum.
Tattersall, I. 1982. *The primates of Madagascar.* New York: Columbia University Press.
Taylor, M. E. 1970. Locomotion in some East African viverrids. *J. Mammal.* 51:42–51.
——— 1976. The functional anatomy of the hindlimb of some African Viverridae (Carnivora). *J. Morph.* 148:227–254.
Thorington, R. W., Jr. 1968. Observations of the tamarin *Saguinus midas. Folia Primatol.* 9:95–98.
Thorington, R. W., Jr., and L. R. Heaney. 1981. Body proportions and gliding adaptations of flying squirrels (Petauristinae). *J. Mammal.* 62:101–114.
Trapp, G. R. 1972. Some anatomical and behavioral adaptations of ringtails, *Bassariscus astutus. J. Mammal.* 53:549–557.
Tuttle, R. H. 1975. Parallelism, brachiation, and hominoid phylogeny. In *Phylogeny of the primates: a multidisciplinary approach,* ed. W. P. Luckett and F. S. Szalay. New York: Plenum.
Vaughan, T. A. 1972. *Mammalogy.* Philadelphia: W. B. Saunders.
Walker, E. P. 1975. *Mammals of the world,* 3rd ed. Baltimore: Johns Hopkins University Press.
Welsch, U., V. Storch, and W. Fuchs. 1974. The fine structure of the digital pads of rhacophorid tree frogs. *Cell and Tissue Res.* 148:407–416.
Whipple, I. L. 1904. The ventral surface of the mammalian chiridium with special reference to the conditions found in man. *Z. Morph. Anthrop.* 7:261–368.
Williams, E. E., and J. A. Peterson. 1982. Convergent and alternative designs in the digital adhesive pads of scincid lizards. *Science* 215:1509–11.
Wimsatt, W. A., and B. Villa. 1970. Locomotor adaptations in the disc-winged bat, *Thyroptera tricolor.* I. Functional organization of the adhesive discs. *Am. J. Anat.* 129:89–119.
Wood Jones, F. 1916. *Arboreal man.* London: E. Arnold.
——— 1924. *The mammals of South Australia.* Pt. 2. *Containing the bandicoots and the herbivorous marsupials.* Adelaide: British Science Guild.
——— 1953. Some readaptations of the mammalian pes in response to arboreal habits. *Proc. Zool. Soc. Lond.* 123:33–41.

6. Digging of Quadrupeds

Ar, A., R. Arieli, and A. Shkolnik. 1977. Blood-gas properties and function in the fossorial mole rat under normal and hypoxic-hypercapnic atmospheric conditions. *Respiration Physiol.* 30:201–218.
Arlton, A. V. 1936. An ecological study of the mole. *J. Mammal.* 17:349–371.
Augee, M. L., R. W. Elsner, B. A. Gooden, and P. R. Wilson. 1971. Respiratory and cardiac responses of a burrowing animal, the echidna. *Respiration Physiol.* 11:327–334.
Barnosky, A. D. 1981. A skeleton of *Mesoscalops* (Mammalia, Insectivora) from the Miocene Deep River Formation, Montana, and a review of the proscalopid moles: evolutionary, functional, and stratigraphic relationships. *J. Vert. Paleontol.* 1:285–339.
——— 1982. Locomotion in moles (Insectivora, Proscalopidae) from the Middle Tertiary of North America. *Science* 216:183–185.
Bateman, J. A. 1959. Laboratory studies of the golden mole and the mole-rat. *African Wild Life* 13:65–71.
Bradley, W. G., and M. K. Yousef. 1975. Thermoregulatory responses in the plains pocket gopher, *Geomys bursarius. Comp. Biochem. Physiol.* 52A:35–38.
Bramble, D. M. 1982. *Scaptochelys:* generic revision and evolution of gopher tortoises. *Copeia* 1982:852–867.
Carlsson, A. 1904. Zur Anatomie des *Notoryctes typhlops. Zool. Jahrb. Abt. Anat. Ontog.* 20:81–122.
Chapman, R. C., and A. F. Bennett. 1975. Physiological correlates of burrowing in rodents. *Comp. Biochem. Physiol.* 52A:599–603.
Chapman, R. N. 1919. A study of the correlation of the pelvic structure and the habits of certain burrowing mammals. *Am. J. Anat.* 25:185–219.
Cox, C. B. 1972. A new digging dicynodont from the Upper Permian of Tanzania. In *Studies in vertebrate evolution,* ed. K. A. Josey and T. S. Kemp. New York: Winchester Press.
Darden, T. R. 1972. Respiratory adaptations of a fossorial mammal, the pocket gopher *(Thomomys bottae). J. Comp. Physiol.* 78:121–137.
Dubost, G. 1968. Les mammifères souterrains. *Rev. Écol. Biol. Sol.* 5:99–197.
Eadie, W. R. 1945. The pelvic girdle of *Parascalops. J. Mammal.* 26:94–95.
Eloff, G. 1951. Orientation in the mole-rat *Cryptomys. Brit. J. Psychol.* 42:134–145.

Emerson, S. B. 1976. Burrowing in frogs. *J. Morph.* 149:437–458.

Gambarian, P. P. 1953. Adaptive specializations of the forelimb of blind mole rats (*Spalax leucodon nehringi* Satunin). (In Russian.) *Zool. Inst. Akad. Nauk Armianskoi SSR* 8:67–125.

——— 1960. Adaptive characteristics of the locomotor organs in burrowing mammals. (In Russian.) *Akad. Nauk Armianskoi SSR, Erevan.*

Genelly, R. E. 1965. Ecology of the common mole-rat *(Cryptomys hottentotus)* in Rhodesia. *J. Mammal.* 46:647–665.

Goldstein, B. 1968. Burrowing mechanisms in some fossorial mammals. Ph.D. diss., University of California at Davis.

——— 1971. Heterogeneity of muscle fibers in some burrowing mammals. *J. Mammal.* 52:515–527.

Gundy, G. C., and C. Z. Wurst. 1976. Parietal eye-pineal morphology in lizards and its physiological implications. *Anat. Record* 185:419–431.

Gupta, B. B. 1966. Fusion of cervical vertebrae in rodents. *Mammalia* 30:25–29.

Hill, J. E. 1937. Morphology of the pocket gopher, mammalian genus *Thomomys. Univ. Calif. Publ. Zool.* 42:81–172.

Hisaw, F. L. 1923. Observations on the burrowing habits of moles *(Scapanus aquaticus machrinoides). J. Mammal.* 4:79–88.

——— 1924. The absorption of the pubic symphysis of the pocket gopher, *Goemys bursarius* (Shaw). *Am. Nat.* 58:93–96.

Howe, D. 1975. Observations on a captive marsupial mole, *Notoryctes typhlops. Australian Mammal.* 1:361–365.

Jarvis, J. U. M. 1978. The energetics of survival in *Heterocephalus glaber* (Rupell) the naked mole rat (Rodentia, Bethyergidae). *Bull. Carnegie Mus. Nat. Hist.* 6:81–87.

Jarvis, J. U. M., and J. B. Sale. 1971. Burrowing and burrow patterns of East African mole-rats *Tachyoryctes, Heliophobius* and *Heterocephalus. J. Zool., Lond.* 163:451–479.

Jenkins, F. A., Jr. 1970. Limb movements in a monotreme *(Tachyglossus aculeatus):* a cineradiographic analysis. *Science* 168:1473–75.

Krapp, F. 1965. Schädel und Kaumuskulatur von *Spalax leucodon* (Nordmann, 1840). *Z. Wiss. Zool.* 173:1–71.

Kriszat, G. 1940a. Untersuchungen zur Sinnesphysiologie, Biologie und Umwelt des Maulwulfs (*Talpa europaea* L.) *Z. Morph. Ökol. Tiere* 36:446–511.

——— 1940b. Die Orientierung im Raume bei *Talpa europaea. Z. Morph. Ökol. Tiere* 36:512–556.

Lechner, A. J. 1976. Respiratory adaptations in burrowing pocket gophers from sea level and high altitude. *J. Appl. Physiol.* 41:168–173.

Lehmann, W. H. 1963. The forelimb architecture of some fossorial rodents. *J. Morph.* 113:59–76.

Manaro, A. J. 1959. Extrusive incisor growth in the rodent genera *Geomys, Peromyscus,* and *Sigmodon. Fla. Acad. Sci. Q. J.* 22:25–31.

McNab, B. K. 1979. The influence of body size on the energetics and distribution of fossorial and burrowing mammals. *Ecol.* 60:1010–21.

——— 1980. Energetics and the limits to a temperate distribution in armadillos. *J. Mammal.* 61:606–627.

Menzies, J. I., and M. J. Tyler. 1977. The systematics and adaptations of some Papuan microhylid frogs which live underground. *J. Zool., Lond.* 183:431–464.

Miles, S. T. 1941. The shoulder anatomy of the armadillo. *J. Mammal.* 22:157–169.

Miller, R. S. 1958. Rate of incisor growth in the mountain pocket gopher. *J. Mammal.* 39:380–385.

Nevo, E. 1979. Adaptive convergence and divergence of subterranean mammals. *An. Rev. Ecol. and Syst.* 10:269–308.

Orcutt, E. E. 1940. Studies on the muscles of the head, neck, and pectoral appendages of *Geomys bursarius. J. Mammal.* 21:37–52.

Pearson, O. P. 1959. Biology of the subterranean rodents, *Ctenomys,* in Peru. *Mem. Mus. Hist. Nat. "Javier Prado"* 9:1–56.

Pevet, P., J. A. Kappers, and E. Nevo. 1976. The pineal gland of the mole-rat (*Spalax ehrenbergi,* Nehring). *Cell and Tissue Res.* 174:1–24.

Puttick, G. M., and J. U. M. Jarvis. 1977. The functional anatomy of the neck and forelimbs of the cape golden mole, *Chrysochloris asiatica* (Lipotyphla, Chrysochloridae). *Zool. Africana* 12:445–458.

Reed, C. A. 1951. Locomotion and appendicular anatomy in three soricid insectivores. *Am. Midland Nat.* 45:513–671.

Rose, K. D., and R. J. Emry. 1983. Extraordinary fossorial adaptations in the Oligocene palaeanodonts *Epoicotherium* and *Xenocranium. J. Morph.* 75:33–56.

Scholander, P. F., L. Irving, and S. W. Grinnell. 1943. Respiration of the armadillo with possible implications as to its burrowing. *J. Cell and Comp. Physiol.* 21:53–63.

Seymour, R. S. 1973. Physiological correlates of forced activity and burrowing in the spadefoot toad, *Scaphiopus hamondii. Copeia* 1973:103–115.

Taylor, B. K. 1978. The anatomy of the forelimb in the anteater *(Tamandua)* and its functional implications. *J. Morph.* 157:347–368.

Vleck, D. 1979. The energy cost of burrowing by the pocket gopher *Thomomys bottae. Physiol. Zool.* 52:122–136.

——— 1981. Burrow structure and foraging costs in the fossorial rodent, *Thomomys bottae. Oecologia* 49:391–396.

Wilson, J. T. 1894. On the myology of *Notoryctes typhlops,* with comparative notes. *Trans. Roy. Soc. So. Australia* 18:3–74.

Yalden, D. W. 1966. The anatomy of mole locomotion. *J. Zool., Lond.* 149:55–64.

7. Swimming

Alexander, R. McN. 1967. *Functional design in fishes.* London: Hutchinson Press.

Blake, R. W. 1979. The energetics of hovering in the mandarin fish *(Synchropus picturatus). J. Exp. Biol.* 82:25–33.

——— 1980. Undulatory median fin propulsion of two teleosts with different modes of life. *Can. J. Zool.* 58:2116–19.

——— 1981. Mechanics of drag-based mechanisms of propulsion in aquatic vertebrates. *Symp. Zool. Soc. Lond.* 48:29–52.

——— 1983. Median and paired fin propulsion. In *Fish Biomechanics,* ed. P. W. Webb and D. Weihs. New York: Praeger.

Breder, C. M. 1926. The locomotion of fishes. *Zoologica* 4:159–297.

Eaton, R. C., R. A. Bombardieri, and D. L. Meyer. 1977. The Mauthner-initiated startle response in teleost fish. *J. Exp. Biol.* 66:65–81.

Gosline, W. R. 1971. *Functional morphology and classification of teleostean fishes.* Honolulu: University of Hawaii Press.

Gray, J. 1933. Studies in animal locomotion. *J. Exp. Biol.* 10:88–104; 386–390; 391–400.

Hunter, J. R., and R. R. Zweifel. 1971. Swimming speed, tail beat frequency, tail beat amplitude, and size in jack mackerel, *Trachurus symmetricus,* and other fishes. *Fishery Bull.* (U.S.) 69:253–266.

Lighthill, M. J. 1975. *Mathematical biofluiddynamics.* Philadelphia: Society for Industrial and Applied Mathematics.

Lindsey, C. C. 1978. Form, function, and locomotory habits in fish. In *Fish physiology,* ed. W. S. Hoar and D. J. Randall, vol. 7. New York: Academic Press.

Pedley, T. J., ed. 1976. *Scale effects in animal locomotion.* New York: Academic Press.

Vogel, S. 1981. *Life in moving fluids.* Boston: Willard Grant Press.

Webb, P. W. 1976. The effect of size on the fast-start performance of rainbow trout, *Salmo gairdneri,* and a consideration of piscivorous predator-prey interactions. *J. Exp. Biol.* 65:157–177.

——— 1978a. Temperature effects on acceleration of rainbow trout, *Salmo gairdneri. J. Fish. Res. Bd. Can.* 35:1417–22.

——— 1978b. Hydrodynamics: non-scombroid fish. In *Fish physiology,* ed. W. S. Hoar and D. J. Randall, vol. 7. New York: Academic Press.

——— 1982a. Locomotor patterns in the evolution of actinopterygian fishes. *Am. Zool.* 22:329–342.

——— 1982b. Fast-start resistance of trout. *J. Exp. Biol.* 96:93–106.

Webb, P. W., and R. S. Keyes. 1983. Swimming kinematics of sharks. *Fishery Bull.* (U.S.) 80:803–812.

Webb, P. W., and J. M. Skadsen. 1979. Reduced skin mass: an adaptation for acceleration in some teleost fish. *Can. J. Zool.* 57:1570–75.

Weihs, D. 1973. The mechanisms of rapid starting of a slender fish. *Biorheol.* 10:343–350.

——— 1981. Effects of swimming path curvature on the energetics of fish motion. *Fishery Bull.* (U.S.) 79:171–176.

Weihs, D., and P. W. Webb. 1983. Optimization of swimming. In *Fish Biomechanics,* ed. P. W. Webb and D. Weihs. New York: Praeger.

Wu, T. Y.-T., C. J. Brokaw, and C. Brennen, eds. 1975. *Swimming and flying in nature,* vol. 2. New York: Plenum Press.

8. Flying, Gliding, and Soaring

Abbott, J. H., and A. E. von Doenhoff. 1949. *Theory of wing section.* New York: McGraw-Hill.

Blick, E. F., D. Watson, G. Belie, and H. Chu. 1974. Bird aerodynamic experiments. In *Swimming and flying in nature,* ed. T. Y.-T. Wu, C. J. Brokaw, and C. Brennen, vol. 2. New York: Plenum Press.

Bramwell, C. D., and G. R. Whitfield. 1974. Biomechanics of *Pteranodon. Phil. Trans. Roy. Soc. Lond. B* 267:503–592.

Duellman, W. E. 1970. The hylid frogs of Middle America. *Monogr. Mus. Nat. Hist. Univ. Kans.,* 1.

Ellington, C. P. 1978. The aerodynamics of normal hovering: three approaches. In *Comparative physiology: water, ions, and fluid mechanics,* ed. K. Schmidt-Nielsen, L. Bolis, and S. H. P. Maddrell. Cambridge: Cambridge University Press.

Feduccia, A., and H. B. Tordoff. 1979. Feathers of *Archaeopteryx:* asymmetric vanes indicate aerodynamic function. *Science* (N.Y.) 203:1021–22.

Feinsinger, P., R. K. Colwell, J. Terborgh, and S. B. Chaplin. 1979. Elevation and the morphology, flight energetics, and foraging ecology of tropical hummingbirds. *Am. Nat.* 113:481–497.
Fisher, H. I. 1957. Bony mechanism of automatic flexion and extension in the pigeon's wing. *Science* 126:446.
Gould, L. L. 1972. Formation flight in the Canada goose *(Branta c. canadensis)*. Masters thesis, University of Rhode Island.
Graham, R. 1934. The silent flight of owls. *J. Roy. Aeronaut. Soc.* 38:837–843.
Greenewalt, C. H. 1962. Dimensional relationships for flying animals. *Smith. Misc. Colls.* 144:1–46.
——— 1975. The flight of birds. *Trans. Am. Phil. Soc.* 65, pt. 4:1–67.
Hankin, E. H. 1913. *Animal flight.* London: Iliffe and Sons.
Heppner, F. H. 1974. Avian flight formation. *Bird-Banding* 45:160–169.
Hertel, H. 1966. *Structure-form-movement.* New York: Reinhold.
Herzog, K. 1968. *Anatomie und Flugbiologie der Vögel.* Stuttgart: Gustav Fisher.
Higdon, J. J. L., and S. Corrsin. 1978. Induced drag of a bird flock. *Am. Nat.* 112:727–744.
Hildebrand, M. 1982. *Analysis of vertebrate structure,* 2d ed. New York: Wiley.
Jepsen, G. L. 1970. Bat origins and evolution. In *Biology of bats,* ed. W. A. Wimsatt, vol. 1. New York: Academic Press.
Kokshaysky, N. V. 1979. Tracing the wake of a flying bird. *Nature* 279:146–148.
Langston, W., Jr. 1981. Pterosaurs. *Sci. Am.* 244:92–102.
Lawson, D. A. 1975. A pterosaur from the latest Cretaceous of West Texas: discovery of the largest flying creature. *Science* 187:947–948.
Lighthill, M. J. 1975. Aerodynamic aspects of animal flight. In *Swimming and flying in nature,* ed. T. Y.-T. Wu, C. J. Brokaw, and C. Brennen, vol. 2. New York: Plenum Press.
——— 1977. Introduction to the scaling of aerial locomotion. In *Scale effects in animal locomotion,* ed. T. J. Pedley. London: Academic Press.
Lissaman, P. B. S., and C. A. Schollenberg. 1970. Formation flight of birds. *Science* 168:1003–5.
McMasters, J. H. 1976. Aerodynamics of the long pterosaur wing. *Science* 191:899.
Nachtigall, W. 1970. Phasenbeziehungen der Flügelschläge von Gansen während de Verbandflugs in Keilformation. *Z. vergl. Physiol.* 67:414–422.
——— 1979. Gleitflug des Flugbeutlers *Petaurus breviceps papuanus.* II. Filmanalysen zur Einstellung von Gleitbahn und Rumpf sowie zur Steuerung des Gleitflugs. *J. Comp. Physiol.* 133:89–95.
Norberg, R. Å. 1981. Why foraging birds in trees should climb and hop upwards rather than downwards. *Ibis* 123:281–288.
Norberg, U. M. 1969. An arrangement giving a stiff leading edge to the hand wing in bats. *J. Mammal.* 50:766–770.
——— 1970. Functional osteology and myology of the wing of *Plecotus auritus* Linnaeus (Chiroptera). *Ark. Zool.* 22:483–543.
——— 1972. Bat wing structures important for aerodynamics and rigidity. (Mammalia, Chiroptera). *Z. Morph. Tiere* 73:45–61.
——— 1976a. Aerodynamics, kinematics, and energetics of horizontal flapping flight in the long-eared bat *Plecotus auritus. J. Exp. Biol.* 65:179–212.
——— 1976b. Aerodynamics of hovering flight in the long-eared bat *Plecotus auritus. J. Exp. Biol.* 65:459–470.
——— 1979. Morphology of the wings, legs and tail of three coniferous forest tits, the goldcrest and the treecreeper in relation to locomotor pattern and feeding station selection. *Phil. Trans. Roy. Soc. Lond. B* 287:131–165.
——— 1981a. Allometry of bat wings and legs and comparison with birds. *Phil. Trans. Roy. Soc. Lond. B* 292:359–398.
——— 1981b. Flight, morphology and the ecological niche in some birds and bats. *Symp. Zool. Soc. Lond.* 48:173–197.
Odum, E. P., C. E. Connell, and H. L. Stoddard. 1961. Flight energy and estimated flight ranges of some migratory birds. *Auk* 78:515–527.
Oehme, H., H. H. Dathe, and U. Kitzler. 1977. Research on biophysics and physiology of bird flight. IV. Flight energetics in birds. *Fortschr. Zool.* 24:257–273.
Pennycuick, C. J. 1969. The mechanics of bird migration. *Ibis* 111:525–556.
——— 1971a. Gliding flight of the white-backed vulture *Gyps africanus. J. Exp. Biol.* 55:13–38.
——— 1971b. Control of gliding angle in Rüppell's griffon vulture *Gyps rüppellii. J. Exp. Biol.* 55:39–46.
——— 1971c. Gliding flight of the dog-faced bat *Rousettus aegyptiacus* observed in a wind tunnel. *J. Exp. Biol.* 55:833–845.
——— 1972a. *Animal flight.* London: Edward Arnold.
——— 1972b. Soaring behaviour of East African birds, observed from a motor-glider. *Ibis* 114:178–218.
——— 1975. Mechanics of flight. In *Avian biology,* ed. D. S. Farner and J. R. King, vol. 5. London: Academic Press.

——— 1982. The flight of petrels and albatrosses (Procellariiformes), observed in south Georgia and its vicinity. *Phil. Trans. Roy. Soc. Lond. B* 300:75–106.

Pennycuick, C. J., and A. Lock. 1976. Elastic energy storage in primary feather shafts. *J. Exp. Biol.* 64:677–689.

Rayner, J. M. V. 1977. The intermittent flight of birds. In *Scale effects of animal locomotion,* ed. T. J. Pedley. London: Academic Press.

——— 1979a. A vortex theory of animal flight. I. The vortex wake of a hovering animal. *J. Fluid Mech.* 91:697–730.

——— 1979b. A vortex theory of animal flight. II. The forward flight of birds. *J. Fluid Mech.* 91:731–763.

——— 1979c. A new approach to animal flight mechanics. *J. Exp. Biol.* 80:17–54.

——— 1981. Flight adaptations in vertebrates. *Symp. Zool. Soc. Lond.* 48:137–172.

Russel, A. P. 1979. The origin of parachuting locomotion in gekkonid lizards (Reptilia, Gekkonidae). *J. Linn. Soc. Zool.* 65:233–249.

Schaefer, G. W. 1975. Dimensional analysis of avian forward flight. (Unpubl.) Symp. on biodynamics of animal locomotion, Cambridge, Eng. Sept. 1975.

Schmidt-Nielsen, K. 1972. Locomotion: energy cost of swimming, flying, and running. *Science* 177:222–228.

Scott, N. J., and A. Starret. 1974. An unusual breeding aggregation of frogs, with notes on the ecology of *Agalychnis spurrelli* (Anura, Hylidae). *Bull. So. Calif. Acad. Sci.* 73:86–94.

Tucker, V. A. 1973. Bird metabolism during flight: evaluation of a theory. *J. Exp. Biol.* 58:689–709.

Tucker, V. A., and G. C. Parrott. 1970. Aerodynamics of gliding flight in a falcon and other birds. *J. Exp. Biol.* 52:345–367.

Weis-Fogh, T. 1965. Elasticity and wing movements in insects. *Proc. 12th Int. Congr. Ent.*:186–188.

——— 1972. Energetics of hovering flight in hummingbirds and in *Drosophila. J. Exp. Biol.* 56:79–104.

——— 1973. Quick estimates of flight fitness in hovering animals, including novel mechanisms for lift production. *J. Exp. Biol.* 59:169–230.

9. Terrestrial Locomotion without Appendages

Bennet, S., T. McConnell and S. L. Trubatch. 1974. Quantitative analysis of the speed of snakes as a function of peg spacing. *J. Exp. Biol.* 60:161–165.

Bogert, C. M. 1947. Rectilinear locomotion in snakes. *Copeia* 1947:253–254.

Brain, C. K. 1960. Observations on the locomotion of the Southwest African adder, *Bitis peringueyi* (Boulenger), with speculations on the origin of sidewinding. *Ann. Transvaal Mus.* 24:19–24.

Carroll, R. L., and P. J. Currie. 1975. Microsaurs as possible apodan ancestors. *J. Linn. Soc. Zool.* 57:229–247.

Chodrow, R. E., and C. R. Taylor. 1973. Energetic cost of limbless locomotion in snakes. *Federation Proc.* 32:422.

Cowles, R. B. 1956. Sidewinding locomotion in snakes. *Copeia* 1956:211–214.

Engelmann, W. E., and F. J. Obst. 1981. *Snakes.* New York: Exeter.

Gans, C. 1962. Terrestrial locomotion without limbs. *Am. Zool.* 2:167–182.

——— 1970. How snakes move. *Sci. Am.* 223:82–96.

——— 1973. Locomotion and burrowing in limbless vertebrates. *Nature* 242:414–415.

——— 1974. *Biomechanics: an approach to vertebrate biology.* Philadelphia: Lippincott.

——— 1975. Tetrapod limblessness: evolution and functional corollaries. *Am. Zool.* 15:455–467.

Gans, C., and D. Baic. 1977. Regional specialization of reptilian scale surfaces: relation of texture and biologic role. *Science* 195:1348–50.

Gans, C., and H. Mendelssohn. 1972. Sidewinding and jumping progression of vipers. In *Toxins of animal and plant origin,* ed. A. deVries and E. Kochva. London: Gordon and Breach.

Gans, C., H. C. Dessauer, and D. Baic. 1978. Axial differences in the musculature of uropeltid snakes: the freight-train approach to burrowing. *Science* 199:189–192.

Gasc, J.-P. 1974. L'interprétation fonctionnelle de l'appareil musculo-squelettique de l'axe vertébrale chez les serpents (Reptilia). *Mém. Mus. Natl. d'Hist. Nat.* sér. A, 83:1–182.

——— 1981. Axial musculature. In *Biology of the Reptilia,* ed. C. Gans, vol. 11. New York: Academic Press.

Gaymer, R. 1971. New method of locomotion in limbless terrestrial vertebrates. *Nature* 234:150–152.

Gray, J. 1946. The mechanism of locomotion in snakes. *J. Exp. Biol.* 23:101–120.

Gray, J., and H. W. Lissmann. 1950. The kinetics of locomotion of the grass-snake. *J. Exp. Biol.* 26:354–367.

Jayne, B. C. 1982. Comparative morphology of the semispinalis-spinalis muscle of snakes and correlations with locomotion and constriction. *J. Morph.* 172:83–96.

Lissmann, H. W. 1950. Rectilinear locomotion in a snake *(Boa occidentalis). J. Exp. Biol.* 26:368–379.

Wake, M. H., and J. Hanken. 1982. Development of the skull of *Dermophis mexicanus* (Amphibia, Gymnophiona), with comments on skull kinesis and amphibian relationships. *J. Morph.* 173:203–223.

10. Energetics and Locomotion

Alexander, R. McN. 1977. Swimming. In *Mechanics and energetics of animal locomotion,* ed. R. McN. Alexander and G. Goldspink. New York: Halsted Press.

Andersen, K. L. 1960. Energy cost of swimming. *Acta Chirurg. Scand.* 253(suppl.):169–174.

Bakker, R. T. 1972. Locomotor energetics of lizards and mammals compared. *Physiol.* 15:76.

Baudinette, R. V., and K. Schmidt-Nielsen. 1974. Energy cost of gliding flight in herring gulls. *Nature* 248:83–84.

Beamish, F. W. H. 1970. Oxygen consumption of largemouth bass, *Micropterus salmoides,* in relation to swimming speed and temperature. *Can. J. Zool.* 48:1221–28.

Benedict, F. G. 1932. *The physiology of large reptiles.* Carnegie Institution of Washington, publ. 425.

Bennett, A. F. 1978. Activity metabolism of the lower vertebrates. *An. Rev. Physiol.* 40:447–469.

——— 1980a. The metabolic foundations of vertebrate behavior. *Biosci.* 30:452–456.

——— 1980b. The thermal dependence of lizard behaviour. *Anim. Behav.* 28:752–762.

Bennett, A. F., and P. Licht. 1972. Anaerobic metabolism during activity in lizards. *J. Comp. Physiol.* 81:277–288.

Bennett, A. F., and J. A. Ruben. 1979. Endothermy and activity in vertebrates. *Science* 206:649–654.

Bernstein, M. H., S. P. Thomas, and K. Schmidt-Nielsen. 1973. Power input during flight of the fish crow, *Corvus ossifragus. J. Exp. Biol.* 58:401–410.

Brett, J. R. 1964. The respiratory metabolism and swimming performance of young sockeye salmon. *J. Fish. Res. B. Can.* 21:1183–1226.

——— 1965. The relation of size to rate of oxygen consumption and sustained swimming speed of sockeye salmon *(Oncorhynchus nerka). J. Fish. Res. B. Can.* 22:1491–1501.

——— 1972. The metabolic demand for oxygen in fish, particularly salmonids, and a comparison with other vertebrates. *Respiration Physiol.* 14:151–170.

Brett, J. R., and N. R. Glass. 1973. Metabolic rates and critical swimming speeds of sockeye salmon *(Oncorhynchus nerka)* in relation to size and temperature. *J. Fish. Res. B. Can.* 30:379–387.

Chowdrow, R. E., and C. R. Taylor. 1973. Energetic cost of limbless locomotion in snakes. *Federation Proc.* 32:422.

Dawson, T. J., and C. R. Taylor. 1973. Energetic cost of locomotion in kangaroos. *Nature* 246:313–314.

Dawson, W. R., and G. A. Bartholomew. 1956. Relation of oxygen consumption to body weight, temperature, and temperature acclimation in lizards *Uta stansburiana* and *Sceloporus occidentalis. Physiol. Zool.* 29:40–51.

Duthie, G. G. 1982. The respiratory metabolism of temperature-adapted flatfish at rest and during swimming activity and the use of anaerobic metabolism at moderate swimming speeds. *J. Exp. Biol.* 97:359–373.

Fedak, M. A., and H. J. Seeherman. 1979. Re-appraisal of energetics of locomotion shows identical cost in bipeds and quadrupeds including ostrich and horse. *Nature* 282:713–716.

Fedak, M. A., B. Pinshow, and K. Schmidt-Nielsen. 1974. Energy cost of bipedal moving. *Am. J. Physiol.* 227:1038–44.

Fry, F. E. J. 1947. Effects of environment on animal activity. *Publ. Ont. Fish. Res. Lab.* 68:1–62.

——— 1957. The aquatic respiration of fish. In *Physiology of fishes,* ed. M. E. Brown, vol. 1. New York: Academic Press.

Gleeson, T. T. 1979. Foraging and transport cost in the Galapagos marine iguana, *Amblyrhynchus cristatus. Physiol. Zool.* 52:549–557.

Greenewalt, C. H. 1975. The flight of birds. *Trans. Am. Physiol. Soc., n.s.* 65:1–67.

Heglund, N. C., N. A. Fedak, C. R. Taylor, and G. A. Cavagna. 1982. Energetics and mechanics of terrestrial locomotion. IV. Total mechanical energy changes as a function of speed and body size in birds and mammals. *J. Exp. Biol.* 97:57–66.

Hemmingsen, A. M. 1960. Energy metabolism as related to body size and respiratory surfaces, and its evolution. *Reports of the Steno Memorial Hospital and Nordisk Insulinlaboratorium* 9:1–110.

Hoar, W. S., and D. Randall. 1978. *Fish physiology,* vol. 7. New York: Academic Press.

Hoyt, D. F., and C. R. Taylor. 1981. Gait and the energetics of locomotion in horses. *Nature* 292:239–240.

John-Alder, H. B., and A. F. Bennett. 1981. Thermal dependence of endurance and locomotory energetics in a lizard. *Am. J. Physiol.* 241 (*Regulatory Integrative and Comparative Physiology* 10):R342–R349.

Lechner, A. J. 1978. The scaling of maximal oxygen consumption and pulmonary dimensions in small mammals. *Respiration Physiol.* 34:29–44.

Paladino, F. V., and J. R. King. 1979. Energetic cost of terrestrial locomotion: biped and quadruped runners compared. *Can. Rev. Biol.* 38:321–323.

Pennycuick, C. J. 1972. Soaring behaviour and performance of some east African birds, observed from a motor-glider. *Ibis* 114:178–218.

——— 1975. Mechanics of flight. In *Avian biology,* ed. D. S. Farner and J. R. King, vol. 5. New York: Academic Press.

Prange, H. D. 1976. Energetics of swimming of a sea turtle. *J. Exp. Biol.* 64:1–12.

Prange, H. D., and K. Schmidt-Nielsen. 1970. The metabolic cost of swimming in ducks. *J. Exp. Biol.* 53:763–777.

Puede, I. G., and F. G. T. Holliday. 1980. The use of a new tilting tunnel respirometer to investigate some aspects of metabolism and swimming activity of the plaice (*Pleuronectes platessa* L.). J. Exp. Biol. 85:295–309.

Rao, G. M. M. 1968. Oxygen consumption of rainbow trout *(Salmo gairdneri)* in relation to activity and salinity. *Can. J. Zool.* 46:781–786.

Rayner, J. M. V. 1977. The intermittent flight of birds. In *Scale effects in animal locomotion,* ed. T. J. Pedley. New York: Academic Press.

——— 1979. A new approach to animal flight energetics. *J. Exp. Biol.* 80:17–54.

Schmidt-Nielsen, K. 1972. Locomotion: energy cost of swimming, flying, and running. *Science* 177:222–228.

Taylor, C. R. 1977. The energetics of terrestrial locomotion and body size in vertebrates. In *Scale effects in animal locomotion,* ed. T. J. Pedley. New York: Academic Press.

——— 1980. Mechanical efficiency of terrestrial locomotion: a useful concept? In *Aspects of animal movement,* ed. H. Y. Elder and E. R. Trueman. New York: Cambridge University Press.

Taylor, C. R., and N. C. Heglund. 1982. Energetics and mechanics of terrestrial locomotion. *An. Rev. Physiol.* 44:97–107.

Taylor, C. R., and V. J. Rowntree. 1973. Running on two or four legs: which consumes more energy? *Science* 179:186–187.

Taylor, C. R., N. C. Heglund, and G. M. O. Maloiy. 1982. Energetics and mechanics of terrestrial locomotion. I. Metabolic energy consumption as a function of speed and body size in birds and mammals. *J. Exp. Biol.* 97:1–21.

Taylor, C. R., K. Schmidt-Nielsen, and J. C. Raab. 1970. Scaling of energetic cost of running to body size in mammals. *Am. J. Physiol.* 219:1104–7.

Taylor, C. R., N. C. Heglund, T. A. McMahon, and T. R. Looney. 1980. Energetic cost of generating muscular force during running. *J. Exp. Biol.* 86:9–18.

Thomas, S. P. 1975. Metabolism during flight in two species of bats, *Phyllostomus hastatus* and *Pteropus gouldii. J. Exp. Biol.* 63:273–293.

Thomas, S. P., and R. A. Suthers. 1972. The physiology and energetics of bat flight. *J. Exp. Biol.* 57:317–335.

Thompson, S. D., R. E. MacMillen, E. M. Burke, and C. R. Taylor. 1980. The energetic cost of bipedal hopping in small mammals. *Nature* 287:223–224.

Torre-Bueno, J. R., and J. Larochelle. 1978. The metabolic cost of flight in unrestrained birds. *J. Exp. Biol.* 75:223–229.

Tucker, V. A. 1968. Respiratory exchange and evaporative water loss in the flying budgerigar. *J. Exp. Biol.* 48:67–87.

——— 1970. Energetic cost of locomotion in animals. *Comp. Biochem. Physiol.* 34:841–846.

——— 1972. Metabolism during flight in the laughing gull. *Am. J. Physiol.* 222:237–245.

——— 1973. Bird metabolism during flight: evaluation of a theory. *J. Exp. Biol.* 58:689–709.

Videler, J. J., and D. Weihs. 1982. Energetic advantages of burst-and-coast swimming of fish at high speeds. *J. Exp. Biol.* 97:169–178.

Vleck, D., T. T. Gleeson and G. A. Bartholomew. 1981. Oxygen consumption during swimming in Galapagos marine iguanas and its ecological correlates. *J. Comp. Physiol.* 141:531–536.

Wardle, C. S. 1975. Limit of fish swimming speed. *Nature* 255:725–727.

——— 1977. Effect of size on the swimming speeds of fish. In *Scale effects in animal locomotion,* ed. T. J. Pedley. New York: Academic Press.

Wardle, C. S., and J. J. Videler. 1980. Fish swimming. In *Aspects of animal movements,* ed. H. Y. Elder and E. R. Trueman. New York: Cambridge University Press.

Webb, P. W. 1975. Hydrodynamics and energetics of fish propulsion. *Bull. Fish. Res. Bd. Can.* 190:1–158.

——— 1977. Effects of size on performance and energetics of fish. In *Scale effects in animal locomotion,* ed. T. J. Pedley. New York: Academic Press.

Weihs, D. 1973a. Hydromechanics of fish schooling. *Nature* 241:290–291.

——— 1973b. Optimal cruising speed for migrating fish. *Nature* 245:48–50.

——— 1974. Energetic advantages of burst swimming of fish. *J. Theor. Biol.* 48:215–229.

——— 1977. Effects of size on sustained swimming speeds of aquatic organisms. In *Scale effects in animal locomotion,* ed. T. J. Pedley. New York: Academic Press.

Wohlschlag, D. E., J. N. Cameron, and J. J. Cech, Jr. 1968. Seasonal changes in the respiratory metabolism of the pinfish *(Lagodon rhomboides). Contrib. Marine Sci.* 13:89–104.

11. Ventilation

Ballintijn, C. M. 1969a. Functional anatomy and movement coordination of the respiratory pump of the carp (*Cyprinus carpio,* L.). *J. Exp. Biol.* 50:547–567.

——— 1969b. Muscle coordination of the respiratory pump of the carp (*Cyprinus carpio,* L.). *J. Exp. Biol.* 50:569–591.

——— 1969c. Movement pattern and efficiency of the respiratory pump of the carp (*Cyprinus carpio,* L.). *J. Exp. Biol.* 50:596–613.

Brackenbury, J. H. 1971. Airflow dynamics in the avian lungs as determined by direct and indirect methods. *Respiration Physiol.* 13:319–329.

Bramble, D. M., and D. R. Carrier. 1983. Running and breathing in mammals. *Science* 219:251–256.

Bretz, W. L., and K. Schmidt-Nielsen. 1971. Bird respiration: flow patterns in the duck lung. *J. Exp. Biol.* 54:103–118.

Campbell, E. J. M. 1958. The respiratory muscles and the mechanics of breathing. London: Lloyd-Luke.

Clark, B. D., C. Gans, and H. I. Rosenberg. 1978. Airflow in snake ventilation. *Respiration Physiol.* 32:207–212.

de Jongh, H. J., and C. Gans. 1969. On the mechanism of respiration in the bullfrog, *Rana catesbeiana:* a reassessment. *J. Morph.* 127:259–290.

De Troyer, A., S. Kelly, and W. A. Zin. 1983. Mechanical action of the intercostal muscles on the ribs. *Science* 220:87–88.

De Troyer, A., M. Sampson, S. Sigrist, and P. T. Macklem. 1981. The diaphragm: two muscles. *Science* 213:237–238.

Duncker, H. R. 1971. The lung air sac system of birds. *Erg. Anat. Entwickl. Gesch.* 45:1–171.

——— 1972. Structure of avian lungs. *Respiration Physiol.* 14:44–63.

Elshoud, G. C. A. 1978. Respiration in the three-spined stickleback, *Gasterosteus aculeatus* L.: an electromyographic approach. *Neth. J. Zool.* 28:524–544.

Fedde, M. R., R. E. Burger, and R. L. Kitchell. 1961. Anatomic and electromyographic studies of the costo-pulmonary muscles in the cock. *Poultry Sci.* 43:1177–84.

Gans, C. 1970. Strategy and sequence in the evolution of the external gas exchangers of ectothermal vertebrates. *Forma et Functio* 3:61–104.

Gans, C., and B. Clark. 1976. Studies on ventilation of *Caiman crocodilus* (Crocodilia, Reptilia). *Respiration Physiol.* 26:285–301.

Gaunt, A. S., and C. Gans. 1969. Mechanics of respiration in the snapping turtle, *Chelydra serpentina* (Linne). *J. Morph.* 128:195–228.

Grigg, G. C. 1970. Water flow through the gills of Port Jackson sharks. *J. Exp. Biol.* 52:565–568.

Gradwell, N. 1971. *Ascaphus* tadpole: experiments on the suction and gill irrigation mechanisms. *Can. J. Zool.* 49:307–332.

——— 1972a. Gill irrigation in *Rana catesbeiana.* I. On the anatomical basis. *Can. J. Zool.* 50:481–499.

——— 1972b. Gill irrigation in *Rana catesbeiana.* II. On the musculoskeletal mechanism. *Can. J. Zool.* 50:501–521.

Heath, A. G. 1973. Ventilatory responses of teleost fish to exercise and thermal stress. *Am. Zool.* 13:491–503.

Holeton, G. F., and D. R. Jones. 1975. Water flow dynamics in the respiratory tract of the carp (*Cyprinus carpio,* L.). *J. Exp. Biol.* 63:537–549.

Holliday, F. G. T., J. H. S. Blaxter, and R. Lasker. 1964. Oxygen uptake of developing eggs and larvae of the herring *(Clupea harengus). J. Mar. Biol. Ass. U.K.* 44:711–723.

Hughes, G. M. 1960. The mechanism of gill ventilation in the dogfish and skate. *J. Exp. Biol.* 37:11–27.

——— 1975. Coughing in the rainbow trout *(Salmo gairdneri)* and the influences of pollutants. *Rev. Suisse Zool.* 83:47–64.

Hughes, G. M., and C. M. Ballintijn. 1965. The muscular basis of the respiratory pumps in the dogfish *(Scyliorhinus canicula). J. Exp. Biol.* 43:363–383.

Hughes, G. M., and M. Morgan. 1973. The structure of fish gills in relation to their respiratory function. *Biol. Rev.* 48:419–475.

Hughes, G. M., and J. L. Roberts. 1970. A study of the effect of temperature changes on the respiratory pumps of the rainbow trout. *J. Exp. Biol.* 52:177–192.

Kramer, D. L. 1978. Ventilation of the respiratory gas bladder in *Hoplerythrinus unitaeniatus* (Pisces, Characoidei, Erythrinidae). *Can. J. Zool.* 56:921–938.

Lauder, G. V., and K. F. Liem. 1983. The evolution and interrelationships of the actinopterygian fishes. *Bull. Mus. Comp. Zool.* 150:95–197.

Liem, K. F. 1963. The comparative osteology and phylogeny of the Anabantoidei (Teleostei, Pisces). *Ill. Biol. Monogr.* 30:1–149.

——— 1980. Air ventilation in advanced teleosts: biomechanical and evolutionary aspects. In *Environmental physiology of fishes,* ed. M. A. Ali. New York: Plenum Press.

——— 1981. Larvae of air-breathing fishes as countercurrent flow devices in hypoxic environments. *Science* 211:1177-79.

Liem, K. F., and J. W. M. Osse. 1975. Biological versatility, evolution, and food resource exploitation in African cichlid fishes. *Am. Zool.* 15:427-454.

McMahon, B. R. 1969. A functional analysis of the aquatic and aerial respiratory movements of an African lungfish *Protopterus aethiopicus,* with reference to the evolution of the lung-ventilation mechanism in vertebrates. *J. Exp. Biol.* 51:407-430.

Muir, B. S., and R. M. Buckley. 1967. Gill ventilation in *Remora remora. Copeia* 1967:581-586.

Osse, J. W. M. 1969. Functional morphology of the head of the perch (*Perca fluviatilis,* L.): an electromyographic study. *Neth. J. Zool.* 19:289-392.

Peters, H. M. 1978. On the mechanism of air ventilation in anabantoids (Pisces, Teleostei). *Zoomorph.* 89:93-123.

Piiper, J., and P. Scheid. 1973. Gas exchange in avian lungs: models and experimental evidence. In *Comparative physiology,* ed. L. Bolis, K. Schmidt-Nielsen, and S. H. P. Maddrell. Amsterdam: North-Holland.

Rahn, H., K. B. Rahn, B. J. Howell, C. Gans, and S. M. Tenney. 1971. Air breathing of the garfish *(Lepisosteus osseus).* Respiration Physiol. 11:285-307.

Rankin, J., and J. A. Dempsey. 1967. Respiratory muscles and the mechanisms of breathing. *Am. J. Phys. Med.* 46:198-244.

Rosenberg, H. I. 1973. Functional anatomy of pulmonary ventilation in the garter snake, *Thamnophis elegans. J. Morph.* 140:171-184.

Scheid, P., and J. Piiper. 1971. Direct measurement of the pathway of respired gas in duck lungs. *Respiration Physiol.* 11:308-314.

Teichmann, H. 1959. Über den Atemmechanismus bei Haifischen und Rochen. *Z. vergl. Physiol.* 41:449-455.

Wassersug, R. J., and K. Hoff. 1979. A comparative study of the buccal pumping mechanism of tadpoles. *Biol. J. Linn. Soc.* 12:225-259.

Whiting, H. P., and Q. Bone. 1980. Ciliary cells in the epidermis of the larval Australian dipnoan, *Neoceratodus. Zool. J. Linn. Soc.* 68:125-137.

12. Aquatic Feeding in Lower Vertebrates

Alexander, R. McN. 1967. *Functional design in fishes.* London: Hutchinson.

——— 1970. Mechanics of the feeding action of various teleost fishes. *J. Zool., Lond.* 162:145-156.

Anker, G. C. 1974. Morphology and kinetics of the head of the stickleback, *Gasterosteus aculeatus. Trans. Zool. Soc. Lond.* 32:311-416.

de Jongh, H. J. 1968. Functional morphology of the jaw apparatus of larval and metamorphosing *Rana temporaria. Neth. J. Zool.* 18:1-103.

Denison, R. H. 1961. Feeding mechanisms of agnatha and early gnathostomes. *Am. Zool.* 1:177-181.

Durbin, A. G., and E. G. Durbin. 1975. Grazing rates of the Atlantic menhaden *Brevoortia tyrannus* as a function of particle size and concentration. *Mar. Biol.* 33:265-277.

Elshoud-Oldenhave, M. J. W. 1979. Prey capture in the pike-perch, *Stizostedion lucioperca* (Teleostei, Percidae): a structural and functional analysis. *Zoomorph.* 93:1-32.

Gans, C. 1969. Comments on inertial feeding. *Copeia* 1969:855-857.

Gordon, K. 1984. Models of tongue movement in the walrus *(Odobenus rosmarus). J. Morph.* 182.

Gosline, W. A. 1971. *Functional morphology and classification of teleostean fishes.* Honolulu: University Press of Hawaii.

Gradwell, N. 1972. Hydrostatic pressures and movements of the lamprey *Petromyzon* during suction, olfaction, and gill ventilation. *Can. J. Zool.* 50:1215-23.

——— 1975. The bearing of filter feeding on the water pumping mechanism of *Xenopus* tadpoles (Anura, Pipidae). *Acta Zool.* 56:119-128.

Gregory, W. K. 1933. Fish skulls: a study in the evolution of natural mechanisms. *Trans. Am. Phil. Soc.* 23:75-481.

Grobecker, D. B., and T. W. Pietsch. 1979. High-speed cinematographic evidence for ultrafast feeding in antennariid anglerfishes. *Science* 205:1161-62.

Hardisty, M. W. 1979. *Biology of the cyclostomes.* London: Chapman and Hall.

Hughes, G. M., and G. Shelton. 1958. The mechanism of gill ventilation in three freshwater teleosts. *J. Exp. Biol.* 35:807-823.

Jenkin, P. M. 1957. The filter-feeding and food of flamingoes (Phaenicopteri). *Phil. Trans. Roy. Soc. Lond. B.* 240:401-493.

Jorgensen, C. B. 1966. *Biology of suspension feeding.* London: Pergamon Press.

LaBarbera, M. 1978. Particle capture by a Pacific brittle star: experimental test of the aerosol suspension feeding model. *Science* 201:1147-49.

Lanyon, L. E. 1976. The measurement of bone strain in vivo. *Acta orth. belg.* 42(suppl. 1):98–108.

Lauder, G. V. 1979. Feeding mechanisms in primitive teleosts and in the halecomorph fish *Amia calva*. *J. Zool., Lond.* 187:543–578.

——— 1980a. The suction feeding mechanism in sunfishes *(Lepomis)*: an experimental analysis. *J. Exp. Biol.* 88:49–72.

——— 1980b. Hydrodynamics of prey capture by teleost fishes. In *Biofluid mechanics*, ed. D. Schenck, vol. 2. New York: Plenum Press.

——— 1980c. The role of the hyoid apparatus in the feeding mechanism of the living coelacanth, *Latimeria chalummae*. *Copeia* 1980:1–9.

——— 1980d. Evolution of the feeding mechanism in primitive actinopterygian fishes: a functional anatomical analysis of *Polypterus*, *Lepisosteus*, and *Amia*. *J. Morph.* 163:283–317.

——— 1982. Patterns of evolution in the feeding mechanism of actinopterygian fishes. *Am. Zool.* 22:275–285.

——— 1983a. Prey capture hydrodynamics in fishes: experimental tests of two models. *J. Exp. Biol.* 104:1–14.

——— 1983b. Functional design and evolution of the pharyngeal jaw apparatus in euteleostean fishes. Zool. J. Linn. Soc. 77:1–38.

Lauder, G. V., and L. E. Lanyon. 1980. Functional anatomy of feeding in the bluegill sunfish, *Lepomis macrochirus*: in vivo measurement of bone strain. *J. Exp. Biol.* 84:33–55.

Lauder, G. V., and K. F. Liem. 1981. Prey capture by *Luciocephalus pulcher*: implications for models of jaw protrusion in teleost fishes. *Env. Biol. Fish.* 6:257–268.

Leong, R. J. H., and C. P. O'Connell. 1969. A laboratory study of particulate and filter feeding of the northern anchovy *(Engraulis mordax)*. *J. Fish. Res. Bd. Can.* 26:557–582.

Liem, K. F. 1967. Functional morphology of the head of the anabantoid teleost fish, *Helostoma temmincki*. *J. Morph.* 121:135–158.

——— 1970. Comparative functional anatomy of the Nandidae. Field. Zool. 56:1–166.

——— 1978. Modulatory multiplicity in the functional repertoire of the feeding mechanism in cichlid fishes. *J. Morph.* 158:323–360.

——— 1980. Acquisition of energy by teleosts: adaptive mechanisms and evolutionary patterns. In *Environmental physiology of fishes*, ed. M. A. Ali. New York: Plenum Press.

Liem, K. F., and P. H. Greenwood. 1981. A functional approach to phylogeny of the pharyngognath teleosts. *Am. Zool.* 21:83–101.

Mallatt, J. 1979. Surface morphology of structures within the pharynx of the larval lamprey *Petromyzon marinus*. *J. Morph.* 162:249–273.

——— 1981. The suspension feeding mechanism of the larval lamprey *Petromyzon marinus*. *J. Zool., Lond.* 194:103–142.

——— 1982. Pumping rates and particle retention efficiencies of the larval lamprey, an unusual suspension feeder. *Biol. Bull.* 163:197–210.

Matthes, E. 1934. Bau und Funktion der Lippensäume wasserlebender Urodelen. *Z. Morph. Ökol. Tiere* 28:155–169.

Moore, J. W., and J. M. Mallatt. 1980. Feeding of larval lamprey. *Can. J. Fish. Aquat. Sci.* 37:1658–64.

Moss, S. A. 1972. The feeding mechanism of sharks of the family Carcharhinidae. *J. Zool., Lond.* 167:423–436.

——— 1979. Feeding mechanisms in sharks. *Am. Zool.* 17:355–364.

Moy-Thomas, J. A., and R. S. Miles. 1971. *Palaeozoic fishes*. Philadelphia: Saunders.

Nelson, G. J. 1969. Gill arches and the phylogeny of fishes with notes on the classification of vertebrates. *Bull. Am. Mus. Nat. Hist.* 141:475–552.

Nyberg, D. W. 1971. Prey capture in the largemouth bass. *Am. Midland Nat.* 86:128–144.

Osse, J. W. M. 1969. Functional morphology of the head of the perch (*Perca fluviatilis* L.): an electromyographic study. *Neth. J. Zool.* 19:289–392.

Osse, J. W. M., and M. Muller. 1980. A model of suction feeding in fishes with some implications for ventilation. In *Environmental physiology of fishes*, ed. M. A. Ali. New York: Plenum Press.

Ozeti, N., and D. B. Wake. 1969. The morphology and evolution of the tongue and associated structures in salamanders and newts (family Salamandridae). *Copeia* 1969:91–123.

Pietsch, T. W. 1978. The feeding mechanism of *Stylephorus chordatus* (Teleostei, Lampridiformes): functional and ecological implications. *Copeia* 1978:255–262.

Randall, D. J. 1971. Respiration. In *The biology of lampreys*, ed. M. W. Hardisty and I. C. Potter. London: Academic Press.

Reynolds, T. L. 1931. Hydrostatics of the suctorial mouth of the lamprey. *Univ. Calif. Publ. Zool.* 37:15–34.

Rosen, R. A., and D. C. Hales. 1981. Feeding of paddlefish, *Polyodon spathula*. *Copeia* 1981:441–455.

Rubenstein, D. I., and M. A. R. Koehl. 1977. The mechanisms of filter feeding: some theoretical considerations. *Am. Nat.* 111:981–994.

Schaeffer, B., and D. E. Rosen. 1961. Major adaptive

levels in the evolution of the actinopterygian feeding mechanism. *Am. Zool.* 1:187–204.

Seale, D. B., and R. J. Wassersug. 1979. Suspension feeding dynamics of anuran larvae related to their functional morphology. *Oecologia* 39:259–272.

Shaffer, H. B., and G. V. Lauder. In press (a). Patterns of variation in ambystomatid salamanders: kinematics of the feeding mechanism. *Evol.*

——— In press (b). Patterns of variation in ambystomatid salamanders: muscle activity during prey capture. *J. Morph.*

Shafland, J. L. 1968. Functional and anatomical convergence in body form and feeding behavior in three diverse species of freshwater bottom dwelling vertebrates. Ph. D. diss. University of Chicago.

Springer, S. 1961. Dynamics of the feeding mechanism of large galeoid sharks. *Am. Zool.* 1:183–185.

Streeter, V. L., and E. B. Wylie. 1979. *Fluid mechanics,* 7th ed. New York: McGraw Hill.

Tchernavin, V. V. 1948. On the mechanical working of the head of bony fishes. *Proc. Zool. Soc. Lond.* 118:129–143.

——— 1953. *The feeding mechanisms of a deep sea fish* Chauliodus sloani *Schneider.* London: British Museum (Nat. Hist.).

Vrba, E. S. 1968. Contributions to the functional morphology of fishes. V. The feeding mechanism of *Elops saurus* Linnaeus. *Zool. Africana* 3:211–236.

Wassersug, R. 1972. The mechanism of ultraplanktonic entrapment in anuran larvae. *J. Morph.* 137:279–288.

——— 1980. Internal oral features of larvae from eight anuran families: functional, systematic, evolutionary and ecological considerations. *Univ. Kans. Mus. Nat. Hist. Misc. Publ.* 68:1–146.

Wassersug, R., and K. Hoff. 1979. A comparative study of the buccal pumping mechanism of tadpoles. *Biol. J. Linn. Soc.* 12:225–259.

Webb, P. W. 1978. Fast-start performance and body form in seven species of teleost fish. *J. Exp. Biol.* 74:211–226.

Weisel, G. F. 1973. Anatomy and histology of the digestive system of the paddlefish *(Polyodon spathula). J. Morph.* 140:243–251.

Winterbottom, R. 1974. A descriptive synonymy of the striated muscles of the Teleostei. *Proc. Acad. Nat. Sci. Philad.* 125:225–317.

Zweers, G. A. 1974. Structure, movement, and myography of the feeding apparatus of the mallard (*Anas platyrhynchos* L.): a study in functional anatomy. *Neth. J. Zool.* 24:323–467.

13. Feeding Mechanisms of Lower Tetrapods

Albright, G. R., and E. M. Nelson. 1959a. Cranial kinetics of the generalized colubrid snake, *Elaphe obsoleta quadrivittata.* I. Descriptive morphology. *J. Morph.* 105:293–241.

——— 1959b. Cranial kinetics of the generalized colubrid snake, *Elaphe obsoleta quadrivittata.* II. Functional morphology. *J. Morph.* 105:241–292.

Alexander, R. McN. 1969. Mechanics of the feeding action of a cyprinid fish. *Proc. Zool. Soc. Lond.* 159:1–15.

Arnold, S. J. 1983. Morphology, performance and fitness. *Am. Zool.* 23:347–361.

Bas, C. 1957. On the relation between the masticatory muscles and the surface of the skull in *Ardea cinerea* (L.). *Proc. Konin. nederl. akad. wet.* 60:480–485.

Beecher, W. J. 1953. Feeding adaptations and systematics in the avian order Piciformes. *J. Wash. Acad. Sci.* 43:293–299.

——— 1962. The bio-mechanics of the bird skull. *Bull. Chicago Acad. Sci.* 11:10–33.

Bemis, W. E., K. Schwenk, and M. H. Wake. 1983. Morphology and function of the feeding apparatus in *Dermophis mexicanus* (Amphibia, Gymnophiona). *Zool. J. Linn. Soc.* 77:75–96.

Bishop, S. C. 1941. *The Salamanders of New York.* New York State Museum, Bulletin 324.

Boag, P. T., and P. R. Grant. 1981. Intense natural selection in a population of Darwin's finches (Geospizinae) in the Galapagos. *Science* 214:82–85.

Boas, J. E. V. 1929. Biologisch-Anatomische Studien über den Hals der Vögel. *D. Kgl. Danske Vidensk. Selsk. Skrifter, Naturvidensk og Mathem.* Afd., 9. Række I. 3:100–222.

Bock, W. J. 1964. Kinetics of the avian skull. *J. Morph.* 114:1–42.

——— 1966. An approach to the functional analysis of bill shape. *Auk* 81:10–51.

Bolt, J. R., and R. J. Wassersug. 1975. Functional morphology of the skull in *Lysorophus:* a snake-like Paleozoic amphibian (Lepospondyli). *Paleobiol.* 3:320–332.

Boltt, R. E., and R. F. Ewer. 1964. The functional anatomy of the head of the puff adder, *Bitis arietans* (Merr.). *J. Morph.* 114:83–106.

Bradley, D. C. 1903. The muscles of mastication and the movements of the skull in lacertilia. *Zool. Jahrb. Abt. Morph.* 18:475–486.

Bramble, D. M. 1973. Media dependent feeding in turtles. *Am. Zool.* 13:1342.

——— 1974. Occurrence and significance of the Os

Transiliens in gopher tortoises. *Copeia* 1974:102–109.
——— 1978. Origin of the mammalian feeding complex: models and mechanisms. *Paleobiol.* 4:271–301.
——— 1980. Feeding in tortoises and mammals: why so similar? *Am. Zool.* 20:931 (abstract).
Burt, W. H. 1930. Adaptive modifications in the woodpeckers. *Univ. Calif. Publ. Zool.* 32:455–524.
Busby, A. 1982. Form and function of the jaw musculature of *Alligator mississipiensis.* Ph.D. diss., University of Chicago.
Camp, C. L. 1923. Classification of the lizards. *Bull. Am. Mus. Nat. Hist.* 48:289–481.
Caple, G., R. P. Balda, and W. R. Willis. 1983. The physics of leaping and the evolution of preflight. *Am. Nat.* 121:455–476.
Carroll, R. L. 1977. The origins of lizards. In *Problems in vertebrate evolution,* ed. S. M. Andrews, R. S. Miles, and A. D. Walker. Linnean Soc. Symp. Ser., 4. London: Academic Press.
Cope, E. D. 1900. The crocodilians, lizards, and snakes of North America. *Rep. U.S. Nat. Mus.* 1898:153–1294.
Crompton, A. W., and N. Hotton III. 1967. Functional morphology of the masticatory apparatus of two dicynodonts (Reptilia, Therapsida). *Postilla* 109:1–51.
Crompton. A. W., P. Cook, K. Hiiemae, and A. J. Thexton. 1975. Movement of the hyoid apparatus during chewing. *Nature* 258:69–70.
Crompton, A. W., A. J. Thexton, P. Parker, and K. Hiiemae. 1977. The activity of the hyoid and jaw muscles during chewing on soft food in the American opossum. In *The biology of marsupials,* ed. D. Gilmore and B. Robinson, vol. 2. London: Macmillan.
Cundall, D. 1983. Activity of head muscles during feeding by snakes: a comparative study. *Am. Zool.* 23:383–396.
Cundall, D., and C. Gans. 1979. Feeding in water snakes: an electromyographic study. *J. Exp. Zool.* 209:189–208.
Cundall, D., and H. Greene. 1982. Evolution of the feeding apparatus in alethinophidian snakes. *Am. Zool.* 22:294.
Dellow, P. G. 1976. The general physiological background of chewing and swallowing. In *Mastication and swallowing: biological and clinical correlates,* ed. B. J. Sessle and A. G. Hannam. Toronto: University of Toronto Press.
Dellow, P. G., and J. P. Lund. 1971. Evidence for central timing of rhythmical mastication. *J. Physiol.* (Lond.) 215:1–13.
Dullemeijer, P. 1956. The functional morphology of the head of the common viper, *Vipera berus* (L.). *Arch. Neerl. Zool.* 11:387–497.
——— 1959. A comparative functional-anatomical study of the heads of some Viperidae. *Morph. Jahrb.* 99:881–985.
Emerson, S. B. 1976. A preliminary report on the superficial throat musculature of the Microhylidae and its possible role in tongue action. *Copeia* 1976:546–551.
——— 1977. Movement of the hyoid in frogs during feeding. *Am. J. Anat.* 149:115–120.
Emerson, S. B., and D. Diehl. 1980. Toe pad morphology and mechanisms of sticking in frogs. *Biol. J. Linn. Soc.* 13:199–216.
Frazzetta, T. H. 1962. A functional consideration of cranial kinesis in lizards. *J. Morph.* 111:287–319.
——— 1966. Morphology and function of the jaw apparatus in *Python sebae* and *Python molurus. J. Morph.* 118:217–296.
——— 1983. Adaptation and function of cranial kinesis in reptiles: a time-motion analysis of feeding in alligator lizards. In *Advances in herpetology and evolutionary biology,* ed. A. G. J. Rhodin and K. Miyata. Cambridge, Mass.: Harvard University Museum of Comparative Zoology.
Gabe, M., and H. St. Girons. 1964. Contribution á l'histologie de *Sphenodon punctatus* Gray. Paris: Edit. Cent. Nat. Rech. Sci.
Gaffney, E. S. 1983. The basicranial articulation of the Triassic turtle, *Proganochelys.* In *Advances in herpetology and evolutionary biology,* ed. A. G. J. Rhodin and K. Miyata. Cambridge, Mass.: Harvard University Museum of Comparative Zoology.
Gans, C. 1952. The functional morphology of the egg-eating adaptations in the snake genus *Dasypeltis. Zoologica* (N.Y.) 37:209–244.
——— 1961a. The feeding mechanism of snakes and its possible evolution. *Am. Zool.* 1:217–227.
——— 1961b. A bullfrog and its prey: a look at the bio-mechanics of jumping. *Nat. Hist.* 70:26–37.
——— 1962. The tongue protrusion mechanism in *Rana catesbeiana. Am. Zool.* 2:524.
——— 1967. The chameleon. *Nat. Hist.* 76:52–59.
——— 1969. Comments on inertial feeding. *Copeia* 1969:855–857.
——— 1974. *Biomechanics.* Philadelphia and Toronto: Lippincott.
——— 1983. Snake feeding strategies and adaptations: conclusion and prognosis. *Am. Zool.* 23:455–460.
Gans, C., and G. Gorniak. 1982a. How does the toad flip its tongue? Test of two hypotheses. *Science* 216:1335–37.

——— 1982b. Functional morphology of lingual protrusion in marine toads *(Bufo marinus). Am. J. Anat.* 163:195–222.

Gans, C., and M. Oshima. 1952. Adaptations for egg eating in the snake *Elaphe climacophora* (Boie). *Am. Mus. Novit.* 1571:1–16.

Gnanamuthu, C. P. 1930. The anatomy and mechanism of the tongue of *Chamaeleon carcaratus* (Merrem). *Proc. Zool. Soc. Lond.* 1:467–486.

Gorniak, G. C., H. I. Rosenberg, and C. Gans. 1982. Mastication in the Tuatara, *Sphenodon punctatus* (Reptilia, Rhyncocephalia): structure and activity of the motor system. *J. Morph.* 171:321–353.

Greene, H. W. 1976. Scale overlap: a directional sign stimulus for prey ingestion by ophiophagous snakes. *Z. Tierpsychol.* 41:113–120.

——— 1982. Dietary and phenotypic diversity in lizards: why are some organisms specialized? In *Environmental adaptation and evolution: a theoretical and empirical approach,* ed. D. Mossakowski and G. Roth. Stuttgart: Fischer.

——— 1983. Dietary correlates of the origin and radiation of snakes. *Am. Zool.* 23:431–441.

Greene, H. W., and G. M. Burghardt. 1978. Behavior and phylogeny: constriction in ancient and modern snakes. *Science* 200:74–77.

Harkness, L. 1977. Chameleons use accommodation cues to judge distance. *Nature* 267:346–349.

Hendricks, H. 1965. Vergleichende Untersuchung des Wiederkauverhaltens. *Biol. Zbl.* 84:651–751.

Hiiemae, K. M. 1978. Mammalian mastication: a review of the activity of the jaw muscles and the movements they produce in chewing. In *Development, function, and evolution of teeth,* ed. P. M. Butler and K. A. Joysey. London: Academic Press.

Hiiemae, K. M., and A. W. Crompton. 1971. A cinefluorographic study of feeding in the American opossum, *Didelphis marsupialis.* In *Dental morphology and evolution,* ed. A. A. Dahlberg. Chicago: University of Chicago Press.

Hiiemae, K. M., A. J. Thexton, and A. W. Crompton. 1979. Intra-oral food transport: the fundamental mechanism of feeding. In *Muscle adaptation in the cranio-facial region,* ed. D. S. Carlson and J. A. McNamara, Jr. Ann Arbor: University of Michigan Center for Human Growth and Development.

Hilyard, N., and H. Biggin. 1977. *Physics for applied biologists.* Baltimore: University Park Press.

Homberger, D. G. 1978. Functional morphology of the parrot tongue *(Psittacus erithacus). Am. Zool.* 18:623.

——— 1980a. Funktionell-morphologische Untersuchungen zur Radiation der Ernährungs-und Trinkmethoden der Papageien (Psittaci). *Bonn. Zool. Mon.* 13:1–192.

——— 1980b. Functional morphology and evolution of the feeding apparatus in parrots, with special reference to the Pesquet's parrot, *Psittrichas fulgidus* (Lesson). In *Conservation of New World parrots,* ed. R. F. Pasquier. Washington, D.C.: Smithsonian Institution Press.

——— 1982. Hydraulic structures in the avian lingual apparatus. *Am. Zool.* 22:943.

——— In press. Functional morphology of the lingual apparatus of the African grey parrot, *Psittacus erithacus* Linné (Aves, Psittacidae): description and theoretical mechanical analysis of a complex system. *Ornithol. Monog.* 38.

Horton, P. 1982. The diversity and systematic significance of anuran tongue musculature. *Copeia* 1982:595–602.

Hutchinson, J. C. D., and W. W. Taylor. 1962. Mechanics of pecking grain. *Proc. 12th World Poultry Cong.* pt. 1:112–116.

Impey, O. R. 1967. Functional aspects of cranial kinetism in the Lacerta. Ph.D. diss., Oxford University.

Iordansky, N. N. 1966. Cranial kinesis in lizards: contribution to the problem of the adaptive significance of skull kinesis. *Zool. Zhur.* 45:1398–1410. (Trans. Smithsonian Herpetological Information Services.)

——— 1970. Structure and biomechanical analysis of functions of the jaw muscles in lizards. *Anat. Anz.* 127:383–413.

Jenkin, P. M. 1957. The filter-feeding and food of flamingoes (Phoenicopteri). *Phil. Trans. Roy. Soc. Lond. B* 240:401–493.

Jenkins, F. A., and G. E. Goslow. 1983. The functional anatomy of the savannah monitor lizard *(Varanus exanthematicus). J. Morph.* 175:195–216.

Jollie, M. 1960. The head skeleton of the lizard. *Acta Zool.* 41:1–64.

Kardong, K. V. 1974. Kinesis of the jaw apparatus during the strike in the cottonmouth snake, *Agkistrodon piscivorus. Forma et Functio* 7:327–354.

——— 1977. Kinesis of the jaw apparatus during swallowing in the cottonmouth snake, *Agkistrodon piscivorus. Copeia* 1977:338–348.

——— 1979. "Protovipers" and the evolution of snake fangs. *Evol.* 33:433–443.

——— 1980. Evolutionary patterns in advanced snakes. *Am. Zool.* 20:269–282.

Kathariner, L. 1900. Die Mechanik des Bisses der Solenoglyphen Giftschlangen. *Biol. Zbl.* 20:45–53.

Kay, R. F., and K. M. Hiiemae. 1974. Jaw movements and tooth use in recent and fossil primates. *Am. J. Phys. Anthrop.* 40:227–256.

Kier, W. M., and K. K. Smith. 1983. Tongues, tentacles and trunks: the biomechanics of movement in muscular-hydrostats. *Am. Zool.* 23:904.

Klauber, L. M. 1939. A statistical study of the rattlesnakes, VI. *Occ. Pap. S. Diego Soc. Nat. Hist.* 30:3–61.

Krebs, U. 1979. Der Dumeril-Waran *(Varanus dumerilii)* ein spezialisierter Krabbenfresser? (Reptilia, Sauria, Varanidae). *Salamandra* 15:146–157.

Larson, J. H., Jr., and D. J. Guthrie. 1976. The feeding system of terrestrial tiger salamanders (*Ambystoma tigrinum melanostictum* Baird). *J. Morph.* 147:137–154.

Lauder, G. V. 1983a. Neuromuscular patterns and the origin of trophic specialization in fishes. *Science* 219:1235–37.

——— 1983b. Functional and morphological bases of trophic specialization in sunfishes (Teleostei, Centrarchidae). *J. Morph.* 178:1–21.

——— In press. Functional morphology of the feeding mechanism in lower vertebrates. In *Functional morphology of vertebrates,* ed. H. R. Duncker and G. Fleischer. New York: Springer-Verlag.

Liem, K. F., and L. S. Kaufman. 1984. Intraspecific macroevolution: functional biology of the polymorphic cichlid species *Cichlasoma minckleyi.* In *Evolutionary biology of species flocks,* ed. A. Echelle and I. Kornfield. Orono: University of Maine Press.

Liem, K. F., H. Marx, and G. B. Rabb. 1971. The viperid snake *Azemiops:* its comparative cephalic musculature and phylogenetic position in relation to Viperinae and Crotalinae. *Field. Zool.* 59:65–126.

Lombard, R. E., and D. B. Wake. 1976. Tongue evolution in the lungless salamanders family Plethodontidae. I. Introduction, theory and a general model of dynamics. *J. Morph.* 148:259–290.

——— 1977. Tongue evolution in the lungless salamanders family Plethodontidae. II. Function and evolutionary diversity. *J. Morph.* 153:39–80.

Lund, J. P. 1976. Evidence for a central neural pattern generator regulating the chewing cycle. In *Mastication,* ed. D. J. Anderson and B. Matthews. Bristol: John Wright.

McDowell, S. B. 1972. The evolution of the tongue of snakes and its bearing on snake origins. *Evol. Biol.* 6:191–273.

——— 1978. Homology mapping of the primitive archosaurian reptile palate on the palate of birds. *Evol. Theory* 4:81–94.

McDowell, S. B., and C. M. Bogert. 1954. The systematic position of *Lanthanotus* and the affinities of the anguinomorphan lizards. *Bull. Am. Mus. Nat. Hist.* 105:1–142.

McLelland, J. 1980. Digestive system. In *Form and function of birds,* ed. A. S. King and J. McLelland, vol. 1. London: Academic Press.

Mudge, G. P. 1903. On the morphology of the tongue of parrots with a classification of the order, based upon the structure of the tongue. *Trans. Zool. Soc. Lond.* 16:211–278.

Ostrom, J. H. 1974. *Archaeopteryx* and the origin of flight. *Q. Rev. Biol.* 49:27–47.

Özeti, N., and D. B. Wake. 1969. The morphology and evolution of the tongue and associated structures in salamanders and newts (family Salamandridae). *Copeia* 1969:91–123.

Panchen, A. L. 1964. The cranial anatomy of two Coal Measure anthracosaurs. *Phil. Trans. Roy. Soc. Lond. B* 247:593–637.

Pianka, E. R. 1966. Convexity, desert lizards and spacial heterogeneity. *Ecol.* 47:1055–59.

Poglayen-Neuwall, I. 1954. Die Kiefermuskulatur der Eidechsen und ihre Innervation. *Z. Wiss. Zool.* 158:79–132.

Pough, F. H., and J. D. Groves. 1983. Specializations of the body form and food habits of snakes. *Am. Zool.* 23:443–454.

Price, I. 1935. Notes on the braincase of *Captorhinus. Proc. Boston Soc. Nat. Hist.* 40:377–386.

Regal, P. J. 1966. Feeding specializations and the classification of terrestrial salamanders. *Evol.* 20:392–407.

Regal, P. J., and C. Gans. 1976. Functional aspects of the evolution of frog tongues. *Evol.* 30:718–734.

Rieppel, O. 1977. Über die Entwicklung des Basicranium bei *Chelydra serpentina* Linnaeus (Chelonia) und *Lacerta sicula* Rafinesque (Lacertilia). *Verhandl. naturf. Ges. Basel* 86:153–170.

——— 1978a. The phylogeny of cranial kinesis in lower vertebrates, with special reference to the Lacertilia. *N. Jahrb. Geol. Palaont.* 156:353–370.

——— 1978b. Streptostyly and muscle function in lizards. *Experientia* 34:776–777.

——— 1980. The evolution of the ophidian feeding system. *Zool. Jahrb. Anat.* 103:551–564.

Robinson, P. L. 1966. The evolution of the Lacertilia. *Colloques Internationaux de Centre National de la Recherche Scientifique* 163:395–407.

Romer, A. S. 1937. The braincase of the carboniferous crossopterygian, *Megalichthys nitidus. Bull. Mus. Comp. Zool.* 82:1–73.

——— 1956. *Osteology of the reptiles.* Chicago: University of Chicago Press.

Rosenberg, H. I., and C. Gans. 1976. Lateral jaw muscles of *Elachistodon westermanni* Reinhardt (Reptilia, Serpentes). *Can. J. Zool.* 54:510–521.

——— 1977. Preliminary analysis of mastication in *Sphenodon punctatus. Am. Zool.* 17:871.

Ruben, J. A. 1977. Morphological correlates of predatory modes in the coachwhip *(Masticophis flagellum)* and the rosy boa *(Lichanura roseofusca). Herpetol.* 33:1–6.

Savitzky, A. H. 1980. The role of venom delivery strategies in snake evolution. *Evol.* 34:1194–1204.

Schwenk, K. 1983. Functional morphology and evolution of the chameleon tongue tip. *Am. Zool.* 23:1028.

——— 1984. Functional and evolutionary morphology of the lizard tongue. Ph.D. diss., University of California at Berkeley.

Severtsov, A. S. 1964. Formation of the tongue in the Hynobiidae. *Dokl. Akad. Nauk SSSR. Biol. Sci. Sect.* 154:34–37 (trans.).

——— 1966. Food-seizing mechanism in Urodela larvae. *Dokl. Akad. Nauk SSSR. Biol. Sci. Sect.* 168:230–233 (trans.).

——— 1971. The mechanism of food capture in the tailed amphibians. *Dokl. Akad. Nauk SSSR. Biol. Sci. Sect.* 197:185–187 (trans.).

Shafland, J. L. 1968. Functional and anatomical convergence in body form and feeding behavior in three diverse species of freshwater bottom dwelling vertebrates. Ph.D. diss., University of Chicago.

Smith, K. K. 1978. Electromyographic and cineradiographic studies of feeding in lizards. *Am. Zool.* 18:623.

——— 1979. The use of the tongue and hyoid apparatus during feeding in lizards. *Am. Zool.* 19:1012.

——— 1980. Mechanical significance of streptostyly in lizards. *Nature* 283:778–779.

——— 1982. An electromyographic study of the function of the jaw adducting muscles in *Varanus exanthematicus* (Varanidae). *J. Morph.* 173:137–158.

——— 1984. The use of the tongue and hyoid apparatus during feeding in lizards (*Ctenosaura similis* and *Tupinambis nigropunctatus*). *J. Zool., Lond.* 202:115–143.

Smith, K. K., and W. L. Hylander. 1982. Strain gage measurement of mesokinetic movement in *Varanus exanthematicus. Am. Zool.* 22:924.

Sondhi, K. C. 1958. The hyoid and associated structures in some Indian reptiles. *Ann. Zool.* 2:155–239.

Sponder, D. and G. V. Lauder. 1981. Terrestrial feeding in the mudskipper *Periophthalmus* (Pisces, Teleostei): a cineradiographic analysis. *J. Zool., Lond.* 193:517–530.

Spring, L. W. 1965. Climbing and pecking adaptations in some North American woodpeckers. *Condor* 67:457–488.

Tanner, W. W., and D. F. Avery. 1982. Buccal floor of reptiles: a summary. *Great Basin Nat.* 42:273–349.

Thexton, A. J. 1973. Some aspects of neurophysiology of dental interest. I. Theories of oral function. *J. Dent.* 2:49–54.

——— 1974. Oral reflexes and neural oscillators. *J. Dent.* 2:137–141.

——— 1976. To what extent is mastication programmed and independent of peripheral feedback? In *Mastication,* ed. D. J. Anderson and B. Matthews. Bristol: John Wright.

Thexton, A. J., K. M. Hiiemae, and A. W. Crompton. 1980. Food consistency and bite size as regulators of jaw movement during feeding in the cat. *J. Neurophysiol.* 44:456–474.

Thexton, A. J., D. B. Wake, and M. H. Wake. 1977. Control of tongue function in the salamander *Bolitoglossa occidentalis. Arch. Oral Biol.* 22:361–366.

Thomson, K. S. 1967. Mechanisms of intracranial kinetics in fossil rhipidistian fishes and their relatives. *J. Linn. Soc. Zool.* 46:223–253.

Throckmorton, G. S. 1976. Oral food processing in two herbivorous lizards, *Iguana iguana* (Iguanidae) and *Uromastix aegyptius* (Agamidae). *J. Morph.* 148:365–390.

——— 1978. Action of the pterygoideus muscle during feeding in the lizard *Uromastix aegyptius* (Agamidae). *Anat. Rec.* 190:217–222.

——— 1980. The chewing cycle in the herbivorous lizard *Uromastix aegyptius* (Agamidae). *Arch. Oral Biol.* 25:225–233.

Throckmorton, G. S., and L. K. Clarke. 1981. Intracranial joint movements in the agamid lizard *Amphibolurus barbatus. J. Exp. Zool.* 216:25–35.

Trueb, L., and C. Gans. 1981. Tongue protrusion in *Rhinophrynus:* a mechanism unique in frogs. *Am. Zool.* 21:970.

——— 1983. Feeding specializations of the Mexican burrowing toad, *Rhinophrynus dorsalis* (Anura, Rhinophrynidae). *J. Zool., Lond.* 199:189–208.

Underwood, G. 1957. On lizards of the family Pygopodidae: a contribution to the morphology and phylogeny of the Squamata. *J. Morph.* 100:207–268.

Versluys, J. 1912. Das Streptostylie-Problem und die Bewegungen im Schädel bei Sauropsiden. *Zool. Jahrb. Anat.* Suppl. 152:545–716.

Wake, D. B. 1966. Comparative osteology and evolution of the lungless salamanders, family Plethodontidae. *Mem. So. Calif. Acad. Sci.* 4:1–111.

——— 1982. Functional and developmental constraints and opportunities in the evolution of feeding systems in urodeles. In *Environmental adaptation and evolution,* ed. D. Mossakowski, and G. Roth. Stuttgart: Fischer.

Wake, M. H., and J. Hanken. 1982. Development of the skull of *Dermophis mexicanus* (Amphibia, Gymno-

phiona), with comments on skull kinesis and amphibian relationships. *J. Morph.* 173:203–223.
White, S. S. 1968. Mechanisms involved in deglutition in *Gallus domesticus*. *J. Anat.* 104:177.
Winokur, R. M. 1974. Adaptive modifications of the buccal mucosae in turtles. *Am. Zool.* 13:1347–48.
Zeigler, H. P., and H. J. Karten. 1973. Brain mechanisms and feeding behavior in the pigeon *(Columba livia)*. II. Analysis of feeding behavior deficits following lesions of quinto-frontal structures. *J. Comp. Neurol.* 152:83–102.
Zeigler, H. P., P. W. Levitt, and R. R. Levine. 1980. Eating in the pigeon *(Columba livia)*: movement patterns, stereotypy and stimulus control. *J. Comp. Physiol. Psychol.* 94:783–794.
Zeigler, H. P., M. Miller, and R. R. Levine. 1975. Trigeminal nerve and eating in the pigeon *(Columba livia)*: neurosensory control of the consummatory responses. *J. Comp. Physiol. Psychol.* 89:845–858.
Ziswiler, V., and D. S. Farner. 1972. Digestion and digestive system. In *Avian biology*, ed. D. S. Farner and J. R. King, vol. 2. New York: Academic Press.
Zoond, A. 1933. The mechanism of projection of the chameleon's tongue. *J. Exp. Biol.* 10:174–185.
Zusi, R. L. 1959. The function of the depressor mandibulae muscle in certain passerine birds. *Auk* 76:537–539.
——— 1967. The role of the depressor mandibulae muscle in kinesis of the avian skull. *Proc. U.S. Nat. Mus.* 123:1–28.
Zweers, G. A. 1974. Structure, movement and myography of the feeding apparatus of the mallard *(Anas platyrhynchos)*. *Neth. J. Zool.* 24:323–467.
——— 1981. Morphology and mechanics of the larynx of the pigeon (*Columba livia* L.): a drillchuck system (Aves). *Zoomorph.* 99:37–69.
——— 1982a. Pecking of the pigeon (*Columba livia* L.). *Behavior* 81:173–230.
——— 1982b,c,d. The feeding system of the pigeon (*Columba livia* L.). In *Advances in Anatomy, Embryology, and Cell Biology*, ed. F. Beck, W. Hild, J. van Limborgh, R. Ortmann, J. E. Pauly, and T. H. Schiebler, vol. 73. Berlin: Springer-Verlag. Pt. I, The lingual apparatus; pt. II, The mouth and pharynx; pt. III, Mechanisms of the feeding system.
Zweers, G. A., A. F. C. Gerritsen, and P. van Kranenburg-Voogd. 1977. Mechanics of feeding of the mallard (*Anas platyrhynchos* L., Aves, Anseriformes). In *Contributions to Vertebrate Evolution*, ed. M. K. Hecht and F. S. Szalay, vol. 3. Basel: Karger.

14. Mastication, Food Transport, and Swallowing

Abd-el-Malek, S. 1955. The part played by the tongue in mastication and deglutition. *J. Anat.* 89:250–254.
Anderson, D. J., and B. Matthews. 1976. *Mastication.* Bristol: Wright.
Ardran, G. M., F. H. H. Kemp, and W. D. L. Ride. 1958. A radiographic analysis of mastication and swallowing in the domestic rabbit. *Proc. Zool. Soc. Lond.* 130:257–274.
Basmajian, J. V., and G. Stecko. 1962. A new bipolar electrode for electromyography. *J. Appl. Physiol.* 17:849.
Bramble, D. 1980. Feeding in tortoises and mammals: why so similar? *Am. Zool.* 20:931 (abstract).
Bremer, F. 1923. Physiologie nerveuse de la mastication chez le chat et lapin. *Arch. int. physiol.* 21:309–352.
Covert, H. H., and R. F. Kay. 1981. Dental microwear and diet: implications for determining the feeding behaviors of extinct primates, with a comment on the dietary pattern of *Sivapithecus*. *Am. J. Phys. Anthrop.* 55:311–316.
Crompton, A. W. 1971. The origin of the tribosphenic molar. In *Early mammals,* ed. D. M. Kermack and K. A. Kermack. London: Academic Press.
Crompton, A. W., and K. M. Hiiemae. 1970. Molar occlusion and mandibular movements during occlusion in the American opossum, *Didelphis marsupialis*. *Zool. J. Linn. Soc.* 49:21–47.
Crompton, A. W., and W. D. Hylander. In press. Changes in mandibular function following the acquisition of a squamosodentary articulation. In *The ecology and biology of mammal-like reptiles,* ed. J. J. Roth, E. C. Roth, P. D. MacLean, and N. Hotton III. Washington, D.C.: Smithsonian Institution.
Crompton, A. W., and F. A. Jenkins. 1979. *The origin of mammals.* In *Mesozoic mammals,* ed. J. A. Lillegraven, Z. Kielan-Jaworowska, and W. A. Clemens. Berkeley: University of California Press.
Crompton, A. W., and Z. Kielan-Jaworowska. 1978. Molar structure and occlusion in cretaceous therian mammals. In *Development, function, and evolution of teeth,* ed. P. M. Butler and K. A. Joysey. New York: Academic Press.
Crompton, A. W., and D. L. Sponder. 1981. The mechanism of food transport. *J. Dent Res.* 60:474 (abstract).
——— 1982. Basic pattern of activity of oral musculature during feeding and swallowing in mammals. *Am. J. Phys. Anthrop.* 57:178 (abstract).
Crompton, A. W., A. J. Thexton, P. Parker, and K. M.

Hiiemae. 1977. The activity of the hyoid and jaw muscles during chewing of soft food in the opossum. In *The biology of marsupials,* ed. B. Stonehouse and D. Gillmore. London: MacMillan.

Edgeworth, F. H. 1935. *The cranial muscles of vertebrates.* Cambridge: Cambridge University Press.

Fish, D. R., and F. C. Mendel. 1982. Mandibular movement patterns relative to food types in common tree shrews *(Tupaia glis). Am. J. Phys. Anthropol.* 58: 1–15.

Franks, H. A., A. W. Crompton, and W. D. Hylander. 1981. Food transport during chewing and swallowing in primate and non-primate herbivores. *J. Dent. Res.* 60:475 (abstract).

——— 1982. A comparative perspective: primate mastication and deglutition. *Am. J. Phys. Anthrop.* 57:189 (abstract).

Goldberg, L. J., and S. H. Chandler. 1981. Evidence for pattern generator control of the effects of spindle afferent input during rhythmical jaw movements. *Can. J. Physiol.* 59:707–712.

Gordon, K. R. 1984a. Microfracture patterns in abrasive wear striations on teeth indicate directionality. *Am. J. Phys. Anthrop.* 63:315–322.

——— 1984b. Models of tongue movement in the walrus *(Odobenus rosmarus). J. Morph.* 182.

Gorniak, G. C. 1977. Feeding in golden hamsters *(Mesocritus auratus). J. Morph.* 154:427–458.

Gorniak, G. C., and C. Gans. 1980. Quantitative assay and electromyograms during mastication in domestic cats. *J. Morph.* 163:253–281.

Greaves, W. S. 1978. The jaw lever system in ungulates: a new model. *J. Zool., Lond.* 184:271–285.

Grine, F. 1981. Trophic differences between "gracile" and "robust" australopithecines: a scanning electronmicroscope analysis of occlusal events. *So. Afr. J. Sci.* 77:203–230.

Herring, S., and R. P. Scapino. 1974. Physiology of feeding in miniature pigs. *J. Morph.* 141:427–460.

Hiiemae, K. M. 1971. The structure and function of the jaw muscles in the rat *(Rattus norvegicus).* III. The mechanics of the muscles. *Zool. J. Linn. Soc.* 50:111–132.

——— 1976. Masticatory movements in primitive mammals. In *Mastication,* ed. D. Anderson and B. Matthews. Bristol: Wright.

——— 1978. Mammalian mastication: a review of activity of jaw muscles and the movements they produce in chewing. In *development, function and evolution of teeth,* ed. P. M. Butler and K. A. Joysey. London: Academic Press.

——— 1984. Functional aspects of primate jaw morphology. In *Food acquisition and processing in primates,* ed. D. J. Chivers, B. A. Wood, and A. Bilsborough. New York: Plenum Press.

Hiiemae, K. M., and Abbas, P. 1981. Viscosity as a regulator of lapping in cats. *J. Dent. Res.* 60:474.

Hiiemae, K. M., and G. M. Ardran. 1968. A cinefluorographic study of mandibular movement during feeding in the rat *Rattus norvegicus. J. Zool. Lond.* 154:139–154.

Hiiemae, K. M., and A. W. Crompton. 1971. A cinefluorographic study of feeding in the American opossum. In *Dental morphology and evolution,* ed. A. A. Dahlberg. Chicago: University of Chicago Press.

Hiiemae, K. M., and F. A. Jenkins, Jr. 1969. The anatomy and internal architecture of the muscles of mastication in the American opossum, *Didelphis marsupialis. Postilla* 140:1–49.

Hiiemae, K. M., and R. F. Kay. 1973. Evolutionary trends in the dynamics of primate mastication. In *Symp. 4th Intl. Congr. Primatol.,* vol. 3, *Craniofacial biology of primates,* ed. M. R. Zingeser. Basel: Karger.

Hiiemae, K. M., A. J. Thexton, and A. W. Crompton. 1978. Intraoral transport: a fundamental mechanism of feeding? In *Muscle adaption in the craniofacial region,* ed. D. Carlson and J. McNamara. Monogr. no. 8., University of Michigan.

Hiiemae, K. M., A. J. Thexton, J. McGarrick, and A. W. Crompton. 1981. The movement of the cat hyoid during feeding. *Arch. Oral Biol.* 26:65–81.

Hylander, W. L. 1975. Human mandible: lever or link? *Am. J. Phys. Anthrop.* 43:227.

——— 1977. In vivo bone strain in the mandible of *Galago crassicaudatus. Am. J. Phys. Anthrop.* 46:309–326.

——— 1979. Mandibular function in *Galago crassicaudatus* and *Macaca fascicularis:* an in vivo approach to stress analysis of the mandible. *J. Morph.* 159:253–296.

——— 1984. Stress and strain in the mandibular symphysis of primates: a test of competing hypotheses. *Am. J. Phys. Anthrop.* 64:1–46.

Hylander, W. L., and A. W. Crompton. 1980. Loading patterns and jaw movement during the masticatory power stroke in macaques. *Am. J. Phys. Anthrop.* 52:239.

Janis, C. 1982. Aspects of the evolution of herbivory in ungulate mammals. Ph.D. diss., Harvard University.

Jenkins, G. N. 1978. *Physiology and biochemistry of the mouth.* Oxford: Blackwell Scientific.

Kallen, F. C., and C. Gans. 1972. Mastication in the little brown bat *Myotis lucifuges. J. Morph.* 136:385–420.

Kay, R. F., and H. H. Covert. 1983. True grit: a microwear experiment. *Am. J. Phys. Anthrop.* 61:33–38.

Kay, R. F., and K. M. Hiiemae. 1974. Jaw movement and tooth use in recent and fossil primates. *Am. J. Phys. Anthrop.* 40:227–256.

Lauder, G. 1979. Feeding mechanics in primitive teleosts and in the halecomorph fish *Amia calva*. *J. Zool., Lond.* 187:543–578.

——— 1982. Patterns of evolution in the feeding mechanisms of Actinopterygian fishes. *Am. Zool.* 22:275–285.

Livingstone, R. M. 1956. Some observations on the natural history of the tongue. *Ann. Roy. Coll. Surg. Eng.* 19:185–200.

Lowe, A. A. 1981. The neural control of tongue movements. *Prog. Neurobiol.* 15:295–344.

Lucas, P. W. 1979. The dental-dietary adaptations of mammals. *N. Jahrb. Geol. Paläontol. Mh.* 8:486–512.

——— 1982. Basic principles of tooth design. In *Teeth: form, function, and evolution,* ed. B. Kurten. New York: Columbia University Press.

Lucas, P. W., and D. A. Luke. 1984. Chewing it over: basic principles of food breakdown. In *Food acquisition and processing in primates,* ed. D. Chivers, B. A. Wook, and A. Bilsborough. New York: Plenum Press.

Luschei, E. S., and L. J. Goldberg. 1981. Neural mechanisms of mandibular control: mastication and voluntary biting. In *Handbook of physiology,* ed. V. B. Brooks, sec. I, *The nervous system,* ed. J. M. Brookhart and V. B. Mountcastle, vol. 2, *Motor control.* Bethesda: American Physiological Society.

McGarick, J., and A. J. Thexton. 1981. A method for a computer-based analysis of jaw and hyoid movement. *Arch. Oral Biol.* 26:247–257.

Miller, A. J. 1982. Deglutition. *Physiol. Rev.* 62:129–184.

Mills, J. R. E. 1967. A comparison of lateral jaw movements in some mammals from wear facets on the teeth. *Arch. Oral Biol.* 12:645–661.

Osborn, J. W., and A. G. S. Lumsden. 1980. An alternative to thegosis and a reexamination of the ways in which mammalian molars work. *N. Jahrb. Geol. Paläontol. Abh.* 156:371–392.

Rensburger, S. 1973. An occlusion model for mastication and dental wear in herbivorous mammals. *J. Palaeon.* 47:515–528.

Ryan, A. S. 1979. Wear striation direction on primate teeth: a scanning electron microscope examination. *Am. J. Phys. Anthrop.* 50:155–161.

Sessle, B. J., and A. G. Hannam. 1976. *Mastication and swallowing: biological and chemical correlates.* Toronto: University of Toronto Press.

Sheine, W. S., and R. F. Kay. 1977. An analysis of chewed food particle size and its relationship to molar structure in the primates *Cheirogaleus medius* and *Galago senegalensis* and the insectivoran *Tupaia glis*. *Am. J. Phys. Anthrop.* 47:15–20.

Smith, K. 1982. An electromyographic study of the function of the jaw adducting muscles in *Varanus exanthematicus* (Varanidae). *J. Morph.* 173:137–158.

Thexton, A. J. 1981. Tongue and hyoid movements in the cat. In *Oral-facial sensory and motor functions,* ed. Y. Kawamura and R. Dubner. Chicago: Quintessence Publ.

Thexton, A. J., K. M. Hiiemae, and A. W. Crompton. 1980. Food consistency and bite size as regulators of mastication in the cat. *J. Neurophysiol.* 44:456–474.

Thexton, A. J., J. McGarrick, K. M. Hiiemae, and A. W. Crompton. 1982. Hyomandibular relationships during feeding in the cat. *Arch. Oral Biol.* 27:793–801.

Throckmorton, G. S. 1976. Oral food processing in two herbivorous lizards, *Iguana iguana* (Iguanidae) and *Uromastix aegyptus* (Agamidae). *J. Morph.* 148:363–390.

——— 1980. The chewing cycle in the herbivorous lizard *Uromastix aegyptus* (Agamidae). *Arch. Oral Biol.* 25:225–233.

Turnbull, W. D. 1970. Mammalian masticatory apparatus. *Field. Geol.* 18:149–356.

von Koenigswald, W. 1982. Enamel structure in the molars of Arvicolidae (Rodentia, Mammalia): a key to functional morphology and phylogeny. In *Teeth: form, function and evolution,* ed. B. Kurten. New York: Columbia University Press.

Walker, A. S., H. N. Hoeck, and L. Perez. 1978. Microwear of mammalian teeth as an indicator of diet. *Science* 201:908–910.

Weijs, W. A. 1975. Mandibular movements of the albino rat during feeding. *J. Morph.* 145:107–124.

Weijs, W. A., and R. Dantuma. 1975. Electromyography and mechanics of mastication in the albino rat. *J. Morph.* 146:1–34.

——— 1981. Functional anatomy of the masticatory apparatus in the rabbit. *Neth. J. Zool.* 31:99–147.

Weijs, W. A., and H. J. de Jongh. 1977. Strain in mandibular alveolar bone during mastication in the rabbit. *Arch. Oral Biol.* 22:667–675.

15. The Octavolateralis System

Baird, I. 1974. Anatomical features of the inner ear in submammalian vertebrates. In *Handbook of sensory physiology,* ed. W. Keidel and W. Neff, vol. 5. New York: Springer-Verlag.

Capranica, R. R. 1976. Morphology and physiology of the auditory system. In *Frog neurobiology,* ed. R. Llinas and W. Prect. Berlin: Springer-Verlag.

Chapman, C. J., and O. Sand. 1974. Field studies of hearing in two species of flatfish, *Pleuronectes platessa* (L.) and *Limanda limanda* (L.) (family Pleuronectidae). *Comp. Biochem. Physiol.* 47A:371–385.

Chung, S. H., A. G. Pettigrew, and M. Anson. 1981. Hearing in the frog: dynamics of the middle ear. *Proc. Roy. Soc. Lond. B* 212:459–485.

Coombs, S., and A. N. Popper. 1979. Hearing differences among Hawaiian squirrelfish (family Holocentridae) related to differences in the peripheral auditory system. *J. Comp. Physiol.* 132A:203–207.

Corwin, J. T. 1981. Audition in elasmobranchs. In *Hearing and sound communication in fishes,* ed. W. N. Tavolga, A. N. Popper, and R. R. Fay. New York: Springer-Verlag.

Crawford, A. C., and R. Fettiplace. 1980. The frequency selectivity of auditory nerve fibres and hair cells in the cochlea of the turtle. *J. Physiol.* 306:79–125.

Dijkgraaf, S. 1963. The functioning and significance of the lateral line organ. *Biol. Rev.* 38:51–105.

Dooling, R. 1973. Behavioral audiometry with the parakeet *(Melopsittacus undulatus). J. Acoust. Soc. Am.* 53:1957–58.

——— 1982. Auditory perception in birds. In *Acoustic communication in birds,* ed. D. E. Koadsma and E. H. Miller, vol. 1. New York: Academic Press.

Dooling, R., J. A. Mulligan, and J. D. Miller. 1971. Auditory sensitivity and song spectrum of the common canary *(Serinus canarius). J. Acoust. Soc. Am.* 50:700–709.

Enger, P. S. 1981. Frequency discrimination in teleosts: central or peripheral? In *Hearing and sound communication in fishes,* ed. W. N. Tavolga, A. N. Popper, and R. R. Fay. New York: Springer-Verlag.

Fay, R. R. 1978. Sound detection and sensory coding by the auditory systems of fishes. In *The behavior of fish and other aquatic animals,* ed. D. I. Mostofsky. New York: Academic Press.

Fay, R. R., and A. N. Popper. 1980. Structure and function in teleost auditory systems. In *Comparative studies of hearing in vertebrates,* ed. A. N. Popper and R. R. Fay. New York: Springer-Verlag.

Feng, A. S. 1981. Directional characteristics of the acoustic receiver of the leopard frog *(Rana pipiens):* a study of eighth nerve auditory responses. *J. Acoust. Soc. Am.* 68:1107–14.

Gourevitch, G. 1980. Directional hearing in terrestrial mammals. In *Comparative studies of hearing in vertebrates,* ed. A. N. Popper and R. R. Fay. New York: Springer-Verlag.

Guinan, J. J., Jr., and W. T. Peake. 1967. Middle-ear characteristics of anesthetized cats. *J. Acoust. Soc. Am.* 41:1237–61.

Henson, O. W., Jr. 1974. Comparative anatomy of the middle ear. In *Handbook of sensory physiology,* ed. W. Keidel and W. Neff, vol. 5. New York: Springer-Verlag.

Hudspeth, A. J. 1983. The hair cells of the inner ear. *Sci. Am.* 248:54–64.

Hudspeth, A. J., and D. P. Corey. 1977. Sensitivity, polarity, and conductance change in the response of vertebrate hair cells to controlled mechanical stimuli. *Proc. Nat. Acad. Sci., U.S.* 74:2407–11.

Iversen, R. T. B. 1967. Response of the yellowfin tuna *(Thunnus albacares)* to underwater sound. In *Marine bio-acoustics,* ed. W. N. Tavolga, vol. 2. Oxford: Pergamon Press.

——— 1969. Auditory thresholds of the scombrid fish *Euthynnus affinis,* with comments on the use of sound in tuna fishing. FAO Conference on Fish Behaviour in Relation to Fishing Techniques and Tactics, *FAO Fisheries rep.* no. 62, 3:849–859.

Johnstone, B. M., and K. J. Taylor. 1971. Physiology of the middle-ear transmission system. *J. Otolaryngol. Soc. Australia* 3:226–228.

Khanna, S. 1983. Interpretation of the sharply tuned basilar membrane responses observed in the cochlea. In *Hearing and other senses: presentations in honor of E. G. Wever,* ed. R. R. Fay and G. Gourevitch. Groton, Conn.: Amphora Press.

Khanna, S., and D. G. B. Leonard. 1982. Basilar membrane tuning in the cat cochlea. *Science* 215:305–306.

Kiang, N. 1975. Stimulus representation in the discharge patterns of auditory neurons. In *The nervous system,* ed. E. L. Eagles. New York: Raven Press.

Knudsen, E. 1980. Sound localization in birds. In *Comparative studies of hearing in vertebrates,* ed. A. N. Popper and R. R. Fay. New York: Springer-Verlag.

Konishi, M. 1973. How the owl tracks its prey. *Am. Sci.* 61:414–424.

Lewis, E. R., E. Leverenz, and H. Koyama, 1982. The tonotopic organization of the bullfrog amphibian papilla, an auditory organ lacking a basilar membrane. *J. Comp. Physiol.* 145:437–445.

Manley, G. 1982. A review of the auditory physiology of the reptiles. In *Progress in sensory physiology,* ed. D. Ottoson, vol. 2. Berlin: Springer-Verlag.

Masterton, B., H. Heffner, and R. Ravizza. 1969. The evolution of human hearing. *J. Acoust. Soc. Am.* 45:966–985.

Miller, M. 1980. The reptilian cochlear duct. In *Comparative studies of hearing in vertebrates,* ed. A. N. Popper and R. R. Fay. New York: Springer-Verlag.

Moffat, A. J. M., and R. R. Capranica. 1978. Middle ear sensitivity in anurans and reptiles measured by light scattering spectroscopy. *J. Comp. Physiol.* 127:97–107.

Platt, C., and A. N. Popper. 1981. Fine structure and function of the ear. In *Hearing and sound communication in fishes,* ed. W. N. Tavolga, A. N. Popper, and R. R. Fay. New York: Springer-Verlag.

Popper, A. N. 1970. Auditory capacities of the Mexican blind cave fish *(Astyanax jordani)* and its eyed ancestor *(Astyanax mexicanus). Anim. Behav.* 18:552–562.

Popper, A. N., and S. Coombs. 1980. Auditory mechanisms in teleost fishes. *Am. Sci.* 68:429–440.

——— 1982. The morphology and evolution of the ear in Actinopterygian fishes. *Am. Zool.* 22:311–328.

Retzius, G. 1881. *Das Gehörorgan der Wirbelthiere,* vol. 1. Stockholm: Samson and Wallin.

Russell, I. J., and P. M. Sellick. 1978. Intracellular studies of hair-cells in the mammalian cochlea. *J. Physiol.* 284:261–290.

Sachs, M. B., N. K. Woolf, and J. M. Sinnott. 1980. Response properties of neurons in the avian auditory system: comparisons with mammalian homologues and considerations of the neural encoding of complex stimuli. In *Comparative studies of hearing in vertebrates,* ed. A. N. Popper and R. R. Fay. New York: Springer-Verlag.

Sand, O. 1981. The lateral-line and sound reception. In *Hearing and sound communication in fishes,* ed. W. N. Tavolga, A. N. Popper, and R. R. Fay. New York: Springer-Verlag.

Saunders, J. C., and S. P. Dear. 1983. Comparative morphology of stereocilia. In *Hearing and other senses: presentations in honor of E. G. Wever,* ed. R. R. Fay and G. Gourevitch. Groton, Conn: Amphora Press.

Schuijf, A., and R. J. A. Buwalda. 1980. Underwater localization: a major problem in fish acoustics. In *Comparative studies of hearing in vertebrates,* ed. A. N. Popper and R. R. Fay. New York: Springer-Verlag.

Schwartzkopff, J. 1973. Mechanoreception. In *Avian biology,* ed. D. S. Farner, J. R. King, and K. C. Parkes, vol. 3. New York: Academic Press.

Smith, C. A. 1975. The inner ear: its embryological development and microstructure. In *The nervous system,* ed. D. B. Tower. New York: Raven Press.

——— 1982. Recent advances in structural correlates of auditory receptors. In *Sensory physiology,* ed. D. Ottoson, vol. 2. New York: Springer-Verlag.

Stebbins, W. C. 1970. Studies of hearing loss in the monkey. In *Animal psychophysics,* ed. W. C. Stebbins. New York: Appleton-Century-Crofts.

——— 1980. The evolution of hearing in mammals. In *Comparative studies of hearing in vertebrates,* ed. A. N. Popper and R. R. Fay. New York: Springer-Verlag.

Takasaka, T., and C. A. Smith. 1971. The structure and innervation of the pigeon's basilar papilla. *J. Ultrastructure Res.* 35:20–65.

Tanaka, K., and C. A. Smith. 1978. Structure of the chicken's inner ear: SEM and TEM study. *Am. J. Anat.* 153:251–272.

Turner, R. G. 1980. Physiology and bioacoustics in reptiles. In *Comparative studies of hearing in vertebrates,* ed. A. N. Popper and R. R. Fay. New York: Springer-Verlag.

Turner, R. G., A. A. Muraski, and D. W. Nielsen. 1981. Cilium length: influence on neural tonotopic organization. *Science* 213:1519–21.

van Bergeijk, W. A. 1967. The evolution of vertebrate hearing. In *Contributions to sensory physiology,* ed. W. D. Neff. New York: Academic Press.

von Békèsy, G. 1960. *Experiments in hearing.* New York: McGraw-Hill.

von Frisch, K. 1936. Über den Gehörsinn der Fische. *Biol. Rev.* 11:210–246.

Weiss, T. F., M. J. Mulroy, R. G. Turner, and C. L. Pike. 1976. Tuning of single fibers in the cochlear nerve of the alligator lizard: relation to receptor morphology. *Brain Res.* 115:71–90.

Wever, E. G. 1949. *Theory of hearing.* New York: Dover.

——— 1967. The tectorial membrane of the lizard ear: species variation. *J. Morph.* 123:355–371.

——— 1970. The lizard ear: Scincidae. *J. Morph.* 133:277–292.

——— 1973. The ear and hearing in the frog, *Rana pipiens. J. Morph.* 114:461–478.

——— 1978. *The reptile ear.* Princeton: Princeton University Press.

——— 1981. The role of the amphibians in the evolution of the vertebrate ear. *Am. J. Otolaryngol.* 2:145–152.

Wever, E. G., and M. Lawrence. 1954. *Physiological acoustics.* Princeton: Princeton University Press.

16. The Vertebrate Eye

Ali, M. A. 1959. The ocular structure, retinomotor and photobehavioral responses of juvenile Pacific salmon. *Can. J. Zool.* 37:965–996.

——— 1975. Retinomotor responses. In *Vision in fishes,* ed. M. A. Ali. New York: Plenum Press.

Ali, M. A., and M. Anctil. 1976. *Retinas of fishes: an atlas.* Heidelberg: Springer-Verlag.

Ali, M. A., F. I. Harosi, and H.-J. Wagner. 1978. Photoreceptors and visual pigments in a cichlid fish, *Nannacara annomala. Sens. Proc.* 2:130–145.

Baylor, D. D., and A. L. Hodgkin. 1973. Detection and resolution of visual stimuli by turtle photoreceptors. *J. Physiol.* (Lond.) 234:163.

Boycott, B. B., and J. E. Dowling. 1969. Organization of the primate retina: light microscopy. *Phil. Trans. Roy. Soc. Lond. B* 225:109–184.

Burkhardt, D. A., J. Gottesman, J. S. Levine, and E. F. MacNichol, Jr. 1983. Cellular mechanisms for color coding in holostean retinas and the evolution of color vision. *Vision Res.* 23:1031–41.

Cajal, S. R. 1933. La rétine des vertèbres. *Trab. Lab. Invest. Biol. Univ. Madrid* 28:1–141. (Trans. and presented as Appendix I in Rodieck, 1973).

Cornsweet, T. N. 1970. *Visual perception.* New York: Academic Press.

Dartnall, H. J. A. 1975. Assessing the fitness of visual pigments for their photic environment. In *Vision in fishes,* ed. M. A. Ali. New York: Plenum Press.

Davson, H. 1980. *Physiology of the eye,* 4th ed. New York: Academic Press.

Daw, N. W. 1973. Neurophysiology of color vision. *Physiol Rev.* 53:571–611.

Dowling, J. E. 1974. Functional organization of the vertebrate retina. In *Retina congress,* ed. R. C. Pruett and D. J. Regan. New York: Prentice Hall.

——— 1979. Information processing by local circuits: the vertebrate retina as a model system. In *The neurosciences: fourth study program,* ed. F. O. Schmidt and F. G. Worden. Cambridge, Mass.: MIT Press.

Harosi, F. I. 1976. Spectral relations of cone pigments in goldfish. *J. Gen. Physiol.* 68:65–80.

Hashimoto, Y., and M. Inokuchi. 1981. Characteristics of second-order neurons in the dace retina: physiological and morphological studies. *Vision Res.* 21:1541–50.

Kaneko, A. 1970. Physiological and morphological identification of horizontal, bipolar, and amacrine cells in goldfish retina. *J. Physiol.* 207:623–633.

Kinney, J. S., S. M. Luria, and D. O. Weitzman. 1967. Visibility of colours underwater. *J. Opt. Soc. Am.* 75:802–809.

Kolb, H. 1970. Organization of the outer plexiform layer of the primate retina: electron microscopy of Golgi-impregnated cells. *Phil. Trans. Roy. Soc. Lond. B.* 258:261–283.

Land, E. H. 1977. The retinex theory of color vision. *Sci. Am.* 237:108–128.

Levine, J. S., and E. F. MacNichol, Jr. 1979. Visual pigments in teleost fishes: effects of habitat, microhabitat, and behavior on visual system evolution. *Sens. Proc.* 3:95–131.

Levine, J. S., P. S. Lobel, and E. F. MacNichol, Jr. 1980. Visual communication in fishes. In *Environmental physiology of fishes,* ed. M. A. Ali. New York: Plenum Press.

Levine, J. S., E. F. MacNichol, Jr., T. Kraft, and B. A. Collins. 1979. Intraretinal distribution of cone pigments in certain teleost fishes. *Science* 204:523–526.

Liebman, P. A., and A. M. Granda. 1971. Microspectrophotometric measurement of visual pigments in two species of turtle, *Pseudemys scripta* and *Chelonia mydas. Vision Res.* 11:105.

Lipetz, L. E. 1978. A model of function at the outer plexiform layer of the Cyprinid retina. In *Frontiers in visual science,* ed. S. J. Cool and E. L. Smith III. New York: Springer-Verlag.

Lythgoe, J. N. 1979. *The ecology of vision.* Oxford: Clarendon Press.

MacNichol, E. F., Jr. 1978. A photon-counting microspectrophotometer for the study of single vertebrate photoreceptor cells. In *Frontiers in visual science,* ed. S. J. Cool and E. L. Smith III. New York: Springer-Verlag.

MacNichol, E. F., Jr., R. Feinberg, and F. I. Harosi. 1973. Colour discrimination processes in the retina. In *Colour 73,* London: Adam Hilger.

Marc, R. E. 1977. Chromatic patterns of cone photoreceptors. *Am. J. Opt. and Physiol. Optics* 54:212–225.

Marc, R. E., and H. G. Sperling. 1976. The chromatic organization of the goldfish cone mosaic. *Vision Res.* 16:1211–24.

——— 1977. Chromatic organization of primate cones. *Science* 196:454–456.

McCann, J. J., S. P. McKee, and T. Taylor. 1976. Quantitative studies in retinex theory: the "color Mondrian" experiments. *Vision Res.* 16:445–458.

McFarland, W. N., and F. W. Munz. 1975a. Pt. II: The photic environment of clear tropical seas during the day. *Vision Res.* 15:1063–70.

——— 1975b. Pt. III: The evolution of photopic visual pigments in fishes. *Vision Res.* 15:1071–80.

Munz, F. W., and W. N. McFarland. 1975. Pt. I: Presumptive cone pigments extracted from tropical marine fishes. *Vision Res.* 15:1045–62.

Naka, K. I., and W. A. H. Rushton. 1966a. An attempt to analyze colour reception by electrophysiology. *J. Physiol.* 185:556–586.

——— 1966b. S-potentials from colour units in the retina of fish (Cyprinidae). *J. Physiol.* 185:536–555.

Newton, I. 1730. *Opticks: or a treatise of the reflections, refractions, inflections and colours of light,* 4th ed. London: Printed for W. Innys.

Northmore, D. P. M., and D. Yager. 1975. Psychophysical methods for investigations of vision in fishes. In *Vision in fishes,* ed. M. A. Ali. New York: Plenum Press.

Richter, A., and E. J. Simon. 1974. Electrical responses of double cones in the turtle retina. *J. Physiol.* 242:673–683.

Rodieck, R. W. 1973. *The vertebrate retina: principles of structure and function.* San Francisco: W. H. Freeman.

Sirovich, L., and I. Abramov. 1977. Photoproducts and pseudopigments. *Vision Res.* 17:5–16.

Sivak, J. G. 1978. A study of vertebrate strategies for vision in air and water. In *Sensory ecology: review and perspectives,* ed. M. A. Ali. New York: Plenum Press.

Smith, R. C., and J. E. Tyler. 1967. Optical properties of natural water. *J. Opt. Soc. Am.* 57:589–601.

Stell, W. K. 1967. The structure and relationships of horizontal cells and photoreceptor-bipolar synaptic complexes in goldfish retina. *Am. J. Anat.* 121:401–423.

Stell, W. K., and F. I. Harosi. 1976. Cone structure and visual pigment content in the retina of the goldfish. *Vision Res.* 16:647–657.

Stell, W. K., and D. O. Lightfoot. 1975. Color-specific interconnections of cones and horizontal cells in the retina of the goldfish. *J. Comp. Neurol.* 159:473–502.

Stell, W. K., A. T. Ishida, and D. O. Lightfoot. 1977. Structural basis for on- and off-center responses in retinal bipolar cells. *Science* 198:1269–71.

Stell, W. K., R. Kretz, and D. O. Lightfoot. 1982. Horizontal cell connectivity in goldfish. In *The S-potential,* ed. B. D. Drujan and M. Laufer. New York: Alan R. Liss.

Svaetichin, G. 1956. Spectral response curves from single cones. *Acta Physiol. Scand.* 39, suppl. 134:17–46.

Svaetichin, G., and E. F. MacNichol, Jr. 1958. Retinal mechanisms for chromatic and achromatic vision. *Ann. N.Y. Acad. Sci.* 74:385–404.

Tartuferi, F. 1887. Sull'anatomia della retina. *Int. Monatss. Anat. Physiol.* 4:421–441.

Tomita, T. 1965. Electrophysiological study of the mechanisms subserving color coding in the fish retina. *Cold Spring Harb. Symp. Quant. Biol.* 30:559–566.

Wagner, H.-J. 1972. Vergleichende Untersuchungen über das Muster der Sehzellen und Horizontalen in der Teleostier-Retina (Pisces). *Z. Morph. Tiere* 72: 77–130.

——— 1978. *Cell types and connectivity patterns in mosaic retina.* Advances in Anatomy, Embryology, and Cell Biology, vol. 55, Fasc. 3. Heidelberg: Springer-Verlag.

Werblin, F. S., and J. E. Dowling. 1969. Organization of the retina of the mudpuppy, *Necturus maculosus.* II. Intracellular recording. *J. Neurophysiol.* 32:339–355.

Walls, G. L. 1942. *The vertebrate eye and its adaptive radiation.* (Cran. Inst. of Sci. Bull., 19.) New York: Hafner, reprinted 1963.

Warwick, R., and P. L. Williams, eds. 1973. *Gray's Anatomy,* 35th Brit. ed. Philadelphia: W. B. Saunders.

Yentsch, L. S., and C. A. Reichert. 1962. The interrelationship between water-soluble yellow substances and chloroplastic pigments in marine algae. *Bot. Marina* 3:65–74.

17. Neural Control of Locomotion

Anderson, O., H. Forssberg, S. Grillner. 1981. Peripheral feedback mechanisms acting on the central pattern generators for locomotion in fish and cat. *Can. J. Physiol. and Pharm.* 59:713–726.

Armstrong, R. B., C. W. Saubert IV, H. J. Seeherman, and C. R. Taylor. 1982. Distribution of fiber types in locomotory muscles of dogs. *Am. J. Anat.* 163:87–98.

Bessou, P., F. Emonot-Denond, and Y. Laporte. 1963. Relation entre la vitesse de conduction des fibres nerveuses motrices et le temps de contraction de leurs unites motrices. *C. R. Acad. Sci.* 256:5625–27.

Binder, M. D., P. Bawa, P. Ruenzel, and E. Henneman. 1983. Does orderly recruitment of motoneurons depend on the existence of different types of motor units? *Neurosci. Lett.* 36:55–58.

Binder, M. D., J. S. Kroin, G. P. Moore, and D. G. Stuart. 1977. The response of Golgi tendon organs to single motor unit contractions. *J. Physiol.* (Lond.) 271:337–349.

Binder, M. D., J. C. Houk, T. R. Nichols, W. Z. Rymer, and D. G. Stuart. 1982. Properties and segmental actions of mammalian muscle receptors: an update. *Fed. Proc.* 41:2907–18.

Bone, Q. 1966. On the function of the two types of myotomal muscle fiber in elasmobranch fish. *J. Mar. Biol. Ass. U.K.* 46:321–349.

Botterman, B. R., M. D. Binder, and D. G. Stuart. 1978. Functional anatomy of the association between motor units and muscle receptors. *Am. Zool.* 18:135–152.

Botterman, B. R., T. M. Hamm, R. M. Reinking, and D. G. Stuart. 1983. Localization of monosynaptic Ia EPSPs in the motor nucleus of the cat biceps femoris muscle. *J. Physiol.* (Lond.) 338:355–377.

Brandle, K., and G. Szekely. 1973. The control of alternating coordination of limb pairs in the newt *(Triturus vulgaris). Brain Behav. Evol.* 8:366–385.

Burke, R. E. 1979. The role of synaptic organization in the control of motor unit activity during movement. In *Reflex control of posture and movement,* ed. R. Granit and O. Pompeiano. Progress in Brain Research, 50. Amsterdam: North-Holland.

——— 1981. Motor units: anatomy, physiology and functional organization. In *Handbook of physiology,* ed. V. B. Brooks, sec. I, *The nervous system,* ed. J. M. Brookhart and V. B. Mountcastle, vol. 2, *Motor control.* Bethesda: American Physiological Society.

Burke, R. E., and V. R. Edgerton. 1975. Motor unit properties and selective involvement in movement. In *Exercise and sport sciences review,* ed. J. H. Wilmore and J. F. Keogh, vol. 3. New York: Academic Press.

Cameron, W. E., M. C. Binder, B. R. Botterman, R. R. Reinking, and D. G. Stuart. 1980. "Sensory partitioning" of cat medial gastrocnemius by its muscle spindles and tendon organs. *J. Neurophysiol.* 46:32–47.

Denny-Brown, D. 1929. On the nature of postural reflexes. *Proc. Roy. Soc. Lond. B.* 104:252–301.

Edwards, J. L. 1976. A comparative study of locomotion in terrestrial salamanders. Ph.D. diss., University of California at Berkeley.

——— The evolution of terrestrial locomotion. In *Major patterns in vertebrate evolution,* ed. M. K. Hecht et al. New York: Plenum Press.

English, A. W. 1979. Interlimb coordination during stepping in the cat: an electromyographic analysis. *J. Neurophysiol.* 42:229–243.

English, A. W., and P. R. Lennard. 1982. Interlimb coordination during stepping in the cat: in-phase stepping and gait transitions. *Brain Res.* 245:353–364.

Freadman, M. A. 1979. Role partitioning of swimming musculature of striped Bass, *Morone saxatilis* Walbaum, and Bluefish, *Pomatomus saltarix* L. *J. Fish Biol.* 15:417–423.

Gambaryan, P. P. 1974. *How mammals run: anatomical adaptations.* New York: Wiley. (Orig. publ. in Russian, 1972.)

Gollnick, P. D., D. Parsons, and C. R. Oakley. 1983. Differentiation of fiber types in skeletal muscle from sequential inactivation of myofibrillar actomyosin ATPase during acid preincubation. *Histochem.* 77: 543–555.

Goldspink, G. 1975. Biochemical energetics for fast and slow muscle. In *Comparative physiology: functional aspects of structural materials.* ed. L. Bolis, S. H. P. Maddrell, and K. Schmidt-Nielson. Amsterdam: North-Holland.

——— 1977. Muscle energetics and animal locomotion. In *Mechanics and energetics of animal locomotion,* ed. R. McN. Alexander and G. Goldspink. London: Chapman and Hall.

Gonyea, W. J., S. A. Marushia, and J. A. Dixon. 1981. Morphological organization and contractile properties of the wrist flexor muscles in the cat. *Anat. Rec.* 189:321–339.

Goslow, G. E., Jr., R. M. Reinking, and D. G. Stuart. 1973a. The cat step cycle: hindlimb joint angles and muscle lengths during unrestrained locomotion. *J. Morph.* 141:1–41.

——— 1973b. Physiological extent, range and rate of muscle stretch for soleus, medial gastrocnemius and tibialis anterior. *Plug. Arch.* 341:77–86.

Graham Brown, T. 1914. On the nature of the fundamental activity of the nervous centers; together with an analysis of the conditioning of rhythmic activity in progression, and a theory of the evolution of function in the nervous system. *J. Physiol.* (Lond.) 48:18–46.

Gray, J. 1968. *Animal locomotion.* London: Weidenfeld and Nicholson.

Grillner, S. 1981. Control of locomotion in bipeds, tetrapods, and fish. In *Handbook of Physiology,* ed. V. B. Brooks, sec. I, *The nervous system,* ed. J. M. Brookhart and V. B. Mountcastle, vol. 2, *Motor control.* Bethesda: American Physiological Society.

Grillner, S., and S. Rossignol. 1978. On the initiation of the swing phase of locomotion in chronic spinal cats. *Brain Res.* 146:269–277.

Hasan, Z., and D. G. Stuart. 1984. Mammalian muscle receptors. In *Handbook of the spinal cord,* ed. R. A. Davidoff, vol. 3. New York: Marcel Dekker.

Henneman, E., G. G. Somjen, and D. O. Carpenter. 1965. Functional significance of cell size in spinal motoneurons. *J. Neurophysiol.* 28:599–620.

Hess, A. 1970. Vertebrate slow muscle. *Physiol. Rev.* 50:40–62.

Jenkins, F. A., Jr., and G. E. Goslow, Jr. 1983. The functional anatomy of the shoulder of the savannah monitor lizard *(Varanus exanthematicus). J. Morph.* 175:195–216.

Landmesser, L. 1976. The development of neural circuits in the limb moving segments of the spinal cord. In *Neural control of locomotion,* ed. R. M. Herman et al. New York: Plenum Press.

Loeb, G. E. 1981. Somatosensory unit input to the spinal cord during normal walking. *Can. J. Physiol. and Pharm.* 59:627–635.

Loeb, G. E., M. J. Bak, and J. Duysens. 1977. Long-term unit recording from somatosensory neurons in the spinal ganglia of the freely walking cat. *Science* 197:1192–1194.

Maeda, N., S. Miyoshi, and H. Toh. 1983. First observation of a muscle spindle fiber in fish. *Nature* 302:61–62.

Maier, A. 1979. Occurrence and distribution of muscle spindles in masticatory and suprahyoid muscles of the rat. *Am. J. Anat.* 155:483–506.

——— 1981. Characteristics of pigeon gastrocnemius and its muscle spindle supply. *Exp. Neur.* 74:892–906.

Matthews, P. B. C. 1972. *Mammalian muscle receptors and their central actions.* London: Arnold.

——— 1981. Muscle spindles: their messages and their fusimotor supply. In *Handbook of physiology,* ed. V. B. Brooks, sec. I, *The nervous system,* ed. J. M. Brookhart and V. B. Mountcastle, vol. 2, *Motor control.* Bethesda: American Physiological Society.

McDonagh, J. C., M. D. Binder, R. M. Reinking, and D. G. Stuart. 1980. A commentary on muscle unit properties in cat hindlimb muscles. *J. Morph.* 166:217–230.

Miller, S., and J. Van Der Berg. 1973. The function of long propriospinal pathways in the coordination of quadrupedal stepping in the cat. In *Control of posture and locomotion,* ed. R. B. Stein et al. New York: Plenum Press.

Nemeth, P. M., H. W. Hoffer, and D. Pette. 1979. Metabolic heterogeneity of muscle fibers classified by myosin ATPase. *Histochem.* 63:191–201.

Nemeth, P. M., D. Pette, and G. Vrbova. 1981. Comparison of enzyme activities among single muscle fibres within defined motor units. *J. Physiol.* (Lond.) 311:489–495.

Ono, R. D. 1982. Structure of tendon organs in fishes of the genus *Polymixia. Zoomorph.* 99:131–144.

Orlovsky, G. N. 1972. The effect of different descending systems on flexor and extensor activity during locomotion. *Brain Res.* 40:359–372.

Peck, A. L., and E. S. Forster, trans. 1937. *Aristotle: parts of animals, movement of animals, progression of animals.* Cambridge, Mass.: Harvard University Press.

Philippson, M. 1905. L'autonomie et la centralisation dans le système nerveux des animaux. *Trav. Lab Physiol. Inst. Solvay* (Bruxelles) 7:1–208.

Prochazka, A., R. A. Westerman, and S. P. Ziccone. 1976. Discharge of single hindlimb afferents in the freely moving cat. *J. Neurophysiol.* 39:1090–1104.

Proske, U. 1981. The Golgi tendon organ. Properties of the receptor and reflex action of impulses arising from tendon organs. In *Neurophysiology,* ed. R. Porter, vol. 4. Intl. Rev. Physiol., 25. Baltimore: University Park Press.

Sacks, R. D., and R. R. Roy. 1982. Architecture of the hindlimb muscles of cats: functional significance. *J. Morph.* 173:185–195.

Sherrington, C. S. 1910. Flexion-reflex of the limb, crossed extension-reflex, and reflex stepping and standing. *J. Physiol.* (Lond.) 40:28–103.

Shik, M. L., and G. N. Orlovsky. 1976. Neurophysiology of locomotor automatism. *Physiol. Rev.* 56:465–501.

Shik, M. L., F. V. Severin, and G. N. Orlovsky. 1966. Control of walking and running by means of electrical stimulation of the mid-brain. *Biophysics* 11:756–765.

Spector, S. A., P. F. Gardiner, R. F. Zernicke, R. R. Roy, and V. R. Edgerton. 1980. Muscle architecture and force-velocity characteristics of cat soleus and medial gastrocnemius: implications for motor control. *J. Neurophysiol.* 44:951–960.

Stein, P. S. G. 1978. Swimming movements elicited by electrical stimulation of the turtle spinal cord: the high spinal preparation. *J. Comp. Physiol.* 124:203–210.

Stephens, J. A., and D. G. Stuart. 1975. The motor units of cat medial gastrocnemius: speed-size relations and their significance for the recruitment order of motor units. *Brain Res.* 91:177–195.

Stephens, J. A., R. M. Reinking, and D. G. Stuart. 1975. The motor units of cat medial gastrocnemius: electrical and mechanical properties as a function of muscle length. *J. Morph.* 146:495–512.

Stuart, D. G., and R. M. Enoka. 1983. Motoneurons, motor units and the size principle. In *The clinical neurosciences,* ed. R. G. Grossman and W. D. Willis, Jr., sec. V, *Neurobiology.* New York: Churchill Livingston.

Stuart, D. G., M. D. Binder, and R. M. Enoka. 1984. Motor unit organization: application of the quadripartite classification scheme to human muscles. In *Peripheral neuropathy,* ed. P. J. Dyck et al. Philadelphia: Saunders.

Stuart, D. G., T. P. Withey, M. C. Wetzel, and G. E. Goslow, Jr. 1973. Time constraints for interlimb coordination in the cat during unrestrained locomotion. In *Control of posture and locomotion,* ed. R. B. Stein et al. New York: Plenum Press.

Sypert, G. W., and J. B. Munson. 1981. Basis of segmental motor control: motoneuron size or motor unit type. *Neurosurg.* 8:608–621.

Szekely, G. 1976. Developmental aspects of locomotion. In *Neural control of locomotion,* ed. R. M. Herman et al. Advances in Behavioral Biology, 18. New York: Plenum Press.

Tracey, D. 1980. Joint receptors and the control of movement. *Tr. Neurosci.* 3:253–255.

Walker, W. F., Jr. 1979. Locomotion. In *Turtles: perspectives and research,* ed. M. Harless and H. Morlock. New York: Wiley.

Walmsley, B., J. A. Hodgson, and R. E. Burke. 1978.

The forces produced by medial gastrocnemius and soleus muscles during locomotion in freely moving cats. *J. Neurophysiol.* 41:1203–16.

Wetzel, M., and D. G. Stuart. 1976. Ensemble characteristics of cat locomotion and its neural control. *Progr. Neurobiol.* 7:1–98.

Willis, W. D., Jr., and R. G. Grossman. 1973. *Medical neurobiology.* St. Louis: C. V. Mosby.

18. Morphology: Current Approaches and Concepts

Alberch, P. 1980. Ontogenesis and morphological diversification. *Am. Zool.* 20:653–667.

Alberch, P., and J. Alberch. 1981. Heterochronic mechanisms of morphological diversification and evolutionary change in the Neotropical salamander *Bolitoglossa occidentalis* (Amphibia, Plethodontidae). *J. Morph.* 167:249–264.

Alberch, P., and E. Gale. 1983. Size dependence during the development of the amphibian foot, colchicine-induced digital loss and reduction. *J. Embrol. Exp. Morph.* 76:177–197.

Alberch, P., S. J. Gould, G. F. Oster, and D. B. Wake. 1979. Size and shape in ontogeny and phylogeny. *Paleobiol.* 5:296–317.

Arnold, S. J. 1983. Morphology, performance and fitness. *Am. Zool.*, 23:347–361.

Boag, P. T., and P. R. Grant. 1981. Intense natural selection in a population of Darwin's finches (Geospizinae) in the Galapagos. *Science* 214:82–85.

Bock, W. J. 1980. The definition and recognition of biological adaptation. *Am. Zool.* 20:217–227.

Bonner, J., ed. 1982. *Evolution and development.* Berlin: Springer-Verlag.

Cherry, L. M., S. M. Case, and A. C. Wilson. 1978. Frog perspective on the morphological difference between humans and chimpanzees. *Science* 200: 209–211.

Cherry, L. M., S. M. Case, J. G. Kunkel, J. S. Wyles, and A. C. Wilson. 1982. Body shape metrics and organismal evolution. *Evol.* 36:914–933.

Dullemeijer, P. 1974. *Concepts and approaches in animal morphology.* Assen, Netherlands: Van Gorcum.

——— 1980. Functional morphology and evolutionary biology. *Acta Biotheor.* 29:151–250.

Eldredge, N., and J. Cracraft. 1980. *Phylogenetic patterns and the evolutionary process.* New York: Columbia University Press.

Fink, W. 1982. The conceptual relationship between ontogeny and phylogeny. *Paleobiol.* 8:254–264.

Gans, C. 1966. Some limitations and approaches to problems in functional anatomy. *Folia Biotheor.* 6:41–50.

——— 1974. Biomechanics: An approach to vertebrate biology. Philadelphia: Lippincott.

Gans, C., and R. G. Northcutt. 1983. Neural crest and the origin of vertebrates: a new head. *Science* 220:268–274.

Goodwin, B. C., N. Holder, and C. G. Wylie, eds. 1983. *Development and evolution.* Cambridge: Cambridge University Press.

Gould, S. J. 1977. *Ontogeny and phylogeny.* Cambridge, Mass.: Harvard University Press.

——— 1980. The evolutionary biology of constraint. *Daedalus* 109:39–52.

Gould, S. J., and R. C. Lewontin. 1979. The spandrels of San Marco and the Panglossian paradigm: a critique of the adaptationist programme. *Proc. Roy. Soc. Lond. B* 205:581–598.

Hall, B. K. 1983. Epigenetic control in development and evolution. In *Development and evolution,* ed. B. C. Goodwin, N. Holder, and C. G. Wylie. Cambridge: Cambridge University Press.

Hennig, W. 1966. *Phylogenetic systematics.* Urbana: University of Illinois Press.

Hickman, C. S. 1980. Gastropod radulae and the assessment of form in evolutionary paleontology. *Paleobiol.* 6:276–294.

Kollar, E. J., and C. Fisher. 1980. Tooth induction in chick epithelium: expression of quiescent genes for enamel synthesis. *Science* 207:993–995.

Lauder, G. V. 1981. Form and function: structural analysis in evolutionary morphology. *Paleobiol.* 7:430–442.

——— 1982a. Historical biology and the problem of design. *J. Theor. Biol.* 97:57–67.

——— 1982b. Introduction. In *Form and function: a contribution to the history of animal morphology,* ed. E. S. Russell. Chicago: University of Chicago Press.

Lauder, G. V., and K. F. Liem. 1983. The evolution and interrelationships of the Actinopterygian fishes. *Bull. Mus. Comp. Zool.,* 150:95–197.

Lewontin, R. C. 1978. Adaptation. *Sci. Am.* 239:212–230.

Liem, K. F. 1980. Adaptive significance of intra- and interspecific differences in the feeding repertoires of cichlid fishes. *Am. Zool.* 20:295–314.

Liem, K. F., and L. S. Kaufman. 1984. Intraspecific macroevolution: functional biology of the polymorphic cichlid species *Cichlasoma minckleyi.* In *Evolutionary biology of species flocks,* ed. A. Echelle and I. Kornfield. Orono: University of Maine Press.

Maderson, P. F. A. 1983. An evolutionary view of epithelial-mesenchymal interactions. In *Epithelial-mesenchymal interactions in development,* ed. R. H. Sawyer and J. F. Fallon. New York: Praeger Scientific.

Mayr, E., and W. B. Provine, eds. 1980. *Evolutionary*

synthesis: perspectives on the unification of biology. Cambridge, Mass.: Harvard University Press.

McKaye, K. R., and A. Marsh. 1983. Food switching by two specialized algae-scraping cichlid fishes in Lake Malawi, Africa. *Oecologia* (Berlin) 56:245–248.

Oster, G. F., and P. Alberch. 1982. Evolution and bifurcation of developmental programs. *Evol.* 36:444–459.

Reif, W.-E. 1982. Functional morphology on the procrustean bed of the neutralism-selectionism debate: notes on the constructional morphology approach. *N. Jahrb. Geol. Paläont. Abh.* 164:46–59.

Ribbink, A. J., B. A. Marsh, A. C. Marsh, A. C. Ribbink, and B. J. Sharp. 1983. A preliminary survey of the cichlid fishes of rocky habitats in Lake Malawi. *So. African J. Zool.* 18:149–310.

Riedl, R. 1978. *Order in living systems.* New York: Wiley.

Rudwick, M. J. S. 1964. The inference of structure from function in fossils. *Brit. J. Phil. Sci.* 15:27–40.

Seilacher, A. 1973. Fabricational noise in adaptive morphology. *Syst. Zool.* 22:451–465.

Vermeij, G. 1973. Adaptation, versatility, and evolution. *Syst. Zool.* 22:466–477.

Wainwright, S. A. 1980. Adaptive materials: a view from the organism. In *The mechanical properties of biological materials,* ed. J. F. V. Vincent and J. D. Currey. London: Society for Experimental Biology.

Wainwright, S. A., W. D. Biggs, J. D. Currey, and J. M. Gosline. 1975. *Mechanical design in organisms.* New York: Halstead.

Wake, D. B. 1966. Comparative osteology and evolution of the lungless salamanders, family Plethodontidae. *Mem. So. Calif. Acad. Sci.* 4:1–111.

——— 1982. Functional and evolutionary morphology. *Persp. Biol. and Med.* 25:603–620.

Wake, D. B., and A. H. Brame, Jr. 1969. Systematics and evolution of neotropical salamanders of the *Bolitoglossa helmrichi* group. *Contrib. Sci. Nat. Hist. Mus. Los Angeles* 175:1–40.

Wake, D. B., G. Roth, and M. H. Wake. 1983. The problem of stasis in organismal evolution. *J. Theor. Biol.* 101:211–224.

Wessells, N. 1982. A catalogue of processes responsible for metazoan morphogenesis. In *Evolution and development,* ed. J. T. Bonner. Berlin: Springer-Verlag.

Wiens, J. A. 1977. On competition and variable environments. *Am. Sci.* 65:590–597.

Wiley, E. O. 1981. *Phylogenetics.* New York: Wiley.

Acknowledgments

1. Functional Adaptation in Skeletal Structures

The authors' work discussed in this chapter has been supported by the Medical Research Council, United Kingdom; the Wellcome Foundation; Arthritis and Rheumatism Council, United Kingdom; and the National Aeronautics and Space Administration. The experiments described were performed at the Department of Anatomy, University of Bristol, United Kingdom; the Concord Field Station, Museum of Comparative Zoology, Harvard University, Cambridge, Massachusetts; and the Department of Anatomy and Cellular Biology, Tufts University, Boston. We are grateful to all our colleagues, without whose assistance much of this work would not have been possible. We should also like to thank John Currey and Geoffery Goldspink for valuable criticism of the manuscript at various stages. Finally, we are grateful to Marie Sollod for typing many drafts of the manuscript.

2. Body Support, Scaling, and Allometry

M. Hildebrand and D. B. Wake read a draft of this chapter and made many useful comments.

3. Walking and Running

A draft of this chapter was reviewed by R. McN. Alexander, W. J. Bock, D. Bramble, G. E. Goslow, Jr., F. A. Jenkins, Jr., S. Rasmussen, and D. B. Wake. Part of the manuscript was read by J. P. Hurley, who assisted with the section on limbs as pendulums. I am indebted to all for important suggestions.

4. Jumping and Leaping

I should like to thank the following persons for allowing me to use their data, published and unpublished, for this chapter: M. Hyde, F. Jouffroy, M. McArdle, J. Nickolai, C. Oxnard, and L. Radinsky. I am also grateful to the curators at the following museums for allowing me access to skeletal material: American Museum of Natural History, Field Museum of Natural History, and the Museum of Comparative Zoology, Harvard University. In addition, D. Bramble, M. Hildebrand, and L. Radinsky provided useful comments on drafts of the manuscript, which was completed in summer 1981.

5. Climbing

I thank M. Hildebrand, W. L. Hylander, R. F. Kay, R. D. E. MacPhee, K. Smith, and D. B. Wake for their help and suggestions. I am particularly indebted to K. Brown for her editorial comments and her invaluable assistance in the production of the manuscript.

6. Digging of Quadrupeds

I am grateful to D. M. Bramble, S. B. Emerson, C. A. Reed, and D. B. Wake for their helpful reviews of a draft of this chapter, and also to A. F. Bennett and K. M. Hiiemae, who read in draft the sections relating to their respective chapters. The curators of the Field Museum of Natural History and the Museum of Vertebrate Zoology, respectively, loaned three skeletons and one alcoholic specimen. The X-ray photograph (Fig.

6-1F) was taken by J. Neves; the remaining photographs were taken by R. W. Kulmann.

7. Swimming

Support for much of the authors' research described here was provided by the National Science Foundation (PWW) and National Research Council of Canada (RWB). We thank R. McN. Alexander, K. F. Liem, and D. Weihs for their comments.

8. Flying, Gliding, and Soaring

I am most grateful to Å. Norberg, C. Pennycuick, J. Rayner, V. Tucker, and P. Webb for reading all or part of an earlier version of the manuscript and for giving many valuable ideas for improving it.

9. Terrestrial Locomotion without Appendages

I should like to thank M. Hildebrand, B. Jayne, A. Savitzky, D. Wake, and M. Wake for their comments. All illustrations were prepared by M. Flanders. Portions of this work were supported by NIH grant NS 16270.

10. Energetics and Locomotion

Many helpful comments on the text were provided by G. A. Bartholomew, T. T. Gleeson, M. Hildebrand, U. M. Norberg, and C. R. Taylor. I thank K. John-Alder for drawing the figures. Financial support for this work was provided by NSF grant PCM81–02331 and NIH grant K04 AM00351.

11. Ventilation

I am greatly indebted to the following persons who have contributed their skills, knowledge, and time: J. Clark, A. Coleman, K. Deyst, C. Fox, C. Gans, K. Hartel, G. V. Lauder, Y. Schulman, W. W. Sung, R. J. Wassersug, and E. Wu.

12. Aquatic Feeding in Lower Vertebrates

I thank D. Bramble and J. Mallatt for making available unpublished information, B. Shaffer for his assistance in filming aquatic salamander feeding, and J. Hives for her assistance in preparing the manuscript. Reviews of this chapter by the editors of the volume are greatly appreciated. Preparation of this paper was supported by grants from the Andrew W. Mellon Foundation, the Block Fund of the University of Chicago, and NSF DEB 81–15048.

13. Feeding Mechanisms of Lower Tetrapods

The following persons have reviewed the manuscript and have offered many useful suggestions, criticisms, and comments: D. Cundall, H. Greene, M. Hildebrand, D. Homberger, G. Lauder, K. Schwenk, K. Smith, G. Throckmorton, and M. Wake. We thank them all. We appreciate as well the important unpublished works made available to us by A. Busby, H. Greene, D. Homberger, W. Hylander, G. Lauder, K. Liem, K. Schwenk, K. Smith, G. Throckmorton, and R. Winokur. We are also indebted to G. Roth for providing the spectacular photograph of salamander tongue projection.

14. Mastication, Food Transport, and Swallowing

This chapter is the result of four or five years' work by H. A. Franks, W. L. Hylander, J. McGarrick, and A. J. Thexton, as well as by us. We are grateful for the support received from the National Institutes of Health (DE-05738) and the National Science Foundation (DEBH 81-01094). Our efforts have focused primarily on food transport and swallowing in the macaque, hyrax, cat, and opossum, and many of the figures used here are based on drawings prepared for various other manuscripts. We thank K. R. Gordon, W. L. Hylander, P. W. Lucas, J. M. Rensberger, A. J. Thexton, and D. B. Wake for their reviews of a draft of this chapter. Without the efforts of Judy Rosovsky, Debra Sponder, students, and other support staff, including Elizabeth Perry, Virginia Withers, Virginia Dwyer, and Sheridan Turner, our task would have been much harder. Laslo Maszoly and Pat Parshall drew the figures.

15. The Octavolateralis System

Writing of the manuscript and portions of the work on fish audition described here were supported by various grants from the National Science Foundation, the National Institutes of Health, and the Office of Naval Research to each of the authors. Both authors are recipients of NIH Research Career Development Awards. We are grateful to R. Dooling, M. Hildebrand, B. Hoxter, E. R. Lewis, W. M. Saidel, and W. C. Stebbins for their constructive comments on our chapter.

16. The Vertebrate Eye

The author would like to acknowledge the assistance of B. A. Collins, E. F. MacNichol, Jr., W. K. Stell, and the editors who gave generously of their time in reviewing earlier drafts of this manuscript. The histological preparations pictured in Figure 16-5 were obtained through the courtesy of B. A. Collins.

17. Neural Control of Locomotion

I am grateful to R. E. Burke, S. Grillner, and D. G. Stuart for the use of unpublished materials. I also wish to thank C. Dragoo for photographic assistance and L. Waller for preparation of the figures. The careful and thoughtful reviews of this chapter by my friends and colleagues saved me from numerous errors and are gratefully acknowledged: C. Dragoo, J. Edwards, B. Foehring, T. Hamm, P. Hannon, J. Hermanson, S. Peters, D. Pierotti, S. Rasmussen, and B. Young. It gives me special pleasure to acknowledge my loyal advisers and good friends, Milton Hildebrand and Douglas Stuart, who not only spent many hours reading and discussing this chapter with me, but instilled in me a love of animal adaptation and neural science. Support for this project was given by the Organized Research Committee, Northern Arizona University.

18. Morphology: Current Approaches and Concepts

We should like to thank P. Alberch, G. V. Lauder, and M. H. Wake for their helpful and critical evaluations.

Contributors

R. McNeill Alexander Department of Pure and Applied Zoology, University of Leeds, Leeds LS2 9JT, United Kingdom

Albert F. Bennett School of Biological Sciences, University of California, Irvine, California 92717

Robert W. Blake Department of Zoology, University of British Columbia, Vancouver, British Columbia V67 1W5, Canada

Dennis M. Bramble Department of Biology, University of Utah, Salt Lake City, Utah 84112

Matt Cartmill Department of Anatomy, Duke University Medical Center, Durham, North Carolina 27710

Alfred W. Crompton Museum of Comparative Zoology, Harvard University, Cambridge, Massachusetts 02138

James L. Edwards Systematic Biology Program, National Science Foundation, Washington, D.C. 20550

Sharon B. Emerson Division of Amphibians and Reptiles, Field Museum of Natural History, Chicago, Illinois 60605

Richard R. Fay Parmly Hearing Institute, Loyola University of Chicago, Chicago, Illinois 60626

George E. Goslow, Jr. Department of Biological Sciences, Northern Arizona University, Flagstaff, Arizona 86011

Karen M. Hiiemae Department of Oral Anatomy, College of Dentistry, University of Illinois, Chicago, Illinois 60612

Milton Hildebrand Department of Zoology, University of California, Davis, California 95616

Lance E. Lanyon Department of Anatomy, Royal Veterinary College, University of London, London NW1, England

George V. Lauder Department of Anatomy, University of Chicago, Chicago, Illinois 60637

Joseph S. Levine Department of Biology, Boston College, Chestnut Hill, Massachusetts 02167

Karel F. Liem Museum of Comparative Zoology, Harvard University, Cambridge, Massachusetts 02138

Ulla M. Norberg Department of Zoology, University of Göteborg, Box 25059, S-400 31 Göteborg, Sweden

Arthur N. Popper Department of Anatomy, Georgetown University Schools of Medicine and Dentistry, Washington, D.C. 20007

Clinton T. Rubin Department of Anatomy, Tufts University Schools of Medicine and Veterinary Medicine, North Grafton, Massachusetts 01536

David B. Wake Museum of Vertebrate Zoology, University of California, Berkeley, California 94720

Paul W. Webb School of Natural Resources, University of Michigan, Ann Arbor, Michigan 48109

Index